CW00815846

Integrated Assessment of Sustainable Energy Systems in China
The China Energy Technology Program

ALLIANCE FOR GLOBAL SUSTAINABILITY BOOKSERIES
SCIENCE AND TECHNOLOGY: TOOLS FOR SUSTAINABLE DEVELOPMENT

VOLUME 4

Series Editor*:** ***Dr. Joanne M. Kauffman
Laboratory for Energy and the Environment
Massachusetts Institute of Technology
1 Amherst St., Room E40-453
Cambridge, Massachusetts 02139 USA
Jmkauffm@mit.edu

Series Advisory Board:

Dr. John H. Gibbons
President, Resource Strategies, The Plains, VA, USA

Professor Atsushi Koma
Vice President, University of Tokyo, Japan

Professor Hiroshi Komiyama
University of Tokyo, Japan

Professor David H. Marks
Massachusetts Institute of Technology, USA

Professor Mario Molina
Massachusetts Insitute of Technology, USA

Dr. Rajendra Pachauri
Director, Tata Energy Research Institute, India

Professor Roland Scholz
Swiss Federal Institute of Technology, Zürich, Switzerland

Dr. Ellen Stechel
Manager, Environmental Programs, Ford Motor Co., USA

Professor Dr. Peter Edwards
Department of Environmental Sciences, Geobotanical Institute, Switzerland

Dr. Julia Carabias
Instituto de Ecología, Universidad Nacional Autónoma de México, México

Aims and Scope of the Series

The aim of this series is to provide timely accounts by authoritative scholars of the results of cutting edge research into emerging barriers to sustainable development, and methodologies and tools to help governments, industry, and civil society overcome them. The work presented in the series will draw mainly on results of the research being carried out in the Alliance for Global Sustainability (AGS).
The level of presentation is for graduate students in natural, social and engineering sciences as well as policy and decision-makers around the world in government, industry and civil society.

Integrated Assessment of Sustainable Energy Systems in China

The China Energy Technology Program

A Framework for Decision Support in the Electric Sector of Shandong Province

Edited by

Baldur Eliasson
ABB Corporate Research,
Baden-Daettwil, Switzerland (retired)

and

Yam Y. Lee
ABB Corporate Research,
Baden-Daettwil, Switzerland (consultant)

KLUWER ACADEMIC PUBLISHERS
DORDRECHT / BOSTON / LONDON

A C.I.P. Catalogue record for this book is available from the Library of Congress.

ISBN 1-4020-1198-9

Published by Kluwer Academic Publishers,
P.O. Box 17, 3300 AA Dordrecht, The Netherlands.

Sold and distributed in North, Central and South America
by Kluwer Academic Publishers,
101 Philip Drive, Norwell, MA 02061, U.S.A.

In all other countries, sold and distributed
by Kluwer Academic Publishers,
P.O. Box 322, 3300 AH Dordrecht, The Netherlands.

Printed on acid-free paper

All Rights Reserved
© 2003 Kluwer Academic Publishers
No part of this work may be reproduced, stored in a retrieval system, or transmitted
in any form or by any means, electronic, mechanical, photocopying, microfilming, recording
or otherwise, without written permission from the Publisher, with the exception
of any material supplied specifically for the purpose of being entered
and executed on a computer system, for exclusive use by the purchaser of the work.

Printed in the Netherlands.

ALLIANCE FOR GLOBAL SUSTAINABILITY

An International Partnership

Alliance for Global Sustainability
International Advisory Board

Interim Chairman:

Mr. Livio D. DeSimone, Chairman of the Board and CEO, 3M (Retired)

AGS University Presidents:

Prof. Olaf Kübler, President, Swiss Federal Institute of Technology, Zürich

Prof. Takeshi Sasaki, President, University of Tokyo

Prof. Jan-Eric Sundgren, President, Chalmers University of Technology

Dr. Charles M. Vest, President, Massachusetts Institute of Technology

Members:

Dr. Markus Bayegan, Technology and Research Director, ABB Ltd

Dr. Thomas Conelly, Chief Science and Technology Officer, DuPont

The Hon. President José-María Figueres Olsen, Managing Director, World Economic Forum

Mr. Hiroaki Fujii, President, The Japan Foundation

Mr. Hiroyuki Fujimura, Chairman of the Board, Ebara Corporation

Mr. Lars Kann-Rasmussen, Director, VKR Holding A/S

Dr. Jill Ker Conway, Chairman, Lend Lease Corporation

Mr. Masatake Matsuda, Chairman, East Japan Railway Company

Mr. Nobuya Minami, President, Tokyo Electric Power Company, Inc.

Mr. Norio Wada, President, Nippon Telegraph and Telephone Corporation (NTT)

Prof. Jakob Nüesch, Member, International Committee of the Red Cross

Mr. Dan Sten Olsson, CEO, Stena AB

Dr. Fred Palensky, Executive Vice President, 3M

Mr. Tei-ichi Sato, Director General, The Japan Society for the Promotion of Science

Sir Nicholas V. (Nick) Scheele, Chief Operating Officer and Director, Ford Motor Company

Dr. Stephan Schmidheiny, President, Avina Foundation

Prof. Francis Waldvogel, President, ETH Board, Switzerland

Dr. Margot Wallström, Member of the European Commission

Prof. Hiroyuki Yoshikawa, President, National Institute of Advanced Industrial Science and Technology

Dr. Hans-Rudolf Zulliger, CEO, Gretag AG

TABLE OF CONTENTS

CHAPTER 3

CHAPTER 4

CHAPTER 5

CHAPTER 7

CHAPTER 8

CHAPTER 9

CHAPTER 10

CHAPTER 11

CHAPTER 12

CHAPTER 13

PREFACE

The purpose of the China Energy Technology Program (CETP) has been to take a holistic view of electricity generation in China with special emphasis on the economic and environmental impact of technology. The program is a collaborative effort involving industry, as leader and manager of the program; researchers from academia and national research institutes; and the stakeholders--users or planners of the electricity market. A power plant, no matter of what kind, is not a solitary unit, which may be considered independently from its environment. Modern society has been built on easy access to power; it is now asking for environmentally clean power. Clean power is the pathway that will lead to sustainability, an extension of the concept of an environmentally clean future. Progress toward a sustainable technical solution must include an evaluation of the environmental, economic, and societal impact of electrical power generation.

To achieve the goals of the CETP program, we wanted to go beyond what had been done in similar programs in the past. Its organizers wanted the program not only to evaluate the overall impact of electricity generation, but also do it in such a way that program results could be applied and adapted to different circumstances, including countries other than China. For this purpose we have developed a methodology and presented it on a DVD disc, which the reader will find inside the back cover of this book. The methodology is conceived to allow insights into the function of the program on various levels.

Very few books have taken such an approach. Even fewer have treated the subject of sustainable power generation in China.

This volume will be especially interesting to decision-makers representing electric utilities, plant suppliers and regulators, engineers and environmental experts, and academics. It will also be valuable to the research community in industrial research organizations, non-governmental organizations, and government laboratories dealing with energy and environmental issue, and academic institutions.

The CETP is a very complex project; it has been divided into 12 different tasks each with a principal investigator. The book summarizes the methodology and findings of the entire project. The chapters are organized to provide background of CETP and its methodology (Chapters 1-2), a description of the nine research tasks and their results (Chapters 3-11), the integration of the tasks and the organization of the DVD (Chapter 12), and conclusions and recommendations (Chapter 13).

The individual chapters describing specific tasks (Chapters 3-11) were written by each principal investigator and the research group engaged in the task under his leadership. The accompanying DVD provides additional detailed information on all tasks performed in this program. The DVD is also an interactive tool in which a user may explore different scenarios within the scope of this study. The wide spectrum of targeted users is reflected in the modular design of the software, which should meet the needs of users having different backgrounds and interests.

The authors are grateful to many people for their contributions to the progress of the CETP. First, we appreciate the management and leadership of the members of the Steering Committee especially its chairman, Dr. Markus Bayegan, Chief Technology Officer of ABB. We would like to thank the members of the

Stakeholders Advisory Group who contributed invaluable input to the technical content and relevance of this research program. We were encouraged by the enthusiastic and strong support of our program by Mr. Huijiong Wang from the China State Council and Mr. Kun Zhang from the State Environmental Protection Administration in Beijing. The Shandong utility SEPRI/SEPCO cooperated closely with us in this project; we especially appreciate the enthusiasm of SEPRI's President Mr. Qingbo Zhao. The members of the Technical Advisory Board, Professor Jefferson Tester from MIT, USA, Professor Wolfgang Kröger from the Paul Scherrer Institute in Switzerland and one of us (BE) provided guidance on the technical aspects of the program and were involved in the review of the technical chapters of the book. We are greatly indebted to them for their help. Special thanks go to the following people who provided tremendous support in the operation of CETP: Mr. Paul Chan, Ms. Jean Gao, Mr. Tianpeng Liu, and Mr. Shiwen Cheng of ABB China, and Mrs. Johanna Nehring, Ms. Daniela Schmied, and Mr. Salvatore Mainardi of ABB Corporate Research in Switzerland.

This volume was written by various authors coming from different backgrounds. Dr. Teresa Hill of Harvard Law School, managing editor of this book had the very difficult and challenging task of organizing the chapters into a coherent whole. We would like to thank her for her excellent efforts. Parts of the book were reviewed by Mr. Zhaoguang Hu from the China State Power Corporation and Professor Alexander Wokaun from the Paul Scherrer Institute in Switzerland. We would like to acknowledge their help. Many thanks go to Ms. Daniela Schmied for helping us prepare the manuscript. We very much appreciate the support of AGS staffs in the course of preparing this book, especially the unwavering support of Dr. Joanne Kauffman, the Chief Editor of the AGS Series. Finally we would like to thank the Publishing Editors of the book, Ms. Nathalie Jacobs and Mr. Arno Schouwenburg from the Kluwer Academic Publishers in the Netherlands, for their excellent cooperation and for making this book possible.

May this book be as much enjoyment for you to read and apply as it was for us to put it together.

Baden-Dättwil September 2002 BALDUR ELIASSON

YAM YEE LEE

FOREWORD I

Although two of the most serious problems facing humanity have been with us for many decades, it is only comparatively recently that they have started to receive the attention they deserve. Perhaps one reason for this is that, in line with contrary nature of problems in many fields, the solution to the first exacerbates the second and vice versa.

Few now doubt that many recent natural disasters around the world are in some way related to the first of these problems: anthropogenic greenhouse gas emission. Reducing such emissions requires a significant shift in thinking in the area of our second problem: the infrastructure and energy needs of the global society, one third of which currently have no access to basic services such as electricity, clean water, telecommunication, etc.

In other words, the world is facing a tremendous challenge: a balancing act which preserves our world for future generations, while, at the same time, meeting the energy demands of a burgeoning world population.

New ideas and technologies are needed to meet this challenge.

Global corporations are often criticized for aggravating global problems. But they can also provide many of the skills and resources that can solve them. ABB, through its activities and investments, is strongly committed to solving global environmental problems and CETP is one example of this.

Although its origins were in 1992-93, the first substantial CETP concepts date from 1997, when ABB began talks with various possible partners.

In 1999 ABB, together with Alliance for Global Sustainability* (AGS), launched its biggest ever study of sustainable development, called the China Energy Technology Program (CETP). The main target for CETP was to enable more sustainable development in the world's largest and fastest growing country - China. Rapid economic growth means China's demand for electricity is soaring. The country has huge coal reserves - but burning coal is one of the most environmentally challenging ways to generate electricity

CETP's aim was to analyze the true, cradle -to-grave impact of a range of power generation options, focusing on the needs of Shandong, a rapidly developing coastal province between Beijing and Shanghai.

The CETP involved some 70 scientists, engineers and academics from three continents, but also customers and consumers of electricity in China.

In this publication the CEPT Project Group presents the results of its work over the past three years. The integrated view presented here considers not only the short-term techno-economic parameters, but also the long-term environmental aspects. It is our hope that the technologies, the methodology and the tools developed under this program can be used by decision-makers in China and elsewhere to make better decisions when it comes to selecting the most appropriate technology for electricity generation.

ABB, as the main fund provider and industrial partner in the program, is proud of the results achieved and has great expectations for their future utilization. Being a world leader in Industrial Automation and Power Technologies, we believe that there are many opportunities for using these results as part of our total efforts in the area of sustainable development. Combined with a wide range of sustainable technologies, products, systems and solutions which we provide to our customers around the world, we believe the results of this program can have a major impact on making this newly dawned century one of sustainable development.

In the case of China, the CETP has already led to some very inspiring conclusions:

- It is feasible, economically and socially justifiably, to generate more electricity with less air pollution
- There are cost-effective methods and technologies which cut air-pollution and which stabilize greenhouse gas emission
- A larger portfolio of products and technologies is needed for appropriate fuel treatment, renewable and distributed power generation, demand side management, etc.

Even though the study concentrated on China as a prominent and important showcase, it has paved the way for a general application in other areas of the world.

I want to congratulate all members of this project for the valuable contribution they have made to a more promising future for mankind.

Zürich September 2002 MARKUS BAYEGAN
Chief Technology Officer, ABB

* The Alliance for Global Sustainability was formed in 1994 by the Massachusetts Institute of Technology, the Swiss Federal Institute of Technology and the University of Tokyo to work on integrated aspects of environmental and sustainability problems through research, education and global outreach.

FOREWORD II

As the twentieth first century dawns, China is a fast growing economy with an annual GDP growth rate of about seven percent. The demand for electricity has increased significantly as the Chinese standard of living has improved. As supply must grow to meet this demand, concern over the environmental impact of power generation also deepens. In recent years, China has applied the principles of sustainable development by implementing various new policies and regulations to accelerate the development of new technology and strengthen scientific management in an effort to control and reduce the environmental impact of the power generation industry.

Three years ago, I became aware of the China Energy Technology Program (CETP) sponsored by ABB. In this program, scientists from several world-renowned universities and three Chinese research organizations have engaged in scientific research activities focusing on the integrated assessment of sustainable energy systems in China. Some 75 research scientists from three continents were involved in this project. It was also a pioneering effort in cooperation between industry and academia. The project was managed and run by the industrial partner and the research was carried out by experts from universities and research organizations.

The interdisciplinary expertise brought to bear on this project was extraordinary. The involvement of stakeholders early in the program was also a very sophisticated and practical approach to ensuring the relevance of the program and its future application. Since the project focuses on Shandong province, it is important that Shandong Electric Power Corporation (SEPCO) was involved; SEPCO personnel have provided very strong support for this research.

The grand scale and pioneering aspects of the project attracted my attention and led to my active participation in CETP as a member of its stakeholder advisory group. During my interaction with the researchers and management of this project, I came to share their strong sense of pride and enthusiasm for the project, and their commitment to its success. Three years of hard work have produced the excellent results presented in this volume and the accompanying DVD.

I firmly believe that the results will have immense impact on the future planning and development of the power industry in China, particularly in Shandong province. The policy makers who read this book and the accompanying DVD will most certainly find them invaluable in developing a sustainable system of energy production that will support China's development goals. I am very glad to have been part of this successful project and look forward to continued international cooperation as the future unfolds.

Beijing November 2002

DINGHUAN SHI
Secretary General
The Ministry of Science and Technology of the People's Republic of China

CHAPTER 1

INTRODUCTION

BALDUR ELIASSON

This book is about energy, special kind of energy--electricity. Electricity is easily generated, easily transported, and easily used and applied in the ordinary households and diverse industries of the world. Its ease of use has made access to electricity grow faster worldwide than the world's population, although one-third is still without a secure supply. Today, most electricity is generated by burning fossil fuels--coal, oil and natural gas--a fact that has raised problems of pollution, but once generated, electricity is clean and nonpolluting.

Progress depends on the use of energy, which can only be applied to human purposes through technology. Technology is the basis upon which progress rests. The program described in this book, the China Energy Technology Program (CETP) was established to address the issues raised by the technologies involved in the electricity generation process.

Concern over pollution produced by energy generation has prompted a shift in emphasis away from the technologies required by large, centrally located power plants to those for smaller and more distributed facilities. Until now most of the world's electricity has been generated in the industrial countries, such as those comprising membership in the OECD. Here again a shift is discernible; the developing countries are building a larger percentage of the world's infrastructure for electrical energy generation, while the percentage represented by construction in the developed world is falling.

If this trend continues, most of the projected increase in the world's greenhouse gas (GHG) emissions will come from developing countries such as China and India. China, the second largest emitter of GHGs, in 1996 emitted 15 percent of the world's carbon dioxide (CO_2). The CO_2 emissions and the percentage continue to grow.[1] The US, the largest emitter, produced around 22 percent of the world's total manmade CO_2 in 1996, and its percentage is falling.

Electricity was first introduced in the last decades of the nineteenth century. There were many competing technologies in the beginning: alternating current versus direct current; high voltage transport of energy versus low voltage transport. By the beginning of the twentieth century, the pattern that would characterize the next 100 years had been stabilized. This standard was characterized by a fairly large power plant, usually burning coal, in one location from which electricity was distributed into an ever-finer net of distribution channels eventually taking the power to homes, factories, or other points of use. This is more or less the way we live with electricity today.

This system has some notable limitations. For example, it is used almost exclusively for stationary purposes--there are very few electric automobiles. A good, portable carrier of electricity, e.g. a good battery, is not available, though recent developments in fuel cell technology might obviate this drawback.

Worldwide electricity use is increasing faster than the world economy as a whole. More and more large plants have been built and the use of fossil fuels is growing at an ever-increasing rate. However, during the latter part of the twentieth century it became evident that many of the gases produced during the combustion processes are causing pollution problems on a regional scale. Others are accumulating in the atmosphere and could possibly affect the global climate. The most serious of all of those emissions is the release of carbon dioxide. Whenever fossil fuels, i.e. gas, oil, or coal, are burned, carbon dioxide is generated and emitted. About half of worldwide CO_2 emissions accumulate in the atmosphere, contributing to global warming.

Large amounts of carbon dioxide occur naturally in nature, in the soil, in the atmosphere, and in the oceans. Anthropogenic emissions are tiny in comparison. However, the values in nature have been constant over a long period of time. The sudden accumulation of manmade emissions in the atmosphere upset this equilibrium. Because atmospheric carbon dioxide blocks the infrared emissions of the Earth, a greater concentration functions to raise global temperatures.

The Earth receives energy from the sun in the form of visible sunlight, and radiates an equal amount of energy back into space in the form of infrared radiation. The more GHGs there are in the atmosphere, the hotter the Earth must be to radiate the amount of energy received from the sun. An increase in temperature in the lower atmosphere--the greenhouse effect--could have very grave consequences for life on Earth. Intensive investigations of this functional relationship began about 100 years ago and it is now fairly well established.

Very few doubts remain that by burning fossil fuels, humanity is affecting the climate through the greenhouse effect. Electricity generation has become part of a larger picture that includes the impacts of fuels, transport, and distribution of this vital part of modern life and development. The electricity production chain is a complex network of natural and constructed phenomena. In Europe, the burning of coal arriving by ship from Australia implicates Europeans in the impacts of mining activity on the other side of the globe.

The objective of the CETP was to try to understand the total cycle of electricity production and use - a "cradle-to-grave" approach. It is vitally important to understand the sources and impacts on nature and human life of all the emissions generated during this cycle.

Only by fully accounting for every part of the process can one hope to calculate its true price in economic terms. In addition to purely economic analysis, environmental impact assessment (EIA) of the production-and-use cycle can suggest the effects of electricity on environmental sustainability—a relatively new scientific and political objective on the global agenda.

What is environmental sustainability? This term was coined to describe a situation in which the world's economy and standard of living continue to improve without leaving any lasting negative effects on the environment, at least not within a time scale of several generations. Simply stated, sustainability means to leave the environment to our children as we received it from our parents. Does sustainability exist? Can it be measured? Isn't environmental impact inevitable whenever electricity is produced?

There is certainly no simple global answer to these questions. Sustainability has different characteristics in different countries. Some aspects of sustainability are easier to manage than others. A more difficult factor may be human vulnerability to environmental disruptions.

The CETP dealt solely with the production of electricity and its environmental impact on nature and people. The goal was to identify technologies for the generation of electricity having the least impact on humans and nature in terms of both health and economic impacts. "What are the emissions associated with digging coal out of the ground and transporting it to a power plant?" is a straightforward question. However, the assessment of the impacts of these emissions on nature and human populations is a very complex inquiry, which requires an ambitious interdisciplinary approach to systemic knowledge.

THE AGS/ABB COLLABORATION

ABB demonstrated its commitment to sustainable development by signing the International Chamber of Commerce Business Charter for Sustainable Development, which developed from Agenda 21, a global action plan adopted by 178 governments at the United Nations 1992 Rio summit. The development and delivery of eco-efficient products and systems with low environmental impact is a cornerstone of the company's strategy and policy towards sustainable development. Other key elements include sharing state-of-the-art technologies with developing countries; contributing to international efforts; and the continuous improvement of ABB's own environmental management performance (Lindhal 1999).

ABB is also one of the initial industrial partners in the Alliance for Global Sustainability (AGS), contributing representation on the AGS International Advisory Board and financial support used by AGS to support its research initiatives. Founded with the support of Stefan Schmidheiny, a Swiss industrialist, AGS is an international partnership of four of the world's leading technical research universities to address complex issues that lie at the intersection of environmental and economic goals. Through the AGS, the Swiss Federal Institute of Technology (ETH), the University of Tokyo (UT), and the Massachusetts Institute of Technology (MIT), Chalmers University of Technology, focus their collective research and educational strength on the set of emerging problems that challenge global sustainability now and in the future. The expressed goal of the AGS is to "build a multidisciplinary international academic forum in partnership with colleagues from other public or private institutions and organizations to advance knowledge and understanding of intelligent alternative practices through research, education, and outreach."

ABB management recognized that the best way to learn to manage the total impact of power plants was to go through one such exercise themselves. Because electricity generation was at the center of ABB's activities in the 1990s, it made sense to apply the concepts of environmental impact to the electricity generation process. It was also evident that ABB should do this in cooperation with experts who had experience in this area. After formulating the first concepts of the Program, ABB leadership determined that the sustainability program of the Alliance for Global Sustainability would be a good home for this kind of inquiry.

Spanning three continents and involving 75 researchers, the CETP project required considerable funds and careful management and coordination. It was clear that such a program could best be served by a new relationship between ABB and AGS that gave greater organizational authority to the industrial partner. Academic partners were initially very skeptical of such a proposal, particularly because of possible danger to academic freedom. However, as consultations progressed, the initial skepticism faded away and the entire CETP research team became very positive about this set-up. Eventually all key participants were ready for this new experiment in industrial/academic collaboration.

THE CETP STORY

The China Energy Technology Program has a fairly long history as an idea. Although the Program focuses on China, it has grown out of efforts of developed countries to learn from the mistakes they made in establishing a dependence on electricity in building their economies. The aim of CETP is to help China avoid these mistakes by considering all options for electrification, especially those that minimize environmental degradation.

For example, it has become increasingly evident that it was not optimal to build isolated power plants in environmentally sensitive places. Analysts now try to see each power plant as one link in a long chain of activities. Environmental scientists emphasized the need to reduce emissions from power plants, especially of small particles, sulfur oxides, and nitrogen oxides.[2]

During the first 100 years of electricity generation, perceived environmental problems were local or regional in nature: One could see the smoking chimney or the dirty washing on the line. Something new happened in the 1980s and 1990s. The public became aware of global environmental problems affecting the world almost synchronously. The ozone hole or the depletion of stratospheric ozone through emissions of chlorofluorocarbons (CFCs) was one such problem.

However, the most serious of these problems was global warming, the warming of the lower part of the atmosphere due to anthropogenic emissions of GHGs, especially carbon dioxide. Forward-thinking designers of power plants realized that in the future questions would be asked and conditions imposed on the emissions of GHGs. Future generations might even demand emission-free power plants. The global warming discussion has led to renewed interest in renewable energies, energies with no net emissions of pollutants, and nuclear energy. Typical of the renewable energy technologies are hydro, solar energy, wind energy, and biomass combustion.

This discussion has also included ideas about how to make fossil fuel power plants more sustainable. Today around 80 percent of the world's total energy comes from the use of fossil fuels; hydro and nuclear contribute about eight percent and the remaining twelve percent comes from the use of non-commercial biomass in developing countries. At present, renewables (excluding hydro) account for only around one percent of the world's installed power of approximately 4000 GW.

China, the most populous country in the world with 20 percent of the world's inhabitants, is rapidly expanding its industry and infrastructure. With electricity generation growing faster than the GDP, China now has installed power in excess of 300 GW, and it is obvious that electricity consumption can only go up. In 2000 the growth rate of electric power installation was on the order of five percent annually. In spite of these impressive numbers, China produces about ten times less electricity per capita than Switzerland.

The main fuel in China is coal, cheaper and more readily available than either natural gas and or oil. Coal combustion is the main contributor to humanly produced emissions of carbon dioxide in the world. How can China provide the electricity it needs for future development, especially electricity produced by burning coal, with the least or, at best, no damage to nature and the inhabitants of the land?

This was the question posed in the China Energy Technology Program. The Program focused on Shandong Province, one of China's 22 provinces. With 12 times the population of Switzerland, Shandong produces about the same amount of electricity annually, most of it from coal. CETP researchers developed suggestions about how modern technologies might improve the electricity supply and make it more environmentally acceptable in the next 20 to 30 years.

Of the world's 6000 million inhabitants, one third, or around 2000 million people, have no access to electricity.[3] All of these people live in the developing countries. They are certainly going to want the same ample and secure supply of electricity that we in the industrial countries enjoy. Of the remaining 4000 million people on Earth, around one fourth live in the industrialized countries; three fourths living in developing countries and those with transitional economies. These 3000 million use five to ten times less electricity than the 1000 million in the industrialized countries, but they will inevitably be using more as they increase their standard of living.

In order to remain meaningful and closely focused on the important questions, CETP adopted a new concept of collaboration between industry and academia. First, the organizers located this program inside the ongoing collaboration and support that ABB gives to the Alliance of Global Sustainability. Through an agreement with AGS some of the money ABB provided as support was redirected to pay for this new three-year program. CETP officially began on May 1, 1999.

Second, the program was run and coordinated by ABB, which employed a separate and full-time Program Manager to run the CETP on a daily basis.

Third, in addition to the academic scientists who bore the brunt of the theoretical work, we engaged from the very beginning, the help of a group of people we called the stakeholders. The stakeholders are the customers of the program, the people who are going to use the results produced or the methodologies defined. They are people familiar with the subject of the program and who have helped keep the program relevant to real needs. The stakeholders have helped ensure that the final products of the program did not end up in a drawer, unseen and unacknowledged.

This threefold approach turned out to be very successful and productive. Through frequent and meetings among the scientists and between the scientists and the stakeholders, it has been possible to maintain an awareness of where the program was moving without losing sight of the initial goals of the program. Feedback from the stakeholders has made it possible to meet the Program's objectives while remaining relevant to their needs.

THE CETP IDEA

The CETP idea is never to work in isolation on a problem in which no one might be interested, or on a solution about which no one ever learns. The idea also involves having the three main actors--industry, academia, and the stakeholders—analyze and manage a problem together.

The three key actors approach problems differently, and herein lies the strength of this method. Industry brings real-world experience to the problem, resources, and connections to the consortium. Academics possess the knowledge and the theoretical and technical overview required to frame an analysis. The stakeholders are the prospective users of the solution.

The success of the process depends on how well it can be made understandable to the stakeholders. If they are disinterested in the solutions found and are not willing to apply them, the whole effort has been in vain. That is why the framers of the Program strongly emphasize interface with the prospective users.

CETP organizers have used two vehicles to communicate research results. One is the book you hold in your hand, the other is a digital video disc (DVD) included with this book. The DVD allows the user to interactively inspect the results and conclusions gained from the Program. The book is organized into chapters each of which corresponds to a specific task in the Program. The chapters contain a relatively detailed description of the approach used and the results gained. The DVD makes possible a more profound and interactive view of the Program.

Chapter 12 includes a description of this part of the work. The DVD enables the user to apply the results of the program to his or her own unique problems. The information is organized on three different levels. The highest level provides an overview and an application of the framework of the Program. The next two levels enable the user to more deeply probe of its results and parameters.

The DVD is a framework for decision making, allowing the user to quickly gain an overview of a complex situation. The user can vary the nature and, to some extent, the location of the technologies he or she might use. The program enables the user to calculate the environmental impact of a chosen solution and the economic consequences. Widespread use of the program to environmentally optimize an approach to electricity generation should, we hope, bring the world community closer to the goal of sustainability.

A LIVING PROGRAM

CETP represents a new approach to the collaboration between academia and industry. This process required a lot of coordination, travel, communication, and patience, but it was ultimately very rewarding in many respects.

One of the main elements of this new approach was a vision of a Living Program. The organizers want the Program to always be receptive to change, and in fact, it was changed and adjusted as it evolved. In the beginning the idea of sustainability and its application to electricity production was new to most of the participants. Furthermore, they were embarking on an endeavor with no script to follow.

The goal was clear: sustainability and an environmentally friendly energy supply. However, Program participants had to create the appropriate approach as they proceeded. This is where the stakeholders played a significant role, as did the frequent and obligatory group meetings among the researchers. As the Program developed a common spirit evolved among the participants, who became highly motivated by sense of a group effort.

THE STRUCTURE OF THE PROGRAM

After some initial complications, the collaboration progressed fairly smoothly as all parties quickly recognized that this effort was more than the sum of its parts. Partners from industry, academia, and the stakeholder group had different and complementary agendas, with few directly opposing elements. It was also important to meet on neutral ground: All participants were encouraged to leave the confines of laboratory or office and meet in the square in front of their buildings. This is indicated in Figure 1.1, which shows symbolically how all the participants meet in front of their home base.

To facilitate communication, group and stakeholder meetings were frequent and mandatory. Participants met at regular intervals of two to three months in different countries to compare notes and ensure that all facets of the Program were moving in the same direction. The meetings offered an opportunity for members of one investigative group to exchange ideas with those pursuing other lines of inquiry and with the stakeholders as well. In the course of the Program, a total of 12 Group Meetings and three special Stakeholder meetings took place. More than half of the Group Meetings and all of the Stakeholder Meetings were held in Beijing and Shandong Province in China.

The CETP Program is divided up into 12 tasks that interact and overlap (see Appendix D). A Principal Investigator runs each task. The Program itself is directed by a Steering Committee consisting of seven members, a secretary, and a financial officer. A Technical Advisory Board (TAB) composed of three members assists the Steering Committee. The TAB's duty is to provide technical and scientific advice and to guarantee the scientific and technical integrity of the Program. A Program Manager facilitates the connections among the various tasks and links the various elements of the Program together.

Figure 1.1 The CETP Forum where Industry and Academia meet with the Stakeholders

A Program Director is the Overall Coordinator of the Program. The Program Director oversees the running of the program in daily consultations with the Program Manager; sets the overall guidelines for the execution of the Program; and maintains links to the Steering Committee and other important institutions outside the Program itself.

Between Group Meetings and Stakeholder Meetings, the 75 program participants are all in contact by e-mail. All important announcements are posted on the Internet, where the program has its own page. Data is exchanged through a database embedded in the Internet page. These World Wide Web connections have turned out to be of extreme importance to this international Program. It is hard to imagine how it could be executed today without the Internet link.

AN IDEA FOR THE FUTURE

The global environmental problems we face today are so immense that no single company, nation, or region can deal with them effectively. Global cooperation must extend from one country to another and between institutes and universities. Researchers speaking many different tongues have to learn to work together and find common goals. This is more easily said than done.

We think the framework we have developed in the CETP can serve as an example for future international cooperation. The experience gained in this program has demonstrated the extreme importance of working together to get as many different viewpoints as possible to bear upon the problem. The grouping of all interested parties into three main groups--industry, academia, and stakeholders—has worked very well.

Program Management devoted to maintaining a common focus is an absolute necessity; without it every effort is doomed. This is an essential function for multidisciplinary research programs in which the final goals might be well defined but the routes to those goals are uncertain. Absent strong program management, researchers from different fields, and indeed, entirely different sectors, are likely to get bogged down in their own specific realm of expertise, and lose sight of the big picture.

A second imperative is to coax researchers out of their labs, out of their familiar surroundings, and strongly encourage—even require--them to meet with their collaborators. It is very important to learn to listen and get to know the viewpoints of other participants. It has always been a problem among the researchers and scientists of this world that the different disciplines speak different languages--English, Chinese, French, and so on. Even when everybody speaks a common language, people from different disciplines and sectors, e.g. industry and academia, use different vocabularies. Even within the academy, it is sometimes hard for the engineer to understand the economist and for the economist to understand the engineer. The best way for these experts to learn from each other is to work closely together towards a common goal.

This is the way investigators must work in the future on global problems such as energy supply and sustainability. Because science and technology are becoming more interdisciplinary, its best practitioners will be those who learn to listen and cooperate. Without interdisciplinary communication, researchers will be incapable of facing the immense tasks before us. This book is but one first step on that journey, but we hope, a step in the right direction.

REFERENCES

Lindahl, G. (1999). An industrialist's perspective on the greenhouse gas challenge. In B. Eliasson, P. Riemer and A. Wokaun, Eds. *Greenhouse Gas Control Technologies*. Oxford. Pergamon.

National Bureau of Statistics. (2001). *China Statistical Yearbook 2000*. Beijing. China Statistics Press.

Sinton, J.E. and Fridley, D.G. (2000). What goes up: recent trends in China's energy consumption. *Energy Policy* 28: 671-687.

NOTES

[1] After 1996, at the beginning of the Asia crisis, China's energy consumption started falling. In 1999 it was estimated to be around four percent lower than in 1996 (Sinton and Fridley 2000). According to official Chinese estimates it was even 12 percent lower in 1999 (National Bureau of Statistics 2001:239). There seem to be many reasons for this drop in energy use, including reduced use of coal and greater power plant efficiencies. The 2000 energy consumption is expected to rise again.

[2] The installation of equipment to reduce these emissions on a large scale had already begun in the 1970s and 1980s, although the first electrostatic precipitators to take out dust were actually introduced in the early 1900s in the Western world.

[3] In China the official number is 60 million people without access to electricity.

CHAPTER 2

GENERAL DESCRIPTION/APPROACH AND METHODOLOGY

BALDUR ELIASSON, YAM Y. LEE AND BINGZHANG XUE

This chapter describes the approach and methodology developed by the organizers of CETP to organize and run an interdisciplinary program, including participants from industry, academia, and the stakeholder community, to analyze the environmental impacts of the generation of electricity in Shandong Province, China. The objective of the Program was to compare different combinations of technologies and policy to identify scenarios feasible for policy makers in Shandong concerned with mitigating the impacts of power generation on the environment and citizens of the region. Ultimately, the idea is to apply such a methodology to any region on earth.

1. GENERAL DESCRIPTION

China's rapid economic growth and development has led to a growing demand for electrical energy. Developing a sustainable energy supply in China is an essential component of global strategies to promote development while mitigating environmental pollution and the potentially disastrous effects of greenhouse gas emissions. The identification, design, and implementation of "sustainable energy systems" must be key elements of any strategy for future economic and energy development.

As mentioned in Chapter 1, ABB's Energy and Global Change Department, in conjunction with the AGS and other world-renowned universities and institutions, has recently complete a three-year program entitled the China Energy Technology Program (CETP) (Eliasson and Lee 1999). The AGS brings together four world-renowned universities: The Federal Institute of Technology (ETH) in Lausanne and Zurich, Switzerland, Massachusetts Institute of Technology (MIT) in Cambridge, USA, the University of Tokyo (UT) in Tokyo, Japan, and Chalmers University of Technology in Gothenburg, Sweden. The goal of the AGS is to advance knowledge and understanding of the intersection of technological, economical and ecological issues and to solve the problems that challenge global sustainability, now and in the future.

The CETP team was comprised about 75 scientists, from four countries on three continents, coming from industry, universities, and other research institutions. Table 2.1 lists the CETP research partners. The Chinese partners in this program, which started on May 1, 1999, played an important role in its development and operation.

Table 2.1 The CETP Research Partners

- Energy and Global Change, ABB Corporate Research, Baden, Switzerland
- ABB China, Beijing, China
- ABB China, Jinan, China
- Energy Research Institute (ERI) of the State Development Planning Commission, Beijing, China
- Massachusetts Institute of Technology (MIT), Cambridge, USA
- Paul Scherrer Institute (PSI), Villigen, Switzerland
- Policy Research Centre for Environment & Economy (PRCEE) of the State Environmental Protection Administration (SEPA), Beijing, China
- Shandong Electric Power Research Institute (SEPRI), of Shandong Electric Power Group Corporation (SEPCO) in Jinan, China
- Swiss Federal Institute of Technology (ETHZ), Zurich, Switzerland
- Swiss Federal Institute of Technology (EPFL), Lausanne, Switzerland
- Tsinghua University (TU), Beijing, China
- University of Tokyo (UT), Tokyo, Japan

The objective of the program was to develop a globally applicable methodology for analyzing the "true," cradle-to-grave impact of electric power generation. Figure 2.1 illustrates the full electricity cycle covered in this study. The CETP took into account the energy technologies and their entire environmental impact, using a case study from China. In order to carry out the complex investigation within a reasonable time frame, one region of China–Shandong Province--was selected for detailed examination. This province was chosen because of the following factors: its independent grid, strong economic growth, diversified energy supply, and availability of data. However, the methodology should be sufficiently general to be applicable to other provinces in China and to other countries.

China, the most populous and third largest country in the world, has seen its economy grow rapidly at an average annual rate of 9.6 percent between 1979 and 2000. During this time Chinese GDP has more than quadrupled.

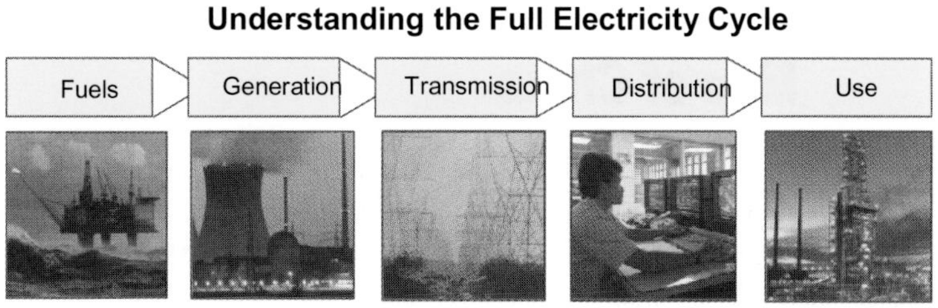

Understanding the Full Electricity Cycle

Fuels Generation Transmission Distribution Use

Figure 2.1 Development of a globally applicable methodology for analyzing the "true" impact of electric power generation requires research into all phases of fuel sourcing, production, and use.

Access to energy is a necessary prerequisite to the economic and social development in all societies. China--the world's second largest energy consumer after the United States—used 1.24 billion tons of coal equivalents (Btce) in 2000. In that year, following 13 consecutive years in which there was a net increase in installed power generation capacity of more than ten GW per year, China had an installed capacity of 319 GW, the second largest in the world after the United States (Electric Power Industry 2001). China is the largest coal producer and consumer in the world, with total production at 1.25 billion tons in 1998, about one fourth of the world's total. In 2000, coal output was reduced to about 900 million tons in an effort to balance supply and demand. However, coal still accounts for about 70 percent of China's primary energy supply. Extensive use of coal is the cause of serious regional pollution, largely in the form of acid rain, and increased global emissions of greenhouse gases.

2. SHANDONG PROVINCE

Shandong, situated on China's eastern coastline, is the country's second most populous province with a population of 88.8 million or 7.1 percent of China's total. By the end of 1999 it had the third highest GDP of all provinces: 766.2 billion Yuan or 9.4 percent of China's total. Figure 2.2 shows a map of China with some statistics for Shandong in 1999. In 1999 Shandong produced 90 Mt of raw coal, 27 Mt of crude oil and 7'333 Mm^3 of natural gas. Raw coal and crude oil are the two main primary energy sources. Figure 2.3 shows the total primary energy supply in Shandong Province from 1949 to 1999. Figure 2.4 shows the trend of total installed power generation capacity and electricity production in Shandong from 1986 to 2000. It illustrates that total installed power generation capacity and electricity production in Shandong have grown steadily since 1986 and reached 20 GW and 110 TWh by the end of 2000. Shandong Electric Power Corporation (SEPCO) owns a majority of the total installed power capacity in Shandong that is connected to the provincial grid.

Figure 2.2 Map of China including statistics on Shandong in 1999

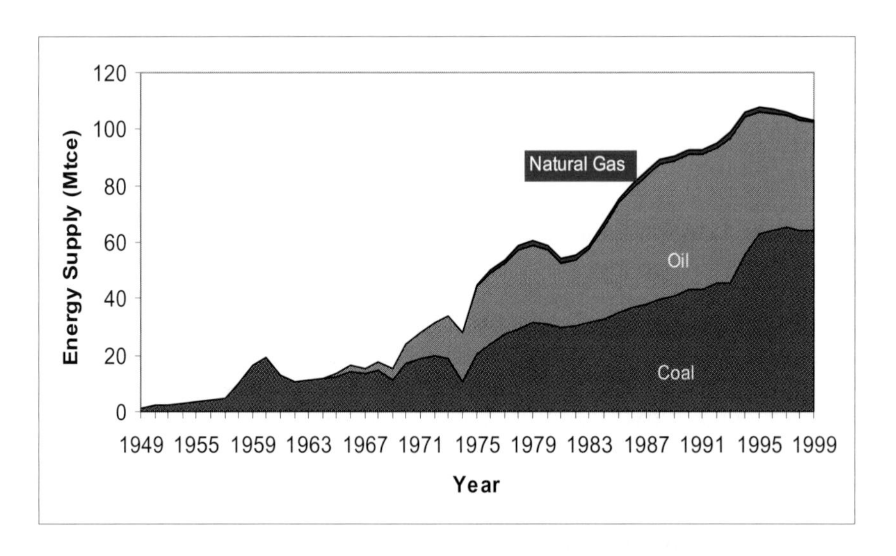

Figure 2.3 Total primary energy supply in Shandong Province from 1949 to 1999

Shandong has limited hydro resources. Total electricity production from this source in Shandong was only six GWh in 1999. The use of renewable energy has attracted more and more attention in recent years because of its potential for low environmental impact and potential for the development of sustainable energy technologies. The average annual solar radiation in Shandong is 4'900-5'400 MJ/m2

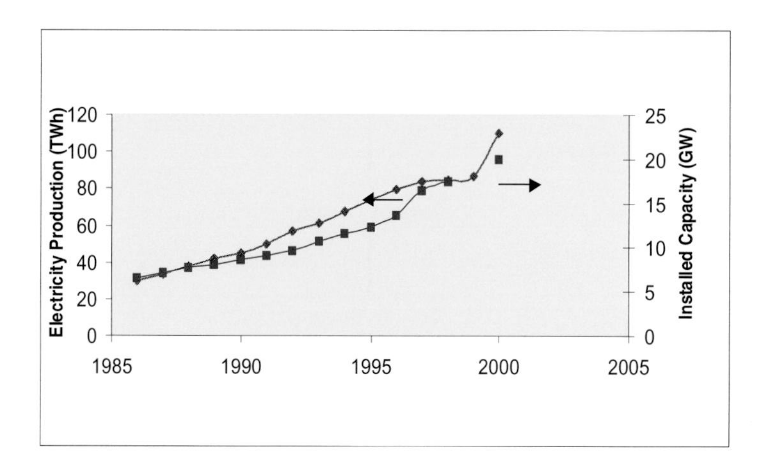

Figure 2.4 Total installed power generation capacity and electricity production in Shandong from 1986 to 2000

(Xue and Eliasson 1999). The sun shines 52 to 65 percent of the year, or between 2'301 and 2'857 hours annually. Solar energy has been used in the form of solar heaters and cookers and solar photovoltaic technology.

China is rich in wind energy that has been used mainly for power generation. The installed capacity of wind power in China reached 375.2 MW at the end of 2000. Presently the main windfarms are in Xinjiang Province, Inner Mongolia Province and Guandong Nan'ao Island. However, the practical wind energy potential in Shandong is limited. Presently, there is only one windfarm on the island of Changdao off the northern coast of the peninsula, with a total capacity of 9 units of 600 kW each.. Shandong has a coastline of 3'121 km with a maximum tidal difference between two to five meters. Fourteen exploitable sites for tidal power stations in Shandong have been identified; they have a theoretical capacity of 460 MW and electricity production of 1.37 TWh per year. Currently recoverable tidal capacity is around 120MW and electricity production is 0.36 TWh per year. In 1978 the third largest tidal power station in China with a capacity of 0.96 MW was commissioned in Rushan, Shandong. Of the 13 thermal springs in Shandong, six have a temperature over 60°C. Annual heat recoverable from geothermal resources is estimated to be 674 TJ, which is equal to 23 ktce.

3. OPERATION OF CETP

CETP was run by a Steering Committee that consisted of seven members, a secretary, and a financial officer. The seven members of the steering committee include three from ABB, three from the AGS, and one from the World Business Council for Sustainable Development. Figure 2.5 shows the CETP organizational structure. A three-member Technical Advisory Board (TAB) assisted the Steering

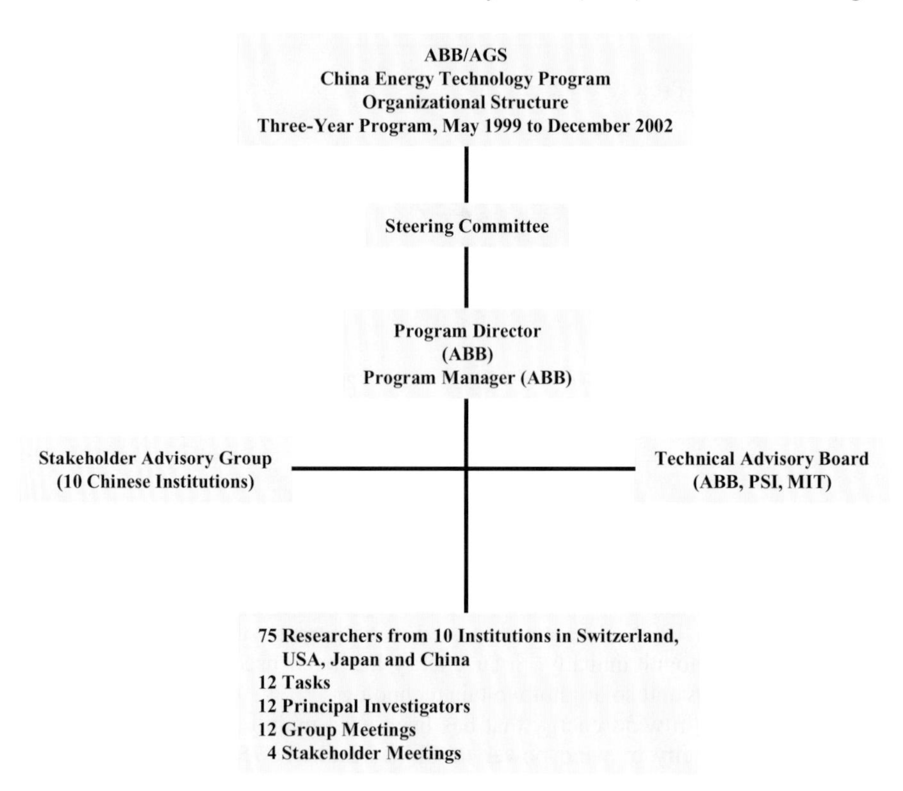

Figure 2.5 CETP organizational structure

Committee. The TAB's duty was to provide technical and scientific advice and make sure the scientific and technical integrity of the Program is guaranteed. A Program Director oversaw the running of the program in daily consultations with a Program Manager; set the overall guidelines for the execution of the program; and maintained links to the Steering Committee and other important institutions outside the program itself. The Program Manager, who also maintained the connections among the various tasks, ran the program on a daily basis. The Program Manager provided the link that tied the various elements of the Program together. The CETP Program was divided up into 12 Tasks. Table 2.2 shows all the tasks and the organizations that were responsible for each. Each program task had a principal investigator, who coordinated and oversaw the task-related activities.

Table 2.2 CETP Tasks and the responsible organizations

Task No.	**Task**	**Responsible Organization(s)**
1	Overall Project Management	ABB
2	Data Collection	ABB
3	Database Development	ABB
4	Demand Forecasting	ERI/EPFL
5	Energy/Economy Modeling	PSI/TU
6	Electric Sector Simulation	MIT/ETHZ
7	Energy Transportation Modeling	U. of Tokyo
8	Life Cycle Assessment	PSI/SEPA
9	Environmental Impact Assessment and External Costs	PSI/SEPA
10	Risk Assessment	PSI/SEPA
11	Integration, Decision Support & DVD	ABB/EPFL/PSI
12	Outreach and Technical Exchange	ABB

ABB together with AGS wanted to break new ground in the area of effective cooperation between industry and academia and between institutions in Europe, USA, and China. The Program rests on three pillars: academia, industry and the stakeholders--the prospective users of the results and methodologies generated by the program. The three-fold nature of the Program has guaranteed its relevance.

The stakeholders are a group of people familiar with the subject of the Program and are also part of the future users of its results and products. Through outreach visits and symposiums organized by CETP project management, the stakeholders gained knowledge of the Program and became active in participating as members of an advisory group facilitating scenario formulation and multi-criteria decision analysis. One of this group's principal tasks was to ensure that the Program focused on the relevant aspects of the problem under study and produced methodologies that meet the needs of Program users. The Chinese stakeholder advisory group includes one representative from each of ten important environmental, academic or governmental institution, or industrial groups, including:

1. The Administrative Center for China's Agenda 21(ACCA21), Beijing
2. Chinese Academy of Sciences, Beijing
3. Development Research Center of State Council, Beijing
4. Environmental Protection Bureau of Shandong Province, Jinan
5. Ministry of Science and Technology, Beijing
6. Shandong Economy and Trade Commission, Jinan
7. Shandong Electric Power Group Corporation, Jinan
8. State Development Planning Commission, Beijing
9. State Environmental Protection Administration, Beijing
10. State Power Corporation of China, Beijing

The members were high-ranking officials in these institutions who had strong backgrounds in science, economics, or industry. The group provided an overview of conditions in China and expertise specific to Shandong. These stakeholder representatives interacted with and conveyed the viewpoints of customers or decision-makers to the research group on issues of government policy, economic development, energy demand, and environmental concerns, and other related issues. They also helped determine criteria selection and weighting factors for the Multi-Criteria Decision Analysis process.

Contact with stakeholders began early in 1999. The first Stakeholders Symposium was held in Jinan in October of that year. In the course of the Program there were five group meetings, symposia, or fora in addition to frequent contact via e-mails, surface mail, or telephone.

The stakeholders provided input to the whole spectrum of the Program, including:

- general comments on CETP,
- Shandong's general needs,
- energy technology choices for Shandong,
- economic considerations,
- environmental issues,
- guidance on scenario formulation (technology and policy) selection,
- Multi Criteria Decision Analysis (MCDA) criteria,
- Weighting factors for economic, social, and environmental issues MCDA and weighting process,
- demand forecasting,
- feedback on the analysis results,
- feedback on the final products of CETP,
- suggestions for application of the CETP results, and
- technology transfer.

The stakeholders provided many valuable insights and observations that helped keep the Program focused and relevant.

4. COMPONENTS OF THE PROGRAM

The 12 Tasks of the CETP, as presented in Table 2.2, above were related as shown in the flowchart in Figure 2.6.

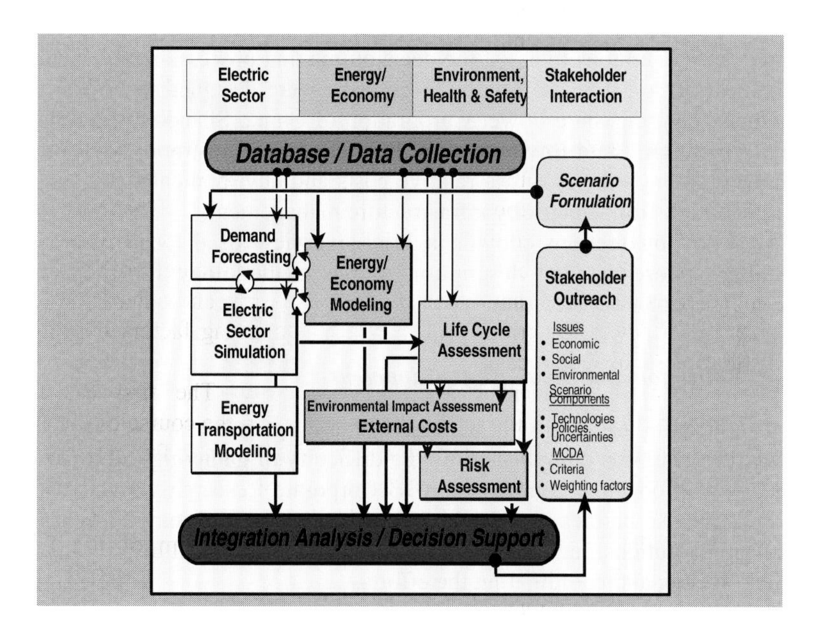

Figure 2.6 The CETP flow chart. A diagram showing the relationship between the various tasks of the program

The four distinctive features of the Program's approach were (ABB/AGS CETP Proposal 1998):

- An integrated methodology involving databases, knowledge bases, and simulation methodologies, covering electric sector supply and demand options, under sustainable and economic driving forces, with technology assessment;
- Development and use of specialized databases and models based on life cycle assessment, health and safety/risk assessment, economic and social

factors, environmental impacts, legal and institutional considerations, technology assessment and operational data;
- Improved use of decision-support techniques, such as Multi-Criteria Decision Analysis; and
- Enlistment of stakeholders in active participation in the decision-making process and in the development of the attributes and energy/electricity mix scenarios.

The objective of the research program was to identify robust portfolio of innovative energy options to assist the decision-making process among stakeholders over the next 20 to 30 years. The research tasks were designed to inform Chinese stakeholders (e.g., the State Power Corporation of China, Shandong Electric Power Group Corporation, State Environmental Protection Administration, and provincial government, among others) of the relative costs and environmental performance of alternative options for "electricity infrastructure management."

The following sections provide a brief overview of the task approach and methodology. More detailed description of the methodology for each task will appear in the chapter devoted to it.

4.1 Data Collection and Database Development

The data collection task was vital because of the Program's emphasis on modeling. However, this task was complicated by the difficulty of gathering information from different sectors on three continents. ABB Corporate Research worked with ABB China and Chinese partner institutions to gather data for the study. Twenty people from seven institutions, industrial and academic, were involved heavily in this task with ABB leading and coordinating the efforts.

These researchers played an important role in coordinating data management for all tasks and interaction between different research groups. A database of accurate technological and economic data is essential to analytic modeling. For this purpose, an advisory group consisting of Chinese experts was specially formed to oversee the reliability of the data. This group ensured data quality through careful review and checking of the consistency of the data. Its task was to evaluate the material and determine which was accurate enough to be used in the program and which was questionable and should be discarded.

In addition to managing the data collection with the help of the Chinese partners, ABB also collected information from the available literature. The brochure entitled *Shandong Energy and Emissions 1999* (Xue and Eliasson 1999) is a product of this effort. It presents a systematic overview of the historic development and the background and boundary conditions of the future prospects for power production in Shandong Province. This publication was a valuable information source for the CETP participants and a broader non-specialized audience. A "China Library," including a large collection of books and documents is located at the ABB Energy and Global Change Department in Switzerland.

Once collected, the data had to be disseminated widely and efficiently among project participants in order to maximize its usefulness. Researchers created an

Internet-based document database of material from multiple sources and made it available to all project tasks. It contains over 400 documents representing the work of dozens of researchers. Along with a database, CETP also developed a website that presents Program research to any interested party at http://www.cetp.ch. It has been operational since May 1999.

4.2 Demand Forecasting

To determine a strategy for energy planning, one must be able to forecast energy and electricity demand. The results from this task served as inputs to the energy and electricity modeling tasks. The principal investigator responsible for this task is the Energy Research Institute (ERI) of the State Development Planning Commission of China. The study uses the MEDEE-S model, designed and implemented in 1983 by the Institute of Energy Policy and Economics (IEPE) at the University of Grenoble, France. This is a long-term (15-20 year), bottom-up final energy demand simulation model for developing countries, and is based on activity level, energy intensity, and energy efficiency. By the end of 1996, 27 countries from four continents had applied the MEDEE-S model.

Since 1987, ERI had used the MEDEE-S model several times to forecast China's energy demand at both the national and provincial levels. Based on the various technological processes and technological levels by sector, the model forecasts final energy demand by end-use as defined by combining techno-economic parameters, social and economical development levels, and energy use equipment in various sectors. The model includes six sectors: industry, construction, transportation, service, agriculture, and household.

Working closely with ERI, EPFL performed electricity demand and load forecasting using a bottom-up approach and the DEMELEC-PRO software. DEMELEC-PRO was developed by the Laboratory of Energy Systems (LASEN) of EPFL in cooperation with power utilities to assess long-term electricity demand using a techno-economic and knowledge-based approach.

The program includes two kinds of variables including scenario and strategic factors. The scenario variables affect the electricity environment and are not dependent on the actors. The strategic variables depend on the electricity consumers and suppliers. The knowledge-based approach (MAGES) makes it possible to control the consistency of the scenarios and strategies through experts' rules. Once consistent strategies and scenarios are established, transparent and well-founded assumptions for the evolution of electricity demand may be derived.

4.3 Energy and Electricity Modeling Tasks

The three modeling activities relating to energy and electricity modeling include energy economy modeling, electric sector simulation, and energy transportation modeling. The approach is to use both optimization and simulation techniques to investigate the energy mix options based on different economic and environmental constraints.

4.3.1 Energy economy modeling

The Energy Economics group of the Paul Scherrer Institute (PSI) in Villigen, Switzerland, in cooperation with the Global Climate Change Institute of Tsinghua University, was responsible for the energy/economy modeling work in the CETP. The primary tool used in this study was the Market Allocation Model (MARKAL) developed at the Brookhaven National Laboratory in the United States. MARKAL is a process-oriented engineering model that describes all energy transformations from primary sources to energy services. It uses as variables the resources of the energy system, the installed capacity of and investments in new technologies, and the flow of energy through the different technologies. Exogenous model parameters are the fuel prices, the availability of resources and the potential for economic growth. Both the supply and demand sides are integrated so supply always adjusts to meet demand.

The optimization capability of the MARKAL model utilized in this task permits the analysis of different objectives or combination of objectives including but not limited to lowest costs and emission targets. PSI has developed a technology database for China, generating scenarios under different energy and environmental policies, and calibrating energy supply model for the reference case. Scenarios are both descriptive (assuming no significant changes in policy) and normative, describing optimal (least cost) allocations of resources under selected environmental and policy constraints.

In addition to using MARKAL to study the Shandong Province, PSI has also developed a multi-regional optimization model called the China Regional Electricity Trade Model (CRETM). The model considers 17 electricity generation technologies utilizing 14 fuel forms in a seven-region model of China's transportation electricity demand and supply system over the period 2000-2030. Shandong is one of the seven regions. Researchers used CRETM to examine a set of scenarios driven by different energy, economic, or environmental factors to quantify related policy implications.

4.3.2 Electric sector simulation

The Electric Sector Simulation task performed by MIT and ETH study complements the Energy Economy Modeling effort described above. This bottom-up, engineering-based modeling of the Chinese electricity sector was performed by the AGREA (Analysis Group for Regional Electricity Alternatives) research group at the MIT Energy Laboratory. Comparative analysis of alternative strategies under various uncertainties is necessary to discover robust, long-term technological strategies and related policies. It is also essential to identify relevant and appropriate performance attributes, and include them in the analytic framework.

During the CETP, MIT investigators sought stakeholders' ideas about which attributes of technologies were attractive and relevant in comparing scenarios to meet short- and long-term performance goals. In evaluating various strategies, a wide range of technologies were being studied. The value and credibility of results from any given technology scenario depends the researchers' ability to reasonably model physical system performance for the sets of technologies and circumstances involved. MIT personnel used a simulation dispatch model (EGEAS) to simulate both technological and operational alternatives.

ETHZ assumed responsibility for creating the electric sector model databases and performing the actual scenario set analysis with MIT. ETHZ also worked closely with PSI on the integration of electric sector simulations into other analytic initiatives such as life cycle assessment.

4.3.3 Energy transportation modeling

The Energy Transportation Model developed at the University of Tokyo was used to simulate the electricity sector for Shandong Province. This study complements the Electric Sector Simulation and the Energy Economy Modeling work. It is also an optimization model but also takes into account the transportation of fossil fuels to the power plants and the electricity transmission networks. It also considered site-specific power generation and demand.

The model represents a fairly simplified trunk power transmission network, which consists of 17 nodes and 25 branches. The nodes correspond to the cities and prefectures in Shandong. The branches are structured with 500kV and 220kV transmission lines. Researchers used this optimization model to study the least-cost expansion planning of the power system in the province. They investigated the configuration of power plants, the electricity generation mix, power plant sites, expansion of transmission lines, coal flows, and SO_2 and CO_2 emission control measures.

4.4 Environmental, Health, and Safety-related Tasks

PSI was responsible for three tasks related to the environment, health, and safety. These include Life Cycle Assessment, Environmental Impact Assessment and External Costs, and Risk Assessment. In carrying out these tasks, PSI worked closely with the Policy Research Center for Environment and Economy of the State Environmental Protection Administration of China.

4.4.1 Life cycle assessment

The Life Cycle Assessment (LCA) methodology was used to develop suitably complete environmental inventories for major present and future energy systems for the generation of electricity in Shandong. All energy systems are described on a cradle-to-grave basis. LCA supplies input to decision-making for energy choices by providing an evaluation of the amounts of pollutants emitted into the environment from the entire chain of each energy system, including fuel mining, processing, power generation, transmission lines, and waste management.

Researchers collected and analyzed as complete as possible sets of technological data for energy and non-energy resources and for main environmental burdens for the current coal chain to establish the reference case. They then collected and analyzed data for future electricity supply systems options (coal, natural gas, nuclear, and wind) included in the scenarios modeled in other CETP tasks. A tool was developed for the calculation of total environmental burdens, including indirect burdens from material and energy requirements throughout the chain.

In performing these tasks, PSI worked closely with the Policy Research Center for Environment and Economy of the State Environmental Protection Administration of China. The results of the LCA also provided inputs to the scenario analysis central to the electric sector simulation task, as well as to the EIA and MCDA tasks.

4.4.2 Environmental impact assessment

Since China's dominant energy source is coal, health effects due to air pollution and acidification stand out as the primary environmental issues to be addressed. Two modeling approaches, i.e. "impact pathway approach" and the Regional Air Pollution Information and Simulation (RAINS) were used in this task:

The "impact pathway approach" was used to estimate health effects and impacts on crops, associated with the various energy chains. These impacts were monetized to obtain external costs of electricity generation by means of current and future technologies. The estimated external costs together with the internal ones give "true" cost of electricity supply. The basic tool used in this analysis is the EcoSense model developed by the University of Stuttgart. It has been applied in many case studies and research projects in Western Europe. The EcoSense multi-source version was extensively adapted to the Chinese conditions for the purpose of this study. In addition, a simplified model AirPacts was calibrated based on the detailed EcoSense analysis.

The RAINS-Asia model was developed by the International Institute for Applied Systems Analysis (IIASA) in Laxenburg, Austria as an analytical tool to help decision-makers analyze future trends in emissions; estimate regional impacts of resulting deposition levels; and evaluate costs and effectiveness of alternative mitigation options. The 2001 version of the RAINS-Asia program was used in this study.

4.4.3 Risk assessment

Electricity systems produce a range of interrelated risks, including possible technology performance, health, environmental, and financial effects which are important when choosing an electricity strategy for a province or country. For the purposes of the CETP, accidental risks were separated from the impacts of normal system operation, such as morbidity or mortality due to pollutant emissions. The objective of this activity was to provide a balanced perspective on the severe accident risks specific to China.

The assessment addressed fossil energy sources such as coal, oil and gas, nuclear power, and hydropower. In addition to the power production (conversion) step in these energy chains, exploration, extraction, transport, processing, storage, and waste disposal were considered whenever necessary. The work focused on the historical experience of accidents and their applicability to China. In this context the most comprehensive data base on severe accidents (ENSAD), recently established by PSI, was further enhanced. The database was extensively used for the evaluation and additional data for China were collected. Apart from the historical data, it is also necessary to employ a probabilistic perspective when analyzing nuclear energy; a simplified Probabilistic Safety Assessment (PSA) was implemented in this task.

4.5 Integration and Decision Support

ABB Corporate Research is responsible for integrating the results from different tasks and for developing a coherent product integrating the contributions of all participating institutions. Integration was a continuous process involving the synchronization of the schedules and progress of the different, interrelated tasks. The integration activities were conducted on various levels and will be detailed in Chapter 12.

The energy economy modeling and the electric sector simulation, including the energy transportation modeling work, generated a large number of options under various scenarios. Even when a conscientious effort is made to use similar inputs and assumptions, the results of these three approaches were each somewhat different, due to the inherent differences in methodology. The big challenge for the CETP was to synthesize a set of coherent conclusions from these three different approaches. Chapter 12 also includes a comparison of the results from these three tasks.

Since the CETP is intended to help in the decision-making process, it is structured in a way that allows scientists to present various factors influencing the decision-making process and their interrelationship. In CETP, the tasks generated a wide spectrum of results of interest to various stakeholders. They could select and utilize various parts or the whole of the CETP results to match their interests. The large portfolio of results gives the decision-makers quantitative understanding of the various potential relationships of interdependent factors.

The interrelationship of the energy-modeling tasks with the LCA and EIA tasks is important in this project. Out of thousands of scenario runs performed in the electric sector simulation task, researchers could choose a set of interests, including weighting factors based on the preferences of stakeholders, to use in a Multi-Criteria Decision Aiding analysis.

4.5.1 Multi-Criteria Decision Aiding (MCDA)

One of the goals of this program was to develop a user-friendly decision support tool to help stakeholders evaluate the options created as various criteria change. Social preferences must be considered as well as technical, ecological, and economic factors. A set of 30 scenarios was selected based on multi-attribute analysis of the electric sector simulations. PSI has provided extensive inputs in terms of aggregated results on LCA-based burdens, environmental impacts, and accident risks. Investigators use a Multi-Criteria Decision Analysis (MCDA) approach to help stakeholders combine the wide range of electric sector, LCA, and risk indicators in support of strategic decisions concerning the sustainable future of the electricity sector in China.

A major advantage of the MCDA method is that it accommodates both quantitative and qualitative criteria in a coherent manner, while allowing for the inherent fuzziness associated with their evaluation. EPFL chose to use the MCDA tool ELECTRE III for its CETP research since the objective of the Program is not to choose a "best" solution but to compare different options from a sustainability standpoint. The results of the MCDA analysis are designed to help stakeholders understand their own strategy preferences, not to enforce consensus.

4.5.2 DVD Presentation

An integrated software tool on DVD was also developed so that stakeholders and users might fully utilize the results obtained from this study. It is included with this book. The DVD contains all the results from the different tasks. PSI was responsible for preparing the DVD, which is based on the contributions from all Principal Investigators and their collaborators. The DVD should not only enhance the communication and comprehension of what has been achieved but also will enable a variety of users to investigate the possibilities opened up by the results of the Program

Those who might use the DVD include Chinese high- and mid-level decision-makers representing electric utilities, plant suppliers and regulators, engineers, environmental experts, and academic researchers. The modular design of the software reflects the anticipated wide spectrum of users' different backgrounds and interests.

The "entrance door" to the DVD is a multi-media introduction and overview of the Program. The rest of the material on the disc ranges from a detailed presentation of all tasks, conclusions and recommendations, to the interactive use of the multi-criteria approach, allowing users to combine the technical knowledge generated by CETP with their own value judgments. The tool also includes some simulation and data-mining capabilities.

To the extent possible, the software is platform-independent. User-friendliness and high quality visualization is assured by use of homogenous graphical user interface (GUI). These requirements can be achieved by the combination of Macromedia's program "Flash" with an open-source web browser such as Netscape Navigator or Internet Explorer.

4.6 Outreach and Technical Exchange

Stakeholder involvement in the Program was essential to ensure that the results are applicable and will be used. Outreach activities were important in soliciting inputs from stakeholders as well as to inform the scientific community and various stakeholders about the progress of the Program. CETP organizers have been quite successful in attracting a lot of public interest and active stakeholder involvement through direct visits, web site visits, presentations, publications, and other public media, including television and the press. The activities and results of CETP were presented at three AGS Annual Meetings: in Boston in 2000, in Lausanne in 2001, and in San Jose, Costa Rica in 2002. Ten CETP publications and three issues of *CETP Newsletter* have also been distributed to a wide audience.

In addition to many visits and meetings to various stakeholders by the program management, a CETP Stakeholder Symposium was held in October, 1999 at Shandong Electric Power Research Institute (SEPRI) in Jinan, in conjunction with our third CETP Group meeting. Of the approximately 50 participants, 40 percent were from foreign countries. The one-day symposium successfully introduced the scope of the CETP to the stakeholders and familiarized them with what they could expect from the Program.

In January 2000 the CETP Forum was held at MIT in Cambridge during the AGS 2000 annual meeting. In March 2000 a stakeholder advisory group consisting of 10 experts from 10 different government organizations and institutions was officially formed in Beijing to work closely with research groups on scenario formulation and multi-criteria decision analysis as well as to provide feedback on the analysis results of this study. In June 2000 and March 2001 two other stakeholder advisory group meetings were held, one each in Qingdao and Beijing. Continuous interaction with the stakeholders helped keep the Program on track and focused on the issues in which stakeholders are interested.

As CETP is a collaborative undertaking, technical exchange among participants is an important part of its activities. More than ten people from the cooperating institutions in China and the other participating countries have exchanged visits of several weeks duration among technical personnel, strengthening the cooperative spirit, enhancing communication, and facilitating technology transfer. Regular CETP group meetings place were held, stimulating interaction among the CETP participants and monitoring the progress of the projects.

5. CONCLUSION

The ABB/AGS China Energy Technology Program (CETP) was a challenging and rewarding interdisciplinary effort to accommodate many important goals within one program. Its objectives were to understand and analyze the economic and environmental impact of energy systems, in this case the production of electricity. The Program took as an ideal a sustainable system of power generation, i.e. one that leaves natural systems as unchanged as possible. Program participants sought results that actual users found feasible, and that could build on infrastructure already in place.

To this end, a wide range of stakeholders from China were invited to participate and comment extensively. In fact, the Program's greatest contribution may be as an experiment in combining the insights and research approaches of investigators from academia, industry, and "stakeholders"—those charged with building and operating energy systems to meet present and future demands. CETP's new contribution to interactive study was to ground academic and industrial cooperation in the expressed needs of stakeholders in the target region, not just at the outset, but throughout the three years of Program research.

CETP research was designed to advance the science of integrated impact assessment of electric power systems and apply it to a region of China, a rapidly developing country and the most populous nation in the world. The methodology developed by CETP is sufficiently general in design to apply to other provinces in China or to other countries. Participants hope that the results of this Program will contribute to the development of a sustainable energy supply in China.

REFERENCES

ABB Website (2002): www.abb.com

CETP Participants (1998) ABB/AGS CETP Proposal

Electric Power Industry in China (2001)

Eliasson, B. and Lee, Y. (1999) China Energy Technology Program, IEA Greenhouse Issues, No 45.

Eliasson, B. and Xue, B. (1997) China Energy and Emissions, ABB Environmental Affairs Brochure, Stockholm, Sweden.

Xue, B. and Eliasson, B. (1999) Shandong Energy and Emissions, ABB Environmental Affairs Brochure, Stockholm, Sweden.

CHAPTER 3

DATA COLLECTION AND DATABASE DEVELOPMENT

BINGZHANG XUE, CHRISTOPHER RUSSO, JIANG MA, QINGBO ZHAO AND YONG XU

Because of the CETP's extensive data requirements, both in quantity and quality, and the large amount of computer modeling in the program, the Data Collection task played a key role in its success. Twenty people from seven institutions, both industrial and academic, were specifically devoted to this task with ABB Corporate Research leading and coordinating the efforts.

The primary goals of the task were to coordinate data collection for all tasks and to function as the communication conduit between tasks and institutions; essentially, the Data Collection task acted as an information clearinghouse for the entire Program. Some of the earliest issues revolved around cultural and language barriers to data exchange. While the data are comprised primarily of numbers, the nuances and subtle points that needed to be communicated presented certain difficulties. Furthermore, a basic understanding of the data requirements and methodology of the different tasks was an essential part of the fundamental task of data collection. The efficacy of the researchers was significantly improved when they knew what they were looking for.

Trust was also an important issue to address; open dialogue and discussion of the tasks and problems confronting us facilitated the opening of doors. In addition to the key function of coordinating data collection for all tasks, the data collection personnel also provided management support and assistance in outreach. This chapter will summarize the efforts and methods used to gather and share data among Program participants, the difficulties and issues involved, and the methods we used to build our Program-wide database and web site.

1. THE DATA COLLECTION TASK

Because of the extensive information requirements of Program participants, reliable raw data were the underpinning of the CETP research efforts. The quality and ease of access to data directly affected not only the speed with which participants were able to perform the research tasks, but also the quality of the final results. The breadth of the Program required us to forge close relationships with Chinese participants in order to find and deliver the required information. We worked closely with Chinese scientists in the areas of energy and environmental science as well as with provincial, national, and governmental stakeholders.

1.1 Work Plan and Guidelines

When the CETP began, the first task of data collection was to identify the data requirements of different tasks. Meetings were held with the principal investigators of different tasks to discuss and refine their data requirements. Due to the large number of data required for some tasks (e.g. Life Cycle Assessment) priorities had to be set to limit the amount of information to be collected.

In order to coordinate the data collection processes, the following guidelines were presented to the research participants at the inception of the Program. We asked them to

- collect the data while still abiding by all legal regulations;
- present data requirements to data collectors and sources as simply and as clearly as possible;
- consult the data collection task information center first before beginning to look for data;
- share data and exchange information between all institutions to avoid duplication of effort; and
- report data collection problems as soon as possible.

1.2 Data Management Strategies

In order to manage the large flow of data and to help organize the attendant database (described later in this section), we developed certain strategies and tenets for Program participants. The management schemes and strategies were presented and discussed before and during the second CETP Group Meeting in June, 1999 in Beijing. The strategies included:

- All CETP participants should contribute to the CETP common database and all tasks should then access information from the same database.
- To facilitate communication between data consumers and producers, data consumers should develop a spreadsheet "template" that can be used by Chinese data providers.
- Responsibility should be divided between researchers in charge of economics, environmental, and electrical unit system data.

Figures 3.1, 3.2 and 3.3 show these three data collection management strategies, data requirements in a unified format, and data collection responsibility assignments, respectively.

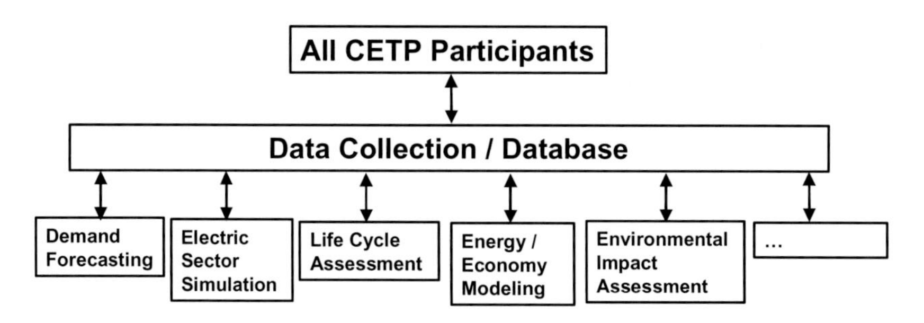

Figure 3.1 Data collection management strategy

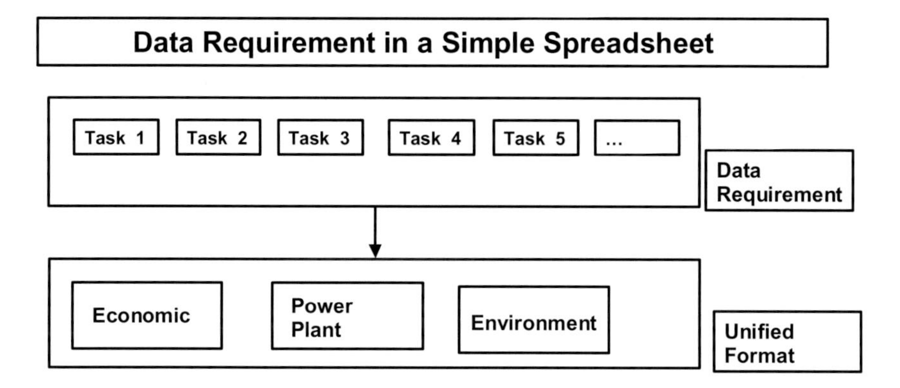

Figure 3.2 Data requirement in a unified format

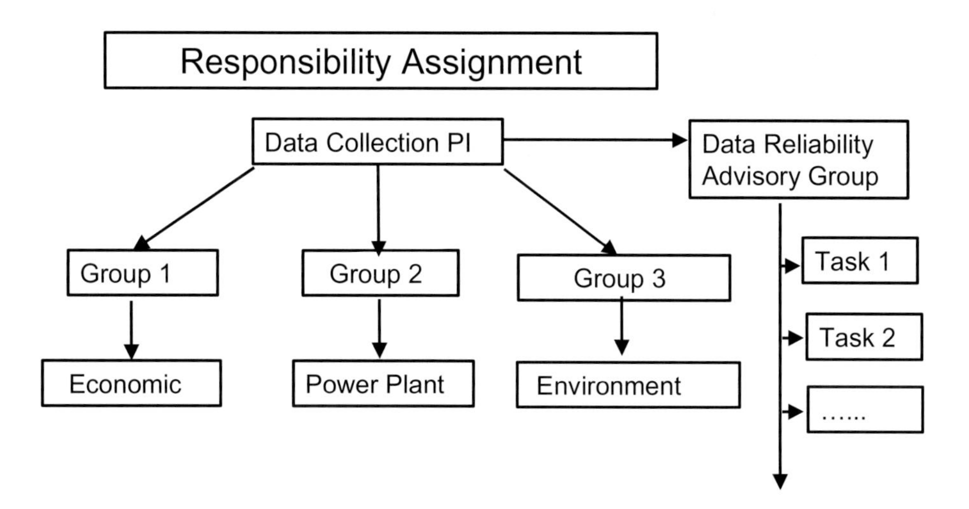

Figure 3.3 Data collection responsibility assignment

The data we needed could only have been found through an organized system of information and literature searches. Unlike information in some western countries, much of the data we needed were not available in published media or did not exist at all. Some data were recorded but not available to the public due to the state and company's confidentiality concerns. This increased the difficulty of data collection tremendously.

Of utmost importance to the research tasks in the Program was the quality of the source data provided. When the data were actually used and shared among different tasks, the consistency of different data and its corroboration by other published sources became more and more important. In fact, the principal challenge faced during the first year of the Program was verifying the source data supplied by Chinese colleagues.

To address this contingency, the Data Reliability Advisory Group was formed in May of 2000. Principal Investigators from the various tasks formed a committee that worked to identify and rectify, if possible, inconsistencies in the data. The Group's task was not necessarily to check every datum available, but rather to provide expert advice and field questions on what researchers thought might be specious information. Members of the group were selected on the basis of their unique knowledge of the industries represented in the source data and their ability to provide informed judgment on the quality of the information.

To better organize data, we prepared data sheets, or templates, according to the data requirements of the tasks, in order to simplify the collection of data for our Chinese colleagues.

One of the most important portions of our job was to communicate data requirements among the researchers and data suppliers. The procedure is shown in Figure 3.4.

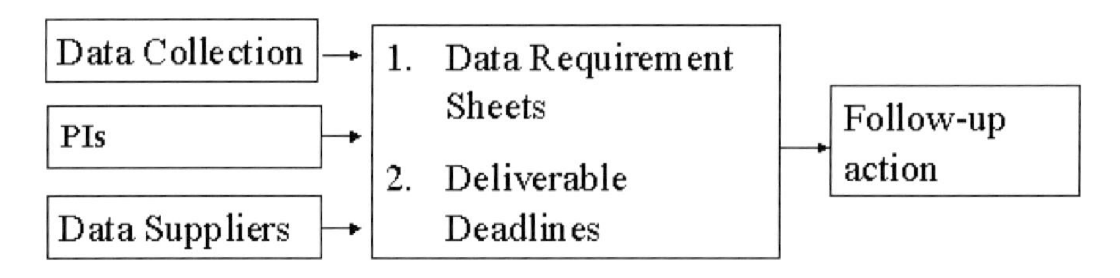

Figure 3.4 Data collection process

Upon delivery of data from our Chinese colleagues to the Program, we passed it on to task researchers and stored copies in the CETP database. We found that storing copies of all data in a central database reduced the amount of data duplication and helped us recognize inconsistencies between data sets more easily. The Data Collection task also acted as a central point for participant feedback, funneling and translating questions, comments, and clarifications between western and Chinese researchers. The data flow between the various elements of the Data Collection task and Program is shown below in Figure 3.5.

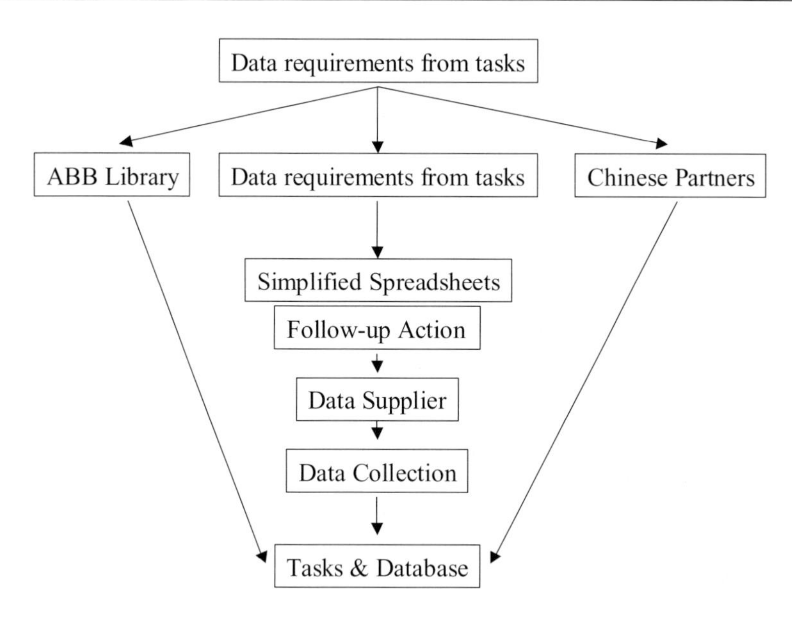

Figure 3.5 Data flow between elements

1.3 Practical Experience

At the outset of the Program, the supply of data was limited in some respects because the mechanism of data collection for the Program was not fully conceived. The process had not yet been fully realized in China, which led to considerable difficulties for researchers at early stages of the Program. In truth, this had an unexpected ancillary benefit: These delays allowed researchers to further refine their respective methodologies in anticipation of receiving data from their Chinese sources. During the early stages, the data supplied were mostly from the ABB Corporate Research Center. For those tasks taking place outside China, individual and pre-established data sources sufficed, while the stream of data from China began to flow gradually.

At first, data requirements were large in scope and very general. Group meetings were often dominated by discussions of Program source data, its quality, or the lack thereof. We attempted to determine more fully the requirements of each task and to adapt our data-gathering strategies to provide the largest quantity and highest quality of information.

When the Shandong Electric Power Research Institute (SEPRI) agreed to participate, methods of data collection were accordingly shifted from consulting public sources to forwarding data requests to partners in Shandong and elsewhere in China. Data requirements for the various tasks were sent in simple spreadsheets to provide a unified format for each data provider according to his or her area(s) of expertise, e.g., emissions, power plants, etc. This significantly eased the burden on the data supplier, as did the decision to make the principal investigator for Data

Collection the primary point of contact for all data requests. As time went on, researchers learned what data were available and adjusted their requests and formats to adapt to the level of information available.

As we were looking for a large number of data spanning disciplines including economics, energy technologies, and environmental science and policy, coordination between the different tasks and data suppliers became very important. Especially difficult to find were performance and environmental data for power plants and transmission network data, which were either not available in the literature or simply did not exist. Patience and great effort were needed to obtain these data. Some data (e.g., for hydroelectric accidents) were unavailable because they were confidential. Some tasks, such as LCA, needed additional information that required special expertise to gather, particularly in collecting data on coal mining.

Facing such challenges, we first analyzed the requirements according to their priority and the task's plan and then collected information step by step according to the task schedule. The two primary tasks faced by Data Collection at the beginning of the Program were to make the data requirements clearly understandable and to organize the formats in which Chinese colleagues should record the data when they found it.

Additionally, language barriers and cultural differences occasionally came between different teams. Many efforts were undertaken to overcome communication barriers and build mutual understanding of eastern and western cultures; many of the western researchers had their first contact with their Chinese counterparts through the Program, and vice versa. Promotion of technical and personnel exchange between Program participants and more than ten group meetings in two years helped build team spirit and facilitate the daunting task of data collection.

1.4 The ABB-China Library

In support of the effort to select a Chinese province for the case study, data collection was underway before CETP officially began. Criteria for the "first round" selection process included: the production of fossil fuel, electricity generation, energy supply, hydro electricity generation, renewable energy potential, pollution, GDP, possible Chinese partners and collaborators, and the location of an ABB office or joint venture for support. Shandong was eventually chosen.

In order to provide an initial source of statistical and energy information about China, large numbers of documents and books were collected by ABB Switzerland. A "China Library" was gradually developed in the Energy and Global Change Department in Switzerland. The primary initial sources of data were the China and Shandong statistical yearbooks; CD-ROMs on Shandong; international publications covering topics such as China's economic development, electric power development, new energy strategies, environmental standard and regulations, and general Chinese energy policies (especially for Shandong). The China Library came to include more than 100 books, journals, maps, articles, and other relevant materials published in various publications inside and outside China and various reports. Approximately 30 percent of these were specific to Shandong Province. These books and publications were extremely useful in familiarizing members of the research teams with the general situation on China. Although the library was limited in size, it provided useful data at the beginning of the Program, especially to the tasks based in Switzerland and the research teams in related countries.

1.5 Other Sources of Information

In addition to building the ABB Library, information from the Internet and newspapers was also collected. When we were able to obtain any useful information or reports related to China's development, especially in the areas of energy, electric power, emissions, etc., we put them into our CETP database or distributed them to all CETP participants directly through e-mail. All of the tasks were actively engaged in searching for useful information and often distributed their data to their colleagues.

1.6 Participation of SEPRI

In the fall of 1999, the Shandong Electric Power Research Institute (SEPRI)[1], a division of the Shandong Electric Power Group Corporation (SEPCO), joined the CETP and began to assist in supplying much-needed information. SEPRI contributed to the Program by

- collecting source data primarily for the Electric Sector Simulation, the Energy-Economy Modeling, and the Energy Transportation Modeling tasks;
- providing input to the modeling of the Shandong power system; and
- providing feedback to results of various tasks in the Program.

Throughout the entire process, SEPRI collaborated with CETP research groups. At the requests of the different tasks, SEPRI collected and gathered essential data related to the electric power system. This included:

- data for different electric power plants in Shandong (installed capacity, numbers of units, and certain parameters, etc.);
- baseline environmental data for Shandong electric power plants;
- detailed and aggregate load data for Shandong; and
- descriptions of the electric power market (including quantities of power generated, used, sold, daily load, and curves of annual continuous load, etc.).

During the Program, there were multiple collaborative discussions about the electric power simulation. Different views and productive ideas were suggested for scenario formulation and preliminary analysis to ensure that the research and its results were as relevant as possible to the development of the Shandong electric power industry. Through its collaboration with the CETP, SEPRI has enhanced its own information database. Personnel exchange during this collaboration has brought new capacity to SEPRI's expertise; the methodologies used in CETP can be further utilized within SEPRI's own research activities.

1.7 Different Stages of the Task

The data collection task was largely completed during the first quarter of 2001. The Data Collection task continued to respond to participant requests for information and verify different data, but the bulk of it had been delivered. We found that as the tasks converged more and more, the exchange of data among participants and tasks increased. As researchers moved from the modeling to the analysis stages of their tasks, the types of data required became more specific and focused. Overall, the different stages of the task can be roughly broken down into the following steps:

Program Stage	Data Type	Primary Sources
Preliminary Stages	Large quantities of general and aggregate data	ABB China Library, Chinese Colleagues and Institutes
Middle Stages	More detailed and task-specific information	SEPRI and Chinese Research counterparts
Integration	Highly specific and focused	Data flow between Task researchers

1.8 Summary

Throughout the entire Program, the Data Collection task collaborated closely with SEPRI, the various tasks, and Chinese institutions. Positive coordination and communication were absolutely critical to ease and reconcile the processes of data collection, distribution, and exchange throughout the CETP program. The lessons learned were the following:

- Providing as much data as possible as early in the project as possible can significantly accelerate the progress of research. Even data that cannot be completely verified is better than none; researchers are often able to use "rough" data to build their methodologies and then refine their work with later adjustments.
- Trying to understand the data requirements of other technical tasks vastly improves the ability of researchers to craft and refine their own information needs.
- A basic understanding of the research methodologies of the tasks considerably simplifies the process of collecting data; knowing how the information will be used allows one to better judge the best source and the accuracy of it.
- Data duplication leads to inefficiencies in the research; a central, open, user-friendly data repository helps to avoid this.
- A central data collection task can improve the interaction between tasks by easing communications and coordination.
- Different researchers with inherently different research areas and methods must be encouraged to share information; data that may appear tangential to one task may be critical to another.

2. DATABASE DEVELOPMENT

The various parts of the CETP rely, to a certain extent, upon the availability and quality of the pertinent data. The value and relevance of task results is also determined by the extent to which data is freely interchanged and shared among researchers. Some of the tasks, in fact, have symbiotic relationships with regard to data. The forecasts of electrical demand, for example, necessarily and naturally lead to the simulation of the electrical system. It became apparent early in the Program that an easy way for users to communicate and share data was absolutely essential to the eventual success of the CETP.

Our primary goals for the portion of the Program were to:

- create a mechanism for project participants to easily exchange source documents;
- create a central repository for all Program-related documents;
- allow participants to access all of our documents for research purposes; and
- allow the documents to act as a historical record of the Program for future researchers to use.

Communication throughout the project has been achieved through, of course, electronic mail. It is difficult to imagine whether such a program could have even been attempted without electronic communication. Our first task, then, was to create electronic mailing lists so news could be distributed to all program participants, about 100 altogether. Through these mailing lists, Program participants have exchanged and broadcast hundreds of messages to the group sharing their research results and questions.

2.1 Lotus Notes Database

To allow users to exchange the data necessary for their respective tasks, we considered several alternatives. Our initial hope was to build a true database that contained numerical and statistical data from the generation side, transmission side, and demand side. We hoped that such a database would be dynamically linked to the various research tools used by program participants who would contribute to it and thereby keep all the source data for all research tasks "synchronized." Unfortunately, we found that the heterogeneity of the data used by different researchers precluded this option; it was simply impossible for us to create one monolithic database that included all the different types of data necessary.

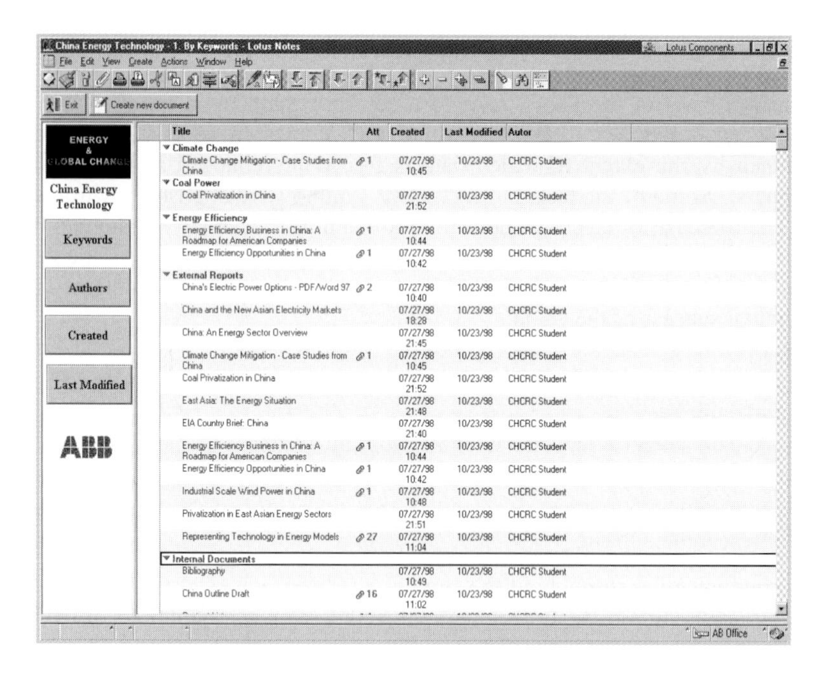

Figure 3.6 Lotus Notes database screen

The initial concepts for the CETP database were presented at the Group Meeting in January of 1999 and refined thereafter. Instead of a true database, we instead proposed and proceeded to develop a "document base", which acted as a repository for different documents generated by the project participants. To create this document base, we used the Notes system from IBM. The Notes system, already the corporate communications system for ABB, allows users in different locations to access the same collection of documents, and to contribute to it through their local servers. These changes are then "replicated" through a wide-area network to other sites, allowing different groups of researchers to synchronize their documents.

To manage the large number of documents generated from the program (reports, presentations, articles, etc.) we categorized and cross-referenced all of them according not only to their task but also to other categories such as author, institution, etc. Additionally, we have version control on our database, so changes in documents over time can be tracked. We also have a full text-search capability to allow users to quickly find the data they are looking for.

For those participants who were not part of ABB, we developed a web-based interface to the system that allows users to access all of the documents through any web browser regardless of their location.

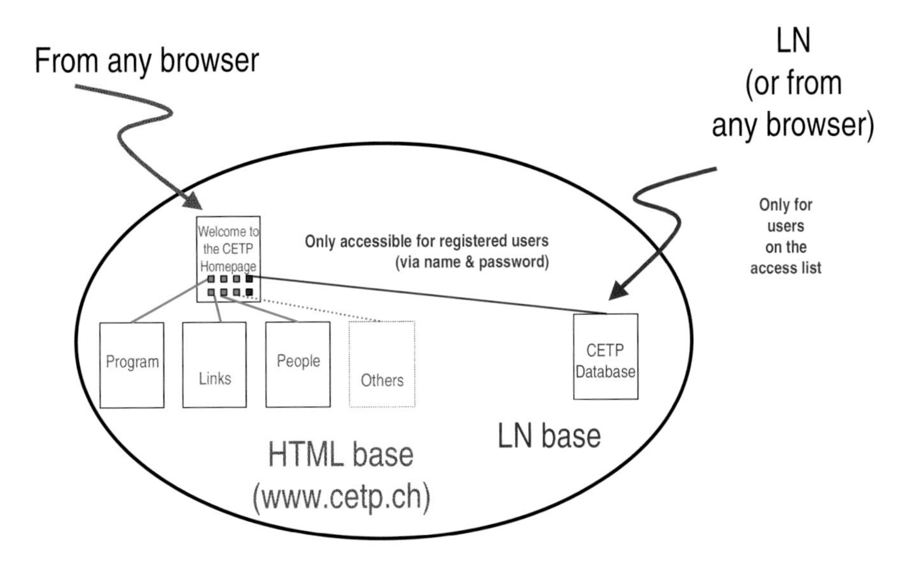

Figure 3.7 Configuration of web interface to Lotus Notes database

The decision to implement a web interface to the Lotus Notes database proved vital to the program. Allowing all participants, not just those from ABB, access to the document database facilitated the transfer of data among participants and allowed all researchers to see what source data the others might be using. Currently, the document database contains over 400 documents, representing not only program research data, but also relevant news articles, meeting notes, program documents, and participant publications.

2.2 CETP Website

In addition to our database, we developed a website, www.cetp.ch, to present our research to the world and act as an outreach mechanism. We have included on the website, along with general information about the Program, multimedia elements such as digital video and downloadable software demos to better communicate our work to the world. The website also has a section linking it to the aforementioned database for Program participants.

The website went online in May 1999 and is accessible at the address www.cetp.ch. Figure 3.8 and Figure 3.9, display a sample page from the website as well as the layout of the pages at the time of publishing.

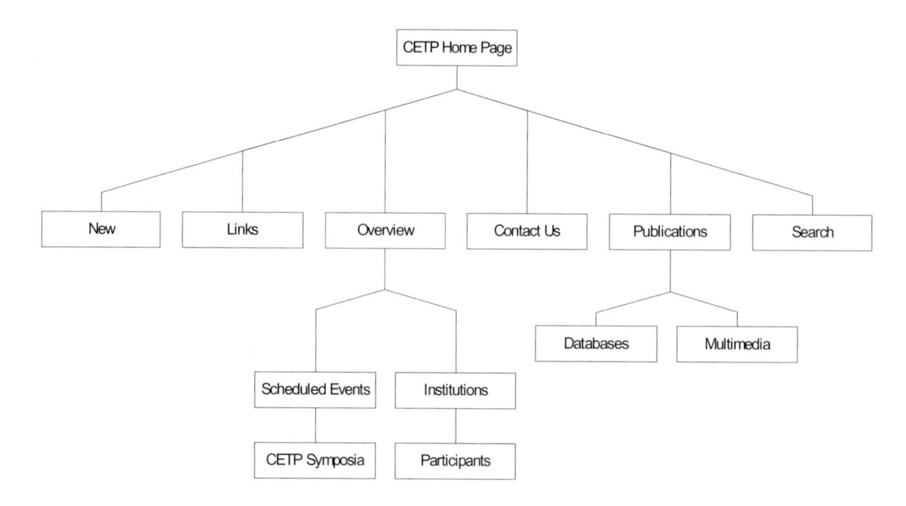

Figure 3.8 CETP website layout

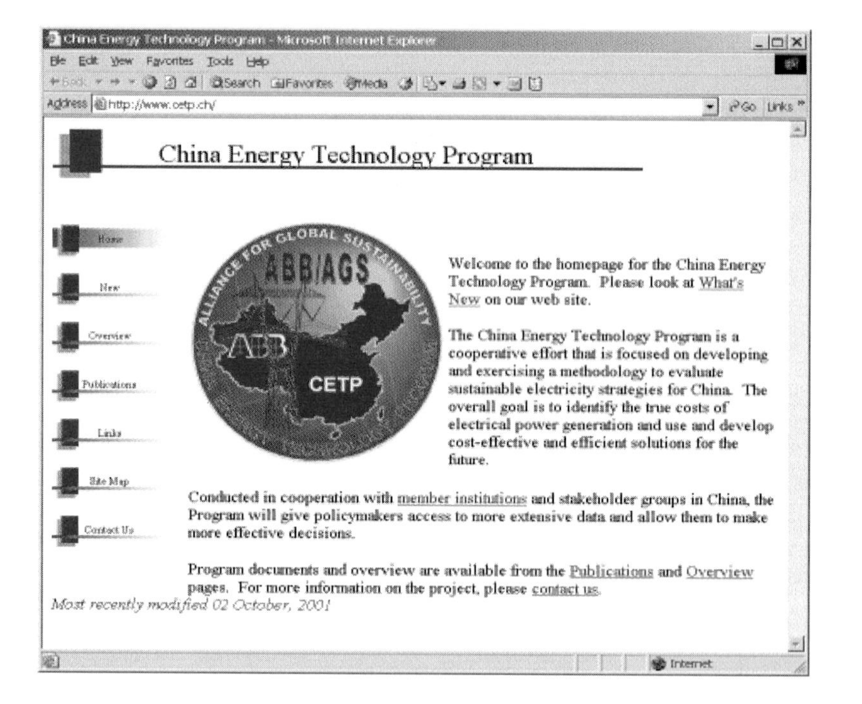

Figure 3.9 CETP website main screen

In addition to various documents and downloadable information, the site also contains multimedia video from the program and contact information for all program participants.

2.3 Conclusions

The database has been, and continues to be, a central element in the operation of the CETP. It would have been impossible to coordinate researchers, engineers, and scholars from multiple continents and countries without modern information technology, and the database allowed a virtual forum in which to share ideas and knowledge. Furthermore, our database[2] and website both provided an important window to the world on our research and progress. In the future, they are will become even more important as a roadmap for those who may follow in our footsteps.

NOTES

[1] SEPRI functions as the research arm of SEPCO and is responsible for technical supervision and service during the course of production and construction and management of the Shandong power system, in addition to unit start-up, commissioning, and performance testing for large-scale thermal-generating units, transmission, and energy conversion projects.

[2] Due to the proprietary nature of the database, it will only be accessible to CETP participants in the future

CHAPTER 4

DEMAND FORECASTING

DADI ZHOU, SHIXIAN GAO,
EDGARD GNANSOUNOU AND JUN DONG

To fully explore the issues of energy demand forecasting in Shandong province, researchers found it necessary to first understand the geography, society, and economy of Shandong province. These are the conditions that determine the region's energy requirements.

1. GEOGRAPHIC, SOCIAL, AND ECONOMIC CHARACTERISTICS

Situated in the lower reaches of the Yellow River and bordered by the Bohai Sea and the Yellow Sea, Shandong is a province in East China with 17 prefectures and cities and 139 counties under its jurisdiction. The province lies between 34°25' and 38°23' north latitude, 114°25' and 112°43' east longitude. Seven hundred km wide from east to west, 420 km long from north to south and 156,700 square km land area, it is the nineteenth largest province in China, making up 1.6 percent of its territory of the country. The province is in the warm temperate zone with a semitropical monsoon climate and an annual average temperature of 11°C-14°C. The annual average precipitation is 550mm-950mm. The frost-free period in the coastal area is more than 180 days while in the inland areas it is more than 220 days.

Facing Japan and the Korean Peninsula to the east, connecting with Hebei Province in the northwest and Henan Province in the south, Shandong's special geological location makes it the crossroads between the Yellow River Economic Belt and the Bohai Ring Economic Area, between North China and East China. It occupies an important position not only in China but also in the East Asia and the Asian-Pacific economic areas.

Shandong is the second most-populated province after Henan province in China. Its population was 88.8 million at the end of 1999, 7.1 percent of the country's total. In 1999, GDP of the province was 766.2 billion yuan, the third largest in China and 9.4 percent of the nation's total. GDP per capita was 8673 yuan, ranking ninth in the country and 2139 yuan higher than the national average. The economic structure of Shandong mirrors that of China as a whole. In 1999, the share of primary, secondary, and tertiary sectors in total GDP was 15.9, 48.4, and 35.7, comparing to 17.7, 49.3, and 33.0 of the country (National Bureau of Statistics, 1999).

2. SHANDONG'S ENERGY PROFILE

Shandong is a province with rich primary energy resources. Coal, oil, and natural gas are relatively abundant and well developed. The proven reserve of primary energy resources (coal, oil and natural gas) is 18 billions toe, or 204 toe per capita.

2.1 Energy and electricity balance

Based on the best available information, the balance of some important types of energy and electricity balances are shown in Table 1. and Table 2. in Appendix F. The energy balance is presented as physical values instead of in standard energy units. Relations among energy types are not included in the energy balance.

In 1999, the electric capacity of the province was 18.24 GW. Electricity production amounted to 91.47 TWh, and consumption equaled 86.32 TWh. Figure 4.1 shows the structure of this consumption.

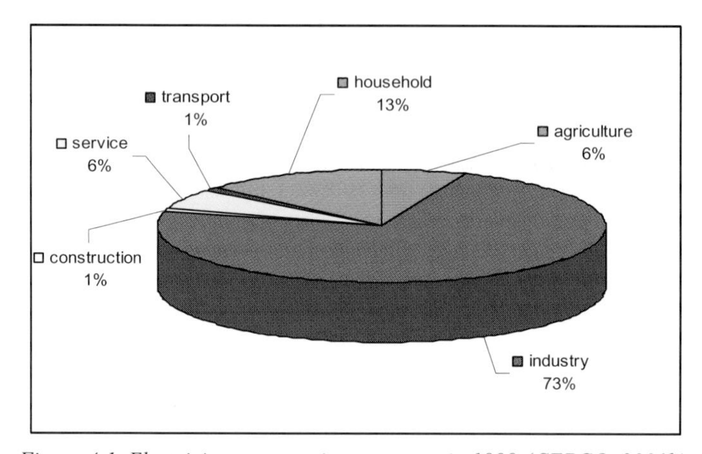

Figure 4.1. Electricity consumption structure in 1999 (SEPCO, 2001b)

2.2 Energy Consumption in China

Research on energy demand in Shandong should be put into the context of China as a whole. This section presents an analysis of the main characteristics of energy consumption at present in China. The three most important characteristics are the impact of transitions in the economic system; the powerful impact of economic factors; and the impact of the concept of sustainable development on the relationships between the economy, energy use, and the environment.

2.2.1 Impact of transitions in the economic system.

There is close relationship between energy demand and economic development. While economic growth is usually coupled with an increase in energy consumption, in some periods of development, it may remain stable or decrease. Since the 1990s, China has been undergoing the transition from a planned to a market economy, the strategic adjustment of the economic structure, and the acceleration of industrialization. In 1991-1995, during the eighth five-year plan (FYP), economic growth slowed and inflation came under control as a new emphasis was place on quality and efficiency. The shortage in the nation's energy supply experienced in the 1980s was relieved.

During the eighth FYP period, average annual GDP growth rate reached 11.6 percent and energy consumption increased by 5.86 percent per year (State Statistics Bureau, 1999). The energy consumption elasticity was 0.51. Unable to meet demand, domestic energy supply became a bottleneck in the nation's growth. In 1996, GDP growth dropped from the peak level of the eighth FYP period of 14.2 percent to 9.6 percent. Meanwhile, the rate of energy growth dropped by one percent compared to 6.9 percent during the most productive year of the eighth FYP (China Energy Statistical Yearbook, 1999). In 1997, GDP growth slowed to 8.8 percent. Energy consumption continued to decrease until 2000.

During the ninth FYP period (1996-2000), the average annual GDP growth rate was 8.3 percent, but energy consumption decreased by 0.46 percent annually. Due to an 8.1 percent annual decrease of energy intensity, 470 Mtoe less was consumed in 1996-2000 than in the previous FYP. As Table 4.1 shows, energy consumption declined with the slackening of growth between 1997 and 1999.

Table 4.1 Growth rates of GDP and energy consumption

Unit: % p.a.

	1996	1997	1998	1999	2000
GDP	9.6	8.8	7.8	7.1	8.0
Energy production	2.8	-0.2	-0.62	-11.3	-0.1
Energy Consumption	5.9	-0.6	-4.3	-1.6	-1.6

Source: National Bureau of Statistics, 2000 and China Energy Statistical Yearbook, 2000

The two main reasons for the decrease in energy consumption may have been improvements in energy efficiency and a shift in the economic structure toward less energy-intensive sectors (World Bank, 1994). In the process of economic reform, the market has been allowed to determine the allocation of resources thus increasing energy efficiency. During the economic transition of the ninth FYP, production in the energy-intensive and low value-added sectors was reduced, while low energy-use and high value-added sectors and products became more important. As a result, energy consumption declined.

The energy industry in China has been undergoing a series of reforms including commercialization, corporation, introduction of market mechanisms, competition, and capital consolidation. The national monopoly in the energy sector is also changing. In a short period, this sector has raised capital from various sources and increased productivity. As consumption has declined, energy supply and demand have come into balance.

Further reform and more open relations with other countries have expanded the scope for procurement; energy may be purchased not only domestically but in international markets as well. Users can compare costs and choose an appropriate consumption pattern based on supplies available in foreign and domestic markets. This will facilitate a shift in energy consumption patterns that will make it possible to use energy economically.

Due to insufficient domestic oil and gas development and production and the strong demand for high quality power, China has increasingly imported energy. It became a net oil importer in 1993 and a net crude oil importer in 1996. According to customs statistics, net oil imports rose from 9.81Mt in 1993 to 33.84Mt in 1997. The amount of imported LPG was less than 1Mt in 1994 but reached 3.35Mt in 1996. In 2000, imported crude oil and oil products amounted to more than 70Mt and 18Mt respectively. Twenty-five percent of oil consumed is imported. Through proper strategies, the importation of energy can facilitate the optimization of the energy consumption structure, environmental protection, and energy efficiency, while still satisfying rapid demand growth. Table 4.2 shows crude oil and oil product imports and exports of China in 1996-2000.

Table 4.2 Oil import and export

unit: Mt

	Export			**Import**			**Total net import**
Year	**Crude oil**	**Oil product**	**Sub-total**	**Crude oil**	**Oil product**	**Sub-total**	
1996	20.33	4.18	24.51	22.62	15.82	38.44	13.93
1997	19.83	5.59	25.42	35.47	23.79	59.26	33.84
1998	15.60	4.36	19.96	27.32	21.74	49.06	29.10
1999	7.17	6.45	13.62	36.62	20.82	57.44	43.82
2000	10.44	8.27	18.71	70.27	18.05	88.32	69.61

Source: China Statistical Yearbook, 2000 and 2001 Newsletter, National Bureau of Statistics

2.2.2 The increasing impact of economic factors on energy consumption

In the past, the energy supply pattern created by planned distribution and national uniform pricing restrained demand, but rigid energy pricing and low energy costs led to inefficiency and waste. The ninth FYP period included extensive energy price reform. In 1998, the State Council announced that the price of crude oil and oil products would henceforth reflect international trends. A preliminary oil price mechanism was established to reflect the natural scarcity of oil and changes in demand and supply. This policy should guide production and rational consumption in enterprises and improve efficiency.

Electricity price reform began in the mid-1980s. Capital was raised through two tiers so that various parties were stimulated to invest in power generation. However, there were side effects such as irrational electricity pricing and multiple fees. High electricity prices constrained electricity consumption. Since 1997, electricity sales had been stagnant. In order to mobilize the power market, the state decided to consolidate electricity prices, to eliminate irrational fees and to accelerate urban and rural grid renovation.

The state also promoted electricity usage. In 2000, consumption increased more than 10 percent over the previous year. Because electric energy is clean, efficient, and high quality, Chinese leadership is encouraging an increase in the share of electricity in energy consumed. The potential is large: Per capita electricity use in China is less than half the world average, and even less in rural areas.

Coal is priced according to the market price except for power-plant coal that is sold at a favorable price. However, since 1997, the price of coal has dropped due to oversupply. For this reason, some enterprises have substituted coal for natural gas, a much higher quality fuel.

2.2.3 Impact of sustainable development on relations between the economy, energy use, and the environment

Sustainable development is one of the basic social and economic development strategies in China in the 21st century. The energy production and utilization required for economic development can cause serious environmental pollution. The development and utilization of high quality energy and energy efficiency improvement are vital in the pursuit of sustainable development.

Since the eighth FYP period, the energy use structure has shifted from coal to high quality energy such as oil and electricity. (See Table 4.3.) The share of natural gas did not increase because domestic production is limited and no imports are available at present. Environmental cost has been considered and people can afford expensive but good quality energy such as oil, gas, and electricity. These forms of energy can be purchased in the market, further enhancing the energy use structure.

Table 4.3 Share of different energy types of total final energy consumption in China

Unit: %

	1991	1996	1998
Coal	65.32	58.70	52.3
Oil	18.00	20.60	26.4
Natural gas	2.44	2.16	2.5
Electricity	9.53	11.60	13.9

Source: National Bureau of Statistics (1991, 1996, 1998) Energy Balance Tables

3. GOALS AND METHODOLOGY

Before presenting the data and the models for demand forecasting, it is necessary to outline the goals and methodology employed in this study.

3.1 Goals

Energy demand forecasting plays a key role in the Chinese planning process. In the planned economy, a supply shortage constrained energy consumption. The amount and type of available fuels and even its supply channels were decided unilaterally by the central government. The influence of the market on energy consumption was small. Energy prices were rigid and unresponsive to demand.

An aggregate approach is used to determine future energy consumption. With the opening of the energy market, both the central planning system and the market system now affect energy consumption. The old method of demand forecasting is no longer suitable to the new situation. It is critical to select an appropriate methodology to forecast the demand of energy.

Electricity is the most versatile and highly valued form of energy; its share in total energy consumption is continuously increasing. Electricity cannot be stored. The simulation and forecast of the electricity market is an essential issue in power system planning. Forecasting can be used to determine the need for additional capacity to generate, transmit, and distribute power. It also plays a major role in policy decisions related to the operation of power systems.

As one of the major tasks in CETP project, the goals of demand forecasting are to analyze the historical and present situation of energy and electricity consumption in Shandong province, to analyze the relationship between energy consumption and social economic development and to make the forecasting on energy and electricity demand taking into account the future economic growth, economic structure shift, energy efficiency improvement. The demand forecasting not only gives a detailed outlook on energy demand in the province but also provides a basis for the other tasks in CETP project.

3.2 Methodology

In this section we consider three methodological perspectives: top-down and bottom-up approaches; optimization and simulation methods; and extrapolation and scenario analysis.

Top-down and bottom-up approaches. Energy demand forecasting can be done from the top down and from the bottom up. In the top-down approach, it is assumed that the market is in long-term balance. The interaction between price, energy demand, energy supply, and investment is simulated under full or partial equilibrium. The final result is the accumulated energy system behavior. At present, China is in a transition from a planned economy to a market economy. The role of the market in balancing supply and demand is becoming apparent gradually. The regularity of economic activities is not yet visible; there is insufficient credible historical data to run the simulation.

In the bottom-up approach, energy demand is disaggregated to a high level of detail. Several important factors such as technical change, energy transformation, market saturation, and structural change are determined mainly by investigation, engineering studies, and expert judgment. This approach does not exclude econometrics. It can adopt the results of econometric model as reference.

In this chapter, the authors have adopted a bottom-up approach. Final energy demand forecasting begins with the end users and takes into account activity level, structural changes, and energy efficiency.

3.2.1 Optimization and simulation methods

The tools for energy demand forecasting fall into two classes: optimization methods and simulation methods.

The optimization method is a planning approach that maximizes or minimizes the targeted functions in order to seek the best system allocation. The success of this method depends on the modeler's knowledge of the energy demand variables and identification of the most important variables. The modeler must also understand other factors constraining the targeted function in the future and their impact on energy demand.

The simulation method is based on energy flow. It calculates energy demand under different assumptions. The accuracy of this method depends on the rationality of the scenarios for energy demand trends and the magnitude of change set by the modeler. If scenarios change, the forecast results will deviate from the actual situation.

3.2.2 Extrapolation and scenario analysis

Energy demand forecast methods can also be classified into extrapolation and scenario approaches. An extrapolation analysis is based on current social economic and energy consumption; the modeler will predict possible future energy consumption according to his perception of future development. The scenario analysis describes future social economic development scenarios based on future patterns of energy consumption.

3.3 Introduction to the Models

In this chapter we use MEDEE-S models and DEMELEC-PRO models to forecast energy demand. The former is used to forecast total future energy demands including different energy types and the latter to forecast electricity demand only. In addition to these two models, a "Modulation" approach is used to forecast peak load and load curve.

3.3.1 MEDEE-S model

MEDEE-S is the abbreviation for French "Modele d'Evaluation de la Demande En Energie." It is a bottom-up model based on economic activity level, structural change, and energy efficiency. The model was initiated by the France Energy Control Agency in 1983 and sponsored by EEC and the French energy ministry. Its design and implementation was organized and carried out by Institute of Energy Policy and Economics of the University of Grenoble (IEPE). Based on the MEDEE2 and MEDEE3 experiences, MEDEE-S was developed to forecast final energy demand in developing countries. Experts and scholars at the Asian Institute of Technology (AIT) are still modifying, improving, and disseminating the model. It has been applied in nearly 30 countries and regions.

The model has the following characteristics:

- Final energy demand forecasting by sector with technological and economic variables;
- Bottom-up final energy demand forecasting;
- Simulation model;
- Flexible structure (flexible choice of sector, base year, forecast year and energy unit. The model consists of a submodel for each sector and 12 optional additional submodels. The final energy demand framework of each sector can be built by choosing the variables; and
- Forecasting made by scenarios (The energy demand forecast process is a scenario analysis process. The three driving factors in the scenarios are activity level, economic structure and energy efficiency).

The typical activity levels in the model are:

- Population and the number of households;
- Value added and production volume;
- Passenger and freight transport;
- Employee number and floor area; and
- Vehicle fleet.

3.3.2 DEMELEC-PRO model

DEMELEC-PRO is a medium- to long-term simulation tool for electricity demand developed by EPFL-LASEN (Gnansounou and Rodriguez, 1999). The tool uses a techno-economic model, and divides consumers into homogeneous categories. For each category, electricity demand is related to a specific demand and a variable measuring the volume of activity for which electricity was consumed, e.g. value added in a productive sector. From its general formulation, the model can be adapted to different systems.

The system studied can be an area inside a country, an entire country, or an area including several countries. The distribution of the consumers into a variable number of homogeneous categories determines the degree of disaggregation of the model. DEMELEC-PRO can allow a vertical disaggregation into four levels; the number of levels can vary from one sector to another. The system is designed to evaluate electricity demand from two components. The first component determines the values of the model's reference variables from the source data. The second component determines the assumptions of evolution of the variables.

The DEMELEC-PRO model is very open to different choices in degrees of disaggregation and in the system configuration of electricity demand evaluation. It allows flexibility in the number of sectors considered, and the degrees of disaggregation for each sector. For example, in DEMELEC-PRO 1.0.1 (Gnansounou and Rodriguez, 1999) there are six sectors: residential, industrial, large-scale services, small-scale services and artisan, public lighting, and other sectors (e.g. agriculture). The residential sector is disaggregated into four levels (area, class of the locality, locality itself, and standard of living) while the industrial sector is divided into two (type of activity, and company). The large-service sector has two levels (type of service and company); the small-service and artisan sector have only one (the sub-sector); public lighting is disaggregated by area, class of locality, and locality itself, and the other sector is classified by sub-sector and company

Based on the value assumptions for the variables for the reference year, DEMELEC-PRO forecasts electricity demand.

3.3.3 "Modulation" approach to load-curve forecasting

A very important element of electricity usage is its time patterns. A "load curve" is a diagram of electricity demand as a function of time. The "modulation" approach involves the use of various "modulation coefficients" to convert annual energy demand to the demand at a particular hour for a particular day of a given week in a year. The modulation coefficients, which may be determined by statistical studies of past annual performance, include average growth during the year, seasonal variations, type of day (weekday or weekend), and hourly variations. The multiplication of all these coefficients provides the mean hourly consumption rate from which the load curve may be derived.

Consumers may also be placed into different categories, each with its own load profile. The future structure of electricity needs may be ascertained in the integrated load curve taking into account changes in consumer demand (Gnansounou, 2000).

3.4 Scope of Demand Forecasting

To determine the scope of demand forecasting it is necessary to group the final energy use sectors; establish time horizons; compare the provincial and national energy demand profiles; and disaggregate demand by sector and by fuel.

The final energy use sectors include agriculture, industry, construction, transportation (including telecommunication), service, and household (See Figure 4.2).

3.4.1 Grouping the sectors for demand forecasting

The industry subsectors are nonferrous metal, textile, chemical, energy and other

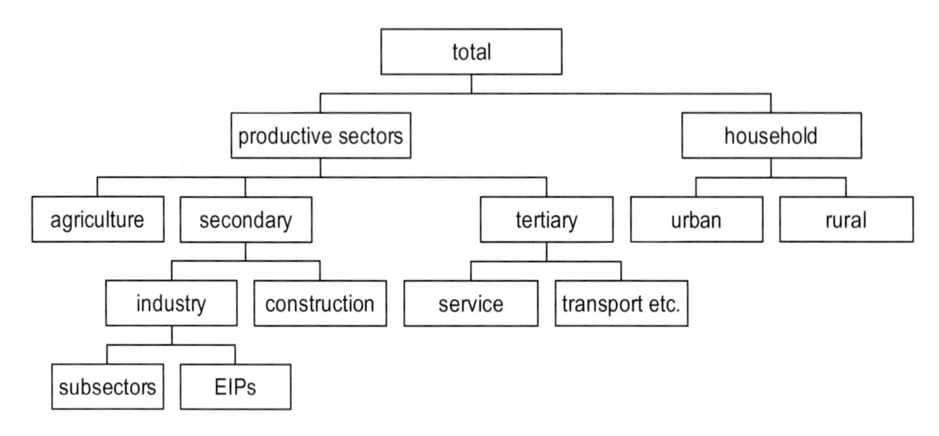

Figure 4.2 Grouping of sectors

industries. In 1995, the energy consumption of these five sub-sectors accounted for 74.6 percent of Shandong's total energy use. Nonferrous metal, textile, chemical and energy industries accounted for 59.7 percent and other industries are 14.9 percent. Energy-intensive products (EIPs) include cement, glass, paper, iron, and steel. In 1995, the energy consumption of these products accounted for 25.4 percent of total industrial energy use in the province (Shandong Statistics Bureau, 1995).

3.4.2 Time horizon

We have taken 1995 as the base year, and forecast demand for 2000, 2005, 2010, 2015, and 2020.

3.4.3 Final energy demand forecasting at the provincial level

Final energy consumption does not include that lost in the transformation of energy from one form to another or the transmission of energy. Final energy demand forecasting of Shandong province is made in the context of the current state and future trends in national and provincial energy consumption.

Comparing energy demand forecasting for Shandong and for China reveals some similarities and conditions specific to the province. Shandong's special circumstances include a rapidly changing activity level during the economic transition period; demand strongly influenced by a few large customers; new energy projects or technical renovation of old facilities; incomplete or scanty basic data and few reference materials. While deficiencies in available data increase the difficulty of demand forecasting, the relatively small number of large customers and concentrated energy-intensive industries make it easier.

3.4.4 Final energy demand forecasting by sector and by fuel
Final energy demand for the province is forecasted by fuels and by sectors. The fuel types are coal, coke, gasoline, diesel, kerosene, aviation kerosene, fuel oil, LPG, gas fuel, and electricity.

3.5 Interaction with Other Partners and Stakeholders

In addition to producing its own results, the Demand Forecasting research group also provides inputs to other CETP tasks such as Energy/Economy Modeling, Electric Sector Simulation and Energy Transportation Modeling (See Figure 4.3).

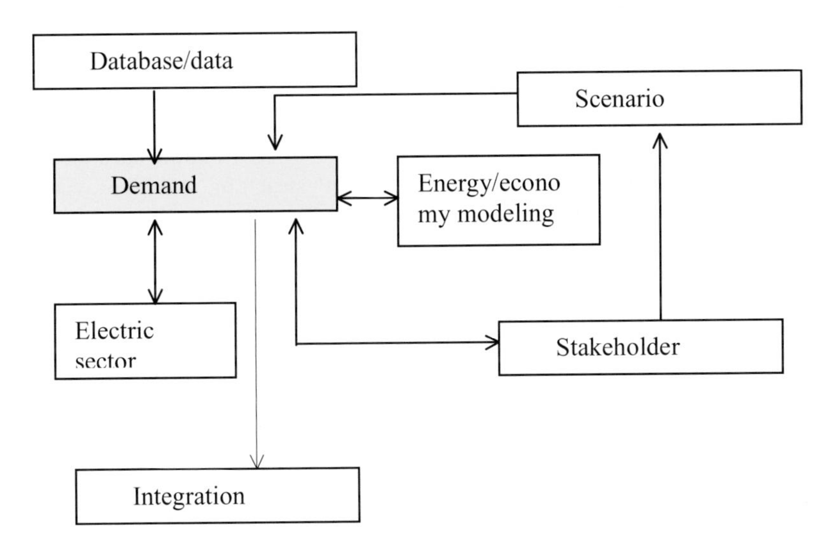

Figure 4.3 Interaction with the other Project partners and the stakeholders

In different stages of the research, - from the data collection to discussions of results, - the demand forecasting group interacted with the stakeholders and other partners in CETP through information exchange, consulting, questionnaires, etc.

4. SCENARIO CONSTRUCTION

The construction of the demand-forecast scenarios required the identification of key scenario-driven factors. These included demographic factors such as total population growth, the distribution of population in urban and rural areas, and the number of households. Macroeconomic factors included GDP growth, increased production on the part of energy intensive end-users, a sectoral analysis of the economic structure of the province, and the energy intensity for each category of end-users. Once these factors were defined, the scenario could be constructed.

To create the scenarios, investigators ran the MEDEE-S and DEMELEC-PRO models using different assumptions about levels of economic activity, economic structure, and energy intensity. In addition to the baseline scenario, three others were built for a low economic growth situation (Lowgdp); a situation in which the economic structure was changing (Structure); and one in which energy efficiency had not been improved (Inefficient). The scenarios differ from the baseline scenario respectively in terms of GDP growth rate, GDP structure, and improvements in energy efficiency (see Table 4.4)

Table 4.4 Factor change in three scenarios compared to a baseline scenario

Scenarios	**Factors in the scenarios**		
	GDP	**GDP Structure**	**Energy Intensity**
Lowgdp	D	S	S
Structure	S	D	S
Inefficient	S	S	D

D: different from baseline scenario
S: same as baseline scenario

4.1 Low GDP Growth Rate Scenario (Lowgdp)

This scenario is characterized by slower GDP growth than the baseline scenario. The difference in growth rates in some industries is shown in Table 4.5.

Table 4.5 A comparison of GDP and EIP output growth rates in Baseline and Lowgdp scenarios

Unit: % p.a.

	Baseline					Lowgdp				
	1995-2000	2000-2005	2005-2010	2010-2015	2015-2020	1995-2000	2000-2005	2005-2010	2010-2015	2015-2020
GDP	10.00	9.00	8.00	7.50	6.50	9.00	8.00	7.00	6.50	5.50
CEMENT	2.00	2.00	1.70	1.70	1.50	1.60	1.00	0.70	0.70	0.50
GLASS	5.50	5.50	4.00	3.00	3.00	5.10	4.50	3.00	2.00	2.00
PAPER	4.92	9.00	7.98	7.51	6.61	4.42	8.00	6.98	6.51	5.61
STEEL	8.69	2.88	3.30	3.50	1.03	8.19	1.88	2.30	2.50	0.03

4.2 Economic Structure Change scenario (Structure)

This scenario is based on the assumption that the service sector's share in GDP is one percentage point higher than in the baseline scenario. The industrial sector share is one percent lower than the baseline scenario. The comparison between Baseline and Structure scenarios is shown in Table 4.6.

Table 4.6 A comparison of GDP structure between Baseline and Structure scenarios

Unit: %

		Baseline					Structure				
	1995	**2000**	**2005**	**2010**	**2015**	**2020**	**2000**	**2005**	**2010**	**2015**	**2020**
Primary sector	20.19	16.50	14.00	12.40	11.50	11.00	16.5	14.00	12.40	11.50	11.00
Secondary sector	47.43	47.50	47.00	45.80	44.50	43.00	46.5	46.00	44.80	43.50	42.00
Industry	42.25	43.80	44.20	43.60	42.80	41.60	42.8	43.20	42.60	41.80	40.60
Construction	5.18	3.70	2.80	2.20	1.70	1.40	3.70	2.80	2.20	1.70	1.40
Tertiary sector	32.38	36.00	39.00	41.80	44.00	46.00	37.00	40.00	42.80	45.00	47.00
Service	26.42	29.50	32.00	34.00	35.50	37.00	30.5	33.00	35.00	36.50	38.00
Transport	5.95	6.50	7.00	7.80	8.50	9.00	6.50	7.00	7.80	8.50	9.00

4.3 Low Energy Efficiency Scenario (Inefficient)

The "Inefficient" scenario was constructed in order to analyze energy demand in the context of insufficient energy-saving investment and the inadequate implementation of energy-saving measures. The comparison of Baseline and Inefficient scenarios is shown in Table 4.7.

Table 4.7 Energy efficiency comparison between Baseline and Inefficient scenarios

		Baseline					Inefficient				
	1995	**2000**	**2005**	**2010**	**2015**	**2020**	**2000**	**2005**	**2010**	**2015**	**2020**
Intensity index for motor fuel	1.000	0.881	0.776	0.700	0.630	0.580	0.904	0.817	0.756	0.698	0.659
Intensity index for electricity	1.000	0.927	0.860	0.810	0.770	0.750	0.951	0.905	0.874	0.852	0.851
Intensity index for thermal use	1.000	1.000	1.000	1.000	1.000	1.000	1.051	1.105	1.161	1.220	1.282

4.4 Additional Scenarios

In addition to the four basic scenarios described above, four additional scenarios are built for electricity demand forecasting. In the additional scenarios, the factors considered are not only economic growth, economic structure and energy efficiency, but also the penetration of electricity use, especially the electrification in service sector. The assumptions on change of electricity intensity are made considering the combined impact of activity shift to less electricity intensive subsectors, improved electricity efficiency and penetration of electricity. The four additional scenarios are: Baseline-E, Lowgdp-E, Structure-E, Inefficient-E. These additional scenarios differ from the others by the high penetration of electricity considered.

5. IMPORTANT ASSUMPTIONS IN ENERGY DEMAND FORECASTING

Energy demand forecasting is based on macroeconomic factors, and facts about the primary sector (agricultural), the industrial, construction, transportation, service, and household sectors. This section also discusses the significance for forecasting of the historical evolution of electricity consumption, and assumptions in electricity demand forecasting in the additional scenarios.

5.1 Macroeconomic Factors

In the final quarter of the twentieth century, the society and economy of Shandong province changed in terms of population and GDP; economic development incentives in Shandong province through institutional reform; adjustments in the economic structure conducive to sustainable economic growth; the promotion of diversified economic ownership and of an investment environment attractive to foreign capital; improvement in rural and urban standards of living; and increased urbanization (SDPC, 1995).

5.1.1 Population and GDP

In Shandong province, total population grew from 76.95 million to 87.05 million in 1985-1995. The rate of population increase was 1.36 percent annually. In urban areas, the total population expanded from 10.17 million to 21.7 million; the average number of persons per household (family size) declined from 3.57 to 3.19. In rural areas, total population grew from 66.78 million to 65.35 million while the average number of persons per household was reduced from 4.72 to 4.07. The growth rate of the population in urban and rural areas in 1985-1995 is shown in Table 4.8.

Table 4.8 Growth rate of population in urban and rural area in 1985-1995

Unit: % p.a.

	Population	**Family size**	**Number of household**
Urban	8.81	-1.26	10.20
Rural	-0.21	-1.55	1.36

Source: Shandong Statistics Bureau (1985-1995), Shandong Statistical Yearbooks

The rate of growth in GDP and changes in the economic structure from 1985 are shown in Table 4.9. In this period, the rapidly developing economy shifted emphasis from primary sector to the industrial sector.

Table 4.9. Historical evolution of GDP and the value added to major sectors

	GDP	**Primary**	**Industry**	**Construction**	**Service**	**Trans.& etc.**
Growth rate of GDP in 1985-1995, % p.a.	11.6	4.7	17.6	10.6	12.4	10.4
GDP share in 1985, %	100	38.9	25.8	5.5	24	5.8
GDP share in 1995, %	100	20.2	42.2	5.2	26.4	6

Source: Shandong Statistics Bureau (1985-1995), Shandong Statistical Yearbooks

5.1.2 Economic development incentives in Shandong province through institutional reform

Since 1978, the province has actively promoted economic reform, initially in rural areas. The old system of ownership was replaced by a system of household co-production contracting system. In the cities, reform focused on the expansion of autonomy for localities and individual enterprises, the adjustment of the ownership structure; the motivation of labor, and economic development. Between 1978 and 1984, the average annual GDP growth rate was 11.1 percent, greatly exceeding the 6.1 percent experienced between 1952 and 1978. The average annual growth rate was 11.5 percent in the primary sector, which remains the major driving force for the provincial economic development.

After 1984, reform efforts prioritized the production and operation mechanisms within individual enterprises, the planning process, taxation, finance, price, wage, foreign trade, and material goods. In this period, China's highly centralized planned system shifted toward one based on commercial and economic forces. After 1992, Shandong province has carried out an even deeper series of reforms in enterprises, taxation, finance, foreign currency exchange and trade, and monetary circulation system. Total industrial output annual growth grew to 20 percent as state ownership dominance yielded to local, cooperative, and private ownership. Currently, the share of state ownership is less than 50 percent. Between 1991 and 1997, the average annual GDP growth rate was 15.4 percent. As commercial markets developed, 95 percent of consumer goods and over 90 percent of investment are privately controlled. The market has begun to determine resource allocation, an important factor in the growing economic prosperity in Shandong province.

5.1.3 Economic structure adjustment and sustainable economic growth

In more than two decades of economic reform, Shandong province has sustained a high level of economic growth. The annual GDP growth rate was 11.1 percent in 1978-1988; 7.8 percent in 1988- 1991; and 15.4 percent in 1991-1997. Quality and efficiency have become important priorities. As the economic structure was adjusted the tertiary sector developed rapidly. Traditional industrial sectors such as machinery and textiles were replaced by emphasis on new materials, microelectronics, and biological engineering.

Another priority is competitiveness of products, expansion of foreign and domestic markets and increases of product's market share. Better management has led to reduced costs and growth in the profit margin for products in the quest for qualitative improvements rather than quantitative growth alone.

5.1.4 Promotion of diversified economic ownership and an investment environment attractive to foreign capital

The policy of encouraging multiple-ownership and introducing foreign capital has produced stable growth in social fixed-asset investment. In 1978, the shares of state-owned, collectively owned, and other forms of ownership were 69.9 percent, 20.1 percent, and 10 percent respectively. After 20 years, by 1997, these shares stood at 43.1 percent, 31.8 percent, and 11.4 percent. The investment environment is improving and a series of preferential policies to encourage foreign investment has been formulated. The field for and scale of foreign investment has expanded continuously, amounting to 13.7 percent of total social fixed assets by 1997. Foreign investment companies are making a significant contribution to economic development in the province. As a main force in foreign trade in Shandong, its share of total provincial exports has increased from 0.04 percent in 1986 to 47.6 percent in 1997.

5.1.5 Improvement in rural and urban standards of living

GDP in the province has grown much faster than population. GDP per capita increased rapidly from 402 yuan in 1980 to 7632 yuan in 1998 (Shandong Statistics Bureau, 1998). Individual income is also increasing dramatically. By 1997, per capita disposable income of urban residents had increased by 14.6 percent annually, 12.3-fold since 1978. The per capita total income of rural residents increased 19-fold, or 17.1 percent annually. The per capita consumption expenditure by urban and rural residents increased 10.9-fold and 16.4-fold respectively between 1978 and 1997.

China's family planning policy has reduced the birth rate from 1.64 percent in 1978 to 1.084 percent in 1997, a factor that reduces employee pensions and household size. The floor space of housing per capita in both urban and rural areas more than doubled from 1978 to 1997. Consumption patterns also changed. The share of household income spent on food decreased and expenditures for durable consumer goods, clothing, and recreation expenditures increased. The Engel's coefficient dropped from 0.577 in 1978 to 0.41 in 1997 in urban areas and from 0.616 to 0.536 in rural areas. Household appliance ownership increased greatly. Refrigerator and air conditioner ownership in urban areas increased from 46.9 percent and 0.23 percent in 1990 to 82.84 percent and 11.35 percent in 1997 respectively. In the same period, the ownership of color TVs, washing machines, and refrigerators in rural areas increased from 6.29 percent, 3.95 percent, and 0.83 percent to 31.43 percent, 13.86 percent, and 13.48 percent respectively (Shandong Statistics Bureau, 1997).

5.1.6 Increased urbanization

The rate of urbanization is an important indicator of the modernization process, and the patterns of energy consumption. In Shandong province, the percentage of urban residents rose from 18.6 in 1990 to 26 in 1999 (Shandong Statistics Bureau, 1999).

5.2 The Outlook for Social and Economic Development

In the tenth FYP (2000-2005), the rate of GDP growth in China is expected to be seven percent annually. The economic structure of China in 2005 is projected to be 13 percent, 51 percent, and 36 percent respectively for primary, secondary, and tertiary sector (SDPC, 2000). GDP growth in Shandong is projected to be a little higher than the average for China.

5.2.1 Basic assumptions for future development trends
Benefiting from economic reform and regional advantages, Shandong province has achieved significant development in the last 20 years, tripling its GDP from 1980 to 2000. However, the following indicators show that the province is still in the preliminary and middle stage of industrialization. In 1998, per capita GDP was 922 US$; the share of the tertiary sector was 34.6 percent; the urbanization rate was 26 percent; food was still the largest part of household expenditure; new and high technology industries accounted for less than 10 percent of industrial output (Shandong Statistics Bureau, 1995-1999).

In the next 20 years, as the economy further expands, the province will enter a new phase of industrialization. According to the international experience, economic expansion and structural adjustment often speed up when the per capita GDP reaches around 1000 US$. In the first decade of the twenty-first century, economic development and structural change should accelerate and then slow down in 2010-2020.

5.2.2 Analysis of economic growth-driven factors
Structural changes will include products and ownership patterns as well as investment and employment structure. All these will directly and indirectly affect future economic growth and efficiency. The current share of the agricultural sector in the province is over 15 percent, compared to five to eight percent in the industrialized countries. There is great potential for the industrialization and modernization of agricultural sector. Within the industrial sector, the share of new and high tech industry should increase from current 8.4 percent of industrial output to 15 percent, comparable to Guangdong, Shanghai, Jiangsu, and Tianjin. The industrial sector will be upgraded through the modification of traditional industries and the fostering of emerging industries.

Market-driven economic reform will optimize resource allocation. This will facilitate capital asset and industrial consolidation and promote profit- and efficiency-oriented production and operations. Effective and rational resource utilization should guarantee sustainable economic development.

The level of development among different regions in the province varies. Decision-makers will consider these regional disparities and pursue a policy of integrated development. The contrast between the economically developed coastal regions and the less developed inland regions creates the opportunity to develop complementary markets among them. A sound integrated regional development policy will create and distribute greater wealth throughout the province.

The population in the province will grow slightly in the immediate future. Between 1995 and 2020, the natural growth rate will be lower than 0.5 percent and the total population will not exceed 100 million. In the urbanization and industrialization process, the urban population will increase steadily. With the rural migration to urban areas, the natural growth rate of rural areas will decrease, not only in its share of total population but also in absolute numbers. The current urbanization rate in Shandong is far below the level of the advanced countries. It is expected that in the next 20 years, the urbanization rate will increase by 20 percentage points. The urban share of the population will have increased from 25 percent in 1995 to 44 percent in 2010 and 50 percent in 2020. The assumptions for the social and economic development in baseline scenario are shown in Table 4.10.

Table 4.10 Social and economical development assumptions in Shandong for baseline scenario

	1995	2000	2005	2010	2015	2020
GDP, Billion Yuan at 1995 price	500.23	805.63	1239.56	1821.32	2614.75	3582.80
GDP breakdown by sector						
Agriculture (%)	20.19	16.50	14.00	12.40	11.50	11.00
Industry (%)	42.25	43.80	44.20	43.60	42.80	41.60
Construction (%)	5.18	3.70	2.80	2.20	1.70	1.40
Service (%)	26.42	29.50	32.00	34.00	35.50	37.00
Transport (%)	5.95	6.50	7.00	7.80	8.50	9.00
Population, Million	87.05	89.6	92.28	95.00	97.00	98.00
Urban Area	*21.76*	*25.98*	*34.61*	*41.80*	*45.59*	*49.00*
Rural Area	*65.29*	*63.62*	*57.68*	*53.20*	*51.41*	*49.00*
Urbanization Rate (%)	25.0	29	37.5	44.0	47.0	50.0

6. THE PRIMARY SECTOR

The outlook for the primary sector and for agriculture, its most important subsector, is mixed. Production will continue to grow as modernization takes effect. However, the growth rate will be lower than that of the secondary and tertiary sectors. The share of agriculture in total GDP will have dropped from 20.19 percent in 1995 to 11 percent in 2020. The share of agricultural product in the primary sector will decrease from the present 50 percent to 33 percent in 2020. Table 4.11 forecasts the value added to the primary sector. Two other primary sub-sectors, livestock and fishing, will expand, increasing their share in total output from the current 47.8 percent to 64.7 percent in 2020. The share of forestry will remain the same.

While the rural population will have dropped from 65.29 million in 1995 to 49 million by 2020, adjustments in plantation structure will increase food production. In the next 20 years, this increase will depend on the application of modern biological engineering technology rather than the expansion of plantation area. In fact, the cropped area will decrease slightly as slope and beach fields are returned to woodland, fisheries, and fruit production. Cotton and vegetable oil production will be produced in these areas. In general, vegetable and fruit farming will develop greatly. The structural adjustment of the economy and China's accession to the WTO will require high quality agricultural products and a larger share of high-value-added, highly processed products.

As high value-added products become a greater part of this sector's output, production volume will increase and energy use per unit product will decrease.

Power and energy consumption in the primary sector will increase with modernization. Currently, field operation is only 50 percent mechanized. The development plan for Shandong province calls for this level to increase to 60 percent, 75 percent, 80 percent, and 90 percent in 2000, 2010, 2015, and 2020 respectively. The associated diesel consumption will also increase. Along with agricultural and general rural economic development, rural transportation needs will grow, demanding more fuel. The rapid growth of fisheries will also require more fuel.

Rural electrification will increase electricity demand for uses such as cropland drainage as pumping in of wetland areas expands and product processing. Agriculture product processing, relying mainly on electricity, will be the main priority in agriculture in the next 20 years.

Table 4.11 Forecasting of primary sector value added

	1995	**2000**	**2005**	**2010**	**2015**	**2020**
Primary sector (BYuan)	101.01	132.93	173.54	225.84	300.70	394.11

7. THE INDUSTRIAL SECTOR

In the current industrialization process, between 1995 and 2005, this sector will develop rapidly. Its share in total GDP will increase. After 2005, industry will maintain a growth rate somewhat lower than the growth of GDP growth in which its share will decrease slightly. In the future, the internal industrial structure will

Table 4.12 Value added structure within the industry sector

Unit: %

	1995	**2000**	**2005**	**2010**	**2015**	**2020**
Subsectors	83.90	82.4	80.80	79.30	77.90	77.60
Chemical	11.00	12.10	12.60	13.40	13.90	15.40
Textile	8.10	7.10	6.70	6.10	5.80	5.60
Nonferrous	1.90	1.20	1.00	0.80	0.70	0.60
Energy	16.40	16.00	15.00	14.00	13.00	12.00
Others	46.50	46.00	45.50	45.00	44.50	44.00
EIPs	16.10	17.6	19.20	20.70	22.10	22.40
EIPs output; Units: Million tonnes for cement, paper, steel; weight cases for glass						
CEMENT	54.97	60.10	67.01	72.90	79.31	85.44
GLASS	13.43	17.55	22.94	27.91	32.36	37.51
PAPER	3.17	4.03	6.20	9.10	13.07	18.00
STEEL	3.89	5.9	6.8	8	9.5	10

change and the share of energy-intensive products will increase. The share of subsector value added will drop. (See Table 4.12)

As the industrial structure changes and technological progress continues, the intensity of industrial energy use will drop steadily at a rate of over seven percent between 1995 and 2005.

7.1 Nonferrous Metals

The nonferrous metals industry in the province is based on aluminum oxides, electrolysis aluminum, and aluminum products. Copper smelting will be limited by depleting resources. There are two aluminum-smelting companies in the province, Shandong Aluminum Company (SAC) and the Pingyin Aluminum Plant.

The priority of structural change in this industry is to accelerate the transition from electrolysis aluminum to deep aluminum product processing; increase the grade of the product; develop new product types; and add value to them. The advantages of combining scientific research and production for multiple aluminum oxide products will be utilized in order to increase market share for these products. Technical renovation and quality improvement will be achieved through increased investment. Policy makers will promote energy conservation, improved efficiency, cost reduction, and comprehensive utilization. In the tenth FYP period, the new small electrolysis aluminum plants will be strictly prohibited and outdated technology will be discarded.

The Pingyin Aluminum Plant is a key enterprise in the province; it uses a 60KA electrolysis compartment—a 1960s technology. In the tenth FYP period, an investment of 420 million yuan will be made to replace this old equipment with an energy-saving 200 KA four-point pre-baking anode electrolysis compartment and construction of two aluminum production lines. The plant's annual capacity will rise to 75,000 tons of electrolysis aluminum and 40,000 tons of aluminum product. New aluminum foil production lines will be constructed to increase the plant's capacity to 5000 tons. Efforts will be made to develop 0.2m/m below refrigerator foil, aluminum belts, aluminum plastic tube and board, etc. in order to increase technology content and value added to these products

Also in the tenth FYP period, the SAC will complete the expansion of Bare AO and aluminum oxide production. The co-production of aluminum, heat, and electricity will be developed. Technical innovations for sintering aluminum oxides will be carried out. The comprehensive utilization of red sludge will also be completed. The annual production capacity will be 800,000 tons of aluminum oxides (including 300,000 tons of multi aluminum oxides), 65,000 tons of electrolysis aluminum, and 45,000 tons of finished aluminum product.

With an emphasis on high value-added product and energy conservation, the energy use per unit product in the nonferrous metal industry will have decreased by 45 percent between 1995 and 2020. The value added to products in 2020 will be 2.22 times the 1995 figures (See Table 4.13) while total energy consumption will stay at about 900ktoe (SDPC, 1995). The share of coal in production of this energy had increased by 2000, compared to 1995 figures, but is now decreasing. By 2010, natural gas will be the major source of fuel. The role of electricity in powering Shandong's industry will increase year by year.

Table 4.13 Trends in the development and energy use for the nonferrous metal industry in Shandong province (SDPC, 1995)

	1995	2000	2005	2010	2015	2020
Value added (million Yuan)	4015	4234	5479	6353	7834	8943
Energy Intensity (toe/million Yuan)	226.46	205.8	158.12	138.42	114.00	102.01

7.2 Textiles

Historically, the light textile industry was well developed in the province, which had served as the production base for this sector in China. In the early period of economic reform, the textile industry once again became a priority, in response to demand for more consumer goods. The textile sector in Shandong relies on light industry.

In recent years, the industrial structure has been adjusted. The share of the textile industry in dropped from 10 percent of GDP in 1993 to eight percent in 1997. In 1998, the total value of textile production in Shandong was 64’055 billion yuan. Shandong produced the highest profits of any province in the country. In terms of total production, Shandong province is the third largest producer of textiles in China, following Jiangsu and Guangdong. Its cotton textile capacity is the first in the country, and wool capacity is 5.42 M spindles, ranking the second in the country. There are 90’000 cloth machines and a crude cloth spindle capacity of 360’000, the nation's second largest. The chemical dying capacity is 500 ktons of fiber, ranking Shandong third in the country.

The textile industry is one of the key energy users in the province. In 1998, its total energy consumption was 1.048 Mtoe, generated by burning 637 ktons of raw coal, 107.8 ktons of gasoline, 32.7 ktons of diesel, 14.6 ktons of fuel oil, 44.5 ktons of LPG, and 850.8 TJ of heat. Ongoing sector reform and structural adjustments have led to greater energy savings. Total energy consumption in the textile industry declined from 2.0512 Mtoe in 1995 to 1.048 Mtoe in 1998. The energy consumption per MYuan of production value decreased from 36 toe in 1995 to 16toe in 1998.

The textile industry has comparative advantages in both domestic and international markets. With 1.2 billion people, China is the largest textile market in the world. Population growth and an improving standard of living create demand for a wide range of textile product types, styles, performance characteristics, and quality. The vast land area of China and its multiple nationalities create diversified textile consumption and a regionalized industrial pattern in the textile industry.

In recent years, international textile products have been used for decorative and industrial purposes. At present, the ratio between clothing, decoration, and industry textile uses in Western Europe and Japan is 46:35:19 and 29:29:40 respectively. In

China, it is 68:22:10. China is currently strengthening its infrastructure. For example, 1000 km of roads and railway are being constructed every year. The Three Gorges project, the Xiaolangdi Hydro Power project, and the south and north water diversion project will create opportunity for the textile industry. Environmental protection, desertification prevention, and control and soil protection will demand large quantities of industrial textiles. Moreover, high efficiency agriculture cannot survive without crop cover material, artificial planting media, insulation material, and drainage pump cloth. Industries such as motors manufacture and paper manufacture also demand large amounts of textiles. It is expected that intensive infrastructure investment and western regional development will require a large amount of decorative and industrial textiles.

The textile industry in Shandong will develop rapidly, as structural change continues. Cotton textile production will be modified, as plants merge and consolidate. The quality and quantity of wool textile production will be upgraded. By 2005, the capacity of each plant will be over 30'000 spindles; such large and medium-scale enterprises can achieve economies of scale and when laid out rationally.

In addition to cotton and wool, new chemical fibers are being developed. Four production bases, one each for Jinan Polyester Fiber, Zibo Qinglun, Weifang Binder, and Yantai New Chemical Fiber will be constructed. Qilu Chemical Fiber, Qingdao Zhongtai, Zibo Xueyin, Weifang Julong, High Density Chemical Fiber, and the Yantai Anlun Companies will become large-scale, highly competitive, and risk-proof enterprises. Research and development of chemical fiber will be enhanced and fiber production modified. The industry will develop and apply new high-tech, highly processed fiber products such as aromatic fiber, carbon fiber, and TENCEL fiber.

It is expected that in 2006-2010, the growth rate of value added in the textile industry will be 5.7 percent annually. Between 2011 and 2020, it will be 5.6 percent. The value added to raw materials by industry will be 36'708 MYuan, 48'440 MYuan, and 83'465 MYuan in 2005, 2010, and 2020 respectively. Table 4.14 shows this development trend and the energy usage of the textile industry from the recent past into the future.

Table 4.14 Developmental trends and energy usage of the textile industry in Shandong

	1995	**2000**	**2005**	**2010**	**2015**	**2020**
Value added (Million Yuan)	17118	25053	36708	48440	64909	83465
Energy intensity (toe/Million Yuan)	119.8	42.09	32.19	27.99	24.49	22.39

7.3 Chemicals

The chemical industry is one of the most important in the province, as Shandong is one of the major chemical producers in China. In 1998, the total production value of the Shandong chemical industry was 64.06 billion yuan, a 4.26 percent increase from 1995, ranking it first among chemical-producing provinces in the country.

The industry is a major user of energy, especially of coal. In 1998, energy consumption amounted to 11.96 Mtoe, nine percent of the energy used in the entire industrial sector. This subsector's use of raw coal, crude oil, and natural gas was 3.38 Mt, 678.9 ktons and 183 Mm^3 respectively, accounting for 10 percent, 87.2 percent and 70 percent of fuel consumed.

In recent years, deeper reforms and structural adjustments have resulted in a significant decrease in the use of energy in the chemical industry. Total energy consumption declined from 11.45 Mtoe in 1995 to 9.07 Mtoe in 1998. Energy consumption per million yuan production value decreased from 179 toe in 1995 to 142 toe in 1998.

According to the provincial tenth FYP and outlook for the next decade of development, the industrial pillars of the province will shift from textile and food to machinery, chemicals, and building materials—capital- and technology-intensive industries. This represents the middle phase of industrialization in which chemicals are produced on a large scale in more technologically sophisticated ways.

In the first decade of the twenty-first century, the production of traditional chemicals is still the most important part of the industry. Rapid development of the national economy will provide a large market for basic chemical raw materials, chemical fertilizers, and petrochemical products. In addition, the international chemical corporations are transferring their manufacture of resource- and labor-intensive and low value-added products to developing countries, creating more opportunities for the chemical industry.

Shandong's petrochemical industry has considerable national stature as resources are available and transportation conditions favorable. The petrochemical industry will develop toward large-scale, technology-intensive facilities. A major series of products will be based on ethylene. Oil refinery equipment will be renovated, and more crude oil imported and processed. It is expected that by 2010, there will be two international level large-scale petrochemical bases. Petrochemicals will be the leading industry in the province.

At present, energy consumption in the chemical industry relies on coal. The "Western Gas to the East" project and imported high quality energy will drive the shift in this industry's energy toward natural gas. Planners project that before 2020, six new or expansion natural gas fired chemical facilities will be constructed. With these plants fully commissioned, the natural gas demand from the industry will be 2.174 billion m^3.

The province will put significant effort into the development of fine chemicals, new chemical materials, biological chemicals, and modern coal-based chemicals. By 2010, the share of fine chemical, high-tech and emerging industries will be 45 percent.

The chemical industry will develop rapidly in the future. Before 2005, due to the large volume of chemicals produced and the transition from extensive to intensive operations, growth will be stable and no higher than ten percent. The growth rate will be nine percent in 2006-2010, and eight percent in 2011-2020. Value added in the chemical industry will amount to 69.03 BYuan, 106.41 BYuan, and 229.53 BYuan in 2005, 2010, and 2020 respectively. The technical renovation, structural adjustment, and use of high quality energy will significantly affect the amount of energy consumed chemical industry. Table 4.15 forecasts trends for the development and future energy use for chemical production.

Table 4.15 Trends in development and energy use in the chemical industry in Shandong

	Unit	1995	2000	2005	2010	2015	2020
Industrial value added	BYuan	23.25	42.70	69.03	106.41	155.56	229.53
Synthesis ammonia	Mt	2.65	4.03	4.33	4.48	4.68	4.88
Fertiliser	Mt	1.98	4.00	4.40	4.80	4.90	5.00
Caustic soda	Mt	0.598	0.90	0.97	1.05	1.17	1.30
Pure soda	Mt	1.187	1.40	1.50	1.60	1.70	1.80
Intensity of chemical industry	toe/ MYuan	489.85	215.89	120.36	72.08	46.89	30.09

7.4 Energy

The production of energy in Shandong includes coal, oil, and gas. The balance between these energy sources will change in decades to come in response to technological development and industrial requirements.

7.4.1 Coal

Energy for Shandong's industries depends on coal, which is abundant and of varied quality in the province. Shandong's coal production is one of the nation's largest. The proven coal reserve in the province is 29.21 billion tons, while the projected reserve at less than 2000 m is 70 billion tons. The accurate reserve is 17.24 billion tons including 15.23 billion tons currently under development and the remaining 2.01 billion tons untouched. The 14 coalfields in the province are national coal and coke coal production bases.

By the end of 1997, there were 723 coalmines in the province with a total designed capacity of 85.58M tons. In 1999, the production capacity was 90M tons. Total asset in state-owned coalmines is over 400 billion Yuan with an annual production value of 14.2 billion Yuan, fifth among the industrial sectors of the province. Its annual profit and taxes paid is 2.1 billion Yuan, second in the province. However, most coalmines in operation are old and declining in production.

Shandong's coal industry faces some important challenges. The "Western Electricity to the East" and "Western Gas to the East" projects, coal and natural gas imports, the exploration and development of offshore oil and gas and nuclear power will affect coal demand in Shandong province and surrounding areas. The development of the western regions and improvements in transportation will benefit western coal and compete with coal production in Shandong. However, the Shandong coal industry will still have the advantages of proximity to markets, a high level of coal technology and equipment, greater efficiency, new production capacity, and a coastal location convenient to external and international markets.

Total coal production in the province will be controlled to keep supply and demand in balance. Twenty-four state-owned coalmines will be closed as will all small township facilities. Exploration for, but no new exploitation of new resources is intended. In accordance with national regulations and coal demand, no new coal wells will be constructed in the immediate future. Any new wells will be larger than 300,000 tons per annum capacity. It is planned that southwest Shandong will be developed as an important energy-export base, especially for coal.

Coal production in the province is expected to reach 90 Mt, 100 Mt, 110 Mt, and 120-125 Mt in 2005, 2010, 2015 and 2020 respectively (SDPC, 1995). In response to demand for high quality coal, the coal washing rate will be increased and reach 30 percent, 40 percent, 50 percent, 60 percent and 80 percent in 2000, 2005, 2010, 2015 and 2020 respectively.

7.4.2 Oil and gas

The nation's second largest oil field—Shengli Oil Field--is in Shandong; it is one of 61 developed oil and gas fields in the province. By the end of 1998, the total proven geological oil and natural gas reserve of the field stands at 3.83 billion tons and 33.09 billion m^3 respectively. The average comprehensive water content of this field is 89.9 percent, and after more than 30 years exploitation, its production is on the wane. Further geological exploration will provide no further resources; the exploitable reserve is in deficit; resource backup is insufficient; and stable production is difficult to maintain. However, further extraction of from old fields and intensive offshore exploration and development will stabilize production for a while. The old wells will mainly rely on water injection extraction. Currently there are 6269 water injection wells with 673,400 m^3 of injected water daily. The accumulated water injection is 3.03 billion m^3; the ratio between water injection and extraction is 0.8.

In 2000 and 2010, the crude oil and natural gas production of the oil fields was and will be 26.75 Mt and 688 Mm^3, 27 Mt and 800 Mm^3 respectively. After 2010, barring any large discovery, oil and gas production will decline.

7.4.3 Electricity

Shandong province has an independent provincial power grid. Since 1990, the electricity industry in the province has developed rapidly. The power generation is only with thermal capacity. Peak load management is difficult. Changdao Island has nine large and advanced wind turbines, each with 600 kW installed capacity. By the end of 2000, the total installed capacity reached 20 GW; and electricity production was 100 TWh. During the ninth FYP period, the commissioned large- and medium-scale unit capacity was 7.95 GW. The total length of 500 kV transmission lines across the province is 1250 km. Shandong was the first province to build a loop grid, consisting of 500 kV transmission lines. The provincial urban and rural grid construction and renovation have been completed.

Due to constraints on coal and water supplies and transportation, it is difficult to site new power plants in the province. The "Western Electricity to the East" project and national grid interconnection will affect the electricity industry. It will develop rapidly, and become the main driver in the growth of available power. It is expected that the total installed capacity and production of power in Shandong will be 33 GW and 170 TWh by 2010 and 40 GW and 217 TWh by 2015.

7.4.4 Forecast of value added in the energy industry

In 1995, value added in the energy industry stood at 34.66 billion Yuan, accounting for 16.4 percent of the industrial sector value added. The share in this figure was coal production (35.22 percent), oil and natural gas production (39.74 percent), electricity and steam and hot water (25.02 percent), and city gas (0.02 percent). In the future, coal, oil, and natural gas production will be constrained as resources dwindle; even now, the Shengli Oil Field is waning. Electricity, steam, and hot water production will be limited by water resources and sites. The national grids will be united, and western power will be transmitted to the east. In general, the growth rate in the energy sector will be limited and its share of value added will decrease. The share in 1995 and 2000 were both 16 percent; in each subsequent five-year increment it will drop one percentage point.

As the share of oil and natural gas available to the energy sector drop dramatically--to 34.5 percent in 2020—and both supply and demand for coal diminish—to 33.5 percent in 2020, reliance on electric power, steam, hot water production and supply will increase to 30.8 percent in 2020. The fastest growth will be seen in city gas production and supply which will together account for 1.2 percent of Shandong's energy utilization in 2020, relying mainly on offshore and imported natural gas (Table 4.16).

7.4.5. Energy conservation

The province attaches great importance to energy efficiency. In the energy sector, technology and management levels are among the country's most advanced. The intensity of energy use—the amount of energy per unit of work performed--is lower than the national average. In 1997, the energy consumption per million Yuan of coal and oil production was 759 toe and 622 toe respectively. The total energy required to produce a million tons of raw coal dropped from 19.7 ktoe to 11.6 ktoe.

In 1995-2000, amount of coal used to produce 1 kWh electricity dropped from 276 goe to 247 goe (Shandong Statistics Bureau, 2000); the average national level in 1995 was 288 goe.

Deep coal extraction, a high level of mechanization, and coal washing will increase energy consumption. Extraction from aging oil fields is more difficult. More and more wells are kept in production by water injection, a process requiring energy. However, technological progress and improved energy efficiency will reduce energy use per unit of product in the power sector. The development of new coalmines in the southeast area of the province will reduce the energy required to produce coal, overall. In general, the energy intensity of the energy industry will not change greatly (see Table 4.17).

Table 4.16 Activity level for each energy industry in the province

Unit: Billion Yuan

	1995		2000		2005		2010		2015		2020	
	Value added	Share %	Value added	Share %	Value added	Share %	Value added	Share %	Value added	Share %	Value added	Share %
Industry	211.3		352.9		547.9		794.1		111.9		149.0	
Energy sector	34.7		56.5		82.2		111.2		145.5		178.9	
The share of energy sector in industry	16.4		16		15		14		13		12	
Coal production	12.2	35.30	19.8	35	28.4	34.5	37.8	34	49.2	33.8	60.0	33.5
Oil and natural gas production	13.8	39.68	22.3	39.47	32.1	39	42.2	38	54.1	37.2	61.7	34.5
Power, steam, hot water production and supply	8.7	24.99	14.4	25.5	21.4	26	30.2	27.2	40.7	28	55.1	30.8
City gas production and supply	0.008	0.02	0.017	0.03	0.411	0.5	0.889	0.8	1.46	1	2.15	1.2

Table 4.17 Energy intensity trends in the energy industry in Shandong

	1995	2000	2005	2010	2015	2020
Value added (BYuan)	34.66	56.46	82.18	111.17	145.49	178.85
Energy consumption (Mtoe)	4.56	7.10	10.53	14.07	18.52	22.51
Energy intensity (Mtoe/BYuan)	0.1317	0.1258	0.1282	0.1266	0.1273	0.1258

7.5 Energy-intensive Products in the Industrial Sector

Energy-intensive products include cement, glass, paper, iron, and steel. For each we review the current status of the industry, its current energy consumption profile, and future energy needs.

7.5.1 Cement

Building material production is important to Shandong. The growth of value added continues to be stable. The main product, cement, ranks the national forefront, accounting for 11 percent of total national production. Per capita cement production is 780 kg, near the level of industrialized countries.

In 1990-1995, regional economies were developing rapidly and a large number of infrastructure projects were underway which resulted in strong growth for the building material industry. The average annual growth rate of the cement industry was 11 percent during this period (Shandong Statistics Bureau, 1995). After 1995, reduced demand and capital circulation began to constrain the construction of many civil and other projects; by 1998, the growth rate of cement production was negative.

Historically, the industry had been dominated by vertical kilns that are not very productive, somewhat unstable, and produce relatively low quality cement. During the 1990s, technologically advanced, less polluting, and energy-saving rotary kilns began to replace vertical kilns and production increased. The production share of higher quality, stable, and energy-efficient rotary-kiln cement is still small, but it increased from seven percent in 1990 to 12 percent in 1995. This figure will increase in the future.

In recent years, all kilns in the province have been renovated to reduce energy consumption and ultimately replace vertical kilns with pre-calcium rotary kilns. Total energy consumption for cement production has been reduced from 140 kgoe/ton to 119 kgoe/ton. In large- and medium-scale cement plants with rotary kilns, it can be reduced to 98 kgoe/ton. It is expected that until 2020, the growth rate of cement production will be around 1.8 percent (see Table 4.18) as advanced, high quality rotary kilns replace all vertical kilns.

Table 4.18 Recent and future cement output in Shandong

	1995	**2000**	**2005**	**2010**	**2015**	**2020**
Output (Mt)	54.97	60.10	67.01	72.90	79.31	85.44

The cement industry is a major user of energy, especially of coal. In 1995, the industry consumed 2.35 Mtoe of this fuel, 85 percent of its total energy consumption. Fuel oil use totaled 0.219 Mtoe, accounting for eight percent. This structure will change somewhat in the next 20 years. Until 2020, coal will remain dominant and the use of fuel oil will not change. The consumption of electricity, LPG, and natural gas will increase significantly.

Due to stricter environmental regulation of fuel combustion equipment and continuous efforts to reduce energy consumption, increased cement production will not require commensurately greater use of fossil fuel; growth will be only in the range of 0.7 to 0.9 percent.

The intensity of energy use (unit of energy per unit of product) in Shandong's cement industry will be lower than the advanced international or even the national average because of statistical deviation: Cement production is calculated on the basis of circulation and stock. In official calculations, the output of county-level and private producers is neglected. (These plants will be converted to clinker production.) Only state-owned and collective cement factories were included in the national and comparative calculations of energy production. To reduce costs, the small plants buy energy through many channels, including from other provinces. In the model, this energy is not included in the circulation market and not accounted for in the statistics. Therefore, the comprehensive energy use per unit of product in the province in the model appears lower than the national average.

The main measures to reduce energy intensity in the cement industry include developing large and medium-size cement plants; shutting down small, highly polluting and energy-using plants; gradually discarding some vertical and dry hollow kilns and renovating mechanized vertical kilns to save energy; and adopting pre-calcium technology. Table 4.19 forecasts fuel use in the production of energy for the cement industry in Shandong.

Table 4.19 Energy forecasting by fuel for cement industry in the province

Unit: ktoe

Year	**1995**	**2000**	**2005**	**2010**	**2015**	**2020**
Total	2789.8	2918.6	3059.5	3205.2	3346.2	3509.0
Coal	2352.5	2414.6	2498.3	2584.7	2645.3	2717.6
Fuel oil	219.3	229.8	242.5	253.6	267.5	282.6
LPG	21.3	32.9	31.9	36.3	50.5	64.7
Natural gas	52.7	65.8	93.0	106.0	133.2	161.8
Electricity	144.0	175.5	193.8	224.6	249.7	282.3

7.5.2 Glass

In recent years, the glass industry in the province has grown rapidly. Production in 1997 was four times that in 1990 (Shandong Statistics Bureau, 1997). Shandong's share in national glass production increased from five percent in 1990 to seven percent in 1995. The expansion and renovation of large and medium-size enterprises are responsible for an in Shandong's share of particularly high quality glass. Before 1990, the up-dragging (65 percent of all glass production) and horizontal dragging techniques were dominant in the industry. In 1995, the share of up-dragging production had increased to 84 percent and horizontal output dropped to 15 percent. In the following year, the float technology was adopted and widely implemented. At present, the shares of floating glass, up-dragging, and horizontal production are is 50 percent, 40 percent, and 10 percent respectively.

The fuels mainly used in the glass industry are coal, fuel oil, and electricity. Electricity use accounts for 85 percent of all energy consumed in the glass-making process. Coal is used in kilns to melt raw material, accounting for 13 percent of the total energy use. Fuel oil, and small amounts of LPG and natural gas are used for kiln firing and to power equipment in factories. The national average comprehensive energy use per case of flat glass, floating glass, up-dragging glass, and horizontal glass is 19 kgoe, 18 kgoe, 23.6 kgoe and 37 kgoe respectively. In Shandong, these figures are 20.7 kgoe, 18 kgoe, 23 kgoe and 38 kgoe respectively. The consumption of coal, fuel oil and electricity per case of flat glass is 18.7 kgoe, 11 kgoe and 4.7 kgoe respectively.

Demand for glass in the province will be large in years to come. The growth rate in this industry is expected to be five percent until 2010, reaching 37.92 M cases. In 1995, production was 13.43 M cases—a figure attained with no increase in the number of plants once the float technique was adopted. The use of the up-dragging and vertical techniques will be completely discarded in 2000 and 2010 respectively, and by 2020 the floating technique will have been adopted throughout the industry. In the next two decades, the two percent energy consumption growth rate in the glass industry is expected to be lower than production growth, due to adoption of large-scale and advanced floating technology. Electricity consumption growth after 2000 will slow slightly.

Coal and fuel oil consumption will fluctuate. Until 2010, the annual growth rate will be three percent, as fuel oil use replaces coal. After 2010, glass production growth will slow, reducing fuel consumption. After 2015, gas will replace fuel oil and total fossil fuel consumption will decrease further.

Nationally, comprehensive energy consumption per glass case will drop to 23 kgoe in 2005 and 20 kgoe in 2010. The share of the glass industry in Shandong's economy will exceed its current national share, and its energy use per unit product will be lower.

The main energy efficiency measures for the glass industry in the province include developing large-scale floating glass technologies; discarding small glass plants; and renovating the existing medium-size and small glass kilns to increase the melting rate.

7.5.3 Paper

Shandong province is one of China's large paper producers. In 1995, paper and paperboard production was 3.17 Mt. In 1997, there were about 500 paper plants in the province with fixed assets of 10 billion yuan, accounting for 12.3 percent of the country's total. In that year, production was 3.42 Mt, highest in the country. In 1998, production declined to 2.79 Mt (Shandong Statistics Bureau, 1998).

With technical progress, product quality is increasing; the distribution of products is improving; and the scale of plants is expanding. At present, enterprises with over 20'000 tons capacity have discarded 1092 round grid machines, renovated others, and introduced long grid machines. Large-scale production accounts for over 60 percent of the total. However, the paper industry of the province faces five main issues.

First, the production of first-class products is restricted by use of a single material. Less than 10 percent of paper is made of wood fiber, lower than the country's average level of 14 percent. New non-wood fiber pulps such as red flax and sesbania pulp are developing slowly. The waste paper recovery and utilization rate is lower than 30 percent and waste paper pulp accounts for less than 20 percent of total pulp in the province. The US waste paper recovery rate is 50 percent.

Second, the types of paper being produced do not fit the needs of the province. Fifty percent of Shandong's paper products are of low and medium quality, while first class paper has to be imported or purchased from other provinces.

Third, plant scale is small, averaging only 5000 tons. The total number of plants is 500. Eighty of them have capacities over 20’000 tons (including 47 plants over 30’000 tons); these account for 70 percent of total capacity. Enterprises of less than 30’000 tons account for 90 percent of the total. The 300 plants under 10’000 tons capacity will be closed.

Fourth, technology is backward, and energy and material consumption is high. The technology of the paper industry in the province is mainly at the level of the 1960s and 1970s. The average international coal consumption per ton of paper is 0.63 toe with 100 tons of water use. In Shandong, these figures are 1.05 toe and 300 tons respectively.

Fifth, environmental pollution is serious. Only large and some medium-size enterprises have installations for alkaline and acidic waste liquid recovery. According to national and provincial regulations, all plants are required to adopt control measures to meet emission standards. This is a heavy burden for medium-size and small enterprises.

In the recent half century, with the world economic development, paper and paperboard production has grown annually by three to five percent, on average. In the developed countries, the paper industry is an important element of economic growth. In Japan, large amounts of raw material is imported to develop the paper industry; Japanese per capita paper use is over 200 kg, a dozen times higher than in China. According to UN Food and Agriculture Program statistics, the paper and paperboard consumption growth rate in developed countries parallels GDP growth rate, while in developing countries consumption growth has exceeded GDP growth.

China is the third largest paper and paperboard producer in the world; half of this total is produced by small plants. The variety, quality, and per capita consumption of paper products are much lower than that of developed countries and the world’s average. In 1995, the world average per capita paper and paperboard use was 50 kg. In the U.S., Asia, and China, these figures were 332 kg, 25 kg, and 16.6 kg respectively. Per capita domestic paper use in the developed countries is 8 kg, but only 0.8 kg in China, and 1 kg in Shandong.

In years to come, Shandong's paper industry will pursue quality improvement, variety expansion, energy consumption reduction, efficiency improvement, pollution reduction, and waste paper recovery. Table 4.20 forecasts these elements of the industry's future.

Table 4.20 Outlook of paper output in Shandong

	1995	2000	2005	2010	2015	2020
Paper output (Mt)	3.17	4.03	6.2	9.1	13.1	18.0
Paper output per capita (t/person)	0.036	0.045	0.067	0.096	0.135	0.184

In 1995, energy use per unit of paper product in Shandong province was 0.26 toe/t. This is lower than the national average of 0.41 toe/t for four reasons. First, the province is first in the nation in paper production. Many large plants with advanced technology and high energy efficiency have been constructed. In 1995, the energy consumption per unit of product was 1.11 toe/t in China but 1.05 toe/t in Shandong. Second, the energy share for related activities is only five percent, lower than in China as a whole. Third, the pulp is mainly imported or raw—no energy is consumed to produce it; the share of paper made from pulp produced in the province is lower than the country average. Based on a 90 percent paper production rate in 1995, 36.6 percent of the paper made in China was processed from domestically produced pulp; in Shandong, this figure was only 16.75 percent. Fourth, because Shandong is a big wheat producer, its machinery pulp is virtually all wheat straw pulp, higher than the country average. The energy use for processing wheat straw pulp is lower than for wood pulp.

Energy consumption per unit product in China is also lower than the international level for three reasons. First, the statistical coverage of machinery paper and paperboard production and energy consumption is different in as calculated in China and internationally. Second, the share of machinery pulp made in China is small; most pulp is imported or raw. Furthermore, the share of wheat straw in the machinery pulp is also high; it costs less energy to pulp this material. The more energy intensive first quality paper is made mostly from imported pulp. Third, paper production consumes less energy in China because little effort is made to recover waste paper or to protect the environment.

Table 4.21 Energy intensity trend of paper industry in Shandong

	1995	2000	2005	2010	2015	2020
Total energy consumption (ktoe)	826.2	934.3	1297.0	1732.7	2273.2	2871.7
Paper and paper board production (Mt)	3.17	4.03	6.20	9.10	13.07	18.00
Energy intensity (ktoe/Mt)	260.6	231.8	209.2	190.4	173.9	159.5

In the future, the scale of paper plants will become larger; technology will be more advanced; paper quality will improve; and energy consumption per unit of product will decrease. Meanwhile, more effort will be made to utilize waste paper and protect the environment. This will require greater energy consumption. Wood pulp production is limited by forestry resources. Such high quality pulp will be imported or processed from fast-growing woods. Thus, for pulp also, energy consumption per unit of product will decline in the future. (See Table 4.21.)

7.5.4 Iron and steel

After 1996 when steel production in China exceeded 100 Mt, the "scale expansion" was replaced by a "structural adjustment" in the iron and steel industry. Priority has shifted from quantitative growth to quality improvement and energy saving and rapid production growth has stabilized. Meanwhile, a structural supply issue emerged: There is an oversupply of ordinary steel product, while first class steel products have to be imported. The ratio of iron and steel in China and in the province was 1.091; the ratio of finished product and steel between China and the province was 0.846. These figures are 5.7 and 9.1 percentage points lower for Shandong than the country average

It is expected that in the next 10 years, the production of ordinary steel will be reduced and high quality production will be increased. Total steel production in 2000, 2010, and 2020 will be 5.9 Mt, 8 Mt, and 10 Mt respectively (SDPC, 1995, see Table 4.22).

Table 4.22 Forecast of iron and steel industry development in the province

	1995	2000	2005	2010	2015	2020
Steel (Mt)	3.89	5.9	6.8	8	9.5	10
Iron and steel ratio	1.17	1	0.934	0.924	0.912	0.9
Rotary furnace/electric furnace	0.816/0.18	0.8/0.2	0.78/0.22	0.76/0.24	0.74/0.26	0.7/0.3
Continuous casting ratio	51.9		80	90	95	100

8. CONSTRUCTION

The construction sector is fundamental to the economy of Shandong. It has had several stages of development. In the first stage the number of enterprises was large, with a single product type, and no comparative advantage among them. A strategic adjustment shifted the sector to knowledge-intensive processes and large and medium-scale operations. Later, the industry was restructured through asset consolidation and capital optimization. Now, corporations own not only building enterprises, but also building material enterprises such as cement, glass, and ceramic plants. The sector is changing from production- to capital-intensive.

The construction sector accounts for two percent of total energy consumption, mainly in the forms of diesel fuel, gasoline, and electricity. The main energy users are civil engineering, currently absorbing 60 percent of all energy in the construction sector, and decoration, in which energy use will grow rapidly in the near future. Diesel fuel and gasoline are used to run construction equipment and vehicles. Electricity, contributing 10 percent of construction energy, is not stable because of unpredictable factors in the operation of equipment and in electricity prices.

The construction sector has seen rapid development in the past ten years. Its GDP growth rate between 1990 and 1995 was 17 percent, dropping to five percent between 1996 and 2000. The reason is that economic development was rapid in the early years and stimulated many civil engineering projects. Because many medium- and small-scale enterprises had low technology and capital content, they had difficulty surviving and developing. The result was low growth in later years. Structural and market adjustment will improve capital and technology and reduce bubble economy effect. The industry should develop steadily in the next 20 years at a rate lower than GDP growth. The average annual growth rate of the industry after 2000 will be 2.6 percent.

In the past ten years, the rapid development of the construction sector drove demand for motor fuel and electricity. The growth rate for motor fuel, mainly diesel, use should decline to some extent as the result of greater energy efficiency. Because the consumption of diesel fuel has not been stable in recent years, it is difficult to forecast future use. As mentioned above, electricity consumption is difficult to forecast for the same reason. In the long term, the energy structure will be further adjusted. Additional equipment will require more electricity in the next 20 years, but the growth rate will be uneven. Electricity consumption changes in each stage of a construction project as different equipment is utilized at each stage.

Small changes will occur in the next 20 years for other energy types. Lesser amounts of coal, used mainly in boilers, will be used after 2010 as the energy structure is adjusted and facilities renovated. Coke consumption is small and stable and will show little growth. Fuel oil consumption will increase slightly in the future. Growth in the use of LPG will be large until 2010. At that point, it will be partly replaced by other high quality energy. It is expected that natural gas consumption will grow significantly in the construction sector after 2010.

The construction sector consumes a great deal of diesel oil, mainly for transportation vehicles. The travel mileage is short but one-way-empty trips are frequent. There is potential to increase energy efficiency in this area. Greater mobility and open-air operations result in more equipment damage, power leakage, and less protection. For these reasons, electricity consumption, though relatively large, is unstable. The efficiency of equipment is low.

9. TRANSPORTATION

Growth in the demand for freight transport is expected slowing down; the elasticity between freight transport and GDP growth will be 0.5-0.6. However, passenger transport will grow rapidly. The transportation infrastructure facilities built in the eighth and ninth FYP period will play an important role in facilitating the mobility of people as they are motivated by the commercialization process and changes in perception. Accelerated urbanization will attract excess labor from rural to urban areas. The elasticity between passenger transport and GDP growth will be 1.0-1.2. Influenced by the world economy, foreign trade transport growth is expected to slow.

In the near future, the structure of freight and passenger transport will change greatly. In the past, energy and raw materials took up a large share of transportation. The transport of coal, oil, metal ores, iron and steel, and mineral construction material accounted for up to 60 percent of total freight volume. However, economic, industrial, and product structure adjustments are changing the transport profile. Changes in the sources and efficiency of energy production will reduce coal demand. Transportation through Qingdao port, Rizhao port, Jiaoji, Heyunshi, Jinghu and Jingjiu railways, the main channels of coal transport, will decrease. The expansion of domestic demand and accelerated urbanization will demand the kind of infrastructure construction that drives iron and steel and mineral building material growth. Transport demand for high value-added products, special goods, and fresh and live agriculture products, will expand though the total will not be large.

The passenger transport subsector will see a significant increase in private trips. This will cause the transport on the Jiaoji railway, which connects major cities to grow considerably. High-speed transport will develop rapidly through the construction of railways, motorway grids, and aviation facilities. The forecast for passenger and freight transport volume is shown in Table 4.23.

Table 4.23 The outlook for traffic in Shandong

	1995	2000	2005	2010	2015	2020
Traffic (Billion Passengerkm)						
Air	1.43	1.93	2.50	3.20	4.00	4.89
PUBLIC TRANSPORT	52.39	88.71	135.25	188.77	250.87	319.86
Rail	17.42	21	26.99	40.52	44.00	47.97
Road	34.74	67.32	107.68	147.54	206.07	270.99
Water	0.23	0.38	0.57	0.70	0.80	0.90
PRIVATE TRANSPORT	12.77	37.94	71.47	105.89	151.59	213.93
TOTAL	66.59	128.58	209.22	297.86	406.46	538.69
Freight Transport, Traffic (Billion tkm)						
Road	26.40	29.01	35.41	42.12	53.29	65.15
Rail	69.86	75	92.50	115.10	140.04	170.17
Waterway	16.40	20.01	21.01	23.50	25.97	30.05
Pipe Line	10.51	12.01	13.51	15.01	16.97	20.00
TOTAL	123.16	136.03	162.43	195.72	236.27	285.37

10. SERVICES

In the last 10 years, the economic structure of the province has been adjusted continuously. The growth rate of value added in services (the tertiary sector) is faster than that of the GDP. In 1995, the value added in the tertiary sector was 161.95

billion yuan, accounting for 32 percent of total provincial GDP (Shandong Statistical Bureau, 1995). In 1990-1995, final energy consumption grew in the tertiary sector, but decreased thereafter due to the development of real estate, financing and insurance, industries with low energy consumption, and to energy conservation measures.

At present, the energy use profile of the tertiary sector energy has some notable characteristics. First, coal is still the dominant fuel. Second, the use of high quality energy is increasing; vehicles used by the finance, insurance, and real estate industries consume about 30 percent of all energy in this sector, but the currently small share of electricity increases annually. While the sector is developing rapidly, its energy demand is stable at about 4.2 – 4.9 Mtoe annually.

Growth in the tertiary sector will be faster than GDP in the next 20 years (SDPC, 1995). The value added and share will be 396.66 billion Yuan (39 percent) in 2005 rising to 1325.64 billion Yuan (46 percent) in 2020. This growth will take place mainly in the subsectors of real estate, financing, information consulting, tourism, and district service.

In the next 20 years, development in the tertiary sector will expand energy consumption but the structure will change dramatically. While still quite large, coal consumption will grow much more slowly than other energy sources. In widely used equipment such as boilers, clean energies will replace coal.

The share of electricity, currently growing faster than any other source of power, will increase greatly, slowed by improvements in energy efficiency. Other important fuels are diesel, fuel oil, and natural gas. It is expected that diesel and fuel oil consumption will exceed coal and share half of the total energy consumed by 2020. Once the infrastructure is in place, natural gas use will grow faster than other energies.

The motor power capacity of the tertiary sector is smaller than in other industries. There are time and seasonal difference for equipment operation and a large potential to save motor energy. Many small coal-fired boilers will be phased out. Boilers using high quality energy will be promoted and district heating will be encouraged.

11. HOUSEHOLDS

Household energy use reflects population, urbanization, economic development level, energy use patterns, energy supply patterns, climate, etc. In 1999, the disposable income of Shandong's urban residents was 5809 Yuan per capita, similar to the country's average of 5854 Yuan. Rural income was 2520 Yuan, slightly higher than the country average of 2210 Yuan. These income gaps were smaller than in other Chinese provinces. Household appliance ownership per 100 households in Shandong and in China in 1995 and 1998 is compared in Table 4.24 (Shandong Statistics Bureau, 1995, 1998, National Bureau of Statistics, 1995, 1998).

Table 4.24 Comparison of household appliance ownership between Shandong and China

Unit: set/100 households

			Washing Machines	**Fans**	**Refrigerator**	**Color TV**	**Air conditioner**
Shandong	**1995**	Urban	88.60	193.42	76.01	90.55	4.07
		Rural	11.40	112.57	6.93	18.14	0.45
	1998	Urban	87.82	196.82	85.83	106.20	15.21
		Rural	13.19	136.64	13.52	35.79	
China	**1995**	Urban	88.97	167.35	66.22	89.79	8.09
		Rural	16.90	88.96	5.15	16.92	
	1998	Urban	90.57	168.37	76.08	105.43	20.01
		Rural	22.81	111.59	9.25	32.59	

Source: China Statistical Yearbook, 1995, 1998 and Shandong Statistical Yearbook, 1995, 1998

The per capita residential energy consumption in the province is considerably lower than the national average. In 1995, it was 37 kgoe, only 40.6 percent of the country average. The gap between consumption in rural and urban areas, - 90 kgoe and 19 kgoe respectively - , is also large (See Table 4.25).

Table 4.25 Comparison of per capita energy consumption between Shandong and China

	Shandong (kgoe)	**China (kgoe)**	**Shandong/ country**
Total	37	91	40.6%
Urban residents	90	167	53.9%
Rural residents	19	59.5	31.9%

In the future, the gap between energy use in rural and urban areas will begin to close. The big cities will use more electricity and piped-in natural gas. Medium cities will make extensive use of LPG, electricity, piped-in natural gas and locally produced gas. Rural and small city residents will use clean coal, electricity and locally produced gas rather than biomass. Electricity demand will grow the fastest, followed by all types of gas. The share of coal and kerosene will decrease.

In the development process, the use of energy for cooking will decrease gradually, while energy used by household appliances will increase. The forecast for appliance ownership is shown in Table 4.26.

Table 4.26 Forecast of household appliance ownership

Unit: Million sets

	1995	2000	2005	2010	2015	2020
Washing machine	7.87	9.76	11.84	14.08	16.57	19.1
Fans	31.25	33.57	35.81	37.94	40.06	41.99
Refrigerator	6.29	7.98	9.73	11.40	13.03	14.38
Color TV	22.13	23.21	24.23	25.18	26.11	26.94
Air conditioner	0.229	0.73	1.85	3.4	5.43	7.85

12. EVOLUTION OF ELECTRICITY CONSUMPTION

By analyzing the growth rate and structure of electricity consumption between 1985 and 2000 (see Table 3. in Appendix F and Figure 4.4), we can see that electricity consumption increased most rapidly in the service and household sectors. The share of these two sectors in total energy consumption also increased. The industrial sector accounted for the largest share of total electricity consumption and also increased rapidly during 1985-1995.

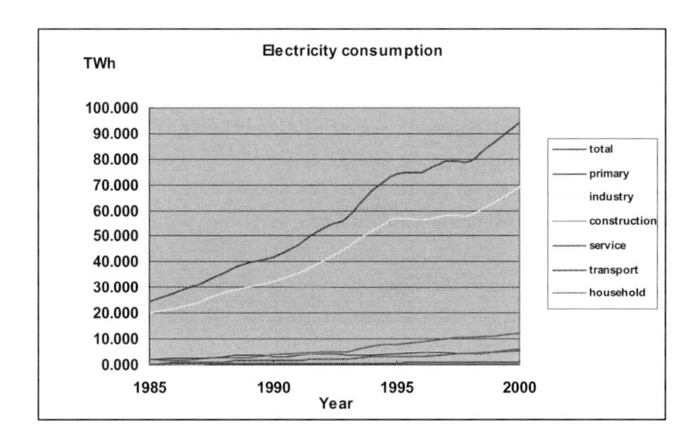

Fig 4.4 Evolution of electricity consumption

Electricity consumption is an important indicator of socioeconomic development. Between 1985 and 1995, consumption per household went from 387.54 kWh to 532.45 kWh in urban households (3.83 percent annually), from 1995 to 2000, the rate of increase was seven percent per annum. In rural areas the 1985-1995 increase went from 61.21 kWh to 269.96 kWh (16.4 percent annually); from 1995-2000, the annual rate of increase dropped to six percent.

As the values in Table 4.27 show, the intensity of electricity use decreased in the secondary sector. This is attributable to efficiency improvement and changes in the economic structure. In this same time period, the intensity of energy use in the primary sector and tertiary sector increased slightly. Because of the electricity consumed in the secondary sector, overall intensity decreased. Figure 4.5 shows the relation between electricity consumption and GDP. Trends during 1995-2000 reflected in Table 4.27 are corroborated in the figure, except that energy intensity in the primary sector intensity did not change.

Table 4.27 Changes in the intensity of electricity use (% p.a.)

	Total	**Primary**	**Industry**	**Construction**	**Service**	**Trans&etc.**
Change of intensity in 1985-1995 (% p.a.)	-0.3	0.9	-5.6	-1.6	2.2	-4.8

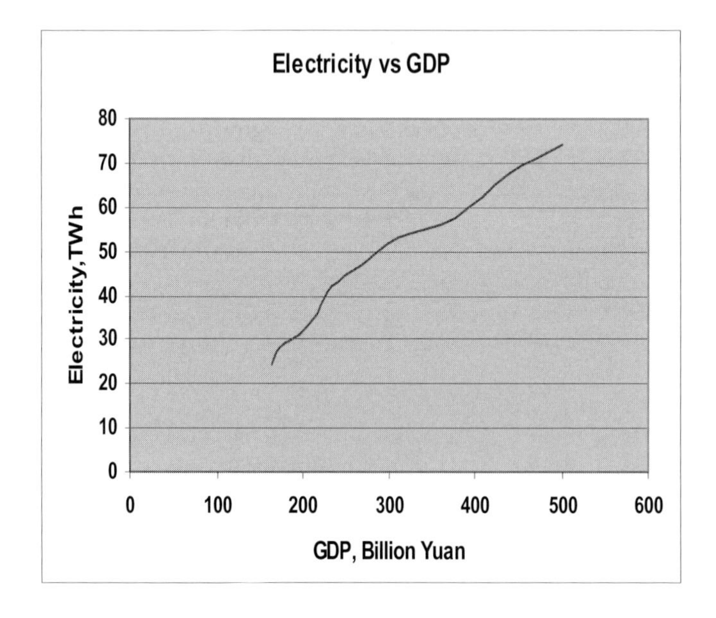

Figure 4.5 Relation between electricity consumption and GDP

Comparing economic development and electricity consumption in China, Shanghai, and Shandong (see Table 4. in Appendix F), we can see that Shandong parallels the economic development, economic structure, and electricity consumption level and structure of China as a whole, while Shanghai is a more developed industrial city. Some basic conclusions can be obtained from the analysis: First, electricity consumption increase will slow down in the future as energy efficiency improves, and as the economic structure shifts towards less electricity-intensive economic activities. Second, although the secondary sector is still the largest consumer of electricity in the province, its share will decrease, and that of the tertiary sector increase. Third, the intensity of electricity use has decreased in the household sector, and slightly increased in the service and industrial sectors, trends that should continue. Finally, the rapid increase in household electricity use will continue in both urban and rural areas.

13. ASSUMPTIONS INCLUDED IN FORECASTING ELECTRICITY DEMAND IN THE ADDITIONAL SCENARIOS

The assumptions for forecasting electricity demand in the additional scenarios are shown in the Tables 5. - 8. in Appendix F. For the household sector, assumptions about total population growth, urbanization rate, family size and growth of average electricity consumption per household in the future are made. For comparison, the historical data from 1985-1995 are also shown in the table (Table 5. in Appendix F).

For the production sectors, the chosen variables are GDP, economic structure, and intensity of electricity use. Table 6. in Appendix F gives the assumptions for these variables in different time periods. It shows four scenarios constructed on the basis of factors parallel to the basic scenarios. The Baseline-E scenario includes the basic-scenario values for GDP, economic structure, and intensity of electricity use. The Lowgdp-E scenario assumes lower GDP growth compared to the baseline scenario. The Structure-E scenario is based on changed economic structures. The Inefficient-E scenario assumes a slower decrease of intensity in all sectors. The assumptions for changes in electricity intensity in industry subsectors and EIPs are shown in Table 7. and Table 8. in Appendix F. Historical data from 1985-1995 were also shown for comparison.

14. FORECASTING RESULTS AND ANALYSIS

This section discusses final energy demand in general, by sector, and by fuel type. The authors then compare demand forecasts for the four scenarios (baseline, lowgdp, structure, and inefficient) in general and by sector.

14.1 Final Energy Demand

Between 1995 and 2020, the average annual growth rate of energy demand will be 2.7 percent. Table 4.28 and Figure 4.6 show the forecast for final energy demand in the baseline scenario. These figures are broken down by sector and by fuel in Table 9. in Appendix F.

Table 4.28 Final energy outlook in Shandong

Unit: Mtoe

	1995	2000	2005	2010	2015	2020
TOTAL	43.55	47.27	56.13	64.48	75.29	84.63

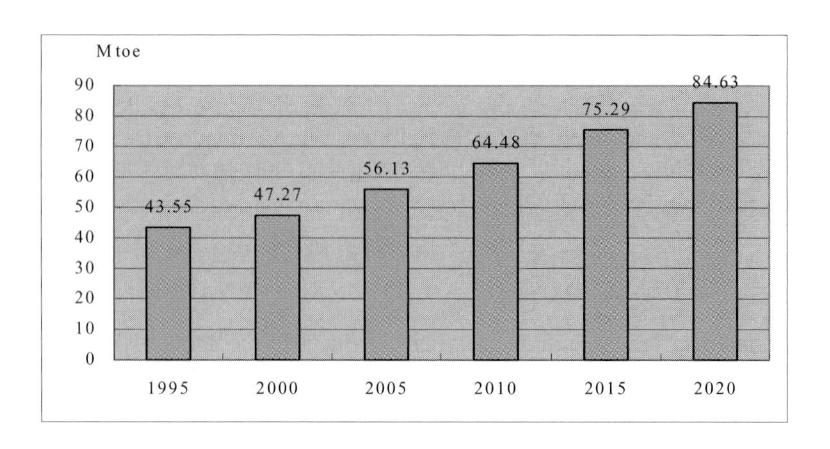

Figure 4.6 Final energy demand

As shown in Table 4.29 and Figure 4.7, energy consumption in the future will shift to high quality fuel. The share of coal will have decreased from 48.66 percent in 1995 to 36.23 percent in 2020. The use of gas and electricity will have increased from 6.02 percent and 14.49 percent in 1995 to 7.36 percent and 26.36 percent resctively by in 2020.

Table 4.29 Forecasted energy breakdown by fuel in Shandong in baseline scenario

unit: %

	1995	2000	2005	2010	2015	2020
Gasoline	4.56	5.19	5.96	6.15	6.27	6.39
Diesel	6.97	7.72	8.15	8.74	8.96	9.48
Jet Fuel	0.16	0.18	0.19	0.20	0.20	0.21
Fuel Oil	10.57	8.32	6.66	5.58	4.90	4.48
Kerosene	0.37	0.27	0.18	0.12	0.08	0.06
LPG	0.97	1.14	1.34	1.47	1.56	1.60
Gas	6.02	6.20	6.42	6.54	6.98	7.36
Coal	48.66	45.42	42.63	40.27	37.94	36.23
Charcoal	5.61	5.87	5.46	5.36	5.71	5.54
Electricity	14.49	17.77	20.88	23.30	25.09	26.36
Motor Fuels	1.62	1.91	2.13	2.27	2.32	2.30

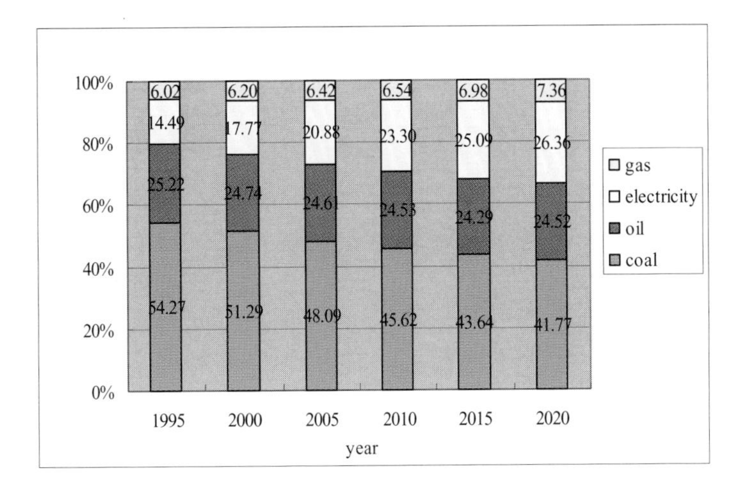

Figure 4.7 Final energy breakdown by fuels

The final energy consumption breakdown by sector will not change greatly. Though industry's share will have dropped from 71.8 percent in 1995 to 66 percent in 2020, this sector will still be the largest consumer of consumer. The share of the transportation sector will increase. Table 4.30 and Figure 4.8 show the structure of energy use forecast energy by sectors.

Table 4.30 Forecast for the energy use structure by sector in the baseline scenario

unit: %

	1995	2000	2005	2010	2015	2020
Agriculture	4.94	6.31	7.36	7.29	6.86	6.55
Industry	71.83	68.48	66.60	65.89	66.23	66.02
Construction	0.35	0.41	0.45	0.51	0.52	0.58
Service	7.01	7.25	6.93	6.86	6.68	6.74
Transport	7.41	8.11	9.09	9.48	9.76	10.26
Household	8.47	9.44	9.57	9.97	9.95	9.84

In the four scenarios, energy growth of the Lowgdp scenario is the lowest, - 6 percent lower than that of the baseline scenario. There is no big difference between the Structure scenario and Baseline scenario. Energy demand in the inefficient scenario is 11.7 percent higher than that of Baseline scenario (See Table 4.31 and Figure 4.9).

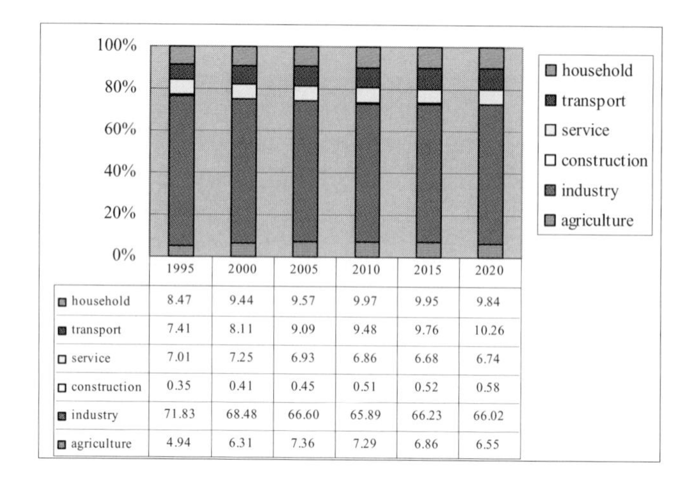

	1995	2000	2005	2010	2015	2020
household	8.47	9.44	9.57	9.97	9.95	9.84
transport	7.41	8.11	9.09	9.48	9.76	10.26
service	7.01	7.25	6.93	6.86	6.68	6.74
construction	0.35	0.41	0.45	0.51	0.52	0.58
industry	71.83	68.48	66.60	65.89	66.23	66.02
agriculture	4.94	6.31	7.36	7.29	6.86	6.55

Figure 4.8 Forecast of final energy structure by sectors in the Baseline scenario

In the four scenarios, energy growth of the Lowgdp scenario is the lowest, - 6 percent lower than that of the baseline scenario. There is no big difference between the Structure scenario and Baseline scenario. Energy demand in the inefficient scenario is 11.7 percent higher than that of Baseline scenario (See Table 4.31 and Figure 4.9).

Table 4.31 Comparison of final energy demand among scenarios

Unit: Mtoe

Scenarios	**1995**	**2000**	**2005**	**2010**	**2015**	**2020**
Baseline	43.55	47.27	56.13	64.48	75.29	84.63
Lowgdp	43.55	46.5	53.05	58.6	65.78	71.11
Structure	43.55	46.83	55.61	63.86	74.53	83.73
Inefficient	43.55	48.35	58.67	68.84	82.17	94.55

The forecasting results for electricity demand in the additional scenarios are shown in Table 10. in Appendix F and Figure 4.10. Figure 4.11 shows the forecast for the electricity demand structure in 2020.

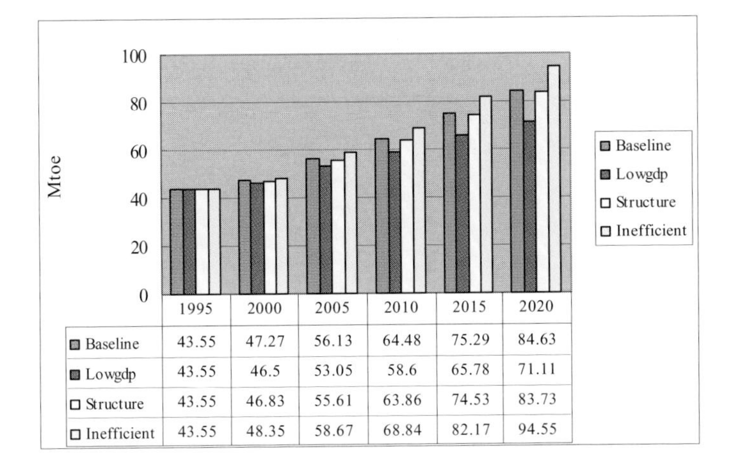

	1995	2000	2005	2010	2015	2020
Baseline	43.55	47.27	56.13	64.48	75.29	84.63
Lowgdp	43.55	46.5	53.05	58.6	65.78	71.11
Structure	43.55	46.83	55.61	63.86	74.53	83.73
Inefficient	43.55	48.35	58.67	68.84	82.17	94.55

Figure 4.9 Comparison of final energy forecast among scenarios

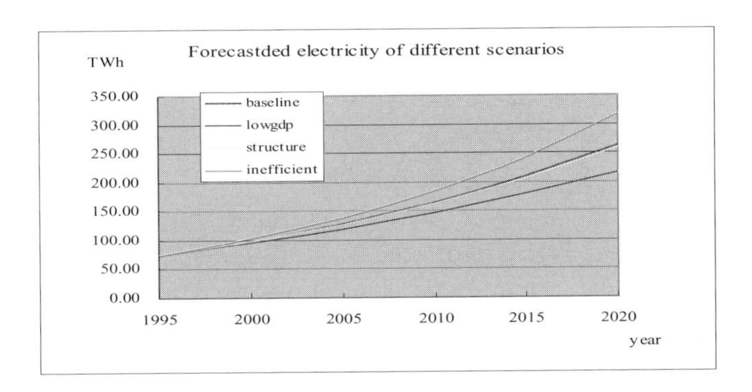

Figure 4.10 Forecast of electricity demand in the different scenarios

As seen in Table 4.32, the growth rate of electricity demand in Shandong will be 4.4 to 6.0 percent each year in different scenarios. For reference, in the tenth FYP, the forecast for electricity consumption growth in China was five percent annually in 2000-2005 (SEPCO, 2000). The IEA perspective on national electricity consumption growth shows a 5.4 percent yearly increase from 1995 to 2020 in China (IEA, 1998).

Table 4.33 compares the sector structure of electricity demand, comparing the forecasts for electricity demand in the baseline and three hypothetical scenarios between 1995 and 2020. The share of the agriculture sector won't change much; consumption by the industry sector will decrease, and the shares of the service and household sectors will increase. Note that the forecasted electricity demand structure in 2000 and the actual structure in 2000 are very similar.

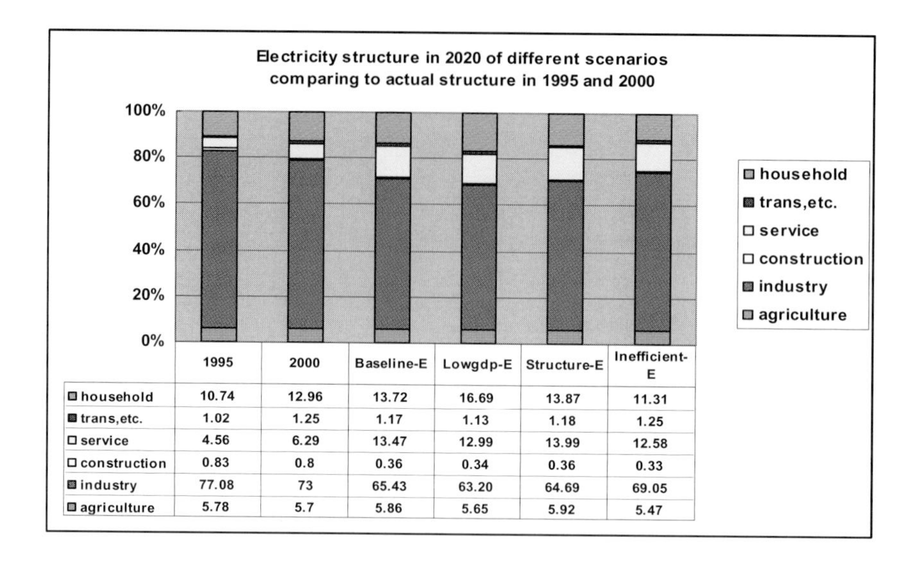

	1995	2000	Baseline-E	Lowgdp-E	Structure-E	Inefficient-E
household	10.74	12.96	13.72	16.69	13.87	11.31
trans,etc.	1.02	1.25	1.17	1.13	1.18	1.25
service	4.56	6.29	13.47	12.99	13.99	12.58
construction	0.83	0.8	0.36	0.34	0.36	0.33
industry	77.08	73	65.43	63.20	64.69	69.05
agriculture	5.78	5.7	5.86	5.65	5.92	5.47

Figure 4.11 Electricity demand structure for different scenarios compared to the actual structure

Table 4.32 Growth rate of electricity demand in 1995-2020

Scenarios	**Growth rate 1995-2020, % p.a.**
Baseline-E	5.2
Lowgdp-E	4.4
Structure-E	5.1
Inefficient-E	6.0

Table 4.33 Electricity structure forecast for different scenarios in 2020

Unit: %

Scenarios	**Agriculture**	**Industry**	**Construction**	**Service**	**Trans,etc.**	**Household**	**Total**
1995	5.78	77.08	0.83	4.56	1.02	10.74	100.00
2000 actual	5.7	73	0.8	6.29	1.25	12.96	100.00
2000 forecasted	5.7	73.93	0.69	6.59	1.03	12.06	100.00
Baseline-E	5.86	65.43	0.36	13.47	1.17	13.72	100.00
Lowgdp-E	5.65	63.20	0.34	12.99	1.13	16.69	100.00
Structure-E	5.92	64.69	0.36	13.99	1.18	13.87	100.00
Inefficient-E	5.47	69.05	0.33	12.58	1.25	11.31	100.00

Table 4.34 shows annual energy consumption per capita and per household in 2020 for the Baseline scenario. The table also gives figures for per capita electricity consumption in Switzerland and other European countries in 1998. Comparatively

speaking, while electricity consumption for individuals and households will increase significantly by 2020 in Shandong, electricity use will , though, still be far below the levels in the developed countries.

Table 4.34 Per capita and household electricity consumption

Country or city	**Per capita electricity consumption (kWh/year)**	**Residential electricity consumption per household (kWh/year, household)**
Shandong 1995	851	348
Shandong 2000	1042	486
Shandong 2020	2632	1127
Switzerland 1998	7484	4940
France 1998	7184	
Austria 1998	6706	
Germany 1998	6162	
England 1998	5874	
Italy 1998	4872	
EU-15 1998	6205	

Source: Shandong Statistics Bureau, this research, and Statistique Suisse de l'Électricité 1999

Table 4.35 shows that the electricity demands forecast in this report are a little higher than those forecast by SEPCO (SEPCO, 2000, 2001a).

Table 4.35 Comparison of CETP and SEPCO electricity demand forecasts

Unit: TWh

	Forecasted electricity demand in 2000,2005,2010,2015 in this report	**Actual consumption in 2000 and demand forecasting of SEPCO in 2005,2010,2020**
2000	98.56	94.59
2005	128.33	122.2
2010	164.06	159.8
2015	208.6	204

Source: SEPCO, 2000, 10th five year plan and outlook of electricity sector in Shandong province

14.2 Forecasting Electric Loads

This section includes analysis of peak loads in general, and by annual and daily needs. The authors lay out the assumptions they have made in building the forecasts (CETP Database, 2000), and estimate the modulation coefficients. Finally, they forecast load curve prospects for 2020.

During the 8th FYP (1991-1995), electricity supply shortages limited increases in consumption, so peak load was determined by the supply capacity. The rapid development of the electricity sector in the ninth FYP (1996-2000), led to a balance between supply and demand; loads increased rapidly and the load factor decreased. In this period, the average peak rates grew by 8.6 percent. The average daily load factor dropped from 88.7 percent (1996), 83.9 percent (1997), and 82 percent in (1998), to 81.2 (1999). The ratio of peak-valley difference to peak load increased from 33 percent in 1996 to 38 percent in 2000 (SEPCO, 2001a, see Table 4.36)

Table 4.36 Peak load and peak-valley difference in 1990-2000

Unit: MW

Year	**1990**	**1995**	**1996**	**1997**	**1998**	**1999**	**2000**
Peak load	6048	8812	9833	10740	11266	12548	12998
Peak-valley load	2030	3004	3286	3986	4012	4700	4938
Peak-valley /peak load (%)	33.6	34.1	33.4	37.1	35.6	37.5	38.0

Source: SEPCO, 2001a

Figure 4.12 shows that before 1997, the monthly peak load did not change much during a year (SEPCO, 2001a). After that, the peak load in summer and winter increased significantly. In Shandong, spring is February through May; summer is June through August; autumn is September to October; and winter is November through January. Normally the annual peak load occurs in winter due to demand for heating, but in 1999, the peak happened in summer due to warm weather. In 2000, the summer peak nearly matched the winter peak (SEPCO, 2001a). Figure 4.13 gives the load duration curve in 1999 (SEPCO, 2001a).

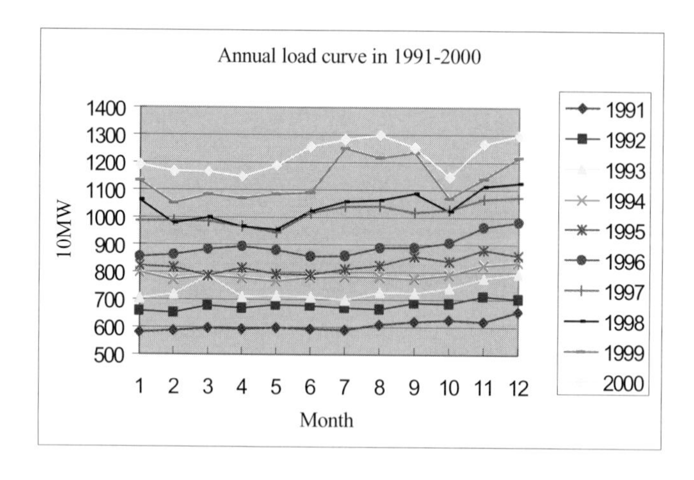

Figure 4.12 Annual load curve in 1991-2000

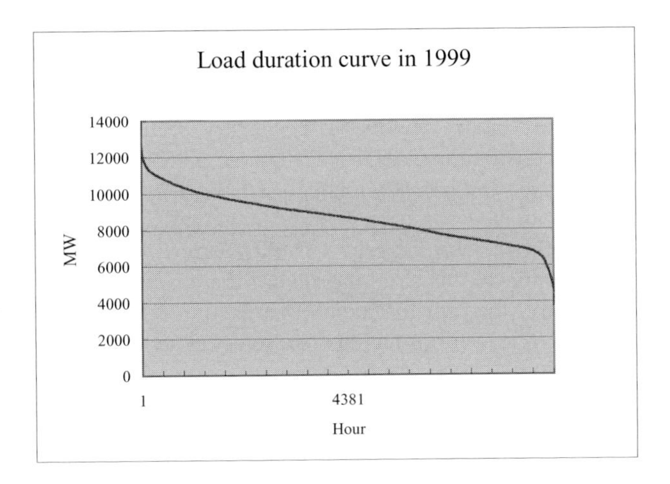

Figure 4.13 Load duration curve in 1999

Figure 4.14 shows the daily load curve for a single day in autumn, winter, spring, and summer in 2000 (SEPCO, 2001a). In the first three seasons, daily load curves show two peak periods in the morning and in the evening. The daily load curve in summer is different from other seasons. The curve remains at a comparatively high level from the morning peak until evening, and the evening peak lasts a long time.

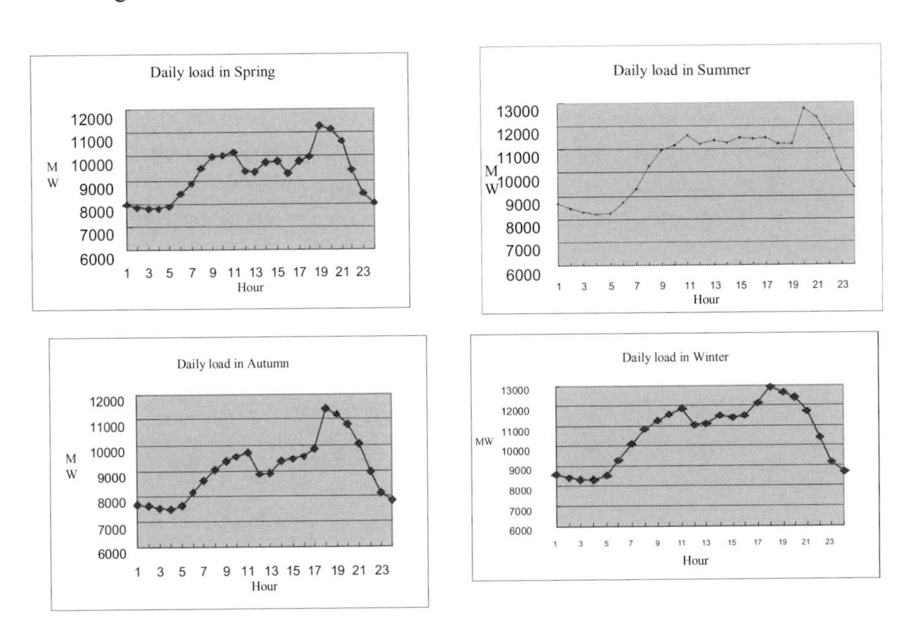

Figure 4.14 Daily load curve in spring, summer, autumn, and winter in 2000

For purposes of analysis, certain assumptions based on available data from the year 2000 have been made in order to forecast electricity load patterns for a baseline scenario and to estimate the modulation coefficients. Consumers are separated into industry, service, and "other" sectors. The industry sector is represented by 33 consumers and service sector includes five; the load pattern of the "other" sector is deduced from the service and industry totals.

The one-day load data for each month sometimes falls on a weekday and sometimes on a weekend. However, it is difficult to distinguish weekday and weekend for each month. In this case, the load of the nearest weekday in the month is taken as the weekday load for the month that has a weekend load data, and vice versa. Several missing bits of information are also estimated from adjacent data.

Load curves are forecast using the modulation approach. The equation is:

$$\mathrm{Psjh} = \frac{\mathrm{ts} * \mathrm{ws} * \mathrm{dsj} * \mathrm{hsjh}}{24 * Ne} * E$$

Psjh:	hourly load for a day in a week
ts:	coefficient of trend for a week
ws:	weekly coefficient, weight of week to yearly average
dsj:	daily coefficient, weight of day to weekly average
hsjh:	hourly coefficient, weight of hour to daily average
Ne:	Number of annual equivalent working days
E:	Forecasted electricity energy demand

The coefficients have been estimated from the load data of 1999 and 2000 for the three different sectors.

The peak load and load curve for the baseline scenario have been forecast using the method mentioned above. The forecast for electricity demand is taken from the Baseline scenario outlined above. It should be noted that according to the basic data, the load is the bus-bar load from SEPCO. Table 4.37 shows the peak load prospect. Figure 4.15 forecasts some examples of daily load curves in 2020 and Figure 4.16 is the load duration curve in 2020.

Table 4.37 Peak load forecasts

	2005	**2010**	**2015**	**2020**
Peak load (MW)	19804	25775	32941	41383
Load factor (%)	74	73	72	72

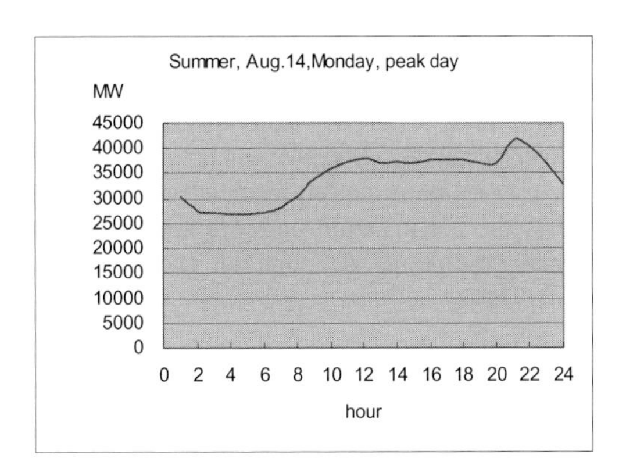

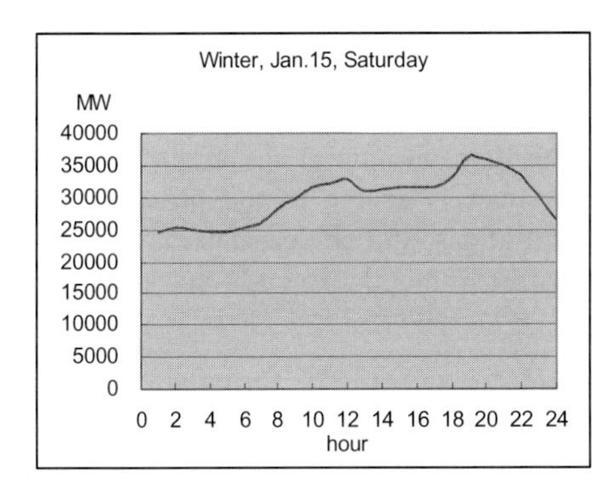

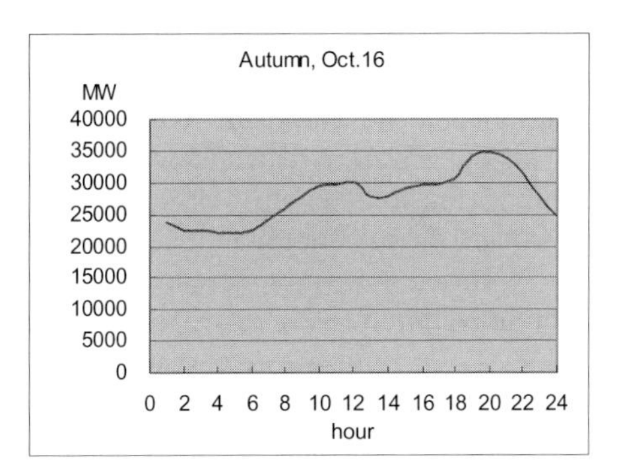

Figure 4.15 Examples of daily load curves forecast for 2020

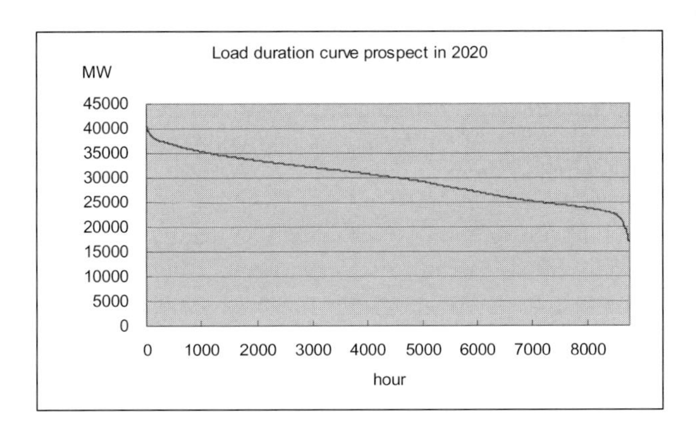

Figure 4.16 Load duration curve prospects in 2020

15. CONCLUSIONS

Several predictions about the demand and supply of energy in Shandong province for the next two decades may be drawn from this study. First, the structure of energy demand will be optimized. Second, energy efficiency will be improved. Third, high quality energy will be imported to meet provincial demand. Fourth, the government will support the construction of more and better energy infrastructure facilities. Each of these predictions depends on the accomplishment of several complex tasks.

15.1 Optimization of Demand

In the future, the demand for higher quality energy sources, such as gas and electricity will grow rapidly. The share of gas and electricity will have increased from 6.02 percent and 14.49 percent in 1995 to 7.36 percent and 26.3 in 2020 respectively. Improved standards of living for Shandong's population and sustainable development require that the final-use energy structure be optimized. This optimization depends on an increase in the share of natural gas, electricity, and oil in the total energy structure. As hydropower is scarce in the province, the primary energy supply will be based on oil, natural gas, and nuclear power.

Based on this research, the authors recommend several measures for optimization:

(1) Stringent environmental laws and regulations should be formulated to limit coal use by final consumers. This use could be gradually phased out in large and medium-size cities and new projects could be designed to depend on natural gas and electricity.
(2) Power sector reform should be carried out and the grid connection between Shandong and other grids explored and studied. The transmission and distribution of electricity, especially in rural areas, should be modified and renovated where necessary. Effective policy and technology should be adopted and power prices reduced.
(3) In the developed rural areas and city suburbs, LPG should be more widely used. The planning of LPG stations and construction of storage facilities should be initiated soon.
(4) The construction of natural gas pipelines and a city network should be sketched out. Active connections should be made between managers at the Shengli Oil Field, Ocean Oil Corporation, China Oil and Natural Gas Corporation, the central government, and foreign and domestic research institutes in planning and developing the natural gas infrastructure.
(5) With the support of the provincial government, tertiary oil extraction technologies should be used more widely, and exploration for oil and natural gas reserves should be stepped up.
(6) The power sector should consider implementing nuclear power and carry out the necessary advance planning and research. The central government should be involved in any nuclear projects.

(7) The province power grid is a thermal grid, which adjusts peak loads with great difficulty. The construction of natural gas power plants should alleviate this weakness.

15.2 Improving Energy Efficiency

Energy efficiency is one of the most important factors affecting final energy demand. For example, if the rate of energy conservation in the industrial sector for the Baseline case in Shandong is only 0.5 percent higher than in the Inefficient scenario, 17.6 percent—10.0 Mtoe of energy--can be saved in 2020.

China is a developing country with a large population and insufficient per capita resources, but its economic development is currently very rapid. . Energy saving and energy efficiency improvement are very important social and economic factors in its future. It is possible that, over time, economic growth in Shandong could impose heavy pressure on resources and the environment. Energy and resource conservation and environmental protection can offset these stresses, which ultimately would retard economic development. Because there is such a big difference in energy utilized in Shandong and in the developed countries, the potential for energy saving is huge. Improved technology is fundamental to advanced energy efficiency, in concert with deepening enterprise reform, energy awareness, and adjustment in the industrial and production structure.

The energy strategy of a country and region depends not only on new strategies in the energy industry, but also on the development and application of energy-efficient technologies in all sectors, for example, equipment manufacturing (World Bank, 1994). At the sector development level, industrial policy and social consumption patterns are important factors to be monitored and shaped with energy conservation in mind.

Clear energy efficiency targets should be formulated for all new facilities, technology, and equipment for the short, middle and long term. From now on, new large-scale projects (expected to be in service for at least 25 years) including industrial projects, commercial and residential buildings, and public infrastructure facilities should be designed to meet the advanced energy efficiency level of the 1990s. Facilities and equipment commissioned before 2010 should reach the international advanced level of the 1990s and beyond. In the next 10 years, equipment such as motors, fans, water pumps, boilers, and vehicles with a service life expectancy of at least 10 years should also meet this standard. It is important to emphasize that energy efficiency should be approached at a systems level, not just at the level of the individual project, equipment, or facility.

The technology and equipment of the large energy intensive industries in the province that is outdated should be discarded and replaced. Experience has demonstrated that partial renovation will not reverse this situation. Furthermore, the growth rate of these industries is dropping below the rate of development for low-energy-use and high value-added industries. If the old-style energy intensive equipment is not replaced, it will continue to account for a large share of total production in the future. It should be replaced with advanced, high efficiency technologies in the next 10 to 20 years.

Several technical measures and policy recommendations would increase energy efficiency in each sector:

(1) Sector and product structures should be adjusted to emphasize extensive product processing and first class products. Research and development are needed to identify new products and increase value added in the industrial process.
(2) Industrial managers should make full use of the scientific research, production, and marketing advantages of the non-ferrous industry; increase market share; encourage investment in technical modifications, identify new techniques; and increase product quality. They should also work to save energy and utilize it efficiently.
(3) Heavily polluting and energy-wasting small cement, glass, electrolysis aluminum plants, and small coal-fired boilers should be closed. Existing medium and small glass kilns should be renovated, and advanced technology and equipment disseminated throughout this industry
(4) Entrepreneurs should develop natural gas chemicals, fine chemicals, and new chemical materials, utilizing new, high technology to support these emerging industries. The introduction of these new enterprises into Shandong's mix of industries will initiate a new direction for economic growth, and help upgrade the entire industrial sector.
(5) China's accession to WTO will affect the chemical industries. The most favorable response would be for each sector to merge and consolidate enterprises; adopt advanced, high, and new technology appropriate to the shift toward large scale, energy and technology intensive operations. These challenges must be met if Shandong's industries are to be competitive in domestic and international markets.

15.3 Importing High Quality Energy

The province is endowed with abundant coal and oil resources. In the future, the demand for high quality energy will increase so rapidly that provincial production will be unable to meet it. The growth of power from coal-fired plants is limited because it is so difficult to site them. Natural gas and electric power should be imported to optimize the energy structure and meet local demand. In addition to Shengli Oil Field, possible natural gas resources include Bohai, Russia, and imported LNG.

The implementation of four policies would help prevent the stunting of industrial growth by an inadequate power supply:

(1) Import power from western China through connection with existing western grids and the construction of direct supply lines and power plants in the west.
(2) Enhance cooperation with the China National Offshore Oil Corporation; promote offshore exploration and development in Bohai Sea; and construct natural gas pipelines in coastal areas. The use of natural gas from Bohai Sea is the easiest of these targets and should begin soon.

(3) Pay close attention to the northeast Asia natural gas pipeline network through which Russian natural gas may be imported and participate in the research needed to complete this connection. Russian natural gas should be introduced into the province as soon as is economically feasible.
(4) Carry out a feasibility study on the importation of LNG and site of terminals.

15.4 Enhancing the Energy Infrastructure

Demand for high quality energy, including natural gas and electricity, will have grown greatly by 2010. To accommodate this demand, government officials must pay attention now to the construction of an adequate energy infrastructure including a natural gas pipeline network system and power grid. Like the central government, the province has attached great importance to the coal supply system and has invested in a major, long-term effort to construct an appropriate transport system, including railways, ports, roads, and water ways). Much has been achieved. At present, the coal market is stagnant, relieving some pressure on this system and its further construction, much of which remains in the preliminary stages.

However, capacity for coal transportation does not meet the demand target of a common coal market. Furthermore, demand for certain types and quality of coal is still constrained by inadequate transportation. Clearly, the construction of transmission and distribution systems for power lags behind the construction of generation capacity. This not only limits power use but also constrains reform in this sector. At present, infrastructure construction is still centered on coal transportation. In the struggle to balance high quality energy supply and demand, managers should emphasize other types of infrastructure.

The new priorities should be: First, build city power, heat, and gas systems to complement urbanization and city construction including some new energy supply facilities to support new industry. In the future, infrastructure construction will be driven by demand for final energy uses. Meanwhile, construction of the rural grid should be advanced. Second, a natural gas and oil pipeline network should be constructed, with a special emphasis on gas. The planning and construction of facilities for the large-scale use of natural gas is the key to economic development in the province for the next 10 to 20 years. The resource is limited only by the infrastructure to exploit it. The construction of the natural gas pipeline should be a first priority for state and local government managers of infrastructure development.

REFERENCES

CETP database (2000), Questionnaire on social economic development and electricity demand in CETP project, March 2000

Gnansounou, E (2000)., Prospective et planification energétique, Cycle postgrade en energie 1998-2000,

Gnansounou, E. and C. Rodriguez, (1999), DEMELEC-PRO1.0.1

IEA (1998), World Energy Outlook 1998 edition, IEA, International Energy Agency

National Bureau of Statistics (1999), China Statistical Yearbook 1999, Beijing, China Statistics Press

National Bureau of Statistics(2000), China Energy Statistical Yearbook 2000, Beijng, China Statistics Press

SEPCO (2000), Tenth five-year plan and outlook of electricity sector in Shandong province by SEPCO, 2000.

SEPCO (2001a), Electricity market analysis report, Research centre of electricity market, SEPCO, 3, 2001.

SEPCO (2001b), Shandong electricity statistics during 1995-2000 from SEPCO

Shandong Statistics Bureau (1995-2000), Shandong Statistical Yearbook 1995-2000, Beijing, China Statistics Press

State Development Planning Commission (SDPC) (1995), Ninth five year plan (1995-2000) and outlook on 2010 of Shandong Development Planning Commission, source: http//www.sdpc.gov.cn:

World Bank (1994), 'Energy efficiency in China: technical and sectoral analyses', World Bank Report 1994.

CHAPTER 5

ENERGY ECONOMY MODELING SCENARIOS FOR CHINA AND SHANDONG

SOCRATES KYPREOS, ROBERT A. KRAKOWSKI, ALEXANDER RÖDER, ZHIHONG WEI AND WENYING CHEN

1. INTRODUCTION

In the year 2000, China generated 1'350 TWh/yr of electricity, representing a 9.5 percent annual rate of increase and an installed capacity of 316 GWe (more than 50 percent of installed capacity in the US). The reduced growth rates over the last few years caused by the Asian financial crisis of 1998 appear to be rebounding to stronger levels and indicate the healthy recovery of the Chinese economy (Zhao 2000). China's GDP rose by 8.2 percent in the first half of 2000, greater growth than in either of the two previous years in their entirety. Even with any eventual tapering of these high GDP growth rates to a still-robust four to five percent a year, the demand for electric power is expected to nearly triple to around 4'000 TWh annually by the year 2030. Significant degradation of already seriously deteriorated environmental (air and water) quality is expected if this growing demand for electrical energy is provided by coal using present technologies without scrubbers at the present level of use.

Figure 5.1 contrasts this large percentage of coal in the primary-energy (PE) menu for China to the energy spectra behind the PE demands for the top seven (total) energy users (Logan, 1999); only the total energy demand for India nearly so dominated by coal. Future scenarios with a reduced reliance on coal for electricity generation, at least on a percentage basis, would require either nuclear or natural gas, combined with decreased energy intensity (ratio of PE demand to GDP, MJ/USD) and increased use of renewable energy sources. Logan (1999) has recently made a case for the latter. Accelerated use of reduced-carbon and renewable energy sources all present unique constraints and deployment thresholds or barriers. Environmental improvements in the face of growing GDP and PE demand will most likely require a combination of all these technologies, along with the implementation of advanced coal technologies in both electric and non-electric sectors.

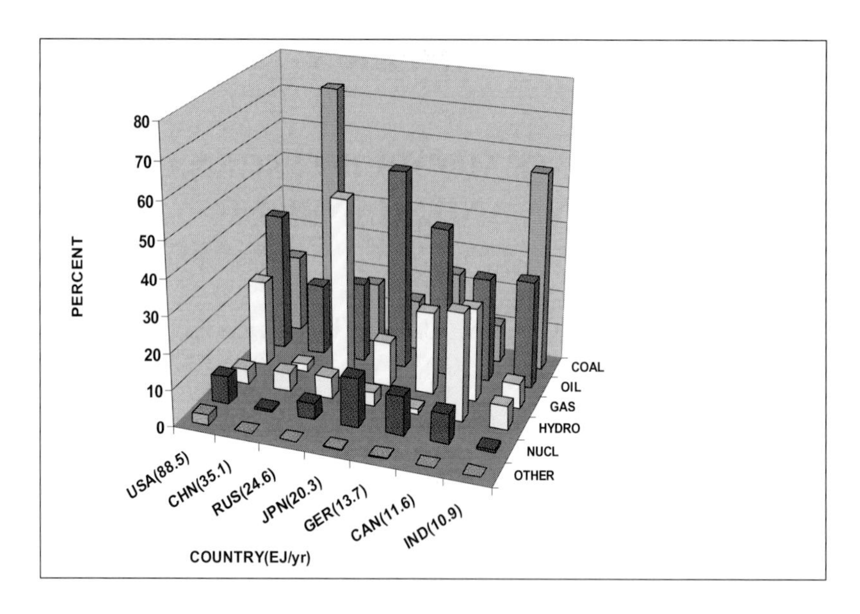

Figure 5.1 Primary energy consumption by fuel from the world's seven largest users (1996).

Assessments of possible energy-economic-environmental (E^3) futures for China have tended to emphasize electricity generation using models that are largely driven by assumed demands for electricity at both regional and national levels. Figure 5.2 includes the results of forecasts from a range of institutions over the last five years for China as a whole; some of these forecasts are based on multiregional electric sector models of China (Chandler, 1998; also reported in Logan, 1998; Kypreos, 2002). Interestingly, the IIASA (earlier) cases (Nakicenovic 1998) tend towards low forecasts, whereas those of Chinese origin favor the higher projections. The CRETM base-case (BHC, see Sec. 3.1.1. for notational explanations) projections are intermediate and somewhat below the PNNL projections (Chandler, 1998; Logan, 1998); the CRETM forecasts are based on GDP projections and assumptions of price and demand elasticities.

The impact of and possible futures resulting from the unique combination of high rates of economic growth and attendant coal-based energy demand in a developing region of high population and currently serious environmental degradation, has been the subject of detailed study at both national (Chandler, 2001; 1998; Zhou, 2000; Levine, 1994; McElroy, 1998; Rogers, 1999) and regional (Zhang, 2001; McElroy, 1998; Baker, 1999; Xue, 1999) levels. The China Energy Technology Project (CETP) built on this literature to formulate a methodology of study, projection, and multi-criteria assessment that can be used by a broad range of industrial, governmental, environmental, and university stakeholders to pursue specific quantitative approaches to understanding mid-term futures (2000-2030) in economic-environmental-energy (E^3) terms for both China and the world. To this end, narrowing of initial scope to a single energy sector, electricity, and to a single province within China, Shandong, was necessary to begin this process.

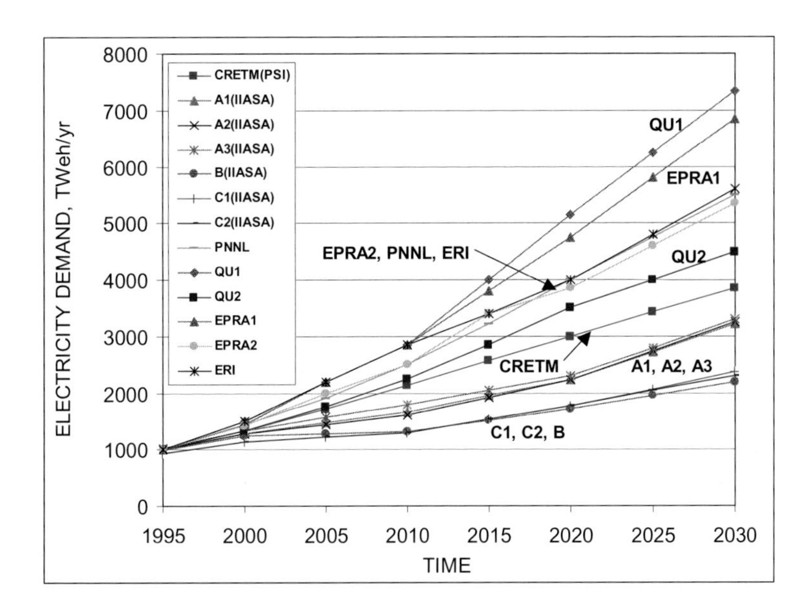

Figure 5.2 Comparison of China electrical energy demand used in this study with projections made for a range of IIASA/WEC scenarios (Nakicenovic 1998); in the parlance of the latter global study, Case A is a "High Growth" scenario; Case B is "Middle Course" scenario, and Case C is an "Ecologically Driven" scenario; Subcases A1, A2, and A3 correspond, respectively, to "conventional and unconventional oil and gas," "coal backstop," and "bio-nuclear," whereas C1 and C2 correspond, respectively to "nuclear phase out" and "new (small) nuclear plants." Also shown in this comparison are projections made from a range of China and US studies, as reported in Chandler (1998). Chandler (1999) adopted the PNNL demand; the ERI demands reflect those made by the Energy Research Institute in China (1995); QU corresponds to projections from Tsinghua University (1994), and EPRA are projections made by the Chinese Electric Power Research Academy (1994). Also shown are base-case (BHC) results from the companion EEM electric-sector model, China Regional Energy Trade Model (CRETM, Kypreos 2002).

In examining options for an E^3 future based on conventional use of coal, CETP used both optimizing (Kypreos, 2002) and simulation (Schenler, 1998) models to assess tradeoffs in the electricity-generation sector for a range of fuel, transport, generation, and distribution options. The optimization models contributing to this overall CETP goal include the seven-region China Regionalized (Electrical) Energy Trade Model, CRETM (Kypreos, 2002), and a single-region MARKAL (market allocation) model (Fishbone, 1981) that has been constructed to describe the Province of Shandong. In the following material, the overall coverage provided by each of the optimization and simulation models that form the basis of the CETP computational modeling hierarchy is elaborated.

The drive to improve the quality of life while maintaining political stability dominates E^3 decisions and directions taken in China today. While technological innovations on both the supply and demand sides of the E^3 equation represent powerful means to ease energy pressures (Ni, 1998), the ways to apply innovation must reflect China's place in the world, and its unique characteristics, including

- large deposits of coal, but limited supplies of petroleum and natural gas;
- shortage of and competition for investment capital;
- difficulty of administering a vast and culturally diffuse territory;
- poorly coordinated distribution systems within this vast territory;
- very large population with commensurately large, but stabilizing growth rates and youthful demographics;
- low per capita energy resources;
- co-existence of modern and less developed social and economic sectors; and
- economic liberalization in selected regions, coupled with cyclic application of political and cultural controls.

The introduction of innovative technologies to accommodate E^3 pressures must ultimately reconcile these unique attributes of a country that for the last two decades has emphasized and achieved significant per capita GDP growth. To this end, the EEM team focused on a) options for generating electricity in China in light of local, regional, and global externalities; b) analyzing the role that advanced technology under different policy assumptions and international co-operation strategies for improving local and regional environments and mitigating global warming; c) sharing experience with Chinese researchers and other model developers in the CETP; d) expanding modeling capabilities; e) contributing to quantitative analyses of the problem; and f) gaining insights into potential developments in China. The EEM team cooperated with researchers at Tsinghua University (TU) and the China Energy Research Institute (Beijing, ERI) in analyzing energy demand and supply for Shandong and China.

This chapter is organized as follows. Section 2 provides a general background of the energy situation in China and Shandong. Section 3 describes the models and the supporting database for both MARKAL and CRETM; the modeling approach for both is based on eight broad-coverage scenarios. These scenarios are evaluated in terms of the MARKAL data for Shandong in Section 4 and for China and Shandong in Section 5, where they are interpreted in the form of an integrated comparison and assessment at both EEM and CETP levels in Section 6; some comparisons also include recent studies performed outside of the CETP. The main findings, general conclusions and recommendations for both policy and model development are reported in Section 7. The audience for these recommendations comprises both CETP stakeholders and others interested in economic and environmental issues in China and Shandong Province specifically.

2. GENERAL BACKGROUND ABOUT CHINA AND SHANDONG PROVINCE

Shandong Province is situated in eastern China midway on an axis defined by Beijing to the north and Shanghai to the south; nearly half of the perimeter of this

620-km X 420-km area is coastal, bounded by the Bohai Sea to the north and the Yellow Sea to the south. The Yellow River (Huang He) flows northeasterly through Shandong Province, near the provincial capital of Jinan and into the Bohai Sea. Shandong ranks twentieth in area among China's 31 provinces, and is rich in minerals (gold, sulfur, diamond, and gypsum reserves are the largest in China) and harvests the largest wheat yields in all of China. Ranking industries are energy, electronics, chemicals, textiles, food processing, and construction materials. The following subsections briefly describe the society and economy, population, economic structure, energy intensity, energy resources, energy uses, and both pollutant and GHG emissions for Shandong Province, and emphasize aspects of these areas that touch upon the EEM modeling task.

2.1 Socio-economic Structure

Shandong Province is one of several bonded free-trade zones along the East Coast. Export-based growth has been driven by a successful mix of foreign investment ($44.9B US in 1997) and the growth of Township and Village Enterprises (T&VEs). Key statistics for Shandong presented in comparison to China as a whole are listed in Table 5.1 which includes a breakdown of 1997 GDP by sector. Key socio-economic trends shaping energy demand are derived directly from Shandong or inferred from comparable trends at the national level. These demand figures are used in the MARKAL model and its Reference Energy System (RES) to assess relationships between possible energy futures, energy-carrier technology spectra, and related emissions impacts.

Shandong Province supports 86.7 million people (7.1 percent of China's population) on an area of 156,700 km^2 (1.6 percent of the land area of China), giving a population density of 564 persons/km^2 and ranking it as the second most populous province in China. A birth rate that is 64 percent of the national average and a somewhat lower death rate yield population growth that is 46 percent of the national average.

Because of Shandong's coastal orientation and its proximity to Japanese and Korean markets, its economy is export-oriented. Products include deals oil, consumer products, machinery, electronics, textiles, and paper. Strong manufacturing and agricultural economic sectors have stimulated the growth of electrical generation capacity. During six years, 1987-1993, 1.2 GWe were added annually 1987-1993 (Baker 1999). As Table 5.1 illustrates, both the per capita GDP and energy use ranks as one of the highest in China.

Table 5.1 Summary Comparison of Key Country and Shandong Statistics, Largely for 1997 (Xue 1999).

Parameter[(a)]	China	Shandong	Prov. Rank	% National
Area, AREA (km^2/1000)	9586	153	18	1.6
Population, POP (10^6)	1236	87.9	2	7.11
Gross Domestic Product, GDP (Yuan/10^9)[(b)]	7693	665	3	8.65
GDP Growth, GGDP (%/a)	8.8 [(c)]	11.2	9	
***per-capita* GDP**, GDPPC (Yuan/*capita*/yr)	6079	4432	9	
o **Primary Energy**, PE(EJ/yr)	37.7	4.70		12.47
o Electrical Generation, EE (TWh/yr)	1105	85.4		7.73
***per-capita* PE Use PEPC** (GJ/*capita*/yr)[(d)]	30.5	30.5	14	
o **Crude Oil** (Mtonne/yr)	162.1	28.0	3	17.3
o **Coal** (Mtonne/yr)	1356	90.9	3	6.7
o **Cement** (Mtonne/yr)	492.6	58.3		11.8
o **Steel** (Mtonne)	107.6	5.23		4.9

(a) largely 1997 data

(b) 8.3 Yuan/US$

(c) 7.8 %/a in 1998, 7.2% in 1999

(d) based on 29.3 GJ/tce

2.3 Structural Changes

At the heart of any Energy-Economy Model (EEM) effort is the demand for final products and services for satisfying either domestic or export needs, as well as the spectrum of energy sources and technologies available to meet these needs under economically and environmentally acceptable conditions. Past performance at both national and provincial levels provide strong guidance for the selection of input required to drive these EEM activities. Table 5.2 gives a "top-level", aggregated sectoral summary of energy requirements for China over the last decades. Most notably from these statistics are: a) the high growth rate in energy demand; and b) the dominance of the industrial sector in driving that demand.

Table 5.2 China Fuel Consumption by Sector, Excluding Biomass (1985-1994) [(a)]

Year		1985	1988	1990	1991	1992	1993	1994
Industry[(b)]	IND	12.04	14.87	15.95	16.85	18.01	19.16	20.74
Agriculture[(c)]	AGR	0.94	1.11	1.16	1.20	1.18	1.13	1.20
Construction	CON	0.31	0.28	0.28	0.31	0.33	0.31	0.31
Transportation	TRA	0.87	1.01	1.06	1.13	1.20	1.32	1.32
Commercial	COM	0.19	0.26	0.28	0.31	0.33	0.45	0.42
Residential	RES	3.14	3.66	3.73	3.78	3.73	3.71	3.63
Other[(d)]	OTH	0.59	0.73	0.83	0.94	1.04	1.27	1.30
TOTAL	TOT	18.08	21.95	23.29	24.50	25.82	27.38	28.96

(a) (Dong, 1998), EJ/yr

(b) Sum of Light Industry plus Heavy Industry.

(c) Farming, Forestry, Animal Husbandry, Fishing, *etc.*

(d) Non-Producing.

2.4 Energy Intensity

The efficiency with which the energy is utilized to produce a given unit of product plays a key role in setting energy requirements and attendant emission rates for a given level of economic activity. Again, at the country level, the energy intensity, EI(MJ/US$), expressed as the ratio of energy required to produce a given contributing unit of GDP for a given economic sector, represents a highly aggregated metric for measuring the economic efficiency of a given sector. Table 5.3 summarizes some of the history of energy demand and energy intensity in China. The energy efficiency improvement with which a unit of GDP (in a given economic sector) is currently produced reflects in large measure the impressive ~50 percent overall decrease in the amount of primary energy needed to produce a given unit of GDP since the late 1970s.

While this growth in efficiency improvement is indisputably impressive, China needs additional improvements in the efficiency with which primary energy is used to heat and transform materials at the industrial level.

2.5 Energy Resources

Ni (1998) has presented a concise and comparative summary of the energy resources in China. This summary includes key elements of China's capacity to provide for its energy future and the form that future may take. Figure 5.3 compares domestic fossil, hydroelectric, and nuclear resources with world resources. Based on this "top-level" comparison at least three inferences can be made: a) China's readily and economically exploitable primary energy resource is coal. b) Hydroelectric

Table 5.3 History of Energy Demand and Energy Intensity for China (Dong, 1998).

Main Fuel Uses in Major (Heavy) Industrial Sectors[a]							
Year	YEAR	**1985**	**1986**	**1987**	**1988**	**1989**	**1990**
Ferrous Metals	FeM	2.14	2.46	2.58	2.75	2.90	3.02
Non-Ferrous Metals	NFeM	0.41	0.44	0.44	0.50	0.53	0.56
Chemicals	CHEM	2.31	2.49	2.81	2.96	3.14	3.14
Construction Materials	CM	2.26	2.49	2.70	2.87	2.93	2.75
Energy Intensity in Major (Heavy) Industrial Sectors[b]							
Ferrous Metals	FeM	406.1	406.1	386.7	372.1	367.2	367.2
Non-Ferrous Metals	NFeM	172.7	167.8	158.1	155.6	153.2	155.6
Chemicals	CHEM	340.5	333.2	313.7	274.8	267.5	243.2
Construction Materials	CM	381.8	352.6	323.4	284.5	257.8	252.9

(a) (Dong, 1998), EJ/yr

(b) MJ/US$; 29.3 GJ/tce, 8.3 Yuan/US$

resources are significant. c) According to present estimates, exploitation of natural gas (NG) resources will require significant exploration and development of endogenous resources, importation, coal conversion, or all three.

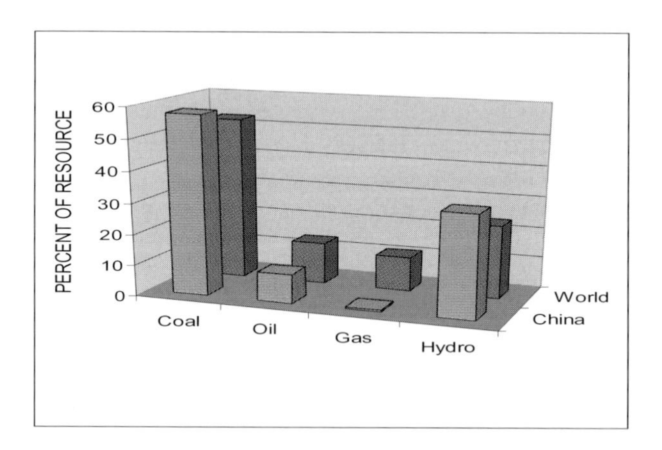

Figure 5.3 Comparison of fossil plus hydroelectric resource distributions between China and the world (Ni, 1998).

2.6 Energy Use

The industrial sector uses most of Shandong's dominant power source, coal. Figure 5.4 compares the primary energy demand in Shandong and in China. As a percentage of PE demand, electricity plays a somewhat reduced role in Shandong as it does in China, the difference being made up by petroleum. At present, the use of natural gas for electricity generation is not important in either China as a whole, or Shandong Province.

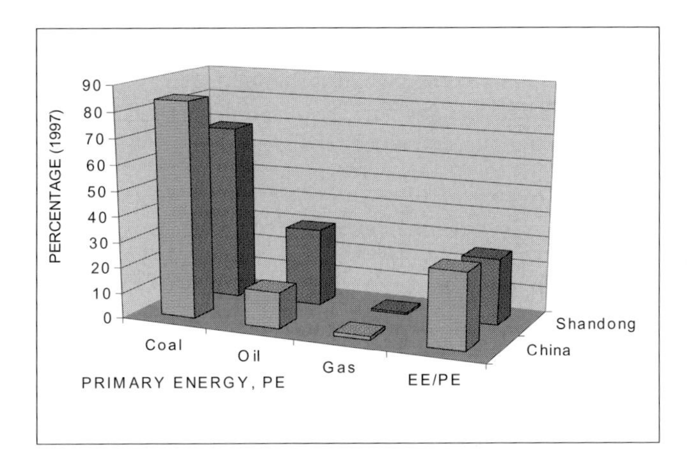

Figure 5.4 Relative comparison of primary energy demand for 1997 in Shandong and China (Xue 1999) and the Electric Energy (EE) to Primary Energy (PE) ratio.

At 85.4 TWh/yr , Shandong in 1997 was the second largest (7.4 percent total) producer of electricity in China. From 1985 to 1990, the rate of electrical generation growth was 6.5 percent annually, increasing to 9.9 percent during 1990-97, and decreasing thereafter. The decrease was felt mostly in the agricultural and industrial (73 percent of generation in 1997) sectors, and reflected the reforming and restructuring of the large state-owned enterprises, and favorable weather conditions.

In 1997 the annual PE consumption in China was 44.5 EJ/yr based on coal (0.92 tce/tonne average, 29.3 GJ/tce), petroleum (42 GJ/toe), and natural gas [36 $EJ/10^{12}m^3$(STP)]. Energy from hydropower (0.67 EJ/yr direct and 1.9 EJ/yr with an 0.35 equivalency factor, in 1997), solar, wind, geothermal, and tidal in 1997 was estimated to be 8.8 EJ/yr, or 16.5 percent of total PE if added thereto. Other estimates suggest new and renewable energy source account for 25 percent of national PE consumption and 47 percent of energy consumption in rural areas.

Three Light-Water Reactors (LWRs) presently form China's relatively small nuclear-energy (NE) fleet, which represented a total capacity of 2.1 GWe in 1997; four other plants are under construction and together will provide a total capacity of 8.75 GWe. With these new plants in operation, the nuclear contribution to China's generation capacity will increase from one percent today to about three percent of total generation capacity if the overall network grows at a five percent annual rate over the next five years. This contribution remains small compared to the world average of 18 percent.

2.7 Gaseous Emissions

Primary concerns about emissions are both global (greenhouse gases (GHGs), primarily CO_2) and local/regional (SO_2, NO_x, and particulates). Figure 5.5 gives the emission history of CO_2, SO_2, and dust for Shandong Province (Xue, 1999).

Carbon emissions in 1997 amounted to 0.89 GtonneC/yr, which represents 13 percent of world emissions. On a per capita basis, atmospheric carbon emissions amounted to 0.71 tonneC/capita for China, compared to 5.3 tonneC/capita for the United States. The economic "flip-side" of this environmental coin is the respective carbon emissions compared to GNP for China and the United States: 1.0 kgC/US$ *versus* 0.2 kgC/US$, using market-valued GDPs.

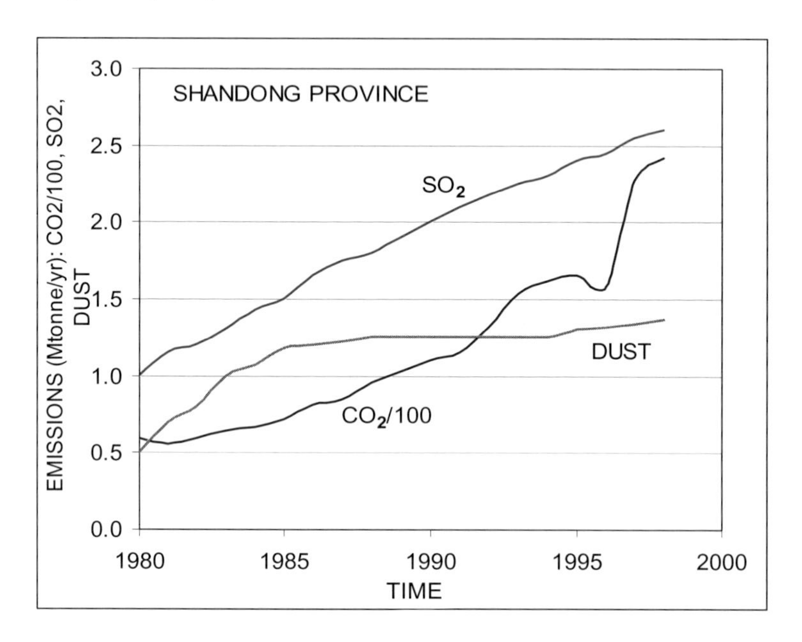

Figure 5.5 Emission history of CO_2 in Mtonne/yr of Carbon, SO_2, and 'Dust' for Shandong Province (Xue, 1999); the last two years for SO_2 and dust are extrapolated.

Whether in terms of population or productivity, the total carbon emissions from both nations are important contributors in driving global environmental impacts beyond any historical experience. Of the 0.54 GtonneC/yr emitted by China in 1994, Shandong Province contributed 7.8 percent, mainly from industrial sources (26 percent), biomass burning and deforestation (24 percent), and electricity generation (22 percent); minor sources of atmospheric CO_2 were residences (three percent), agriculture (two percent), and transportation (two percent).

Sulfur dioxide emissions from industrial wastewater in 1997 amounted to 29.7 MtonneSO_2/yr from China; Shandong Province contributed 3.6 percent to that total. Power plants (>6 MWe) contributed 7 MtonneSO_2/yr to the atmosphere in that year.

Fly ash and dust from power plants (>6MWe) in 1997 amounted to 3.5 Mtonne/yr, or 30 percent of the total emissions of this pollutant. Industrial dust emissions in 1997 amounted to 6.7 Mtonne/yr, of which Shandong Province contributed 4.9 percent.

3. ENERGY AND ECONOMY MODELING (EEM) APPROACH

A central question and goal for CETP is how to capture the trends, practices, and realities described above in a modeling system that can meaningfully project possible futures. These projections must contain sufficient detail to reflect reality while remaining usable by decision-makers in as they create policies that steer towards a future of sustained national growth and individual well-being for China's 1.3 billion citizens. Following a general overview of the overall CETP modeling task, this section first describes the basis, scope, and limitations of the MARKAL model developed for Shandong Province and then the seven-region (one of which is Shandong Province), electric-sector LP CRETM (China Regional Energy Trade Model).

3.1 Overview

The CETP has identified a number of modeling tasks and approaches to projecting energy futures for China in general, and electrical energy requirements for Shandong Province specifically. These modeling approaches to understanding E^3 interactions are distributed between optimization and simulation models. Both of these approaches fit into a multi-criteria decision analysis package. (See Figure 2.7 in Chapter 2) CETP's E^3 modeling focuses on the electrical energy sector of Shandong in both the optimization (EEM task) and the simulation tasks (ESS, see Chapter 6). However, the scope of the EEM task includes the full energy sector for the province alone and the electric sector of the multi-regional expanse of China (CRETM; Kypreos, 2002).

This MARKAL-Shandong/CRETM optimization-modeling capability is complemented by a single-region Shandong electric-sector Energy Transportation Model (ETM; Chapter 7), which also uses an optimization strategy to model the Shandong electrical sector. It includes a geographical resolution including the transportation of fuel and transmission of electricity within the Shandong electrical-generation network. While pursuing similar goals, the optimization modeling approaches pursued under both EEM and ETM tasks differ from that of the electrical-energy simulation (ESS) task; both these differences and similarities are elaborated in the following subsections. All of these modeling tasks are based on analyses of relatively surprise-free "visions of the future" quantified in terms of specific scenarios. The specific elements and limitations of Scenario Analysis are also elaborated below.

3.1.1 Optimization models (EEM, ETM)

Figure 5.6 illustrates the essential characteristics of an optimization modeling approach that focuses only on those cases that yield constrained minimization of an objective function, such as the total present-value (discounted) cost of energy generation over the time frame of the computational scenario. Both optimization and simulation models are "bottom-up" engineering/technology approaches to finding a "trade-off frontier" in, for example, cost *versus* emission "phase space;" both express results in terms of scenarios.

The scenario analysis approach followed by the Energy Systems Analyses group of PSI and by other energy and environmental analyses around the world uses engineering and economic optimization techniques. Following (Murray 1998), Figure 5.6 illustrates the general logical flow of inputs, constraints, and results for the optimization approach to E^3 modeling. The optimization methodology uses integrated resource planning to links the "bottom-up" energy model to "top-down" economic models to simulate economic growth and demands for energy services. This coupling is provided through the use of energy-price and energy-income (GDP) elasticities.

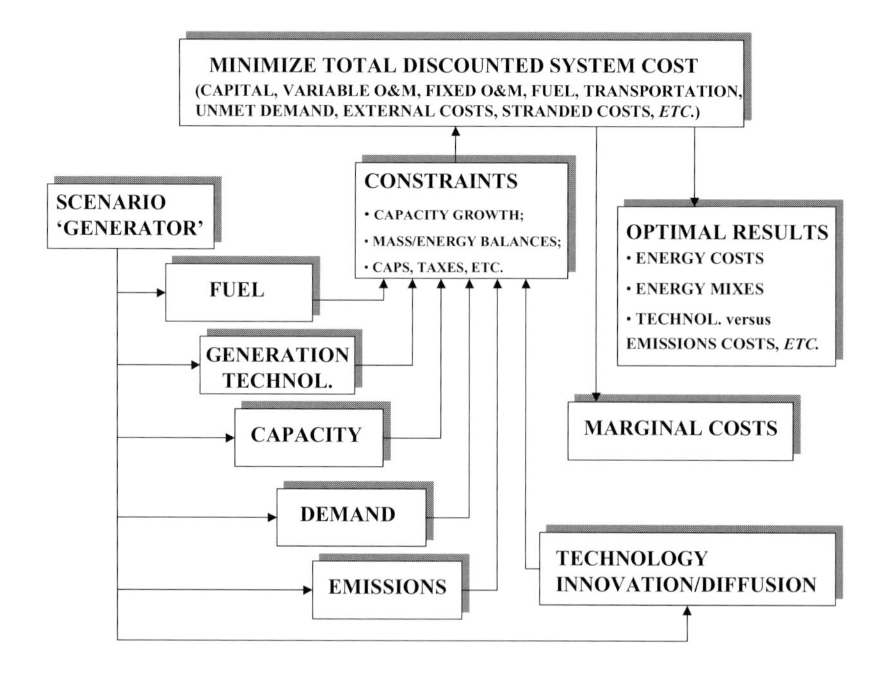

Figure 5.6 Typical inputs, constraints, and output from an optimization model; Sample constraints are defined for each of the classes listed above.

Decision making under uncertainties related to future economic developments and environmental/energy policy can also be performed by applying stochastic analyses with hedging options (Kann, 2000); this extension, however, has not been made in the CETP study. Although the methodology is based on optimization techniques, "tradeoff analyses" between the system costs and other considerations related to security of supplies and to specific elements of environmental policies being implemented play a central role in assessing the policy implications emerging from these optimization models.

In summary, the Energy-Economy Modeling (EEM) approach defines the allocation of scarce resources by minimizing production cost (or maximizes the welfare of a county or region, quantified by a utility function), under a set of constraints describing the energy and economic system under consideration. Welfare or production cost can be defined to either include or exclude externalities associated with environmental costs that might be included in a stakeholder's decision making. The database of the model includes the technological options available, associated characteristics and costs, available resources, the cost of extraction and transformation, and/or the price of imports. Decisions on technology implementation are taken by minimizing cost (or maximizing some kind of welfare function); they are not exogenous.

Policy option scenarios may include or exclude technologies such as nuclear or wind; the importation of gas through pipeline or electricity through wires; or LNG harbors. The final implementation of these options is a question of economic competitiveness.

In policy development, the advantages and disadvantages of these options are usually analyzed separately, and recommendations derived from the study. Every assumption, including decisions on technological implementation are generally made endogenously to the model results, and, as such, are not considered as scenario parameters.

3.1.2 Scenario Paradigm and Attributes

The optimization models in this study examine scenarios defined in terms of *technological, economic*, and *environmental* attributes. The technologies studied include (advanced) coal, gas, nuclear, and renewable generation technologies. The introduction of a technology into a given scenario establishes specific operational and economic parameters (primarily cost, efficiency, emission rates, and introduction/deployment rates). Economic attributes of each scenario include assumptions and uncertainties related primarily to fuel prices (which include both resource and policy-driven clean-up considerations); energy demand (*e.g.*, primarily through GDP growth and attendant price and income elasticities); and discount rates. Environmental scenario attributes include a choice between imposed emission limits or caps *versus* taxes to control SO_2 (local, regional) or CO_2 (countrywide, global) emissions. While the allocation of caps *versus* taxes is sometimes related to particular economic structures, examples of the use of both instruments within the same economic paradigm can be cited.

3.2 MARKAL-Shandong; Model Description

A Linear-Programming (LP) model, MARKAL (Market Allocation; Fishbone, 1981) optimizes the Reference Energy System (RES) of a country or region in terms of its fuel sources (SRC: imports, mining exports, renewable); energy technologies (TCH: processing, conversion, demand-related); and final-energy (FE) demands (DM) or demands for energy services.

As illustrated in Figure 5.7, a SRC-TCH-DM system may be evaluated over multiple time periods and is driven by assumptions about the time horizon of final-energy demands for current economic sectors (industry I, residential R, transportation T, commercial services S, and agriculture A). The MARKAL model expresses the energy flows and transformation needed to meet these exogenous FE demands in terms of sets of energy carriers, {ENT}, and sets of energy technologies, {TCH}. Within the Figure 5.7 framework, each of the supply, processing (non-generation transformation), and generation elements within each set compete to meet

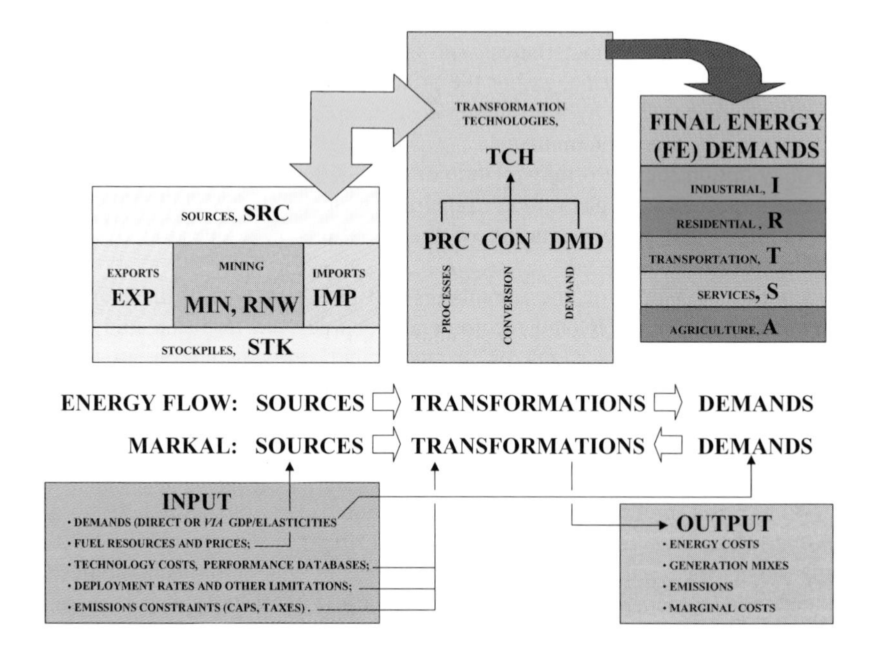

Figure 5.7 Energy flows within MARKAL (Fishbone, 1981) showing the connectivity between sources, technologies, and demands, as well as typical input and outputs.

the exogenous FE sectoral market demands (I, R, S, T, A) and exogenously imposed constraints (*e.g.*, mass and energy balance, technology deployment constraints, emission limits, *etc.*) to minimize an objective function (*e.g.*, total system cost discounted over the specified time period).

Sources of energy required to meet FE demand through the constrained technology set are all one-way flows from imports (IMP), exports (EXP), mining (MIN), renewable (RNW), and stockpiles (STK). The energy-conversion

technology set, {TCH}, divides into: a) processes {PRC} that convert one energy carrier into another, independent of load; b) load-dependent conversions {CON} of specific energy carriers into electricity and/or process heat to meet specific FE demands; and c) demand technologies {DMD} that consume energy carriers to meet demands for energy services.
The RES for a given region or country can be as complex as the modeler desires. The RES constructed for Shandong MARKAL models is relatively simple. It focuses primarily on the dominant industrial and residential sectors, with the remaining transportation, service (commercial), and agricultural sector modeled at a relatively high level of aggregation.

3.2.1. Scenario Specifications
The MARKAL-Shandong LP optimization model was used to generate relatively "surprise-free" futures for the province and for China. These scenarios show relatively few high-level attributes. While scenario definition and designation differences between CRETM and MARKAL models are inevitable, a special effort was made to maintain a high degree of uniformity to facilitate the comparison of results generated from each (Section 5.3). Table 5-4 lists the key elements used to define scenarios for both the CRETM and the MARKAL-Shandong optimization modeling tools. The scenarios reflect a given set of attributes including energy demand, prices, technology (market) penetration, and the environmental attributes (C, S, or E, as given in Table 5.4). These include:

- High (H) *versus* Low (L) energy-demand developments, as reflected either directly by separated demand estimates, as in the case of MARKAL-Shandong (see also Chapter 4), or by assumed growths in productivity (GDP) and the use of price and income elasticities in CRETM (Kypreos, 2002);
- High (H), Medium (M), or Low (L) long-term and inflation-free discount rates used to determine the present value of all energy costs pertinent to a given model (*e.g.*, electricity costs in the case of CRETM, total energy costs in the case of MARKAL-Shandong); The reference value was 10 percent per annum.
- Nominal and Constant (C) *versus* High (H) and increasing (doubling over the next two decades) fossil-fuel prices;
- High (H) *versus* Low (L) success rates for advanced electricity generation technologies measured in terms of market penetration, either explicitly, as in the case of nuclear energy, or by using endogenous technology in the models;
- Environmental policy, as reflected in C, S, and E attributes, are listed in Table 5.4, along with the ways in which these three scenario drivers alter related goals symbiotically.

Lastly, the actual (computational) drivers behind the implementation of any one of these scenarios [*e.g.*, S-BHHC = Shandong(S) – BAU Policy (B) High Demand (H) High Discount Rate (H) Constant Fuel Prices (C)] are implemented through the following variables or procedures:

- expected economic and population growth rates;
- income and price elasticities used to couple economic growth and demand for both total energy and electricity consumption;
- structural (Sector) shifts in the overall economy that affect the level and nature of energy demand;
- real (inflation-free) discount rates and investment pay-back times;
- fuel availability and prices for both domestic and imported sources;
- energy and environmental policy options and instruments (caps, taxes, advanced technologies having reduced emissions, etc.).

Table 5.4 Essential Scenario-Defining Elements Used in the Optimization Models

Scenario Designation	**Scenario Identifier**	**Scenario Attribute Definition**
REG	S	Regions: S = Shandong Province for MARKAL; No index for the seven-region China (CRETM)
BAU	B	BAU = Business-As-Usual economic, emissions, technology, *etc.* conditions and/or policy
CO_2	C	Emphasizes control and trade of CO_2 emissions, with a China policy that is more open to international cooperation and trade, with watchwords being technology transfer, lower interest rates, higher infrastructure investments in energy and environmental control, *etc.*
SO_2	S	An emphasis is placed on atmospheric sulfur emissions through caps or taxes based on both local and regional targets and the recognition and incorporation of external costs.
EMI	E (= C + S)	Places and emphasis on both sulfur and carbon emission controls to examining synergism and secondary benefits that might arise from a range of policy instruments (*e.g.*, caps *versus* taxes *versus* investments in advanced reduced-emissions technologies.

To clarify the discussion we omit the letters for the discount rate and the penetration rate of new technologies; all scenarios analyzed in this chapter have a high discount rate and high penetration rates of new technologies. However these scenario attributes belong to an optimization model that includes exogenous energy demand. In the case of the MARKAL-based models, the demand for end-use or Useful Energy (UE) services must be specified. In an ideal situation, one should specify only useful demands or demands for energy services, e.g. passenger km/yr and tonne km/yr for transportation; heating and cooling requirements for commercial or domestic uses; and specific industrial demands, e.g., how many tonnes of oil are required to produce a tonne of product (steel, cement, paper). For the MARKAL-Shandong model, the mix of FE demands provided by ERI required considerable interpretation and interpolation before it could be used in MARKAL, as is described in the following subsection.

3.2.2 Energy Demand interface with MARKAL

The first systematic specifications projected for energy demand in Shandong Province and an improved understanding of the forces driving this demand using the MEDEE-S bottom-up model have been applied by ERI (see Chapter 4). Based on these projections, an interface model of demands for energy services coupled to the supply model MARKAL was developed at Paul Scherrer Institute (PSI). ERI added information and background data to accomplish this interface task. This information was either available as underlying assumptions in MEDEE-S, or extra efforts were made to collect or derive it.

The interface model coupling ERI end-use energy demand and MARKAL specifies both the demand for energy services under BAU (B) scenario conditions and the implicit income elasticities for these conditions. The income elasticities together with an exogenous set of price elasticities for energy services then allow alternative scenarios to be determined directly in MARKAL using the partial equilibrium version of the model. The following paragraphs describe the MEDEE-S demand structure, and the creation of an Interface Model to implement the ERI demand projections in MARKAL.

The MARKAL models require exogenous input of all end-use demands for each time interval being included in the optimization. These demands provided in UE units or indicators (e.g., tonne km/yr for freight transport, or m^2 of heated or illuminated household area in rural and urban settings, etc.) together with the efficiency factors to convert these demands to FE(PJ/yr) demands enable the full optimization potential of MARKAL. The Interface Model had to be developed to accept these ERI-generated FE demand projections, to test these projections for self-consistency and "reasonableness," and to generate input files that MARKAL could use.

The MEDEE-S structure is as follows: The Agricultural subsector includes all mining and fishing as well as farming. The Industry sector includes energy intensive subsectors, energy intensive products, and the construction sector. The tertiary sector includes Commercial (Services) and Transportation. The Households sector was partitioned to Residential Urban and Rural. The task of the Integration Model is to take these demand projections reported by ERI for Shandong and convert them into a form that could be used by the MARKAL optimization model.

The ERI group considered four demand scenarios: Baseline, low-GDP, reduced efficiency (inefficiency), and changed sectoral structure. Most of the studies with MARKAL-Shandong focused on the ERI Baseline and Low-GDP case scenarios while the Interface model was defined to facilitate alternative scenario and sensitivity analyses. The integrated impact of the four ERI scenarios on projected

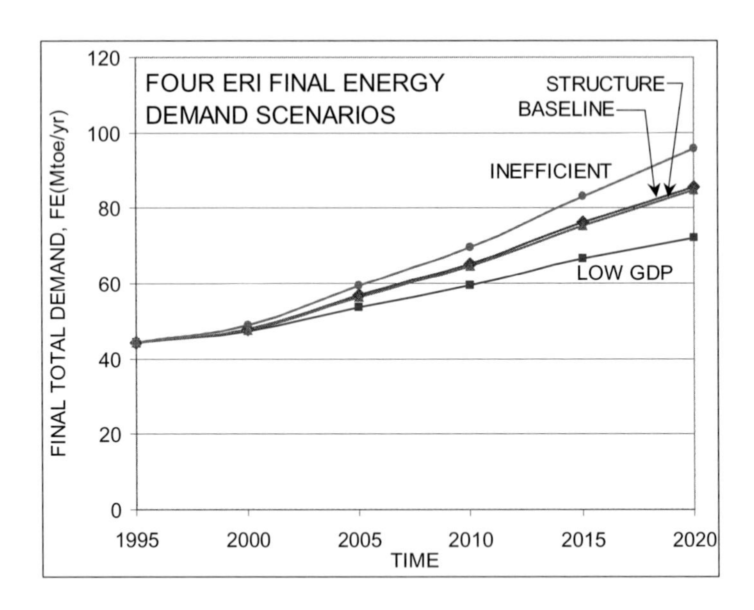

Figure 5.8 Differences between integrated FE demand among the four ERI scenarios considered (Baseline, low-GDP, Structure, and Inefficient) for Shandong Province; the MARKAL-Shandong used primarily the Baseline scenario of ERI.

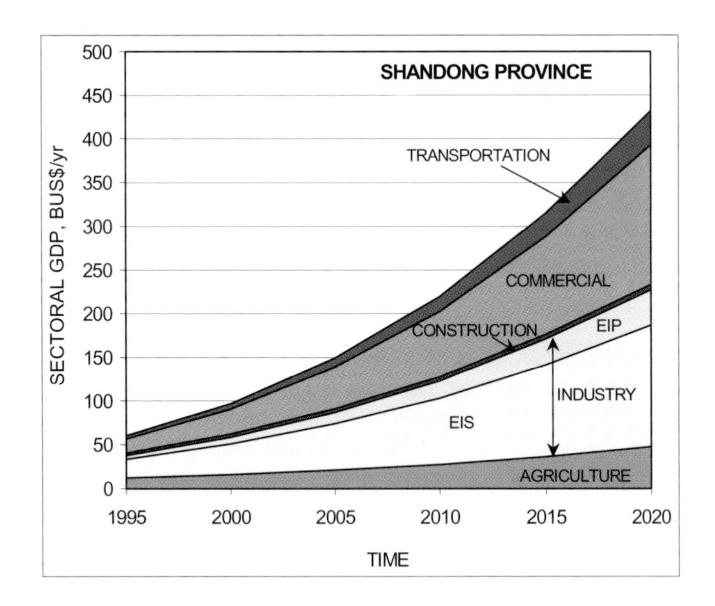

Figure 5.9 Variation of the GDP structure used by ERI to generate MEDEE-S Final-Energy (FE) demand projections. EIP: Energy-intensive products, EIS: Energy-intensive sectors.

FE demand is illustrated in Figure 5.8, with the expected results: Lower GDP growth translates into lower FE demand; less efficient use of UE leads to higher FE; the sectoral restructuring at the level implemented by ERI would have little impact on the Baseline case, the focus of the present study. The assumption of (decreasing) energy intensity for each subsector plays an important role in setting the FE demands and, thereby, the MARKAL results. Finally, based on ERI results a sample of GDP divisions is given in Figure 5.9.

The high level of economic growth for Shandong Province and the associated structural changes assumed by ERI emphasize a shift in the economy away from agriculture and industry towards a greater reliance on the service sector. These shifts are shown explicitly in Table 5.5. Another important shift for the CETP study relates to the implicit assumptions on the income elasticity for electricity compared to the elasticity assumed for other fossil fuels. The level of income elasticity in Shandong is below historical values for China.

The energy-to-GDP ratio and the income elasticity are low, (logarithmic energy-to-GDP ratio, ELASGDP in Equation 5.1, below) which implies very efficient energy use in the future. In any case, the income elasticity for energy and electricity shown in Figure 5.11 reflect the known characteristics of the Chinese energy system. It should be noted that the income elasticity values shown in Figure 5.11 reflect all potential changes (e.g., efficiency improvement, price effects, and the structural changes) but are low in relation to international experience.

Table 5.5 Shifts in Sectoral Structure and GDP Assumed as a Basis for the ERI Final-Energy Projections (see also Chapter 4).

Year	1995	2000	2010	2020
GDP (billion Yuen)	500	806	1820	3580
Population (million)	87	89.6	95	98
GDP/cap(US$/cap)	692.4	1083.8	2308.2	4401.3
GDP Structure (%)				
o Primary Industry[(a)]	20.2	16.5	12.4	11
o Secondary Industry[(b)]	47.4	47.5	45.8	43
o Tertiary Industry[(c)]	32.5	29.5	41.8	46

[(a)] Primary Industry = Mining + Agriculture

[(b)] Secondary Industry = EIP + EIS + Construction

[(c)] Tertiary Industry = Commercial + Transportation

The vital statistics assumed in making the ERI projections for Shandong Province are summarized in Figure 5.10. The increase in energy efficiency is indicated by energy intensity, EIFE(MJ/US$), based on the FE demand for the Baseline ERI projections.

Periodic changes in the shares of energy use among sectors, as computed from the MEDEE-S data provided by ERI, are irregular. At the beginning of the time horizon, energy use in the Industrial sector assumes a low growth rate of less than one percent, but later increasing. Nevertheless, the income elasticities, i.e., the relative change in energy use per unit change of the value added, is ~0.4, while the income elasticity for electricity is ~0.65.

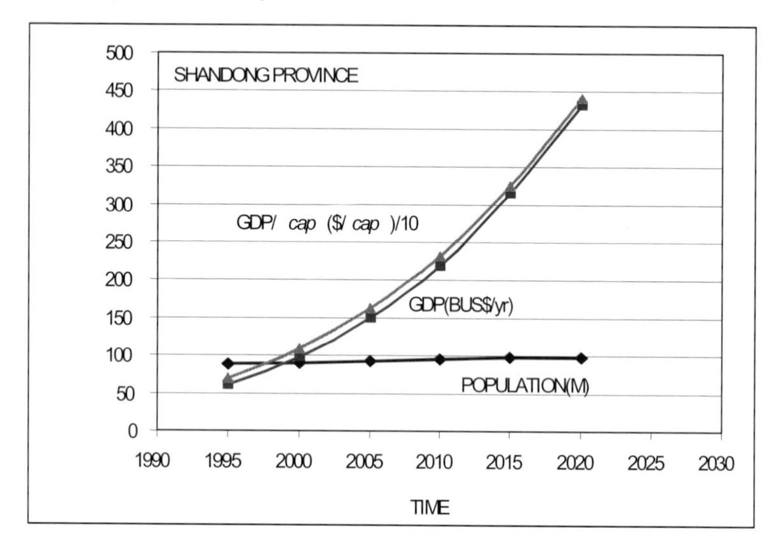

Figure 5.10 GDP, population, and per capita GDP growths assumed by ERI for Shandong Province in providing the FE and UE inputs to MARKAL-Shandong.

In addition to developing end-use demand projection for the BAU (B) cases being considered by MARKAL-Shandong (e.g., ERI's Baseline scenario reported above), ERI has generated projections based on alternative assumptions about economic growth; structural changes in the sectors; and a range of assumptions dealing with efficiency improvement. The values reported here reflect most potential changes (e.g., efficiency improvement, price effects, and the structural changes). However, in many cases they differ from international experience and some, e.g. low-income elasticity and high rates of decrease in EIFE, require greater justification.

The complete Reference Energy System (RES) that results from this melding of the final demand structure with the detail required in the standard MARKAL model is given in Figure 5.12 in which each of these sectors is briefly described. The specification of income elasticities, ELASGDP., is based on the demand projections of ERI and Equation 5.1, and assumes constant prices over time. The traditional way to specify elasticity values is to define the time series of relevant variables and

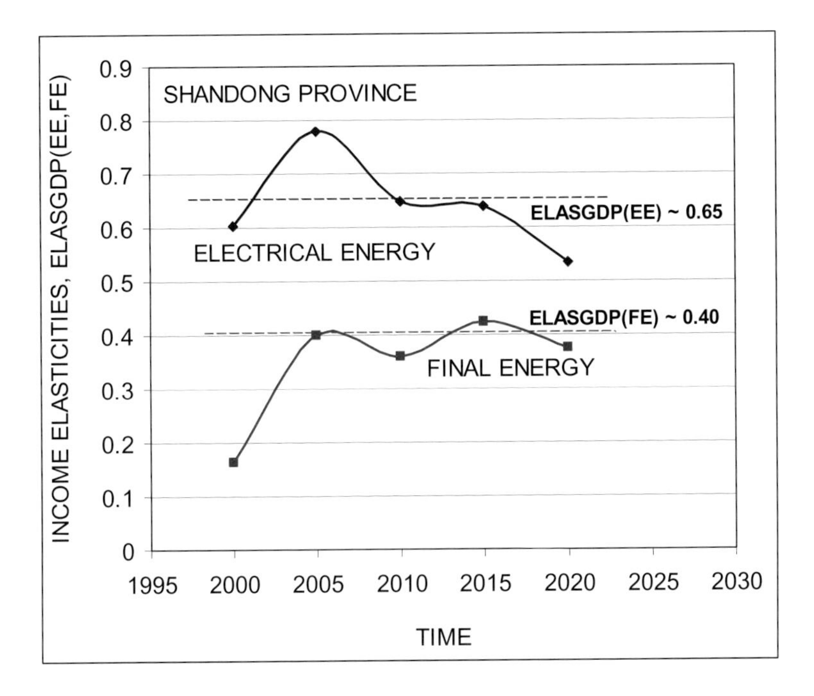

Figure 5.11 Income elasticities computed from ERI projections used to drive the MARKAL-Shandong model.

perform econometric analysis. Here, the use of implicit elasticities associated with ERI assumptions and results simultaneously aid in understanding and validating underlying assumptions.

$$D_t / D_{t-1} = (Y_t / Y_{t-1})^{ELASGDP} * (P_t / P_{t-1})^{-ELASPRC} \qquad (Eq.\ 5.1)$$

where Y_t is the GNP at time t, and P_t is the price associated with items aggregated or classified into the particular demand sector or subsector. This relationship can also be used to adjust the demand projections, assuming other elasticities and/or other GDP growth rates and prices. These adjustments can be made either in the Interface Model directly or in the partial-equilibrium version of the MARKAL model that generates internally the marginal costs of demands for energy services and endogenously adjusts the demand level applying the relation above. Cross-price elasticities are not defined in this work.

The Interface Model that makes the connection between MARKAL-Shandong and the MEDEE-S-generated demand estimates of ERI also uses the technology database of TU, as well as the ERI input to the EEM task. Other inputs include technologies for power generation; systems for removing sulfur and nitrogen oxides; space-heating systems; industrial boilers; and data for transportation systems (e.g., cars and trucks for both private and public use and for freight or passenger cargo).

The Interface Model begins with the specification of FE and/or UE demand categories from MEDEE-S from ERI. Demands are grouped together in the so-called "bottom-up" categories, in which a set of representative end-use technologies and associated fuels are described in terms of market shares using MARKAL. For the rest of the demand categories, for which detailed technical information is not available, markets shares are simulated using the "top-down" (econometric) approach explained above.

Final-energy demands in these categories are specified from the ERI reference development or Baseline scenario, together with the implicit income (and price) elasticity per demand category. Price assumptions other than those valid for the baseline scenario, or price changes resulting from the imposition of taxes or environmental constraints, could modify the reference demands, depending on the level of price elasticity assumed for Shandong Province. The sectoral structure of these FE demand categories is shown in Figure5.12. Finally, the modified values for elasticities together with alternative growth assumptions can be used to define alternative demand projections. The results of scenario analyses using the above-described MARKAL-Shandong and Interface Models are presented in Section 4.1.

3.3 The China Regional Electricity Trade Model (CRETM)

Like MARKAL, the CRETM is an optimization model based on Linear Programming (LP) methods. While limited to the electrical energy sector, the main attraction of CRETM compared to full-energy-sector models like MARKAL (Fishbone, 1981; Kypreos 1996; Goldstein, 1995), is its ability to follow interregional allocation and transport of fuels within the seven-region model of China (Murray, 1998). Like the earlier version of the Harvard University energy model, the PSI-modified CRETM model minimizes the total cost of the electrical system, including all costs associated with generation per se (e.g., capital, fixed, and variable costs), coal cleaning, interregional fuel transport, inter- and intraregional power transmission. When so constrained, it ignores external costs attributable to the emission of both pollutants (mainly SO_2 and NO_x) and of greenhouse gases (GHG, mainly CO_2).

A still earlier version of the Harvard University electrical-energy model, first developed and used to model a six-region China, was restricted to coal-fired stations and a much shorter time horizon. CRETM examines a full-range of electrical generation options. This full-spectrum generation model covers a 30-year time scale, similar to that used by PNNL (Chandler, 1998), from which much of the data used herein derives. Several PSI modifications to the Harvard model contributed to the development of the CRETM:

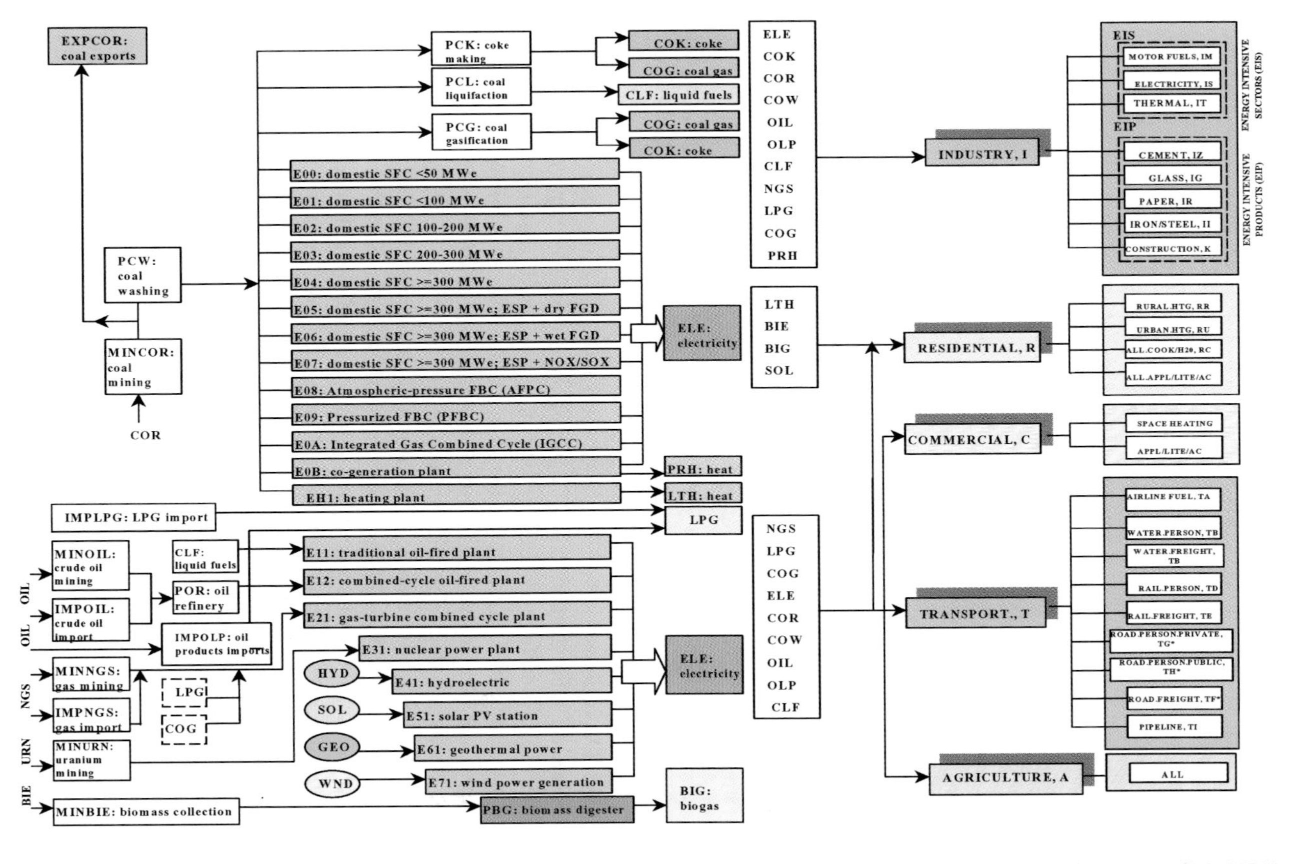

Figure 5.12. Reference Energy System (RES) used in MARKAL-Shandong, illustrating connection and flows from the PE/SE parts of MARKAL to a simplified FE/UE demand structure that interfaces with the ERI projections for Shandong Province.

- Shandong Province is explicitly included.
- The time horizon is extended to the year 2030 and becomes a scenario-dependent input.
- Electricity Demand is not fixed, but is now price (e.g., marginal cost) and income elastic.
- Resource, imports, and transportation costs are time dependent.
- Transportation costs are redefined based on mode, distance, and mass.
- Explicit variables for investment and capacity constraints have been introduced.
- Imports to explicit regions by transport mode and regional balances are now included.
- Technology market penetration and growth rates for expanding technologies are described by explicit constraints.
- Emission of SO_2, NO_x, and CO_2 are limited either by direct taxation or explicitly constrained (regional or country-wide).
- Endogenous technological learning is included so that costs are reduced in accordance with accumulated installed capacity of a given technology;
- Growth of generation capacity is decoupled from growth in electricity generation per se and forced by a capacity constraint to estimate the impacts of peaking management.
- Finally, Scenario Generation and Results Report Generation are broadened and redefined to create directly comparative graphs and tables.

The CRETM is comprised of four basic relationships: a) the *objective function* (present value of total system costs); b) *mass conservation* relationships that balance the flows of fuel masses across processes and regions; c) *energy conservation* relationships that connect the flow of fuel masses to energy generation and transport; and d) a *set of constraints* that establish bounds on both energy and mass flows, levels at which specific generation technologies can be deployed, and emission rates. These latter endogenous constraints combine with controls enforced through specific exogenous inputs (e.g., discount rates, demand growths, GDP growths, fuel and capital unit costs, *etc.*) to create the various scenarios and used to define and/or bracket possible E^3 (energy-economic-environmental) futures for Shandong Province in the context of China.

The computational algorithm used to evaluate CRETM, like that under which MARKAL operates, is based on a set structure established by the GAMS operating system (Brookes, 1998; ILOG, 1997). Within the GAMS framework, technologies, fuels, regions, etc. are described in terms of *sets*. The four essential elements of CRETM described above are evaluated in terms of a set of generation technologies that are driven by a set of fuels, which is a subset of fuel-related commodities and electricity. These fuel and technology sets are combined within the GAMS architecture to describe the energy and material flows within and between the seven China regions.

The CRETM describes an electrical energy sector model that cost-optimizes on the basis of 17 generation technologies; 16 fuel forms (including renewable energy);

both domestic and foreign fuel sources (for oil, gas, and uranium); and four transportation modalities. It can search over eight time periods into the future, examine energy and material exchanges within and between seven regions, and use an object function composed of eight cost categories (domestic fuel, imported fuel, coal cleaning, transportation, possible emission taxes on three species, capital costs, fixed generation costs, and variable generation costs). A product of these six dimensions (generation technologies, fuel commodities, temporal, regional, fuel transport, and cost categories) yields about 500,000 possibilities, all of which will deliver a single economic optimum given a particular set of constraints.

Even for a relatively simplified model, the size of the task can become daunting, particularly when ten or 15 optimizing scenarios are superimposed. Furthermore it is necessary to understand the effects of deviations or departures from any "optimum" that is reported, as well as the impact of changing the objective function, on possible conclusions and recommendations. The results from CRETM presented in the following begin a quantification of that impact.

3.3.1 Scenario Specifications

Scenario definitions and specification used in CRETM parallel as much as possible those described in Section 3.2.1. for MARKAL-Shandong. Use of the baseline demand for electrical energy reported before is given a "high" or "H" designation, nominal or "constant" (C) oil and gas prices, and the input values given in detail in Appendix A of Kypreos (2002), results in the BHHC scenario; the added “H” indicates a high discount rate, which for most of the CRETM optimizations was assumed to be 10%/yr. For nominal and constant (C) fuel prices, the time-independent unit costs given in (Kypreos, 2002) are use to generate a baseline BHHC-B scenario (referring to the base-case designations: Business as Usual=B, high demand=H, high discount rate=H, constant fossil-fuel prices=C, and base-case technology costs=B).

The rest of this chapter will use a simplified notation that allows a systematic distinction of all scenarios presented in detail. In this notation, the first letter stands for the environmental policy, the second one for the demand, and the third one for the assumed level of fossil fuel prices. In this notation, the base case BHHC-B becomes BHC.

In addition to the BHC base case, seven additional main policy scenarios have been defined. The scenario BHH is the basic case, with "high" overall demand and "high" and increasing fuel costs, wherein time-dependent unit-fuel-cost multipliers are applied to the Scenario BHC values (see also Kypreos [2002]). As seen in Table 5.6, scenarios CHC and CHH include a carbon-emissions cap; scenarios SHC and SHH include a sulfur cap; and scenarios EHC and EHH include the imposition of both. All the scenarios assume a high real discount rate of 10%.

Table 5.6 Carbon and Sulfur Emission Limits on the Eight Scenarios.

Scenarios	**CHC, CHH EHC, EHH**	**SHC, SHH EHC, EHH**
Time, t	CO_2TARGET(t) MtonneCO_2/yr	SO_2TARGET(t) MtonneSO_2/yr
1995	Unconstrained	Unconstrained
2000	1300	10
2005	1500	12
2010	2000	13
2015	1800	14
2020	1600	14
2025	1500	13
2030	1500	12

3.3.2 Electricity Demand

As explained in the previous section, the exogenous demand for electrical energy for each of the seven China regions being modeled is fixed at a value termed H for "high". This designator is relative, particularly when compared to other projections and used in China electrical-sector modeling exercises (Murray, 1998; Chandler, 1998; Dadi, 2000). In CRETM, the demand for electrical energy in period t for region i, $DELEC(i,t)$, is determined from the following relationship:

$$DELEC(i,t) = DELEC(i,t-1)\frac{[1+GGDP(i,t)]^{ELAGDP*YEARPP}}{[1+GPRICE(i,t)]^{ELAPR(t)*YEARPP}} \quad \text{(Eq. 5.2)}$$

The exogenous growth rates for fuel prices and GDP are $GPRICE(i,t)$ and $GGDP(i,t)$, respectively; the respective elasticities are $ELASPR(t)$ and $ELAGDP$; and $YEARPP$ is the number of years per period. The growth in regional GDP is given below. The coefficients used are reported in Table 5.7.

$$GDP(i,t) = GDP(i,t-1) * [1+GGDP(i,t)]^{YEARPP} \quad \text{(Eq. 5.3)}$$

Table 5.7 Summary of Regional GDP Growth Rates and Income and (Electricity) Price Elasticities Used in this Study (assumed the same for all seven CRETM regions).

Time Period	**GDP Growth** Rate, %/yr	**Income** Elasticity	**Price** Elasticity
1995-2000	9	0.65	0
2000-2005	8	0.65	0
2005-2010	7	0.65	0
2010-2015	6	0.65	0.1
2015-2020	5	0.65	0.1
2020-2025	4.5	0.65	0.2
2025-2030	4	0.65	0.2

Total GDP(B$/yr) over all seven regions along with the resulting time-dependence of country-wide electricity demand and the Electrical Energy Intensity, EEI(MJ/$) for the BHC scenario are shown on Figures 5.13 and 5.14. The key assumptions about cost and technology underlying the BHC scenario used to generate the results are summarized in Table 5.14. Lastly, Figure 5.2 gives a comparison of Chinese electrical-energy demand used in this study with projections made from a range of Chinese, IIASA, and US studies.

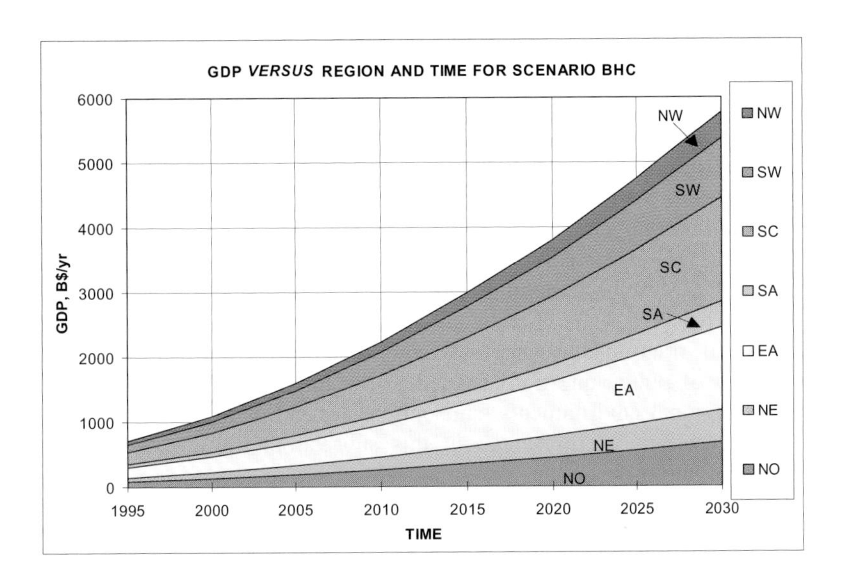

Figure 5.13 Regional GDP versus *time for the BHC scenario used in CRETM. All CRETM results presented herein are based on electricity demands derived from GDP growth vis-à-vis Equation 5.2). There are some differences between the demands for Shandong and those projected from MARKAL-Shandong using the exogenous final end-use energy demands projected by ERI, but, as shown in Figure 5.75, these differences fall into a reasonable range.*

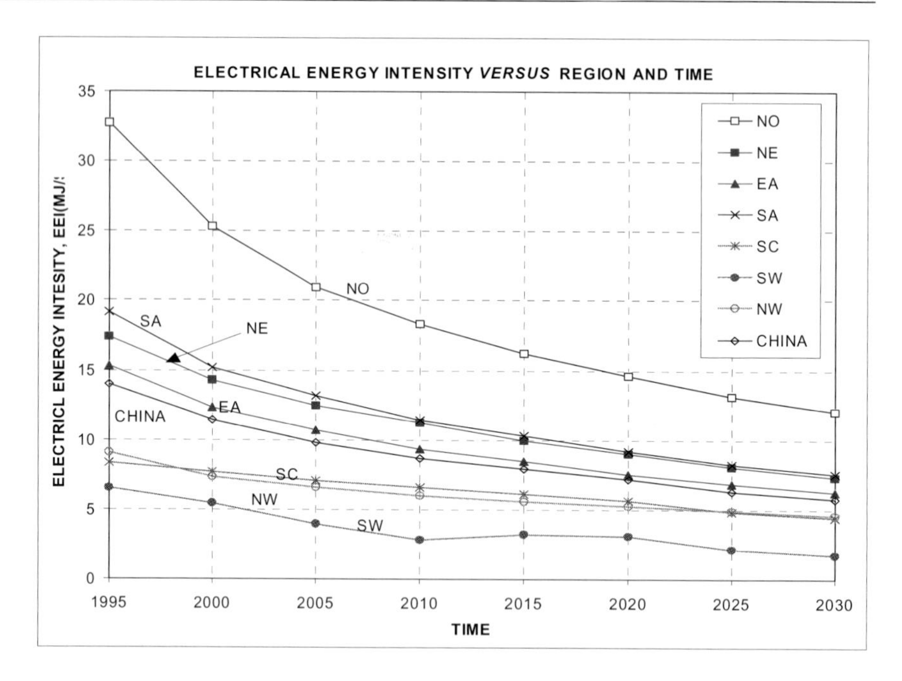

Figure 5.14 Regional Electrical Energy Intensity versus *time for BHC scenario.*

In the last few years, electricity-demand projections for China have varied widely. More recent projections based primarily on the strong recent performance of the Chinese economy reflect greater optimism. The BHC scenario-demand adopted for the CRETM studies generally falls in the mid-range of these past and more recent (optimistic) projections, as is illustrated in Figure 5.2.

3.3.3. Optional Endogenous Modeling

The CRETM is a regionalized state-of-the-art optimization model able to assess options for electricity generation in China. Some new model options define equilibrium between demand and supply; endogenous technological learning; and a multiregional trade of energy fuels and electricity. These features allow the examination of policy questions related to the improvement of energy efficiency, local and regional environments, and options for mitigating global warming. The new model options are presented in this subsection; the results of a sensitivity analysis applying these options are given below.

Lately, advanced endogenous learning models have been integrated into energy optimization models to determine the cost of technology development using the concept of learning curves. In these models, the cost of a technology is not defined as an exogenous trend over time, but instead as a function of experience accumulated as people learn how to fabricate and use the technology. A typical learning curve describes the cost of a technology as a function of cumulative installed capacity. Unit costs (e.g., \$/We) for the technologies in question are reduced at a constant fraction for every doubling of cumulative installations. This empirical relation represents "learning-by-doing" and "learning-by-using" a technology. Learning curves characteristically assume that the greatest cost reductions occur at the outset, and then tail off as the amount of installed capacity increases. The cost-reduction relationship appears as a straight line on double logarithmic scales[1].

The EEM group at PSI integrated this option in CRETM. This capability broadens insights into the future electricity markets in China; helps pinpoint the cost reduction for sulfur and carbon control, and views research and development as a policy option for improving technology performance.

Advanced economic equilibrium models introduce price feedback in the evolving electric markets. The CRETM was modified[2] to introduce partial equilibrium and achieve the aims of CETP. The basic idea behind this is that demand adjusts to consumer choices through price changes. Demand feedback is based on relative changes in the marginal generating costs of electricity in each region and the level of price elasticity. Consumers gradually become active players in the optimizations that establish minimum-cost fuel and generation mixes; market reforms occur in the electricity sector, and these behaviors are quantified for electricity services through price elasticity. Since price elasticity in China is not well known, near-scenarios begin with low (conservative) values. Further-term scenarios (2020-30) assume a move toward international market behavior.

It is well known that China has insufficient generation capacity for peak load management. Coal-burning units provide the main supply of power, supplemented with small diesel-burning engines. Algorithms for peak management should be developed for CRETM, as well as for other optimization models, like MARKAL. One way to extend the model is to include seasonal balances and load profiles. This is possible in the MARKAL model, but it does not offer the level of detail afforded in a simulation model. This option, however, would complicate the CRETM, and, furthermore, requires exact load characteristics.

Instead, the peaking model adopted in CRETM uses a simplifying approach to define peak power demand. First, the average load factor, as defined in Chinese compilations of energy statistics is used to determine the total capacity for the starting year. The model differentiates then between energy demand growth and peaking requirements. Multiplying the demand growth by time-dependent peaking factors yields the required growth. Extra constraints force the construction of new capacity to satisfy peak demand. Sections four and five, below, detail model results for the MARKAL for Shandong and CRETM for China and Shandong.

4. MODEL RESULTS: MARKAL

The primary and final energy use in Shandong shown in Figs. 5.15 and 5.16 illustrate the dominance of coal for the SBLC baseline scenario. These SBLC results are based on ERI's projection of final (the low GDP growth case discussed above) as extrapolated by the EEM team for the year 2030 using the MARKAL model of Shandong to describe the energy system. This description is based on a set of assumptions related to resource availability and prices, a given level of technological development, and environmental policy. In the SBLC case, environmental policy is assumed not to change, but other scenarios include policy variants. The baseline scenario, therefore, is by no means the most probable. The assumption of low income elasticity in Shandong causes final and primary energy demand to double that of 1995, while economic output is assumed to increase twice as rapidly as energy demand.

An increase in the share of electricity in the final energy demand corresponds to the implicit assumptions of ERI concerning changes in income elasticity favoring the substitution of electricity for the burning of fossil fuels. The income elasticity for electricity use is 0.65 (*versus* 0.85 for China), while the overall final energy demand elasticity was taken to be 0.4. As previously noted, electricity is produced mainly by coal, which explains the primary-energy distribution shown in Figure 5.16, scenario SBLC. Advanced coal power generation systems appear only by the end of the time horizon under prevailing baseline case conditions.

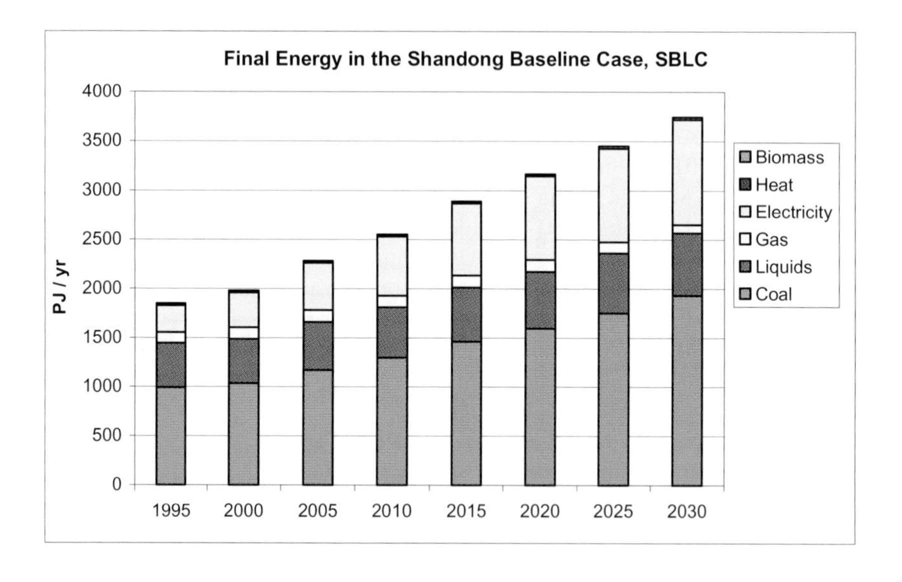

Figure 5.15 Final energy by fuel in the low-demand, constant-price, baseline (SBLC) scenario computed with the MARKAL model.

Figure 5.16 defines the primary energy profiles under different environmental policy scenarios. These include the sulfur control case (SSLC) in which sulfur emissions are stabilized at present levels; the carbon control case in which carbon emissions are stabilized at 150 Mtonne/yr (SCLC scenario) in the last decade; and the combination of both policies (SELC). Finally, for comparison purposes, a case requiring a sulfur tax of 1500 \$/tonne SO_2 (special sulfur-tax case SSLC-tax) is shown.

Systems using gas, nuclear energy, or wind are not competitive with coal under nominal prices and present environmental policy, though wind power is included in the solution at a minimum capacity. However, a radical change in the primary-energy supply distribution in favor of gas and nuclear energy would occur should a carbon cap (SCLC) be imposed, and gas would be much more widely used should in response to a tax on sulfur emissions (special sulfur tax case SSLC-tax).

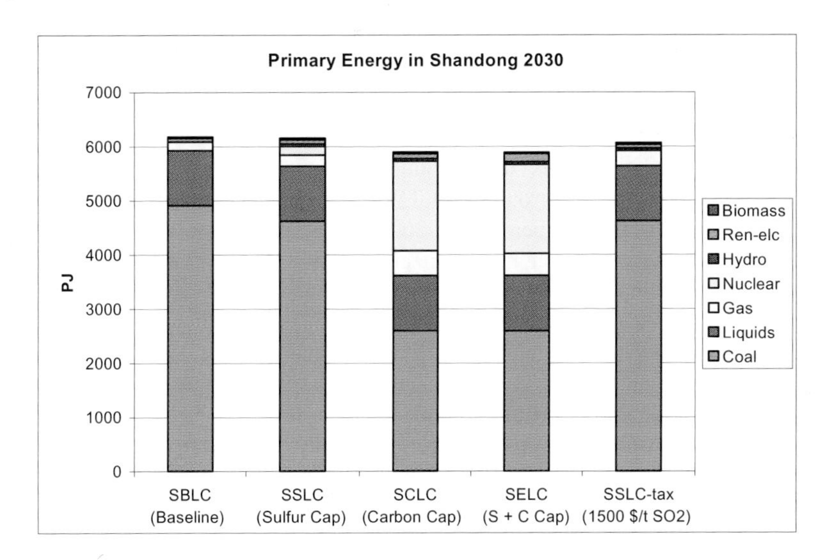

Figure 5.16 Primary energy by fuel in the low-demand, constant-price cases (S(B,S,C,E)LC, for a range of special caps and sulfur tax cases) computed with the MARKAL model.

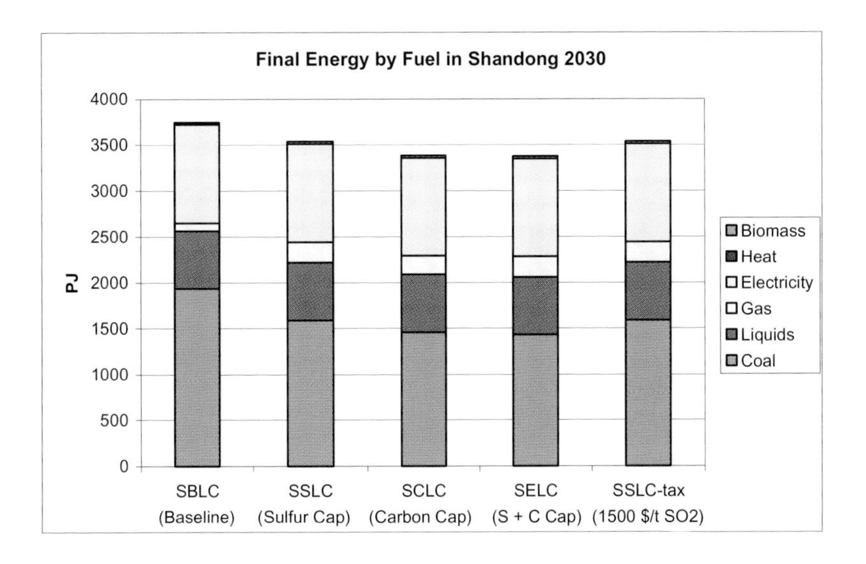

Figure 5.17 Shandong-2030; final energy by fuel, low-demand, constant-price scenario.

Figure 5.17 shows the final energy profiles for these cases. The main conclusions are:

- Final energy demand responds to policy changes. Demand is reduced under a carbon constraint that drives costs for energy services higher and favors efficiency improvement across all sectors of the energy system
- Gas and electricity will be substituted for coal in response to a sulfur control policy, and should a carbon control policy be adopted, gas will be substituted for coal at a dramatic rate.

Figure 5.18 breaks down electricity generation by different technologies in different policy regimes. Because seven percent of electric power is lost in transmission and distribution, the amount of electricity generated exceeds final demand, reaching values of around 330 TWh/yr. Final demand is higher by a few percentage points (e.g., 2.5 percent in 2020) than that projected by ERI because MARKAL favors substitution based on electricity. The ranking of systems like advanced coal technology *versus* scrubbers in the cases of sulfur control shows interesting behavior. First, without active policy limiting sulfur emissions, conventional coal technology will prevail (as in scenario SBLC). Second, scrubbers and/or advanced coal technology based on AFBC and IGCC will be introduced at different levels, depending on the amount of taxes imposed. Finally, a carbon tax makes the options of nuclear power, natural gas, and wind competitive. In the case of controls on both sulfur and carbon (scenario SELC), advanced coal systems only are introduced, which indicates that scrubbers are less competitive. However, in the case of high sulfur taxes (special sulfur-tax scenario SSLC-tax) with no carbon tax or cap, the contribution of coal plants with scrubbers becomes significant. The market will be shared between conventional coal systems with scrubbers and advanced coal technology.

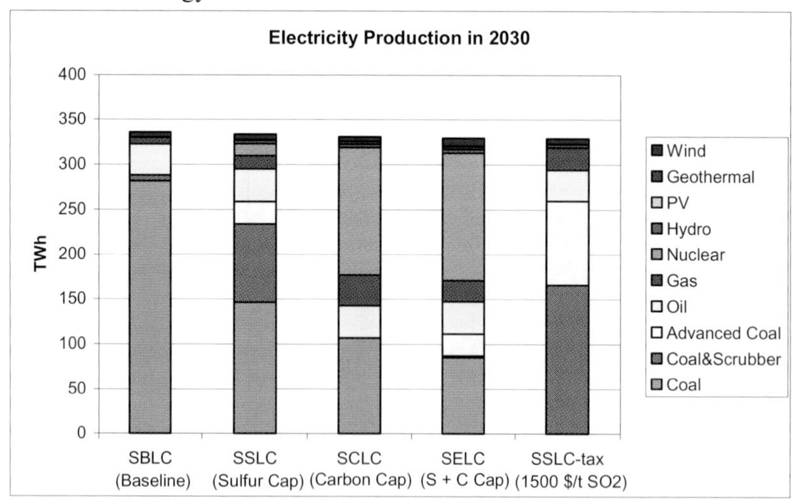

Figure 5.18A. Shandong-2030; Electricity generation by fuel and technology in the low-demand, constant-price scenario [mainly S(B,S,C,E)LC].

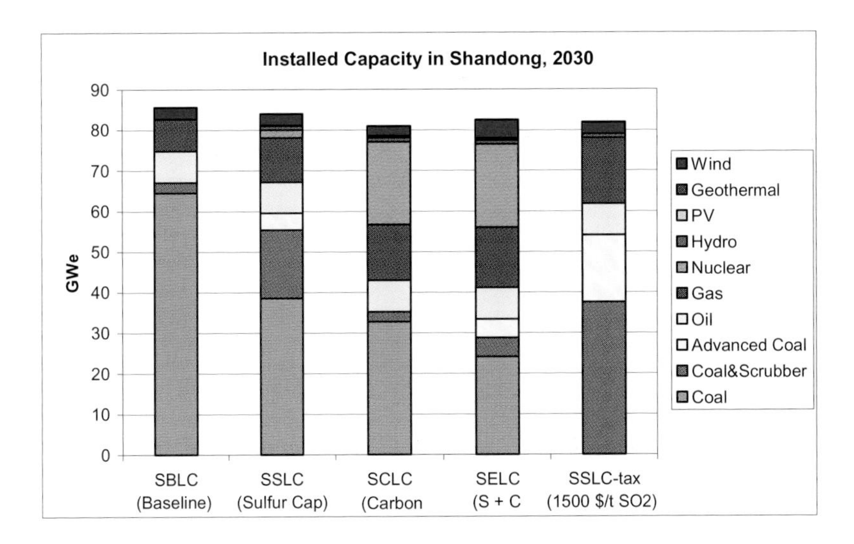

Figure 5.18B Shandong-2030; Installed capacity by technology in GWe, in the low-demand, constant-price scenario [mainly S(B,S,C,E)LC].

A policy to stabilize sulfur emissions at present levels will stimulate a moderate increase—2.11 percent--in cumulative and discounted energy system costs. Carbon stabilization to 150 Mtonne/yr (i.e., 11 percent below the reference case at the year 2030) would require similar financial resources (1.34 percent). Pursuing both policies results in a cost increase of 2.94 percent (less than the arithmetic sum of the cost of each policy) related to the secondary benefits of carbon control policy.

The real discount rate of 10%/yr is high compared to the industrialized world, but it may be realistic in the case of China. Such high discount rates eventually distort the impact of the cost of environmental policies, since it lowers or devalues all costs incurred towards the end of a time horizon. The undiscounted cost increase in the last period is high. Cost differences caused by environmental policy for China approaches levels of 5%/yr for sulfur control and 13%/yr for carbon control.

Figure 5.19 explains the level of SO_2 emissions in Shandong in the year 2030 for different scenarios. All of the policies discussed above lead to significant reductions. The most significant SO_2 emission reduction takes place in the electricity sector (~90 percent) under a high sulfur tax rate, while the overall reduction obtained is ~45 percent compared to present emission levels.

4.1 Sulfur Taxes and Caps at Constant Fossil-Fuel Prices (SSLC)

The discounted system cost increase, including sulfur and/or carbon tax revenue, is moderate; changes range from 0.2 percent, at 500 \$/tonneSO_2, to 5.1 percent at 1500 \$/tonneSO_2. Figure 5.20 gives the market share of conventional and advanced power generation systems of Shandong for the year 2030 as a function of a sulfur tax. In this scenario, scrubbers are less competitive than advanced coal technology represented by AFBC and IGCC generation technologies. All of these results correspond to nominal prices for fossil fuels.

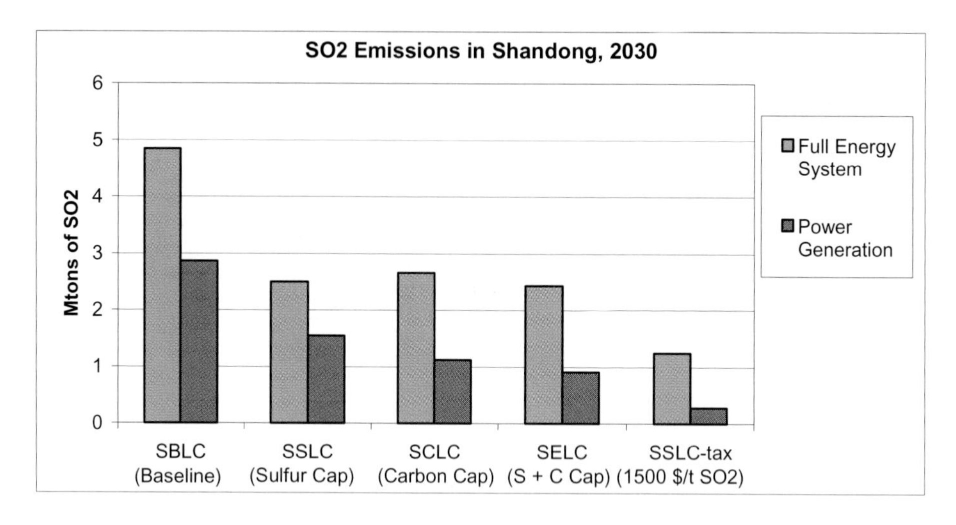

Figure 5.19 Shandong-2030; SO_2 emissions in the low-demand, constant-price scenario as a function of environmental policy.

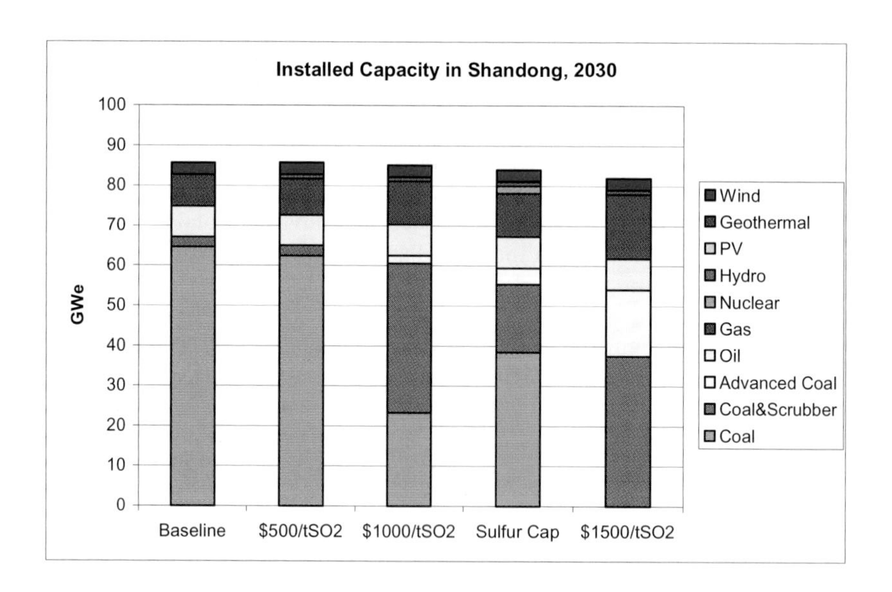

Figure 5.20 Shandong-2030; Installed capacity by technology in the low-demand, constant-price scenario under sulfur tax or caps.

4.2 Carbon Tax and Caps at Constant Fossil-Fuel Prices (SCLC)

The last results of the parametric analysis refer to carbon control policies. The findings describe the change in systems costs as a function of taxes (or marginal control costs) is discussed and show the market share of conventional and advanced power generation systems in Shandong for the year 2030. As substitutes for coal, nuclear energy and wind power technologies are introduced at high system cost and at high marginal control costs. Stabilizing emissions at annual levels of 150 MtonneC/yr increases cumulative costs by 1.3 percent, while the cost increase from a combined policy for sulfur and carbon would be three percent. Figure 5.21 shows that in both cases the carbon tax reaches around 100$/tonne of carbon, which allows for a significant deployment of nuclear energy (*e.g.*, 20.3 GWe by the year 2030).

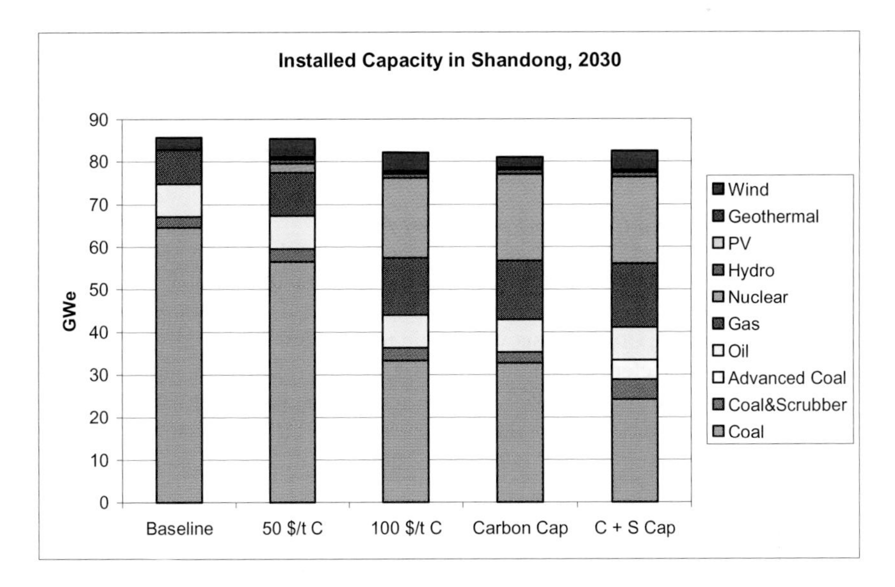

Figure 5.21 Shandong-2030; Installed capacity by technology in the low-demand, constant-price scenario under carbon taxes or caps.

4.3 Low Demand and High (Increasing) Fossil-Fuel Prices (SBLH)

Table 5.8 presents the same scenarios in terms of increased system cost versus emission reduction; all scenarios refer to high (increasing) fossil-fuel prices. The constant-price cases is based on 30$/tonne(coal); 2.7$/GJ for Chinese gas; and 3-4$/GJ for gas imported *via* pipeline or LNG, respectively. The case involving high fossil-fuel prices assumes that they will double (linearly) by the end of the time horizon (2030) leading to higher energy system costs. The relative energy system cost change in this case is lower than those reported for the constant fossil-fuel price scenario because of the high absolute cost level associated with the baseline (SBLH) scenario. High prices for fossil fuels make carbon reduction economically more attractive. Apart from this behavior, the general trend is the same as in the case with constant nominal prices.

Table 5.8 Shandong 1995-2030; Discounted Cumulative Cost Increase versus Cumulative Emission Reduction relative to the Baseline Case, in Percent.

Indicator	**SBLH**	**SSLH**	**SCLH**	**SELH**
	Baseline	Sulfur Cap	Carbon Cap	S&C Cap
Disc. System Cost	0	1.90	1.08	2.60
Total CO_2	0	0.38	9.06	9.39
CO_2 from Power Gen.	0	5.72	19.85	22.87
Total NO_x	0	26.33	13.59	28.89
Total SO_2	0	28.47	13.18	29.70
SO_2 from Power Gen.	0	34.07	21.47	43.20

4.4 High Demand and Constant Fossil-Fuel Prices

The last result presented from the MARKAL-Shandong model is a series of graphical scenario comparisons for the high-demand projection in Shandong Province. This sensitivity analysis was made to see how the model deals with very high growth in projected demand. The parametric results are shown using the format adopted in the previous subsections and comparisons are made of primary, and final, electricity production, generating capacity, and emissions. Conclusions for technology penetration and costs as a function of environmental policy are compared to the baseline SBLC scenario.

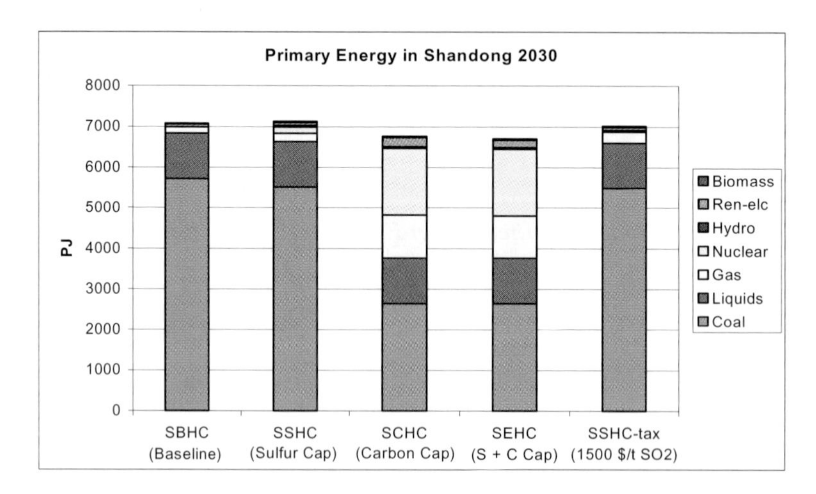

Figure 5.22 Shandong-2030: primary energy by fuel; high-demand, constant-prices.

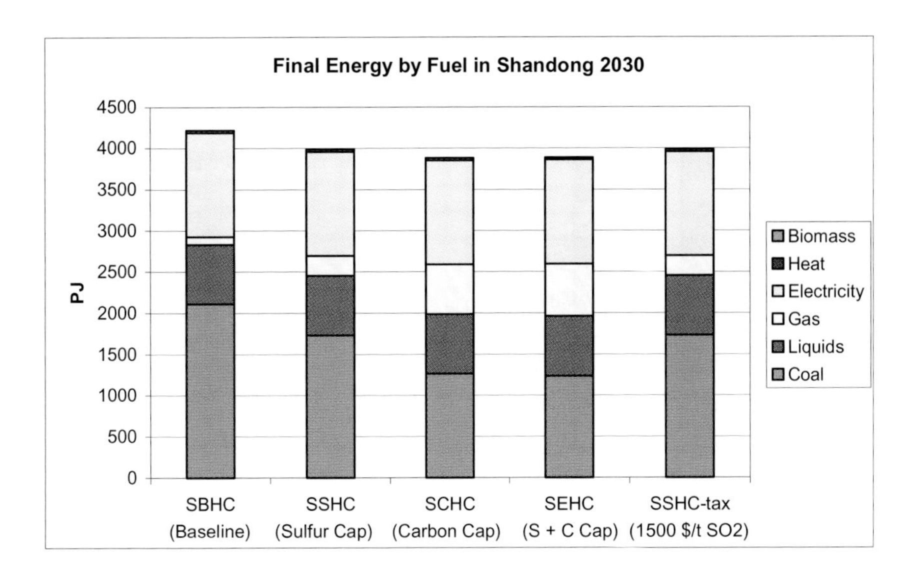

Figure 5.23 Shandong-2030: final energy by fuel; high-demand, constant-prices.

Figures 5.22 and 5.23 show the primary and final energy use projected for Shandong in the year 2030. Primary energy consumption increases to around 7,000 PJ/yr; final energy use is on the order of 4,000 PJ/yr, indicating an overall energy sector efficiency of 58 percent. The large difference between primary and final energy use reflects the poor conversion efficiency of the power generation system and the energy sector as a whole. The structure of the energy mix remains mainly the same, however. Coal use still accounts for the differences between the high- and low-demand cases. Coal also functions as a buffer against variations in demand for electricity when it falls below the SBLC case. Total installed capacity for power generation reflects the same relationship, as shown in Figure 5.24; In this case it rises to 100 GWe, versus ~80 GWe for the low-demand case in the year 2030. The absolute contribution of nuclear energy remains the same.

In the high-demand case, fixed tax rates for sulfur and carbon emissions are higher (cf. Figure 5.25. Both types of emissions are proportional to the use of coal. At a sulfur tax of 1500 $/tonne SO_2, (e.g., in the SSHC-tax special scenario), sulfur emissions are reduced to 60 percent of present emission levels. On the other hand, for fixed emission caps the marginal cost is higher; carbon taxes, for instance, reach 128 $/tonneC, versus 100 $/tonneC in the low-demand case.

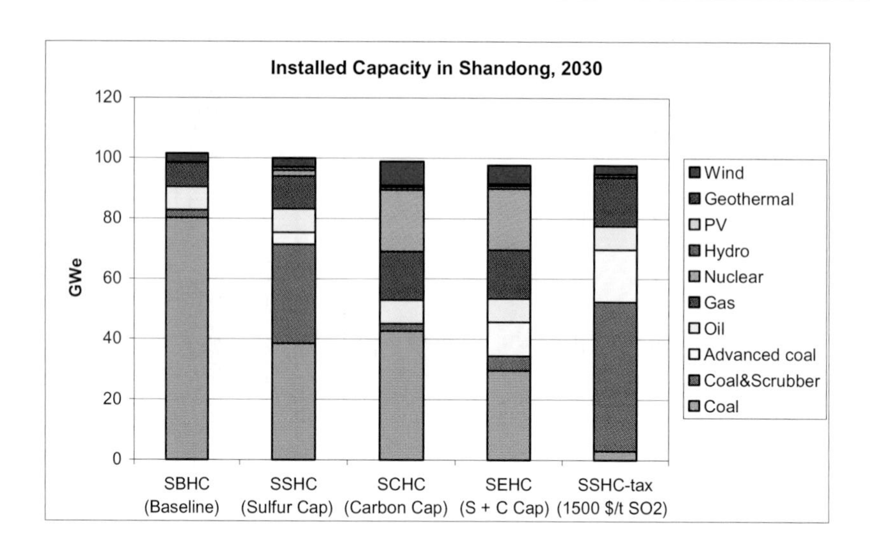

Figure 5.24 Shandong-2030: Installed capacity by technology in GWe in the high-demand, constant-price scenario under carbon taxes or caps.

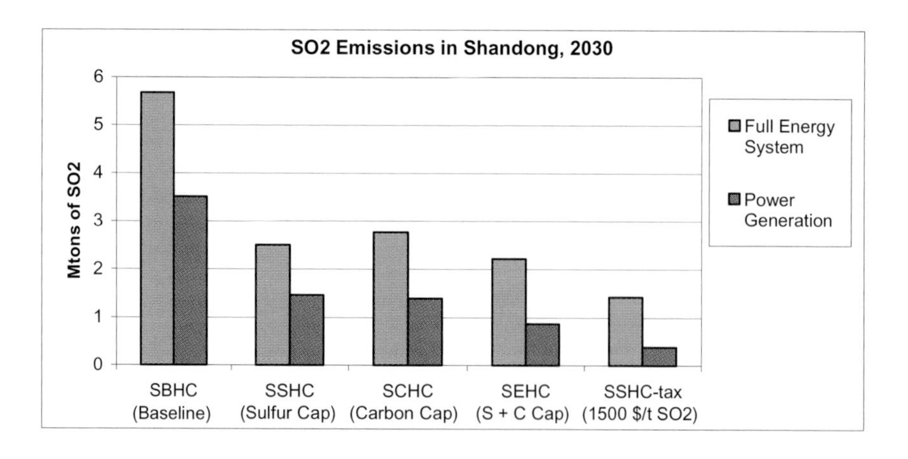

Figure 5.25 Shandong-2030: sulfur emissions in the high-demand, constant-price scenario under sulfur taxes or caps; the marginal cost is 820 \$\tonne$SO_2$ in SSHC scenario (stabilizations at present levels) and 1500 \$/tonne SO_2 in the SSHC-tax special sulfur tax case.

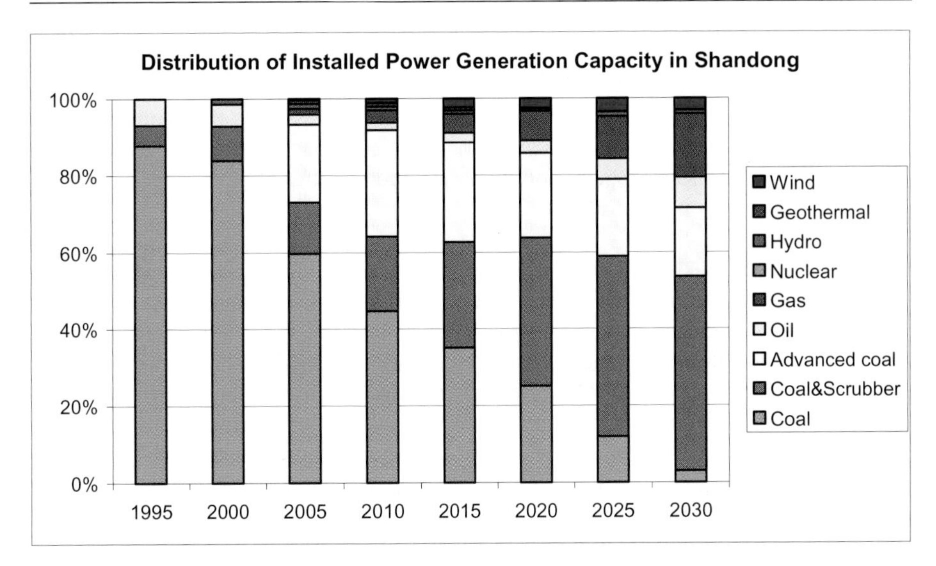

Figure 5.26 Shandong-2030: distribution of technology installations over time for power generation in Shandong under a tax of 1500 \$/tonne SO_2 introduced in the year 2005.

Finally, Figure 5.26 shows the dynamics in capacity installations for power generation in Shandong under a sulfur tax of 1,500 \$/tonneSO_2 tax. Conventional coal power stations under these imposed conditions are displaced gradually by systems using scrubbers and advanced coal-burning technologies. Natural gas CC systems are competitive under a sulfur tax and constant prices while nuclear energy is not. The discount rate for these cases is 10%/yr. Sensitivity for this rate will be performed using the CRETM model for the whole China, with Shandong Province being one of seven regions.

4.5 MARKAL-Specific Interim Technology Assessment

Based on the results emerging from the MARKAL-Shandong modeling effort, investigators have been able to make an assessment of technology implementation for Shandong.

Coal: Pulverized coal power technologies will continue to play a dominant role for new power generation in Shandong. Chinese traditional coal power stations have average capital costs of about 600 \$/kWe without sulfur control technology. The introduction of flue gas desulfurization (FGD) is feasible in the presence moderate emission control taxes, and advanced systems can be implemented at higher tax levels or higher cost for clean coal and transportation. These extra costs could make other generation technologies, like nuclear energy, more competitive.

IGCC: Integrated Gasification Combined Cycle (IGCC) systems that first gasify coal to generate electricity by using gas turbines emit fewer particulate and sulfur oxides and have less carbon emissions per unit of electrical energy generated, because of their high efficiency. Final specific investment costs for IGCC systems fabricated in China would be around 1000 $/kWe (*i.e.*, less than the capital costs for nuclear power, but still more costly than other coal based options), although the competitiveness of such depends on the magnitude of these highly uncertain install costs.

CCGT: The market share for Combined-Cycle Gas-Turbine (CCGT) units operating on natural gas could increase if they were manufactured in China and if natural gas were available at reasonable prices (less than 2.5 US$/GJ). An increase in gas supplies would require advanced exploration and extraction technologies or greater use of coal bed methane (CBM). Gas could be piped in from Siberia or imported as LNG. Again, the delivered cost of natural gas is subject to major uncertainties. Furthermore, the annualized infrastructure cost of natural gas equals approximately the present-day price of coal in China. Because China prefers to use natural gas in other economic sectors to improve environmental quality, managers might consider combining users from other sectors with the power generation sector, in its overall energy strategy. The consumption of large quantities of natural gas for power generation increases the short-term viability of the risky and capital-intensive investments needed to make natural-gas pipeline systems an operational reality.

Nuclear Energy (NE): Shandong Province is one of the candidate regions for deployment of nuclear energy (NE). NE total capital costs over the next decades should fall below 1,500 $/kWe when the technology is fabricated in China. Given low total specific capital cost, reduced construction times, and low interest rates for capital, nuclear power could become competitive against advanced fossil systems. NE is especially attractive if emission control policies targeting coal are adopted in China, and when fossil fuel prices increase in international markets.

Wind: Most of China's quality wind sites are in Inner Mongolia, Xinjiang, and southeastern coastal regions. Current international capital costs for 750-kilowatt turbine units are below 1,000 $/kWe. (MARKAL-Shandong) studies assumed limited wind potential (10 GWe) in the region. However, it also was assumed that some wind power would be developed regardless of the economics of the system because international organizations that advocate this particular renewable energy source will support research and development.

Non-Power Generation Technology: Although the cost data used for Shandong are China-specific and not sufficiently precise for policy conclusions, a few general remarks can be made. First, future transportation systems in China and Shandong will rely on advanced gasoline vehicles with catalytic converters and diesel trucks, which will later be supplanted by compressed natural gas systems in cities when moderate sulfur and nitrogen oxide emissions constraints are imposed. Second, heating systems based on clean fuels like natural gas and district heating will gradually replace the coal and biomass systems that now dominate markets in both rural and urban regions. Finally, AFBC and scrubbers will be introduced to control emissions from relatively large boilers, even when sulfur and carbon constraints are relatively moderate.

5. CRETM FOR BOTH CHINA AND SHANDONG PROVINCE

CRETM produces results for both Shandong and for the other regions of China. This section presents the results of scenarios comparable to those of MARKAL; parametric studies of the effects of exogenous energy demand, the discount rate, and technology costs; the impact of endogenous factors on the base case; and concludes on the main results from the CRETM study. The canonical discount rate used to generate all scenarios is high (H) e.g., *10%/yr*. For this reason the corresponding scenario index is dropped.

5.1 Scenario Results

This sub-section reporting CRETM results is organized as follows. An early emphasis has been placed on: a) the driver for electricity demand; b) description of key results from the baseline case BHC; c) a summary of key results (e.g., energy mixes, emission rates, etc.) from the eight main scenarios analyzed above; d) a limited set of single-point parametric studies that explore sensitivities around the BHC scenario; and e) a summary of results generated by assuming endogenous learning and partial equilibrium. Approximately 1,000 parameters must be fixed to evaluate a given scenario using CRETM so that a single optimized energy mix is delivered as a function of time for the seven regions of China.

Key technological, economic, and operational parameters for the 17 electricity generation technologies are summarized on Table 5.9. Except where explicitly stated, these parameters remain fixed for the scenario studies presented herein, with the full report of input, equations and results described in Kypreos (2002). Key results reported by CRETM, upon minimization of the total Net Present Value (NPV) of all electricity generation costs over the period of the analysis (1995-2030, in five-year time intervals), include:

- primary fuel use; electricity production by fuel; installed generation capacity;
- local, regional, and countrywide atmospheric emissions;
- technology- and region-dependent marginal costs of electricity for specified emission constraints (*e.g.*, effective [minimum] taxes for emissions control, \$/tonneCO_2, etc.);
- capital investment and annual O&M charges for electricity by technology and region;
- capital investment and annual O&M charges for transport across China; and
- amount of fuel and electricity transported by a given mode (e.g., road, rail, ship, wire).

Table 5.9 Summary of Key Technology, Economic, and Operational Parameters Fixed for the Scenario Analyses (Reported Herein Using CRETM)

Generation Technology	Fuel	ESP	Scrub.	Construction Time	Operational Lifetime	Economic Lifetime	UnitTotal Cost	Variable O&M Cost	Fixed O&M Cost	Conversion Efficiency	Annual Load Factor
				yr	yr	yr	$/kWe	M$/TWh	%UTC/yr	%	hr/yr(%)
Domestic small (>100 MWe) plant	coal			2	20	15	676	5	3	27	4800 (55%)
Domestic medium (100-200 MWe) plant	coal			3	30	20	650	4	3	29	5000 (57%)
Domestic medium (300 MWe) plant	coal	x		3	30	20	600	3.5	3	36	5200 (59%)
Domestic medium (300 MWe) plant	coal	x	x	3	30	20	890	4.5	3	35	5200 (59%)
Domestic large (300-600 MWe) ss plant	coal	x	x	3	30	20	1100	4.5	3	42	5200 (59%)
Foreign atm. fluid.-bed combst. plant	coal			3	30	20	900	8.5	3	38	6000 (69%)
Foreign integ. gasif.comb.-cycle plant	coal			2	20	15	1000	4.5	3	43	6000 (69%)
Pressurized fluidized-bed combst plant	coal			3	30	20	1100	4	3	45	6000 (69%)
Traditional oil-fired plant	oil			2	20	15	530	2.8	3	35	4800 (55%)
Combined-cycle oil-fired plant	oil			2.5	20	15	500	2.8	3	42	5600 (64%)
Combined-cycle gas-fired plant	gas			2	20	15	530	8	3	45	6000 (69%)
Adv. gas-turbine comb.-cycle plant	gas			2.5	20	15	500	8	3	55	6000 (69%)
Nuclear plant	U			5	30	20	1500 + 200*	4 + 5	1.6	33	7000 (80%)
Hydroelectric plant				8	50	40	1200	1	1.5	33	3800 (43%)
Wind plant				1	20	15	1200*	2	1.5	33	2600 (30%)
Solar photovoltaic plant				2	25	20	10000*	0.07	1.5	33	2600 (30%)
Geothermal plant				1	15	12	2000	0.05	1.5	33	3200 (37%)

*Initial capital cost. See (Kypreos, 2002) for the evolution of specific capital costs for these technologies.

5.1.1. Base Case: High Demand and Constant Fossil-Fuel Prices (BHC)
Table 5.10 lists key integrated results reported by CRETM for the BHC scenario, which serves as a "point-of-departure" for subsequent scenario investigations and single-point parametric studies. The electricity system cost for BHC amounts to 3.31 percent of the cumulative GDP. Please note that in contrast to the MARKAL analyses the base case exhibits "high" demand.

Table 5.10 Integrated Results for the BHC Scenario.

Parameter	**Value**
Total Discounted Energy Cost, ENC(B$)	478
Total Undiscounted Energy Cost, B$	3,247
Total Electrical Demand, TWh	94,967
Average (Undiscounted) COE, mill/kWeh	34.2
Total Discounted GDP, B$	14,379
Total Discounted ENC/GDP, %	3.31
Total Integrated CO_2 Emissions, GtonneCO_2	60.9
Specific CO_2 Emissions, MtonneCO_2/TWeh	0.64
Total SO_2 Emissions, MtonneSO_2	632
Specific SO_2 Emissions, ktonneSO_2/TWeh	6.7

The countrywide electrical generation mix for China over the 1995-2030-study period needed to meet the baseline demand for these BHC scenario conditions is shown in Figure 5.27. Figure 5.28 gives the evolution over time of regional distribution of total electrical generation. Total emissions of CO_2, SO_2, and NO_x are given in Figure 5.29. The time-dependence of the specific emission rates (tonneXO_2/TWeh) for China is shown in Figure 5.30. Most notable in the BHC electrical-generation mix are: a) the nominally constant contribution of hydroelectricity, with some growth after 2020; b) the miniscule contribution of nuclear; and c) the growing dependence on gas and oil; domestic coal remains the dominant fuel, albeit through the growing use of clean-coal technologies (not explicitly shown in the relatively aggregated Figure 5.27). Kypreos (2002) shows that modest decreases in the capital cost of nuclear can dramatically increase its contribution to the electrical-generation mix.

5.1.2 High Demand and High (Increasing) Fossil-Fuel Prices (BHH)
The impact of increasing the price of fuel by the time- and region-dependent factors (Kypreos, 2002) on the countrywide energy electrical generation mix for China is illustrated for the BHH scenario in Figure 5.31. The impact of these fuel-price increases on CO_2, SO_2 and NO_x emissions is depicted relative to the BHC baseline scenario on Figure 5.32. A comparison of the BHC and BHH scenarios shows that

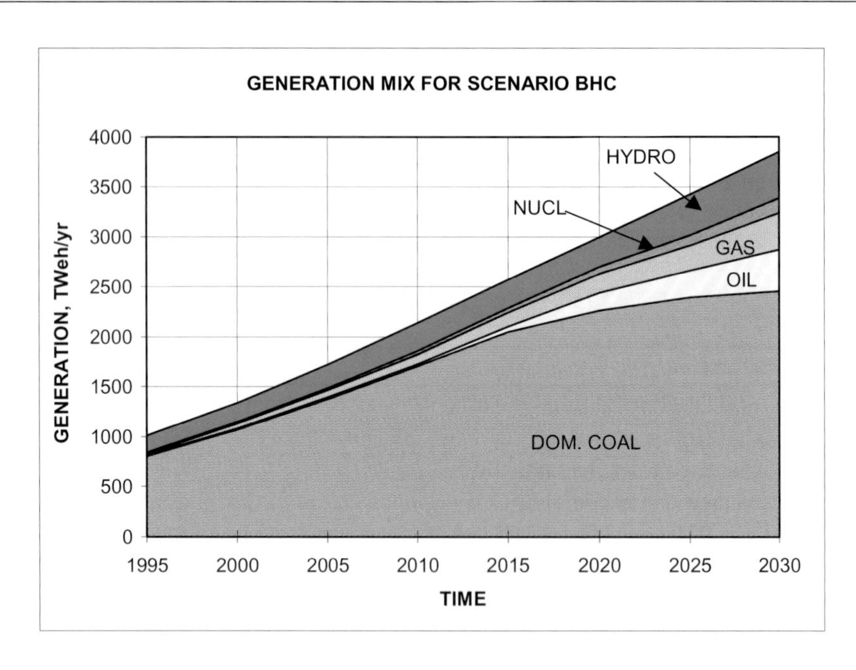

Figure 5.27 Countrywide electrical generation mix for the BHC scenario in China.

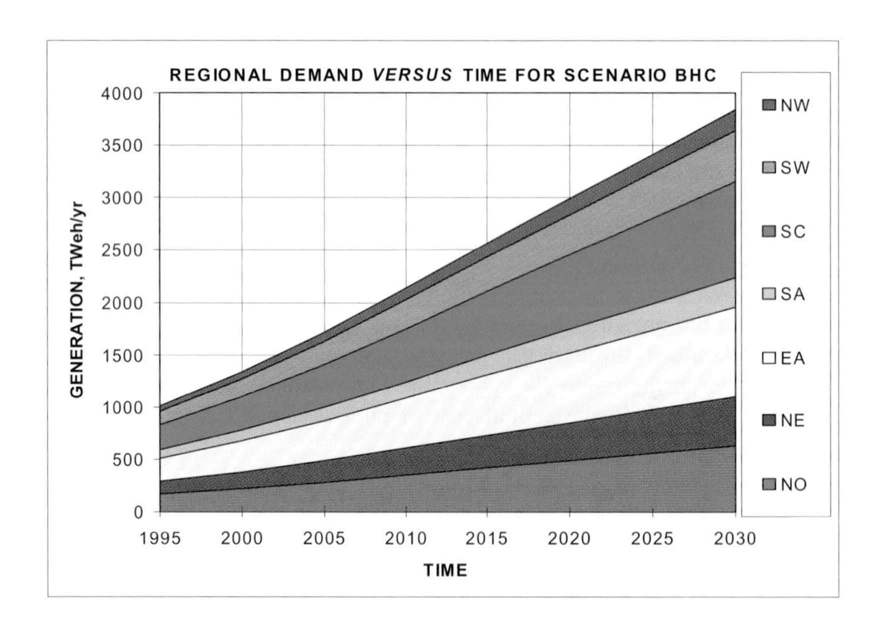

Figure 5.28 Regional distribution of electrical generation for the BHC scenario in China.

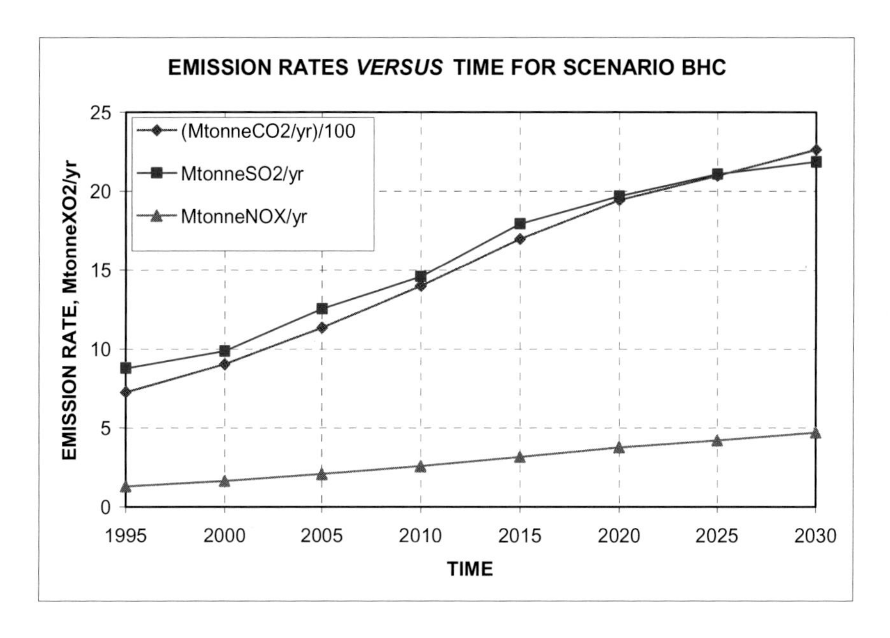

Figure 5.29 Total CO_2, SO_2, and NO_x emission rate for the BHC scenario in China.

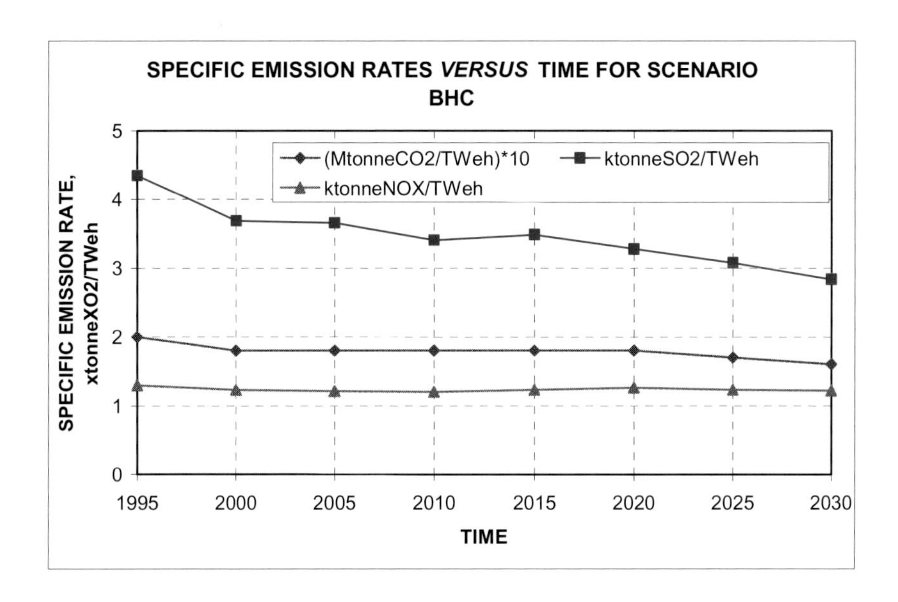

Figure 5.30 Specific CO_2, SO_2, and NO_x emission rates for the BHC scenario in China.

the kind of change in price structure involved in the transition from BHC to BHH transition is accompanied by a major switch to advanced coal technologies; the introduction of some renewable generation in the form of wind; and a modest increase in the use of nuclear energy. As seen from Figure 5.32, modest (~15-20 percent) decreases in CO_2 and NO_x emissions in the year 2030 result from these price increases, but SO_2 emissions in 2030 decrease by nearly 40 percent as the use of "Domestic Coal" generation technologies decreases. In assessing these results, the absence of price-economy (GDP) feedback in CRETM must be recognized, but this feedback is expected to be of secondary importance. Price effects on demand are much more significant.

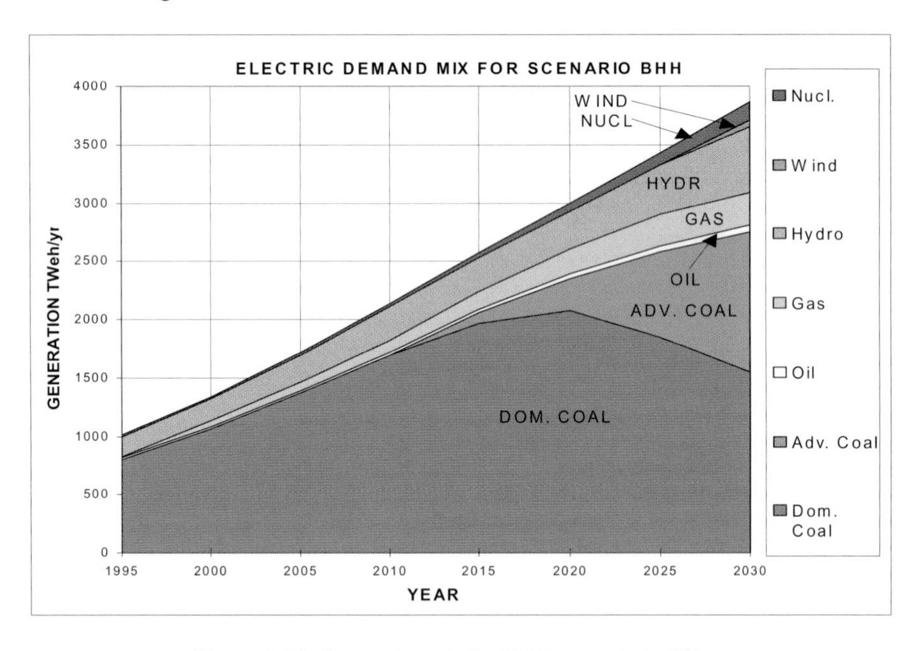

Figure 5.31 Generation mix for BHH scenario in China.

5.1.3 Carbon-Caps at Different Fossil-Fuel Price Levels (CHC, CHH)

Carbon emission limits were applied as indicated in Table 5.11 for both the "H" and the "C" fuel-price conditions, with both demand and discount rate remaining in the "H" mode; scenarios CHC and CHH result. The CO_2 emission constraints were applied separately and in conjunction with those for SO_2 emissions. Figures. 5.33 and 5.34 show evolution of the electrical generation mix for both the CHC and the CHH scenarios. The comparison of CO_2, SO_2, and NO_x emissions for the CHC scenario with those for the BHC base-line scenario is given in Figure 5.35. The marginal cost of CO_2 emissions is given as a function of time in Figure 5.36, which also compares these results with those from the other carbon-constraining scenarios.

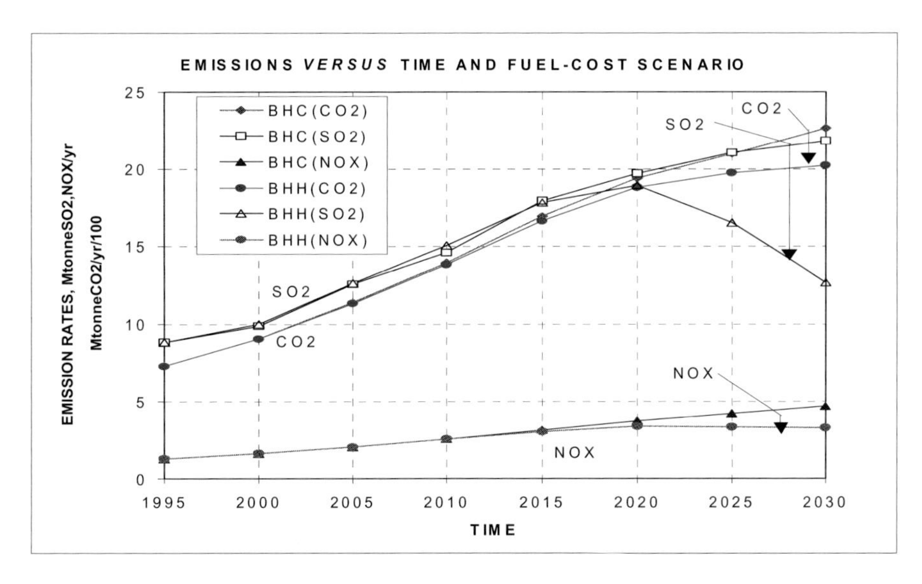

Figure 5.32 Comparison of emission rates for CO_2, SO_2, and NO_x as the price of fuel is increased from scenario BHC to BHH in China.

This figure also shows that for the constant prices (CHC and EHC) the marginal costs are higher because more coal-fired technologies must be replaced than is necessary in the CHH or EHH cases.

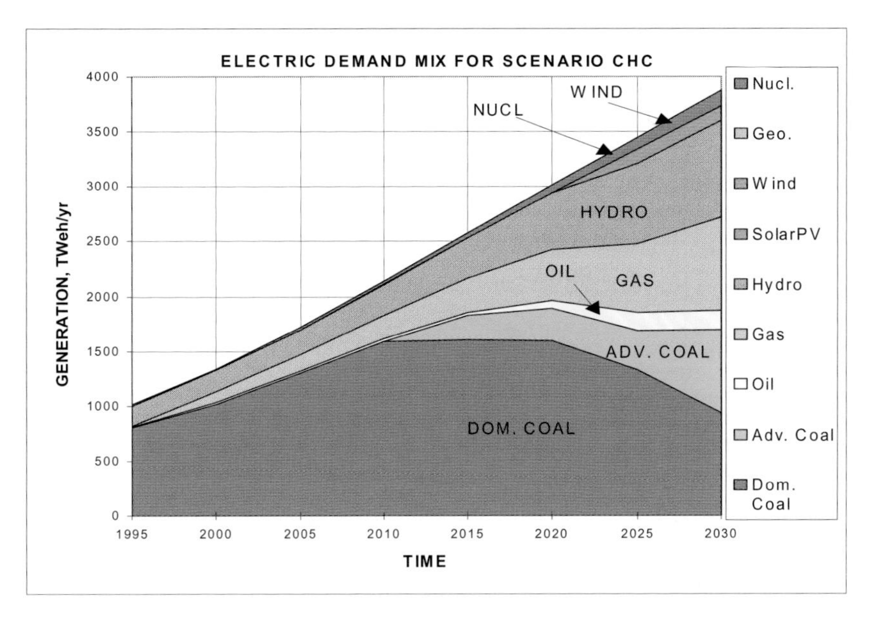

Figure 5.33 Generation mix for CHC scenario: (C)arbon constraint with (H)igh demand and (C)onstant fuel prices in China.

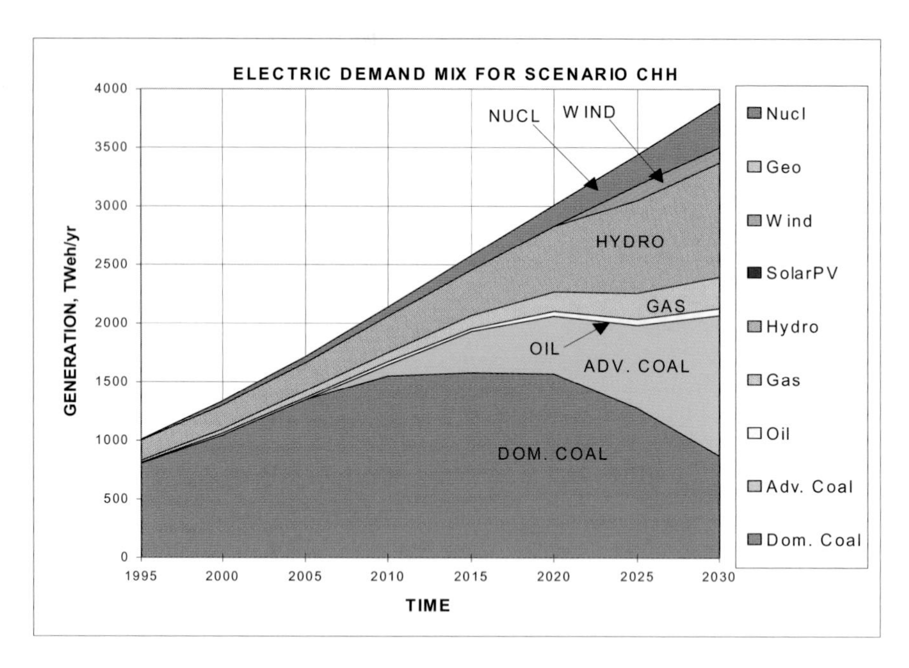

Figure 5.34 Generation mix for CHH scenario: (C)arbon constraint with (H)igh demand and (H)igh fuel prices in China.

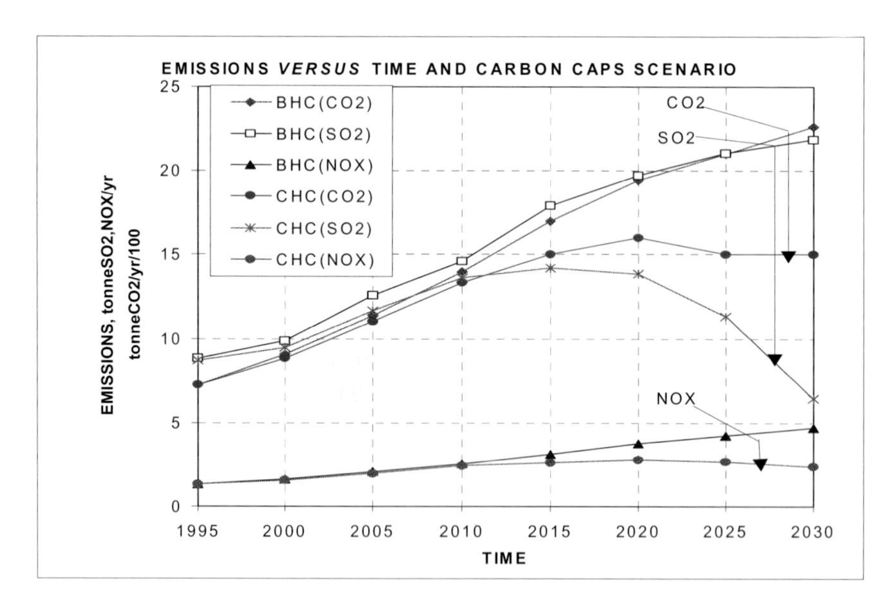

Figure 5.35 Comparison of CO_2, SO_2, and NO_x emission rates between scenarios BHC, BHC and CHC as a function of time in China.

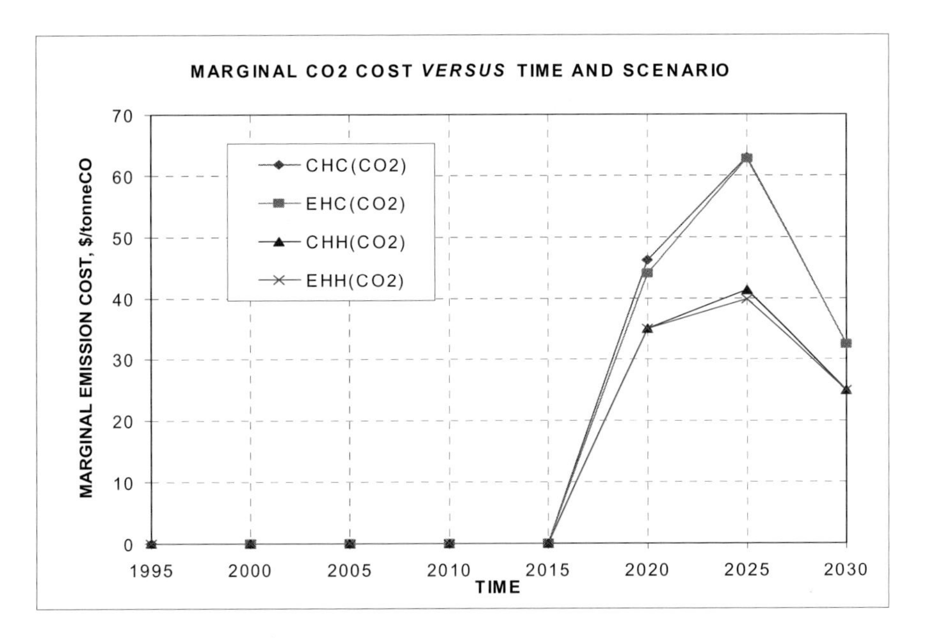

Figure 5.36 Time evolution of marginal costs for CO_2 emissions for the range of carbon-constrained scenarios considered.

5.1.4 Sulfur-Caps at Different Fossil-Fuel Price Levels (SHC, SHH)

As seen in the above presentation of carbon-constrained results, Figures 5.37 to 5.40 show the electric-generation mixes reported by CRETM for the sulfur-constrained scenarios with H and C fuel-price conditions imposed, and with both being based on the "high" (H) discount rate (10%/yr) and demand assumptions. Figures 5.37 and 5.38 summarize the electrical-energy mix for SHC and SHH scenarios. Figure 5.39 compares emission rates of CO_2, SO_2, and NO_x for the SHC and baseline BHC scenarios, and Figure 5.40 gives the time dependence of all marginal costs associated with the sulfur-constrained cases.

5.1.5. Carbon + Sulfur-Caps at Different Fossil-Fuel Price Levels (EHC, EHH)

Finally, Figures 5.41 and 5.42 demonstrate the evolution of the electrical generation mix for China if both of the previously reported carbon and sulfur emission constraints are applied simultaneously for the "high" (H) discount rate assumption and both (oil, coal and gas) fuel-price (H and C) scenarios. The comparison of emission rates for the "constant" (C) fuel-price condition with emissions for the baseline BHC scenario is given in Figure 5.43. These marginal costs are equivalent to the "taxes" needed to achieve (within the confines and limitation of CRETM) the corresponding emission rates.

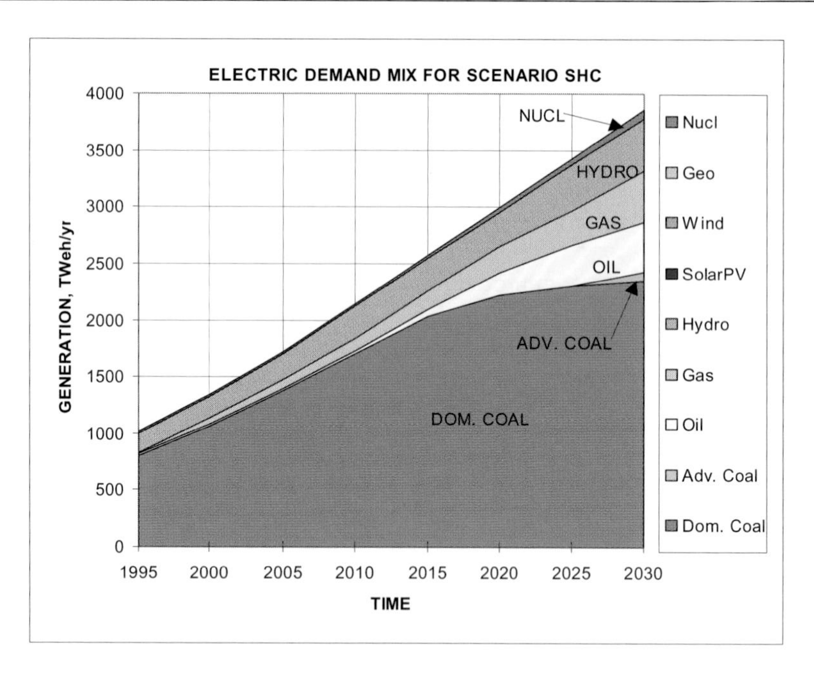

Figure 5.37 Generation mix for SHC scenario: (S)ulfur constraint with (H)igh demand and (C)onstant fuel prices in China.

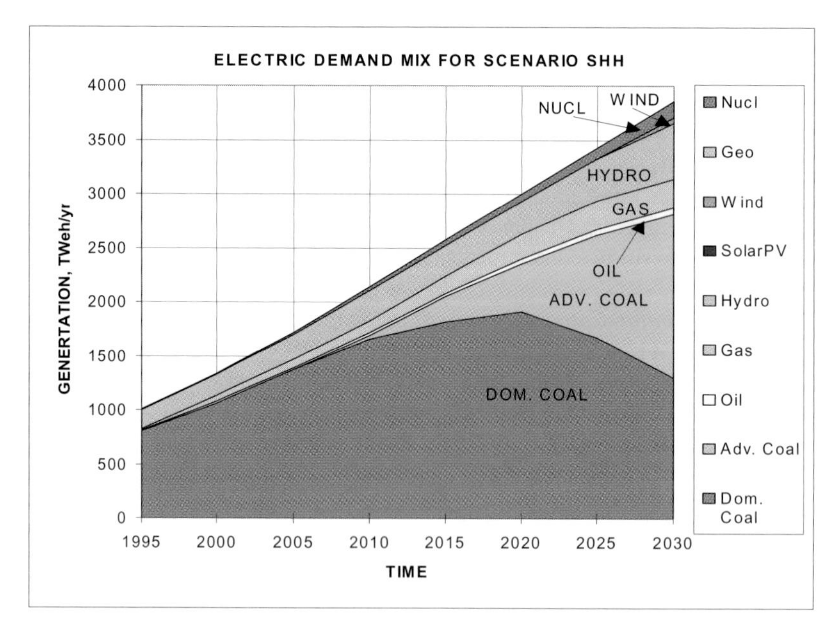

Figure 5.38 Generation mix for SHH scenario: (S)ulfur constraint with (H)igh demand and (H)igh fuel prices in China.

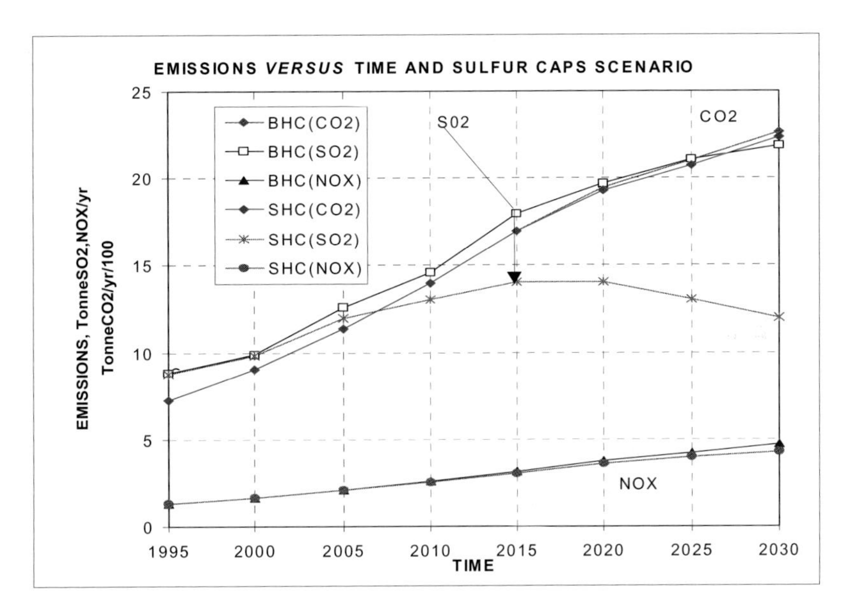

Figure 5.39 Comparison of CO_2, SO_2, and NO_x emission rates between scenarios BHC and SHC as a function of time.

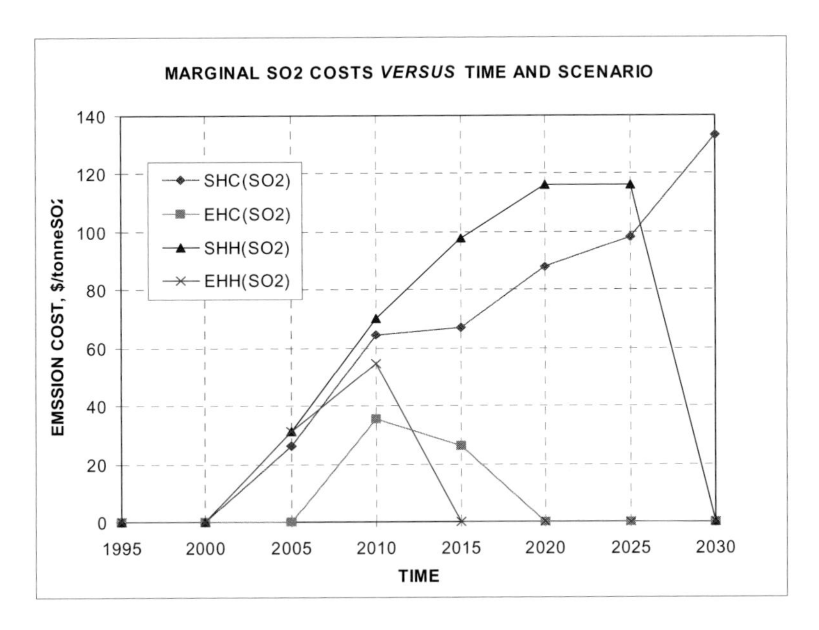

Figure 5.40 Time evolution of marginal costs for SO_2 emissions for the range of sulfur-constrained scenarios considered.

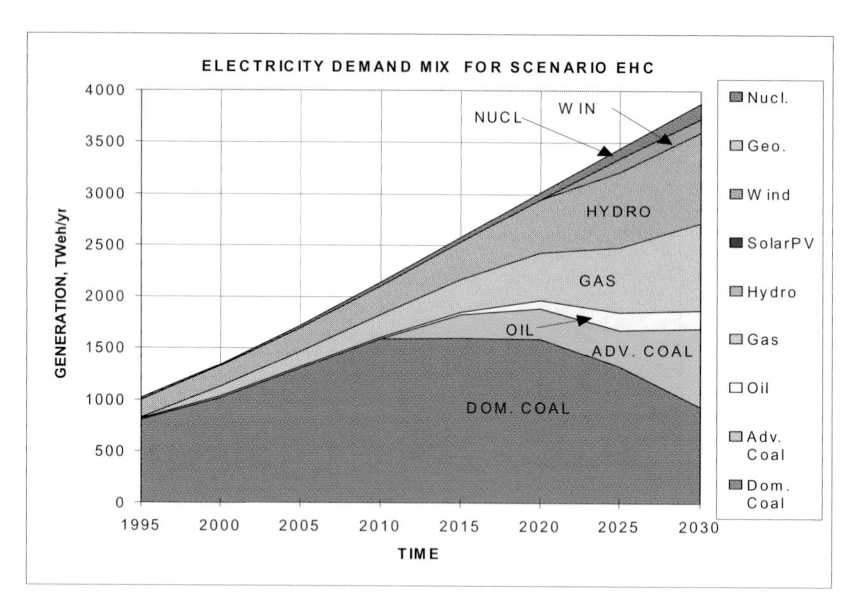

Figure 5.41 Generation mix for EHC scenario: Both sulfur and carbon emission constraints are applied (E = S + C) with (H)igh demand and (C)onstant fuel prices in China.

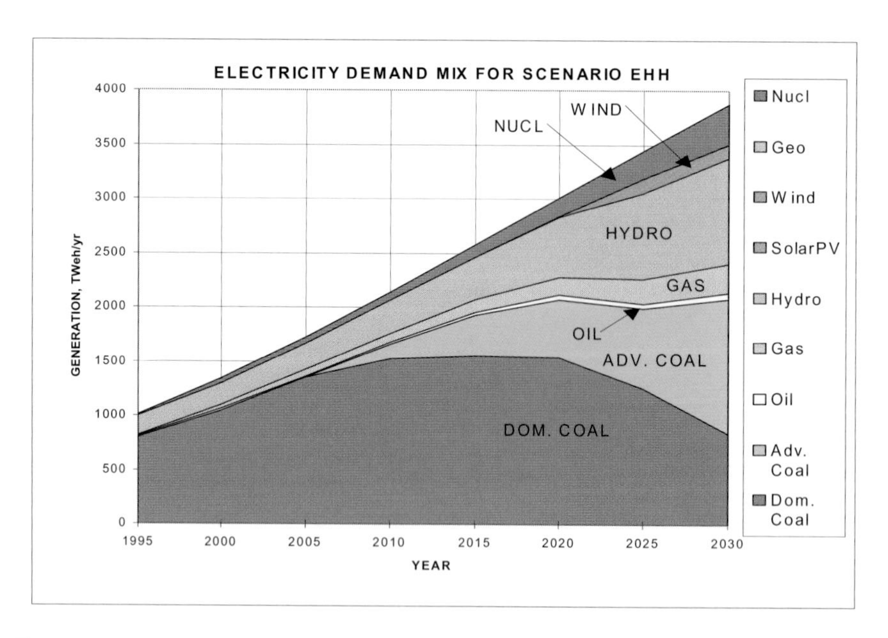

Figure 5.42 Generation mix for EHH scenario: Both sulfur and carbon emission constraints are applied (E = S + C) with (H)igh demand and (H)igh fuel prices in China.

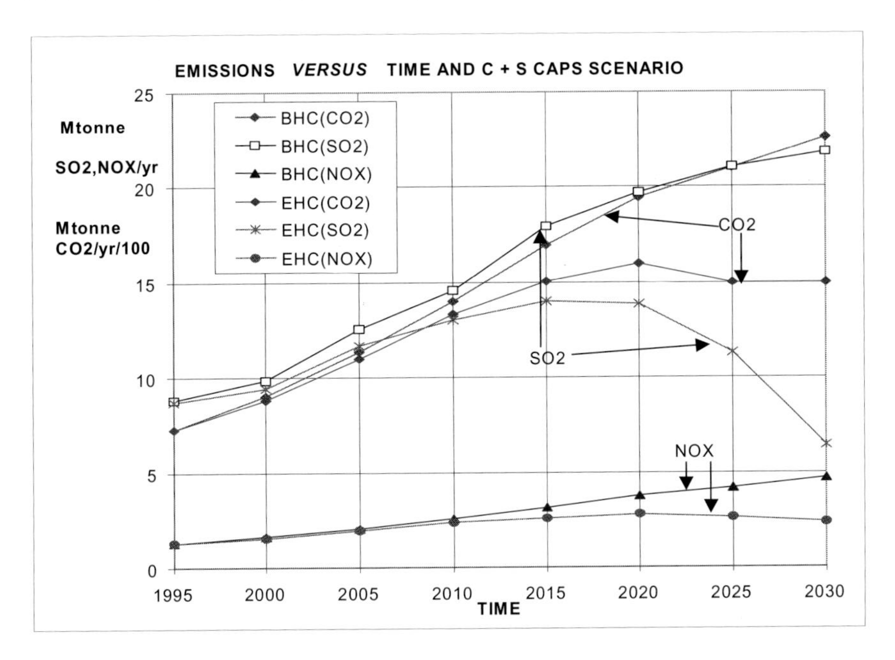

Figure 5.43 Comparison of CO_2, SO_2, and NO_x emission rates between scenarios BHC and SHC as a function of time.

5.2 Parametric Sensitivity Studies

The eight scenarios--one baseline and seven (coal, oil and gas) fuel-cost and emission variations--provide a broad range of possible (and relatively "surprise-free") futures for China's electric generation system over the next three decades, though many of the "fixed" scenario attributes can powerfully affect the magnitude and the structure of those possible futures. To advance understanding of the role played by some of these key exogenous (to the LP model) variables, a series of single-point parametric studies were conducted. These studies varied: a) the energy demand driven by the assumed GDP growth rate described by Equations 5.2 and 5.3; b) the rate at which expenditures are discounted to determine the present-value of total (electrical) energy costs to be minimized (10%/yr for the present computations, which contain no inflationary factors); and c) the costs of key, non-carbon-emitting technologies. Key results from these single-point parameter variations are reported below, with more extensive results being given in Kypreos (2002).

5.2.1 Exogenous Energy Demand

The growth rate of electrical-energy demand in China assumed for the BHC baseline scenario falls in the mid-range of recently reported projections. While the ERI projections, at least for Shandong Province, will eventually provide a common basis for all components of the CETP, investigations so far using the CRETM considered a range of demands that was both above (HH) and below (L, LL, and LLL) the baseline "high" (H) demands used to generate the BHC and derivative scenarios.

In this parametric study, an enormous GDP variation between 1,500 and 10,000 US dollars per capita, for 2030 was assumed to identify the associated growths in electricity demand over the five cases considered. The baseline BHC and all of its seven derivative scenarios were based on a "high" (H) GDP and associated energy demands. The impact of varying this demand on the country-wide generation mix shows variations between an electric-generation system for fueled almost entirely by hydropower and domestic coal (LL demand) and a system (HH case) in which domestic coal is displaced by gas and oil, with some growth in nuclear energy and renewable energy sources. The latter regime requires more thorough analysis.

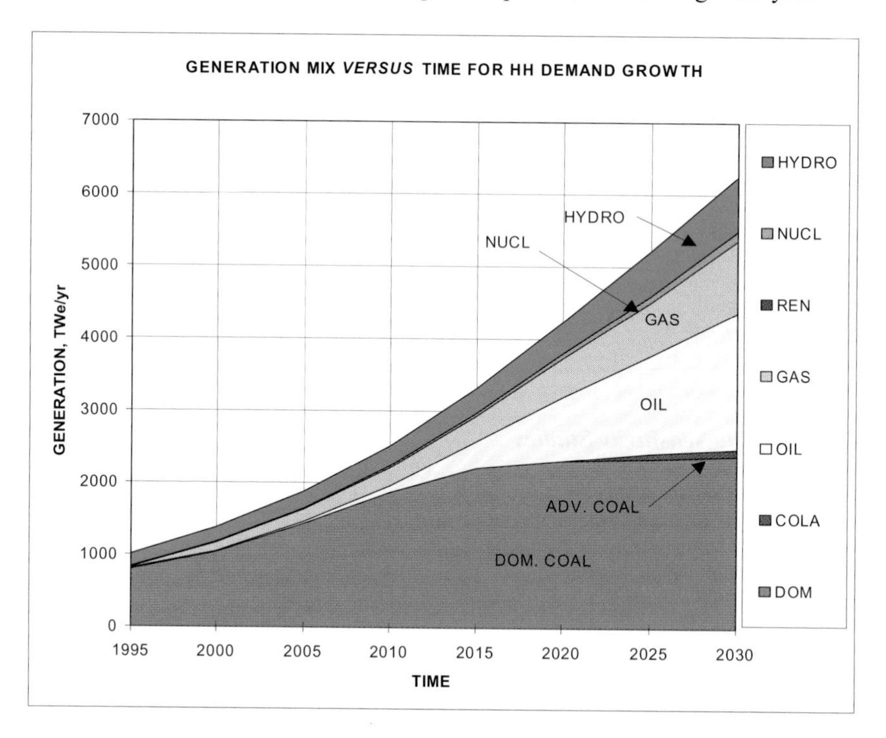

Figure 5.44 Countrywide mix of electric generation technologies for electric energy demands that are above (case HH in Figure 5.24) those used to generate the baseline BHC scenario.

As has been discussed, the optimization and simulation approaches to modeling complex energy systems depend on significantly different approaches and levels of intuitive judgment.

Simulation modeling typically examines thousands (if not tens of thousands) of "scenarios," and uses experience and judgment to identify those that either might "appeal" to a given set of stakeholders, or represent an acceptable compromise among different stakeholders. This process typically divides the task into ranges or sets of possible *futures* to be attained by a set of *strategies*, with typically the sum *futures* + *strategies* = *scenarios* (Schenler, 1998). In simplified (two-dimensional) form, a cost-*versus*-emissions "phase space" can be constructed in which the *simulation* processes collect various possible scenarios from which some kind of operational "frontier" might form, suggesting a region of "best" solutions.

Optimization modeling focuses on constrained, cost-based examination of a more limited number of economically optimized scenarios. The process describes a limited set of optimized scenarios residing near some kind of operational frontier of the kind that *simulation modeling* identifies through stakeholder experience, opinion, or (in the case of multi-stakeholder groups) compromise. A properly constrained optimization model takes into account all these combinations of possible solutions and selects only those that residing on the optimality front.

New approaches are being developed (Kann 2000) to enable optimization models to perform analyses sensitive to data uncertainties (distributions) using Monte Carlo sampling techniques, thereby leading to the distribution of possible values of the objective function. This approach will allow decisions to be made under conditions of uncertainty including risk aversion among targeted stakeholder groups.

The limited number of parametric studies reported here can be expressed in a way (*e.g.*, cost *versus* emissions) that combines the insights provided by both approaches in the search for feasible approaches to developing electric generation technologies needed to meet the energy demands of China and Shandong Province. Figure 5.45 correlates total (integrated) CO_2 and SO_2 emissions with the present value of total energy costs (in China and Shandong, respectively) for the case in which variations are driven by the range of rates for energy demand. Finally, figure 5.45 shows the exogenous rate of electrical energy demand used to correlate total cost (ENC), generation (GEN), average unit costs (<COE>), and total integrated emissions of CO_2 and SO_2. This correlation is shown on Figure 5.46. It is interesting to note that while all extrinsic parameters evaluated over the 1995-2030 time frame of the optimization grow, as expected, with increased demand, the average cost of energy remains relatively insensitive to it, even showing a slight minimum at a discount rate of ~4%/yr.

5.2.2. Discount Rate

The canonical discount rate used to generate all scenarios is 10%/yr (no inflation was considered in the study), and is expected to have the greatest impact on those technologies with high capital costs and long construction times (e.g., present-day nuclear energy). To develop an improved appreciation of the influence that the discount rate has on the optimal energy mixes for both China and for Shandong Province, the discount rate was varied over the range {0,0.15}.

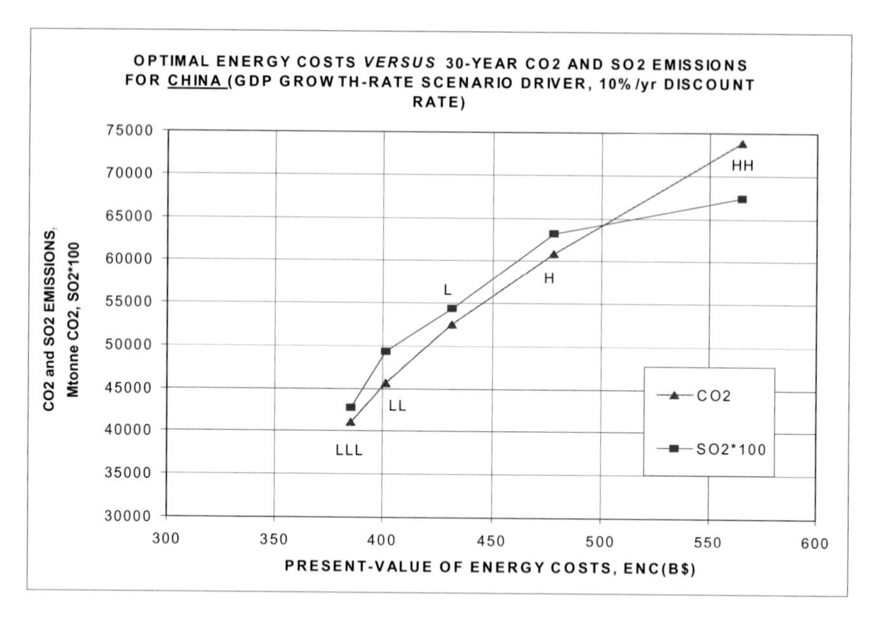

Figure 5.45 Correlation of integrated CO_2 and SO_2 emissions over the period 1995-2030 with present value of total energy costs for a range of demand scenarios.

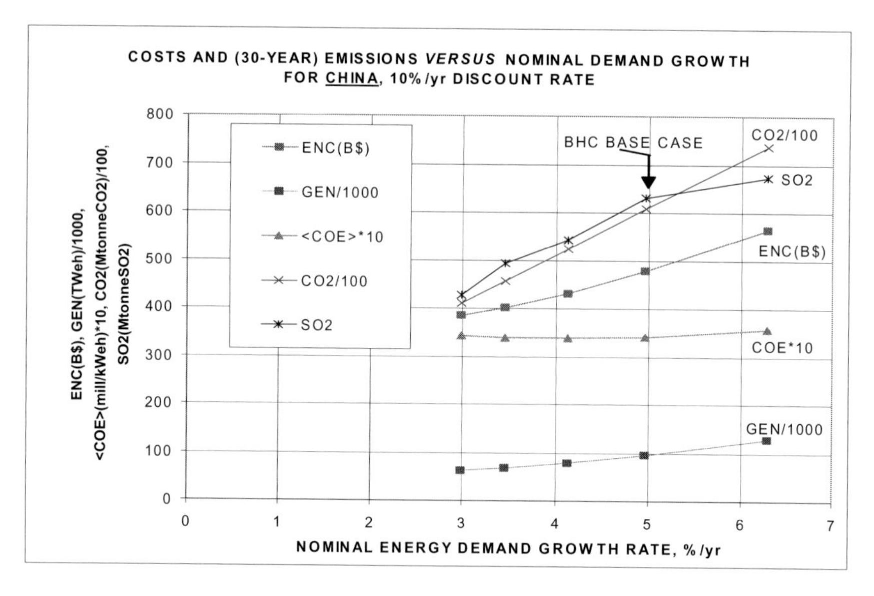

Figure 5.46 Summary of dependence of key integrated variables on the growth rate of electricity demand for China: ENC(B$) = present-value of total energy generation over 1995-2030; GEN(TWh) = total generation over 1995-2030; <COE>(mill/kWeh) =average (undiscounted) cost of electricity; CO_2($MtonneCO_2$) and SO_2($MtonneSO_2$) = total integrated emissions of these respective gases.

Taking conditions that are integrated over the computational period, Figure 5.47 gives the dependence of total (discounted) energy costs, *ENC(B$)*, average energy costs, *<COE>(mills/kWeh),* and total CO_2 and SO_2 emissions on the discount rate over the range indicated for both China and for Shandong Province. While total electrical-energy costs, *ENC(B$)*, expectedly decreases as discount rate increases, the ratio of total undiscounted cost to total energy generation, *<COE>,* increases because of increasing total capital charges and the interest during construction.

Figure 5.47 shows only the weakest of integrated 30-year CO_2 and SO_2 emissions correlation with the discount rate, for reasons discussed below. When the present value of total energy costs are expressed as a function of the present value of total GDP taken over the 30-year computational period, this percentage increases with increasing discount, as is indicated in Figure 5. 48; this indicates a nominal slope of *0.22 %*. This result assumes no feedback between the exogenously varied discount rate and the economy, as modeled through the GDP and the assumed connection with energy demand. Increased discount rates in the model will increase the capital cost charges and the stabilized cost of electricity. In the case under investigation, doubling the interest on capital from 5 %/yr 10 %/yr increases power generation cost by 22 percent. This increase occurs because the structure of power generation in China depends mainly on coal. At very high capital charges, the optimization approach slightly changes the structure of power production, but coal dominates as the primary fuel for electricity generation.

In the spirit of relating total costs to emissions in a way that connects with the simulation (ESS) modeling efforts, Figure 5.49 illustrates a correlation that results when total integrated emissions of CO_2 and SO_2 over the period 1995-2030 are plotted against the present value of total energy costs, *ENC*, for both China and Shandong Province, as driven by the variations in discount rate. Figure 5.50 gives a similar correlation of CO_2 and SO_2 emissions with average energy costs, *<COE>*.

The results given in both Figure 5.49 and 5.51 are presented as "scatter plots" that might result from uncertainties in discount rates, although lines are used to connect points within respective sets. On these scatter plots the direction of decreasing *ENC(B$)* correlates with decreases in the discount rate, whereas the direction of increasing *<COE>* correlates with an increasing discount rate (Figure 5.47). As noted previously, Figure 5.47 shows only a weak correlation between integrated 30-year CO_2 and SO_2 emissions and the discount rate.

This behavior is replicated in the emissions versus cost correlation (Figures 5.49 and 5.50), with weak negative correlation of emissions with *ENC* and weak positive correlation with *ENC* possibly indicated at the level. Generally, for low discount rates the advanced coal, nuclear, and (to some extent) hydroelectric technologies are more readily introduced and CO_2 emissions decrease (again, the direction of the decreasing discount rate in Figure 5.49 is that of increased present-value of total electric energy costs, *ENC*.

The share fractions of key generation technologies in the last period (2030), as well as emission rates, are shown on Figures 5.51 and 5.52, as a function of the discount rate; this figure illustrates the technology mix and emissions for both BHC and CHC scenarios. The reason for the previously noted weak or nonexistent correlation of both CO_2 and SO_2 emissions with the discount rate reflects the dominance of domestic coal-fired technologies for all ranges. Increased discount rates generally push advanced (costly) technologies like nuclear and advanced coal (COLA) out of the market for the capital (in the case of nuclear and renewable energy sources, including hydroelectric) costs assumed. Very low discount rates favor nuclear over advanced coal, but their relative (30-year) market shares reverse at low-to-intermediate discount rates; for all intents and purposes they are eliminated in the presence of the 10% 1/yr baseline or above discount rates. Very high discount rates will give oil a dominant market share for the relatively small fractions of total electrical energy provided by fossil fuel sources other than domestic coal, which dominates the market.

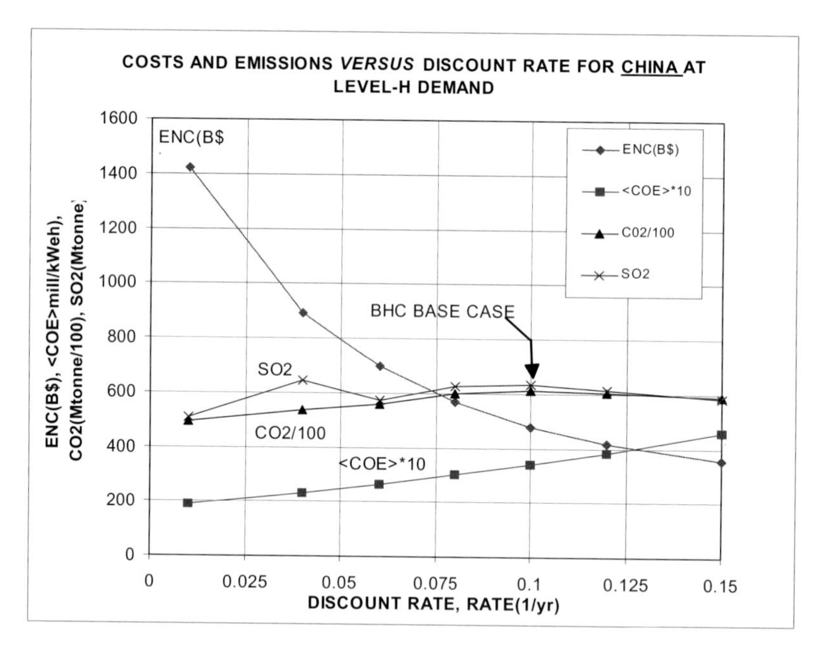

Figure 5.47 Dependence of present-value of total energy costs, ENC(B\$); average cost of electricity, <COE>(mill/kWeh); and the total integrated (1995-2030) emissions for both CO_2 and SO_2 on discount rate for China.

The simplified (direct discounting) approach used to generate these results is not entirely correct in the sense that in reality GDP growth rates and capital discount rates are correlated. However, these second order effects are not taken into account in CRETM.

Lastly, for the baseline capital costs assumed, renewable energy sources (mainly solar PV and wind) remain minor players in electrical-energy mixes modeled for China. These single-point parameter studies have all been conducted without CO_2 and/or SO_2 emission constraints (caps or taxes) imposed. The interactions between emission constraints and discount rates remains to be examined.

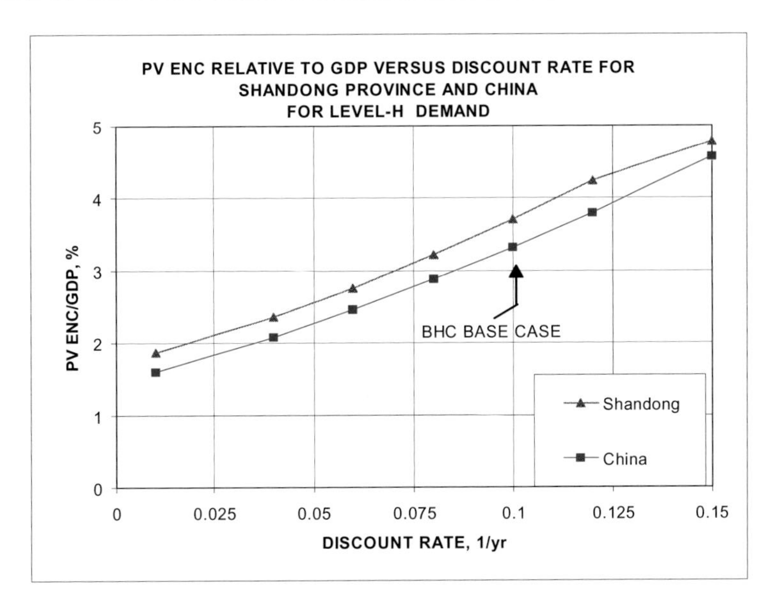

Figure 5.48 Dependence of the present value of total energy cost, ENC, as a percent of the present value of total (integrated and discounted over the computational time frame) GDP on discount rate for China (and Shandong Province).

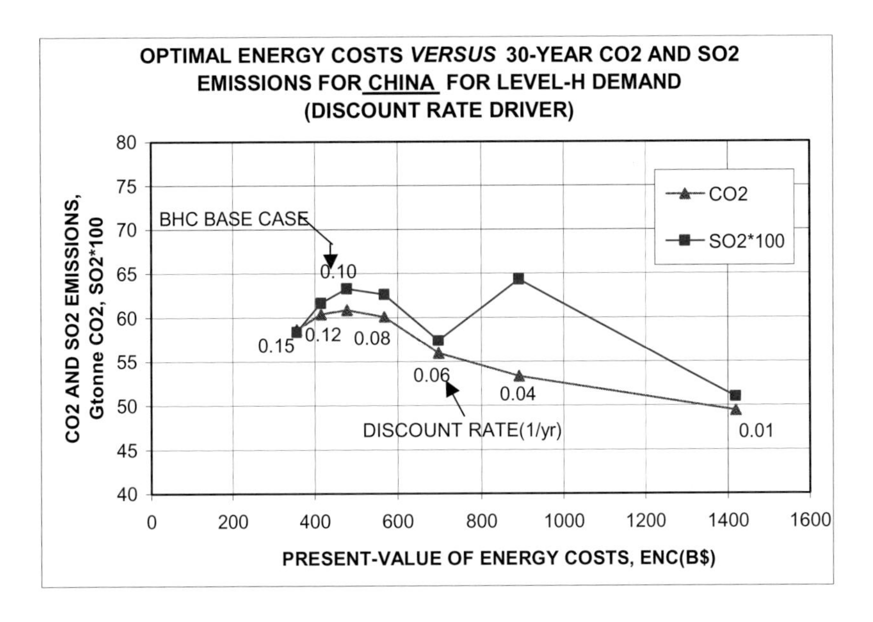

Figure 5.49 Dependence of total integrated (1995-2030) CO_2 and SO_2 emissions on present-value of total energy costs, ENC(B$), as the discount rate is varied.

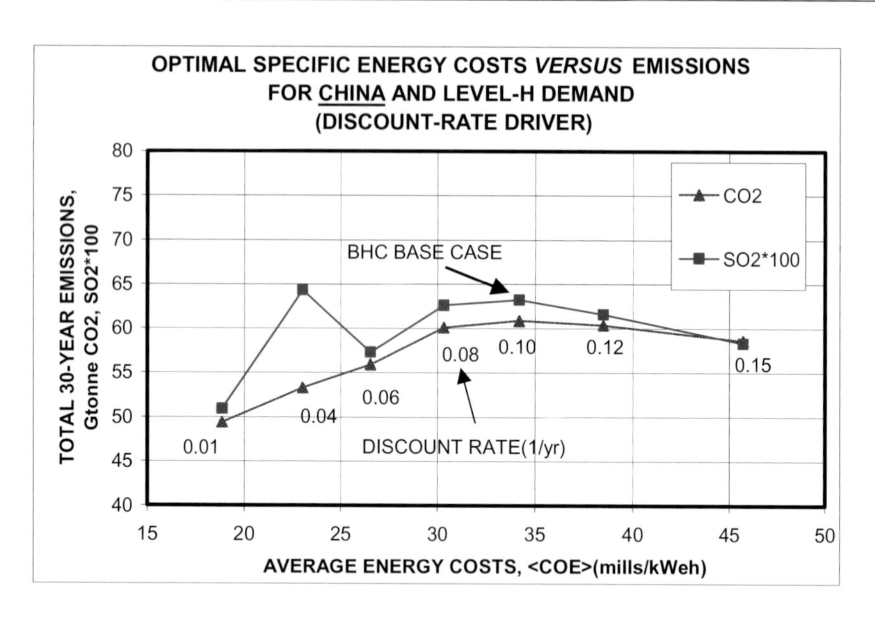

Figure 5.50 Dependence of total integrated (1995-2030) CO_2 and SO_2 emissions on average cost of energy, <COE>(mill/kWeh), as the discount rate varied

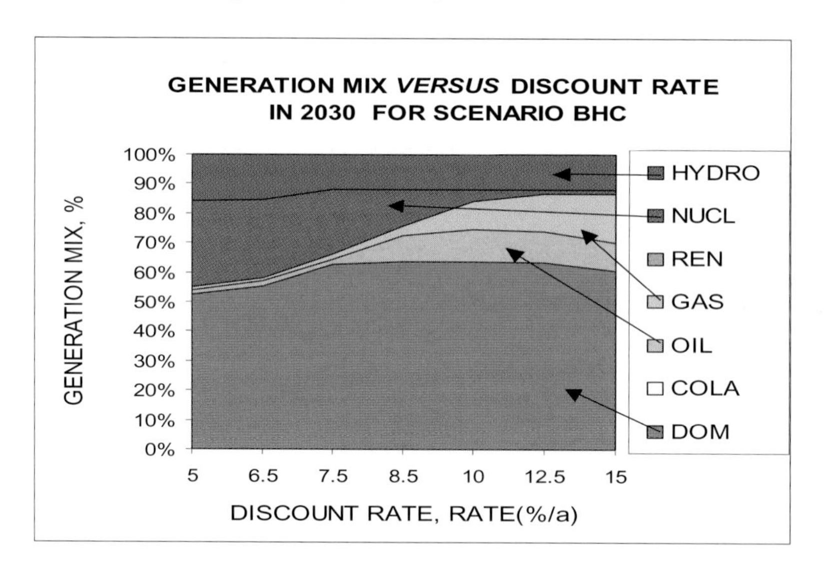

Figure 5.51 Dependence of share fractions of key generation technologies and emission rates in the last computational period (2030) on discount rate for BHC base-line scenario.

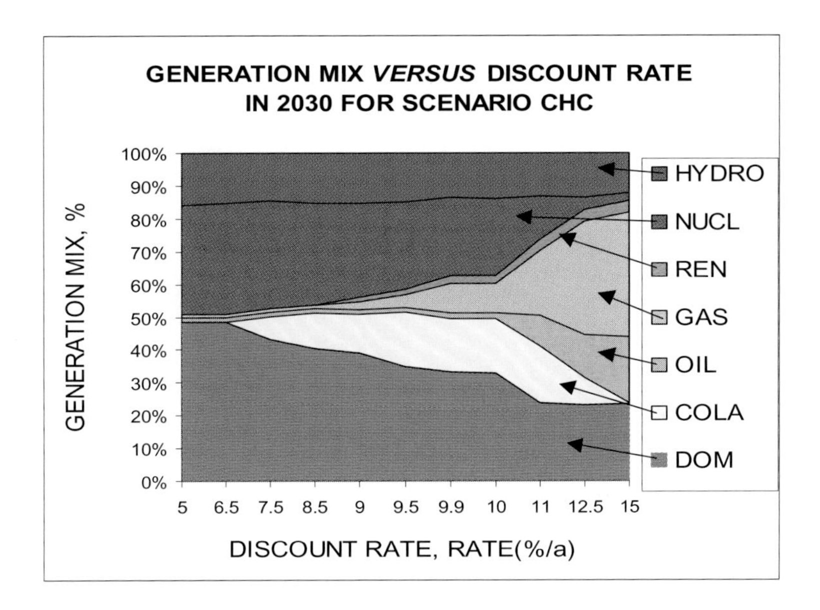

Figure 5.52 Dependence of share fractions of key generation technologies and emission rates in the last computational period (2030) on discount rate for CHC (carbon caps) scenario.

5.2.3. Technology Costs

The cost of any given generation technology takes the form of capital, fixed and variable operating costs, fuel, and decommissioning and decontamination charges (D&D) incurred so far only for nuclear, although long-term use of coal portends unique heavy-metal contamination issues charges. Charges related to waste disposition will also be incurred, but so far such costs pertain only to nuclear technology, and have been internalized primarily through the fuel charge. Given the key, largely exogenous (in the case of CRETM), economic drivers (e.g., GDP growth, related growth in energy demand both in kind and in magnitude: "learning-by-doing" cost reductions; and the intensity with which energy is used to provide certain services, etc.), these technology costs are by far the strongest determinants of cost-optimized generation mixes *versus* time.

For this reason, the sensitivity of the baseline BHC-optimized generation mixes to technology cost variations has been examined and reported in Kypreos (2002). Cost changes were affecting nuclear energy or "learning-by-doing" improvements were considered simultaneously for a series of technologies. Figure 5.53 shows the penetration of nuclear energy in the BHC case in the presence of a moderate capital cost reduction and carbon tax rate of 3 \$/tonne CO_2/5yr. Kypreos (2002) reports on 19 different combinations of costs and carbon and sulfur taxes.

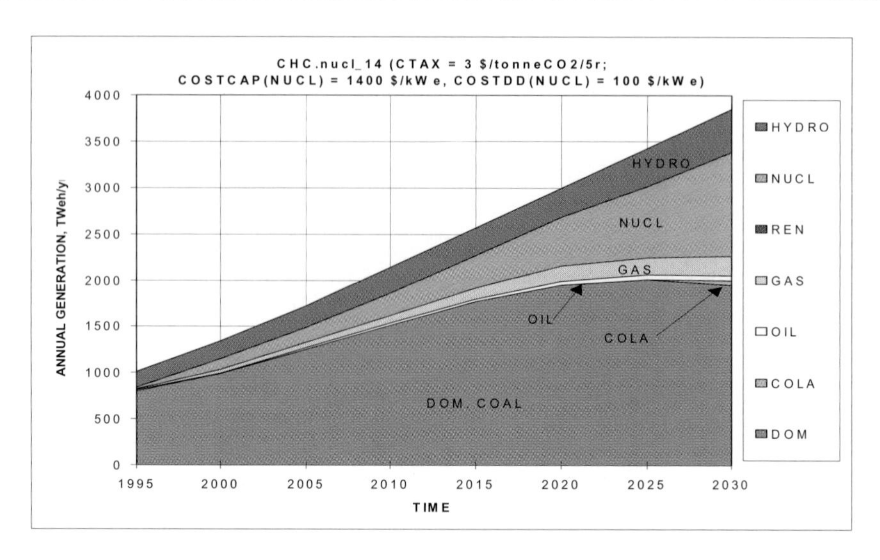

Figure 5.53 Time evolution of generation technology mix for a carbon tax imposed *at a linear rate CTAX = 3 \$/tonneCO$_2$/5yr for [COSTCAP(NUCL) = 1500 \$/kWe, COSTDD(NUCL) = 150 \$/kWe.*

5.3 Base Case with Optional Endogenous Modeling

This section presents selected results of a sensitivity analysis of these exogenous effects. The first examples examine the baseline BHC scenario in the presence of endogenous-learning, partial-equilibrium, and the peaking-constraint options. The partial-equilibrium problem is solved directly using a linear approximation, without the need in using an iterative approach. Only the shadow prices for the starting year and each region should be known. The endogenous-learning formulation creates a non-convex and non-linear problem not solvable by plain optimizers. It has been formulated as a Mixed Integer Programming approximation (Barreto, 1999) and requires few minutes of CPU time to solve. The model has a size of approximately 7500 rows and 34000 variables.

This section presents scenario-based results that illustrate a range of possible futures for electricity generation in China, over the next three decades, with and without considering the effects of endogenous learning in the implementation of new technologies. Following a period in which foreign technology will be imported, Chinese R&DD will incorporate local experience, producing improved technology for Chinese uses, which should reduce the investment cost of systems built in the country. Such "learning by doing" effects should affect IGCC systems, supercritical coal, nuclear, combined cycle gas turbines; wind turbines and solar-PV.

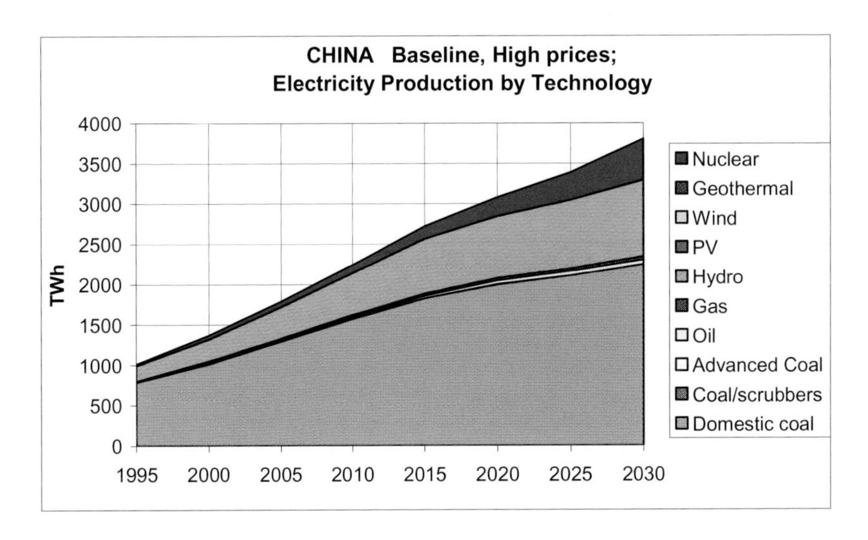

Figure 5.54 Electricity generation in the Business as Usual (BaU) development is based mainly on domestic coal technology, hydropower and nuclear energy. Discount rate is now 5%/yr that makes nuclear energy more competitive than in the BHC cases.

Table 5.11 describes the technical and economic characteristics of generation systems used in China. Capital costs are assumed to be 25-30 percent less than in the industrialized countries, but the systems will operate two to four percent less efficiently. To examine ETL impacts, investigators adopted the CRETM baseline BHC scenario with a 5%/yr discount rate rather than the nominal 10%/yr used throughout the study. The generation mix displayed in Figure 5-54 shows that this lower discount-rate case (assuming endogenous technological learning for all competing systems) allows the early and significant introduction of nuclear power.

Figures 5.54 and 5.55, show that even in the BHC baseline scenario (with discount rate of 5%/yr) the diffusion of scrubbers and advanced coal technology into the markets is stimulated by R&DD investments that reduce their costs. Normally, sulfur removal options and advanced IGCC technology are introduced when policies favor regional and global environments. However, in this case cost reductions due to R&DD and accumulated experience are sufficient to enable advanced, more efficient technology to penetrate the market.

Table 5.11 Techno-Economic Characteristics of Systems Analyzed for China Based on Conservative Learning Data[(a)].

Generation Technology	**Capital cost**	**Variable O&M**	**Fixed O&M**	**Efficiency**	**Learning rate**	**Capital floor cost**
	$/kW	M$/TWh	%CC/a	%	-	$/kW
Domestic small coal plant	676	5	3	27	1	676
Domestic 100-200 MW coal plant	650	4	3	29	1	650
Domestic 300 MW coal plant with ESP	600	3.5	3	36	1	600
Domestic 300 MW coal plant, ESP&SCR	790	4.5	3	35	0.97	400
Supercritical 600 MW coal, ESP&SCR	950	4.5	3	42	0.97	450
Foreign AFBC coal plant	900	8.5	3	38	1	900
Foreign IGCC coal plant	1100	4	3	43	0.96	700
Foreign PFBC coal plant	1100	4	3	43	1	1100
Traditional oil-fired plant	530	2.8	3	35	1	530
Combined-cycle oil-fired plant	500	2.8	3	42	1	500
Combined-cycle gas-fired plant	530	8	3	45	1	530
Advanced gas-turbine CC plant	550	8	3	55	0.97	300
Nuclear plant	1500	5	1.6	33	0.98	1000
Hydroelectric plant	1200	1	1.5	33	1	1200
Wind plant	1200	2	1.5	33	0.90	500
Solar photovoltaic plant	6000	0.07	1.5	33	0.90	1000
Geothermal plant	2000	0.05	1.5	33	1	2000

[(a)] The penetration of technologies in the CRETM model is limited by constraints on the regional availability of endogenous resources, transported fuels and renewables as well as market penetration constraints applying maximum growth rates *per* region and technology. Thus, when the best option reaches the boundary of capacity growth, the second best becomes the marginal technology introduced in the solution.

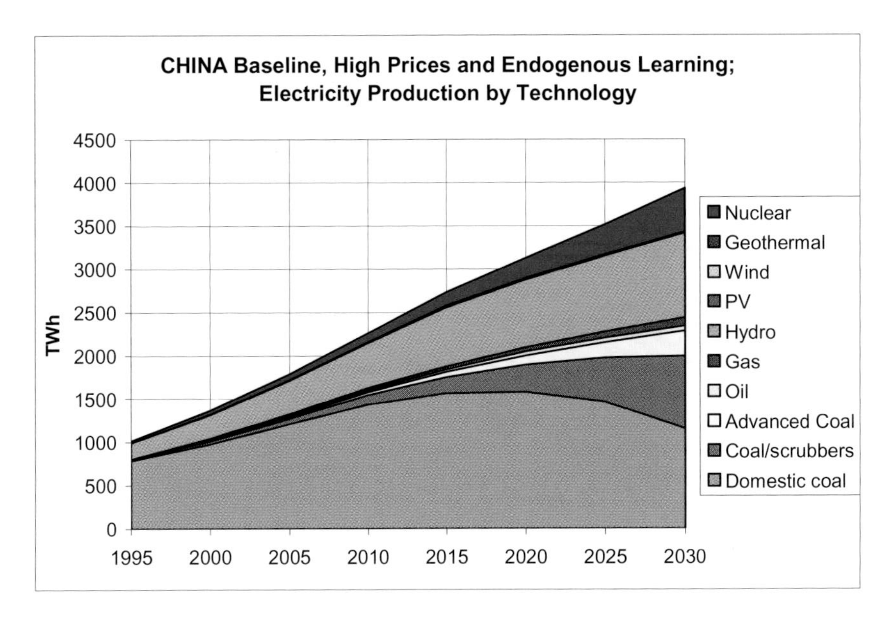

Figure 5.55 Electricity generation in the Business as Usual (B) development is based again on domestic coal technology, hydropower and nuclear energy but the penetration of scrubbers and advanced coal technology is higher. The real discount rate is now 5%/yr instead of the (H)igh 10%/yr used in all other CRETM analyses.

5.2 Main Results from CRETM

While neither the model nor this investigation is comprehensive enough to warrant detailed analyses, these seven scenario results compared to the eighth baseline (BHC) scenario gives some sense of the cost *versus* emissions trade-off associated with each approach. Figure 5.56 summarizes the CO_2 and SO_2 emission rates for all eight scenarios. Figures 5.57 and 5.58 present the discount percentage changes (again, all scenarios assumed a 10%/yr discount rate except those specifically reporting the single-point impact of increasing and decreasing it) in energy costs computed by CRETM compared to the percentage change in total CO_2 or SO_2 accumulated emission over the 1995-2030 time frame. Generally, it appears that increasingly higher coal, oil and gas prices (BHH) reduce sulfur emission rates compared to the BHC lower fuel-cost scenario as these fuels are substituted for coal and advanced coal technologies are introduced. The introduction of carbon constraints alone (as in ,e.g., the CHH, CHC scenarios) reduces sulfur emissions as a secondary benefit.

Turning to CRETM-Specific Interim Technology Assessment, we will examine the penetration of energy generation technologies under sulfur and carbon control policies are first without and then with endogenous learning imposed. The scenarios

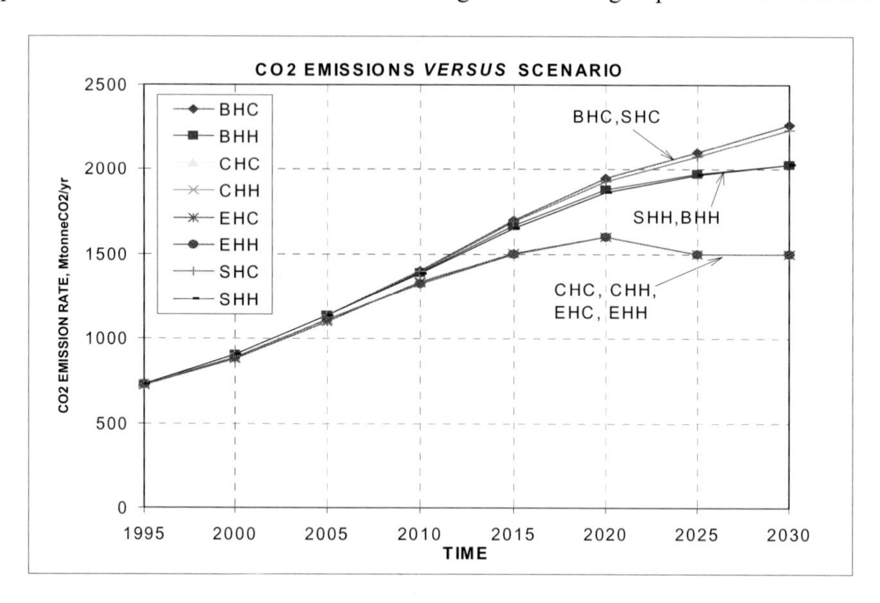

Figure 5.56A Summary of CO_2 emission rates for the eight scenarios.

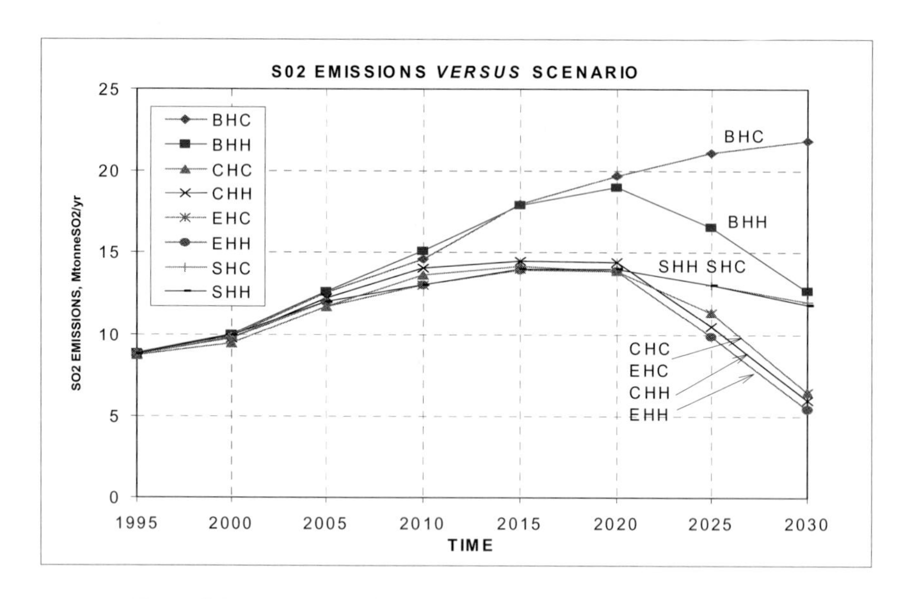

Figure 5.56B Summary of SO_2 emission rates for the eight scenarios.

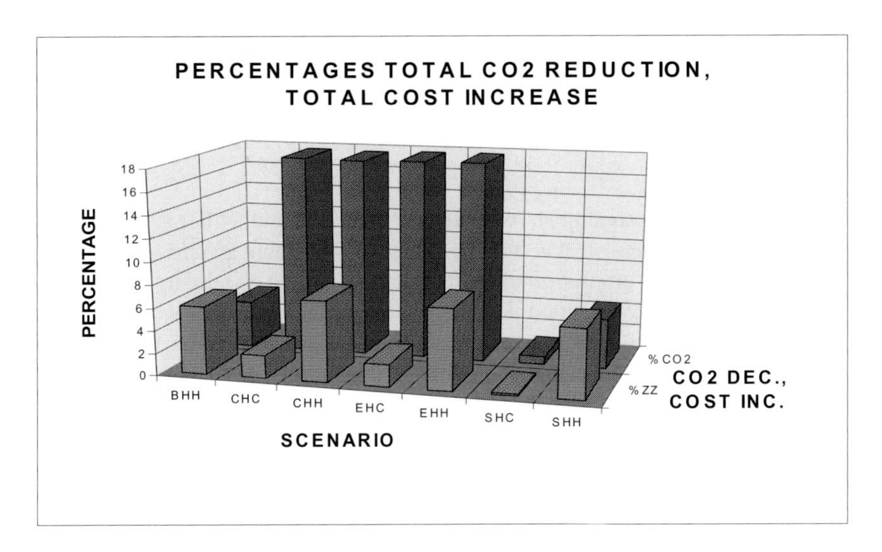

Figure 5.57 Percentage change in total (1995-2030) CO_2 emissions (decrease) in relationship to percentage increase in present-value of total energy costs ZZ, (10%/year discount rate over the period 1995-2030) in China for seven scenarios given in relationship to the baseline BHC scenario.

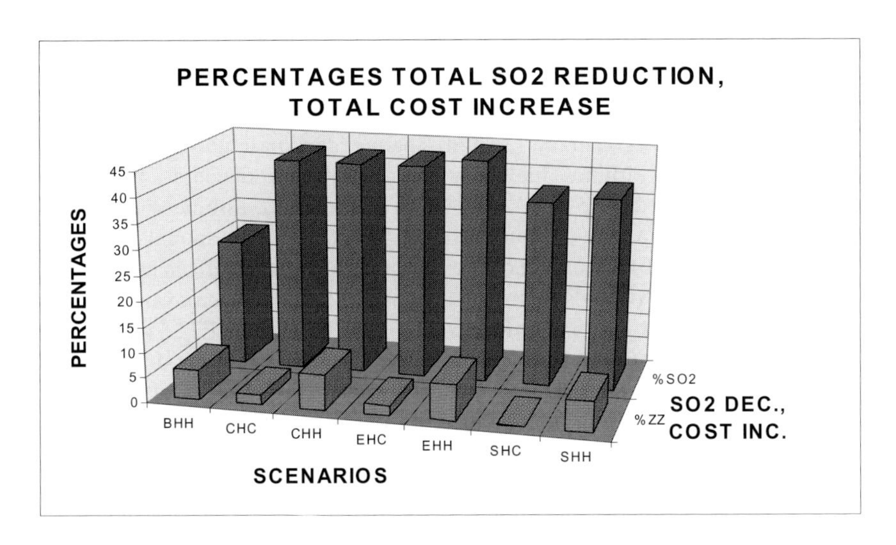

Figure 5.58 Percentage change in total (1995-2030) SO_2 emissions (decrease) in relationship to percentage increase in present-value of total energy costs ZZ, (10%/year discount rate over the period 1995-2030) in China for seven scenarios given in relationship to the baseline BHC scenario.

with endogenous learning all occur in the presence of high prices for fossil fuels. In order to distinguish them from the same cases without endogenous learning the price indicator is replaced by an "E", so BHE means the BHH case (no emission constraints, high demand, high prices) with endogenous learning, etc.

Investigators reached some important conclusions related to the new ETL options of CRETM:

- As a general statement, we can say that demand will respond to higher production costs. It has been assumed that an active market liberalization policy based on marginal cost pricing will be gradually introduced in China. Under these conditions, the sulfur and/or carbon constraints increase the marginal costs of power generation, and consumers will react to higher prices by adjusting their `electricity consumption. Demand and supply equilibrium shows that without endogenous learning, relative costs changes will respond to variations in power generation between 3,800 TWh and 3,400 TWh. Under ETL, investment and generation cost will be reduced; demand for the reference case will approach 4,000 TWh.
- Endogenous technological learning favors scrubbers and advanced coal technology while gas CC and wind turbines will contribute to power generation in the last decades.
- Endogenous learning first improves the performance of scrubbers and makes sulfur control easier; it also reduces the cost of carbon control. In the cases without endogenous learning, the shadow prices for carbon control would be reduced from $110 per tonneC to around 60 $/tonneC because of cost reductions induced by learning in systems other than coal. The carbon tax in the learning case corresponds to a nearly two- to three-fold increase in the price of coal.
- The peaking constraint would change the baseline BHC results if conventional turbines fueled by oil/gas are introduced to address peak energy demand. However, as seen in Figure 5.60, turbines introduced for peak management would not produce any significant amount of electricity (see Figure 5.61), which indicates a minimum amount of gas use.
- The exchange of fuels by mode of transport has also been examined (Figure 5.62). Pipeline gas was introduced only in the case of ETL, which reduces the exchange of electricity by wire. The results here are not detailed enough to support any policy recommendations about the merits of gas-pipelines versus wire transmission of electricity. The uncertainty in these results suggests the need for a more detailed study of the problem.
- The last observation concerns a comparison of fuels by mode of transport. CRETM estimates of regional trade in fuels and electricity includes the costs of transport and availability of resources. Changes in the amount of coal transported across regions with and without ETL considerations have been presented. Pipeline gas was introduced as an alternative only in the case of endogenous learning (GCC becomes more competitive in some regions); in this case it acted to reduce the exchange of electricity by wire.

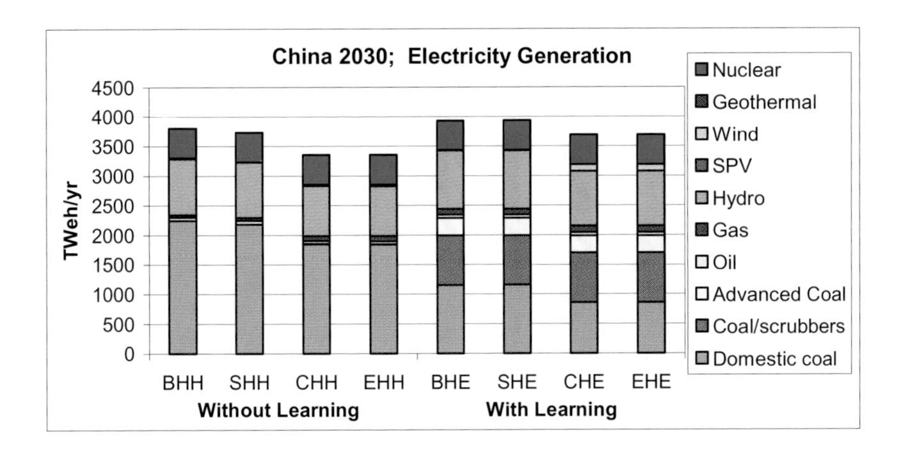

Figure 5.59 Electricity generation in the baseline and the environmental benign policies with and without endogenous learning considerations. Coal systems with scrubbers, advanced coal technologies and hydropower are the pillars of electricity supply in China. Nuclear energy, gas CC and wind become important under learning considerations. But, again learning needs R&DD policy support for technologies to follow their learning curves. The discount rate is 5%/yr, compared to the (H)igh rate of 10%/yr used in most of the CRETM reported herein.

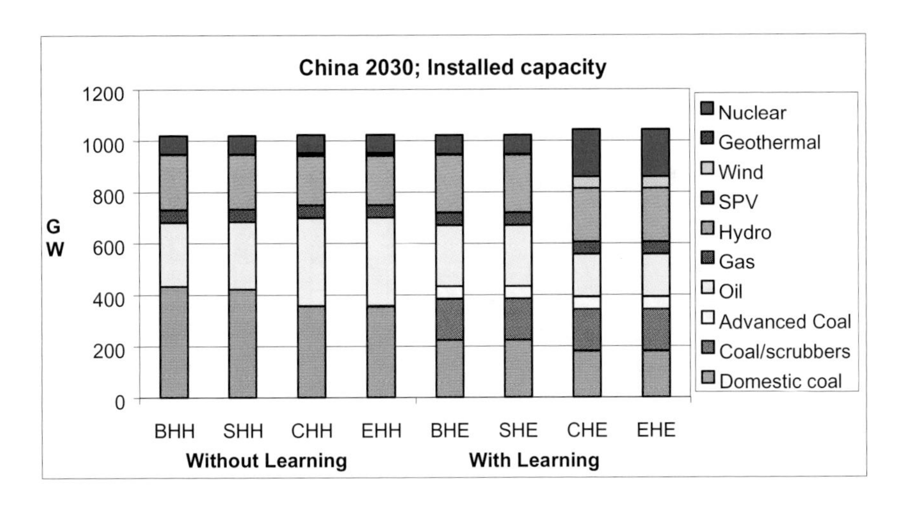

Figure 5.60 Electrical generation capacity in the baseline and the environmental benign policies with and without endogenous learning considerations.

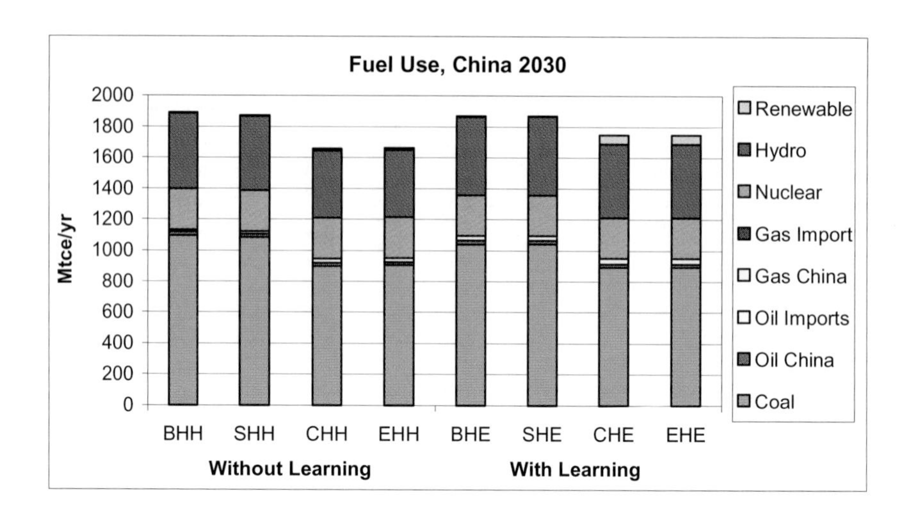

Figure 5.61 Fuel use for electricity generation in the environmental benign policies with and without endogenous learning considerations. Coal, hydropower and nuclear energy are the pillars of electricity supply in China. Only wind contribution is increased with learning.

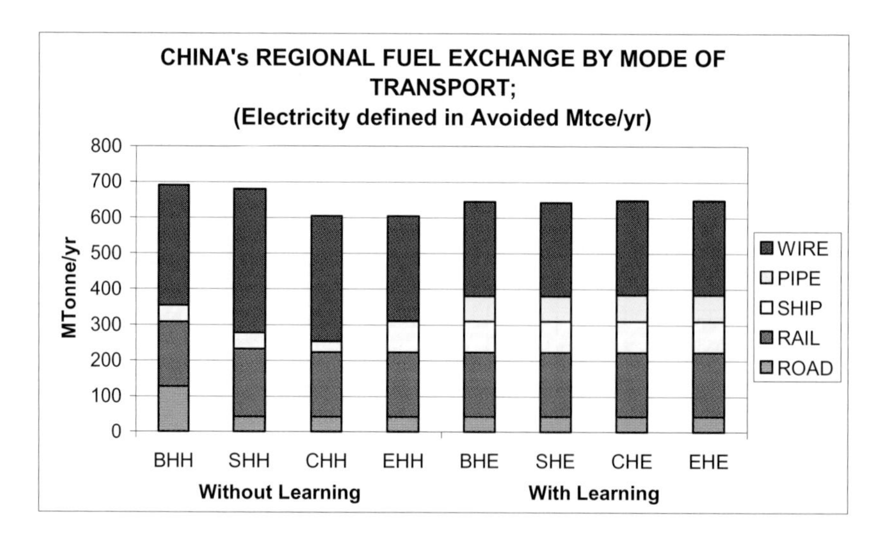

Figure 5.62 The regional fuel exchanges in China is large and reach the limits of capacity for the railroad systems of China. Transmissions of electricity by wire, gas by pipe and ship transport for coal are alternatives to release pressure on the overloaded railroad system. ***Whatever the solution to the problem, investment requirements for infrastructures in China are again enormous.***

The most robust conclusion derived from this ETL analysis is that China should place great importance on establishing an efficient R&DD policy and international networks for technology diffusion. This policy would stimulate the development of a robust and flexible portfolio of alternative technologies for power generation and would be a hedge against future uncertainties. If the ETL performance simulated in the analyses can be maintained in the future, many of the environmental concerns and the quest for energy security will be resolved to the benefit of China and the industrialized nations. What the model cannot identify is the level (construed in terms of, *e.g.*, percent GDP) of R&DD expenditure at which these benefits will emerge and accrue.

6. INTEGRATION OF RESULTS

As is illustrated in Figures 2.7 and 5.63, CETP has included a wide range of modeling efforts and approaches. Many of the resources devoted to these modeling efforts were directed toward the development of individual tools for generating policy-related tradeoffs associated with the EEM, ESS, ETM, LCA, EIA, and RA tasks. As illustrated in Figure 2.7, some of these individual modeling efforts were designed to provide input to the others; this was achieved with varying degrees of success. Although considerable effort was made early in the Project to assure a reasonable level of commonality among data bases and assumptions behind key drivers, the comparability of these (e.g., demand projections; time frames; unit technology costs and the cost of fuels that drive them; levels of discounting and inflation; degrees to which existing technologies were included in estimates of total electrical energy costs; specific emission rates and related fuel qualities, etc.) was not perfect. Residual differences still remained by the time results were being generated and shared among tasks. This is to be expected, given the large uncertainties associated with development in China, but it can lead to differences in results for modeling activities that share common goals, but use different approaches. The results generated by, most notably, the EEM, ESS, and ETM tasks pursued common goals but used different methods, modeling scope, and less than perfectly common databases.

These differences notwithstanding, comparisons of key results from the EEM, ESS, and ETM tasks remain a crucial component in CETP's final policy recommendations to the stakeholders, Though imperfect, the comparisons made in this section are as accurate as is possible and reasonable. Figure 5.63 shows these comparisons from three perspectives on the EEM task: a) comparisons between EEM results (primarily costs, emissions, technology and fuel mixes, etc.) and related studies made outside of CETP; b) comparison of EEM results with those produced from other tasks within CETP; and c) comparison of EEM results from models used within the EEM task.

In making these comparisons within the EEM task (*e.g.*, CRETM *versus* MARKAL), results from the range of scenarios (Table 5.8) are aggregated into correlation or (scenario-driven) scatter plots of average energy costs, <COE>(mill/kWeh) versus total emissions for both CO_2 and SO_2, and for fuel and generation-technology mixes versus time for the BHC base case at both Shandong and China regional levels. Behind each set of models and the regions covered are assumptions related to electricity demand projections; these are summarized in Figures 5.67 and 5.75 for the CRETM and MARKAL tools. CRETM electricity demands are generated directly from GDP growth projections and related price and income elasticities (see Equation 5.1). The electricity demands from MARKAL result from exogenous input for final energy or end-use energy demands.

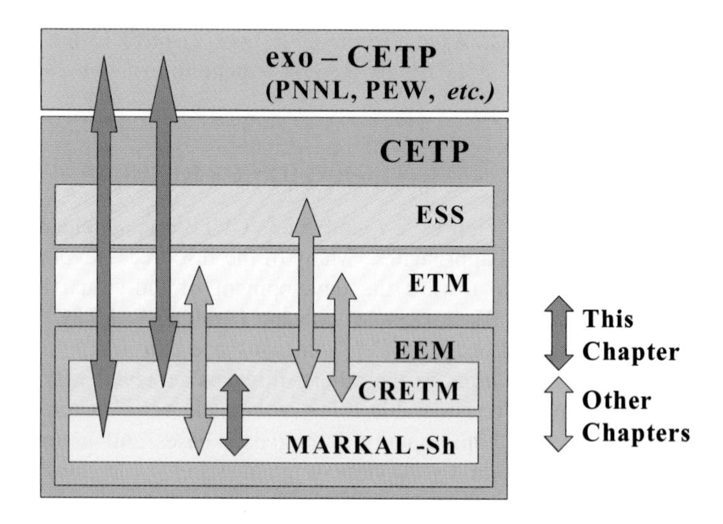

Figure 5.63 Spectrum of CETP results comparison levels, showing those of concern to the present section.

This section concludes with a brief synthesis of results and recommendations. The discussion of ultimate recommendations has been broadly divided into a) *Policy and Planning Implications* of these model results; and b) *Future Work and Model Advancement*. At the present stage of the EEM activities, the policy and planning component has been sketched out on the basis of EEM results only. Following a summary and synthesis of results, recommendations for future model development are given. This second set of recommendations, projected provisionally, still requires integration with other relevant parts of CETP.

6.1 Syntheses

In this section, we present the CRETM results first for China and for Shandong Province and the MARKAL results for Shandong specifically.

6.1.1 China

The focus of the CRETM results (Kypreos, 2002) has been on China countrywide, with an aim towards explaining the bases, workings, capabilities, and limitations of the China Regional Energy Transport Model (CRETM). Table 5.12 summarizes the integrated emissions results, as well as impacts on the total present value of energy costs, ENC. The average cost of energy, <COE>(mill/kWeh) reported in Table 5.12 is determined by dividing the total electrical energy generation over the period 1995-2030 by the total undiscounted energy costs. It is also given on this table, along with other incremental parameters.

Table 5.12 Summary of Integrated Emission and Cost Impacts for the Eight Scenarios Examined for China.

SCENARIO	**ENC M\$**	**TOTAL CO_2 MtonneCO_2**	**TOTAL SO_2 MtonneSO_2**	**<COE> mill/kWeh**	**ENC/ $C0_2$ \$/tonneCO_2**	**ENC/ $S0_2$ \$/tonneSO_2**
BHC	478393	60874	632	34.2	0	0
BHH	507053	58544	563	38.2	12.29	411.9
CHC	484860	50714	446	35.9	0.64	34.7
CHH	512321	50778	450	38.9	3.36	186.5
SHC	478990	60469	483	34.3	1.47	4.0
SHH	507502	58274	482	38.2	11.19	194.3
EHC	487878	50714	442	35.9	0.93	49.8
EHH	512362	50727	433	39.0	3.35	170.4

At the country-wide level, the cost *versus* emissions impacts of the eight scenarios are best shown relative to the baseline BHC case. Figure 5.64 shows these results for both CO_2 and SO_2. As noted previously, the main difference between these eight scenarios is the fuel cost. The scenario points in Figure 5.64 have been artificially connected to highlight those high fuel costs *versus* those with constant fuel cost. The relationships between some scenarios on this cost-emissions "phase space" may be distorted by the fact that the relative cost changes refer to discounted costs at a substantial (10%/yr) rate. To eliminate this distortion, Figure 5.65 is included to show the impacts of trades between emissions and costs using undiscounted total energy costs (1995-2030). Meeting scenarios with carbon emission constraints at the Shandong Province level under minimum total cost conditions requires the import of electrical energy in the out years of the optimization (Kypreos, 2002).

The single-point parametric studies are comprised of 19 nuclear variations, six discount rate variations and five demand growth rate cases; together with the BHC scenario, about which the variations where conducted, a total of 31 ENC-minimizing "scenarios" result. Figure 5.66 captures them on an energy-emissions "scatter" plot. The average energy cost *<COE>(mill/kWeh)* was selected for the cost dimension, and the total, integrated CO_2 emissions M_{CO2}(MtonneCO_2) was selected for the emissions dimension. Since most of the 31 cases represented in Figure 5.66 were evaluated for a discount rate of 10 %/yr, the corresponding discount-rate variation curve included on Figure 5.66 provides an approximate scale whereon the center of the remaining cluster of optimized points would shift if the corresponding value of discount rate was increased or decreased above the BHC value of 10%/yr. This cluster of points represents a kind of minimum ENC-M_{CO2} "frontier." Under the ideal conditions described by CRETM, this frontier limits access to the lower regions of this *<COE>*-M_{CO2} phase space. As for most economic or trade-off frontiers of this kind, the realities of the technological, social, and cultural (Huntington, 1996) world can limit actually achievable cost-emission goals to some distance above this idealized trade-off frontier.

Once the identity of each cluster of points is identified, as shown on Figure 5.66, the trends become more-or-less self evident: a) the CTAX points expectedly move to higher <COE> and lower M_{CO2} regions as CTAX(\$/tonneCO_2/5yr) is increased; b) modest decreases in the cost of nuclear power can push the frontier to desirable regions of lower <COE> and lower M_{CO2}, albeit with a rough model of the nuclear fuel cycle; c) the availability of moderately cheaper nuclear energy in combination with an effective carbon tax (CTAX = 3 \$/tonneCO_2/5yr, COSTCAP(NUCL) = 1,400 \$/kWe) pushes back the cost-emissions frontier without increasing the average cost of energy; d) the application of SO_2 limits alone can push back the frontier without significant increases in the average cost of electricity; e) small percentage changes in the discount rate used has a large impact on the vertical (<COE>) position of the cost-emission trade-off frontier. These findings may require more extensive analysis, but the cost-emissions trade-off frontier of the kind suggested in Figure 5.66 can provide an excellent means for joining, interpreting, and utilizing EEM and EES results for policy-making purposes.

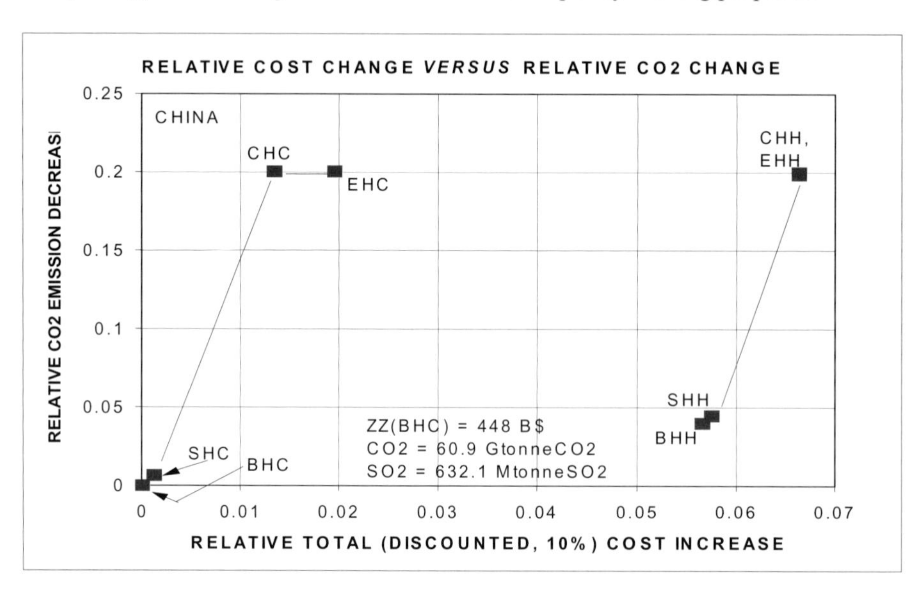

Figure 5.64A Relative discounted total energy cost versus *CO_2 emissions impacts for the eight scenarios examined for China.*

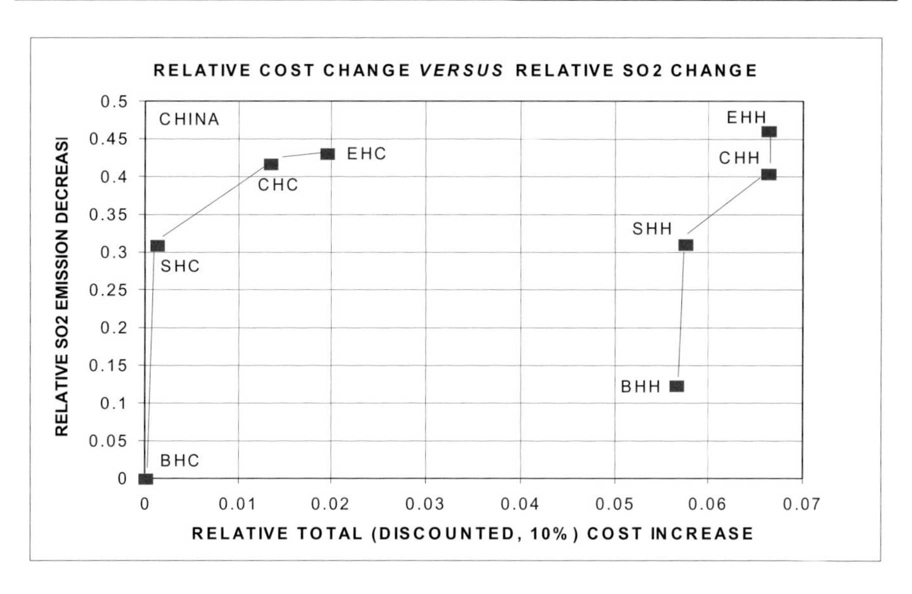

Figure 5.64B Relative discounted total energy cost versus *SO_2 emissions impacts for the eight scenarios examined for China*

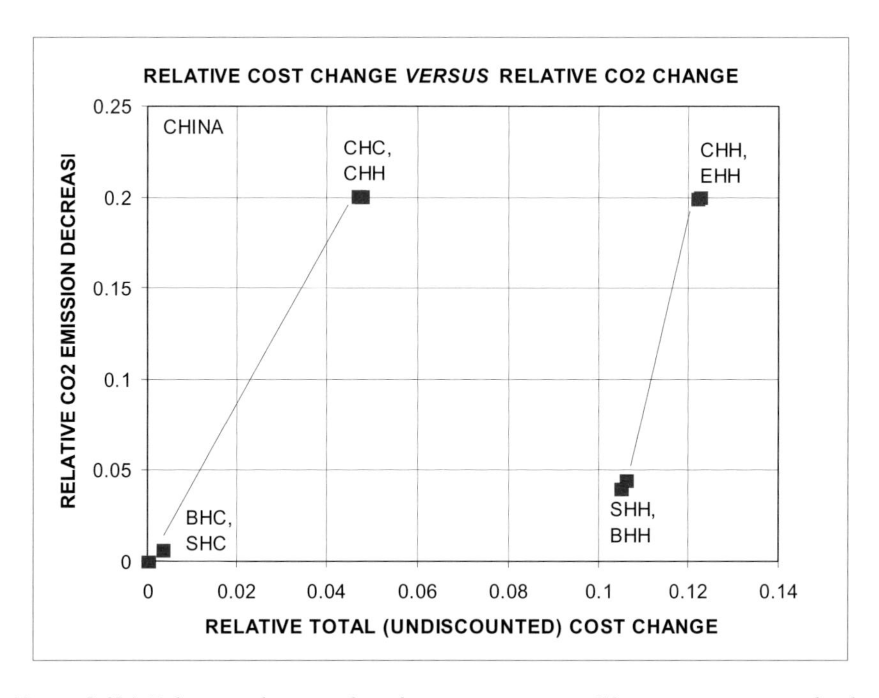

Figure 5.65A Relative undiscounted total energy cost versus *CO_2 emissions impacts for the eight scenarios examined or China.*

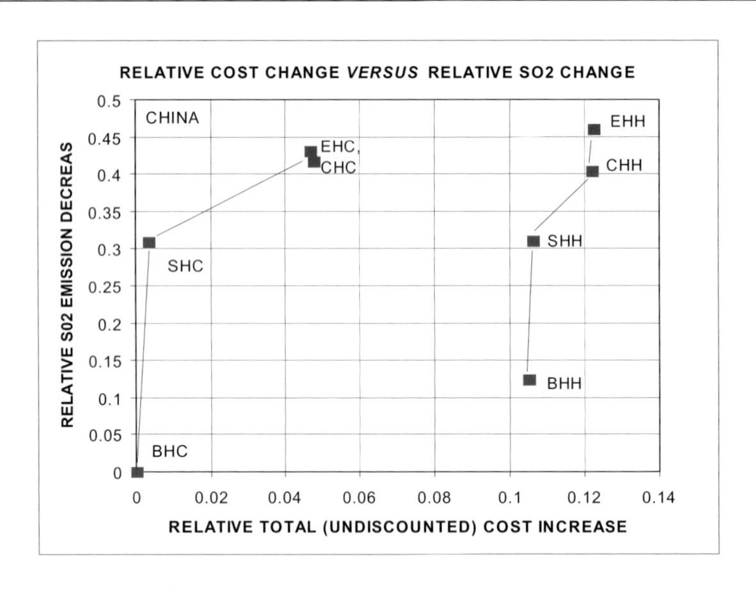

Figure 5.65B Relative undiscounted total energy cost versus *SO_2 emissions impacts for the eight scenarios examined for China*

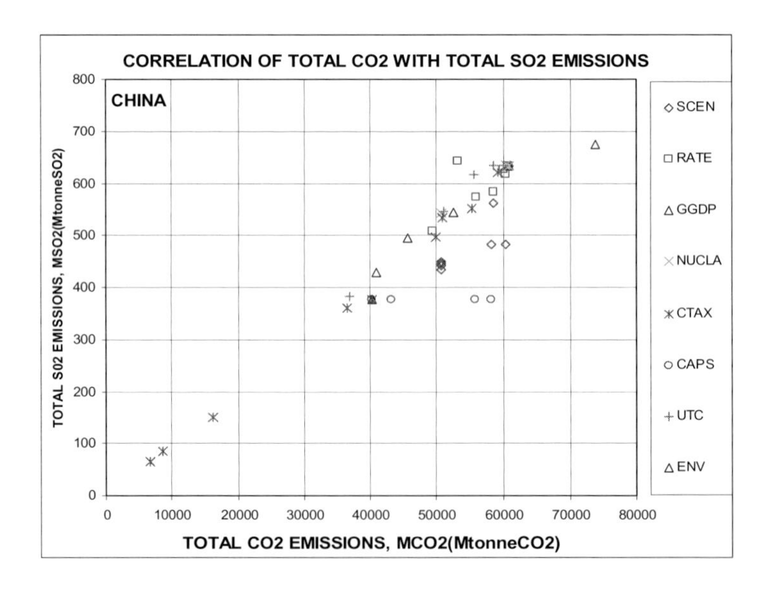

Figure 5.66A Collection of 21 <COE> - MCO_2 couplets from the single-point parametric variations, indicating a minimum-ENC "frontier" for the BHC parameters and variations thereon, for China.

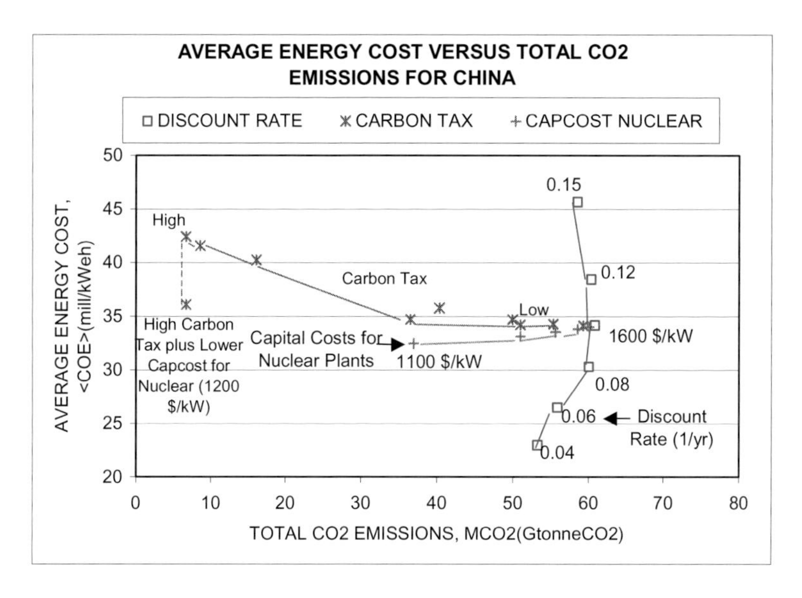

Figure 5.66B Collection of 44 MSO_2-MCO_2 couplets from the single-point parametric variations, indicating a minimum-ENC "frontier" for the BHC parameters and variations thereon for China.

6.1.2 Shandong.

We now turn to the CRETM results specific to Shandong Province, the subject of CETP. Table 5.13 recapitulates the generation mixes and emission rates of the eight wide-ranging scenarios adopted, and Figure 5.67 gives the baseline country-wide demand for electric energy used in each. A wide range of options and opinions can be found on this key issue, as is seen in Figure 5.2. National electrical-energy demand is calculated from the regional GDP growths and the algorithm produced by Equation 5.2. Figure 5.67 also shows the country-wide demand for electricity as projected from the MARKAL-China model (Chen, 2000). Figure 5.68 gives the assumed exogenous growth in GDP and its electricity demands. (Recent projections for Shandong Province made by the Energy Research Institute may be found in Chapter 4).

The six frames in Figure 5.69 give the evolving generation mix for Shandong Province for the scenarios listed in Table 5.13. The CO_2 and SO_2 emission rates associated with each scenario are depicted in Figure 5.70. Generally, the richness of the technology mixes and options for Shandong increases as it is optimized under variations and combinations of the two main scenario attributes: fuel costs and emission caps. Common to all eight scenarios is the early, designed phase-out of small domestic coal plants (*domsml*, Table 5.9), followed by a phased departure of medium (*dommed*) and large (*dombig*) coal plants lacking emissions cleanup technologies.

Table 5.13 Summary of (Economic, Technological, Environmental) Attributes for Eight Scenarios.

Scenario(a)	**Price(b)**	**Caps(c)**
BHC	C(constant)	
BHH	H(increasing)	
CHC	C(constant)	C
CHH	H(increasing)	C
SHC	C(constant)	S
SHH	H(increasing)	S
EHC	C(constant)	C + S
EHH	H(increasing)	C + S

(a) B = baseline; H = high; C = constant or carbon; S = sulfur; E = C + S
Order: Emissions-Demand-Price
Demand is High in all Scenarios
(b) C = Table 20; H = Table 22; Appendix A (Kypreos, 2002)
(c) see Table 5-6

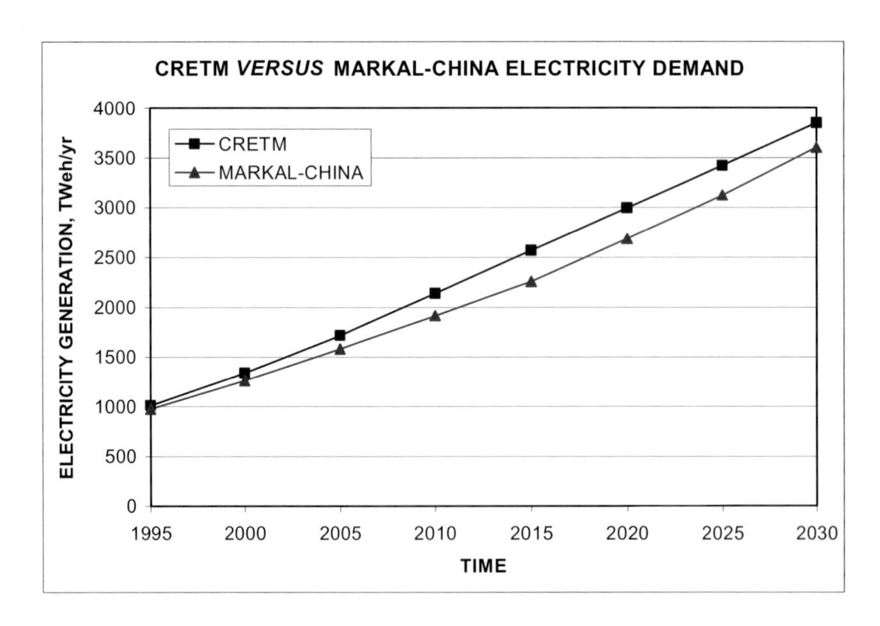

Figure 5.67 Countrywide China electrical energy demand used to drive CRETM [Eq. (5-2)], and comparison with projection from MARKAL-CN model (Chen, 2000).

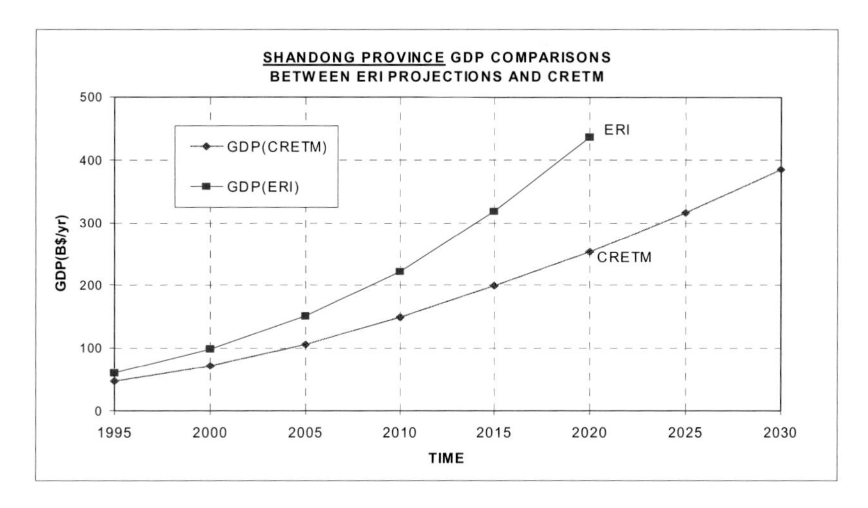

Figure 5.68A GDP projections for Shandong Province used to generate the results in this report, as well as more recent projections made by the Energy Research Institute (cf. Chapter 4).

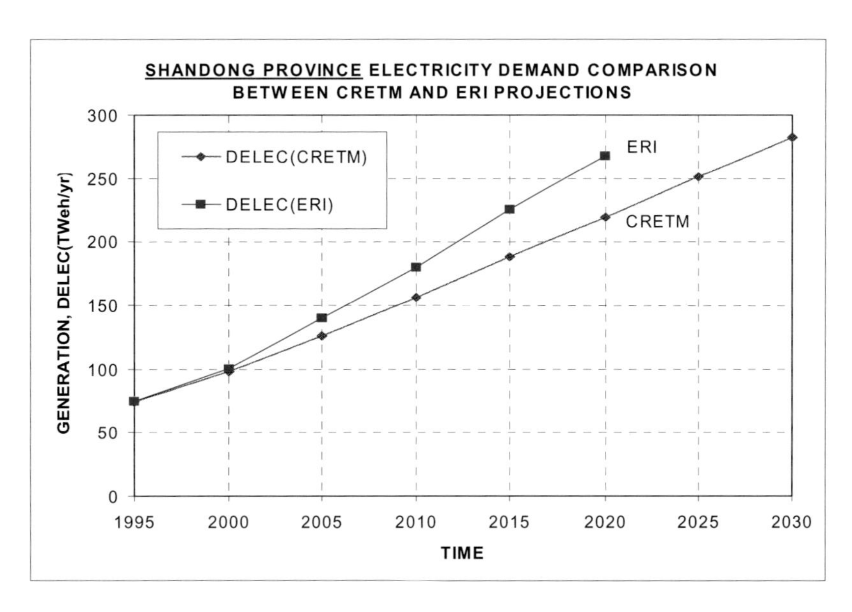

Figure 5.68B Electricity demand projections for Shandong Province used to generate the results in this report, as derived from Eq. (5-2), as well as more recent projections made by the Energy Research Institute (cf. Chapter 4).

Depending on the cost of fuel and the level and nature of emissions caps, large coal-fired plans with electrostatic precipitators (*domesp*) will become a significant part of the generation mix. Because of the high cost of scrubbers and competition from advanced coal technologies, large domestic coal-fired plants with flue-gas scrubbers (*domscb*) will not be part of the mix. The most significant variations in the mix shown in Figure 5.69 generally appear first along lines indicating whether a "constant" (C) or a "high" (H) fuel cost is exogenously enforced.

Secondly, for a given fuel-cost scenario, the generation mix varies according to the kind (e.g., C, S, or E = C + S) of caps enforced. While integrated gas combined cycle (*igcc*) plants under high-fuel-cost scenarios take an important market share, as do advanced gas combined-cycle plants (*gascca)*, combined-cycle plants fueled by oil barely appear in the generation mix. For all cases in which carbon emission caps are enforced (e.g., scenarios CHC, CHH, EHC, EHH), transfer of electricity generation (trans) from outside of Shandong Province is an optimal choice in order to satisfy the exogenous demand. In one case (EHH), for the years between ~2010 and 2020, Shandong Province would be able to reverse this flow and to export electrical energy (Figure 5.69F).

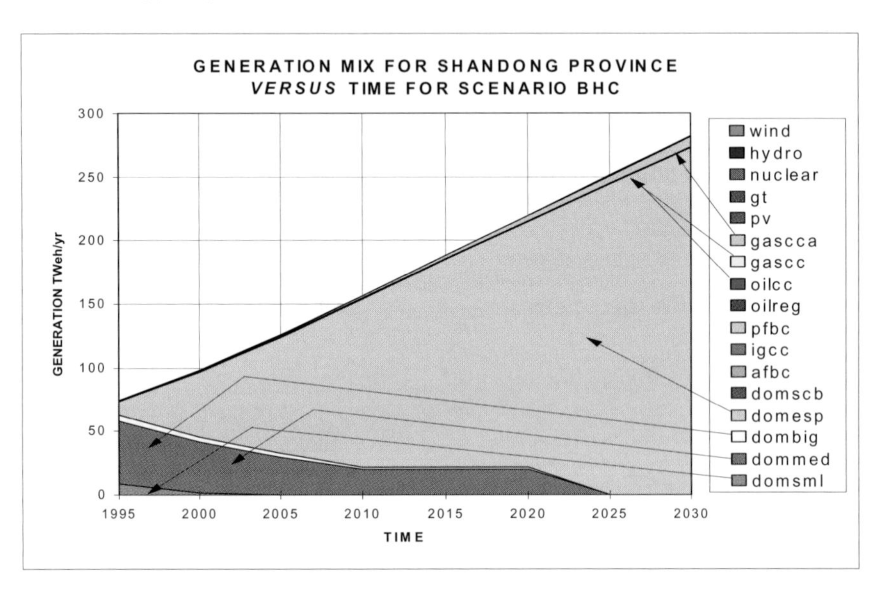

Figure 5.69A Generation mix for scenario BHC [Baseline (no CO_2 or SO_2 emission constraints), (H)igh demand, (C)onstant fuel costs].

Figure 5.70 gives a composite summary of CO_2 and SO_2 emission rates for the eight scenarios. Generally, for a given fuel-cost scenario attribute (H or C), imposition of an SO_2 emissions constraint leads to little benefit in the form of reduced CO_2 emissions. Figure 5.71 shows undiscounted cost versus emission impacts relative to the baseline BHC case for Shandong province. A striking--and at first sight puzzling--result is that under the assumptions of increasing fossil prices

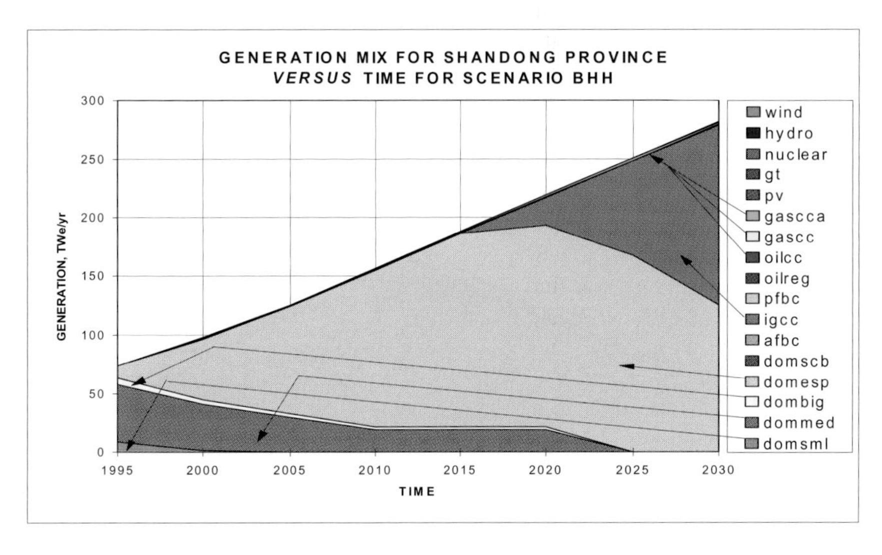

Figure 5.69B Generation mix for scenario BHH [Baseline (no CO_2 or SO_2 emission constraints), (H)igh Demand, (H)igh (increasing) fuel costs].

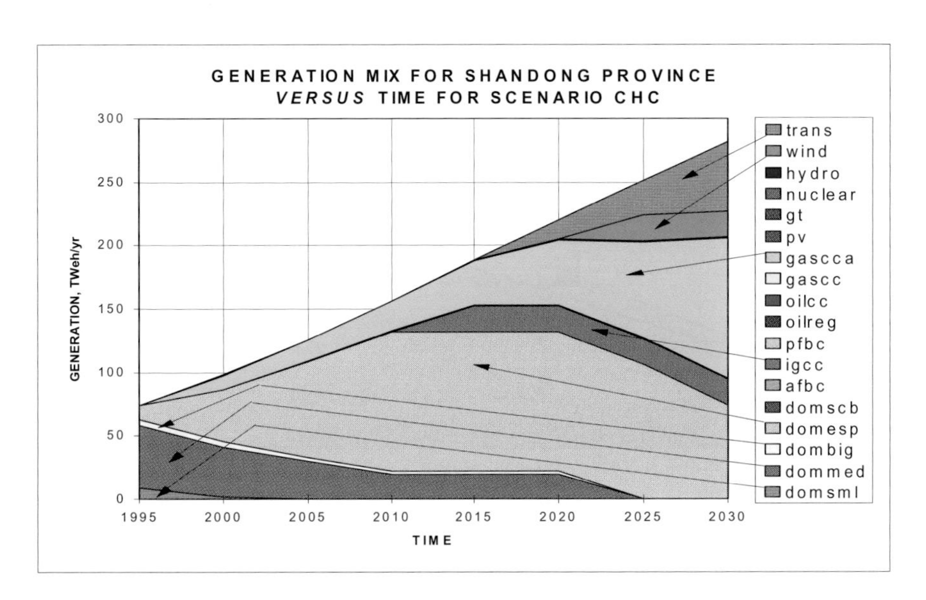

Figure 5.69C Generation mix for scenario CHC [CO_2 emission constraints via *caps), (H)igh demand, (C)onstant fuel costs].*

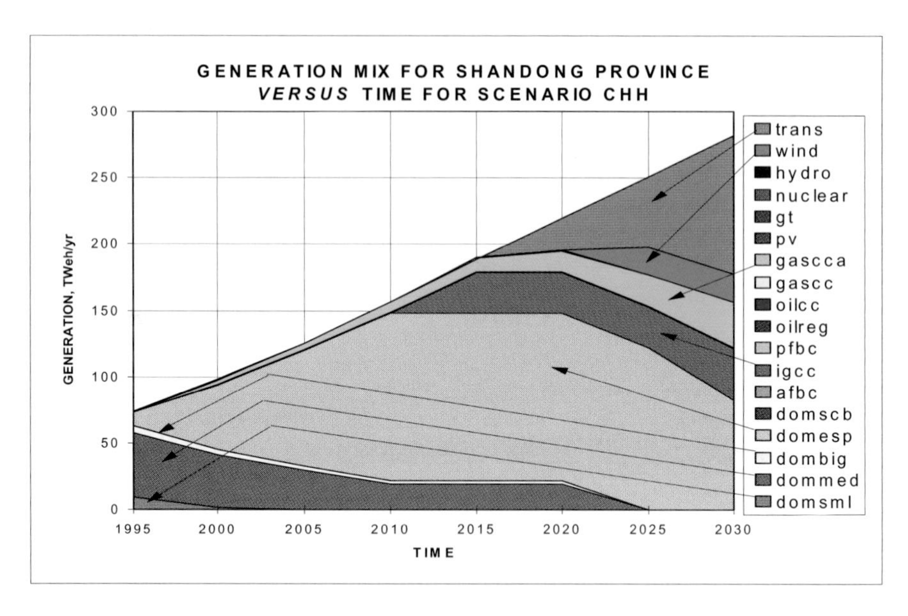

Figure 5.69D Generation mix for scenario CHH [CO_2 emission constraints via *caps), (H)igh demand, (H)igh (increasing) fuel costs].*

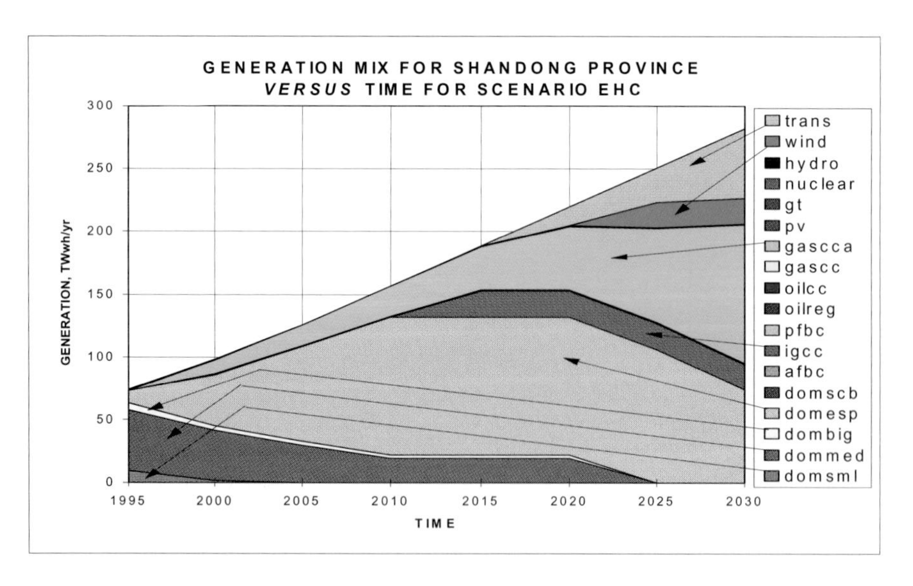

Figure 5.69E Generation mix for scenario EHC [CO_2 + SO_2 emission constraints via *caps), (H)igh demand, (C)onstant fuel costs].*

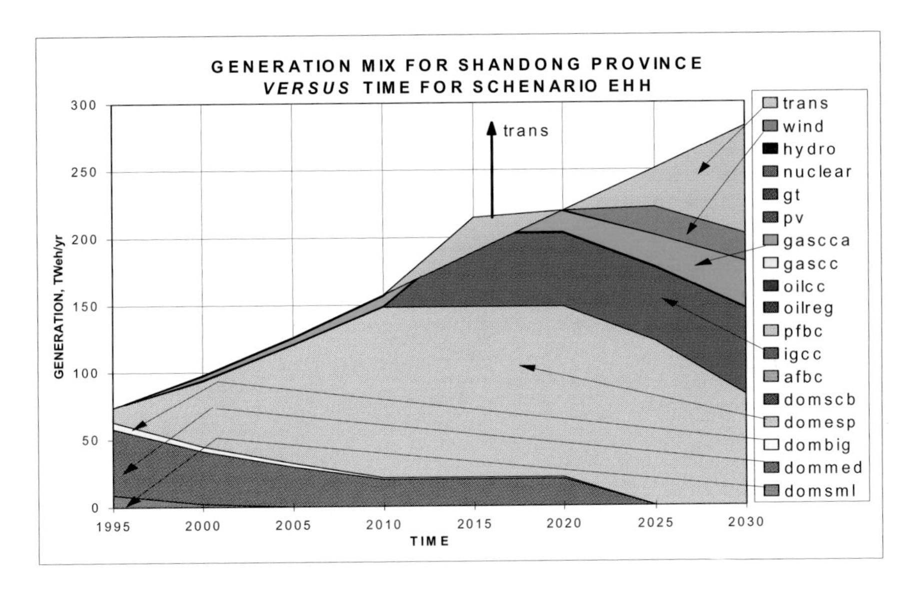

Figure 5.69F Generation mix for scenario EHH [CO_2 + SO_2 (E = S + C) emission constraints via *caps), (H)igh demand, (H)igh (increasing) fuel costs.*

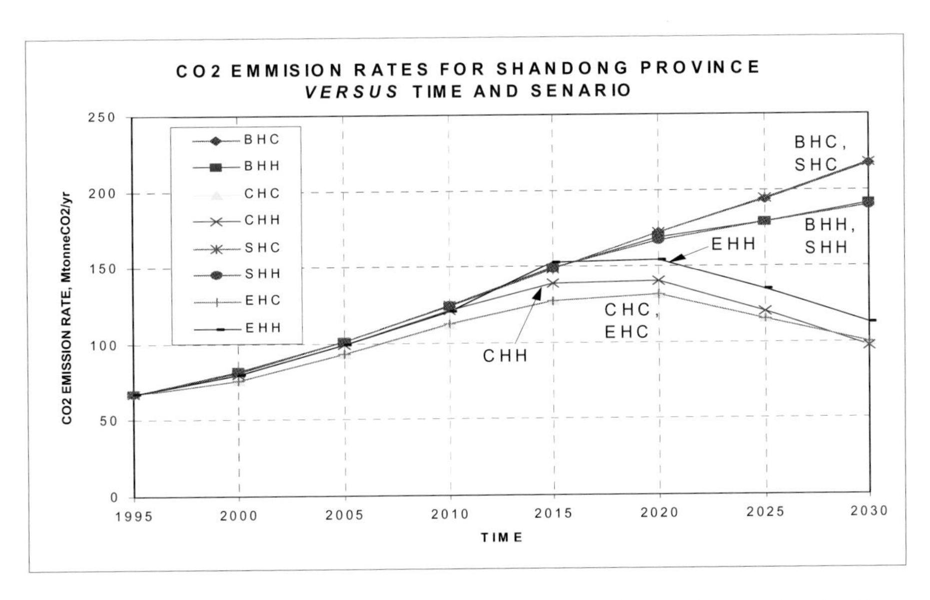

Figure 5.70A Comparison of CO_2 emission rates for each of the eight scenarios listed in Table 5.18 and described in Figure 5.69.

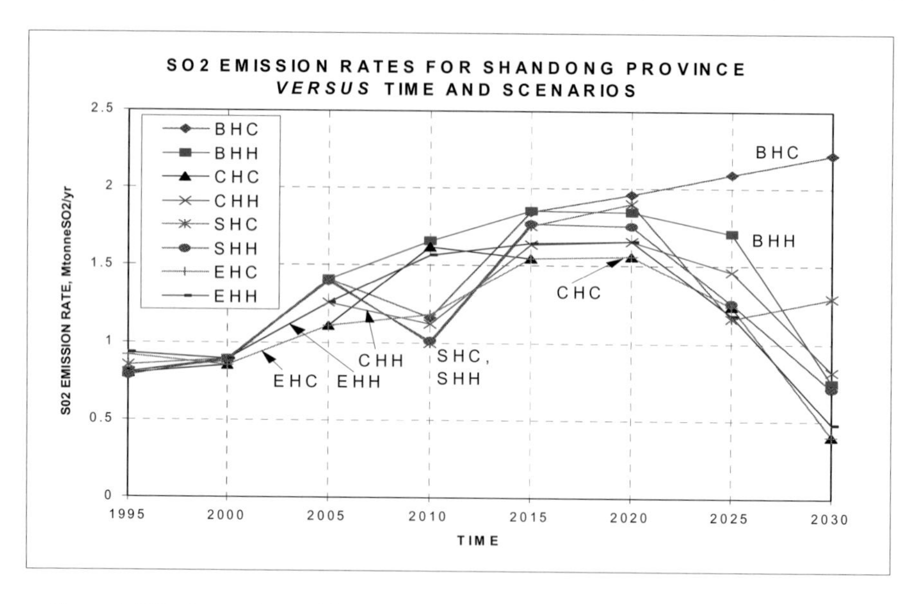

Figure 5.70B Comparison of SO_2 emission rates for each of the eight scenarios listed in Table 5.18 and described in Figure 5.69.

some of the constrained scenarios (namely SHH and EHH) seem to be less costly than the unconstrained case (BHH). However, the model optimizes total cost for China (and not for Shandong province): On a country-wide level these scenarios are more expensive. (Figure 5.65 shows the same results for China).

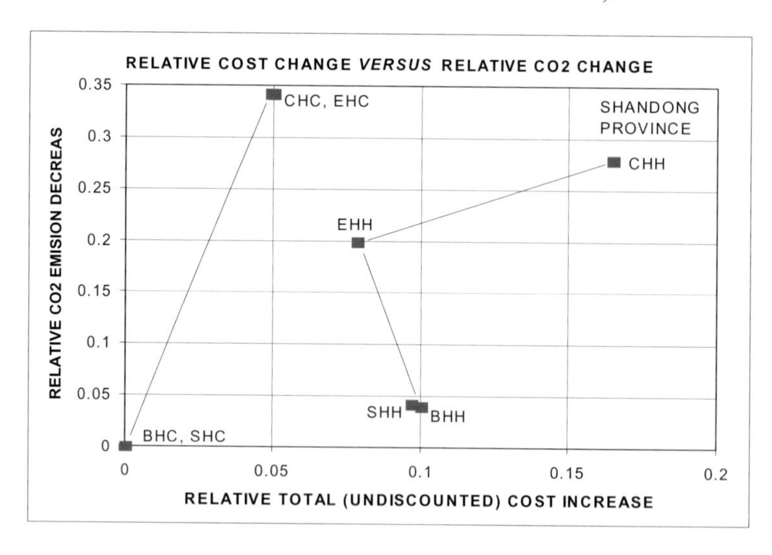

Figure 5.71A Relative undiscounted total energy cost versus *CO_2 emissions impacts for the eight scenarios examined: Shandong Province.*

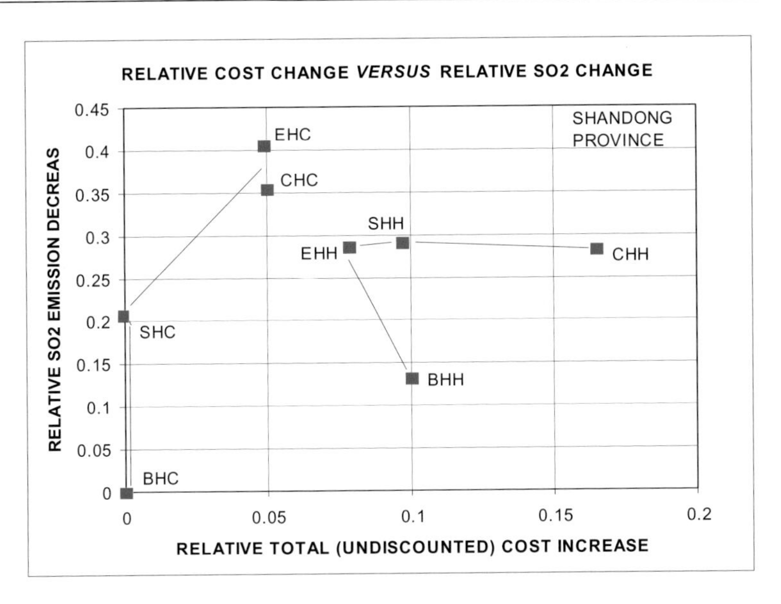

Figure 5.71B Relative undiscounted total energy cost versus *SO_2 emissions impacts for the eight scenarios examined: Shandong Province.*

6.2 MARKAL Results for Shandong

In parallel to the results given from CRETM scenario variations, Figure 5-72 gives similar correlations of CO_2 and SO_2 emissions with average (undiscounted) cost of electricity as projected from the MARKAL full-energy-sector model. The new and unique information that a total-energy-sector model like brings to the CETP, which has focused on the electrical generation sector, is relative importance of key emissions associated with electricity generation compared to those emanating from the total energy system. The plot of CO_2 emissions over SO_2 emissions shows that for the majority of cases these emissions are proportional to each other. Only in cases with a sulfur cap (circles, filled squares) or high taxes on SO_2 emissions (diagonal cross, open squares) the reduction of SO_2 emissions is significantly more than proportional. Figure 5.73 gives the fraction of CO_2 and SO_2 emissions incurred over the 1995-2030 computational timeframe for the range of scenarios considered in applying MARKAL to Shandong Province.

Note:
The following legend has been used for all frames in Figure 5-72:

The colors denominate demand and price levels for fossil fuels, whereas the symbols stand for different environmental policies:

■ Low Demand, High Fossil-Fuel Prices

■ Low Demand, Constant Fossil-Fuel Prices

■ High Demand, High Fossil-Fuel Prices

■ High Demand, Constant Fossil-Fuel Prices

▲ Baseline

■ S Cap

✱ S Tax 500 $/ t SO2

× S Tax 1000 $/ t SO2

□ S Tax 1500 $/ t SO2

◆ C Cap

\+ C Tax 50 $/t C

◇ C Tax 100 $/t C

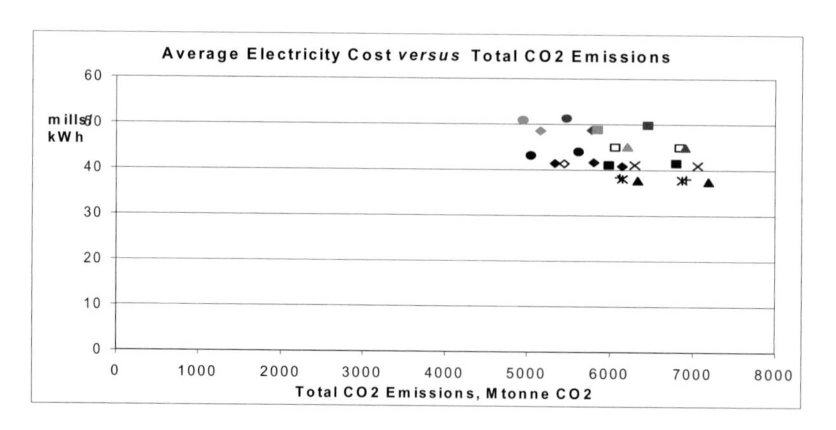

Figure 5.72A Average energy costs versus total CO_2 emissions for Shandong Province based on MARKAL.

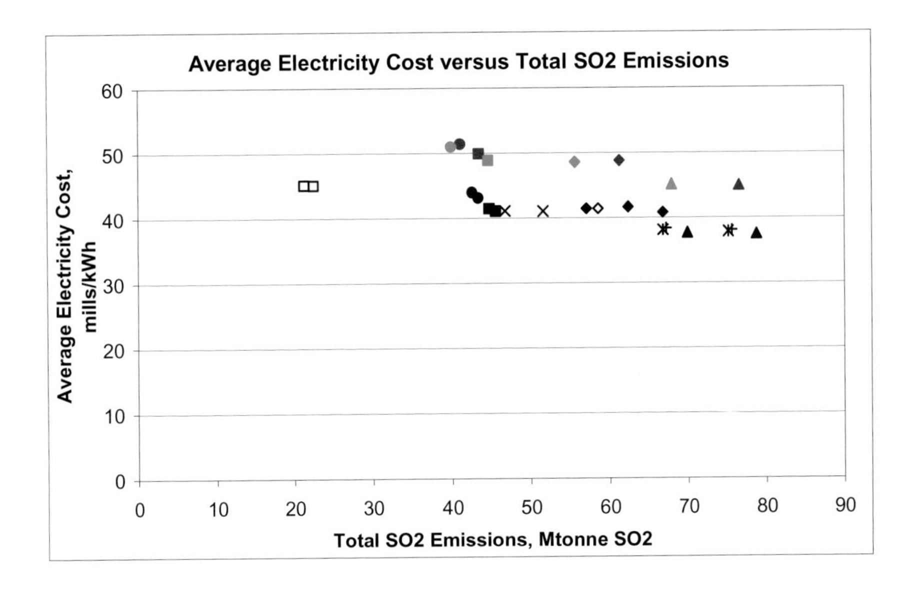

Figure 5.72B Average energy costs versus *total SO_2 emissions for Shandong Province based on MARKAL.*

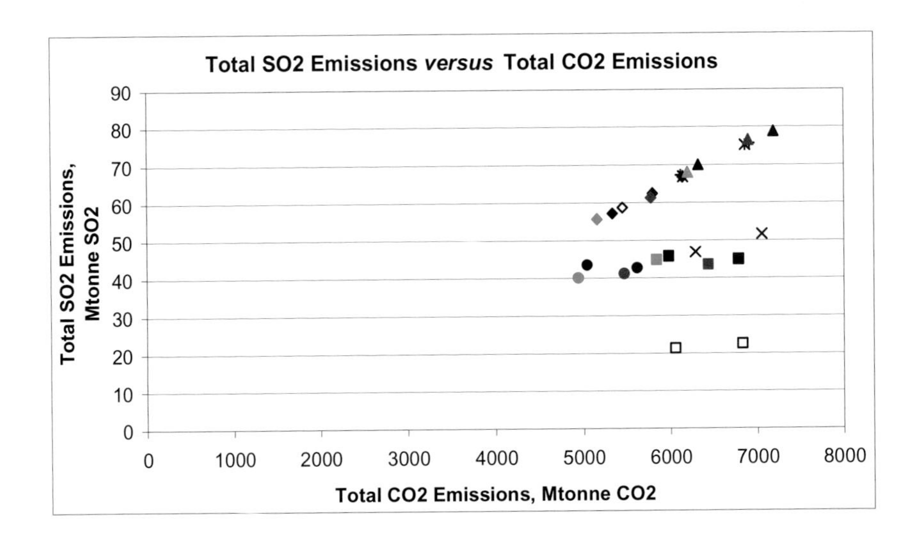

Figure 5.72C Correlation of total SO_2 emissions with total CO_2 emissions for Shandong Province based on MARKAL.

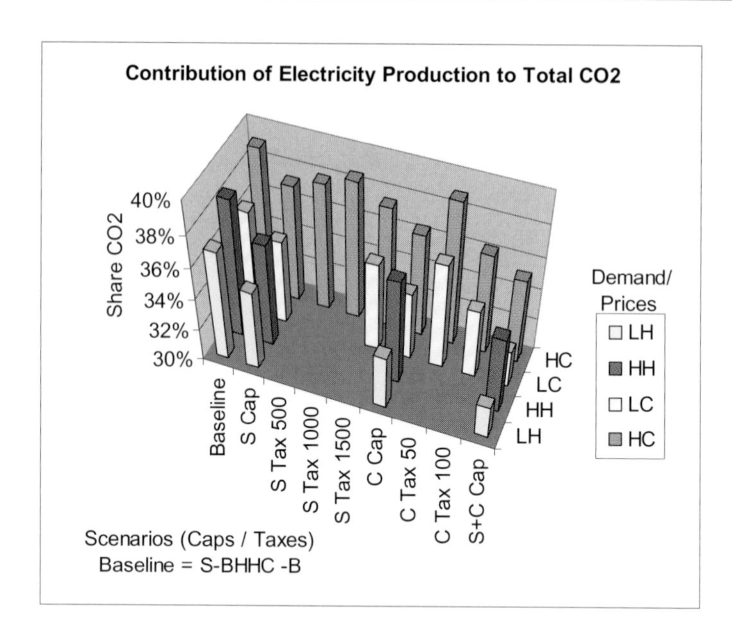

Figure 5.73A. Total CO_2 emissions from electricity generation in Shandong as a fraction of total emissions over the time period 1995-2030 for a range of policy-driven scenarios, as projected by MARKAL.

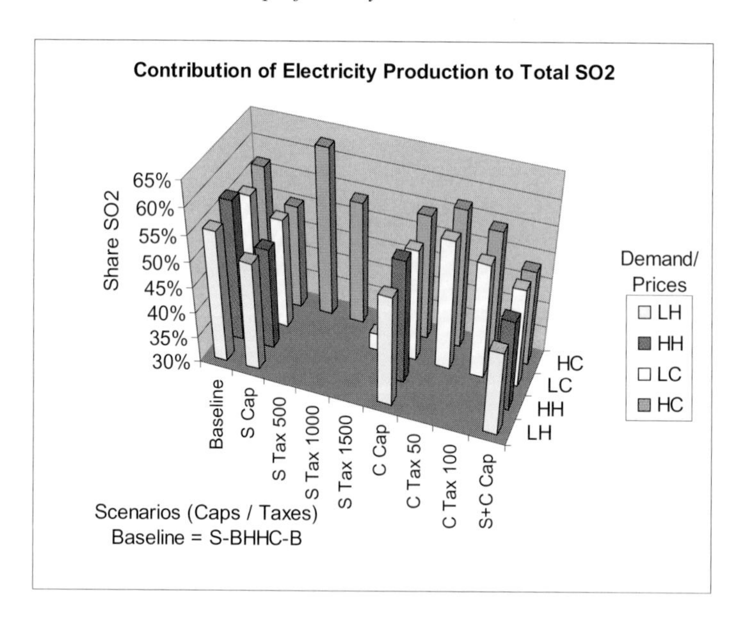

Figure 5.73B Total SO_2 emissions from electricity generation in Shandong as a fraction of total emissions over the time period 1995-2030 for a range of policy-driven scenarios, as projected by MARKAL.

6.3 Comparison with Previous Studies

Comparisons of projections made by other investigations carry with them the risk of generating false confidence. However, studies of similar detail and fidelity have been reported and policy recommendations issued there from. For that reason, leaders of the Project thought it was appropriate to compare their findings with those of the key forerunner studies--that conducted by the Battelle Memorial Institute (Chandler, 1998) and reported under the auspices of the Pew Foundation (Zhou, 2000). The two frames of Figure 5.74 give, respectively, the country-wide demands for electrical energy in each study for its respective baseline case and the resulting CO_2 emissions. Though they use different time scales (2015 for the PEW study, 2030 for the present study), these emission rates track both qualitatively and quantitatively the assumed demand, indicating that the energy mixes and specific emissions are in reasonable agreement.

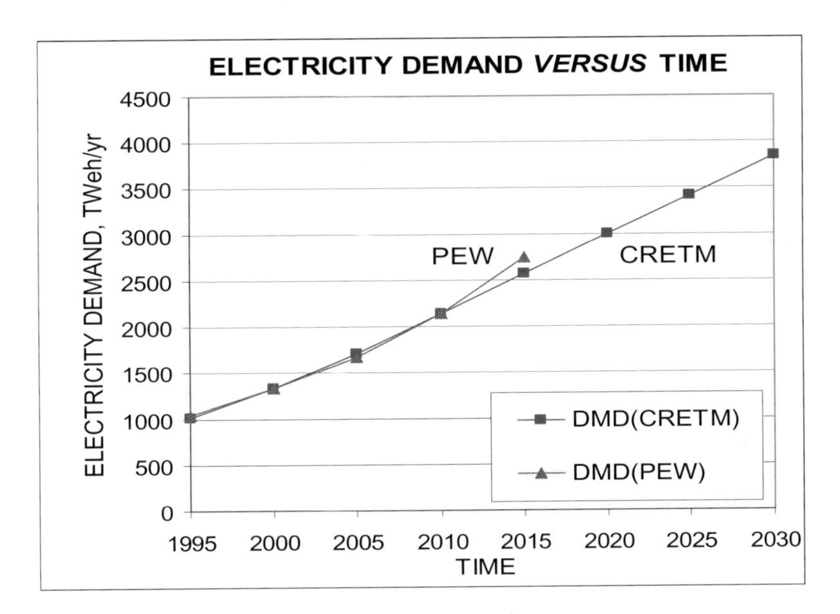

Figure 5.74A Comparison of base-case demand for electrical energy in China used by present study (CRETM) and the PEW study (Dadi, 2000).

6.4 Comparison with Results from within EEM Task (MARKAL and CRETM)

Figure 5.75 shows electricity demand in Shandong as projected in the CRETM and MARKAL models. Figure 5.76 compares the generation mix for Shandong Province for the BHHC scenario in a histogram. The MARKAL results give a richer mixture of fuel options for electricity generation, although both MARKAL and CRETM suggest the dominance of coal over next three decades.

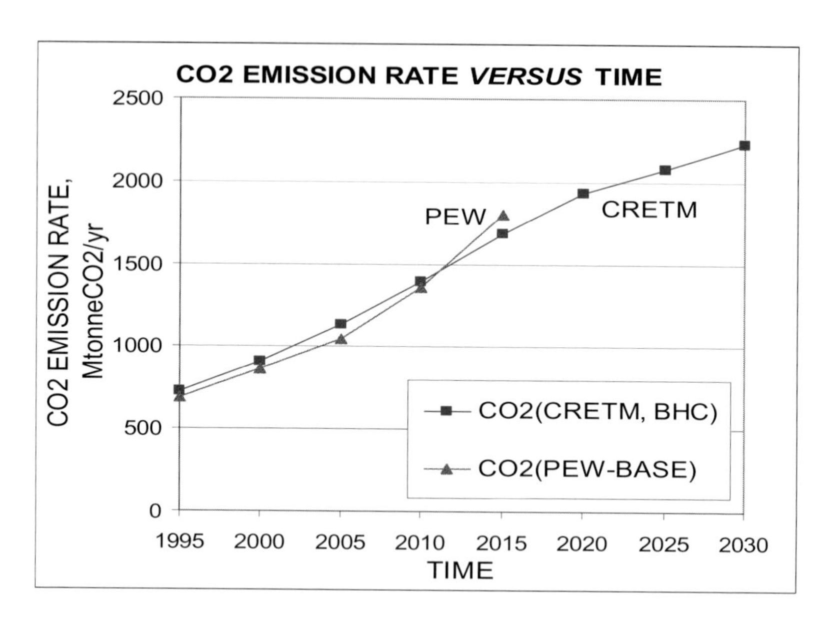

Figure 5.74B Comparison of CO_2 emission rates for base-case conditions between present study (CRETM) and the PEW study (Dadi, 2000).

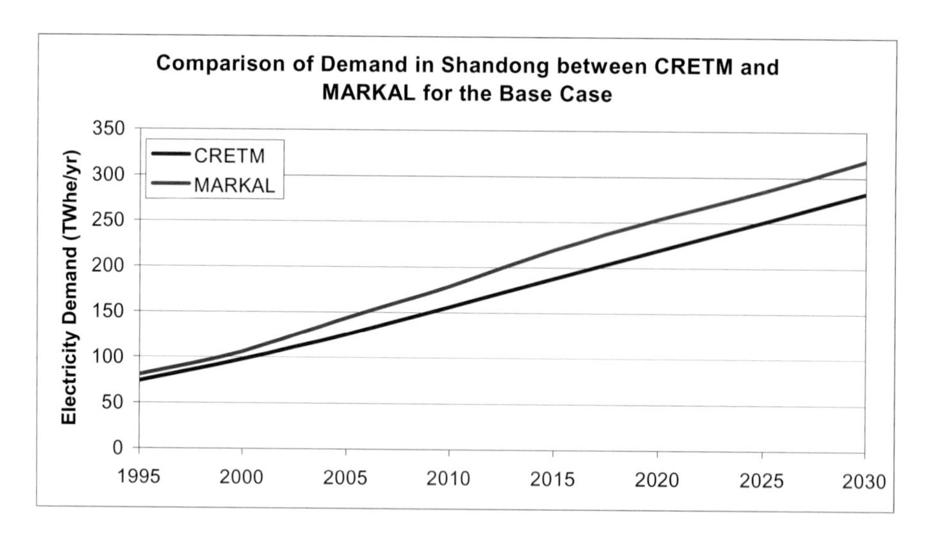

Figure 5.75 Comparison of electricity demands used to drive the CRETM and MARKAL models for Shandong Province. As discussed in the beginning of Sec. 5., MARKAL is actually driven by demands for final energy services, and the electricity demand projected above result from theses assumptions use for final and end-use energy demand.

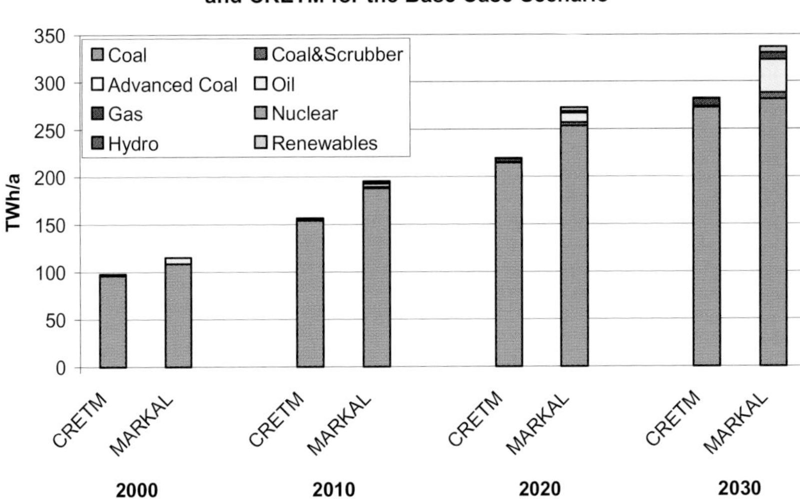

Figure 5.76 Comparison of generation mixes for the base case projected from the CRETM and the MARKAL models: histogram charts.

Figure 5.77 integrates the CRETM and MARKAL results for a wide range of policy and technology-driven scenarios on the basis of average (undiscounted) electrical energy costs *versus* combined CO_2 and SO_2 emissions for the period 1995-2030. The correlative CO_2 *versus* SO_2 plots are also shown. One can easily see that the unconstrained cases, i.e. the base cases in both models, are close together. With constraints, the cost increase in the MARKAL model is higher than in CRETM. This can be explained at least partly by the fact that in CRETM the reduction of emissions in Shandong is influenced by reductions in other regions as well.

7. FINDINGS, CONCLUSIONS, AND RECOMMENDATIONS

In this section, we bring the data above to bear on the actual process of policy making for Shandong's energy needs in the foreseeable future. How might they be met with the greatest efficiency for this vibrant and growing economy at the least cost for the environment? In this section we call out the most significant findings; draw some conclusions about the energy future of the province; and distill some recommendations for policy makers.

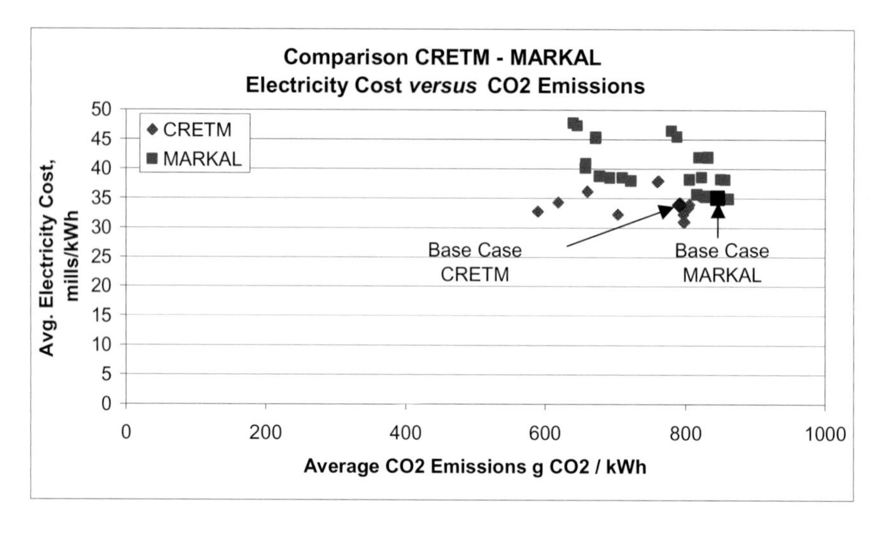

Figure 5.77A Comparison of CRETM and MARKAL multi-scenario results on the basis of average (undiscounted) electrical energy costs and total CO_2 emissions. The base cases of both models are highlighted.

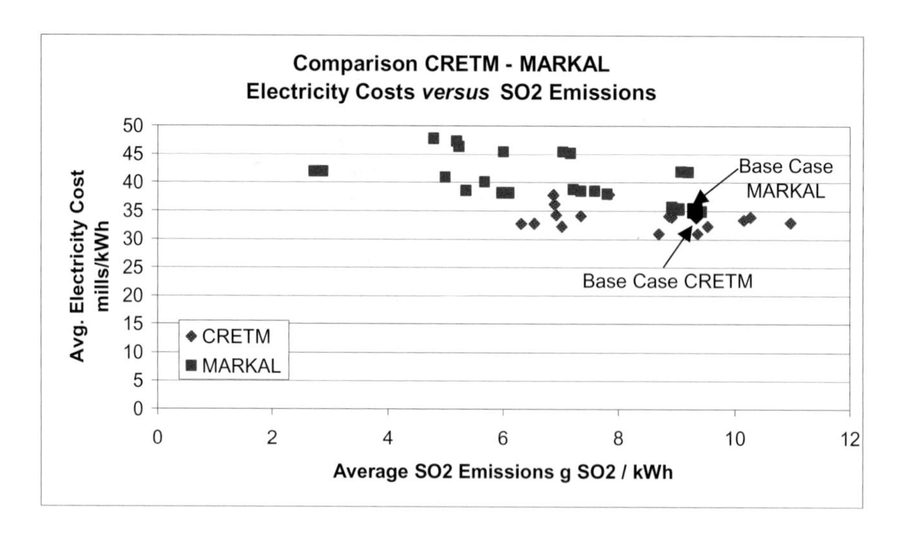

Figure 5.77B Comparison of CRETM and MARKAL multi-scenario results on the basis of average (undiscounted) electrical energy costs and total SO_2 emissions. The base cases of both models are highlighted.

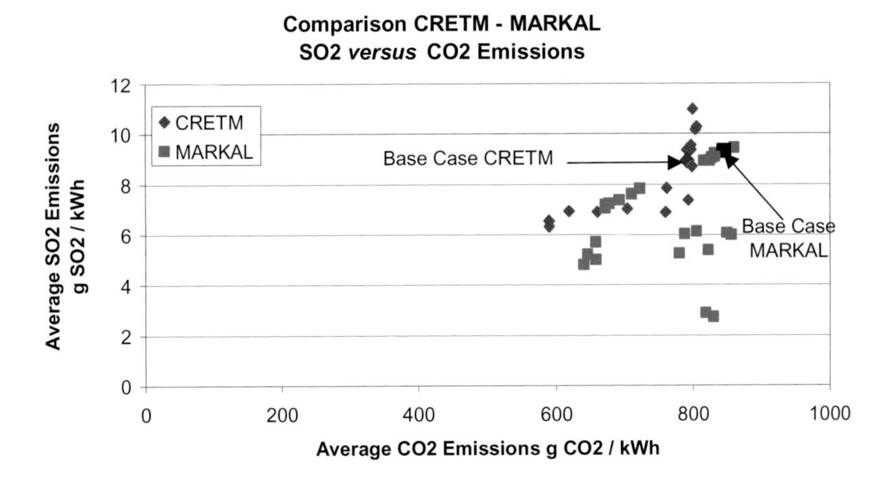

Figure 5.78C Comparison of CRETM and MARKAL multi-scenario results on the basis of correlation of total CO_2 emissions with total SO_2 emissions. The base cases of both models are highlighted.

7.1 Findings for Technologies and Emissions

The implications of the data are most pronounced for eight areas of inquiry:

1. Increased Power Demand: The demand for power in China will increase four-fold by 2030. This demand will reach, and maybe go above, the 4,000 terawatt-hours (TWh) by 2030, which represents a four-fold increase from 1995 levels of consumption. Similar levels of growth are expected in Shandong Province.

2. Pollution Control: According to a report compiled at the U.S. Embassy in Beijing (2000) and based on estimates of various Chinese and Western specialists, pollution costs China anywhere from three to eight percent of GDP annually. Air pollution and acid rain damages human health and buildings, and acidifies soil and lakes. Ecological damage is estimated to cost another potential five to 14 percent of the annual GDP. Even at the lower bounds of these estimates, environmental damage roughly cancels annual economic growth. When damages associated with sulfur emissions are internalized, sulfur control clearly becomes a cost-optimal policy. The CETP study concludes that installing sulfur-control equipment in new plants or introducing advanced coal technologies would be cheaper, by several factors, than the cost of pollution. Initiating policies with region-specific emission caps and/or a sulfur permit system, across regions and sectors, is the most efficient way to control future emissions in China and Shandong Province.

3. Clean Coal Technologies: Both CRETM and MARKAL have identified promising solutions for efficient power generation and control of sulfur emissions in China. These models indicate that conversion to clean technologies should begin by reducing local emissions through improving the performance of pulverized coal systems and introducing coal washing and/or sulfur scrubbing. Meanwhile, China takes the next step by building advanced coal systems like IGCC power plants and eventually supercritical steam coal. Furthermore, it is prudent to concentrate activities on the development of the systems that could solve many problems simultaneously, (i.e., efficiency improvement and sulfur control); IGCC is such a technology for China.

4. Chinese-Manufactured Technology: Advanced coal-fired systems can be competitive with conventional Chinese coal technology if manufactured in China or under conditions of high coal prices and transportation cost. Development sensitive to Chinese needs will be most effective in improving efficiency and will stimulate the transition to clean coal technology use.

5. Diversified Fuels and Technologies: It is essential to either diversify supply options or establish a rich portfolio of technologies. Oil and gas supply options must be improved to reduce dependency on coal, especially in the coastal provinces. The most promising substitutes for coal are advanced-gas combined-cycle systems plus an appropriate complement of large-scale hydropower, nuclear energy, and renewable wind- and small hydropower sources. A policy of supply diversification coincides well with international efforts to reduce carbon emissions in order to mitigate global warming. It provides an important argument in support of China's requests for technology transfer via the establishment of international networks and cooperation with the industrialized world.

6. Competitive Nuclear Power: Nuclear power can be competitive if reactors are fabricated in China, the construction time is under five years, and capital costs are at or below $1,500 per kilowatt for the high discount rates assumed (10%/yr). Even at moderate tax levels, nuclear energy is competitive at a five percent discount rate (or higher when regional and global externalities are addressed). This conclusion differs from other studies, in which researchers have overestimated fuel cycle costs (Chandler, 1998). The importance of nuclear energy becomes apparent only under a carbon control policy. Without a significant share of nuclear energy, the marginal costs of carbon control in China will cost more than can reasonably be paid. This would powerfully undermine the hope that China will participate effectively in international protocols for carbon emissions control

7. Natural Gas Markets: The development of natural gas markets could start with enhanced exploration and domestic production, extensive use of coal-bed methane resources, and, finally, with increased LNG imports. Competition against coal is not an easy undertaken. It is likely that such imports would first affect the coastal provinces; pipeline gas imports from Siberia and Kazakhstan could follow. Conditions for natural gas need to be improved to encourage exploration and the development of markets via marginal cost pricing. GCC power generation systems should also be encouraged, as their operation will help mitigate the high risk faced by investors in LNG port facilities and/or pipeline networks. While other users, e.g. providers of residential heat, need more time to build networks of consumers, GCC plants are able to consume large quantities of gas right away.

8. Commitment to RD&D Technology Development: China needs to establish robust research, development, and demonstration (RD&D) programs for advanced coal power generation with scrubbers and for alternative technologies such as wind turbines, IGCC and gasification processes, and nuclear fuel cycles for the fabrication of sensitive alloys. Improved RD&D is at best implemented through cooperation and joint capital investments with the institutions of industrialized countries.

7.2 Conclusions

Results from studies using the CRETM (seven-region optimization model of China's nation-wide electric sector, including Shandong Province)) and MARKAL (single-region optimization model of the total energy system for Shandong Province, including the electric sector) are based on the set of technical and economic data, price assumptions, and information about the availability of fossil fuel (indigenous and imported) combined with projections of the technical potential and performance of renewable energy technology. Despite the uncertainties that characterize these data and, therefore, cloud the results (e.g. optimal energy mixes, fuel compositions, emission rates, costs), it is possible to formulate some useful conclusions.

1. China will continue to rely on coal for power production, independent of environmental policies.
2. Chinese RD&D for advanced generation technology, along with integrated foreign investments, can improve energy efficiency and, through reduced emissions, the environment.
3. SO_2 pollution can be reduced at moderate cost by introducing scrubbers and/or advanced coal technology; some reductions in CO_2 automatically accompany SO_2 emission reductions.
4. Significant carbon emission reduction needs a considerable investments in reduced- or carbon-free generation technologies; this policy will improve local environments through reduces SO_2 emissions (secondary benefits).
5. The cumulative discounted cost of building a more sustainable path for electricity generation in China increases by two to six percent between 2000 and 2030 for the high discount rates assumed (10%/yr). However, these added costs are less than that of the damage that will be caused by pollution in China during these years.
6. The non-discounted cost differences that appear in the scenarios are high; they will and reach levels of around 20 to 30 percent by 2030.
7. Transmitting electricity between regions within China is a viable economic option, according to the CRETM analyses. Interregional transmitting also reduces local pollution.

7.3 Recommendations

Given these findings and conclusions, it is possible to recommend several options for policymaking and the advancement of modeling for energy systems.

Suggestions for important directions in formulating long-term energy policy for China include:

1. Improvements in the performance of clean coal technology;
2. Introduction of scrubbers and IGCC systems;
3. Diversification of fuel supply to include a larger component of gas, nuclear energy, wind, and small hydroelectric power;
4. Continued reforms to encourage the use of market forces, especially in the gas and electricity sector;
5. Introduction of demand-side management and conservation options as complementary means of stimulating the development of new supply technologies for the needs of an economically growing China and for maintaining its excellent record of efficiency improvement;
6. Participation in international protocols, which like Kyoto, commit signers to reducing carbon emissions (when China attains sufficient economic development) to facilitate technology transfer, international R&DD cooperation, and CDM projects.

The findings have implications for model development, which will facilitate more finely tuned projections of energy supply and demand in the Chinese context. On the basis of the development and evaluations experience so far accumulated by CRETM as part of CETP, important limitations and areas of improvement have surfaced. In addition to crucial interface areas between EEM, ESM, LCA, and MCDA modeling activities, important issues and areas of integration within the EEM task itself have emerged. Finally, within the CRETM itself, a number of developments and enhancements were identified as important work for the future. Clearly, the extensive effort needed to accomplish these tasks and goals cannot be accomplished within the temporal and funding limitations that presently define the CETP. However, it is useful to identify the work remaining to be done, whether or not it is accomplished within the Project, even if that work may not be made under the present CETP. Future work to refine current models might include further investigation of:

1. Fuel Transportation: The model presently includes flows of material based on modal flow-variables based on official Chinese statistics and the assumption that changes occur smoothly over time; capacities variables per se and load factors are not modeled. In any case, since it is not possible to model demand and supply for all products transported in China, constraints on the maximum load imposed by the transportation of fuels are legitimate.

2. Electricity Transmission: The CRETM transports energy by wire across regions under the assumption of average pricing and losses per kilometer and smooth growth.

3. Technology Diffusion: Advanced "learning-by-doing" R&D models need to be integrated into the determination of technology costs. While some progress has been made in this important area, the model used to achieve this goal requires better collaboration with Chinese institutions.

4. Partial Equilibrium: To endogenize electricity demand, advanced partial economic equilibrium models have been applied as a function of price feedback under conditions of an evolving electric generation energy mix; price elasticities for Chinese conditions must be elaborated in cooperation with Chinese Institutes.

5. Nuclear Energy: The accuracy of current models should be enhanced, particularly in relation to a) the dependence of fuel price on resource depletion, the flows of dangerous (environmentally and in terms of proliferation) materials for a given set of nuclear-fuel-cycle (NFC) assumptions; and b) the introduction of newer technologies to deal with (perceived or real) proliferation, waste, and/or safety issues so that these might reflected more accurately in all cost components (e.g., capital, variable, fixed, decommissioning, transmutation, waste disposal, etc.) (Trellue, 2000; Brogli, 2001). The growth-rate constraints now enforced (8%/yr, or ~50 percent increase in each five-year period) needs to be reconciled with both technical and financial limits. The capability to include advanced NFC modeling methods and data bases exist, but the resources within the present CETP are not sufficient to apply them.

6. External Costs: Environmental externalities estimated in the CETP project (Chapter 9) should be explicitly included in the objective function used in CRETM. For that reason particulate material and NO_x that are not yet treated, at either source or consequence levels, must be included. Knowing the damage cost as a non-linear function of emission levels can help integrate the results of the Environmental Impact Assessment group into CRETM analyses.

7. Other Renewable Energy Sources: Constraints on the availability of land, competition, and “footprint” issues, and factors in the distribution, peaking impacts, capacity, and storage of renewable energy sources require better modeling methods. More realistic assumptions need to be determined. MARKAL-style balances would be useful here.

8. Hydroelectric: Cost and licensing issues related to big versus small hydroelectric generation need to be addressed, particularly as evolving regional and cultural differences evolve in response to, e.g., environmental, economic (longer construction periods), and adverse public opinion. For example, the trend in South America is towards the construction of smaller hydroelectric installations, while China is pursuing a large-dam policy. Up to now, China’s economy has been only affected by global trends and protests on the part of populations affected by construction have not been a strong feature of the culture.

The CRETM offers the advantages of a fast-running, relatively transparent E3 optimization model that describes a regionalized China. Its main limitation is the narrowness of the energy sector examined--electricity production and consumption only. Nevertheless, within this limitation, province-specific analyses can be conducted and expressed in a country-wide context. With the improvements suggested above, the CRETM can be applied with great policy and planning utility to a range of Chinese provinces, once the experience of the Shandong study has been gathered, integrated, and advertised. Furthermore, through the application of user-friendly graphical-user interfaces (GUI), the CRETM can be put at the disposal of policy makers through a PC/laptop platform. Through this technology, this flexible and standardized modeling tool can provide information and policy analyses directly to provincial stakeholders anywhere in China. While many of the tasks listed above are crucial to a more effective application of these model results to the generation of useful energy policy, most of these items fall outside the budget presently allocated to the EEM task.

The MARKAL model can be reformulated for a more detailed and regionalized analysis of the Chinese energy system, as the CRETM might be for the electricity sector. To do this one needs a better cooperation with the Chinese providers of sources of data. The work done for Shandong, especially the interface model between the MEDEE-S and MARKAL, can be easily extended to include all the Chinese regions. This accomplished, the availability of fuel resources and infrastructure, renewables, demands, and technology performance can be specified on a regional scale. The MARKAL family of models could play an important role in analyzing the development of a growing China on a regional scale and derive conclusions at the country-wide level. ERI and QU could play a key role by regionalizing demand projections and specifying technologies now in place and recommended for future installations for China.

REFERENCES

Baker, J. B. (1999), *China Energy Technology Program*, Alliance for Global Sustainability report, Massachusetts Institute of Technology Energy Laboratory, Cambridge, MA (June, 1999).

Barreto, L. and S. Kypreos, (1999), *Technological Change in energy Models: Experience and Scenario Analysis with MARKAL and the ERIS Model Prototype*, Paul Scherrer Institute report Nr. 99-08 (September 1999).

Brookes, A., D. Kendrick, A. Meeraus, R. Raman, (1998), *GAMS: A User's Guide*, GAMS Development Corp., Washington, DC (December 1998)[1]

Brogli, R. and R. Krakowski (2001) Proliferation and the Civilian Nuclear Fuel Cycle: Towards a Simplified Recipe o Measure Proliferation Risk, Paul Scherrer Institute report Nr. 01-10 (August 2001).

Chandler, W., Guo Yuan, Jeffery Logan, and Zhou Dadi (1998), *China's Electric Power Options: An Analysis of Economic and Environmental Costs*, Battelle Pacific Northwest Laboratory report PNWD-2433 (June, 1998).

Chen, W. (2000), private communication, Paul Scherrer Institute (March, 2000).

Dong, Fang, Li Ping, D. Lew. D. M. Kammen, and R. Wilson (1998), Strategic Options for Reducing CO_2 in China: Improving Energy Efficiency and Using Alternatives to Fossil Fuel, in *Energizing China: Reconciling Environmental Protection and Economic Growth*, edited by M. B. McElroy, C. P. Nielsen, and P. Lydon (Eds.), Harvard University Press, Cambridge MA , p. 119-166.

Fishbone,. G., and J. Abilock (1981), *MARKAL, A Linear-Programming Model for Energy Systems Analysis: Technical Description of the BNL Version*, Energy Research, **5**, 353-375 (1981).

Goldstein, G. A. (1995), *MARCAL-MACRO: A Methodology for Informed Energy, Economic, and Environmental Decision Making*, Brookhaven National Laboratory report BNL-61832 (May 16, 1995).

Huntington, S. P. (1996), *The Clash of Civilizations and the Remaking of World Order*, Touchstone Books, London .

ILOG, Inc. (1997), *Using the CPLEX Callable Library, Including Using the CPLEX Base System with CPLEX Barrier and Mixed Integer Solver Options*, Version 5.0 Incline Village, NV .

Kann, A., and J. P. Weyant, (2000): Approaches for Performing Uncertainty Analysis in Large-Scale Energy/Economic Policy Models, *Environment Modeling and Assessment*, **5**, 29-46 .

Kypreos, S. (1996), *The MARKAL-MACRO Model and the Climate Change*, Paul Scherrer Institute report 96-14.

Kypreos, S. and R. Krakowski (2002), *Electrical-Generation Scenarios for China*, Paul Scherrer Institute report Nr. 02-08 .

Levibe, M. D., F. Fang, J. Sinton, X. Xi. N. Martin (1994), The China Project: Energy Efficiency and Policy Studies, *Energy Analysis Program 1993 Annual Report*, Lawrence Berkeley Laboratory report LBL-35240.

Logan, J., and Zhang, J. (1998). Powering Non-Nuclear Growth in China with Natural Gas and Renewable Energy Technologies. China Environment Series 2, Summer 1998. Retrieved from: http://ecsp.si.edu/ecsplib.nsf/6b5e482eec6e8a27852565d1000e1a4c/54b8301f5e346cf08525664b007 16a08?OpenDocument

Logan, J. S. and Dongkun L., (1999) *Natural Gas and China's Environment*, Presentation at the IEA-China Natural Gas Industry Conference, Beijing.

McElroy, M. B., C. P. Nielsen, and P. Lydon (Eds.) (1998), *Energizing China: Reconciling Environmental Protection and Economic Growth*, Harvard University Press .

Murray, E. and P. Rogers (1998), Living with Coal: Coal-Based Technology Options for China's Electricity Power Generating Sector, in *Energizing China: Reconciling Environmental Protection and Economic Growth*, edited by M. B. McElroy, C. P. Nielsen, and P. Lydon (Eds.), Harvard University Press, Cambridge MA , p. 167-200.

Nakicenovic, N., A. Gruebler, and A. McDonald, (1998), *Global Energy Perspectives*, International Institute for Applied Analysis and World Energy Council report, Cambridge University Press Cambridge, UK .

Ni, Weidou and Nien, Dak Ze (1998), Energy Supply and Development in China, in *Energizing China: Reconciling Environmental Protection and Economic Growth*, edited by M. B. McElroy, C. P. Nielsen, and P. Lydon (Eds.), Harvard University Press, Cambridge MA , pp.67-118.

OECD/IEA (2000), *China's Worldwide Quest for Energy Security.*

Rogers, P. (1999), China Energy and Environmental Planning Model, Harvard University, Division of Engineering and Applied Sciences, personal communication .

Schenler, W., A., V. Gheorghe, S. Connors (1998), P.-A. Haldi, and S. Hirschberg, *Strategic Electric Sector Assessment Methodology Under Sustainable Conditions: A Swiss Case Study*, Alliance for Global Sustainability report .

Trellue, H. and R. A. Krakowski (2000), *A Multi-Objective Nuclear Fuel Cycle Optimization Model: Interim Results*, Los Alamos National Laboratory document (in preparation).

Xue, B. and Eliasson B., (1999), ABB Corporate Research Ltd. Report, *China (Shandong Province): Energy and Emissions, Status, Statistics, and Prospects.*

ZHANG, C. , M. M. May, and T. C. Heller, (2001), Impact of Global Warming on Development and Structural Changes in the Electricity Sector of Guangdong Province, China, Energy Policy, **29**, 179-203

ZHAO, S. (2000), Electricity Lights 10% Growth, *China Daily* ,13 August 2000.

ZHOU, D., Yuan, G., Yingyi, S.,Chandler, W., and Logan, J., (2000), *Developing Countries and Global Climate Change: Electric Power Options in China*, Pew Center on Global Climate Change Report.

NOTES

[1] The expression for the specific cost as a function of cumulative capacity, i.e., the learning curve, is defined as: $SC(C) = a * C^{-b}$; With SC being the specific capital cost; C the cumulative capacity; b the learning index; and a the specific capital cost of the first unit.

[2] Demands (relative to the reference case) depend on the relative economic growth and the relative change in the marginal cost of electricity: $D = Y^{a} * P^{-e}$; with a and e being the income and price elasticity, respectively.

CHAPTER 6

ELECTRIC SECTOR SIMULATION: A TRADEOFF ANALYSIS OF SHANDONG PROVINCE'S ELECTRIC SERVICE OPTIONS

STEPHEN R. CONNORS, WARREN W. SCHENLER
CHIA-CHIN CHENG, CHRISTOPHER J. HANSEN,
AND ADRIAN V. GHEORGHE

1. INTRODUCTION

Three key aspects in transitioning to a sustainable energy future are technology development, deployment and use. This part of the China Energy Technology Program uses electric industry simulation models to explore the deployment and use of numerous electricity supply and end-use options, which could provide Shandong Province with cheaper and cleaner electric service. To do this the Electric Sector Simulation (ESS) team uses the Scenario-Based Multi-Attribute Tradeoff Analysis approach developed at MIT's Laboratory for Energy and the Environment, and employed in previous Alliance for Global Sustainability projects (Schenler, et. al, 2000 and 2002). With the assistance of the analysis team, the tradeoff analysis approach allows stakeholders to look at a broad range of options and uncertainties and compare the performance of the resulting strategies. This "inclusive"–and therefore extensive–analytic approach was developed to help multi-stakeholder groups jointly evaluate combinations of options that meet their collective interests.

As well as being complex, future energy infrastructures must be designed for a highly uncertain future. Good short-term forecasts, especially for an export-oriented province like Shandong, are extremely difficult to make; long-term ones are impossible. They are also dangerous to bet one's long term infrastructure decisions on. By looking at the performance of multiple-options across multiple uncertainties, tradeoff analysis helps identify robust long-term strategies, which perform well across a range of futures. By looking at common aspects of the better performing strategies, Chinese decision-makers can determine which actions to take.

Based upon the input and interests of the CETP's Stakeholder Advisory Group, the ESS team constructed a set of over a thousand strategies and looked at their costs, investment requirements, fuel consumption and power plant emissions for eighteen combinations of electricity demand growth and possible future fuel costs.

The results are dramatic. Shandong Province has numerous opportunities to reduce its pollutant emissions at little or no impact on cost – relative to its historical course of action. These "cheap and clean" strategies require a rethinking of the problem, where old and new generation, and the growth in the province's need for electric service are managed in a coordinated fashion. Several of these options are currently being pursued by the Chinese government, however national initiatives to reform the power sector may weaken the government's ability to direct technology choice. Whether centrally planned or coordinated via market forces, to manage China's demand for electricity in an environmentally responsible manner Chinese decision-makers will need to "manage" their infrastructure. The results presented in this chapter hopefully offer some useful insights on what course to follow.

The first part of this chapter describes the tradeoff analysis approach and construction of the electric sector scenarios for Shandong Province. It then looks at which types of options were most successful at reducing pollutant emissions and costs, and why. After looking at how these options perform as part of an integrated strategy, twelve of them are then selected for more detailed analysis of their life cycle and environmental impacts and acceptability to Chinese decision-makers.

2. SCENARIO-BASED MULTI-ATTRIBUTE TRADEOFF ANALYSIS

The Electric Sector Simulation task employs the Scenario-Based Multi-Attribute Tradeoff Analysis approach developed at MIT. In the early 1980s, researchers at the then Energy Laboratory at MIT employed a multi-scenario approach to evaluate alternative sites for a capacity constrained New York City. Faced with a capacity shortage and regulatory stalemate in New England in the late 1980's, the MIT team extended and refined the multi-scenario approach, using it to facilitate a discussion among stakeholders in the New England electric policy debate. In addition to helping electric utilities, economic and environmental regulators, large customers, and consumer and environmental advocates discuss issues outside of regulatory proceedings, the research team was able to identify new opportunities for dealing with the region's long-term challenges under shifting utility and environmental policies. Since then it has been used in various locales, and extended further in Alliance for Global Sustainability projects looking at China's, Mexico's and Europe's energy alternatives.

2.1 Policy Relevant Research – Stakeholders and Scenarios

The tradeoff analysis approach was designed to conduct "policy relevant" research. The tradeoff analysis framework has been designed around a stakeholder dialogue, whereby the analysis team constructs a set of scenarios, and the attributes which measure a scenario's performance, based upon its discussions with regional stakeholders. The analysis team discusses with the stakeholder advisory group how it has "packaged" the group's collective interests and concerns into a scenario set, how it will be analyzing the scenarios, and the key assumptions they are making regarding the cost and performance of technologies, changes in electricity demand, fuel costs, regulations, etc., over time. The analysis team then analyzes the performance of the scenarios. The scenario team then presents the outcomes to the advisory group, and based upon a discussion of the results, a new revised and refined scenario set is developed which builds upon the knowledge generated in the previous set, and includes new topics, options, and uncertainties that may have arisen in discussion with the stakeholders.

During the course of its eight year run, the New England Project explored a wide, evolving range of issues and options, such as increasing demand for natural gas as a power plant fuel, the comparative performance of conservation, peak load management, wind and solar options, the costs and emissions implications of repowering old power plant sites, the emissions impacts of electric vehicle fleets, and sub-regional impacts of a cap and trade system for summertime nitrogen oxides (NO_x) reductions.

Figure 6.1 shows the general structure of the tradeoff approach with its iterative design and evaluation of scenarios in support of a discussion among diverse stakeholders. With competition in the electric industry, the number and types of stakeholders has expanded. In general, a simulation model is used, in this case an industry standard production costing model developed under sponsorship of EPRI in the United States.

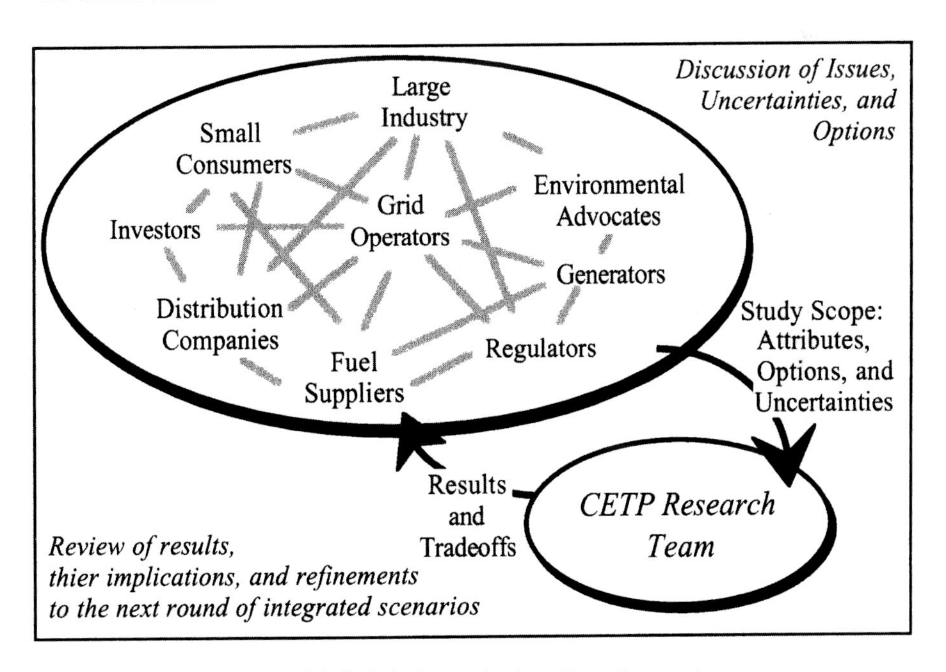

Figure 6.1 Stakeholder – Analysis Team Interactions

In order to maintain the integrity of the stakeholder dialogue, each stakeholder's concerns and interests, as well as favorite options–good or bad–need to be included in the scenario set. As will be shown later in this chapter, this leads to rather large scenario sets. For this a simulation model, usually a bottom-up engineering model, is used. Simulation models generally run faster than optimization models, since they perform fewer iterations before coming up with a result, allowing for a greater number of scenarios to be analyzed. In the electric power industry, many optimization models also choose which future technologies to build, and may be less credible with some stakeholders if they fail to select their "favorite" technologies. This can be compensated for in the simulation approach by adding more scenarios. Finally, if the optimization model uses a "utility maximization" approach, derived from the utility functions, or preferences of a user, the dialogue process can get bogged down. The "utility functions" of differing stakeholders cannot reasonably be combined into a single utility function for society, and even if they could the results from those simulations may not be credible to some stakeholders as the integrated utility function does not sufficiently reflect their interests. Furthermore, the intent of the tradeoff analysis approach is to help stakeholders identify novel strategies. Thus, their perspectives, and therefore preferences regarding certain technologies and policies will change over time making the original utility functions obsolete.

2.2 Tradeoff Analysis

As mentioned previously, one of the key uses of the tradeoff analysis approach is to identify robust strategies under uncertainty. Most optimization models implicitly "believe" the forecast to which they are optimizing the power system. Such models are best used for near-term, tactical assessments where deviations from the forecast will not be large. While more sophisticated optimization models may perform perturbations around a forecast, they have difficulty with "surprise," or "noisy" uncertainties, such as the Asian economic crisis, global conflict-oil price interactions, or a "step change" in regulations.

Figure 6.2 shows how the tradeoff analysis approach helps identify superior and inferior strategies. Here strategies are being evaluated for cost and emissions, usually one pollutant at a time. The group to the right reflects the initial set of scenarios performed for the stakeholder advisory group. By looking at the position of the scenarios, it is easy to identify strategies along the tradeoff frontier for which there are no strategies that are lower in *both* cost and emissions. Those on or near the tradeoff frontier are referred to as superior strategies, and those well back of the frontier are inferior or dominated strategies. Strategies not on the frontier for one pair of axis attributes may appear on the frontier for another pairing, or under another set of uncertainties, so care should used when identifying "reasonably good" strategies. Furthermore, strategies that may look just okay from a techno-economic viewpoint, through discussion with the stakeholders, may turn out to be better once political and implementation related factors are considered.

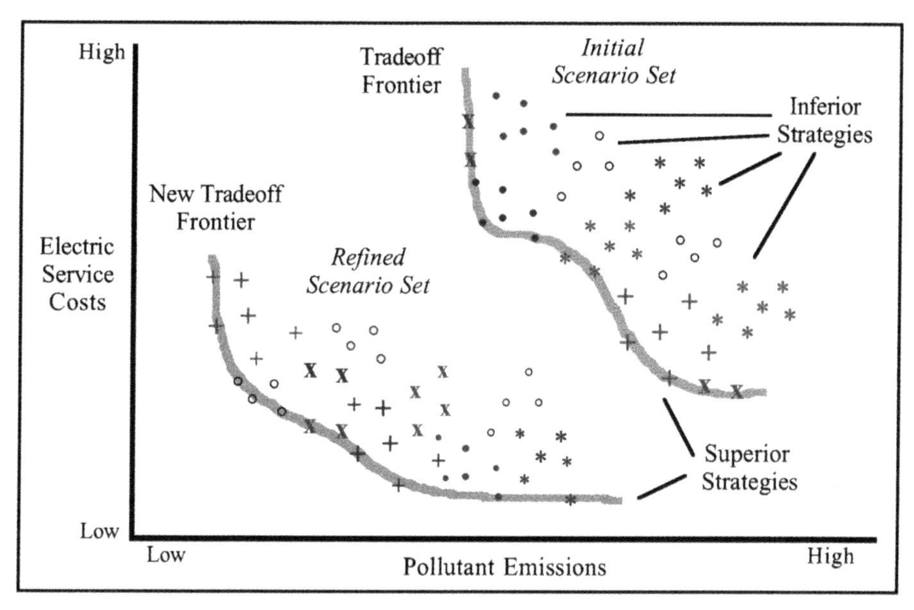

Figure 6.2 Identifying Superior Strategies

From this initial scenario set a second set of refined strategies is then developed and analyzed. Dropping the poor performing options or combinations of options from the scenario set, and introducing (hopefully) new and more refined options which moves the scenario set in the cheap-clean direction. Successive scenario sets can be analyzed until the group believes that all of the major opportunities for refinement, or additional issue exploration, have been evaluated.

Another important aspect of the tradeoff frontier is its shape. For the initial scenario set there are only a few strategies on the low-cost, higher emissions end of the tradeoff frontier. However for the refined scenario set, there are substantial slightly higher cost, substantially lower emissions strategies populating the frontier before it "turns up." Such transition points are important for stakeholders to know about, as decisions pushing emissions reductions beyond that point do so at considerable cost.

Two additional features of the tradeoff analysis approach should be noted. First, it is generally very difficult to give technologically prescriptive recommendations when looking at how to evolve an infrastructure out several decades. However, by looking at which strategies are consistently far from the tradeoff frontiers, researchers can inform stakeholders on what *not* to do. Second, in addition to identifying consistently bad combinations of options, the tradeoff plots show the overall range within which costs, emissions, reliability and other attributes can change.

Although stakeholders often have differing preferences, say for reducing costs before emissions, or visa versa, they generally agree that both should be reduced if possible. The tradeoff analysis approach allows them to see which combinations of options perform best. Stakeholders can then discuss which strategies on the tradeoff frontier might be the better long-term solution, rather than argue over favorite but inferior strategies. In contrast, optimization approaches by definition are intended to show the best one or two strategies, within the confines of the model's structure, leaving little bargaining room among stakeholders. The tradeoff analysis approach however informs stakeholders about the entire "option space," not just the portion of the tradeoff frontier intersecting a stakeholder's principal interests (utility function), or best guess circumstances (forecast).

Figure 6.3 shows the other dimension of the tradeoff analysis approach, identifying robust strategies. Here we see the performance of a group of superior strategies for three futures. Careful examination of the Figure shows that some strategies are on the frontier for one or two of three futures, but for the futures where they are not on the frontier, they are well back. Similarly, there are several strategies that are not on the tradeoff frontier in each case, but are nevertheless close to it in all instances.

So far we have been discussing the tradeoff analysis approach without really defining its components. Figure 6.4 illustrates the components, or "building blocks" of a scenario. In our nomenclature a strategy is comprised of multiple options, often grouped into option-sets targeting some aspect of the system, old power plants for instance. Likewise individual uncertainties are combined into futures. A scenario is then the combination of one specific strategy with one unique future.

Figure 6.4 illustrates a scenario set comprised of three option-sets, with a total of nine individual options that must be selected. If there is one alternative plus a reference option for each of these nine, then this set of strategies would be a combination of 2^9 options, or 512 unique strategies. Often however the number is smaller since options are linked, with one only being possible if another is selected, such as subsidizing natural gas costs only when natural gas power plants are built. A similar calculation for the uncertainties yields 2^5 or 32 futures. Combined there would be a little over sixteen thousand unique scenarios. Here the iterative nature of the tradeoff analysis approach helps deal with the combinatorial aspects of scenario formation. Certain options can be "held back," with stakeholder permission, to the next iteration, thereby making the process more manageable.

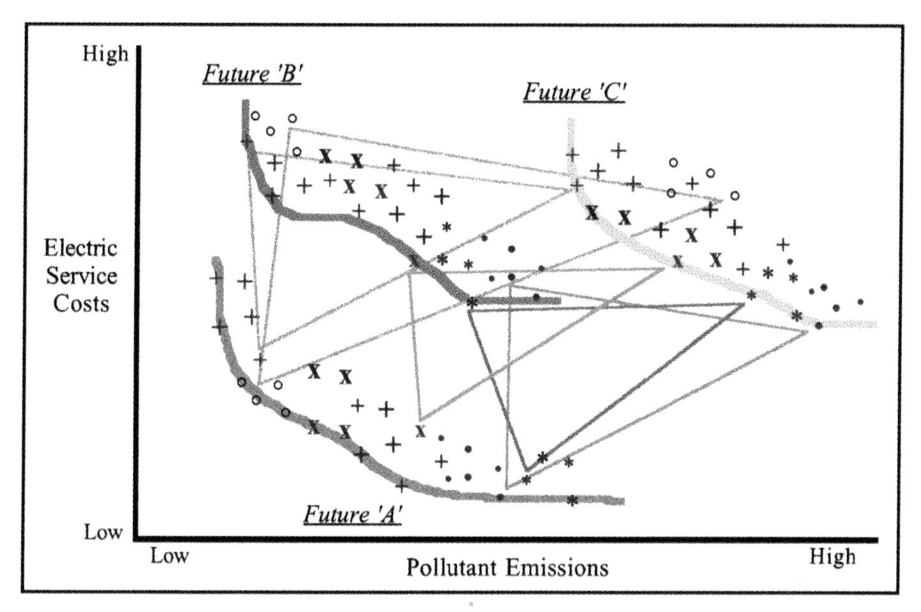

Figure 6.3. Identifying Robust Strategies

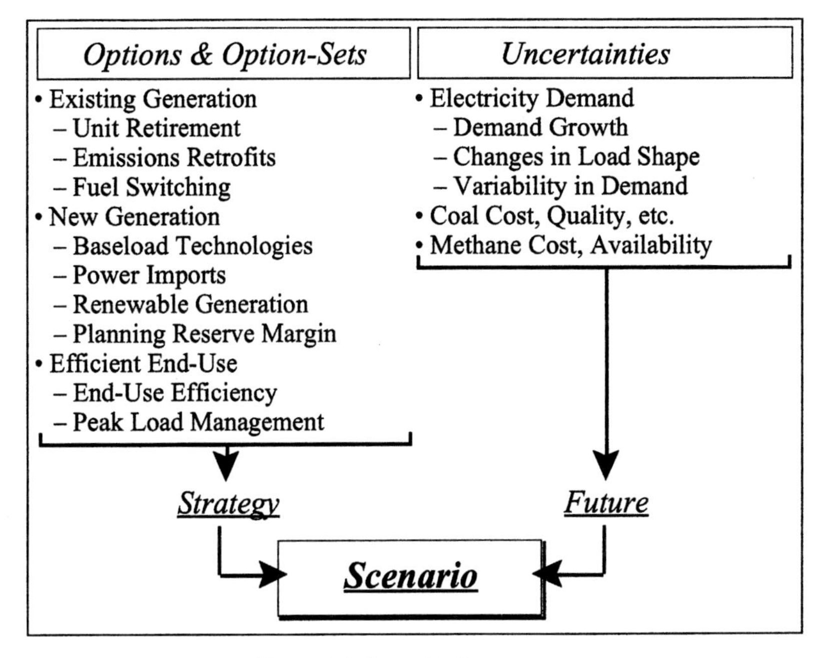

Figure 6.4 Scenario Components

Attributes are the metrics by which strategies are evaluated, becoming the axis of the tradeoff plots. Computer software allows analysts to perform real-time evaluation of tradeoff results with stakeholders, changing the attributes of the tradeoff plots, focusing in on specific groups of strategies, and then looking at specific scenarios in greater detail. For the Shandong scenario sets, the ESS analysis team set up the modeling to automatically calculate over two hundred attributes. While only a dozen or so of these were used in the tradeoff analysis, the remainder serve to help understand why the scenarios performed the way they did in terms of capacity utilization, fuel consumption, or peak load growth, for example.

As will be shown below, from a computational viewpoint the scenario is the unit of analysis. However, from a conceptual viewpoint, the strategy is the unit of observation. Through the iterative analysis of scenarios, stakeholders, with the analysis team's assistance, can design multi-option strategies that perform well across multiple futures.

One final aspect of the scenario-based multi-attribute tradeoff analysis approach addresses the issue of near-term actions in long-term infrastructure management. Strategies commonly look at which options a region should use to evolve its infrastructure out over several decades. Understanding the long-term path, and the actions needed to realize it over those intervening years is a highly valuable piece of knowledge. However decision-makers rarely lock-in such future decisions. They do not need to, and different decisions may need to be made as the future unfolds different from what was anticipated or hoped for.

One possible solution is to have stakeholders look at what common features, or sets of options the best strategies have. If there are similarities, then these can be considered robust options, and stakeholders, even if they cannot agree on an entire strategy, can at least agree that these options should be pursued. The differences among superior strategies reflect options that decision-makers should take steps to develop, but withhold implementation of until the right circumstances occur. In this light, a robust strategy is comprised of *both* robust and flexible options, and by recognizing what they are, a near-term tactical plan can be devised from several long-term strategies.

2.3 Shandong Building Blocks

Previous tradeoff analysis on electric systems in other regions have shown that there are many factors that must be considered in the construction of useful, informative scenarios. It is important to include options that impact nearly every aspect of the infrastructure. Figure 6.5 illustrates the diversity components for which options should at least be considered. Here we see infrastructure components for both the supply (generation) and use of electricity, including the existing infrastructure. Not pictured, but also potentially important are fuel and cooling water supplies, and regulatory and fiscal policies related to the investment in and use of various elements of the infrastructure.

The arrows between components indicate that a decision targeting one component often influences the behavior of other components. One example that will reappear below is energy conservation. By promoting new "clean" demand, fewer new "clean" generators are required. However, new "clean" generators displace the operation of old "dirty" generators. Strategies should not only be constructed to uncover these interactions, but address them. In this example, an option turning an old "dirty" generator into an old "clean" generator by installing a scrubber could be considered, in addition to other options targeting the deployment and use of new electricity supplies and end-uses.

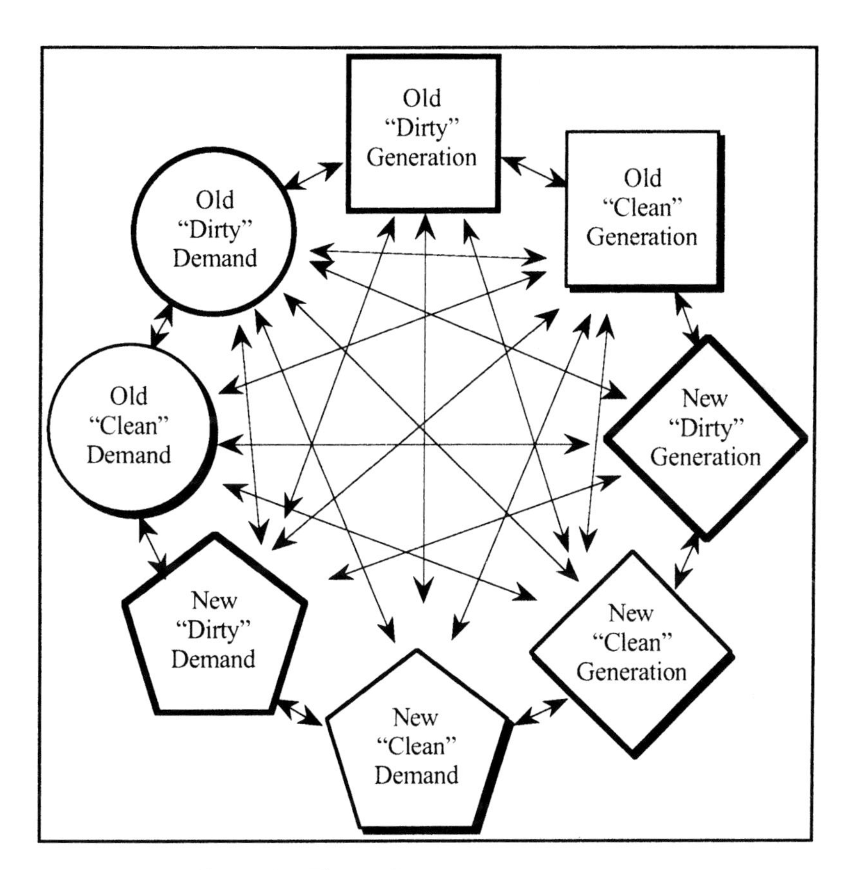

Figure 6.5 Electric Sector Resource Components

Table 6.1 shows the options sets and uncertainties incorporated into the Electric Sector Simulation group's Shandong scenario sets. Results reported here represent the third scenario set analyzed in the project as various options were dropped, refined and reintroduced into the set. Beside each category of option or uncertainty is the number of alternatives evaluated.

As the table illustrates, it is very easy to get a great many scenarios. Good organizational skills, reasonable simulation times, and computer graphics has made scenario-based tradeoff analysis a useful planning and educational tool. The following sections show how individual scenarios were analyzed, and the key assumptions used in putting together the scenarios for Shandong Province.

Table 6.1 Option-Sets and Uncertainties

Strategy Components	
Option Sets	*No. of Options*
Existing Generation Options	
Retire Additional Existing Units	2
Emissions Retrofit Existing Units with Sulfur FGD	2
Switch to Prepared Coal	3
New Generation Options	
Mix of Future Generation Technologies in Shandong	7
Extra-Provincial "By-Wire" Generation	2
Demand-Side Options	
Peak Load Management	2
End-Use Efficiency/Conservation	3
No. of Strategies	1,008
Future Components	
Uncertainty Sets	*No. of Uncertainties*
Economy/Demand Growth	
Demand for Electrical Energy	3
Fuel Cost and Availability	
Delivered Cost of Steam Coal	3
Delivered Cost of Pipeline Natural Gas	2
No. of Futures	18
Total No. Scenarios	18,144

3. ELECTRIC SECTOR SIMULATION

At the core of the tradeoff analysis approach is the analysis of individual scenarios. The above discussion focused on the “framework” for analysis, built around the dialogue with stakeholders. Figure 6.6 shows the overall process for how scenarios were analyzed on behalf of the CETP’s Stakeholder Advisory Group. The scenarios outlined in Table 6.1 require changes to the model inputs, lower electricity demand resulting from end-use efficiency programs for example. In addition, various options may alter the parameters by which the model runs, a cap on emissions for example that will shift the utilization of system resources. After this “data management” and option construction phase is completed, the scenarios are ready to run. Two core models comprise the ESS’s analytic implementation of the tradeoff analysis approach. The Pre-Specified Pathway program takes load growth uncertainties, power plant retirements and preferences for new generation and chooses how many of each technology get built and when. This and other parameters are then passed to the power system simulator. In this project we used the EPRI EGEAS™ program, which given a set of loads, generators, and various costs and operational constraints determines the relative "dispatch" of the generators. These two models are explained in greater detail below.

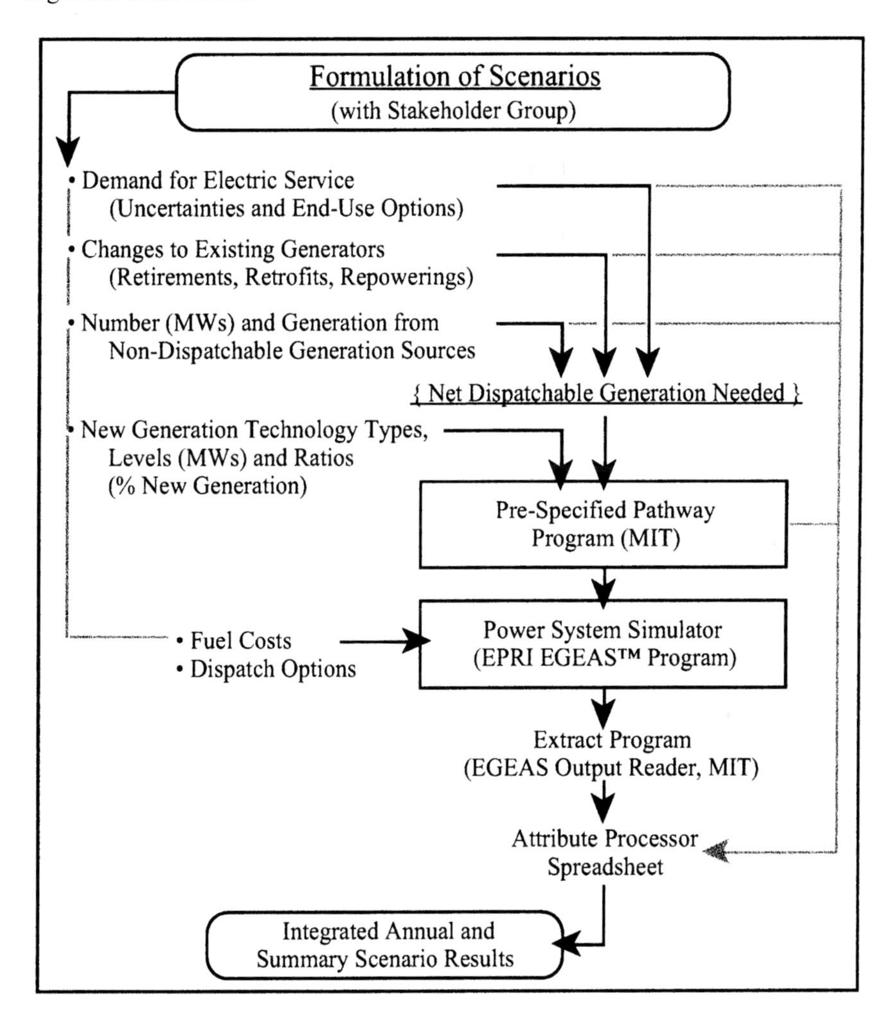

Figure 6.6 Analyzing a Single Scenario

Another program reads the output from the EGEAS simulation, making a condensed table of the simulator's results. These results are then dropped into the "Attribute Processor" spreadsheet, which recombines the simulator's output with other information such as the savings and costs of implementing end-use efficiency programs. The attribute processor also performs some additional calculations, such as the calculation of transmission and distribution costs as a function of electricity throughput, and finally calculates the scenario specific stream of attributes, which are then appended to the results from other scenarios and used to perform tradeoff analysis. The attribute processor also holds the year by year results whereby analysts and decision-makers can review in detail "what happened" in a specific scenario that makes its costs and emissions change. A more complete description of this methodology can be found in the SESAMS report on a previous AGS planning project done for Switzerland (Schenler, et. al., 2000).

3.1 Simulating Growth in the Electric Sector

Growth in the demand for electric service, and the construction of generators to meet that demand is inherently lumpy. These year-to-year variations in load and capacity growth can be attributed to changes in the economy and the weather, and due to the time it takes to permit and build new power plants. If short of capacity, smaller, shorter lead-time generators may be built. Similarly, if there is excess capacity, power plants under consideration may be canceled or postponed.

Early in the development of the tradeoff analysis approach it became apparent that a more "human" capacity expansion model was needed to produce more "realistic" trajectories showing the addition of new generators. While simulation programs like EGEAS had capacity expansion modules, it was found that they were "too good," and allowed the addition of new generators to meet demand, irrespective of their construction lead times. Such "optimal capacity expansion" modules also tend to build power plants to the exact forecast of load growth, when in fact power system planners must "guess" future capacity needs based upon past experience and current system conditions. To address these realities MIT researchers developed an "imperfect planning" program called the Pre-Specified Pathway (PSP) program. Figure 6.7 shows the PSP's overall logic.

Starting with known changes to the systems fleet of generators, including pre-determined unit retirements and power plants permitted and under construction, the program guesses the future need for capacity based upon past trends in load growth and the desired capacity reserve margin – the amount of extra installed capacity over anticipated peak load to account for unit outages and uncertainty in forecasting. The model then steps through the study period, in this case 2000 through 2024, and commits future units based upon the strategy's preferred mix of new technologies, the near term capacity situation, and the different lead times of those technologies.

Deviations from the preferred mix of technologies are allowed in order to meet anticipated near term capacity shortages. Technologies are given both a permitting and construction lead time, so that if excess generation has been ordered but has not "broken ground" it can be cancelled or deferred. The PSP model's output is the schedule of new technologies that are built over the scenario's time horizon and fed to the EGEAS simulator. When compared to historical demand growth and capacity expansion, the PSP provides a realistic trajectory of capacity additions, given the different lead times of technologies and the "noisiness" of electricity demand growth.

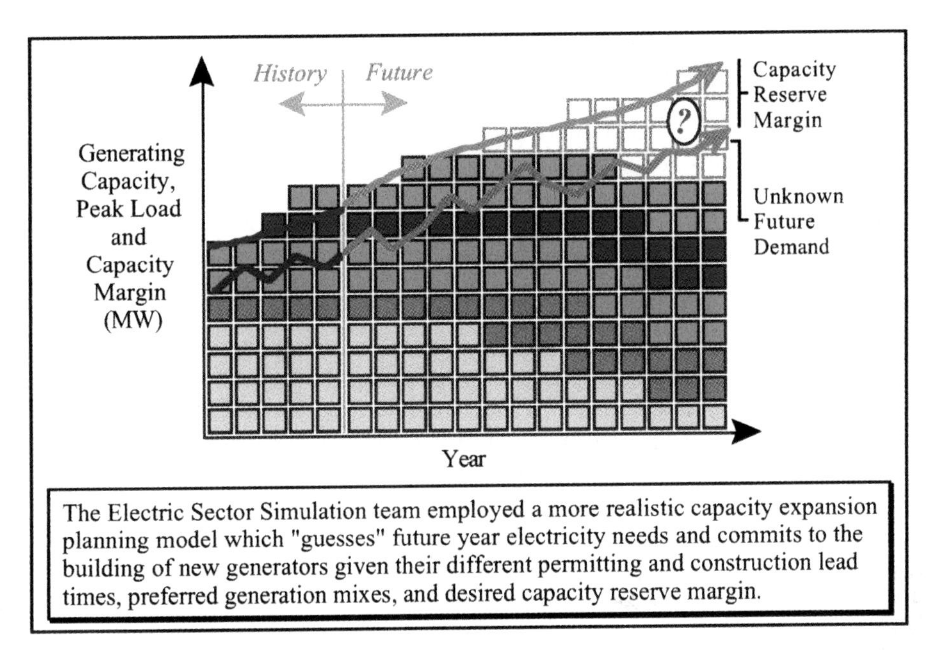

Figure 6.7 Sub-Optimum Capacity Expansion

3.2 Simulating Power System Operation

The simulation model used was the commercial program EGEAS (Electric Generation Expansion Analysis System). EGEAS was developed in the early 1980s by MIT and Stone & Webster (now a division of the Shaw Group) under contract with EPRI. The model offers numerous features including subperiod modeling, energy constrained unit dispatch, statistical modeling of non-dispatchable technologies, emissions constraints and shadow pricing, reliability analysis, sensitivity analysis and optimized expansion.

EGEAS is fed the output from the PSP program, as well as fuel costs and electricity demand uncertainties. It then decides which units run, and for how many hours, based upon system conditions, operational constraints and the units' relative operating costs. It also calculates the capital costs associated with the addition of new generators. This is critically important for estimating the costs and emissions impacts of changes to electricity demand and the mix of generation units. As will be shown below, if a new, cleaner power plant is added to the system it can displace more generation than an existing unit of similar size, by running at a higher capacity factor. Likewise, if for a given year the unit's operational cost are slightly higher, a clean unit may be used for fewer hours resulting in a smaller, or negative, impact on system emissions. The primary outputs of the model are generation utilization, fuel, operation and maintenance (O&M) and investment costs, fuel consumption and pollutant emissions. Other factors such as system reliability, aggregate generation efficiencies and other numerical results are also available.

4. OVERVIEW OF THE ESS SCENARIOS

Until now we have discussed the methodological approach and principal analytic components of the Electric Sector Simulation research, with only minor comments on how it was applied to Shandong Province. This section presents the scenarios and their key assumptions in detail before exploring their combined impacts on Shandong's electric sector.

Figure 6.8 shows a map of Shandong Province and the location of its larger, grid connected power plants. The circles indicating the location of generators are sized to their 1998 annual output if they are older (red) units. Recently completed generation plants (orange) and those under construction (blue) are shown with their circles sized to a 65% annual capacity factor, except for the Taian pumped storage power plant (15%).

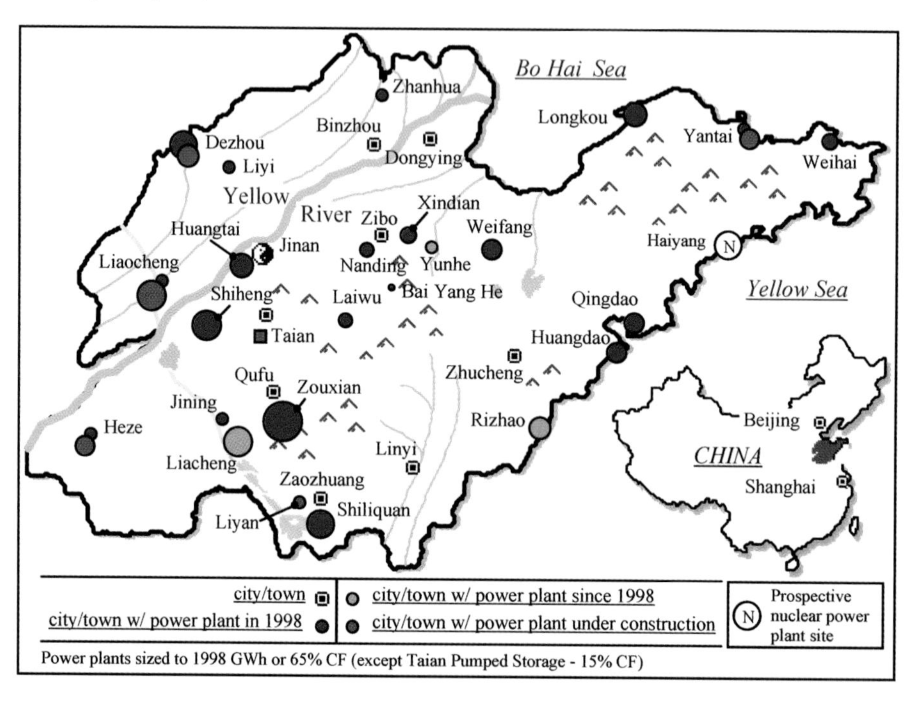

Figure 6.8 Shandong Province's Principal Power Plants

Shandong is one of China's most highly populated and economically productive provinces, with a large export trade to Japan, Korea and elsewhere. It sits on China's northeastern seacoast, southeast of Beijing, between Tianjin and Shanghai.

Shandong's population has increased to over 93 million people (2002), 10 million more than in 1990 (World Gazetteer, 2002). Shandong covers roughly 155 thousand square kilometers with the Yellow River (Huang He) river valley and delta being the principal geographic features in the West, with a mountain range extending from the South Central portion of the province to the Shandong Peninsula in the Northeast. Despite this mountainous area, Shandong's hydropower potential is considered very low (Xue and Eliasson, 2000). The province is roughly 600 km from East to West, and 400 km from North to South.

Generation in Shandong is produced almost exclusively with coal mined within the province, but also shipped by rail, and rail and ship from Shanxi and other provinces to the Northeast. Only an exceeding small portion of the province's generation comes from oil-fired generation or hydropower. The Shandong Electric Power Group Corporation (SEPCO) manages generator dispatch and transmission system operations. Shandong is China's largest stand-alone provincial network, although current plans for power sector reform in China will integrate it into the North China Grid Company over time. (Connors, et. al., 2002)

4.1 Attributes

The Electric Sector Simulation research team set up the Attribute Processor to automatically calculate over 240 attributes. A detailed list of them can be found in the report Shandong, China Electric Sector Simulation Assumption Book (Connors et. al., 2002). Table 6.2 shows the various classes of attributes that the team calculated.

Not included in the above list, but available for calculation were system efficiency (heat rate) and reliability attributes. Attributes generally fall into two categories, decision and descriptive attributes. Decision attributes are analogous to criteria, and measure performance relative to the stakeholders' interests. Table 6.3 shows the key decision attributes that will be used in presenting electric sector results in this chapter. (A different set of criteria were developed for the MCDA analysis, incorporating results from the life cycle, environmental impact, and risk and safety assessment tasks.) When performing tradeoff analysis it is handy to have attributes available that describe how many of which technologies were present, and the degree to which they were used. During the presentation of results to stakeholders, this helps explain why costs or emissions are higher for some strategies than others.

Table 6.2 Classes of Attributes

Cost Attributes

Regional Costs: Present value cost of supplying electric service to Shandong Province.

Unit Costs: Average cost on a per kWh basis of supplying electricity and electric service to Shandong Province.

Component Costs: Present value cost of capital expenditures, fuel and operation and maintenance costs, demand-side expenditures, etc.

Emissions and Effluent Attributes

Power Plant Stack Emissions: Sulfur Dioxide (SO_2), Particulate Emissions (PM10), Nitrogen Oxides (NO_x) and Carbon Dioxide (CO_2).

Solid Wastes and Consumables: Flyash and scrubber sorbent use.

Water Consumption: Cooling water not returned/evaporated, and boiler make-up water.

Electricity Demand and Generation Attributes

Electricity Demand and Growth: Cumulative and end-year (2024) electricity demand (generation and sales–GWh, and peak loads–MW), including growth rates and end-use impacts.

Generation by Type and Location: Annual and cumulative generation broken down by old versus new source, technology and fuel type, both installed MWs and GWh utilization.

Fuel Transport and Consumption by Fuel Type and Source: Breakdowns on the level and growth in fuel consumption by type, source and mode of transport.

Within the electric sector simulation's scenario tradeoff analysis only direct costs and emissions from the sector were calculated. The life cycle and environmental impact assessment teams build upon these results adding emission from the entire fuel chain and external costs.

As shown in Table 6.1, the final large scenario set presented here was the third iteration of scenarios. The differences between the initial and final scenario sets focused mainly on the number and timing of additional unit retirements, several alternative mixes of new generation technologies involving the choice of clean coal technologies and the operational modes and location of natural gas-fired combined cycle generation, as well as several alternative coal and natural-gas fuel cost uncertainties. These are also presented in Connors, et. al. (2002).

Table 6.3 Key Decision Attributes for Tradeoff Analysis

Cost Attributes

Regional Cost of Electric Service :

(Net Present Value – Billion of 1999 Yuan (¥))

The direct cost of providing electric service to Shandong Province for the entire study period (2000-2024 inclusive), including non-utility costs such as those for end-use programs. Expressed as the present value of the twenty-five year cost stream using a discount rate of 10% (including an average inflation of ≈ 4%). It includes cost recovery for existing as well as new generation, as well as costs for transmission and distribution and other electric company operations costs.

Average Unit Cost of Electricity:

(Yuan/kWh– electricity sales)

Average of each year's cost of providing electric service (inflation adjusted), divided by electricity sales in that year. It represent the normalized *rate* for electric service in Shandong.

Average Unit Cost of Electric Service:

(Yuan/kWh – electric service)

Average of each year's cost of providing electric service (inflation adjusted), divided by the amount of electricity *service* provided in that year. The amount of electric service is in any given year is equal to the electricity sales plus the electricity consumption avoided by end-use efficiency programs. A measure of the average electricity *bill* for electric service in Shandong. As such, the Unit Cost of Electric Service is the more appropriate normalized metric for electric service costs, and is the one that appears in most tables and figures.

(A uniform long-term exchange rate of 8 ¥ per U.S. Dollar was used in this analysis.)

Emissions Attributes:

Cumulative Power Plant Sulfur Dioxide Emissions:

(Millions of Tonnes of SO_2, (Mt))

The sum of all power plant SO_2 emissions from 2000-2024. Sulfur Dioxide contributes to acid deposition, and is a precursor of fine particulates (PM2.5).

Cumulative Power Plant Particulate Emissions:

(Millions of Tonnes of PM10, (Mt))

The sum of all power plant PM10 emissions from 2000-2024. Particulate emissions contribute to diminished visibility and respiratory and other health effects.

Table 6.3 Key Decision Attributes for Tradeoff Analysis (cont.)

Cumulative Power Plant Nitrogen Oxides:

(Millions of Tonnes of NO_x, (Mt))

The sum of all power plant NO_x emissions from 2000-2024. Nitrogen Oxides contribute to acid deposition, and are a precursor of photochemical smog (ozone) and fine particulates (PM2.5).

Cumulative Power Plant Carbon Dioxide:

(Billions of Tonnes of CO_2, (Gt))

The sum of all power plant CO_2 emissions from 2000-2024. Carbon Dioxide is the principal greenhouse gas resulting from fossil fuel combustion.

(Costs and emissions will be shown as annual numbers when performance trajectories are present in addition to tradeoff results.)

5. SHANDONG UNCERTAINTIES AND FUTURES

For this analysis the ESS research team looked at how changes in electricity demand growth and coal and natural gas prices affect the performance of strategies. Table 6.4 lists the individual uncertainties incorporated into the analysis, including their letter codes by which the combined futures are identified, and employed in the analysis. The individual uncertainties are presented below.

Table 6.4 Shandong Electric Sector Uncertainties and Futures

Future Components			
Uncertainty Set	Uncertainty	Code	No.
Economy/Demand Growth			
Demand for Electrical Energy (Shandong Grid)			
	Slow Demand Growth (<4%/yr)	T	3
	Moderate Demand Growth (>5%/yr)	**F**	
	Stong Demand Economy (>7%/yr)	S	
Fuel Cost and Availability			
Delivered Cost of Steam Coal			
	Business as Usual Coal	**I**	3
	Mechanized Mining Lowers Coal Costs	U	
	Transportation Investment Raises Costs	A	
Delivered Cost of Pipeline Natural Gas			
	Base Natural Gas Costs ('26/GJ - 1999)	**B**	2
	Lower Natural Gas Costs ('15/GJ - 1999)	F	
(Ref. Future: FIB)		No. of Futures	18

Combined, the three electricity demand uncertainties, three coal cost uncertainties and the two natural gas cost uncertainties yield eighteen unique futures. For this analysis we have selected the future 'FIB' as the "reference future," and most results present here will refer to this combination of load growth and fuel costs. Other combinations will cut across one or more uncertainties, with the series TAF, FIB, and SUB reflecting the "best" and "worst" futures from an emissions standpoint. TAF where electricity demand growth is low, coal costs high and natural gas costs low. SUB in contrast has electricity demand continuing to grow rapidly, with coal costs declining in constant terms.

5.1 Demand for Electrical Energy

One of the largest uncertainties facing Shandong's electric sector is the rate at which the demand for electrical energy, or electric service, grows. Growth in the demand for electrical energy (TWh per year) influences the number and hours of operation of generators, which impacts costs and emissions, and the need for new power plants. Growth in annual peak load (maximum MW of demand for the year) impacts the number and size of power plants needed to meet demand, and influences costs and emissions through investment requirements and its impacts on the generation mix and its use.

Figure 6.9 shows how the demand for electrical energy has grown in the recent past, with a projection based upon a curve fit of the past. Presented is "busbar demand" in terawatt-hours (TWh). This reflects generator output to the grid, prior to transmission and distribution (T&D) losses. In our analysis we assumed T&D losses of 6.2%, based upon 1999 electricity generation and sales as reported by SEPCO. This is considerably lower than those of say the United States (≈ 10%), but considering the high level of industrial electricity demand it was considered reasonable. Electricity sales ("meter load") reflect the amount of electricity delivered to customers *after* T&D losses, and is used for unit cost (per kWh) calculations.

Statistics for Shandong electricity demand can reflect the entire province's electricity demand, or just the load served by the provincial grid. The difference (≈ 15 TWh) reflects self-generation by some industrials and municipal electric companies. The ESS scenarios analyze the Shandong Grid, primarily because information on non-grid controlled generators and loads was unavailable. As most new generators and loads are connected to the provincial grid, the 15 TWh difference between the provincial and grid demand was held constant throughout the study period. Table 6.5 shows the historical growth rates compiled from the available information at both the provincial and grid level (Connors et. al., 2002). Note the high rates of growth, and large swings in year-to-year growth rates.

The other thing illustrated in Figure 6.9 is the variability in year-to-year growth. Note the drop in electricity demand from 1997 to 1998 due to the Asian Economic Crisis. ESS electricity demand uncertainties (which are not forecasts) also reflect this year-to-year variability or "noise," based upon the past variability. Figure 6.9 shows the "noise" superimposed on the long-term trend. This influences both the capacity planning in the PSP program, and the capacity margin in any given year and how that impacts power plant use.

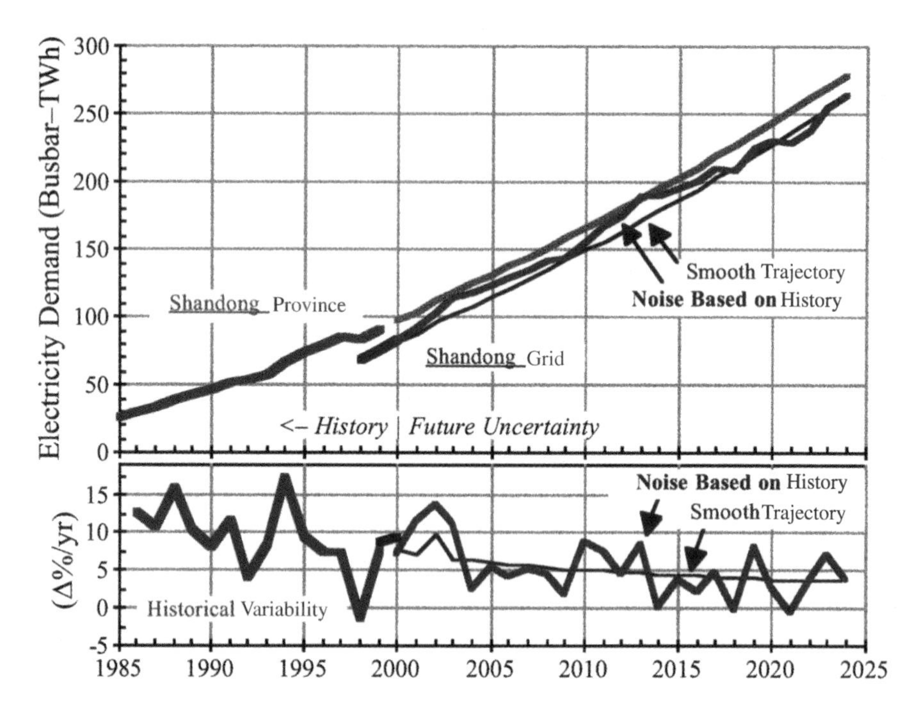

Figure 6.9 Historical and Future Electricity Demand, Shandong

Table 6.5 Historical Demand Growth, Shandong

	Electricity Demand		Peak Load
Annual Statistics	Province	Grid	Grid
1998	84.33	69.61	11.27
1999	90.90	75.68	12.55
2000	–	82.72	13.00
	(Busbar–TWh)		(Busbar–GW)

Rates of Growth	Province (1985-99)	Grid (1998-00)	Grid (1996-00)
"Long-Term" Growth Rate	9.16	9.01	7.23
Minimum Annual Change	-1.43	8.72	-4.44
Maximum Annual Change	17.53	9.30	19.90
	(Δ%/yr)		(Δ%/yr)

Not only is Shandong electricity growing fast, but its peak demand is growing even faster. This can be seen in Figure 6.10, which shows hourly electricity demand for three years. It can be seen that peak loads occur in the evening and during summer heat waves, driven by residential and commercial demand for electricity. As Shandong's population becomes more affluent, this trend is likely to accelerate, putting greater pressure on the entire system, and especially the need for new power plants. Note the low load periods in February associated with the Chinese Lunar New Year.

These observations provide the source information for the Electric Sector Simulation's electricity demand uncertainties. The long-term growth trends are shown in Table 6.6, and are derived from different curve fitting approaches to the province's historical demand, except for the Strong/High demand uncertainty that was capped at 7%/yr. When the non-grid demand for electricity is subtracted out, Shandong Grid growth rates become higher.

Table 6.6 Electricity Demand Growth Rate Uncertainties

	Electricity Demand Uncertainty		
	Slow (T)	Moderate (F)	Strong (S)
	Electricity Demand (Busbar–GWh)		
Shandong Province	3.43	4.58	7.00
Shandong Grid			
Historical Noise	3.89	5.11	7.66
Smooth	3.89	5.12	7.65
	Peak Load (Busbar–MW)		
Shandong Grid			
Historical Noise	4.19	5.58	8.36
Smooth	4.20	5.59	8.36
	(Δ%/yr)		

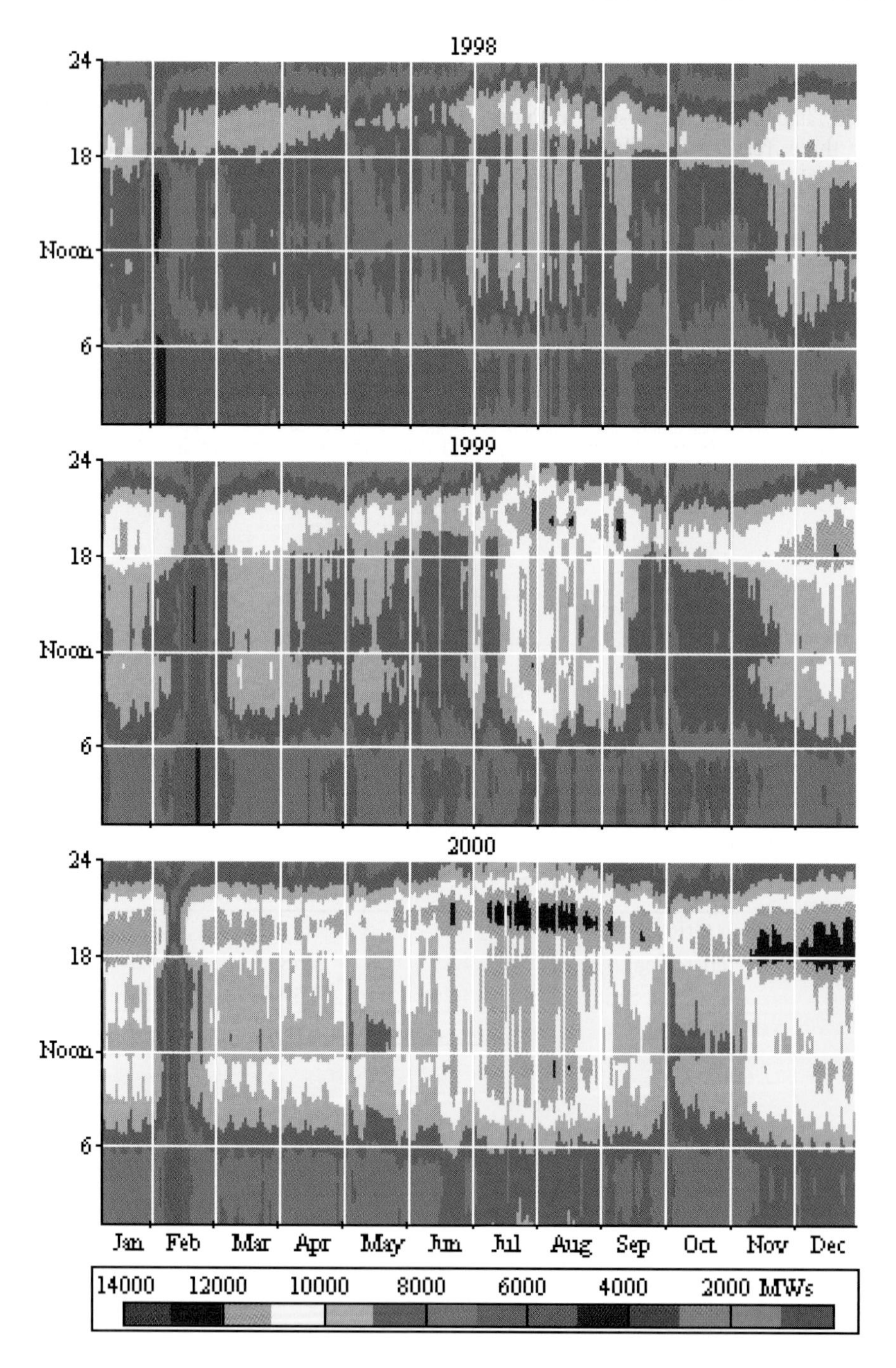

Figure 6.10 Hourly Busbar Electricity Demand, Shandong Grid

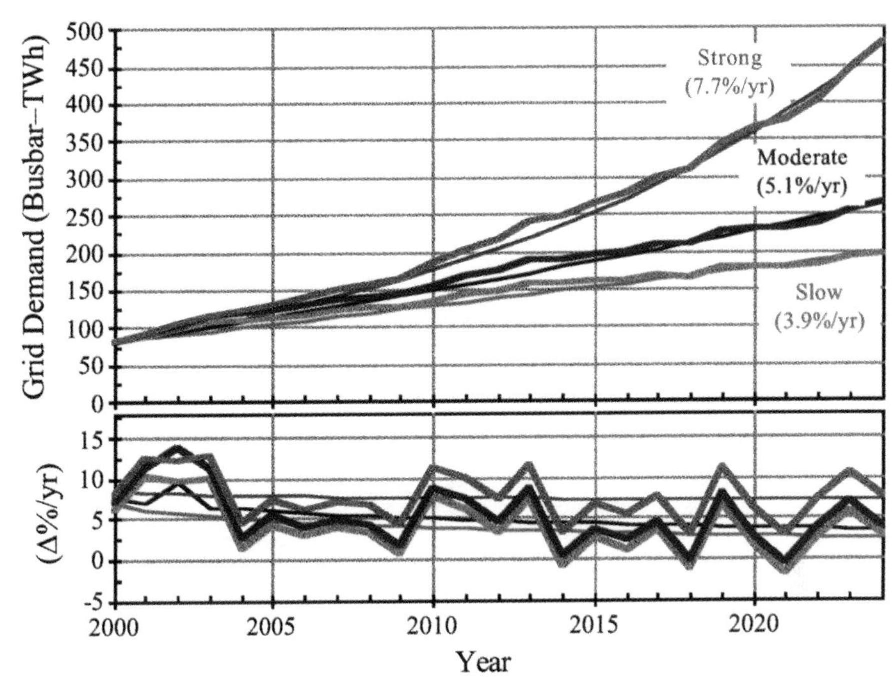

Figure 6.11 Electricity Demand Uncertainties

These ranges of electricity demand growth are analogous to those in the Energy Demand Forecasting chapter. The principal difference is that the ESS uncertainties have separate electricity demand and peak load growth trajectories, with peak load growing faster than annual electricity demand, indexed to the growth of residential and commercial electricity demand which contributes more to peak load than industrial end-uses. Figures 6.11 and 6.12 show the trajectories for electricity demand and peak load growth for the Slow, Moderate and Strong electricity demand uncertainties. Both the base smooth, and noisy trajectories for each TWh and GW trajectory are shown.

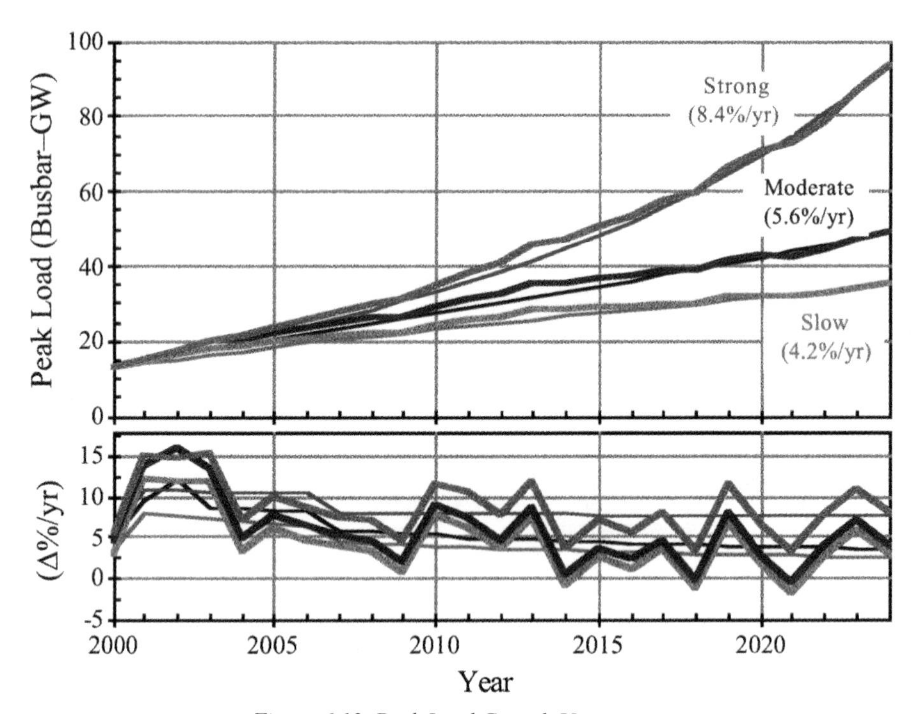

Figure 6.12 Peak Load Growth Uncertainties

5.2 Steam Coal Cost Uncertainties

As coal is the principal fuel for generating electricity in Shandong, and most of China, its delivered price to generators within the province is a major concern, and therefore uncertainty. Forecasts for the cost of delivered steam coal were unavailable, so these uncertainties are based upon some broader assumptions regarding the evolution of the Chinese coal industry. Figure 6.13 shows the cost trajectories for two types of coal, one from Shandong (SD) transported by rail, and another from Shanxi (SX) transported by rail and ship. We assumed all coal mined in Shandong is distributed within the province by rail. Coal brought into the province from Shanxi and other provinces to the Northwest is transported two ways. Inland power plants in the Western parts of the province receive their coal directly by rail, however those power plants using non-Shandong coal on the coast receive their fuel by ship. This coal is transported by rail to ports such as Qinhuangdao, where is loaded on ships for final delivery. As is shown in the figures, these increased transportation costs are reflected in the delivered cost of the coal. In our analysis, based upon the advice of Chinese research colleagues, *all* new coal-fired generation gets its coal from outside of Shandong Province.

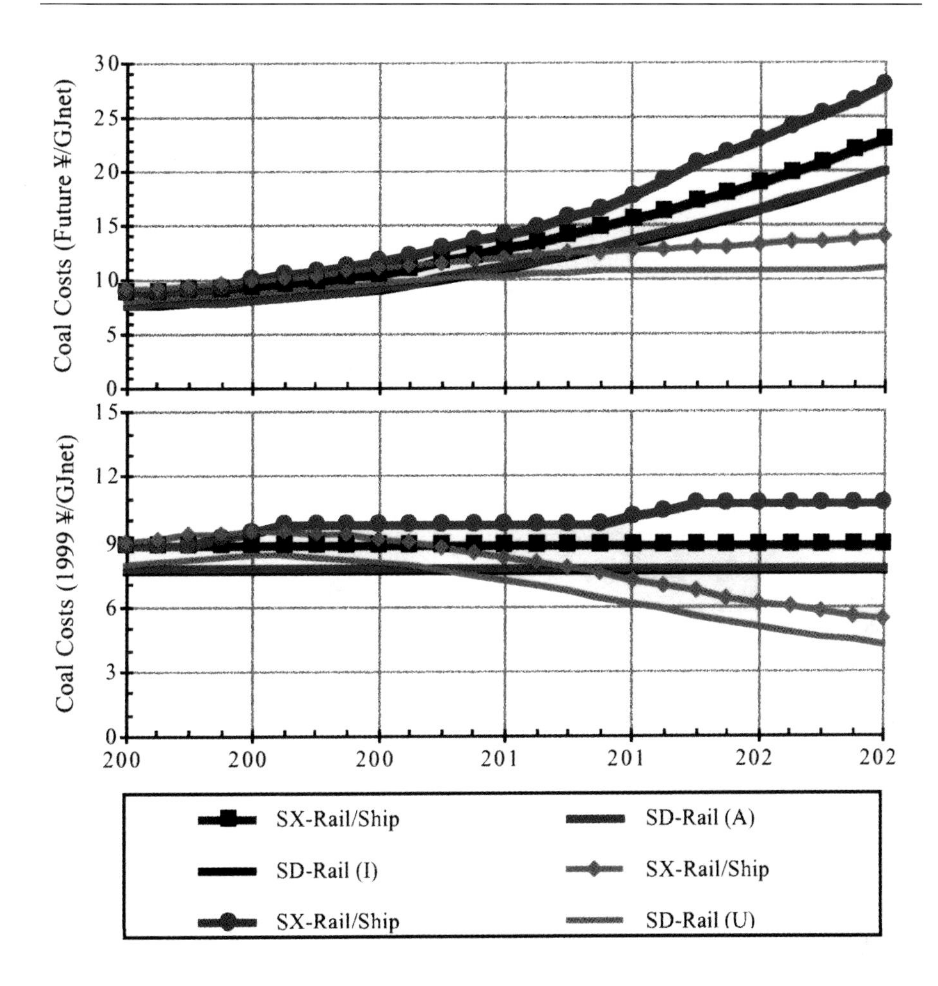

Figure 6.13 Coal Cost Uncertainties

The top portion of the graph shows coal costs in Future (current) Yuan, while the bottom shows coal costs in Base Year 1999 (constant) Yuan. The "Business as Usual" (I) coal cost uncertainty simply escalates coal costs with inflation. However, international coal statistics show that coal costs have declined dramatically over the past several decades, primarily due to increased mechanization of coal mining. The Mechanized Mining – Productive (U) coal cost uncertainty reflects this trend. A large portion of rail transport capacity in China is used for transporting coal, so the third coal cost uncertainty reflects a stressed or aggravated coal transport situation (A), where the need to invest in more rail transportation capacity is reflected in the delivered price of coal shipped from outside the province increases, first in 2005 and then again in 2016.

5.3 Natural Gas Cost Uncertainty

Although there are no significant natural gas supplies to Shandong at present, that does not mean it may not be an important fuel for power generation in the future. As several of our strategies call for natural gas-fired generation, we included a natural gas cost uncertainty. The base cost for natural gas was assumed to be 26¥ per GJnet (Gigajoule – Lower Heating Value or "net") and escalated through time as shown in Figure 6.14. A lower natural gas cost beginning at 15¥/GJnet is the second uncertainty, and reflects either a drop in cost due to country-wide investment in gas transportation infrastructure, subsidization, or a combination of both. For comparison purposes, Figure 6.14 also shows Business as Usual uncertainty coal costs. Like the coal uncertainty graph future/current fuel costs are shown on top and base year-1999/constant costs are shown on the bottom. For both natural gas cost uncertainties, costs growth with inflation until 2008 and then are slightly higher than inflation. As the gray areas in the figure indicate, we assumed that power production from baseload natural-gas fired generation did could not occur until 2015, due to the time it takes to extend the pipeline infrastructure into the province in sufficient quantities to supply such generation.

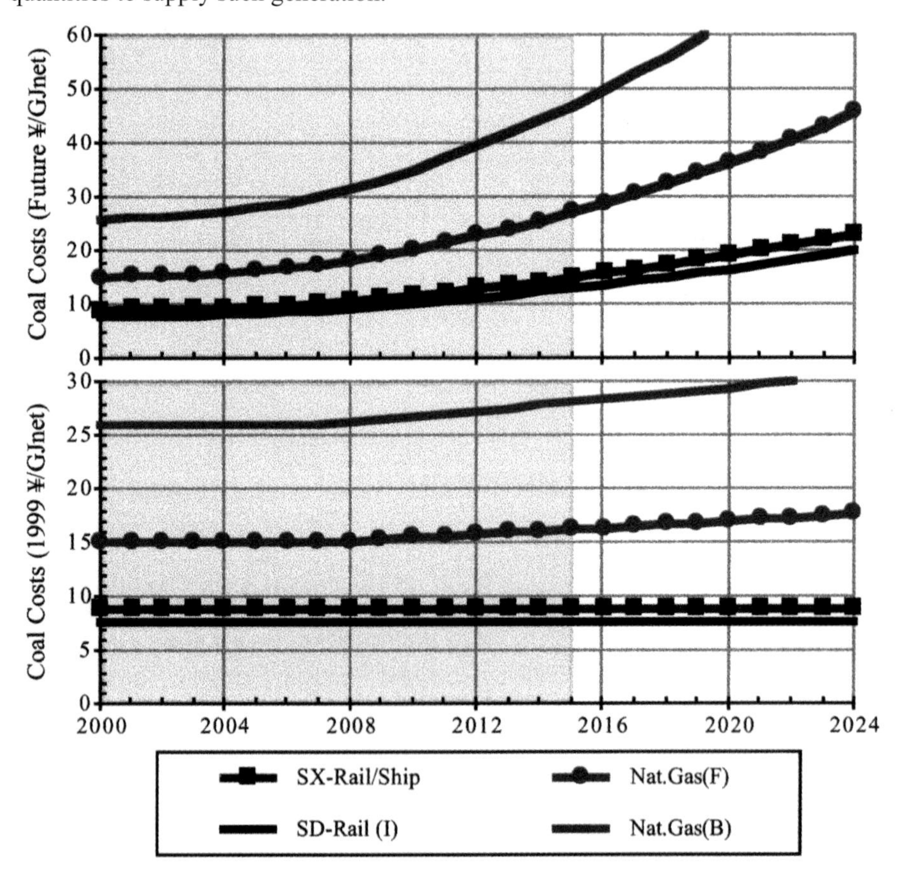

Figure 6.14 Natural Gas Cost Uncertainties

6. SHANDONG OPTIONS AND STRATEGIES

Strategies for Shandong's electric power sector were grouped into three categories, Existing Generation, New Generation and End-Use. Table 6.7 shows these categories, the options-sets within each of them, and the options and their letter codes that were combined to form 1008 strategies. Highlighted in each option-set is each group's reference option. Strung together they form the strategy "BOC-CONPAS."

Table 6.7 Shandong Electric Sector Options and Strategies

Strategy Components			
Option Set	Option	Code	No.
Existing Generation Options			
Retire Additional Existing Units			
	Baseline Retirements 50 MW and Under by 2003	**B**	2
	Retire Select Units by 2008 and at 35 Years	D	
Retrofit Existing Units with Sulfur FGD			
	None beyond Planned	**O**	2
	Retrofit Select Units	U	
Switch to Prepared Coal (Dry Processing)			
	No Switch to Prepared Coal	**C**	3
	Switch only Existing Coal Units	X	
	Switch All Conventional Coal Units	P	
New Generation Options			
Mix of Future Generation Technologies in Shandong			
	Conventional Coal with FGD & ESP	**C**	7
Conv. Coal plus...	AFBC Beginning 2010	F	
	IGCC Beginning 2012	L	
	Nat. Gas Combined Cycle Beginning 2015	M	
	Nuclear Beginning 2010	N	
	Nuclear and Natural Gas	D	
	Nuclear, Nat. Gas and IGCC	T	
Extra-Provincial Generation			
	No "Generation by Wire"	**O**	2
	Natural Gas from the West Beginning 2010	A	
End-Use Options			
Peak Load Management/Peaking Generation			
	Nat. Gas Peaking Turbines Beginning 2008	**P**	2
	Reduce Need for Peakers via Peak Load Mgt.	L	
End-Use Efficiency/Conservation			
	Current Efficiency Standards	**S**	3
Moderate Efforts	10% Cumulative Reduction	M	
Aggressive Efforts	20% Cumulative Reduction	G	
(Ref. Strategy: BOC-CONPAS)		No. of Strategies	1,008

This "reference strategy" will be used as the basis for comparisons, and represents a "static technology" combination of options, not a "Business as Usual" forecast. With China's ongoing reform of the power sector, and its ascension to the World Trade Organization, which will affect both the economy, and access to technology, a "Business as Usual" strategy would be meaningless. After discussing some of the crosscutting assumptions in the analysis of the ESS strategies, each set of options is presented.

Table 6.8. Key Modeling Assumptions

Power System Modeling Assumptions:

Study Period: 2000 to 2024, inclusive.

Transmission and Distribution Losses:

Annual average of 6.2% between power plant output to grid (busbar generation) and end-use consumption (meter load) based on 1999 busbar generation of 75.73 TWh and electricity sales of 71.04 TWh. (SEPCO, 2000)

Planning Reserve Margin

A planning reserve margin of 20% was assumed for all scenarios. This represents the amount of extra generating capacity the power system would like to have in place, relative to annual peak load, to account for scheduled and unscheduled generator outages, and changes in the actual demand for electricity due to extreme weather and short-term changes in the economy.

Coal Supplies

For existing power plants, no changes from the current sources of coal were assumed. New generation was assumed to get their coal from Shanxi province, with inland units receiving it by rail, and coastal units receiving it by rail then ship.

Natural Gas Supplies

It was assumed that gas-fired peaking units could be supplied by associated natural gas from the Bo Hai oil fields beginning in 2008, however baseload natural gas combined-cycle units would have to wait for sufficient pipeline capacity bringing methane from the west or north to reach Shandong Province. Commissioning of gas combined-cycle generation was therefore prohibited before 2015 to reflect this availability constraint.

Cost Modeling Assumptions

Base Year for Costs: 1999

Discount Rate: 10%

A discount rate of 10%, including inflation was used to calculate the net present value of all cost streams in the results presented here

Table 6.8 Key Modeling Assumptions (cont.)

Borrowing Rate / Weighted Average Cost of Capital: 7%/year

By looking at the financial statements of Chinese independent power producers, a Weighted Average Cost of Capital (WACC) of 7% was assumed for future power plant capital expenditures, including inflation. This reflects roughly 75% of the financing coming from loans (debt) at 5%/yr., and 25% coming from the issuance of stock at 13%/yr. (Connors, et. al., 2002).

Average Rate for Delivered Electric Power, 1999: 0.45 Yuan/kWh

In 1999, SEPCO's revenue from electricity sales totaled 32.07 billion Yuan, on sales of 71.04 TWh. This yields an average rate of 0.45 ¥/kWh. This number is used as to determine the balance of costs required beyond generation expenditures to provide power to Shandong grid customers (SEPCO, 2000).

6.1 Baseline Assumptions

Although over a thousand multi-option strategies were analyzed for Shandong Province, there are still many common assumptions that cross all strategies. One that has been discussed already is transmission and distribution losses. Another was that all new coal-fired generation is supplied by coal mined outside Shandong Province, even if retirements of existing generators theoretically free up local mining capacity. Table 6.8 presents these key crosscutting assumptions and their derivation.

One of the principal challenges of the Electric Sector Simulation team was the estimation of transmission, distribution and other costs to add to the costs of generation, which together represent the cost consumers pay for electric service. This was achieved by taking the electricity sales in the first year of the study period (2000) and multiplying it by 0.45 ¥/kWh to get the total revenue, and therefore revenue requirements, to provide the province with electric service. Power plant fuel and operating costs, plus new generation capital costs for 2000 obtained from the simulation model were then subtracted from 2000 revenue requirements. This difference represents the annual expenditure for the debt service on existing generators; maintenance and expansion of the transmission and distribution system; and corporate general and administrative (G&A) costs. 50% of this difference was allocated to debt service of existing generators, which declined to zero over the course of the twenty-five year study period. Another 35% of the difference was allocated to transmission and distribution. This number was divided by 2000 sales to get a base Yuan/kWh T&D expenditure, which was then used to calculate future year annual T&D expenditures. The remaining 15% was handled in a similar fashion to calculate future G&A costs.

As the choice of new generating technologies is normally a high profile electric sector decision these options are covered next, followed by options targeting existing generators and fuel quality, and the consumption of electricity.

6.2 New Generation Options

The modeling of new generation options is a two-step process. First, what technologies should be included, and when might they first come on line in Shandong Province? Second, what combinations of generation technologies should be examined? Tables 6.9 and 6.10 show the key performance and cost assumptions for the new generation technologies analyzed in the ESS's large scenario sets. These include new subcritical pulverized coal power plants, with and without flue gas desulfurization, two clean coal technologies, natural gas combined-cycle power plants and nuclear power. The two clean coal technologies were atmospheric fluidized bed combustion (AFBC) and integrated gasification combined-cycle (IGCC). Additional technologies such as windpower, advanced modular high temperature nuclear generation and waste-to-energy were examined as sensitivity analyses and are reported on later, and in Hansen (2002). Details on all these technologies can be found in Connors et. al. (2002).

Table 6.9 shows the general performance characteristics of the technologies including size, thermal efficiency and percent reduction from uncontrolled emissions rates resulting from the installation of pollution control equipment. Two basic assumptions are made regarding the cost and performance of new generators, depending on whether they are located on the coast or inland. Coastal units are assumed to have Once Through Cooling where cooling water is returned to the sea, while inland power plants are modeled with evaporative or wet cooling systems. The cost of the cooling water return loop makes the coastal units slightly more expensive than the inland units with cooling towers. However, coastal units are slightly more efficient due to the smaller energy requirements of Once Through Cooling. All nuclear units were modeled as coastal units, while all natural gas combined-cycle units were modeled as inland units, proximate to pipeline natural gas supplies.

All conventional coal and fluidized bed units are assumed to have electrostatic precipitators (ESP) to control particulate emissions. Most large existing power plants in Shandong currently have ESPs or some other form of particulate control. We assumed the efficiency of ESPs to be 95%, due to the high ash content of Chinese coals. All new fossil-fueled generators are also assumed to have combustion modifications to reduce nitrogen oxide emissions, but no flue gas treatment for NO_x.

While only a few Shandong coal units currently have flue gas treatment to capture sulfur dioxide, conventional coal units were all modeled with one type or another of flue gas desulfurization technology (FGD). Inland units employed the more common wet flue gas desulfurization, which uses lime or limestone as a sorbent. Coastal units however were assumed employ a seawater scrubber. Both are explained further in Connors et.al. (2002), with the principal differences being the increased capital cost and auxiliary power consumption of wet scrubbers. Sulfur removal efficiencies were assumed to be 90%, again due to the high ash content of Chinese coals. Emissions requirements for new power plants in China call for flue gas sulfur controls on new units unless less than 1% sulfur coals are used. Based upon conversations with Chinese colleagues, the ESS team chose to model FGD on all new conventional coal fired power plants, since this provides plant operators greater flexibility with extra-provincial fuel suppliers. The impact this has on baseline sulfur emissions is shown later.

Table 6.9 New Generation Technology Characteristics

Generation Technology	*Unit Size*	*Thermal Efficiency*	*Emissions Removal Efficiency* SO2	PM10	NOx
Conventional Coal – Pulverized Coal, Subcritical Boilers (Conv.Coal)					
Coastal Locations – Once Through Cooling					
No Desulfurization	300	36.0		95.0	50.0
	600	37.0		95.0	50.0
Sea Water Scrubbers	300	35.0	90.0	95.0	50.0
(OS)	600	36.0	90.0	95.0	50.0
Inland Locations – Wet/Evaporative Cooling					
No Desulfurization	300	35.5		95.0	50.0
	600	36.5		95.0	50.0
Wet Scrubbers	300	34.5	90.0	95.0	50.0
(WW)	600	35.5	90.0	95.0	50.0
Atmospheric Fluidized Bed Combustion (AFBC)					
Coastal – Once Through	300	38.0	95.0	99.0	73.8
Inland – Wet Cooling	300	37.5	95.0	99.0	73.8
Integrated Gasification Combined Cycle (IGCC)					
Coastal – Once Through	500	45.0	99.0	*n/a*	69.8
Inland – Wet Cooling	500	44.5	99.0	*n/a*	69.8
Natural Gas Fired Combustion Turbines (CT, Peaking)					
Closed Loop Cooling	155	38.0	*n/a*	*n/a*	70.0
Natural Gas Fired Combined Cycle (NGCC)					
Wet/Evaporative	250	57.5	*n/a*	*n/a*	70.0
Cooling	500	57.5	*n/a*	*n/a*	70.0
	750	57.5	*n/a*	*n/a*	70.0
Nuclear – Advanced Light Water Reactors (ALWR)					
Coastal – Once Through	1000	33.0	*n/a*	*n/a*	*n/a*
	(MW, Busbar)	(%, LHV)	(% reduced from uncontrolled)		

Cost assumptions were taken from the literature (Connors et. al., 2002), and reflect adjustments for domestic production of components and Chinese labor rates. The OS and WW associated with the conventional coal units in the Tables 6.9 and

6.10 refer to Once Through Cooling and Sea Water Scrubbers (OS) for coastal units, and Wet Cooling and Wet Scrubbers (WW) for inland units. Differential impacts on these units' operation and maintenance costs were also assumed. Table 6.10 also presents the key availability metrics of annual scheduled maintenance in weeks per year, and the equivalent forced outage rate in percent of annual operation. The permitting and construction lead-time associated with each technology is also provided.

Table 6.10 New Generation Technology Costs and Availability

Generation Technology	*Capital Cost*	*Fixed O&M*	*Var. O&M*	*Main – tenance*	*Outage Rate*	*Lead Time*
Conventional Coal – Pulverized Coal, Subcritical Boilers (Conv.Coal)						
Coastal Locations – Once Through Cooling						
No Desulfurization	600	20	1.0	7	5	5
	550	18	1.0	8	5	6
Sea Water Scrubbers	624	22	2.0	7	5	5
(OS)	574	20	2.0	8	5	6
Inland Locations – Wet/Evaporative Cooling						
No Desulfurization	588	21	1.0	7	5	5
	540	19	1.0	8	5	6
Wet Scrubbers	660	23	4.0	7	5	5
(WW)	610	20	4.0	8	5	6
Atmospheric Fluidized Bed Combustion (AFBC)						
Coastal – Once Through	900	30	4.0	5	5	5
Inland – Wet Cooling	880	31	4.0	5	5	5
Integrated Gasification Combined-Cycle (IGCC)						
Coastal – Once Through	1200	30	1.0	5	8	6
Inland – Wet Cooling	1200	31	1.0	5	8	6
Natural Gas Fired Combustion Turbines (CT, Peaking)						
Closed Loop Cooling	400	1	3.0	1	8	3
Natural Gas Fired Combined Cycle (NGCC)						
Wet/Evaporative	600	14	0.5	3	5	4
Cooling	600	13	0.5	3	5	5
	600	12	0.5	3	5	6
Nuclear – Advanced Light Water Reactors (ALWR)						
Coastal – Once Through	1400	42	0.5	4	5	8
	(Overnight, $99/kW)	($/kW-yr)	($/MWh)	(Wks)	(%)	(Yrs)

Table 6.11 shows how these individual generation technologies were combined into portfolios, or generation mixes. In all cases conventional coal units are built in the early years of the study period, with the other technologies coming on line as

they become available. The conventional coal mix ('C') continues to build pulverized coal throughout the twenty-five years, and for comparison purposes is considered the "reference" generation mix option. The following four (F–ABFC, L-IGCC, M-Methane and N-Nuclear) add one new generation technology in addition to pulverized coal. The final two mixes, 'D' and 'T' combine conventional coal with natural gas and nuclear, and those three plus IGCC.

Except for nuclear, each technology's contribution to the mix is expressed as the percent of "new" megawatts built. For nuclear we assumed one new nuclear unit was added every other year beginning in 2010, resulting in a total of eight nuclear power units. This was independent of the load growth uncertainty. For the two clean coal mixes (F and L), no conventional coal generation was built after 2017. It should be noted that "new" power plants can be either "replacement" or "additional" capacity, replacing the megawatts from retired power plants or meeting the growth in demand. This allows for modernization of power generation beyond the need to meet just load growth. When peak load management programs are part of a strategy's broader mix of options, a slightly lower ratio of peaking combustion turbines was used.

Table 6.11 New Generation Technology Mixes

		Peaking		*Baseload Generation*					
New Generation Technology Mixes		No LM	Load Mgt.	Conv. Coal: Coastal	Conv. Coal: Inland	Clean Coal: AFBC	Clean Coal: IGCC	Nat. Gas: NGCC	Nuclear: ALWR
First Year Avail.:		*2008*		*2000*		*2010*	*2012*	*2015*	*2010*
C	Conventional	5		50	45				
	Coal		3	50	47				
F	Clean Coal –	5		35	30	30			
	AFBC		3	35	32	30			
L	Clean Coal –	5		35	30		30		
	IGCC		3	35	32		30		
M	Natural Gas –	5		25	20			50	
	NGCC		3	25	22			50	
N	Nuclear –	5		50	45				*8 GW*
	ALWR		3	50	47				*8 GW*
D	Nat. Gas &	5		25	20			50	*8 GW*
	Nuclear		3	25	22			50	*8 GW*
T	Nat.Gas, IGCC	5		25	20		25	25	*8 GW*
	& Nuclear		3	25	22		25	25	*8 GW*
		(Percent of New MWs)							(GW)

Superimposed on the generation mixes was a "generation by wire" option. This assumed a firm purchase of natural gas fired generation from a province or provinces to the west. Starting with 500 MWs of must run generation in 2010 this option added an additional 500 MW for each of the following nine years. An additional T&D loss of 5% was added to account for long distance transmission. Coal-by-wire and hydro-by-wire were also considered. While the "gas-by-wire" option substantially reduced SO_2, NO_x and CO_2 emissions, due to its expensive fuel and must-run formulation it cost substantially more, therefore it will not be discussed in detail in the results section of this chapter.

6.3 Existing Generation Options

There were three option-sets targeting existing generation. First was the additional retirement of old power plants. Second was retrofitting select existing units with FGD. Third was assuming the processing of coal at or near mines to reduce ash content, and that coal's use in either existing or all coal-fired generation.

6.3.1 Retire Additional Existing Units

Current Chinese policy requires old power plants 50 MW or smaller in size to be turned off after 2003, and this is reflected in the ESS's reference strategy – BOC-CONPAS. The other assumption in the reference strategy was that units are not otherwise retired unless there was a firm retirement year specified. Before this final scenario set, several approaches to additional retirements were tested, which are shown in Figure 6.15.

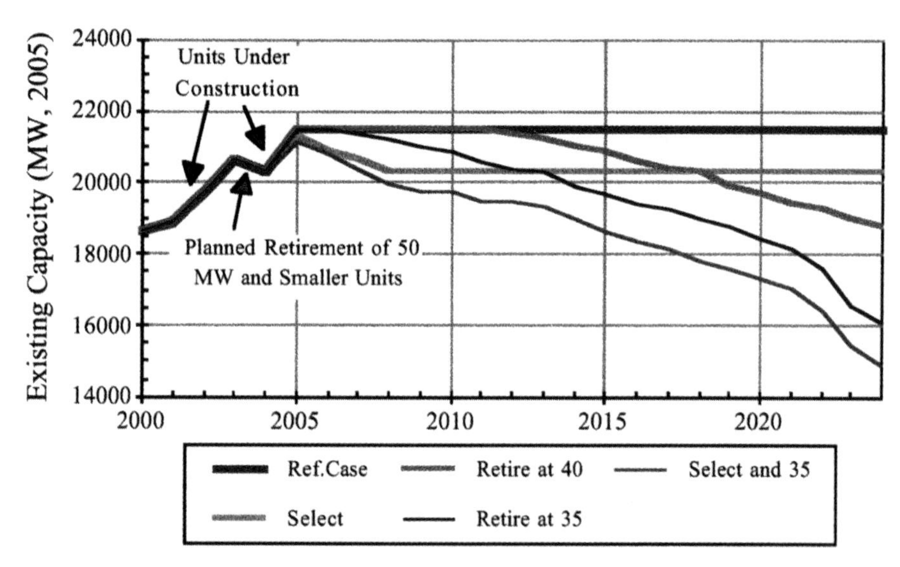

Figure 6.15 Changes to Existing Capacity

The first was "select" retirements, where based upon annual output, numerous units were considered for retirement if they operated few hours but had large emissions. Here nine units, totaling 1175 MW were retired over the years 2006 to 2008. Another 600 MW of municipal generating units were retired between 2006 and 2016 in this option a well. Finally, firm retirement dates were considered at 35 and 40 years from unit start date. The Select plus Retire at 35 year option (D) results in 30% reduction from the Reference retirements option (B). No extra decommissioning costs were assumed for these options, nor was any site value ascribed if the location became the site for a new generation unit.

6.3.2 Retrofit Existing Units with Sulfur FGD

The second existing unit set of options was to install flue gas desulfurization units on existing power plants. Some FGD retrofits have already been planned or completed and are incorporated into the Reference Case. Additional candidates for FGD retrofit were units that had relatively high capacity factors but high emissions rates. These totaled 14 units with a total of 3670 MWs, with the retrofits phased in between 2004 and 2007.

6.3.3 Switch to Prepared Coal

During the course of formulating the initial scenario set it became apparent that the high ash content of Chinese coals has a substantial impact on the operational and emissions performance of Chinese power plants. Therefore a set of options looking at mine-based treatment of coals to reduce their ash content were developed. As some sulfur is bound up in the ash, sulfur reductions also occur. Due to water availability issues in coal mining regions, only dry preparation techniques, consisting primarily of sorting and screening raw coal, were considered. The assumptions regarding coal preparation are presented in Connors et. al. (2002). Table 6.12 shows examples of how dry preparation of coal impacted it chemical composition, energy content and cost assuming a preparation cost of five Yuan per tonne.

Table 6.12 Composition and Cost Impacts of Coal Preparation

Select Bituminous Coals *by Source, Transport and Preparation*		Ash Content	Total Sulfur	Total Carbon	Energy Content	1999 Coal Cost Energy	Mass
		(weight %)			(GJn/t)	(Yuan/GJnet)	(Yuan/t)
Shandong by Rail	*Raw*	22.39	0.75	43.22	22.50	7.74	174.2
Low Sulfur	*– Prepared*	10.47	0.65	50.61	24.66	7.83	193.2
	(Δ%)	(53.2)	(14.0)	17.1	9.6	1.2	10.9
Shandong by Rail	*Raw*	29.09	1.25	39.81	21.50	7.76	166.8
Medium Sulfur	*– Prepared*	16.00	1.08	48.07	23.92	8.07	193.1
	(Δ%)	(45.0)	(14.0)	20.8	11.3	4.1	15.8
Shanxi by Rail	*Raw*	15.70	0.75	46.64	23.50	8.37	196.7
Low Sulfur	*– Prepared*	10.47	0.65	50.61	24.66	8.44	208.2
	(Δ%)	(33.3)	(14.0)	8.5	5.0	0.8	5.8
Shanxi by Rail/Ship	*Raw*	15.70	1.25	46.64	23.50	8.80	206.7
Medium Sulfur	*– Prepared*	10.47	1.08	50.61	24.66	8.85	218.2
	(Δ%)	(33.3)	(14.0)	8.5	5.0	0.6	5.6
Shanxi by Rail/Ship	*Raw*	29.09	1.75	39.81	21.50	8.92	191.8
Medium Sulfur	*– Prepared*	16.00	1.51	48.07	23.92	8.62	206.2
	(Δ%)	(45.0)	(14.0)	20.8	11.3	(3.3)	7.6

Operational impacts of "switching" to prepared coals in pulverized coal units included an increase in ESP removal efficiency from 95% to 97%. The availability of old generators was also assumed to improve with scheduled maintenance dropping from 10 weeks per year down to 8, as well as a reduction in the unit's equivalent forced outage rate from 8% to 5%, essentially the same as for new conventional coal units.

This option assumed that prepared coal was used instead of raw coal for the entire study period, an overestimation of the degree of actual coal switching that could actually occur. Two levels of switching to prepared coals were assumed. First, in all existing conventional coal units (X), and second, in both existing and all new coal-fired power plants (P), including clean coal technologies.

6.4 Demand-Side Management Options

The third category of options was that aimed at the demand for electricity, commonly referred to as demand-side management (DSM). It is also the category for which there was the least amount of information upon which to base assumptions. The peak load management (PLM) and end-use efficiency (EUE) options shown here should therefore be considered theoretical. Even so, they do allow the determination of the benefits of achieving the assumed levels of conservation and peak load reduction, identifying the avoided operating and investment costs in delivering power, and its associated stack emissions, and therefore what the province should be "willing-to-pay" for DSM.

6.4.1 Peak Load Management

As shown above in the demand uncertainty section, annual peak load is assumed to grow faster than annual electrical energy. In the peak load management option peak load is assumed to grow at the *same rate* as annual electrical energy. Peak load management is modeled as a pure load shift with no impact on the annual demand for electrical energy (TWh per year). The cost of peak load management was assumed to be 120 Yuan per kW-yr ($15/kW-yr) from the no Peak Load Management level of demand, and increased with inflation. Figure 6.16 shows the peak load impacts of this option across the three load growth uncertainties. As you can see, peak load growth is substantially different across the three load growth uncertainties. The 2.6, 5.2 and 14.4 GW peak load reductions associated with the Slow, Moderate and Strong electricity demand uncertainties must be increased by 20% to get the GWs of avoided generation investment.

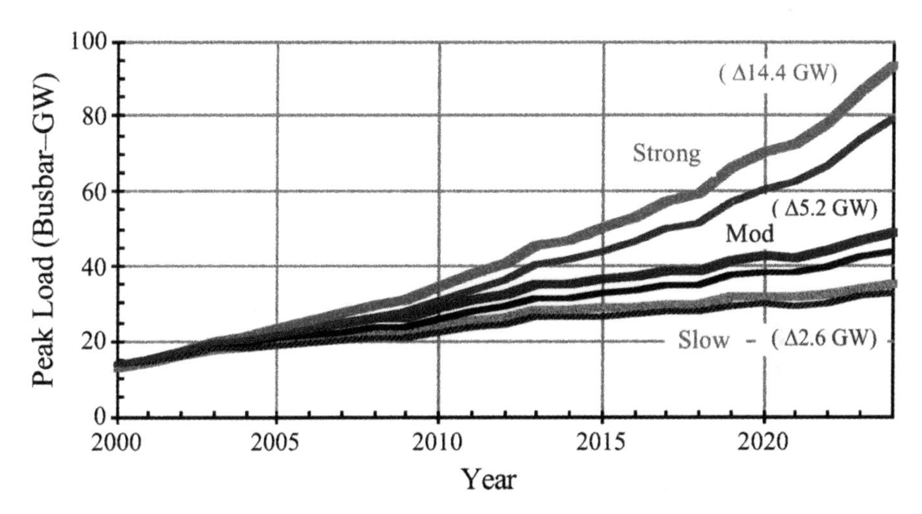

Figure 6.16 Peak Load Management Impact on Annual Peak Load

6.4.2 End-Use Efficiency

The second DSM option was end-use efficiency. Here we assume that the deployment of efficiency end-use technologies achieve Moderate 10% (M) and Aggressive 20% (G) reductions in cumulative electricity sales over the twenty-five year period, relative to the no-DSM option "current standards" (S). These saving are phased in over time by end use sector. Figure 6.17 shows these reductions.

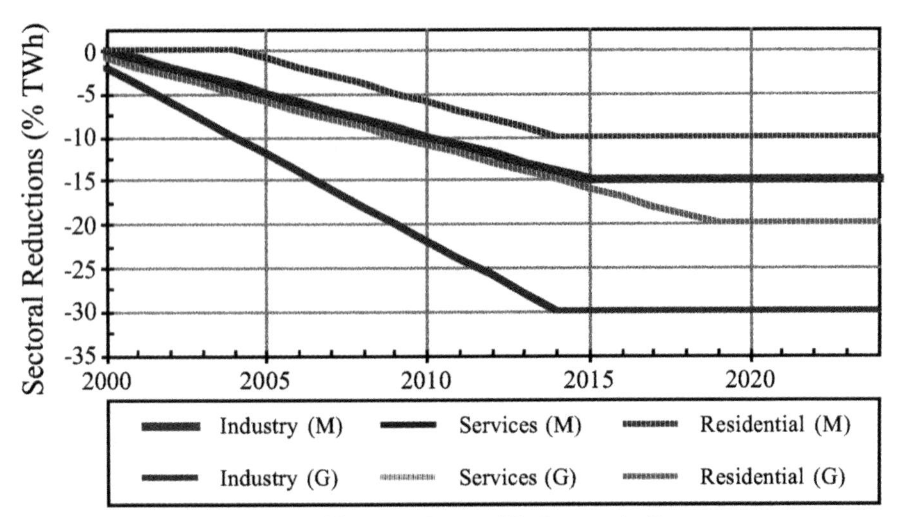

Figure 6.17 Electricity Demand Reductions by Sector (Reduction from Current Standards option)

Most reductions are attained in the industrial sector. In the Moderate case, efficiency savings start in the first year and achieve a level of 15% by 2015. Percent reductions in the services and residential sectors are identical and begin in 2005, leveling out at 10% in 2014. Industry sector efficiency gains in the Aggressive option are double those of the Moderate Option. For the services and residential sectors, the doubling to 20% is phased in over a longer time period, from 2000 to 2019.

The impact these two levels of end-use efficiency have on total demand for electricity is show in Figure 6.18, for each of the three load growth uncertainties. Moderate load growth with aggressive efficiency programs, has about the same electricity demand as slow load growth with no end-use efficiency. Figure 6.10 showed that peak electricity demand in Shandong is strongly influenced by residential demand for electricity. We modeled growth in residential demand as a having greater impact on peak load growth. Therefore, reductions in electricity demand from that and other sectors also reduces peak load growth. These impacts are shown in Figure 6.19.

Table 6.13 Combined DSM MW Impacts

	Demand-Side Option Impacts		
	Slow (T)	Moderate (F)	Strong (S)
	Growth in Electricity Demand (Busbar-GWh)		
No DSM	3.89	5.11	7.66
20% End-Use Efficiency	3.31	4.55	7.08
10% End-Use Efficiency	2.62	3.87	6.40
	Growth in Peak Load (Busbar-MW)		
No DSM	4.19	5.58	8.36
Peak Load Mgt.	3.89	5.12	7.66
20% End-Use Efficiency	3.65	5.05	7.82
10% End-Use Efficiency	3.01	4.41	7.19
20% EUE & PLM	3.30	4.54	7.07
10% EUE & PLM	2.62	3.87	6.40
	(Long-Term Growth Rate – Δ%/yr)		
	Electricity Demand Reductions (Busbar-GWh)		
20% End-Use Efficiency	-13.25		-12.76
10% End-Use Efficiency	-26.51	-26.01	-25.52
	(Δ% GWh in 2024 from No-DSM)		
	Peak Load Reductions (Busbar-MW)		
Peak Load Mgt.	-2635	-5248	-14377
20% End-Use Efficiency	-4447	-6047	-11277
10% End-Use Efficiency	-8894	-12094	-22554
20% EUE & PLM	-6951	-10940	-24502
10% EUE & PLM	-11268	-16631	-34628
	(ΔMW in 2024 from No-DSM)		

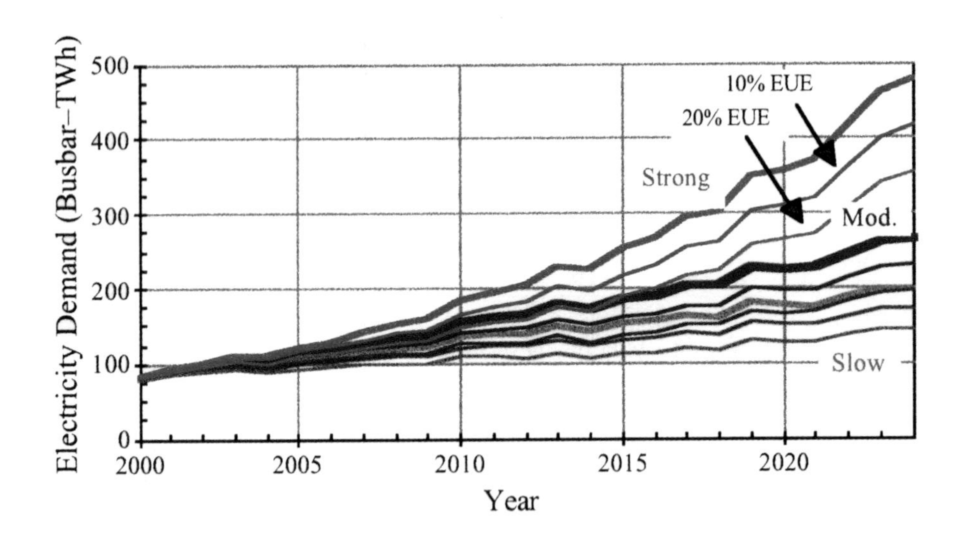

Figure 6.18 End-Use Efficiency Impacts on Electricity Demand

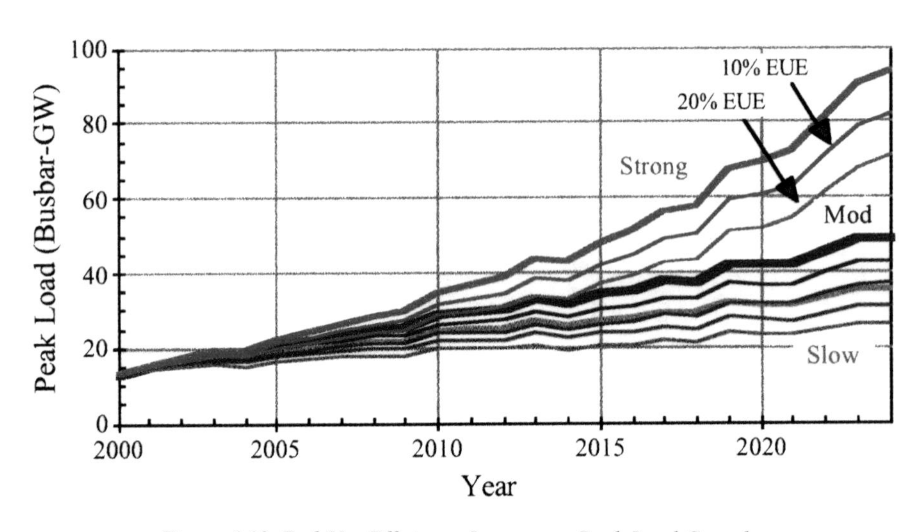

Figure 6.19 End-Use Efficiency Impacts on Peak Load Growth

Both DSM options, peak load management and end-use efficiency can work together. Table 6.13 shows the separate and combined impacts of the DSM options across all three load growth uncertainties as changes to long-term growth rates in electrical energy and peak load growth, as well as the percent reduction in electricity demand and peak load in 2024.

Like peak load management, we assumed a cost for implementing end-use efficiency programs. DSM cost assumptions were derived from Yang and Lau (1999). For the Moderate EUE option the cost of end-use efficiency was assumed to be 0.10 ¥/kWh for industry, 0.15 for services and 0.20 for residential applications. For the Aggressive option, 0.15, 0.20 and 0.25 ¥/kWh costs were used to reflect that lower cost applications had been used up. These DSM implementation costs escalated with inflation. Much greater detail on these and all the other options can be found in " LFEE Reports".

7. THE REFERENCE STRATEGY AND THE IMPACT OF GROWTH AND FUEL COST UNCERTAINTIES

The above sections gave a quick overview of the various options used to construct the strategies for Shandong Province's electric sector. Each of the two new generation options is coupled to each of the three existing generation and then the two demand side options, for a total of 1008 unique strategies. Additional sensitivity analyses were done in addition to these, such as the reference strategy without FGD. Each strategy is in turn coupled with a future comprised of an electricity demand, and a coal and a natural gas cost uncertainty. In this chapter many results will be shown in comparison to the reference strategy (BOC-CONPAS), and future (FIB). So, before jumping into the broad set of results, we will review the performance of this scenario.

Figure 6.20 shows the capacity expansion, capacity utilization and costs of supplying Shandong grid customers electric service for the entire twenty-five year study period, for the FIB future. The top graph also shows the growth in peak load and the capacity target, which is the peak load plus the planning reserve margin of 20%. Except for some combustion turbines built for peak generation beginning 2008, all new generation is subcritical pulverized coal with flue gas desulfurization. Due to long lead times of five and six years for the coal units, some years are over and under the capacity target. The middle graph shows how the simulation model dispatches the available generation. Old generation is grouped by location within the province, and new generation by is grouped by technology type. These designations were used to help calculate population exposures for the environmental impact assessment. The third plot shows the costs of operating the system, beginning with generation and T&D capital costs on the bottom, general and administrative costs, and then operation and maintenance and fuel costs for generation. Costs are shown in base year Yuan (1999), and grow with the level electricity demand. Unit costs will be shown later.

These plots show the true value of a simulation approach. As new coal generation enters the system it displaces most of the generation by older power plants, since the new units are more efficient and cheaper to operate. The impact this has on emissions are shown in Figure 6.21. Here SO_2, PM10, NO_x and CO_2 emissions are plotted. Old generation has been aggregated into emissions from old-small, old-large and new generation.

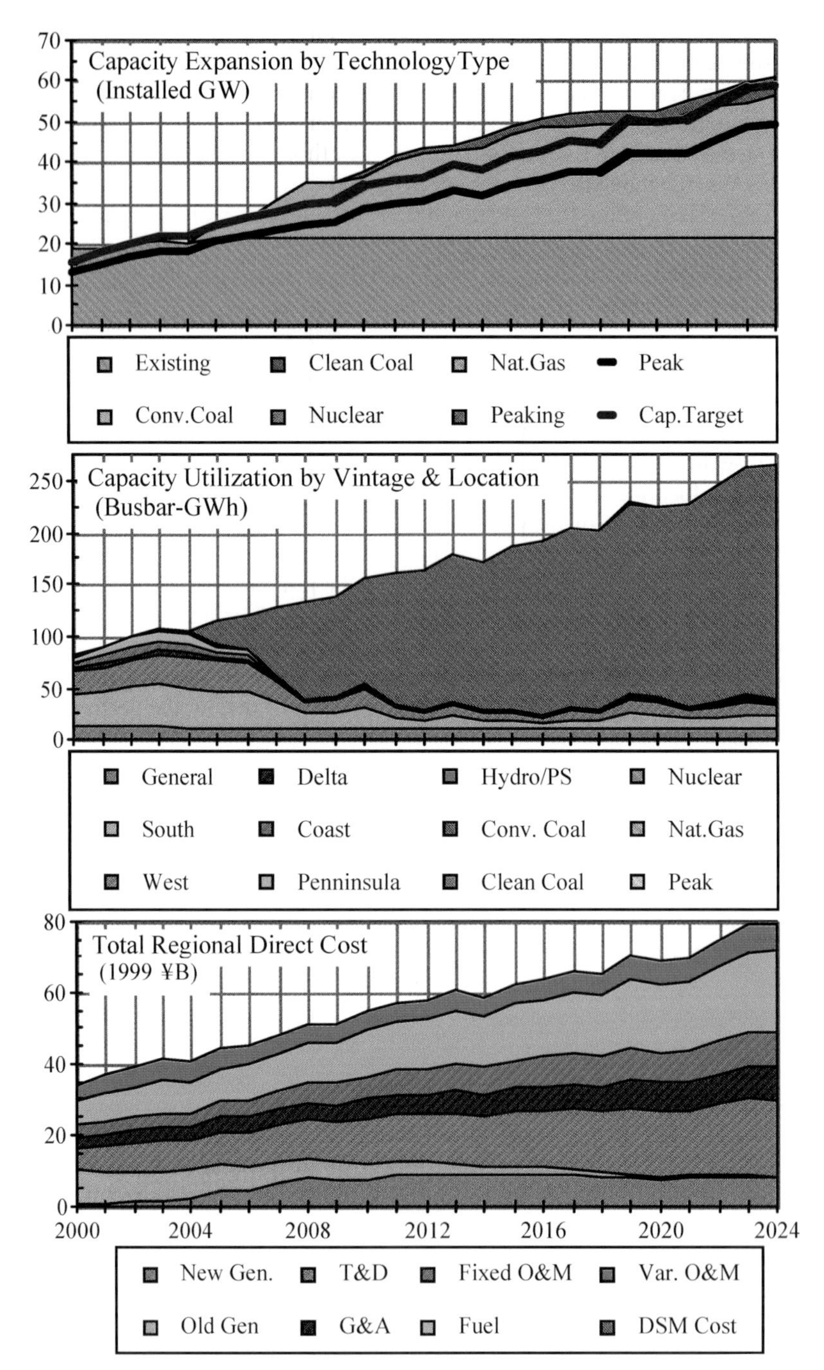

Figure 6.20 Reference Scenario Technical and Cost Performance (BOC-CONPAS-FIB)

The emissions results show several things. First is that the current policy of retiring old units 50 megawatts or smaller is very sound from an emissions viewpoint. However, emissions from small units continue to contribute substantially to particulate emissions. Both SO_2 and particulate emissions drop in the first third of the study period and then increase gradually, even though annual electricity demand more than triples. NO_x and CO_2 emissions however continue to grow.

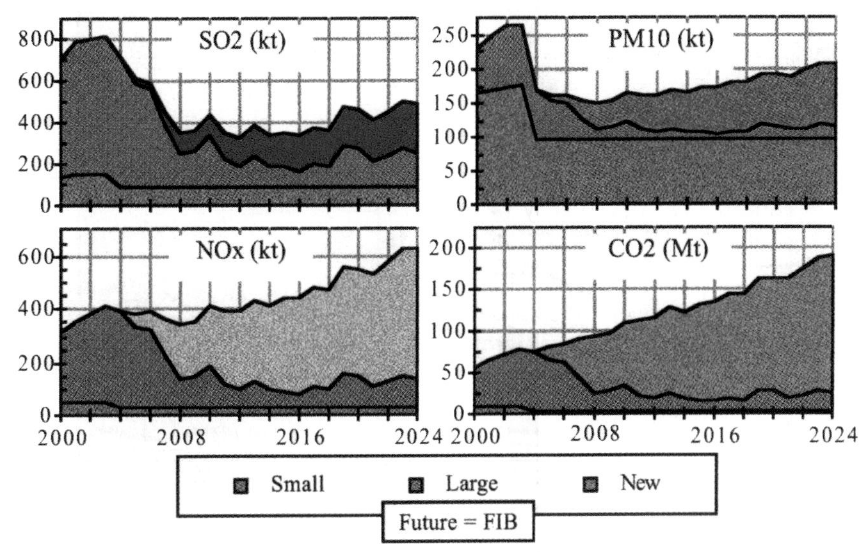

Figure 6.21 Reference Scenario Emissions Performance (BOC-CONPAS-FIB)

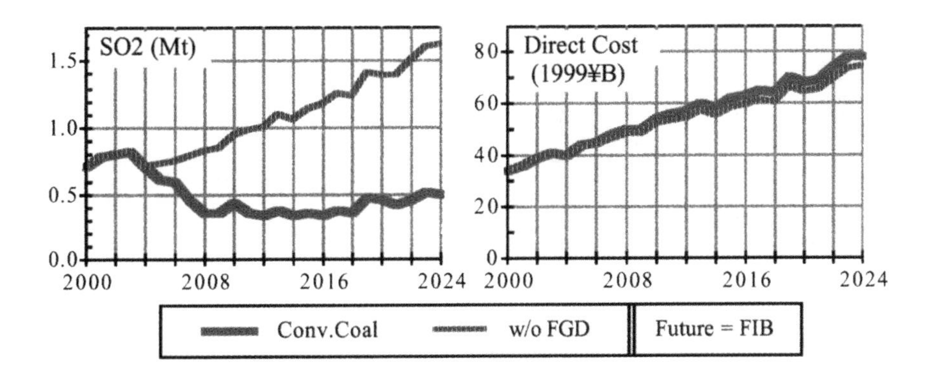

Figure 6.22 Impact of the FGD Assumption

Figure 6.22 shows how the assumption that new conventional coal units use FGD instead of low sulfur coal impacts SO_2 emissions and costs. While the cost of the reference case without FGD is slightly lower, due to the reduced cost of non-FGD generators, and their slightly better efficiency, sulfur emissions grow substantially. Particulate, NO_x and CO_2 emissions were effectively the same with and without FGD. Therefore, assuming FGD on new conventional coal units is the better option of the two allowed in by China's current sulfur reduction policy.

How does the reference case perform across uncertainties? Figure 6.23 shows how the reference strategy performs for the slow, moderate and strong load growth uncertainties, retaining the reference uncertainties for coal and natural gas costs (_IB). Displayed are installed capacity, the *unit* cost of electric service, and the four principal air emissions.

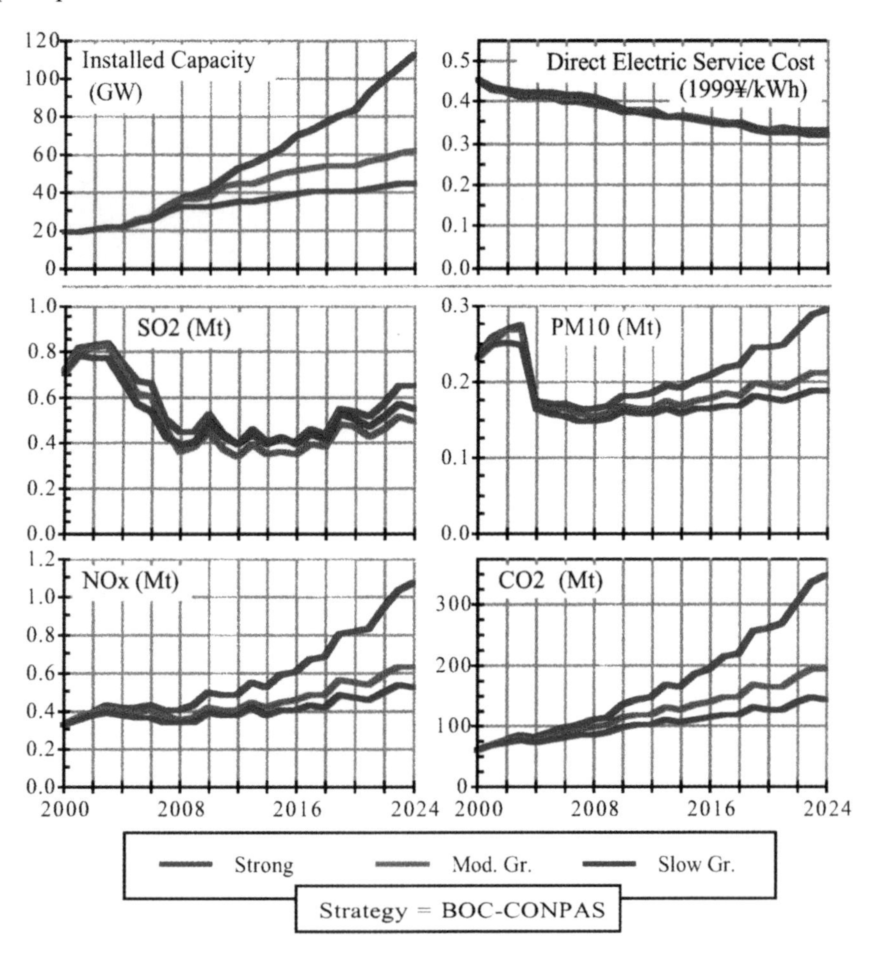

Figure 6.23 Impact of Load Growth on Capacity, Costs and Emissions

While capacity grows, cost to the consumer remains about the same, as the increased costs are spread over a greater number of kilowatt-hours. Total direct costs continue to grow as in the previous graph. Sulfur emissions stay about the same as well, as the operation of high SO_2 emitting older units cannot increase by too much. Close examination of Figure 6.23 shows that SO_2 emissions are higher in the slow growth case compared to moderate growth, highlighting the dynamics between old and new generation. There is greater emissions growth in the other three pollutants, especially CO_2 and NO_x, whose emissions do not benefit from the retirement of older smaller units at the end of 2003.

In the following sections we examine how the choice of options around the reference case impact emissions and costs, then we look at those options in combination.

8. PERFORMANCE OF ESS SCENARIOS

Before delving into the performance of individual options and strategies it is important to understand the range over which costs and emissions change due to both options and uncertainties. Figure 6.24 shows costs versus emissions for three of the eighteen futures analyzed.

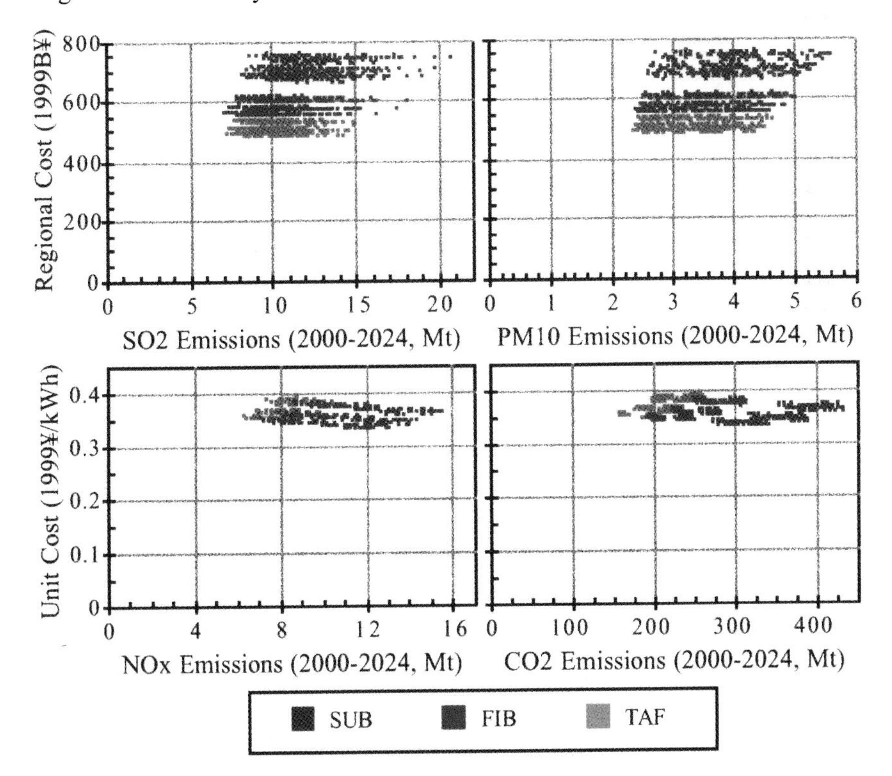

Figure 6.24 Range of Variation in Costs and Emissions Across Futures

All the strategies depicted are without gas-by-wire for the reference future FIB, and two other futures selected so that emissions would be either high or low. The low emissions future is where electricity demand is slow, coal prices high and natural gas prices low so that generation shifts to units with higher efficiency or other fuels (TAF). In contrast, the high emissions future is where growth in electricity demand is high, coal costs low, and natural gas costs high (SUB). The top two plots show cumulative SO_2 and PM10 emissions versus the present value of total direct regional costs, while the bottom two show NO_x and CO_2 against the average unit cost of electric service. Each plot has a zero-zero origin so that the true relative movement in costs and emissions can be evaluated.

Significant in Figure 6.24 is the overlap in the emissions and unit costs across the futures. For regional costs there is almost no overlap on costs, due to the scale effects of the different load growth rates and the amount of electricity being produced. Normalizing this in the lower two graphs eliminates the overlap, and actually makes the unit cost for strong demand slightly lower since growth in demand is slightly faster than growth in expenditures. This trend should be treated with caution however, as the costing, especially of non-generation activities, was calculated using a benchmarking technique rather than with confidential utility cost information. Ranges from highest to lowest unit costs within these three futures 11%, and 16% maximum to minimum cost across the three futures.

More significant perhaps is the lack of overlap along the emissions axes. The overlap among SO_2, PM10 and NO_x emissions is substantial. CO_2 emissions overlap to a lesser degree. The ranges are considerably higher than those for cost; roughly 52-61% reductions for SO_2, 49-54% for PM10, 39-43% for NO_x and 37-41% for CO_2. This implies two important lessons. First, there is more opportunity to reduce emissions than there are costs. Second, that while ranges in costs are less important in a design-oriented tradeoff analysis, the design element in crafting scenarios will have a very large impact on the emissions performance of the strategies. With these thoughts in mind, we next look at the performance of the individual classes of options and their cumulative and annual impacts on costs and emissions.

8.1 New Generation Options

How did the seven combinations of conventional coal, clean coal, natural gas and nuclear generation technologies plus the gas-by-wire option perform? Table 6.14 shows how these options performed relative to the reference strategy for the moderate growth FIB future, without any other option choices. The reference case was the cheapest of the eight strategies shown, with the AFBC and gas-by-wire strategies being the most expensive. In the AFBC case, these clean coal units had roughly the same efficiency as the pulverized coal units, and so did not provide any operational cost savings to offset their higher capital cost. In the gas-by-wire case, must-run dispatch with a high fuel cost and additional transmission losses, makes it the most expensive new supply alternative.

Focusing on emissions performance, the combinations with IGCC and nuclear performed best, with the nuclear combinations providing substantial reductions in

CO_2 as well. The NGCC and Nuclear and NGCC options had relatively poor environmental performance in the FIB future, with substantial increases in SO_2 emissions. Although natural gas units were built, they operated very few hours due to the high cost of natural gas relative to coal. The emissions increases occur because older, higher emissions power plants are not displaced as they are with the other options. In the lower natural gas cost futures, this does not occur as much.

Table 6.14 Cost and Emissions Performance of New Generation Technology Options

New Generation Options	Electric Service Direct Costs		Power Plant Stack Emissions			
	Regional	Unit	SO2	PM10	NOx	CO2
Reference	601.0	0.373	12.34	4.69	11.05	3.01
AFBC-2010	610.8	0.379	11.91	4.64	9.91	2.96
IGCC-2012	606.5	0.377	11.31	4.28	9.79	2.84
NGCC-2015	602.5	0.374	15.98	4.87	12.00	2.99
Nuclear-2010	604.5	0.375	11.26	4.47	9.81	2.64
Nuclear & NGCC	603.9	0.375	14.15	4.60	10.66	2.65
Nuc., NGCC & IGCC	608.4	0.378	11.30	4.18	9.06	2.52
NGCC by Wire	612.6	0.381	11.38	4.50	10.00	2.85
	(NPV BYuan)	(Yuan/kWh)	(Mt)	(Mt)	(Mt)	(Gt)
Percent Change from Reference - Conventional Coal with FGD						
AFBC-2010	1.6	1.7	(3.5)	(1.0)	(10.3)	(1.4)
IGCC-2012	0.9	0.9	(8.4)	(8.7)	(11.4)	(5.5)
NGCC-2015	0.2	0.3	29.6	3.9	8.6	(0.5)
Nuclear-2010	0.6	0.6	(8.7)	(4.7)	(11.2)	(12.0)
Nuclear & NGCC	0.5	0.5	14.7	(1.9)	(3.5)	(11.8)
Nuc., NGCC & IGCC	1.2	1.3	(8.4)	(10.9)	(18.0)	(16.3)
NGCC by Wire	1.9	2.0	(7.7)	(4.0)	(9.5)	(5.1)
(Future = FIB)	(Δ%)	(Δ%)	(Δ%)	(Δ%)	(Δ%)	(Δ%)

Figure 6.25 shows how all but the gas-by-wire options perform on an annual basis. Apparent are the similarities in performance for the amount of installed capacity and regional direct costs. Another important observation is that substantial costs and emissions impacts only occur as the new technologies come on-line in sufficient numbers. As seen before, SO_2 and PM10 emissions drop as planned retirements of old, smaller generators occur. Differences in CO_2 emissions are directly linked to those options including combinations of nuclear and IGCC. Here the failure of natural gas options to displace older generation and reduce SO_2, PM10 and NO_x emissions is apparent. Furthermore, the strategies with nuclear and/or IGCC are able to sustain, to some degree, SO_2 and PM10 reductions over time.

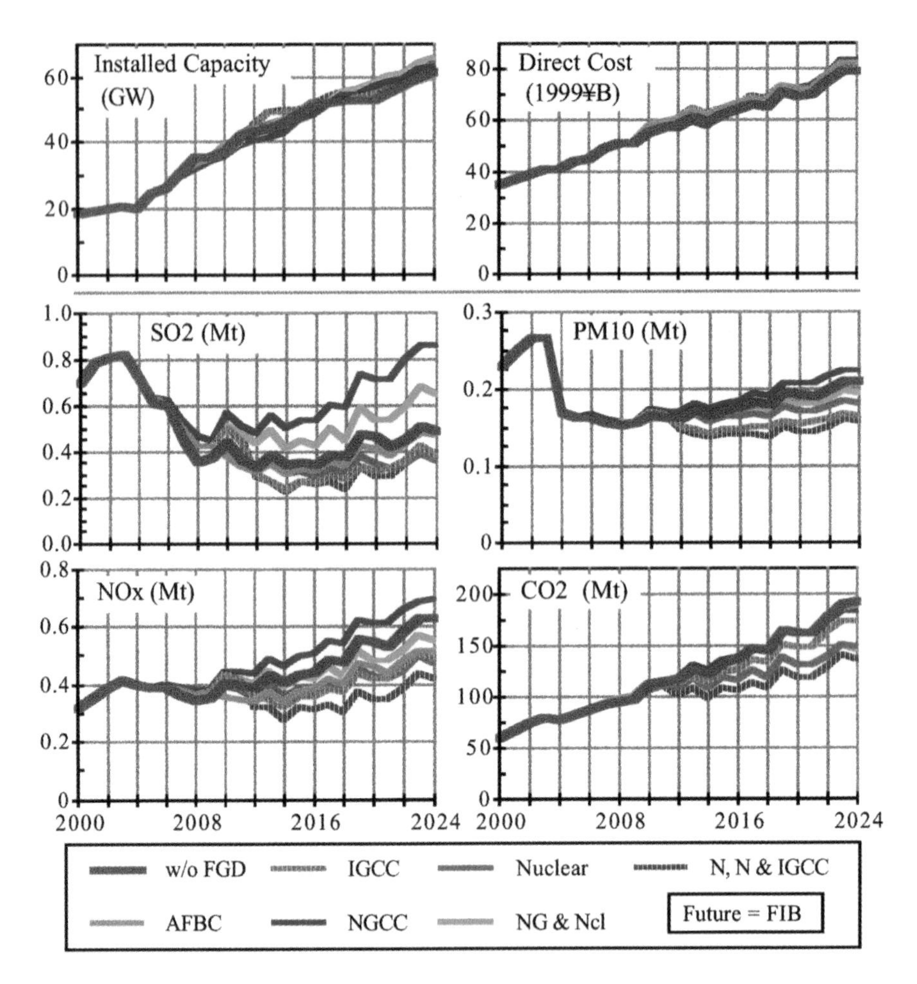

Figure 6.25 Annual Performance of Generation Mix Options

How well does the selection of generation technology perform as we add back in the other options? The 5 to 20% reductions in emission is only part of the overall 40-55% range shown in Figure 6.24. Figure 6.26 shows all of the 1008 strategies for the FIB future, keyed by choice of new generation technology mix. Highlighted are the positions of the reference strategy (BOC-CONPAS), and the least-cost (BOX-CONLAG) strategy that defines the low-cost, high emissions end of the tradeoff frontier for the FIB future. As can be seen, the choice of new generating technologies is not the principal driver for reducing SO_2 and PM10 emissions, as there is substantial overlap in the generation technology clusters. Some differentiation does occur in the cost versus CO_2 plot with the options including nuclear clustered to the left, although other options are exerting significant influence here as well. We explore what these are in the next sections.

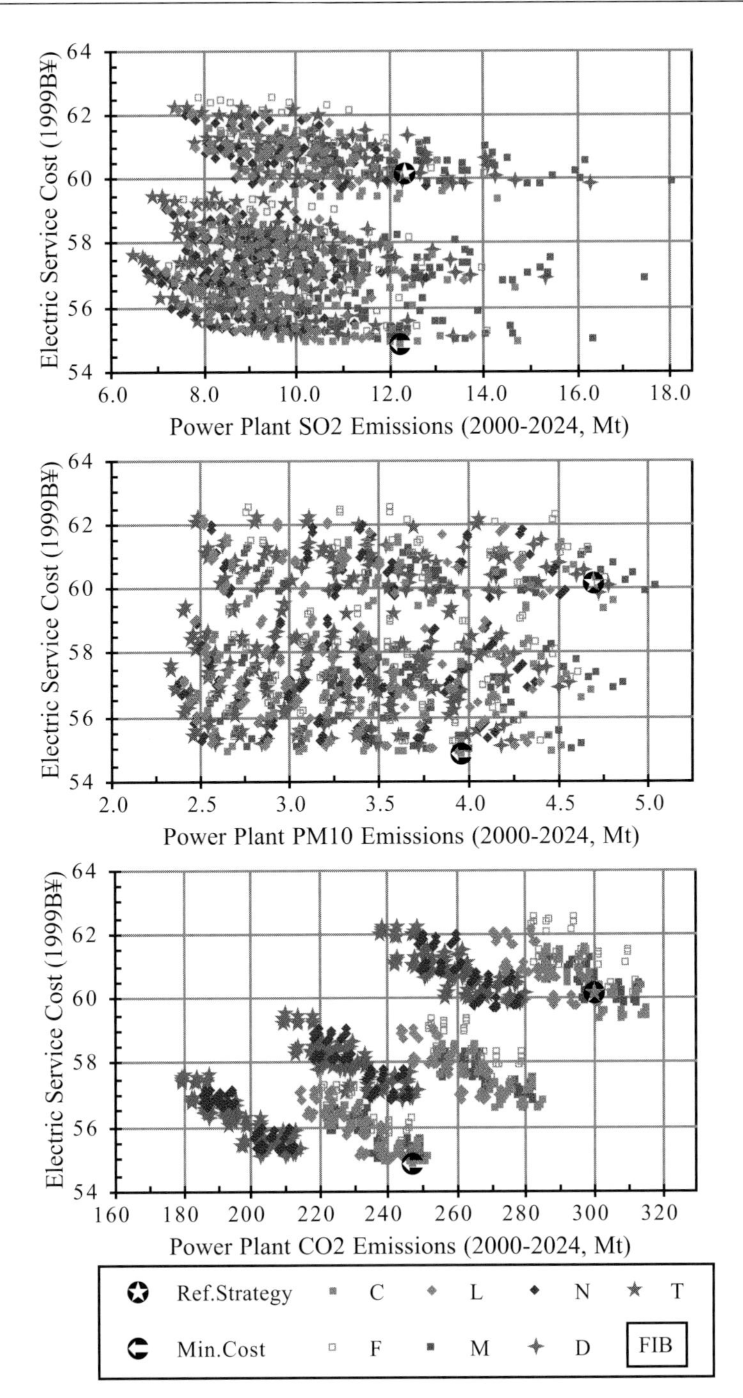

Figure 6.26 Comparative Performance of FIB Strategies Keyed by Future Generation Mix

8.2 Existing Generation Options

Existing generation options include the select retirement of old power plants, as well as the "forced" retirement of units after 35 years of operation, plus select scrubber retrofits, and the use of lower-ash, lower-sulfur prepared coals. The switch to prepared coal option is a hybrid option as it was also applied to new coal-fired generation in the "switch all" formulation. When these options were combined, units scheduled for early retirement were not given scrubbers. Similar to the presentation of new generation technology option results, Table 6.15 and Figures 6.27 and 6.28 show the performance of these options relative to the reference case, and across all options.

Table 6.15 Cost and Emissions Performance of Existing Generation Options

Existing Generation Options	Electric Service Direct Costs		Power Plant Stack Emissions			
	Regional	Unit	SO2	PM10	NOx	CO2
Reference	601.0	0.373	12.34	4.69	11.05	3.01
Retire Select	601.9	0.374	10.87	3.78	10.73	3.01
FGD Retrofit Some	602.6	0.374	11.73	4.71	11.03	3.01
Switch Existing	600.5	0.373	10.80	4.35	11.05	3.05
Switch All Conv. Coal	601.1	0.373	10.27	3.65	11.05	3.12
Retire & Sw. Exist	601.4	0.373	9.54	3.49	10.73	3.06
Retire, FGD & Sw. All	603.9	0.375	8.68	2.74	10.72	3.14
	(NPV BYuan)	(Yuan/kWh)	(Mt)	(Mt)	(Mt)	(Gt)
Percent Change from Ref. - No Retirements, Retrofits or Cleaner Coals						
Retire Select	0.1	0.2	(11.9)	(19.5)	(2.9)	0.3
FGD Retrofit Some	0.3	0.3	(4.9)	0.4	(0.1)	0.2
Switch Existing	(0.1)	(0.1)	(12.5)	(7.2)	0.0	1.6
Switch All Conv. Coal	0.0	0.0	(16.8)	(22.2)	0.0	3.9
Retire & Sw. Exist	0.1	0.1	(22.7)	(25.6)	(2.9)	1.7
Retire, FGD & Sw. All	0.5	0.5	(29.7)	(41.7)	(3.0)	4.4
(Future = FIB)	(Δ%)	(Δ%)	(Δ%)	(Δ%)	(Δ%)	(Δ%)

Compared to the reference strategy, all cost slightly more, except for the "switch existing" option. In this case, making the older units slightly more expensive shifted even more generation to the newer, more efficient generators. Reasons for the cost increases are the additional cost of the prepared coal, and replacement capacity (retire) and FGD systems (retrofit) investment costs. Given the uncertainties in the cost assumptions and how they propagate through time, these options are effectively the same cost.

Overall the emissions reductions from these options are more than double those from the choice of new generation technology, except for CO_2, for little or no cost impact. When the options are combined, emissions reductions are greater still. The increase in CO_2 is attributed to the retirement of cogenerators, which were not only efficient from an electricity generation viewpoint, but also assumed to be industrial thermal-following units and therefore modeled on a "must-run" basis.

Figure 6.27 shows the year-to-year performance of the existing generation options. As can be seen overall installed capacity and direct costs are essentially the same, as are NO_x and CO_2 emissions. Early reduction in SO_2 and PM10 come from switching to prepared coal. Whether such a degree of fuel switching could actually occur needs to be explored. The switch and retire combinations achieve the largest reductions, and are roughly half the year 2000's SO_2 and PM10 emissions.

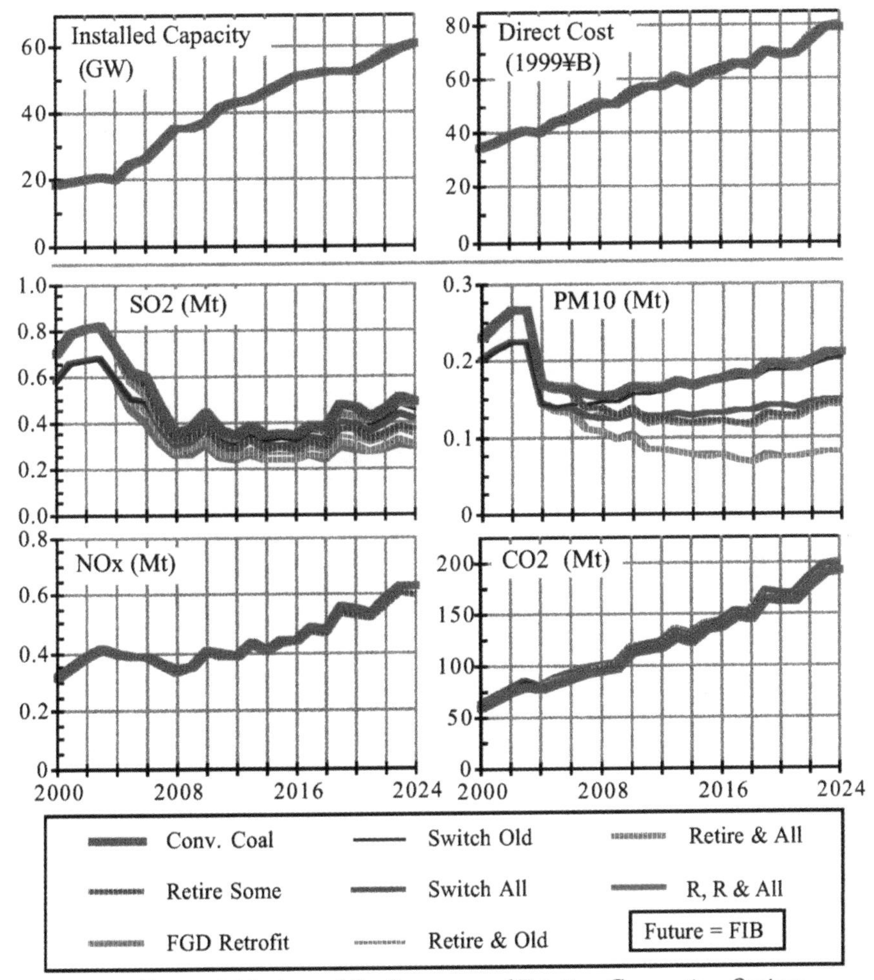

Figure 6.27 Annual Performance of Existing Generation Options

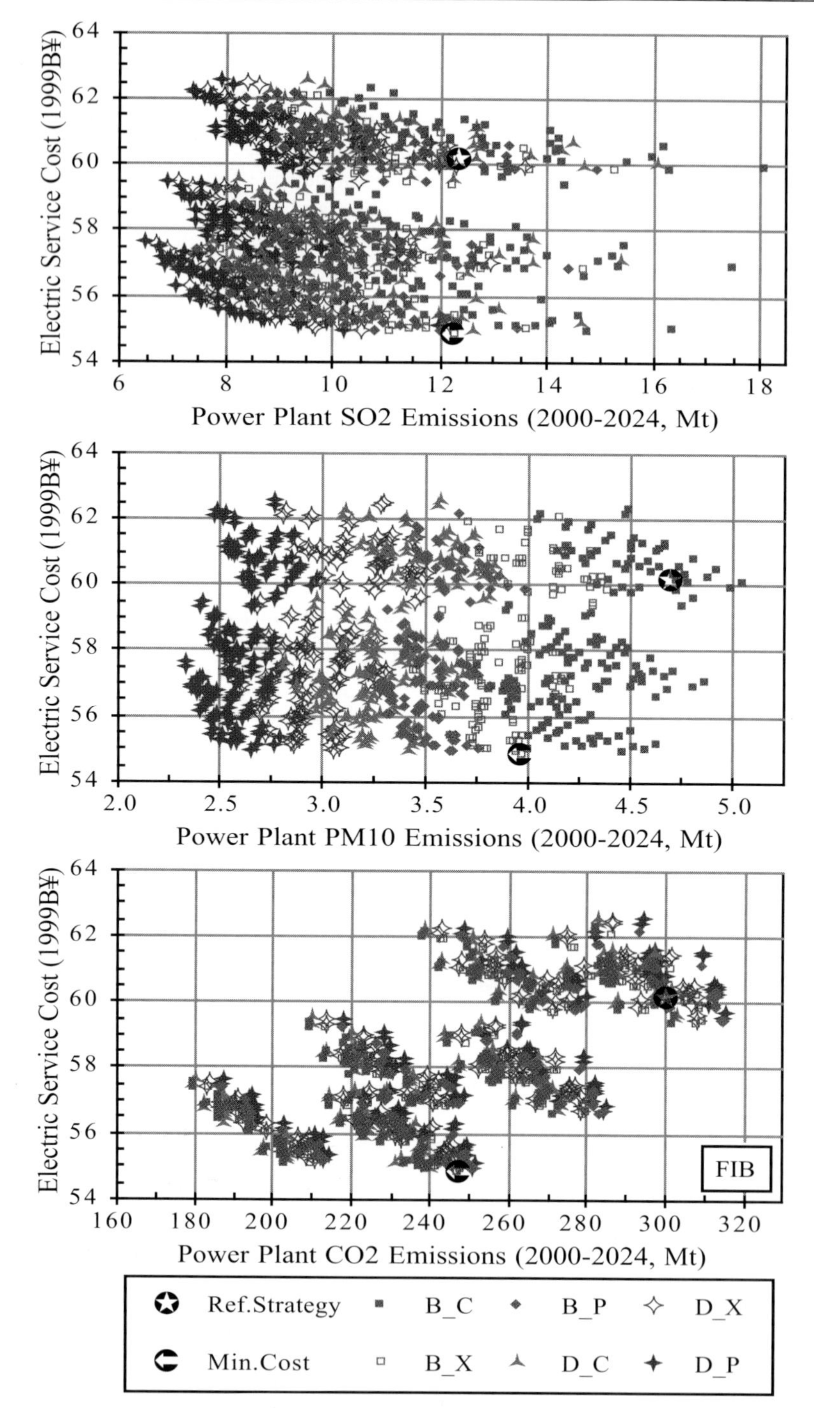

Figure 6.28 Comparative Performance of FIB Strategies Keyed by Existing Generation Options

Figure 6.28 shows the performance of these options along with all the other choices. Strategies are grouped by their combination of retirement (B, D) and switching (C, X, P) options. The impact this has on SO_2, and especially PM10 emissions is readily apparent.

8.3 Demand-Side Options

To look at the avoided costs and emissions of alternate types and levels of demand-side management, the electric sector strategies included 10% (M, Moderate) and 20% (G, Aggressive) reductions in total electricity demand, compared with current efficiency standards (S) option. These reductions were phased in over time, and contributed to reductions in peak load as well. In addition to end-use efficiency programs, the impact of peak load reductions were also analyzed (P–No Peak Management, L-Load Management). Table 6.16 shows the impacts of these end-use efficiency options working alone and together with peak load management.

Table 6.16 Cost and Emissions Performance of Demand-Side Options

Demand-Side Options	Electric Service Direct Costs		Power Plant Stack Emissions			
	Regional	Unit	SO2	PM10	NOx	CO2
Reference	601.0	0.373	12.34	4.69	11.05	3.01
Peak Load Mgt.	593.8	0.369	14.32	4.76	11.69	3.01
Moderate Efficiency	571.1	0.354	12.16	4.51	10.22	2.70
Moderate & Load Mgt.	565.9	0.351	14.71	4.62	11.04	2.71
Aggressive Efficiency	552.7	0.342	12.08	4.31	9.39	2.37
Aggressive & Load Mgt.	548.7	0.340	14.74	4.45	10.22	2.39
	(NPV BYuan)	(Yuan/kWh)	(Mt)	(Mt)	(Mt)	(Gt)
Percent Change from Ref.- Current Eff. Standards and No Peak Load Mgt.						
Peak Load Mgt.	(1.2)	(1.2)	16.1	1.4	5.8	0.3
Moderate Efficiency	(5.0)	(5.1)	(1.4)	(3.8)	(7.5)	(10.2)
Moderate & Load Mgt.	(5.8)	(6.0)	19.2	(1.4)	(0.1)	(9.7)
Aggressive Efficiency	(8.0)	(8.2)	(2.1)	(8.0)	(15.0)	(21.0)
Aggressive & Load Mgt.	(8.7)	(8.9)	19.5	(5.0)	(7.5)	(20.5)
(Future = FIB)	(Δ%)	(Δ%)	(Δ%)	(Δ%)	(Δ%)	(Δ%)

Again the dynamic among electricity demand, and old and new generation is apparent. The peak load management option, where peak load grows at the same rate as annual electricity demand, avoids the need for new power plants, but not the need for additional generation. While this saves considerable cost, it means that there are fewer new generators to displace the hours of operation of the older dirtier units. Costs go down, but emissions go up. End-use efficiency in contrast achieves both reductions in costs and emissions. Enough generation is avoided such that the increased use of older generators still results in a net reduction in emissions.

When both end-use efficiency and peak load management are pursued together, emissions still go down for all pollutants except sulfur dioxide. This is apparent when the annual changes in installed generation, costs and emissions are examined in Figure 6.29. Since both old and new conventional coal-fired generation have roughly equivalent conversion efficiencies and particulate controls, there is little divergence between end-use efficiency and peak load management programs' emissions. The difference between new and old generation–without flue gas desulfurization–however makes a large difference in annual SO_2 emissions.

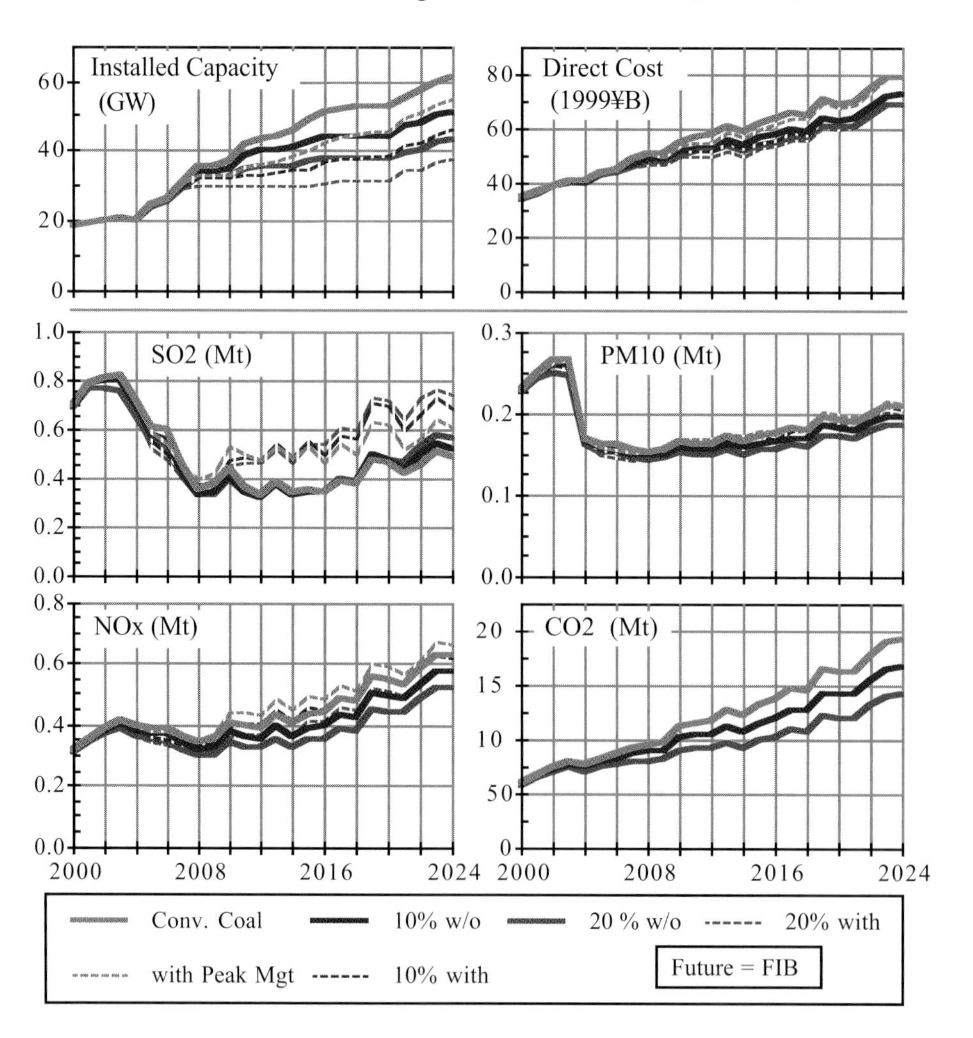

Figure 6.29 Annual Performance of Demand-Side Options

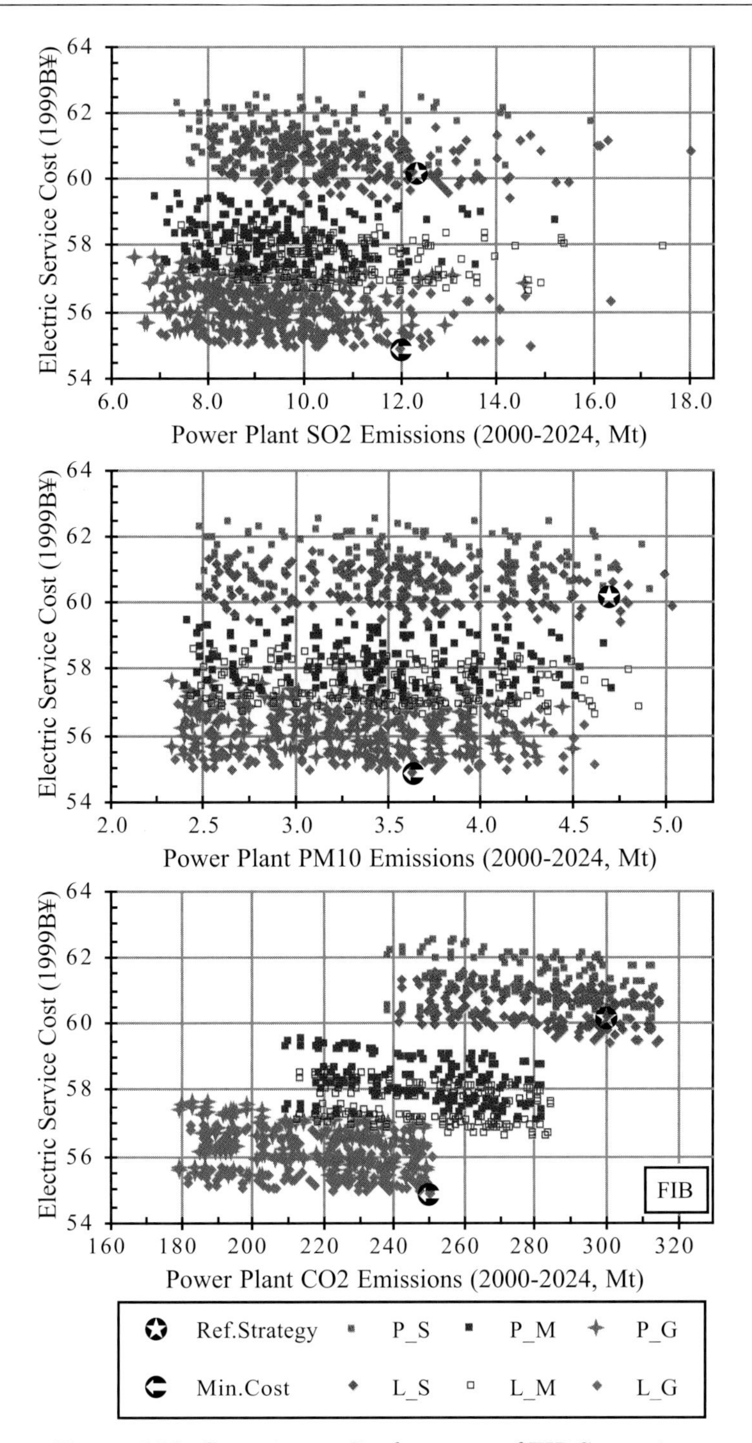

Figure 6.30 Comparative Performance of FIB Strategies Keyed by Demand-Side Options

Figure 6.30 shows how the DSM options perform when all the other options are in force. Here we see dramatic reductions in costs and CO_2 emissions, but like the choice of new generation technologies, only minor reductions in SO_2 and particulate emissions. Like the 10% and 20% reductions in electricity demand, the costs of implementing DSM are to a large degree hypothetical, and need to be refined. However, a quick sensitivity analysis, where the cost of implementing end-use efficiency was doubled, made these strategies roughly the same cost as the no EUE strategies, but still with substantially lower PM10, NO_x and CO_2 emissions.

8.4 Integrated Supply and Demand-Side Strategies

As the above sections showed, phasing out or cleaning up older generation and the fuels it uses leads to large reductions in particulate and sulfur dioxide emissions. End-use efficiency and investment in higher efficiency and lower carbon generation technologies reduced carbon dioxide. End-use efficiency and peak load management were the lower cost strategies. How do these options perform in combination? Table 6.17 shows the performance of strategies that retire select units and units as they reach thirty-five years of operation (D), use of prepared coal in all coal fired generation (P), plus peak load management (L) and aggressive (20%) end-use efficiency programs (G). These "DPLG" strategies are shown for conventional coal, IGCC, nuclear, nuclear and natural gas, and nuclear, natural gas and IGCC new generation mixes. As can be seen, all offer substantial cost and emissions reductions, compared to the reference case.

Table 6.17 Cost and Emissions Performance of Integrated Strategies

	Electric Service		Power Plant			
Integrated Demand-Side &		Direct Costs		Stack Emissions		
Existing Unit Strategies	Regional	Unit	SO2	PM10	NOx	CO2
Reference	601.0	0.373	12.34	4.69	11.05	3.01
with "DPLG"	549.0	0.340	10.23	2.66	9.73	2.51
plus IGCC-2012	551.0	0.341	9.41	2.56	8.96	2.42
plus Nuclear-2010	552.5	0.342	8.66	2.51	8.19	2.12
plus Nuclear & NGCC	551.4	0.342	9.93	2.58	8.70	2.13
plus Nuc., NGCC & IGCC	554.0	0.343	8.32	2.46	7.74	2.06
	(NPV BYuan)	(Yuan/kWh)	(Mt)	(Mt)	(Mt)	(Gt)
Percent Change from Reference - BOC-CONPAS						
with "DPLG"	(8.7)	(8.8)	(17.1)	(43.4)	(11.9)	(16.6)
plus IGCC-2012	(8.3)	(8.5)	(23.8)	(45.4)	(18.9)	(19.3)
plus Nuclear-2010	(8.1)	(8.2)	(29.8)	(46.6)	(25.8)	(29.4)
plus Nuclear & NGCC	(8.2)	(8.4)	(19.5)	(45.0)	(21.3)	(29.0)
plus Nuc., NGCC & IGCC	(7.8)	(8.0)	(32.6)	(47.6)	(29.9)	(31.3)
(Future = FIB)	(Δ%)	(Δ%)	(Δ%)	(Δ%)	(Δ%)	(Δ%)

Figure 6.31 shows the annual performance of these six strategies. By bundling the best performing individual options from each of the new generation, old generation, and end-use classes of options, substantial and sustained reductions in all emissions are achieved. By concurrently promoting DSM and the renewal of the province's fleet of generators, the problem of aggressive DSM contributing to the increased use of old, dirty generators is avoided. If newer, cleaner generation technologies are selected in addition to this, then emissions are lower still.

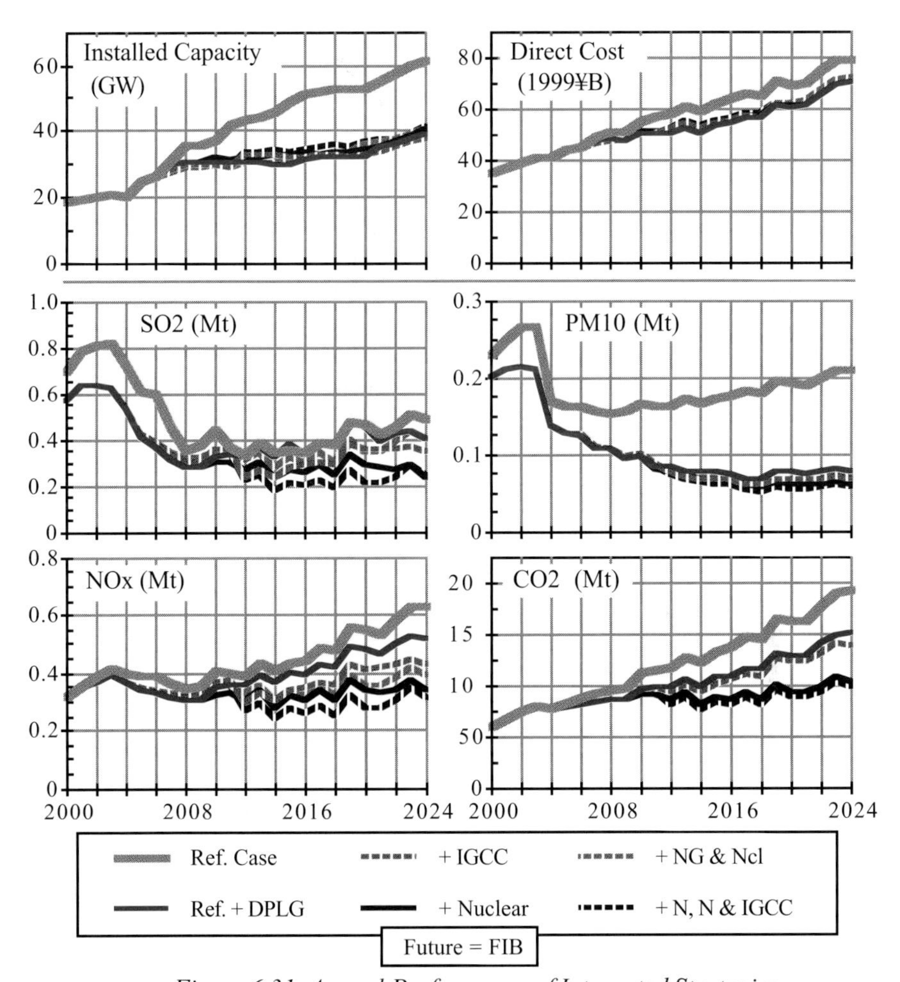

Figure 6.31 Annual Performance of Integrated Strategies

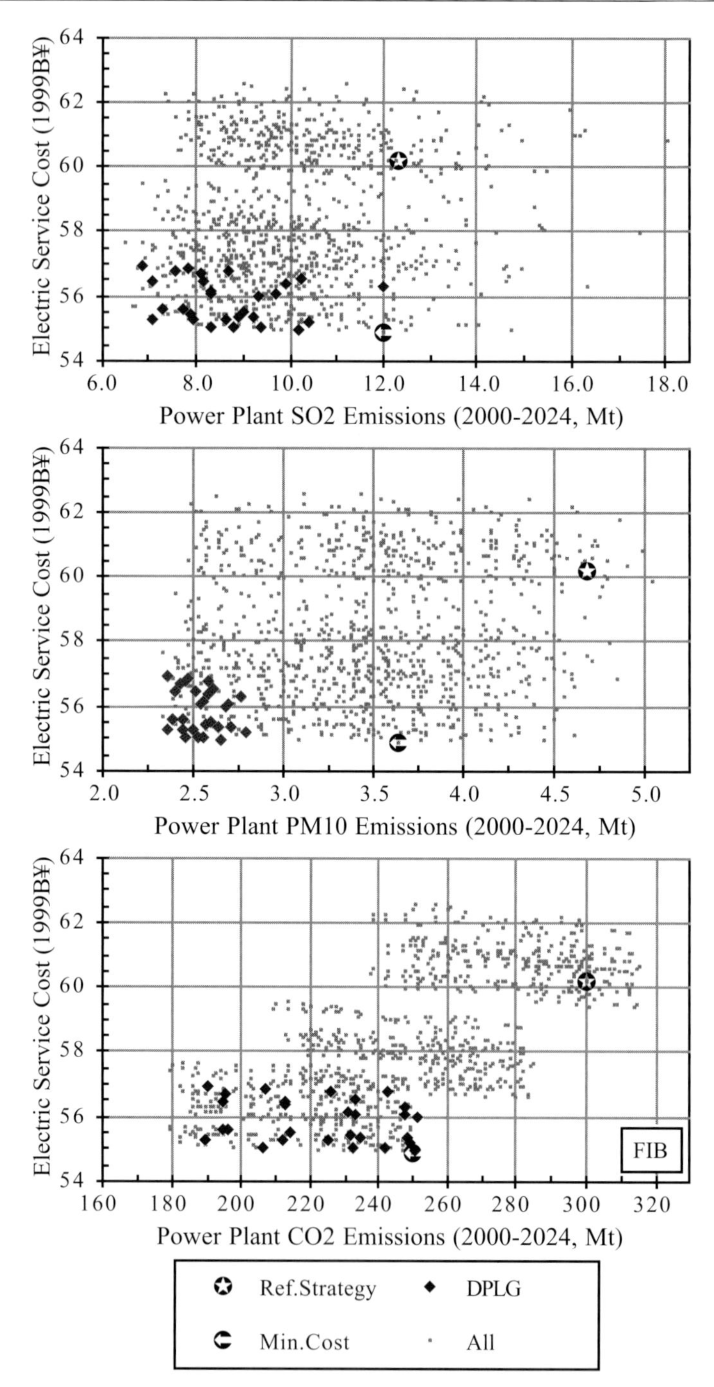

Figure 6.32 Comparative Performance of Select FIB Strategies Keyed by Strategies with Retirements, Prepared Coal, Peak Load Management and Aggressive End-Use Efficiency

Figure 6.32 shows how seven of the twenty-eight "DPLG" strategies perform in comparison to the others. These seven include all of the new generation mixes, but not the gas-by-wire and FGD retrofit options. These are at or near the bend in the frontiers for all three cost-emissions tradeoff plots. Strategies with even lower sulfur emissions include the FGD retrofit and No Peak Load management options. Lower CO_2 strategies either do not use as much prepared coal, or retire as many old units.

8.5 Selection of Strategies for Further Analysis

The current scope of the electric sector simulation task was to identify the direct cost and power plant emissions characteristics of a broad mix of options, combined into multi-option strategies. A subset of twelve of these strategies were then selected for further analysis in the life-cycle analysis (LCA), environmental impact assessment (EIA), and multi-criteria decision-aiding (MCDA) tasks. These "MCDA Strategies" are shown in Table 6.18. Rather than just selecting the reference case and the tradeoff frontier strategies, such as those just presented, a cross section of the scenario set was chosen. This was done to respect the uncertain feasibility and implementability of certain options, which includes not only their technology performance and cost, but also their timing. A cross section also helps decision-makers determine the consequences of "half measure" strategies, where some but not all of the better options are implemented.

Table 6.18 The MCDA Strategies

MCDA Strategies	Retire More	Retrofit FGD	Prep. Coal	New Gen.	FGD on New	Peak Mgt	End-Use Eff.
(1) BOC-CENPAS	–	–	–	C	No	*No*	–
(2) BOC-CONPAS	–	–	–	C	Yes	*No*	–
(3) BOX-CONPAM	–	–	Exist.	C	Yes	*No*	10%
(4) DOX-CONLAG	Yes	–	Exist.	C	Yes	Yes	20%
(5) BOX-LONLAM	–	–	Exist.	L	Yes	Yes	10%
(6) DOX-MONLAM	Yes	–	Exist.	M	Yes	Yes	10%
(7) BOC-NONLAS	–	–	–	N	Yes	Yes	20%
(8) BOX-NONLAM	–	–	Exist.	N	Yes	Yes	10%
(9) BOX-NONLAG	–	–	Exist.	N	Yes	Yes	20%
(10) DOX-TONLAG	Yes	–	Exist.	T	Yes	Yes	20%
(11) DUX-DONLAG	Yes	Yes	Exist.	D	Yes	Yes	20%
(12) DUX-TONPAS	Yes	Yes	Exist.	T	Yes	No	–

The MCDA scenarios include the reference case (2) and also the same strategy without flue gas desulfurization (1) on new conventional coal-fired units. The remaining ten strategies reflect alternate combinations of options that clean up electricity supplies, reduce the demand for electricity, or both. The next strategy (3) builds upon the reference case, BOC-CONPAS, by using prepared coal in existing units only, and achieving a 10% reduction in electricity demand without a separate peak load management program. For the MCDA strategies, use of prepared coal was restricted to existing generators due to uncertainties regarding how large and fast coal preparation technologies could be deployed in the coal mining sector. Also, using prepared coal in existing units only makes these generators slightly more expensive to operate relative to new coal units, thereby achieving an additional shift in dispatch to newer, cleaner generators. The next strategy (4) adds retirement of old units, 20% end-use efficiency and peak load management. This strategy will be shown in detail below, and has been named "*Conventional Coal Plus*", for descriptive purposes, as it aggressively pursues cleaning up older generation and implementing DSM, but continues to rely upon conventional coal-fired technologies for the production of electric power. Then next two strategies (5, 6) pursue moderate EUE and PLM, but choose coal-gasification or natural gas combined-cycle generation in addition to conventional coal. The strategy with natural gas combined cycle also retires old units to address the fact that natural gas units may be "underutilized" if natural gas costs remain high relative to coal costs. The next three strategies (7, 8, 9) look at coal plus nuclear with peak load management, and different combinations of fuel switching and end-use efficiency. Strategies 10 and 11 reflect full spectrum strategies, with retirements, fuel switching, 20% EUE and peak load management. Strategy 10 has nuclear, IGCC and natural gas generation. Below we refer to this strategy as the "*Modernization*" strategy. Strategy 11 has only nuclear and natural gas, but retrofits select existing generation with desulfurization equipment. The final MCDA strategy (12) focuses only on the supply side, with retirements, FGD retrofits, fuel switching as well as nuclear, IGCC and natural gas generation, but no end-use efficiency or peak load management. Below this is referred to as the *"Clean Supply"* strategy.

Table 6.19 and Figure 6.33 show the costs and emissions for the twelve MCDA strategies for the FIB future. Highlighted are the reference strategy and the three "named" strategies mentioned above. Note that sulfur emissions of the no FGD version of reference case are well outside the plot area in Figure 6.33. While cumulative SO_2 emissions of this strategy are more than double those of the reference case with FGD, the other emissions of the no FGD strategy are slightly lower. This is due to the dispatch effect, where new conventional coal units without FGD displace more older generation, since they are even less expensive to operate, as well as more efficient than new units with FGD.

Figure 6.34 shows the installed capacity, annual cost, SO_2, PM10, NO_x and CO_2 emissions for the reference case (1, BOC-CONPAS), Clean Supply (12, DUX-TONPAS), Conventional Coal Plus (4, DOX-CONLAG), and Modernization (10, DOX-TONLAG) strategies. There are several fundamental dynamics illustrated by this figure. First is that growth in electricity production, here indicated by the installed capacity, is *not* a proxy for changes in pollutant emissions. While the two high DSM strategies (Conventional Coal Plus and Modernization) avoid significant investment in new generation, the Conventional Coal Plus strategy has SO_2 and NO_x

emissions roughly equivalent with those of the reference case in later years. In contrast, the Clean Supply and Modernization strategies, by pursuing a future mix of generating technologies including nuclear, IGCC and NGCC, have the lowest SO_2, PM10 and NO_x emissions, even though the Clean Supply strategy has no DSM. It is only the Modernization strategy, by including options for old dirty generation, new cleaner generation technologies, *and* the growth in electricity demand, that achieves superior performance for costs and *all* emissions.

Table 6.19 Cost and Emissions Performance of the MCDA Strategies

MCDA Strategy	Electric Service Direct Costs: Regional	Electric Service Direct Costs: Unit	Power Plant Stack Emissions: SO2	PM10	NOx	CO2
BOC-CENPAS	578.5	0.359	26.59	4.66	10.89	2.95
BOC-CONPAS	601.0	0.373	12.34	4.69	11.05	3.01
BOX-CONPAM	570.7	0.354	10.54	4.16	10.22	2.75
DOX-CONLAG	548.6	0.340	10.54	3.06	9.73	2.47
BOX-LONLAM	568.1	0.352	11.40	3.90	10.14	2.68
DOX-MONLAM	569.6	0.353	12.93	3.23	11.18	2.77
BOC-NONLAS	597.0	0.371	12.83	4.50	10.32	2.65
BOX-NONLAM	568.4	0.352	10.85	3.93	9.55	2.41
BOX-NONLAG	552.6	0.342	10.22	3.72	8.52	2.08
DOX-TONLAG	553.7	0.343	8.53	2.73	7.74	2.03
DUX-DONLAG	552.7	0.343	9.25	2.88	8.70	2.12
DUX-TONPAS	610.7	0.379	8.16	2.94	8.75	2.57
	(NPV BYuan)	(Yuan/kWh)	(Mt)	(Mt)	(Mt)	(Gt)
Percent Change from BOC-CONPAS						
BOC-CENPAS	(3.7)	(3.8)	115.6	(0.6)	(1.5)	(1.8)
BOX-CONPAM	(5.0)	(5.2)	(14.5)	(11.3)	(7.5)	(8.5)
DOX-CONLAG	(8.7)	(8.9)	(14.6)	(34.7)	(11.9)	(17.9)
BOX-LONLAM	(5.5)	(5.6)	(7.6)	(16.8)	(8.3)	(10.7)
DOX-MONLAM	(5.2)	(5.3)	4.8	(31.1)	1.2	(7.7)
BOC-NONLAS	(0.7)	(0.7)	4.0	(4.0)	(6.5)	(11.7)
BOX-NONLAM	(5.4)	(5.6)	(12.0)	(16.3)	(13.6)	(19.8)
BOX-NONLAG	(8.1)	(8.2)	(17.2)	(20.7)	(22.9)	(30.8)
DOX-TONLAG	(7.9)	(8.0)	(30.9)	(41.9)	(29.9)	(32.5)
DUX-DONLAG	(8.0)	(8.2)	(25.0)	(38.6)	(21.3)	(29.6)
DUX-TONPAS	1.6	1.7	(33.9)	(37.4)	(20.8)	(14.5)
(Future = FIB)	(Δ%)	(Δ%)	(Δ%)	(Δ%)	(Δ%)	(Δ%)

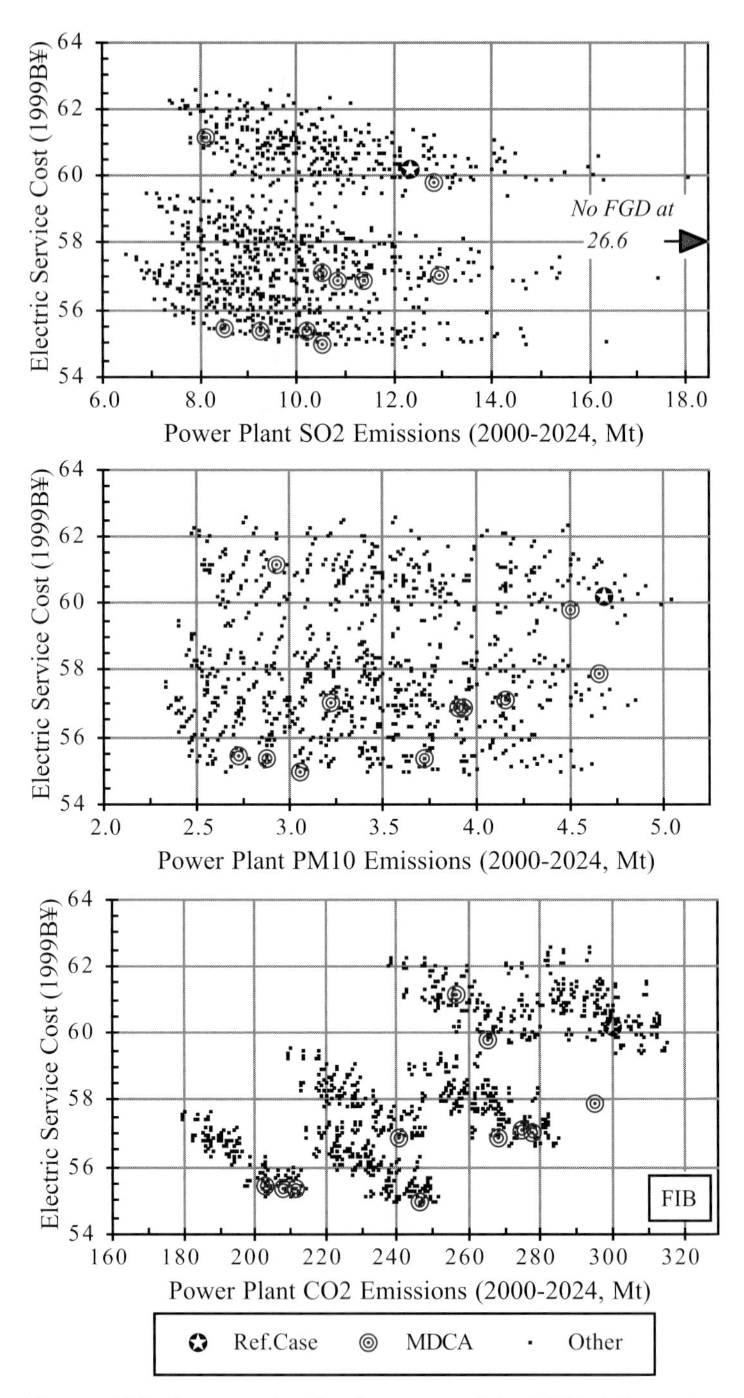

Figure 6.33 Comparative Performance of the MCDA Strategies for the FIB Future

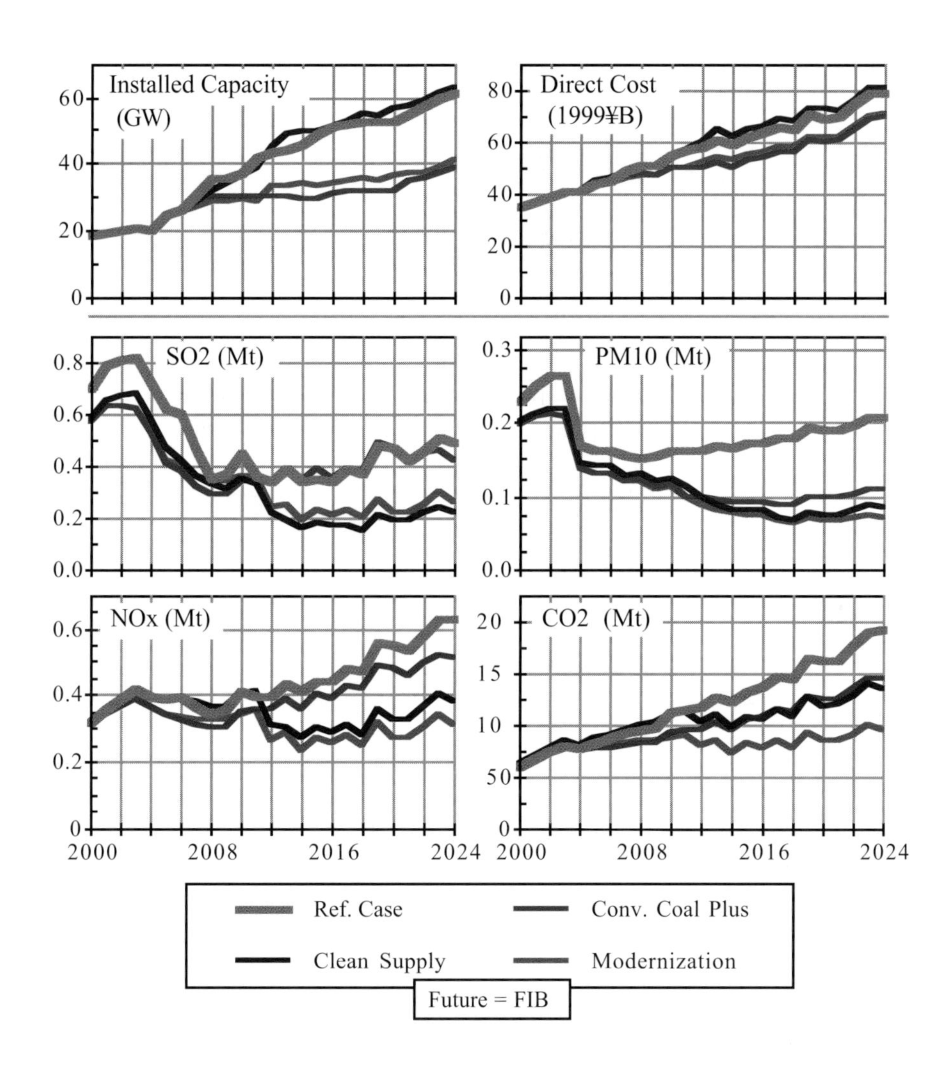

Figure 6.34 Annual Performance of Select MCDA Strategies

To better understand the dynamics behind these trends Figures 6.35 and 6.36 show capacity additions, utilization, and costs for the same four strategies. Figures 6.37 and 6.38 do the same for power plant emissions.

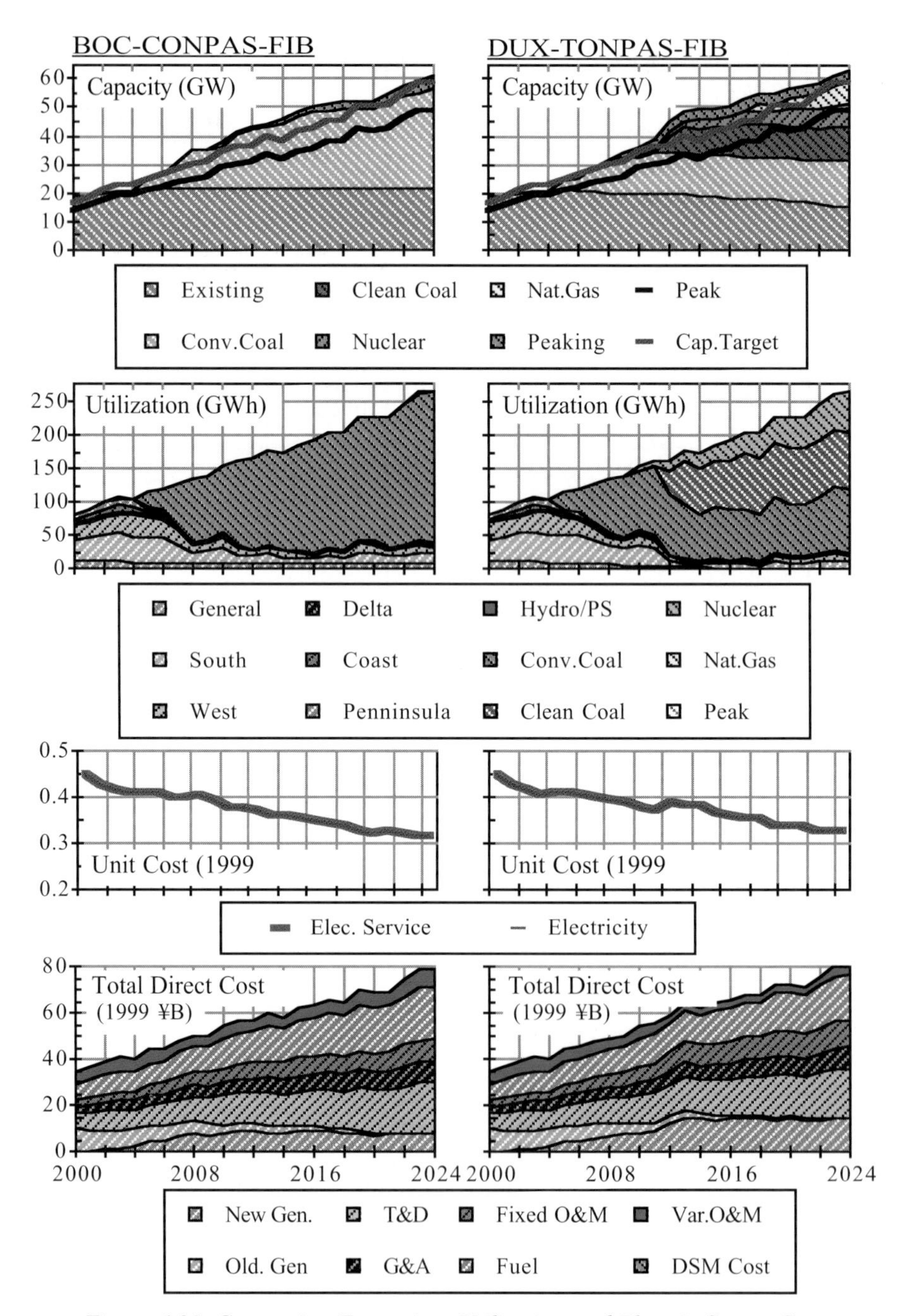

Figure 6.35 Generation Expansion, Utilization and Electric Sector Costs for the Reference Case and Clean Supply Strategies

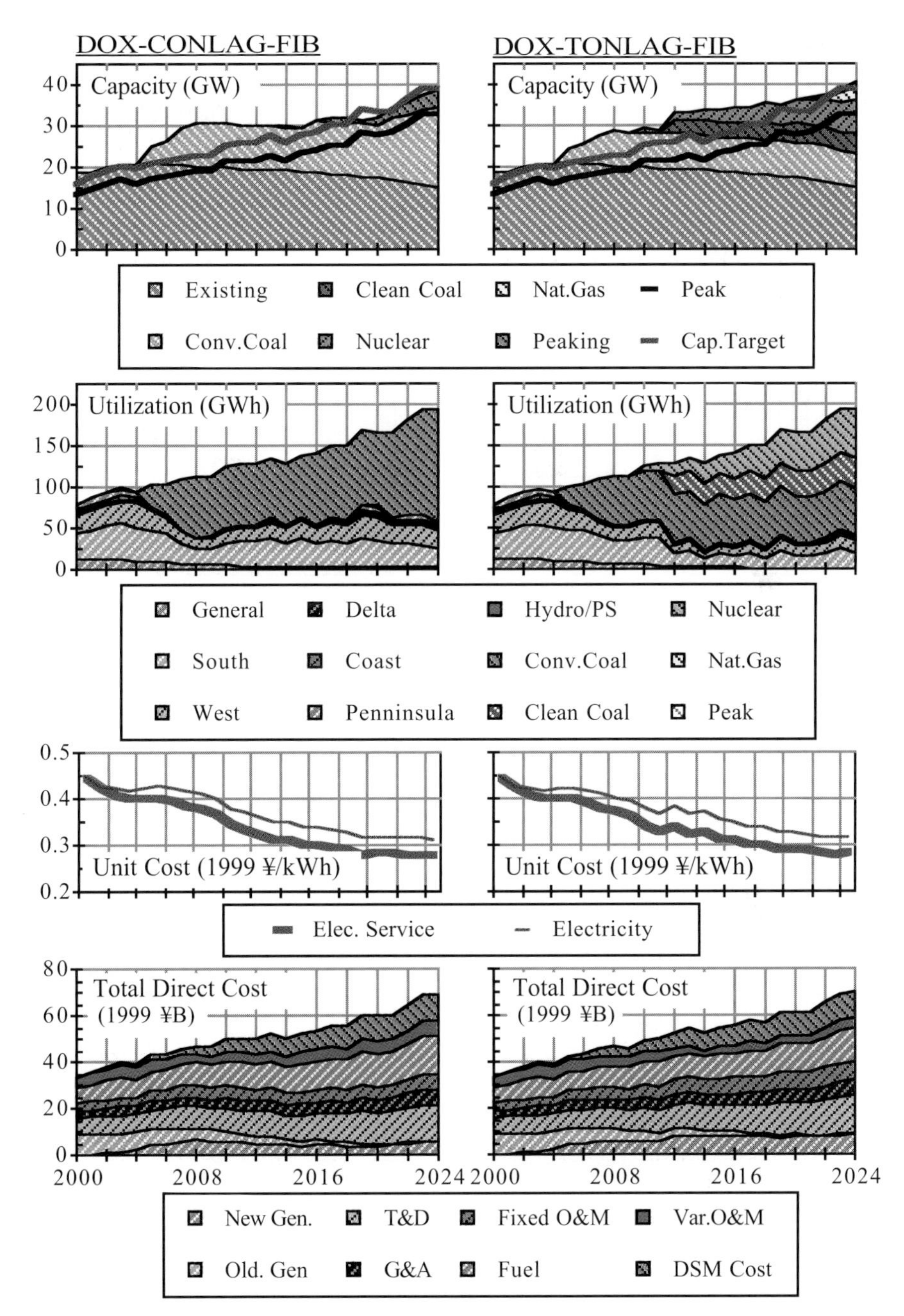

Figure 6.36 Generation Expansion, Utilization and Electric Sector Costs for the Conventional Coal Plus and Modernization Strategies

Here the dispatch effect is clearly evident. In the Reference Case and Clean Supply strategies, new generation displaces nearly all of the generation from older units. In the Conventional Coal Plus and Modernization strategies, while total generation is substantially lower, much more of it comes from older units, even after many of the dirtier ones have been retired. This is especially true for the Conventional Coal Plus strategy. Expect for SO_2 emissions, even in this case, there is enough a reduction in all generation to offset the increased emissions from the older units. Of the four, only the Modernization strategy sustains the reductions of all four emissions.

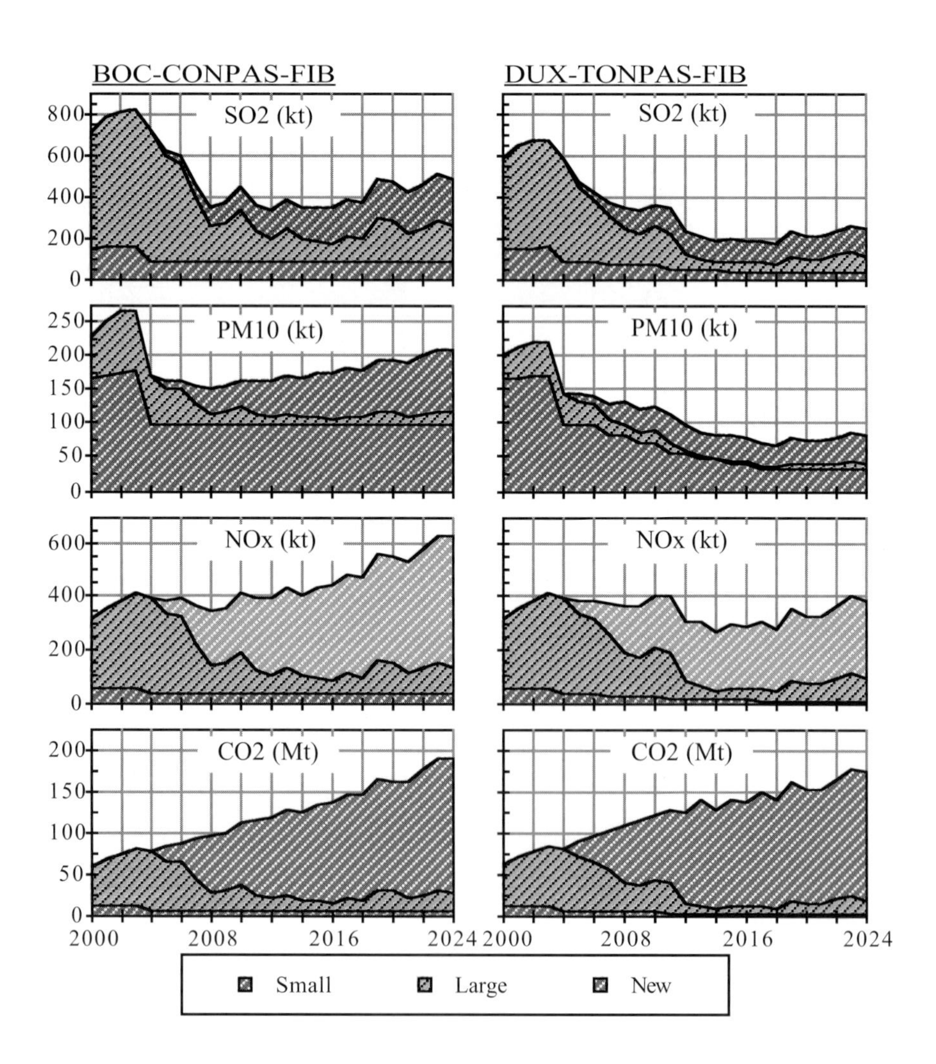

Figure 6.37 Annual Pollutant Emissions by Size and Vintage of Generation for the Reference Case and Clean Supply Strategies

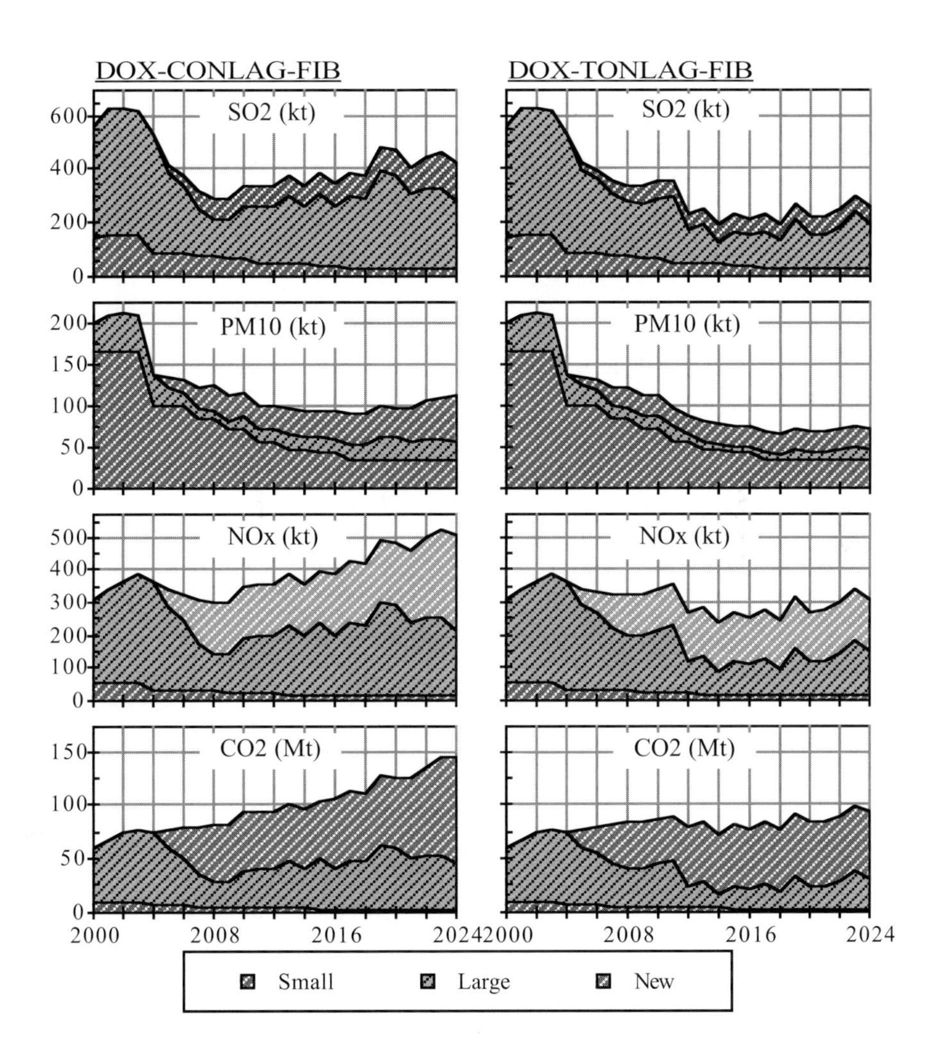

Figure 6.38 Annual Pollutant Emissions by Size and Vintage of Generation for the Conventional Coal Plus and Modernization Strategies

How robust are these strategies? Figure 6.39 shows the twelve MCDA strategies plotted for all eighteen futures. Highlighted are the four named strategies shown in detail above. The lines circumscribe the "performance envelope" of these strategies across the different load growths and fuel costs, and give us a sense of how their costs and emissions change across with changes in electricity demand and fuel costs.

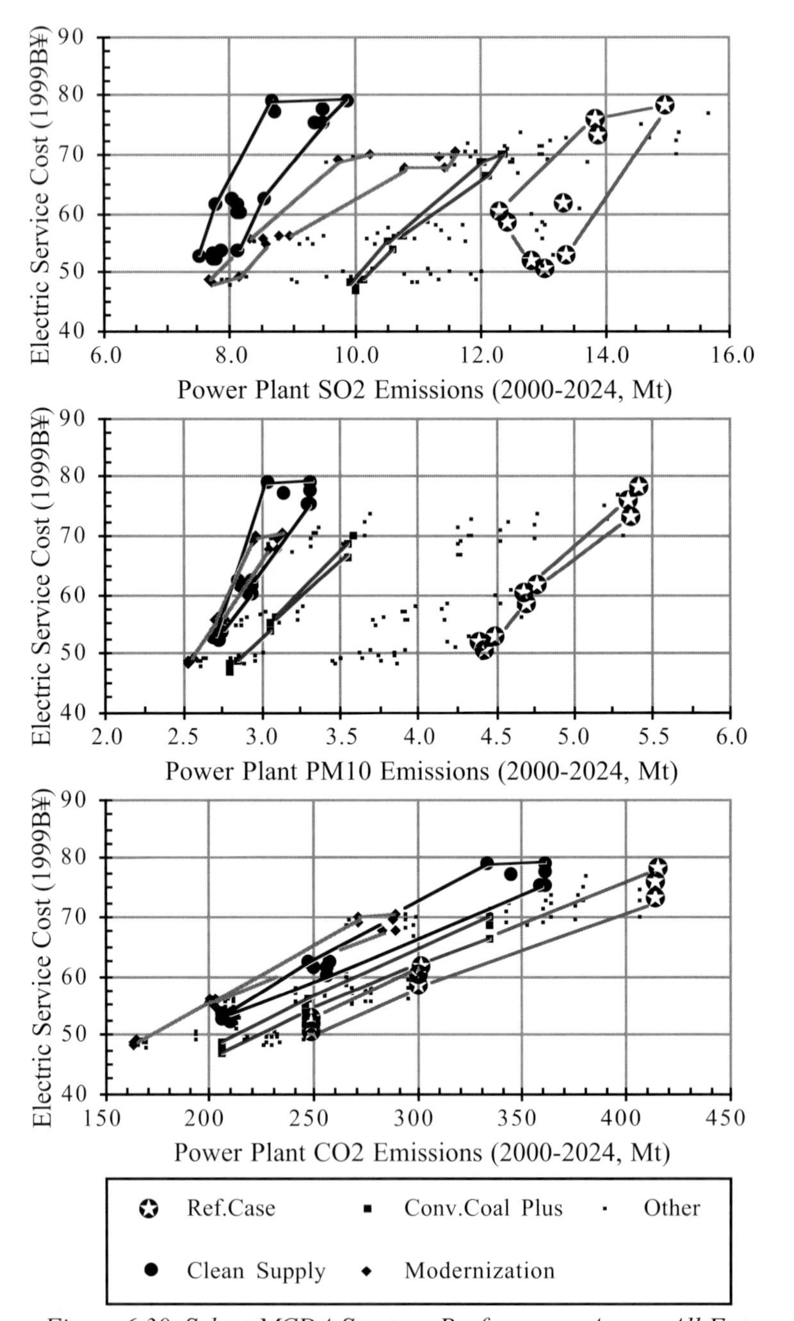

Figure 6.39 Select MCDA Strategy Performance Across All Futures

Several things are apparent from Figure 6.39. First is that the high DSM Conventional Coal Plus and Modernization strategies' costs are less sensitive to an increase in electricity demand. This is due, in part, to the fact that the 20% reductions in electricity demand from efficiency programs "scale up" as electricity demand increases. This is in no way a problem. In fact, it may be argued that rapidly rising demand for electric service provides greater opportunities for the deployment of more efficient electrical appliances, since the increased affluence, and therefore purchasing power, may increase the turnover of refrigerators, lamps and air conditioners.

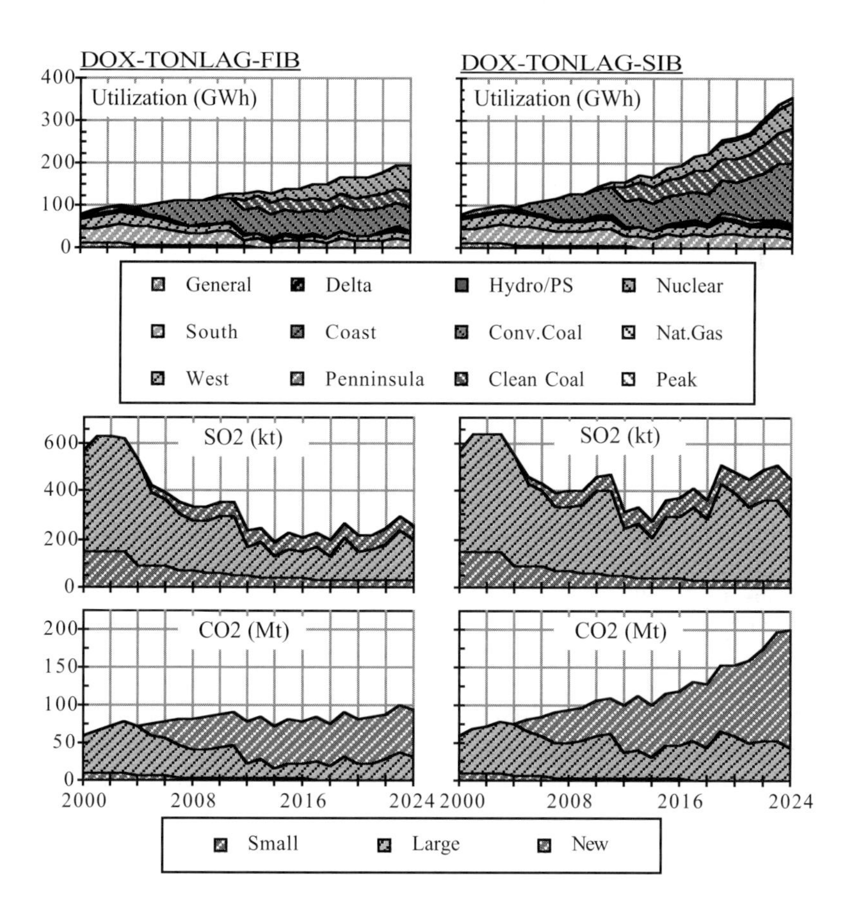

Figure 6.40 Performance of the Modernization Strategy Under Moderate and Strong Load Growth Uncertainties

Also in Figure 6.39, we see that the Modernization strategy's SO_2 emissions increase more than the other strategies'. There are two reasons for this. First is that the nuclear option is fixed at eight 1000 MW power plants whether load grows fast or not. This means that the emissions displacement potential of the nuclear option, as it was modeled, does not scale with the increase in demand. Second is that if load grows faster than new capacity additions, then older generation will be used more. Figure 6.40 shows these impacts on capacity utilization, SO_2 and CO_2 emissions for the Modernization strategy under moderate and strong electricity demand growth.

9. EXTENDING THE RANGE OF OPTIONS

Even though a broad range of options were incorporated into the full scenario set run by Electric Sector Simulation analysis team, from a sustainable energy viewpoint, there are some omissions. Most noticeable is that of renewable energy sources. As mentioned early in the chapter, hydropower potential in the province is severely limited. Although one of China's major rivers, the Yellow River, flows through the province, it cannot be considered a possible source of hydropower. Not only is the topography rather flat as the Yellow River approaches its delta, but upstream utilization of the water often prevents it from reaching the Bo Hai Sea at all. In 1997, the lower reaches of the Yellow River ran dry for 226 days (Kirby, 1999). Since the first time the Yellow River ran dry in 1972, the event has reoccurred roughly four out of every five years (Liu, 2001). The demand for what water resources there are is intense, whether for irrigation, industry, power production or petroleum refining (Singer, 1998). Even now, officials are still considering additional diversions of river water to some of Shandong's population centers (Qingdao, Yantai, Weihai) far away from the river (China Daily, 2002). It was with this knowledge in mind that the Electric Sector Simulation team placed such an emphasis differentiating new inland and coastal generators.

Other prospective renewable resources are wind and sun. These options were not included in the Electric Sector Simulation's scenario set since adequate resource data was not available. The cost and quantity of generation from renewable technologies is directly attributable to the size and dynamics of the renewable resource (McGowan and Connors, 2000).

Dong et. al. suggest that substantial wind resources may be present in Shandong. (1998). At the end of 2000 there was roughly five and a half megawatts of wind generating capacity installed in the province, primarily on the northeast coast including islands. The Chinese government is actively pursing this option, with both small and large wind turbine technologies, and is considering numerous mechanisms for providing incentives–including wind resource concessions–to those wishing to develop wind farms. (Raufer and Yang, 2002)

Even so, the location, quantity, and seasonal and daily distribution of wind needs to be known in order to estimate whether wind farms along Shandong's mountains and coasts can displace both the investment in, and the use of, fossil generation, either old and new. Several dozen additional scenarios were run looking at three levels of windpower deployment. 1500 MW of onshore wind was analyzed, phased in from 2005 to 2019, as was 3000 MW of offshore wind, deployed from 2010 through 2019, and then the two together. Onshore windpower was given a 25%

capacity factor, while offshore wind was assumed to have a 35% capacity factor (Hansen, 2002). Both represent fair wind regimes given current technology costs.

The performance is these strategies are reported in Hansen (2002). In general they perform like the DSM strategies above. As a non-dispatchable resource, conventional generation sees the impact of wind and other uncontrollable generation resources as a reduction in demand that power system operators normally use dispatchable generation to meet. As such, the amount of emissions reduction is scaled to amount of non-dispatchable generation deployed. Whether this generation avoids the need for new dispatchable generation, or displaces the operation of old versus new controllable generation, can only be determined with a solid understanding of not only the renewable resource, but how it matches the demand for electricity as well. The historical hourly demand for 1998 through 2000, shown earlier in Figure 6.10, suggests that Shandong's electricity demand has strong late day and summertime peaks. This shows promise for both wind and solar, but only a detailed study of these renewable resources can show for sure whether they are both well timed, and exist in sufficient enough quantities, in order to have a large potential impact. Both wind and solar technologies are progressing rapidly on both a cost and performance basis, especially offshore wind. Both should be considered for a future detailed study.

Also examined as additional technological sensitivities were waste-to-energy and advanced nuclear technologies. A long-standing issue, when looking at energy from municipal solid waste in developing and emerging economies is its energy content. Such waste in developing countries is often high in inorganic materials, making it a poor fuel. Food wastes are also poor energy candidates for conventional "mass burn" waste-to-energy technologies, due primarily to their moisture content. Hansen therefore looked at performance of strategies including methane production from bio-reactor type landfills. Again, the contribution to emissions reduction on a provincial level is a function of how much of the waste "resource" there is that can used to generate fuel for power production. As an alternative to the 1000 MW advanced light water reactors included in the scenarios above, modular high temperature gas reactors were also examined. These HTGR, or pebble-bed, reactors are still under development, but if successfully commercialized have numerous advantages in cost, modularity (113 MW), ease of deployment, and the utilization of uranium fuels. These also are discussed in Hansen (2002).

10. ENCOMPASSING GREATER UNCERTAINTIES

One of the reasons Shandong Province was selected for study is that the province and its electric grid are geographically and institutionally the same, making interactions with stakeholders, the search for pertinent information, and its incorporation into a coherent study, that much easier. As power exchanges with neighboring provinces have historically been small, it also makes setting the "boundary" of the study far easier.

However, China is currently in the process of reforming its electric sector. One of the largest changes that power sector reform is expected to bring is that of regionalizing the country's power grids, and separating generation–ownership-wise–

from transmission and distribution, and inviting more investment by independent power producers (Hillis, 2002). Also anticipated is a consolidation of existing state owned generation companies, including SEPCO (China Business, 2002).

In addition to policies directly affecting the structure of the electric sector are changes in the fuel sector and the economy in general. The strategies examined here assumed unconstrained access to new, advanced technologies, including the fuels they consume, and that these will be available at reasonable cost, including financing. With China's recent admission to the World Trade Organization (WTO), these seem reasonable assumptions. However WTO ascension also raises issues regarding the degree of subsidization–especially borrowing rates–that China may make in certain sectors, increased scrutiny in licensing and the use of intellectual property, as well as greater harmonization of rules and regulations, including accounting and environmental performance. Barker (2001) explored several of these topics using the ESS scenarios as a baseline, and the results indicate that these may influence the cost and choice of technologies in different ways. Government policies regarding other infrastructure investments, fuel and water supplies for example, all suggest that there are a broad range of uncertainties that need to be considered as both Shandong and China examine their long-term electric power choices.

11. CONCLUSIONS

This analysis shows that Shandong Province has numerous opportunities to simultaneously meet its future electricity needs, and substantially reduce both criteria pollutants and greenhouse gas emissions (CO_2). However, in order to achieve these combined goals, it must embark on a strategy that targets the emissions from older, dirtier power plants, and balances this with a well coordinated effort that manages the growth in electricity demand, and introduces new generation technologies. Furthermore, Shandong may be able to implement such a coordinated strategy at equivalent or even lower direct costs than other alternatives. While the future costs of generation, fuels and DSM are quite uncertain, the general trends in relative costs should hold true.

Specifically, the feasibility of providing lower-ash, lower-sulfur prepared coals should be explored in earnest. This not only lowers the pollution from coal combustion, but also extends the capabilities of the coal transportation infrastructure by having more fuel energy transported per tonne of coal shipped. Lower ash coal also improves the performance of conventional power plants by reducing the slagging of boilers, thereby extending the period between scheduled maintenance and increasing unit availability.

The select retirement or retrofitting of smaller, dirtier, older generation should also be considered. Historical emissions data show that there are some units that emit very high levels of SO_2 and particulates, even though they only see limited use on an annual basis. Older generation units are usually quite small ($\approx$ 100 MW), compared with the much larger ones (300 – 1000 MW) being built today. Given the continued rapid expansion of power generation in the province, the "extra" investment required to replace these units does not appear provide any significant cost pressure to providing the people of Shandong with electric service.

While continued use of pulverized coal, but with flue gas desulfurization, looks cost-effective, the development of newer, higher efficiency, lower emissions technologies and fuel supplies looks like a solid long-term option. China is gaining considerable experience in the area of nuclear. Greater expertise in the development, deployment and integration of other technologies such as clean coal–especially IGCC, natural gas combined-cycle, and windpower should also be pursued. What contribution niche technologies such as waste-to-energy, cogeneration, biomass and solar may make should also be explored.

Finally, great attention should be paid to the demand side. While "moderate" growth of 5% per year in electricity demand is much smaller than what the province has experienced over the past quarter-century, it will still result in electricity demand tripling over the next twenty-five years. With growing urban populations, and individual purchasing power, growth in peak electricity demand may be even larger. Opportunities for using electricity more wisely need to be identified, and policies ensuring their timely deployment implemented.

The above scenarios, formulated in conjunction with the CETP's Stakeholder Advisory Group, indicate that the greatest benefit to the province will come when all three elements of these robust strategies; managing old generation and the fuels it uses; promoting the efficient use of electricity; and encouraging investment in advanced generation technologies, are deployed in concert.

REFERENCES

Barker, J. (2001). *Planning for economically and environmentally sound electricity in Shandong Province, China.* Massachusetts Institute of Technology. Department of Urban Studies and Planning. Masters of City Planning. May 2001.

China Business (2002). *China to merge 5 independent power generators with 9 listed firms.* China Business. 15 November 2002.

China Daily (2002). *"Yellow River Water to Flow into 3 Cities."* China Daily. 29 July 2002. www.china.org.cn/english/China/37999.htm, accessed October 2002.

Connors, S. and C. C. Cheng, C. Hansen and J. Barker (2002). *Shandong, China Electric Sector Simulation Assumptions Book.* Massachusetts Institute of Technology, Laboratory for Energy and the Environment, LFEE Report No. 2002-002 RP.

Dong, F., D. Lew, P. Li, D. Kammen and R. Wilson (1998). Strategic Options for Reducing CO_2 in China: Improving Energy Efficiency and Using Alternatives to Fossil Fuels. In *Energizing China: Reconciling Environmental Protection and Economic Growth.* eds. M. McElroy, C. Nielsen and P. Lydon. Harvard University Press.

Hansen, C. (2002). *Evaluation of Renewable and Advanced Electricity Generation Alternatives for Shandong Province, China.* Massachusetts Institute of Technology. Technology and Policy Program. Masters Thesis. September 2002.

Hillis, S. (2002). *China cabinet approves power reform plan*. Reuters. 23 October 2002. biz.yahoo.com/rf/021023/energy_china_1.html, accessed November 2002.

Kirby, A. (1999). *Half the World's Rivers at Risk*. BBC News, 29 November 1999. News.bbc.co.uk/1/hi/sci/tech/538457.stm, accessed October 2002.

Liu, C. (2001). *Water quantity and quality management regarding eco-environment: a case of China's Yellow River*, at SCOPE, Scientific Committee on Problems of the Environment, XIth General Assembly and Scientific Symposia. 24-28 September 2001. www.scope-germany.uni-bremen.de/scope_ga/ga_lui.html, accessed October 2002.

McGowan, J. and S. Connors (2000). Windpower: A Turn of the Century Review. Annual Review of Energy and the Environment. Vol. 25, pp. 147-97.

Raufer, R. and S. Wang (2002). *Wind Resource Concession Approach in China*. IEEE Power Engineering Review, Vol. 22, No. 9. pp. 12-15.

Schenler, W. and A. Gheorghe, S. Connors, P. A. Haldi, S. Hirschberg, C. Frei, L. Beurskens and N. Zhuikova (2002). *Strategic Electric Sector Assessment Methodology under Sustainability Conditions: Two Swiss Case Studies on Stranded Costs and Environmental Externality Dispatch*. A Research Report for the AGS SESAMS Project 1998-1999.

Schenler, W. and A. Gheorghe, S. Hirschberg, P. A. Haldi and S. Connors (2002). Strategic electric sector assessment methodology under sustainability conditions: a Swiss case study on the costs of CO_2 emissions reductions. *International Journal of Sustainable Development*, Vol.5, No. 4, pp. 7-63.

SEPCO – Shandong Electric Power Group Corporation (2000). SEPCO Annual Report 1999. www.sepco.com.cn/sepco_web/eindex.htm, accessed March 2000.

Singer, R. (1998). *"China's Yellow River, Now a Trickle, Poses New Threat."* The Philadelphia Inquirer. www.pewfellowships.org/stories/china/yellow_river.html, accessed October 2002.

World Gazetteer (2002). www.world-gazetteer.com, accessed November 2002.

Xue B. and B. Eliasson (2000). Shandong Province Energy and Emissions: Status, Statistics and Prospects, 1999. ABB Energy and Global Change.

Yang and Lau (1999). Demand-Side Management (DSM) and Its Applications. China Electricity Power Press. (in Chinese).

CHAPTER 7

ENERGY TRANSPORTATION MODELING

KENJI YAMAJI AND TAKEO IMANAKA

1. INTRODUCTION

Energy systems depend on transportation to deliver fuel to power plants and to convey energy commodities to their consumers. For this reason, the development of the infrastructure for energy transportation is an essential part of energy planning. The authors of this chapter used a modeling approach in preparing an analysis of electric power system in Shandong province, taking account of the fuel transportation and the power transmission. This is the third approach for energy modeling effort in CETP, which complements the two approaches described in Chapter 5 and Chapter 6.

Shandong Province is one of the largest coal producers in China, and has little in the way of hydropower resources. Electricity production has been almost entirely dependent on coal-fired plants. The major suppliers of fuel for power generation are local coal mines, but Shanxi Province, also endowed with large amounts of coal, supplies Shandong as well. Sulfur content, a key factor in determining SO_x emissions from coal-fired generating plants, is generally higher in the local product than in Shanxi coal. Because the energy supply in Shandong Province is strongly dependent on coal, , the most pressing environmental concern in this area is the control of sulfur emissions; emissions of nitric oxides and carbon dioxide will also be subject to control some time in the future.

A complicating factor is that Shandong Province must meet its dynamic electricity demand growth in an economically efficient manner under the tight constraints of environmental regulation. To meet sulfur emission targets, decision makers planning Shandong's future electric power system have an array of alternatives from which to choose. They could introduce more environmentally sound coal-firing technologies including a flue gas desulfurization (FDG) system; increase the imports of the low-sulfur coal from Shanxi province; increase the use of domestic low sulfur coal by modifying the transmission network; or introduce more advanced technologies such as natural-gas-fired or nuclear power plants.

To analyze the impact of these alternatives on the Shandong electric power system, we developed an engineering-based, bottom-up mathematical optimization model of the electric power system. The Energy Transportation Model (ETM) is designed to show how to plan an energy system for the province at the least cost under a specified condition. In ETM, fuel transportation for power generating plants and the generation and transmission of electricity are modeled taking the alternatives mentioned above into account.

The model analyses presented in this chapter suggest several possible futures for the Shandong electric power system under a range of specified conditions. The results should not be considered predictions but interpreted as alternative futures among which planners may select in developing an economically efficient and environmentally sound power system in Shandong Province.

The remainder of this chapter consists of the three sections. Section 2 describes the outline, formulation, and supporting data for the model. In Section 3, we describe seven cases and present results for then as calculated using the model. Section 4 concludes the chapter.

2. MODEL DESCRIPTION

In this section, the model used to suggest options for Shandong's power system is described, followed by a description of model assumptions and data setting.

2.1 Outline of the Model

The Energy Transportation Model (ETM) is a linear optimization model designed to seek the least-cost plan for the power system in Shandong province through an engineering approach. The model is formulated to describe the structure of the energy system as associated with a specific scenario, such as the presence of strong environmental policy.

The scope of the model is indicated in Figure 7.1, which includes coal transportation for the generating plants and the generation and transmission of electricity, taking account of various types of fuel, a number of coal mines, and various power-generating options including those for environmental pollution control. The trunk power transmission network is defined geographically in conjunction with 17 cities in Shandong Province, each of which has specific power-generating plants and electricity demands. The coal transportation routes are also defined including the relevant cities and coal mines. The power system is modeled as an independent system inside the province: Power interchanges between Shandong province and other provinces do not fall within the model. The only exception is that the coal mines in Shanxi province and the transportation of coal from Shanxi to Shandong province are included.

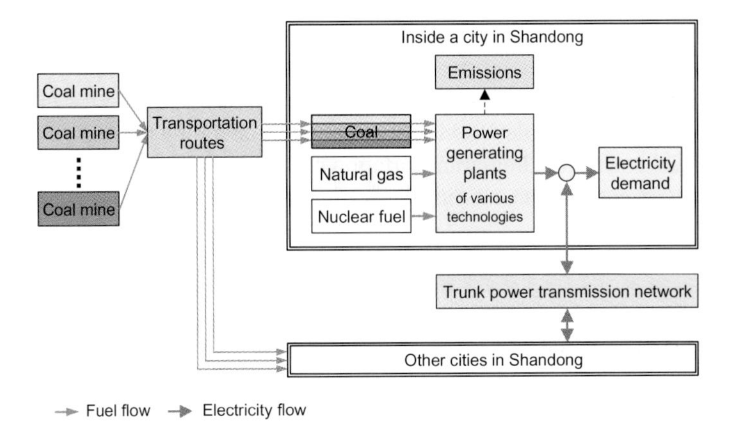

Figure 7.1 Modeling Scope

ETM suggests a least-cost plan for the power system under a specified scenario in terms of:

- Expansion planning of power generating plants in conjunction with their siting and the technology utilized;
- Operation of the power generating plants and electricity generation mix;
- Locational fuel consumption. The model defines the source of the coal utilized at respective coal-fired plants, thus identifies the coal flow from production to consumption;
- Locational SO_x emissions and CO_2 emissions.
- Expansion planning of the transmission lines and the network power flow;
- Total system cost with its specification; and
- Shadow prices for the emission constraints -- the theoretical sulfur or carbon taxes related to the emission caps --, when assumed.

The geographical resolution of the modeling is one of the essential characteristics of ETM, together with handling of energy transportation. The model has sufficient flexibility to address a variety of scenarios such as the imposition of environmental policies or anticipated states of the energy supply.

2.2 Formulation

As described above, the ETM is a linear optimization model: decision variables which represent the system configuration and operation are determined endogenously in the model as the result of optimizing the objective function under the constraints, which define the system under consideration.

The objective function of ETM is to minimize the discounted total cost of the expenditures distributed over the time horizon. This cost includes:

- capital recovery costs of power-generating plants;
- fixed and variable O&M costs of power-generating plants,
- fuel costs including transportation costs;
- fixed cost of transmission lines. (including capital recovery cost and O&M cost); and
- tax charged on emissions (when assumed.)

Table 7.1 describes the decision variables, and major constraints in ETM.

Table 7.1. Model formulation (Xue and Eliasson, 2000)

Decision variables	
Installation capacity of generating plants	by generating technologies, periods, and nodes
Modification capacity of coal-fired generating plants*	by plants, and periods
Outputs of generating plants	by plants, fuel (for coal-fired plants), time zone, and periods
Installation capacity of transmission lines	by transmission routes, periods

Constraints	
Power system	Supply-demand balance
Power generation	Sufficient Capacity installation to cover the demand
	The maximum output and operating hours
	Fuel availability (depends on the scenario)
Power transmission	The maximum power flow on transmission lines
	Physics of network power flow (: power flow equations, which define the relationship between plants' outputs and power flows on the transmission lines)
Emission	Emission caps

*To retrofit the FGD system

To formulate the model by linear programming techniques, the decision variables must be able to accept continuous values, and the objective function and constraints need to be formulated linearly. For example, the installation capacity of generating plants and transmission lines takes a discrete value based on the unit capacity in the real world; in the model however, they accept continuous values.. A number of assumptions are associated with the input data for formulating the model as well. Information about the assumptions is described in the following section. For a detailed mathematical formulation of ETM, see Appendix G.

2.3 Input data and assumptions

ETM simulates the power system planning after 2000 through 2024 at five-year intervals: with decisions in 2005, 2010, 2015, and 2020. The power system in 2000 is simulated independently through a modified version of ETM, which defines the initial status for planning the system. The objective function of this model is to minimize the fuel cost of the existing generating plants plus the annual expenditure for the transmission lines in 2000. The installation of generating plants is not considered and the source of coal used at the existing plants is exogenously determined by input data described later in the chapter. The results of this model define the operation of the power system, associated fuel consumption and emissions, and the capacity of transmission lines in 2000.

Four major assumptions and settings for data are described below. These include the power transmission network, electricity demand, power generation technologies, and fuel.

2.3.1. Power transmission network.

Power transmission network is modeled simply, as a representation of the trunk power transmission system. The modeled network consists of 17 nodes and 25 branches. The nodes correspond to the cities in Shandong Province and represent the access points of electricity demands and generating plants in the cities. The branches are the transmission lines connecting the nodes, which are structured with 500kV and 220kV transmission lines. The model does not include the lower voltage transmission system and the distribution system (Figure 7.2).

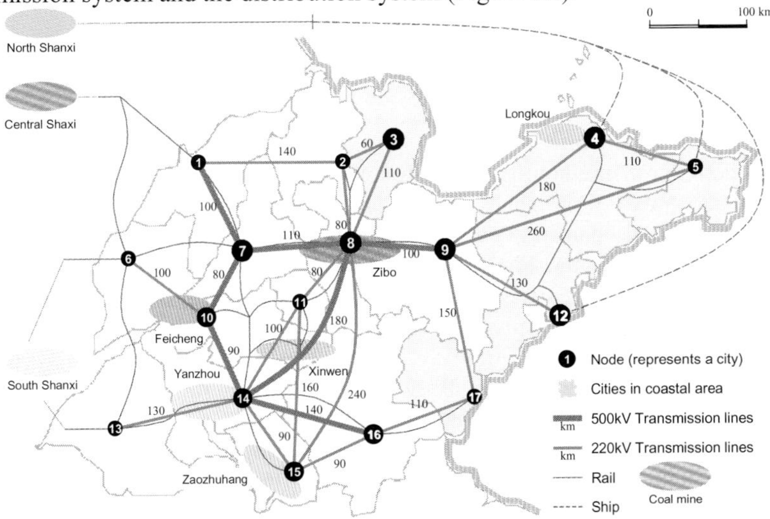

Figure 7. 2 Power transmission network model(Xue and Eliasson, 2000; SEPCO 2000)

The reinforcement of the power transmission network is modeled in terms of the capacity expansion of the existing branches; construction of the transmission lines on new transmission routes is not taken into account in this model. This is a requisite assumption for formulating ETM by a linear-programming technique, comparable to the assumption noted above that the expansion capacity of transmission lines can take continuous value. In addition, the following assumptions are made for the same purpose:

- A DC approximation can be applied to calculate the power flow on transmission lines, which implies that losses in the modeled trunk transmission network are negligible.
- Line reactance (Chugoku Furyoku 1997) will not change even if the capacity of the transmission line changes. The reactance of each branch is set equivalent to the reactance of two circuits in parallel.

The model also assumes that no transmission line (capacity) will be abandoned during the planning periods. The data for modeling the network are shown in Table 7.2.

Table 7. 2 Transmission Data (Aratame; Chandler et. al.,1998; Imanaka and Yamaji,1999)

voltage (kV)	**500**	**220**
line reactance (%/km [1000MVA base]) [1]	0.1	0.8
maximum operation rate at peak (%) *	50	
construction cost ($/kW/100km)	22.25	44.5
annual fixed charge rate (%)	11	

* taking the outage at one of two circuits into account

2.3.2. Electricity demand

In consideration of the hourly and seasonally varying electricity demand, a year is divided into three time zones based on the demand level. In addition, the instantaneous peak demand, which determines the required system capacity is considered independently. Figure 7.3 shows the assumed load duration curve for Shandong province. The system peak load (L_0 in the figure) is calculated from the peak demand and the transmission and distribution (T&D) loss in Table 7.3. The peak demand is assumed to grow as indicated in Table 7.3, which is originally derived from Demand Forecasting Task in CETP. The duration curve also changes in conjunction with demand growth, resulting in the lower load factor. As defined here, T&D loss occurs as power flows through distribution lines and through lower voltage transmission not included in the modeled transmission network. The total capacity of the generating plants must cover the system peak load (kW) taking required capacity reserve margin into account. However, electricity generation (kWh) in an instantaneous peak time zone is not counted in the model.

The load at each node is set by assigning the system load in proportion to the demand share in Table 7.4. Thus the shape of the load curve at each node is assumed to be uniform.

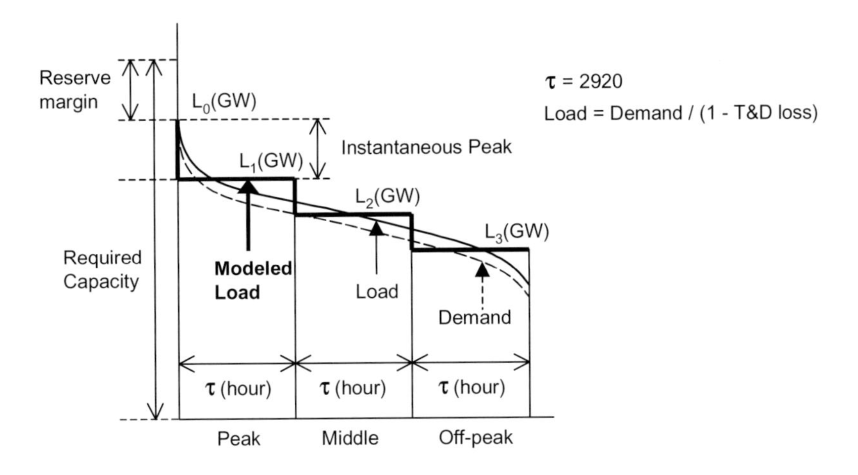

Figure 7. 3 Model load duration curve

Table 7.3 Electricity demand assumptions (CETP database, 2001; Kaigai-denryoku, 1997)

	2000	2005	2010	2015	2020
Peak demand (GW)	13.95	21.74	29.77	38.1	47.22
Load factor (%)	70	64	62	62	62
L_1 / L_0	0.8	0.75	0.74		
L_2 / L_0	0.7	0.64	0.62		
L_3 / L_0	0.6	0.53	0.5		
T&D loss (%)	8				
Required reserve margin (% of system peak load)	20				

2.3.3. Power Generation Technologies

Table 7.5 shows the power generation technologies considered in ETM. As with the transmission lines, the expansion capacity of generating plants is also assumed to accept a continuous value in the model. The installation sites of the generating plants are represented in terms of the connecting nodes, which correspond to the cities. The cities are grouped into two areas with regard to the generating technologies -- the coastal area and the other area. (See Figure 7.2).

Table 7.4 Demand Share (Chugoku Furyoku,1997)

node	city	kWh (%)	node	city	kWh (%)
1	**Dezhou**	3.0	10	**Tai'an**	4.9
2	**Binzhou**	3.5	11	**Laiwu**	2.3
3	**Dongying**	11.5	12	**Qingdao**	10.0
4	**Yantai**	9.3	13	**Heze**	2.0
5	**Weihai**	3.1	14	**Jining**	5.9
6	**Liaocheng**	2.8	15	**Zaozhuang**	5.4
7	**Jinan**	9.7	16	**Linyi**	5.9
8	**Zibo**	10.3	17	**Rizhao**	1.6
9	**Weifang**	8.8	Total		100

Advanced light water reactors (ALWR) require large amounts of water for cooling, hence, these plants can only be installed in the cities in the coastal area of the model. For this technology, installation capacity is also limited to under 1000MW per period (five years) per city. The value of is assumed to be the capacity of one standard-size unit, because the construction of a nuclear power plant is much more demanding than for a thermal power plant.

The cost of flue gas desulfurization (FGD) systems also depends on the location. The FGD system is the technology attached to a power plant to reduce sulfur emissions. This option may be applied to both existing and new pulverized-coal-fired facilities, assuming the same installation in both cases. However, a less expensive FGD system —the seawater scrubber--can be used only in coastal areas. Coal for coal-fired power plants including Atmospheric Fluidized-Bed (AFB) and Integrated Gas Combined Cycle (IGCC), may come from various coal mines as will be described below. In ETM, coal-fired power plants can change or mix them without extra cost.

Basically, there is no limitation on where or how much new capacity could be installed for any type of generating plant when the technologies are available, except for the capacity is mentioned above and the plants under construction or planned. However, constraints would be imposed on plant installation in conjunction with the scenarios. Plants under construction or planned (Table 7.6) are represented by the lower bounds for plant installation capacity.

As a consequence of the assumed lifetimes in Table 7.7, no newly installed generating plant will be abandoned during the planning periods. However, plants existing in the year 2000 will be assumed to be phased out as described in Table 7.8. All existing plants assumed to be without FGD, however, may be retrofitted as the new pulverized coal-fired plants will be.. The coal sources for existing plants indicated in Table 7.8 are those for 2000. These are used to specify the system configuration and SO_x emissions in 2000 as the initial status for the system planning as described above. The sources of coal used at the existing plants can also be changed in the planning periods after 2005.

Table 7.5 Power generation technologies (CETP database, 2001)

	Invest-ment ($/kW)	Fixed O&M* cost ($/kW/yr.)	Variable O&M* cost ($/MWh)	Net Thermal Efficiency (%, LHV basis)	desulfur-ization Efficiency (%)	Available year	NO_x emission (kg-NO_x / GJ by Fuel)
Coal							
Pulverized Coal-fired (PC)	550	18	1	37	-	2000	0.214
+ Flue Gas Desulfurization (FGD) system							
Wet Scrubber	+ 70	+2	+3	-1	90	-	-
Sea Water Scrubber	+ 24	+2	+1	-1	90	-	-
Atmospheric Fluidized-Bed combustion (AFB)	880	31	4	38	95	2000	0.1121
Integrated Gasification Combined Cycle (IGCC)	1200	31	1	45	99	2000	0.1292
Natural Gas							
Combustion Turbine (CT)	400	1	3	38	-	2010	0.0003
Combined Cycle Gas Turbine (CCGT)	600	13	0.5	58		2015	0.0003
Nuclear							
Advanced Light Water Reactor (ALWR)	1400	42	0.5	33	-	2010	-

The generating costs counted in the model, besides fuel cost, are the annual recovery cost of investment, and fixed and variable O&M cost (see Table 7.5, Table 7.7, Table 7.9). However, for existing plants, the annual recovery cost of investment and the fixed O&M cost are not included in the objective function of ETM except for those of retrofitted FGD systems, for they do not affect the results. Closing of the generating plants before the end of the lifetime is not modeled.

Table 7.6 Plants under construction or planned(CETP database, 2001)

node	1	4	6	13	14	15
plant type	PC w/ FGD					PC w/o FGD
total capacity (MW)	660	600	1200	600	600	260
on-line year in ETM	2005					

* PC: Pulverized Coal-fired, FGD: Flue Gas Desulfurization system

Table 7.7 Common assumptions for generating plants (CETP database, 2001)

maximum capacity factor (%)	85
financial lifetime (yr.): y	20
plant lifetime (yr.)	30
discount rate (%): R	10
annual recovery factor (%)*	11.75

*= R /(1- (1+R)-y)

Table 7.8 Existing Plants (CETP database, 2001)

node		Plant No.	Plant capacity (MW) in the year					Net thermal efficiency (%)	Source of coal in 2000
No.	City		**2000**	**2005**	**2010**	**2015**	**2020**		
1	**Dezhou**	1	1200	1200	1200	1200	1200	37	Central Shanxi
		2	100	0	0	0	0	31	Central Shanxi
2	**Binzhou**	1	350	250	0	0	0	31	Zibo
3	**Dongying**	1	600	400	0	0	0	31	Zibo
4	**Yantai**	1	200	200	200	200	200	34	North Shanxi
		2	800	800	800	800	400	37	Longkou
		3	200	200	200	0	0	36	Longkou
		4	250	0	0	0	0	31	Zibo
5	**Weihai**	1	250	250	250	250	250	36	North Shanxi
		2	600	600	600	600	600	37	North Shanxi
		3	100	0	0	0	0	31	Zibo

6	**Liaocheng**	1	100	100	100	100	50	32	South Shanxi
		2	200	200	200	200	200	33	South Shanxi
		3	100	0	0	0	0	31	Feicheng
7	**Jinan**	1	600	600	600	600	0	37	Central Shanxi
		2	200	200	200	0	0	34	Zibo
		3	300	100	0	0	0	31	Zibo
8	**Zibo**	1	390	390	390	390	390	37	Zibo
		2	1000	600	0	0	0	31	Zibo
9	**Weifang**	1	300	300	300	300	300	37	Central Shanxi
		2	300	300	300	300	300	37	Zibo
		3	150	0	0	0	0	31	Zibo
10	**Tai'an**	1	1200	1200	1200	1200	600	37	Feicheng
		2	250	200	0	0	0	31	Feicheng
11	**Laiwu**	1	375	0	0	0	0	35	Xinwen
		2	100	0	0	0	0	31	Xinwen
12	**Qingdao**	1	600	600	600	600	600	37	Zibo
		2	670	670	545	420	0	35	Central Shanxi
		3	290	200	0	0	0	31	Zibo
13	**Heze**	1	250	250	250	250	250	35	South Shanxi
		2	100	0	0	0	0	31	South Shanxi
14	**Jining**	1	1800	1800	1800	1800	1800	37	Yanzhou
		2	1200	1200	1200	900	0	36	Yanzhou
		3	200	100	0	0	0	34	Yanzhou
		4	200	0	0	0	0	32	Yanzhou
15	**Zaozhuhang**	1	1225	1225	975	600	600	36	Zaozhuhang
		2	200	200	0	0	0	31	Zaozhuhang
16	**Linyi**	1	100	100	100	100	100	32	Central Shanxi
		2	250	250	250	250	250	35	Central Shanxi
		3	100	0	0	0	0	31	Yanzhou
17	**Rizhao**	1	100	0	0	0	0	31	Yanzhou
		2	700	700	700	700	700	37	Yanzhou
Total			18200	15385	12960	11760	8790		

Table 7.9 Assumptions about existing plants (CETP database, 2001)

NO_x Emission (kg/GJ by Fuel)	0.214
Variable O&M cost ($/MWh)	1

2.3.4. Fuel

The types of fuel taken into account in ETM are coal, natural gas, and nuclear. Among the fuels considered, coal is handled in particular detail by modeling a number of coal mines and transportation routes from the mines to generating plants, i.e. nodes, as shown in Figure 7.4. Transportation routes determine the distances between coal production sites and nodes (Table 7.10), and then the transportation costs from the mines to the nodes. . It must be noted that the capacity of the transportation routes is not specified in ETM -- there is no constraint on the amount of transportation, although the limitations of the transport system is a substantial issue in the development of China's energy system. The construction of new routes is not taken into account in the model either. The transportation routes models correspond to those that currently exist.

Nine coal mines are modeled in ETM, six of which are in Shandong, and three in Shanxi (see Figure 7.4). For the sake of convenience,, coal from each production site is grouped into "Domestic" coal or "Import" coal. The former indicates coal from mines in Shandong Province, and the latter indicates that from other sources (Shanxi). . The coal from each mine has different heat values, the sulfur content, and production costs (Table 7.11, Table 7.12). SO_x emissions from coal-fired plants are determined by the sulfur content of the coal used in conjunction with each generating plant's desulfurization efficiency.

The price of coal at each node equals transportation cost plus the production cost of raw coal (Table 7.12). Even though the assumed transportation distance is set zero, transportation will be required except for the mine-mouth generating plants. Thus, the basic fee for transportation is generally charged except for mine-mouth facilities. The existing mine-mouth plants assumed in ETM are shown in Table 7.8, which uses the coal from Longkou; its total capacity is 1000MW. However, the availability of Longkou coal is limited in the model, reflecting the limited production capacity of the mine. The production capacity of Longkou mine is assumed to remain the maximum generation of the existing mine-mouth plant. Except for the Longkou mine, the production capacity of each mine is not considered in ETM.

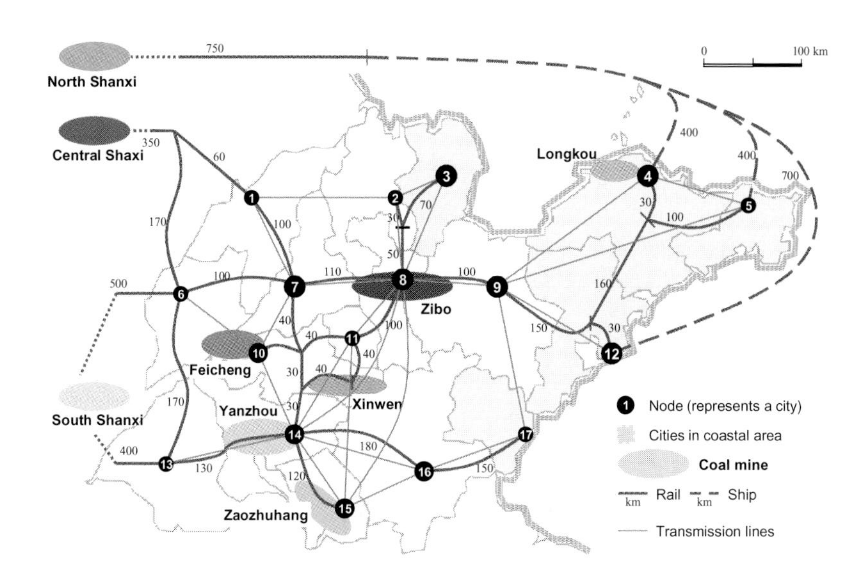

Figure 7.4 Coal production sites and transportation routes (CETP database, 2001)

Table 7.10 Coal transportation distances (km) (CETP database, 2001)

node	Domestic coal						Import coal		
	Zibo	Xinwen	Zaozhuhang	Feicheng	Yanzhou	Longkou	North	Central	South
1	210	230	340	200	220	-	-	410	700
2	80	220	400	260	280	-	-	700	790
3	120	260	440	300	320	-	-	740	830
4	440	580	760	620	640	0	750 (400)*	1060	1150
5	510	650	830	690	710	-	750 (400)*	1130	1220
6	210	230	340	200	220	-	-	520	500
7	110	130	240	100	120	-	-	510	600
8	0	140	320	180	200	-	-	620	710
9	100	240	420	280	300	-	-	720	810
10	180	110	220	0	100	-	-	610	620
11	100	40	220	80	100	-	-	610	620
12	280	420	600	460	480	-	750 (700)*	900	990
13	330	200	250	230	130	-	-	690	400
14	200	70	120	100	0	-	-	630	520

15	320	190	0	220	120	-	-	750	640
16	380	250	300	280	180	-	-	810	700
17	530	400	450	430	330	-	-	960	850

*(transportation distance by ship)

Table 7.11 Coal characteristics (CETP database, 2001)

	Coal mine	sulfur (%)	GJ/t (LHV)
Domestic	Zibo	1.5	23.0
	Xinwen	1.2	
	Zaozhuang	1.0	
	Feicheng	1.0	
	Yanzhou	0.8	
	Longkou	0.5	
Import	North Shanxi	0.8	22.2
	Central Shanxi	1.1	
	South Shanxi	0.5	

Table 7.12 Cost assumptions about coal (Chandler et. al. 1998; CETP database, 2001)

Production Cost ($/t)	
Domestic	22.5
Import	17.5
Transportation Cost	
basic fee($/t)	1.25
per distance ($/t/100km) by Rail	1.38
per distance ($/t/100km) by Ship	0.42
carbon emission coefficient (t-C/TOE)	1.08

The facts assumed in the model about natural gas and nuclear fuels are shown in Table 7.13. The costs for both natural gas and nuclear fuel are considered to be uniform in the entire province; and no limitation is imposed on their supply.

Table 7.13 Assumptions for natural gas and nuclear fuel (CETP database, 2001; Handbook of Energy and Economics in Japan, 2001)

	Natural Gas	**Nuclear**
Fuel price (\$/GJ)	3.25	0.6
Carbon emission coefficient (t-C/TOE)	0.64	0
Sulfur emission coefficient (kg-SO_2/GJ by Fuel)	0.0003	0

3. SCENARIO ANALYSIS WITH ETM

This section presents results from seven cases calculated with ETM to analyze the Shandong power system under typical scenarios from an environmental point of view. The cases include a business-as-usual case, a variety of SO_x control cases, a CO_2 control case, and a case of simultaneous control of SO_x and CO_2. Results of the analysis are first described in terms of emissions, then for each scenario, and lastly for costs.

3.1 Scenario Settings

Seven cases were calculated with ETM:
1. Business-As-Usual (BAU) Case. This scenario includes no environmental constraints or policy. This case serves as a reference for comparison with other cases. Actually China and Shandong Province are likely to enforce environmental control measures, so the business-as-usual label might not be appropriate.

2. TAX Case. In this case, a charge is imposed on SO_2 emissions in particular cities. China has begun a trial implementation of emission charges for SO_2 emissions in the Acid Rain Control Zone and SO_2 Control Zone. We have assumed that 10 of the 17 cities in Shandong are included in the Control Zone (see Figure 7.5). The charge rate is 0.2 yuan/kg-SO_2 (\$50/t-S). The TAX case simulates this trial policy.
3. SO_x Control Zone (SCZ) Case.

This case simulates a situation in which the SO_2 Control Zone policy mentioned above is enforced comprehensively. According to recent information, in the Control Zone flue gas desulfurization (FGD) system, or measures that can produce the same SO_2 reduction effects, must be installed in the existing thermal power plants burning coal with a sulfur content of more than one percent by 2010. Newly built plants burning coal with a sulfur content of over one percent must install an FGD system.

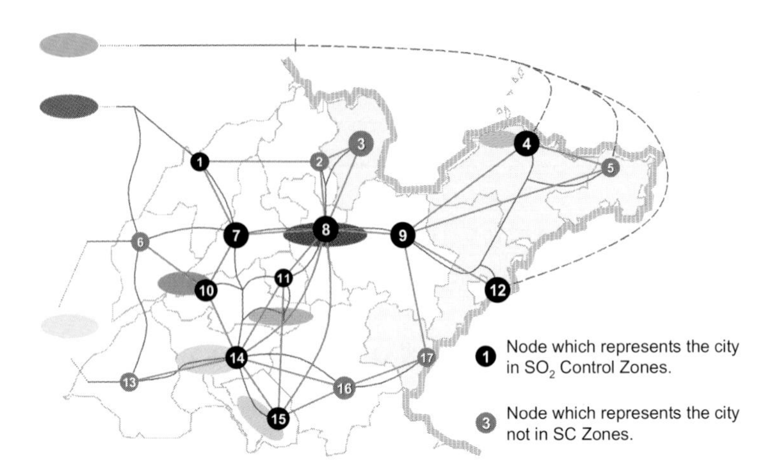

Figure 7.5. SO_x Control Zones

The sulfur tax mentioned in the TAX case is also imposed on SO_x emissions. In the SCZ case to be simulated, in addition to the taxation, the FGD system must be installed in existing plants by 2010 as well as in the newly built pulverized-coal-fired plants, if they utilize coal with one percent sulfur content or greater.

4. Total Sulfur Control (TSC) Case. In this case, total sulfur emissions in the province are bounded to fall below existing conditions in 2000. SO_2 emissions cause damage not only provincially but also locally, so this case might not be appropriate in considering the local environment.

5. Local Sulfur Control (LSC) Case. In this case, sulfur emissions at each city must fall below what they were in 2000, representing local quantitative regulation of SO_2.

6. CO_2 Control (CC) Case. Here, provincial carbon emissions are limited to less than half of that allowed in the year 2020 in the BAU case.

7. CO_2 and Local SO_x Control (CLSC) Case. In this case, SO_x and CO_2 emissions are simultaneously controlled equivalent to the LSC case and the CC case respectively:

- Sulfur emissions at each node, corresponding to cities, must fall below the level existing in 2000.
- Provincial carbon emissions are limited to less than half of the level existing in 2020 in the BAU case.

3.2 Results

ETM analysis of each of the scenarios listed above yields information on emissions, the optimal configuration and operation, and the associated cost for the Shandong electric power system in the presence of the constraints defined for each scenario.

3.2.1 Emissions

Figures 7.6, 7.7, 7-8, and 7.9 present the total emissions of SO_x for the province and the locales; and total CO_2 and NO_x for the province. In the BAU case, provincial SO_x emissions in 2020 reach a level over three times those in 2000 and would increase significantly in some cities with large demands. These results would not change very much if a sulfur tax were imposed on the SO_x as it now has on a trial basis in the SO_x Control Zones. (See the TAX case.) Even in the SCZ case, wherein this policy is implemented comprehensively by requiring FGD systems or the use of low-sulfur coal in addition to taxation, the suppression effect on total provincial SO_x emissions is modest; emissions still increase up to 2.5 times as much as they were in 2000. Moreover, SO_x emissions increase greatly in excess of those in the BAU case in some cities outside the SO_x Control Zones, even though within the zones, they might decrease.

The story is different for those cases involving quantitative SO_x emission control. Though the emissions are controlled not locally but provincially in the TSC case, in no city did emissions noticeably exceed 2000 levels. And in the CC case, even though there is no SO_x regulation at all, emissions are lower than those in the BAU or SCZ cases.

Like SO_x emissions, 2020 CO_2 emissions in the province almost triple those of 2000 in the BAU case. In the cases with SO_x emission control alone, CO_2 emissions appear to increase slightly more than that in the BAU case. Relations between SO_2 and CO_2 controls are asymmetric: CO_2 reduction measures lead to SO_2 reduction, but SO_2 controls do not suppress CO_2; in fact they increase them slightly.

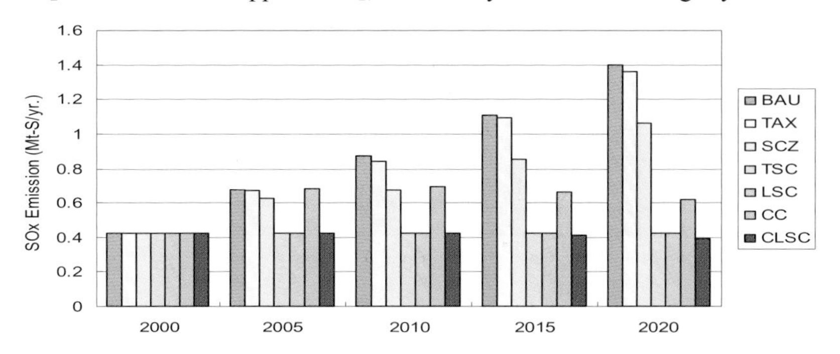

* BAU: Business As Usual case, SCZ: SOx Control Zones case, TSC: Total SO_x Control case, LSC: Local SO_x Control case, CC: CO_2 Control case, CLSC: CO_2 and SO_x Control case

Figure 7.6 Provincial total SOx emissions

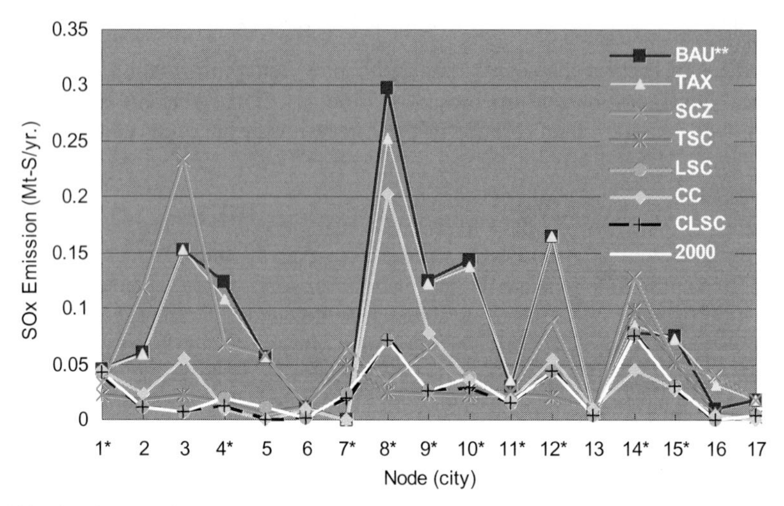

* Cities in SOx control zones.

** BAU: Business As Usual case, SCZ: SOx Control Zones case, TSC: Total SO_x Control case, LSC: Local SO_x Control case, CC: CO_2 Control case, CLSC: CO_2 and SO_x Control case, 2000: the existing conditions in 2000.

Figure 7.7 Local SO_x emissions in 2020

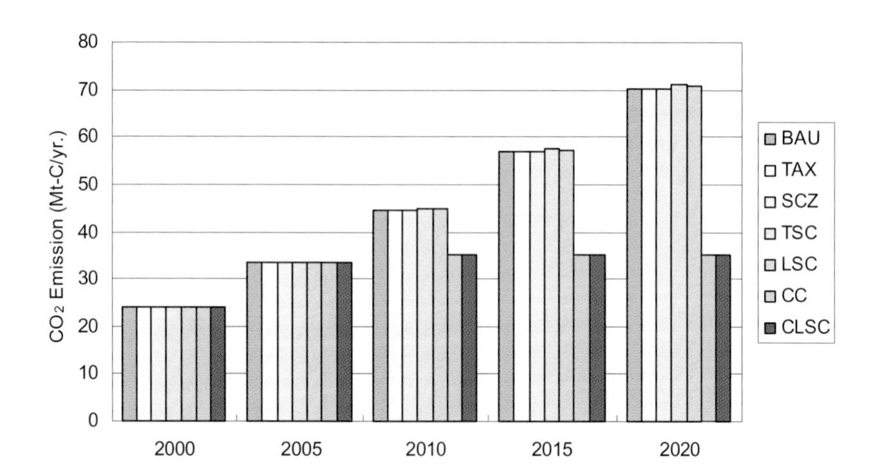

* BAU: Business As Usual case, SCZ: SOx Control Zones case, TSC: Total SO_x Control case, LSC: Local SO_x Control case, CC: CO_2 Control case, CLSC: CO_2 and SO_x Control case

Figure 7.8 Provincial CO_2 emissions

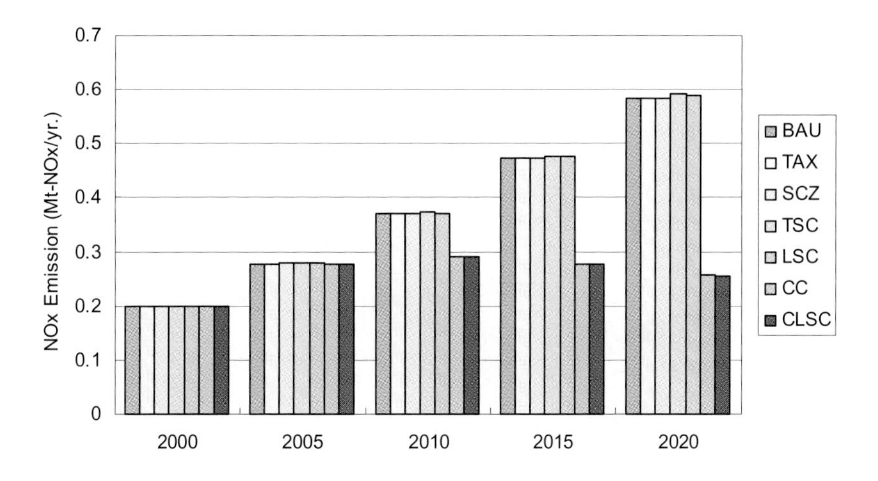

* BAU: Business As Usual case, SCZ: SOx Control Zones case, TSC: Total SO_x Control case, LSC: Local SO_x Control case, CC: CO_2 Control case, CLSC: CO_2 and SO_x Control case

Figure 7.9. Provincial NO_x emissions

3.2.2. Analysis of results for each case

In this section the impacts of the scenarios for the Shandong power system are analyzed, based on the results of different configurations of generating plants, generation mix, fuel consumption, and location.

1. The Business-As-Usual (BAU) Case. In the BAU case, with no environmental constraints or policy, the pulverized coal-fired unit without a flue gas desulfurization (FGD) system is the least-cost generating option. (Some newly installed pulverized coal-fired plants have FGD systems, which are exogenously determined to reflect actual planning commitments already in place.) The utilization rate of pulverized coal-fired plants with FGD systems is kept low to avoid the additional energy consumption and operating cost of FGD technology. Combustion turbine plants are installed as reserve generating capacity and for instantaneous peak requirements, because their fixed cost is low, though the cost of fuel--natural gas--is high. In this model, the electrical energy output for instantaneous peak is not counted.

The fuel consumed in the BAU case is coal only, most of it produced in Shandong Province. The coal from Shanxi Province is not economical because of the cost for long-distance transportation, except in a few cities relatively close to Shanxi. The major coal source in this case is the Zibo mine which produces the highest sulfur coal in this model. This result is strongly shaped by the mine's location. Zibo is the closest mine to many cities having high demands for energy.

As it turned out, SOx and CO_2 emissions increase greatly, especially in cities with large demand, as mentioned in the previous section.

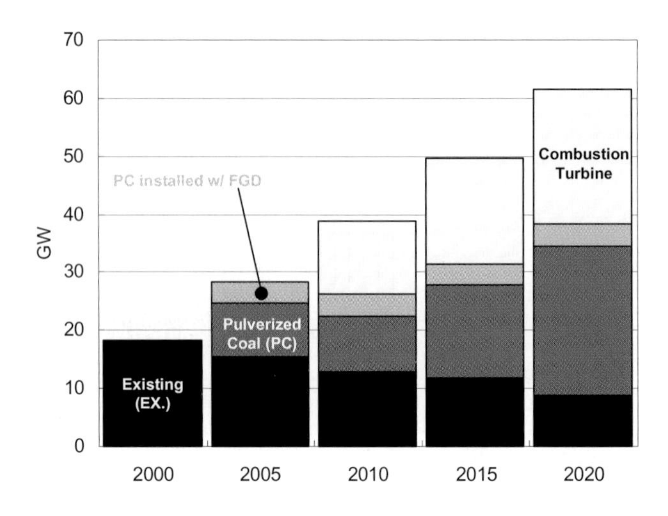

Figure 7.10 Configuration of generating plants in the BAU case

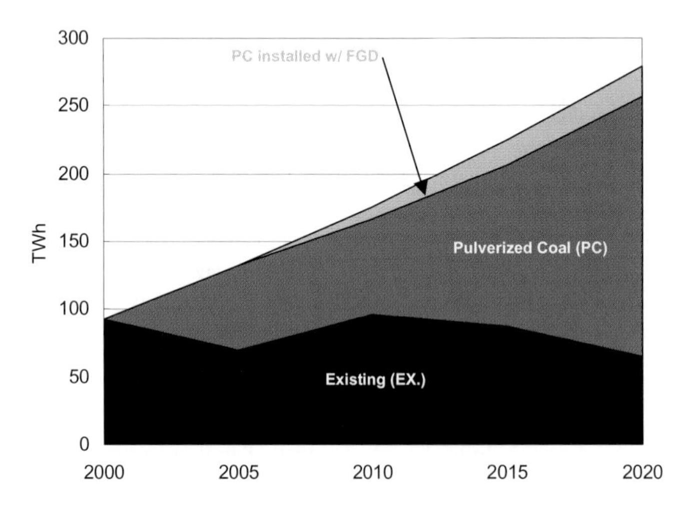

Figure 7.11 Generation mix in the BAU case

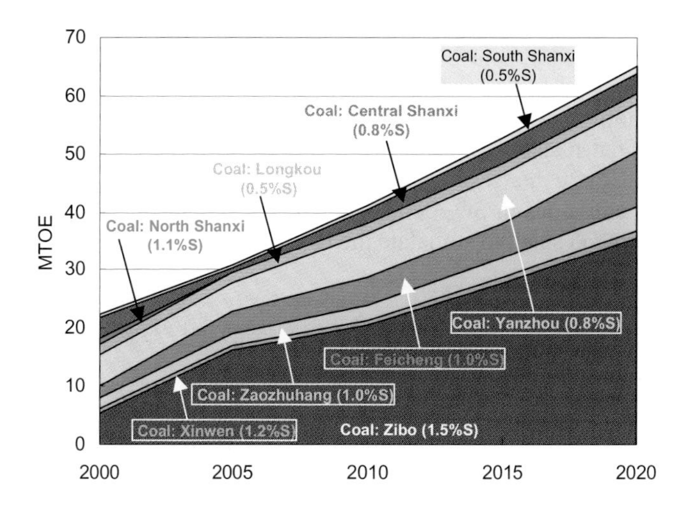

Figure 7.12 Provincial fuel consumption in the BAU case

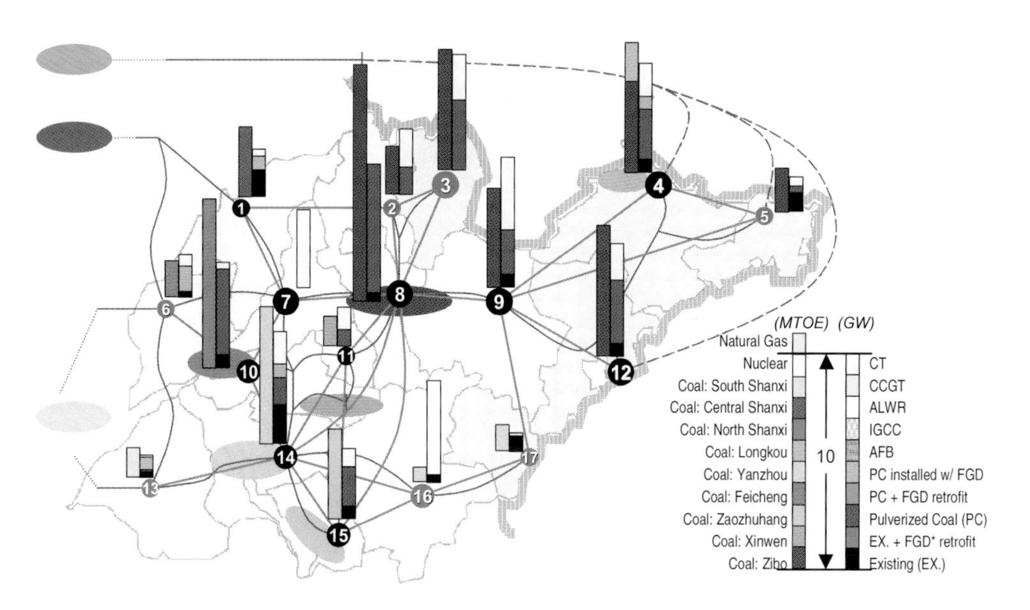

*CT: Combustion Turbine, CCGT: Combined Cycle Gas Turbine, AFB: Advanced Light Water Reactor, IGCC: Integrated Gasification Combined Cycle, AFB: Atmospheric Fluidized-Bed, FGD: Flue Gas Desulfurization system

Figure 7.13 Plant location and local fuel consumption in 2020 in the BAU case

2. The TAX Case. The SO_2 emission charge rate assumed in the TAX case is 0.2 yuan/kg-SO_2 in SO_x control zones, in which 10 of 17 cities are included. The results of this case are almost the same as those of the Business-As-Usual case: Plants burn relatively high sulfur coal, most of which is produced in Shandong Province at pulverized coal-fired plants without a flue gas desulfurization system. Provincial SO_x emissions in 2020 increase to about three times the 2000 emissions. The reduction of SO_x emissions in SO_x control zones is imperceptible. The charge rate, 0.2 yuan/kg-SO_2, is too small to stimulate the reduction of SO_2 emissions.

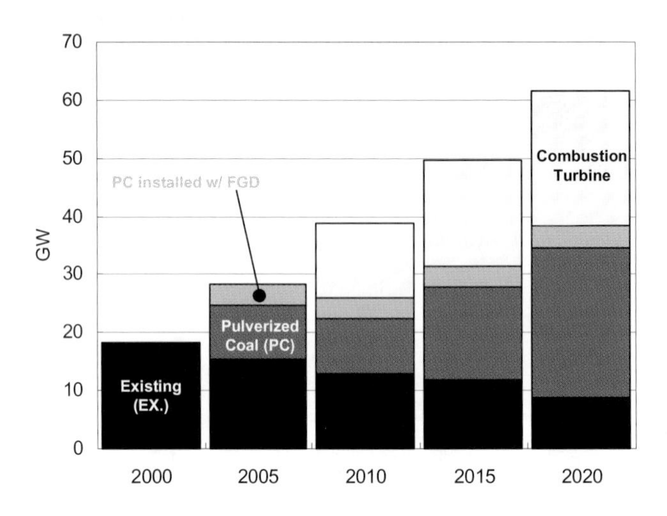

Figure 7.14 Configuration of generating plants in the TAX case

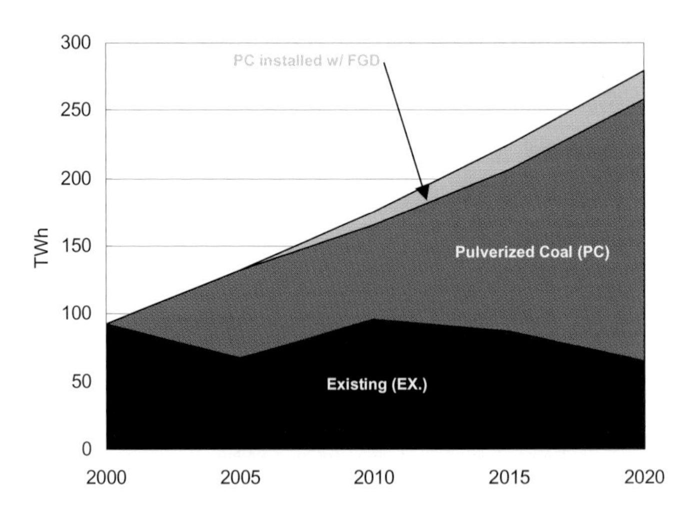

Figure 7.15 Generation mix in TAX case

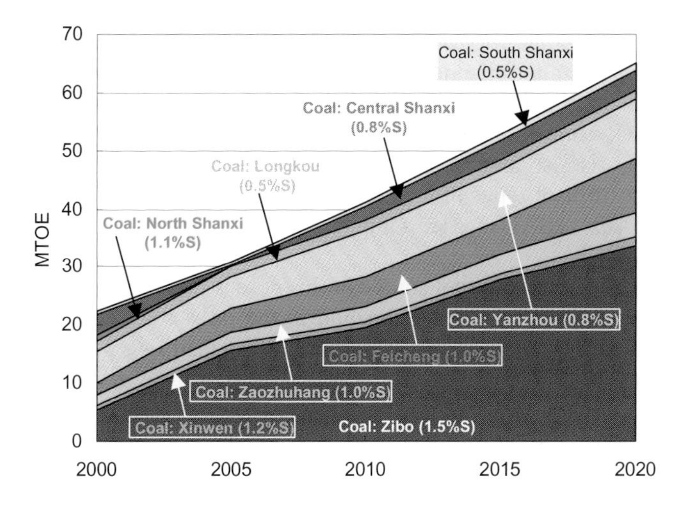

Figure 7.16 Provincial fuel consumption in the TAX case

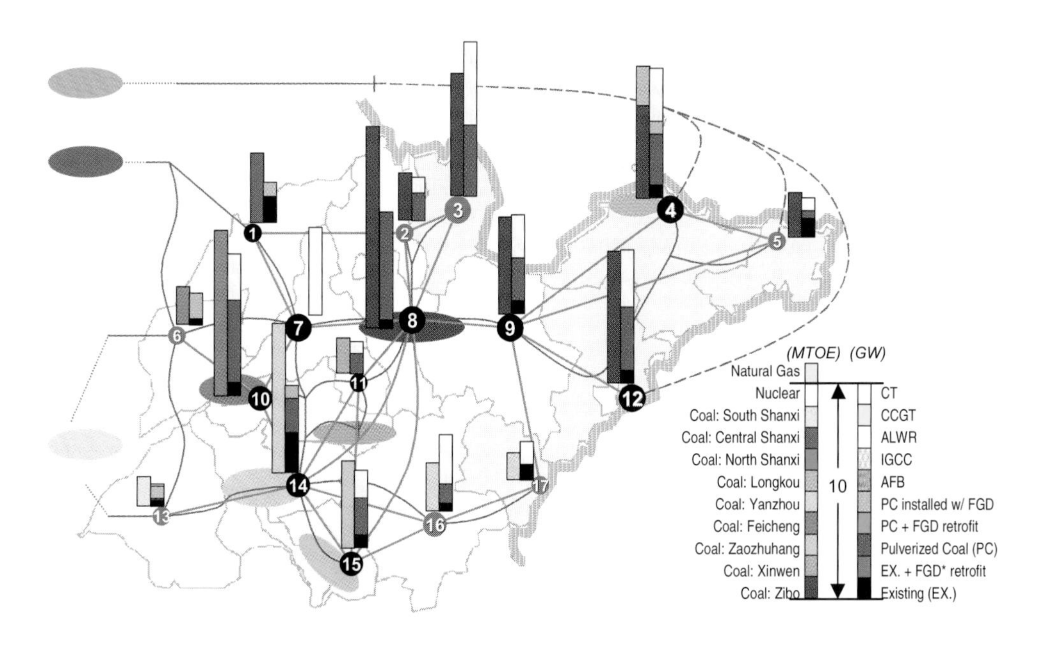

*CT: Combustion Turbine, CCGT: Combined Cycle Gas Turbine, AFB: Advanced Light Water Reactor, IGCC: Integrated Gasification Combined Cycle, AFB: Atmospheric Fluidized-Bed, FGD: Flue Gas Desulfurization system

Figure 7.17 Plant location and local fuel consumption in 2020 in the TAX case

3. The SO_x Control Zone (SCZ) Case. In this case, flue gas desulfurization (FGD) systems must be installed in existing coal-fired plants after 2010 and in newly built pulverized coal-fired plants if they burn coal with a sulfur content of one percent or more in the SO_x control zones mentioned above. The SO_2 emission charge rate of 0.2 yuan/kg-SO_2 is also assumed in SO_x control zones in this case. The provincial generation mix of plant types in the SCZ case is not much different from that in the Business-As-Usual (BAU) case. Only exogenously planned plants have FGD systems.

However, the dominant fuel source in this case is the Yanzhou mine in place of the Zibo mine in the BAU case. Yanzhou is the only coal mine assumed in the model to produce coal with a sulfur content of less than one percent in Shandong province. Yanzhou coal is used at coal-fired plants in the cities in SO_x Control Zones, therefore these facilities are not required to install FGD systems under the control regulation.

In comparison to the BAU case, the capacities of the generating plant in each city also change. Among the cities in SO_x control zones, both fuel consumption and generating plant capacity decrease, especially in the cities having coal mines (node 8, 10), except for the city including the Yanzhou mine (node 14) where they increase. However, among the cities which are not designated as SO_x Control Zones such as Binzhou city (node 2) and Dongying city (node 3), there are significant increases in both the installation capacity of pulverized coal-fired plants without FGD systems and coal consumption. The fuel utilized in these cities is the high-sulfur coal from the Zibo mine.

Thus, SO_x emissions increase greatly over those in the BAU case in those cities outside of the SO_x Control Zones, though decrease in those within the zones in this case. As a result, total SO_x emission of the province increase up to about 2.5 times the 2000 level, though they remain less than in the BAU case.

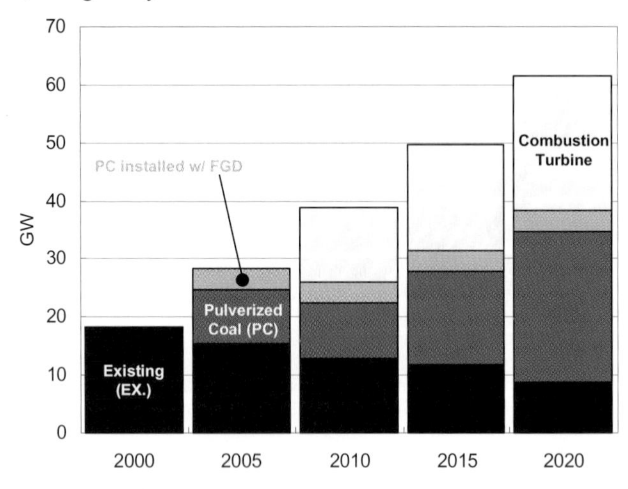

Figure 7.18 Configuration of generating plants in the SCZ case

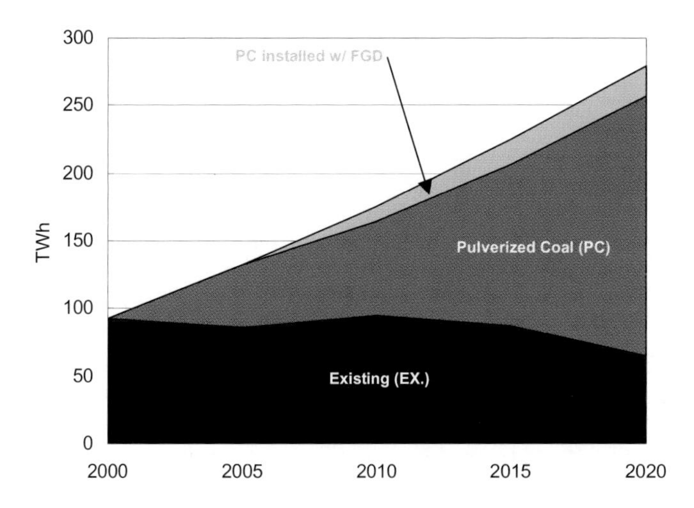

Figure 7.19 Generation mix in the SCZ case

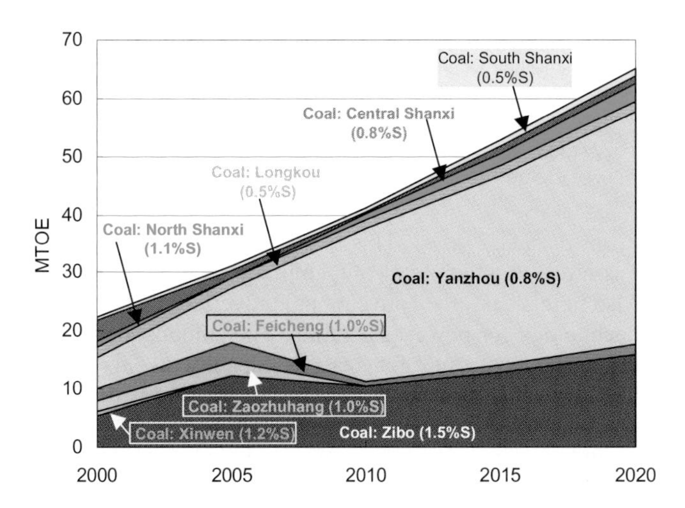

Figure 7.20 Provincial fuel consumption in the SCZ case

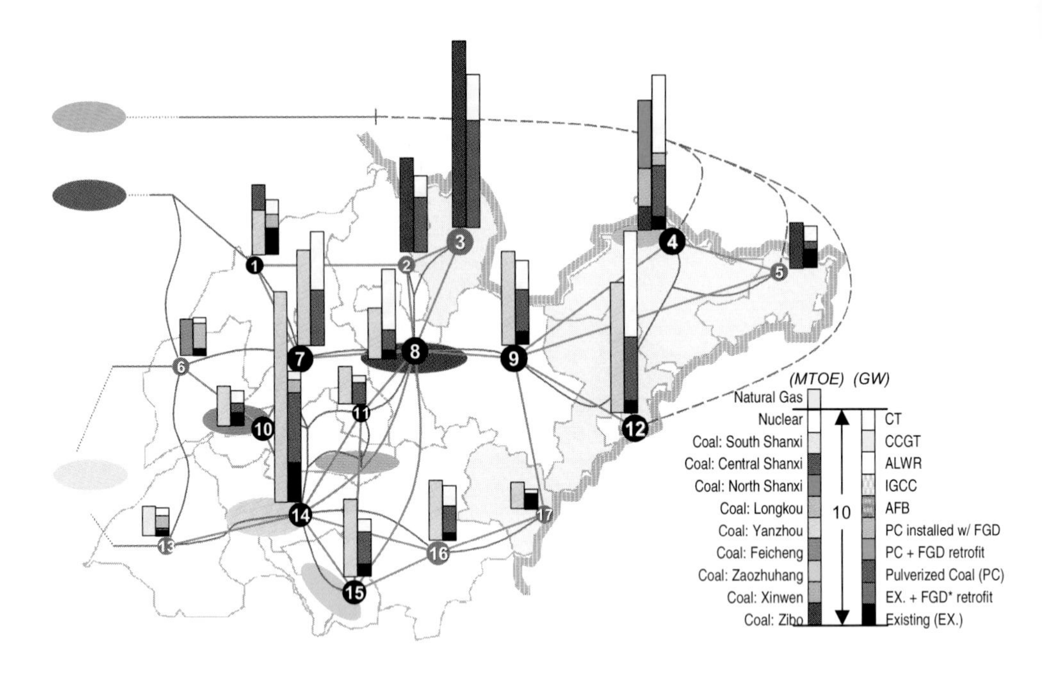

*CT: Combustion Turbine, CCGT: Combined Cycle Gas Turbine, AFB: Advanced Light Water Reactor, IGCC: Integrated Gasification Combined Cycle, AFB: Atmospheric Fluidized-Bed, FGD: Flue Gas Desulfurization system

Figure 7.21 Plant location and local fuel consumption in 2020 in the SCZ case

4. Total SO_x Control (TSC) Case. This is one of three cases involving quantitative SO_2 emission control, in which total SO_x emission of the province must fall below 2000 level. Also in this case, the conventional pulverized coal-fired plant is selected as the major technology for power generation. The combustion turbine is used only for the instantaneous peak and for reserve capacity as in the previous cases. However, in contrast to the previous cases, installed flue gas desulfurization (FGD) system exceeds exogenously committed capacity, being installed in some of the existing plants and in over half of the newly installed pulverized coal-fired plants. Except for the exogenously committed ones, these FGD systems are installed in plants in coastal area. This means that the FGD technology utilized is the low-cost one--the seawater scrubber available only in the coastal area.

The fuel supply in this case is less dependent on low-sulfur domestic coal than in the SOx Control Zones (SCZ) case, but more so than in the Business-As-Usual (BAU) case. In the latter periods simulated, the low-sulfur coals from Shanxi Province comprise large shares of the supply. These lower-sulfur coal are used at the plants without a FGD system, and the high-sulfur coal produced in Shandong is consumed at the plants with FGD systems.

Though there are some cities where SO_x emissions are greater than those in 2000, the rise in each city is less than in the BAU or SCZ cases. CO_2 emissions grow slightly over those in the BAU case because the expanded utilization of FGD systems lowers the net thermal efficiency of a power plant.

In this case, the theoretical sulfur taxes to stabilize the SO_x emissions at the constraint level is calculated, that is shadow prices for those constraints. The shadow price ranges from \$390/t-S in 2005 to \$1100/t-S in 2020 (undiscounted). The results confirm that the tax rate of \$50/t-S (0.2 yuan/kg-SO_2 assumed in the TAX and SCZ cases is insufficient to control SO_x emissions.

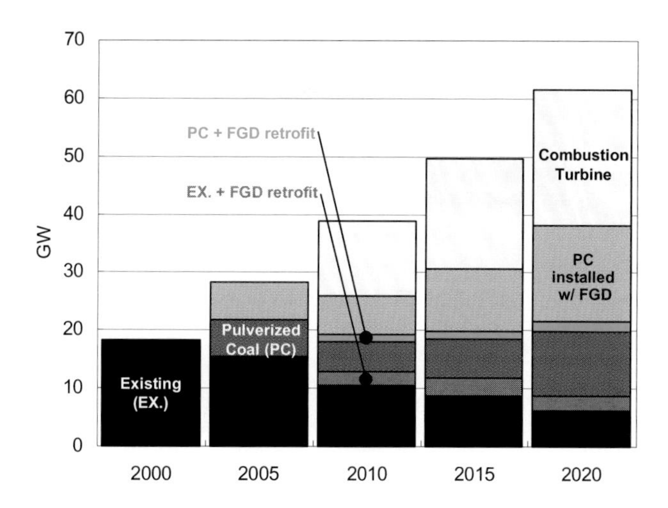

Figure 7.22 Configuration of generating plants in the TSC case

5. Local SO_x Control (LSC) Case. The results of this case suggest measures to keep SO_x emissions below the 2000 level in each city. The regulation for SO_x emissions here is more stringent than in the Total SO_x Control (TSC) case with respect to the flexibility of locating the emission source. Even under tighter regulation, the only plants installed burn pulverized coal, except for combustion turbines. Furthermore,, flue gas desulfurization (FGD) systems are installed only in the coastal areas: only the low-cost type is utilized.. Moreover, the capacity of the FGD systems becomes less than the capacity seen in the TSC case. However the lowest-sulfur coal produced in Shanxi Province provides the larger share of the coal supply.

This can be explained as follows: In some of the coastal cities where low-cost (seawater) FGD systems are available and, in the TSC case, installed SO_x emissions in 2000 are so low that they can only be reduced by cutting the amount of power generated. For this reason, the installed capacity of FGD systems will also decrease

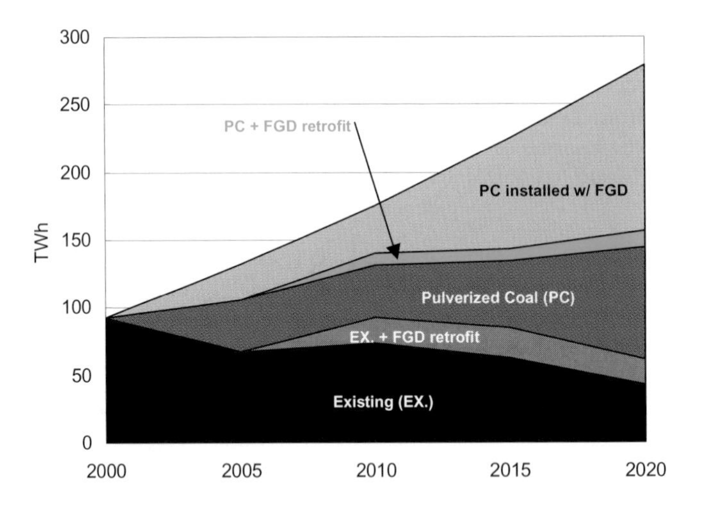

Figure 7.23 Generation mix in the TSC case

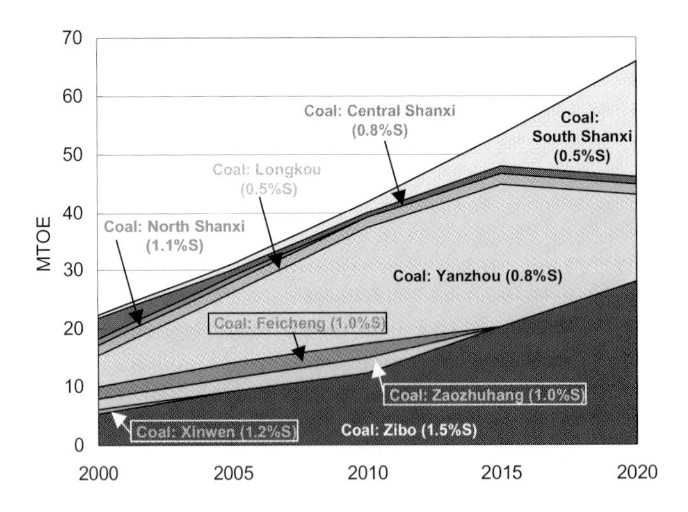

Figure 7.24 Provincial fuel consumption in the TSC case

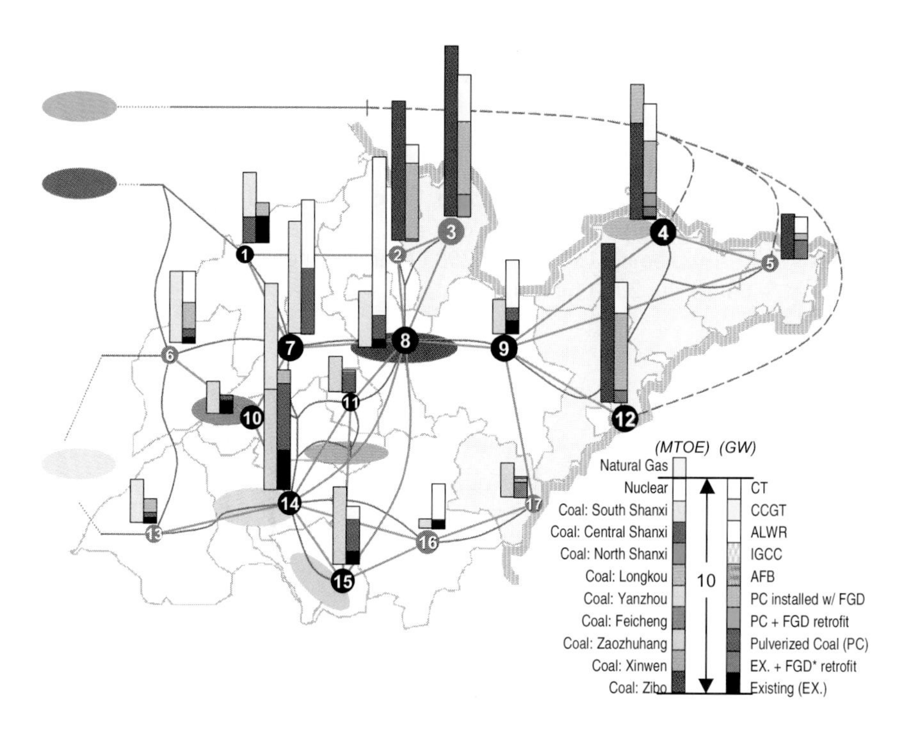

*CT: Combustion Turbine, CCGT: Combined Cycle Gas Turbine, AFB: Advanced Light Water Reactor, IGCC: Integrated Gasification Combined Cycle, AFB: Atmospheric Fluidized-Bed, FGD: Flue Gas Desulfurization system

Figure 7.25 Plant location and local fuel consumption in 2020 in the TSC case

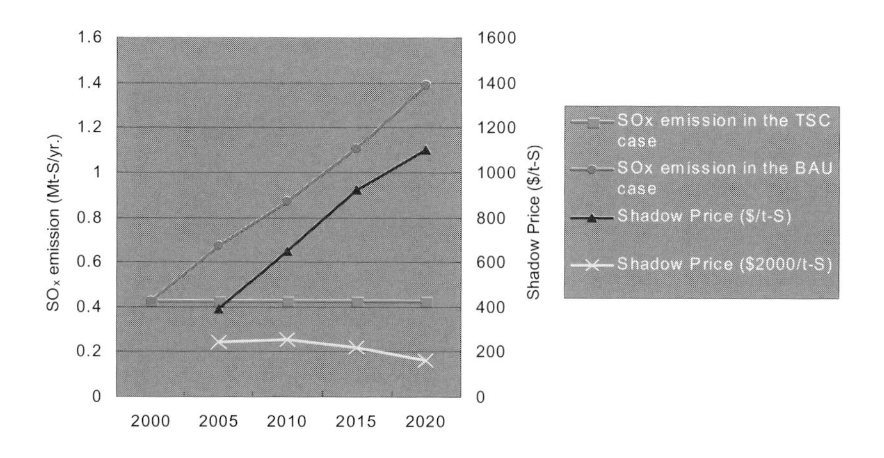

Figure 7.26 Theoretical sulfur tax in the TSC case

along with the generating plants. In other cities where low-cost FGD is unavailable, SO_x reduction can be achieved for lower cost by shifting to lower-sulfur coal rather than installing FGD technology. As a result, SO_x emissions control in each city is neither more nor less effective than in 2000.

In this case, a theoretical sulfur tax -- a shadow price -- for stabilizing the SO_x emission at the local level is calculated for each city. The theoretical sulfur tax would range extremely widely between cities, as well as between periods: from about \$300/S-t to about \$1300/S-t in 2005 (undiscounted); and from about \$350/S-t to about \$4000/S-t in 2020 (undiscounted).

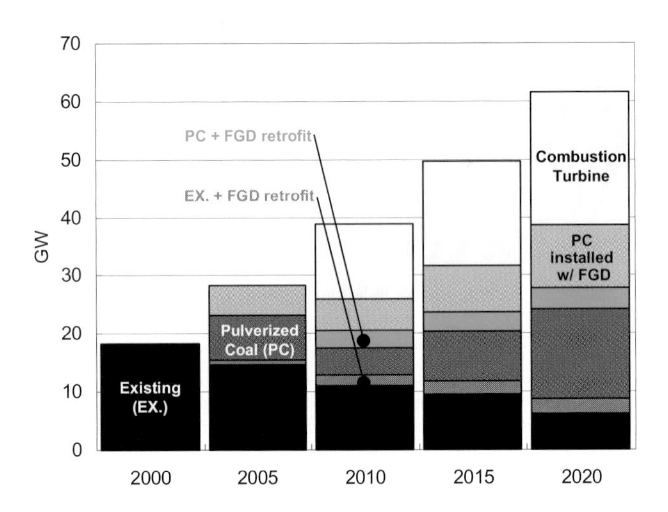

Figure 7. 27 Configuration of provincial generating plants in the LSC case

6. CO_2 Control (CC) Case. In this case, in which CO_2 emissions are kept below 50 percent of 2020 emissions in the Business-As-Usual (BAU) case, nuclear power plants (Advanced Light Water Reactor: ALWR) and Combined Cycle Gas Turbine (CCGT) plants are utilized to reduce them. The model assumes that nuclear power plants will be installed only in coastal areas and its installation capacity is limited to under1000MW per city per period. , CCGT plants however, can be installed anywhere in the province once the technology becomes available and the natural gas price is assumed to be identical in any city. The simulation reflects the installation of nuclear power plants almost to the maximum capacity specified by the assumption. These results indicate that the utilization of nuclear power is more economical in reducing CO_2 emissions than either the use of natural gas by CCGT plants or the use of clean coal technology such as Atmospheric Fluidized Bed (AFB) and Integrated Gasification Combined Cycle (IGCC).

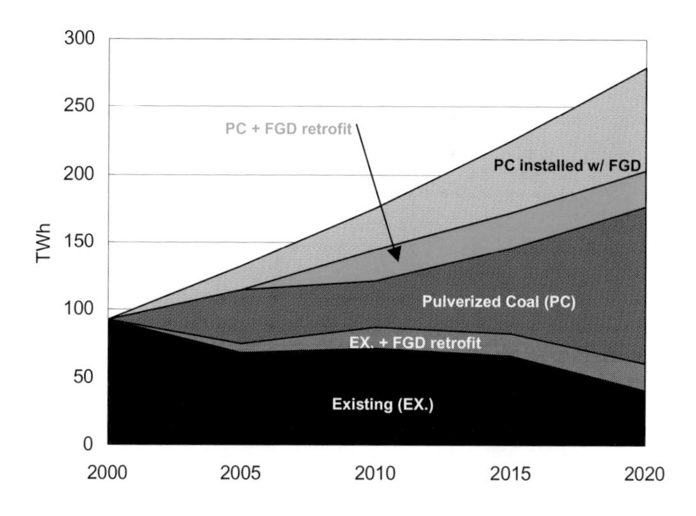

Figure 7. 28 Generation mix in the LSC case

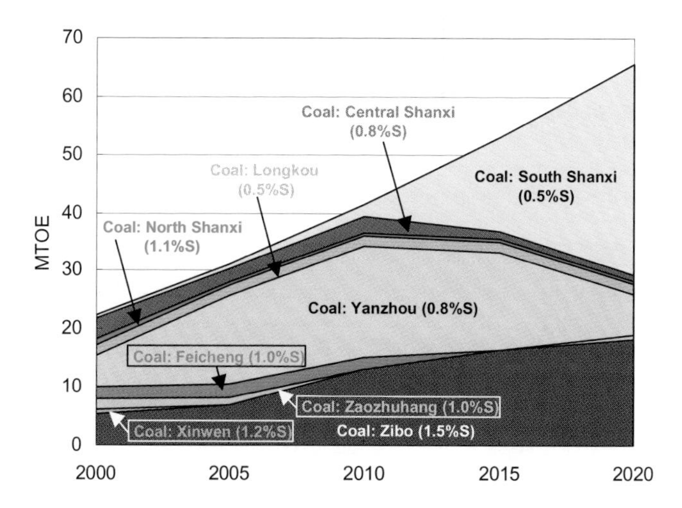

Figure 7. 29 Provincial fuel consumption in the LSC case

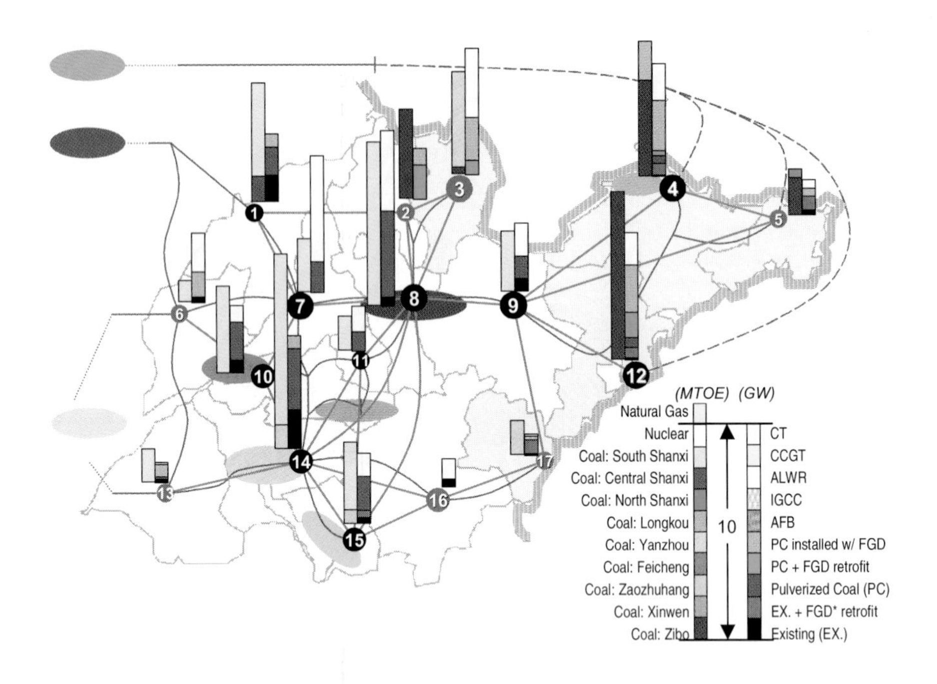

*CT: Combustion Turbine, CCGT: Combined Cycle Gas Turbine, AFB: Advanced Light Water Reactor, IGCC: Integrated Gasification Combined Cycle, AFB: Atmospheric Fluidized-Bed, FGD: Flue Gas Desulfurization system

Figure 7. 30 Plant location and local fuel consumption in 2020 in the LSC case

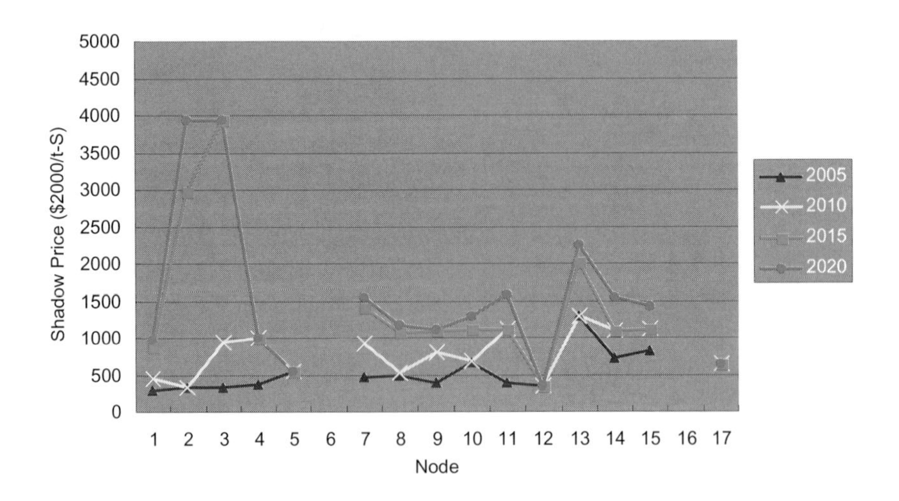

Figure 7.31 Theoretical sulfur tax in the LSC case

Although the use of nuclear power plants and CCGT plants also reduce SO_x emissions, the total for the province in this case increases over the 2000 figure by up to around 1.5 times. Moreover, in some cities where large demands exist, SO_x emissions greatly exceed those of 2000 because high-sulfur coal is burned without a FGD system.

The theoretical carbon tax obtained as a shadow price related to the CO_2 emission constraints ranges from about \$60 to \$80 (undiscounted).

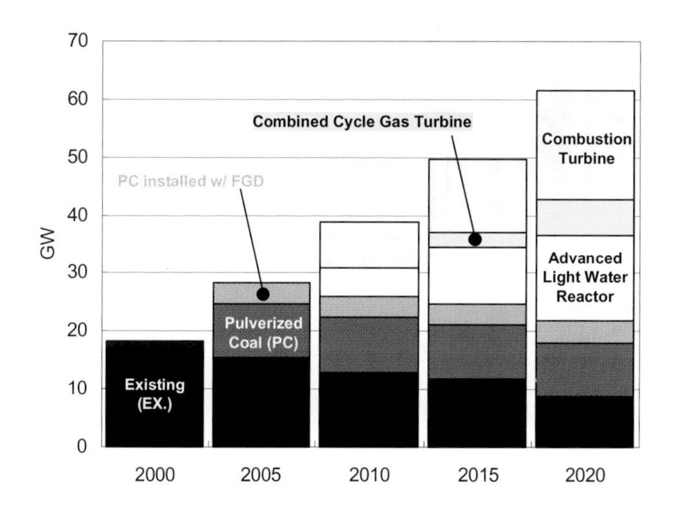

Figure 7.32 Configuration of generating plants in the CC case

7. CO_2 and Local SO_x Control (CLSC) Case. In the CLSC case, both local SO_x emissions and provincial CO_2 emission are controlled as they are in the Local SO_x Control (LSC) case and the CO_2 Control (CC) case respectively: SO_x emissions in each city are kept below the 2000 level; provincial CO_2 emissions are kept below 50 percent of the 2020 level specified in the BAU case.

In this case, nuclear power plants (Advanced Light Water Reactor: ALWR) and Combined Cycle Gas Turbine (CCGT) plants would be installed to reduce CO_2 as they are in the CC case. SO_x reduction measures, however, are not as substantial as they are in the LSC case: The lowest-sulfur coal produced in Shanxi is utilized to a lesser extent; this coal comprises a large share of the fuel supply in the LSC case; and the installation capacity of the FGD system is minimal. These results are attributed to the utilization of nuclear power and CCGT plants, which significantly reduces SO_x emissions.

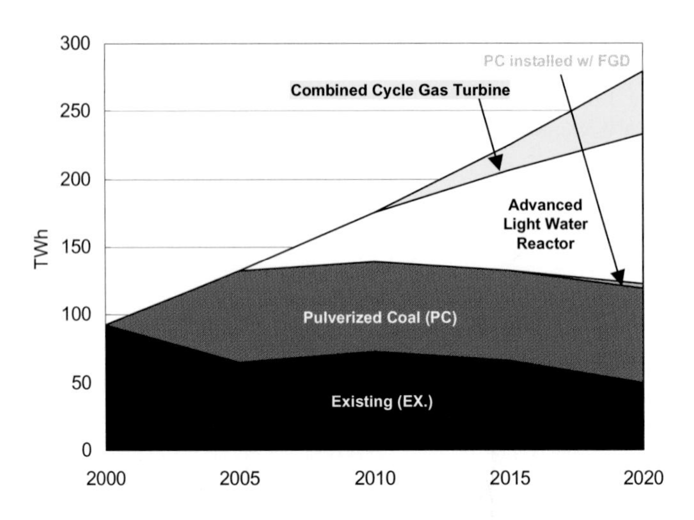

Figure 7.33 Generation mix in the CC case

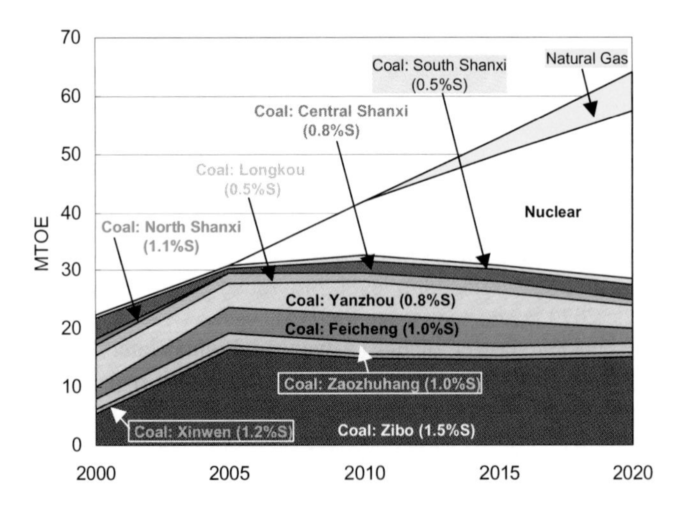

Figure 7.34 Provincial fuel consumption in the CC case

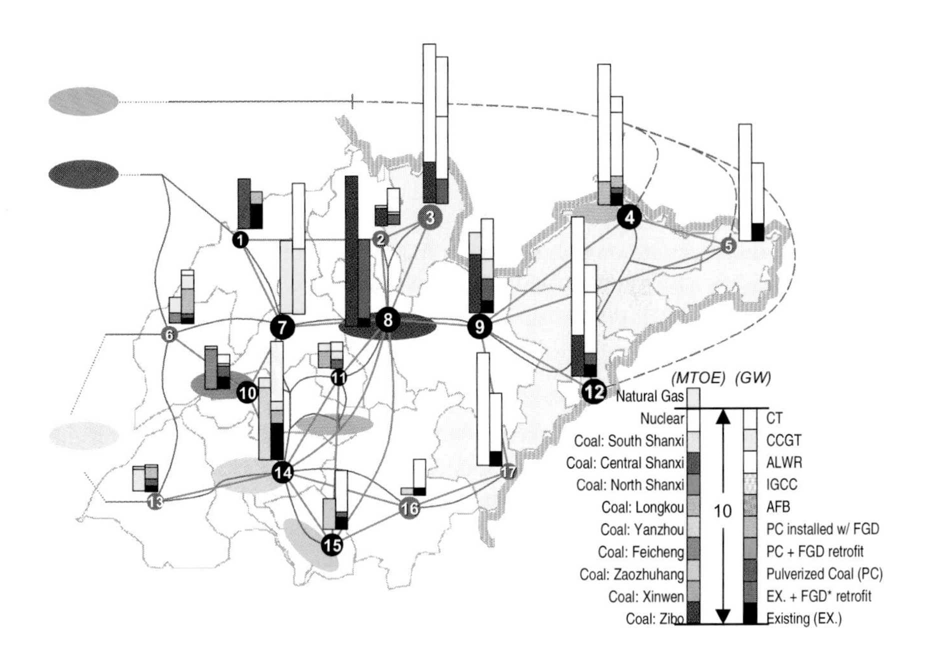

*CT: Combustion Turbine, CCGT: Combined Cycle Gas Turbine, AFB: Advanced Light Water Reactor, IGCC: Integrated Gasification Combined Cycle, AFB: Atmospheric Fluidized-Bed, FGD: Flue Gas Desulfurization system

Figure 7.35 Plant location and local fuel consumption in 2020 in the CC case

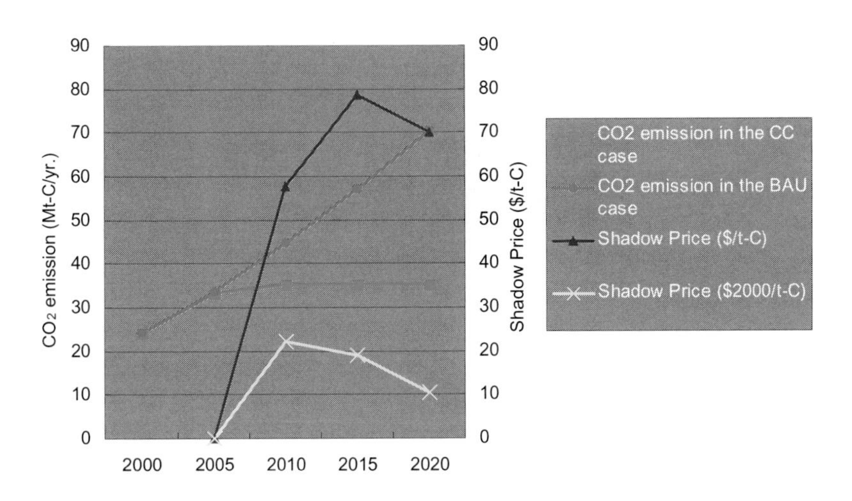

Figure 7.36 Theoretical carbon tax in the CC case

In this case, almost all the coal is supplied by Shandong mines, sulfur content of which are higher than that in the LSC case on the average. Nevertheless, there are some cities where SO_x emissions remain less than the emission control level.

The theoretical sulfur tax in each city, which equals to the SO_x reduction marginal cost, ranges from about \$300/S-t to about \$1300/S-t in 2005 (undiscounted); and from about \$0/S-t to about \$1300/S-t in 2020 (undiscounted). The theoretical taxes calculated in this case are generally lower than those in the LSC case in the latter periods simulated. Since the SO_2 reduction measures are less substantial on account of CO_2 reduction measures in this case, the marginal costs to reduce the SO_2 emissions turned out to be lower. The theoretical tax of \$0/S-t in a city implies that the regulation is not needed to hold emissions under 2000 levels in that city at that period. Emissions levels fall below control levels there.

Also, the theoretical carbon tax derived from this case is lower than the tax found in the CC case, which ranges about \$50 to \$70 (undiscounted). The theoretical tax is equal to the marginal CO_2 reduction cost, which equals the cost of replacing coal-fired plants with nuclear power or CCGT. In the CLSC case, coal-fired plants with SO_2 reduction measures are replaced by nuclear power plants or CCGT facilities. This contrasts to the CC case, in which the less expensive coal-fired plants without SO_2 reduction measures are replaced. Thus, the additional cost--the theoretical carbon tax-- in the CLSC case is lower than that in the CC case.

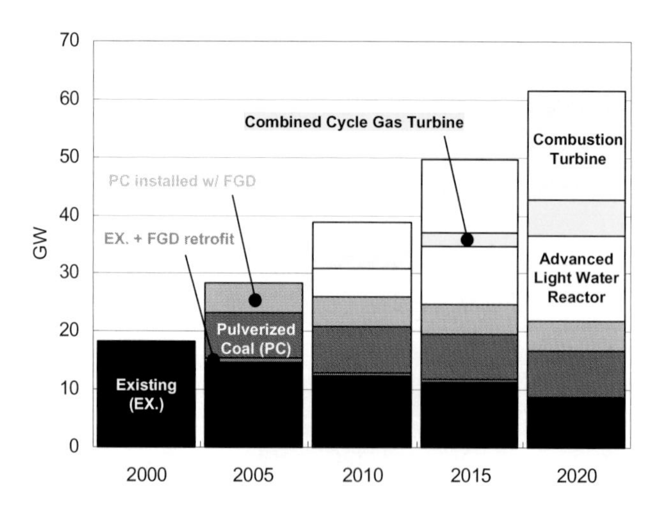

Figure 7.37 Configuration of generating plants in the CLSC case

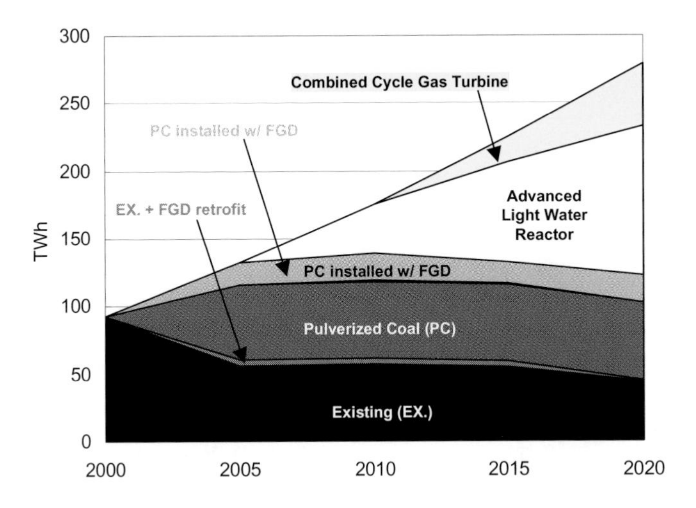

Figure 7.38 Generation mix in the CLSC case

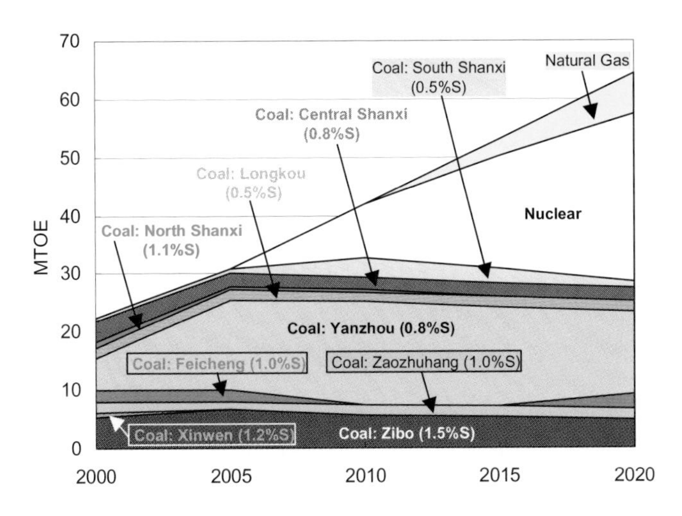

Figure 7.39 Provincial fuel consumption in the CLSC case

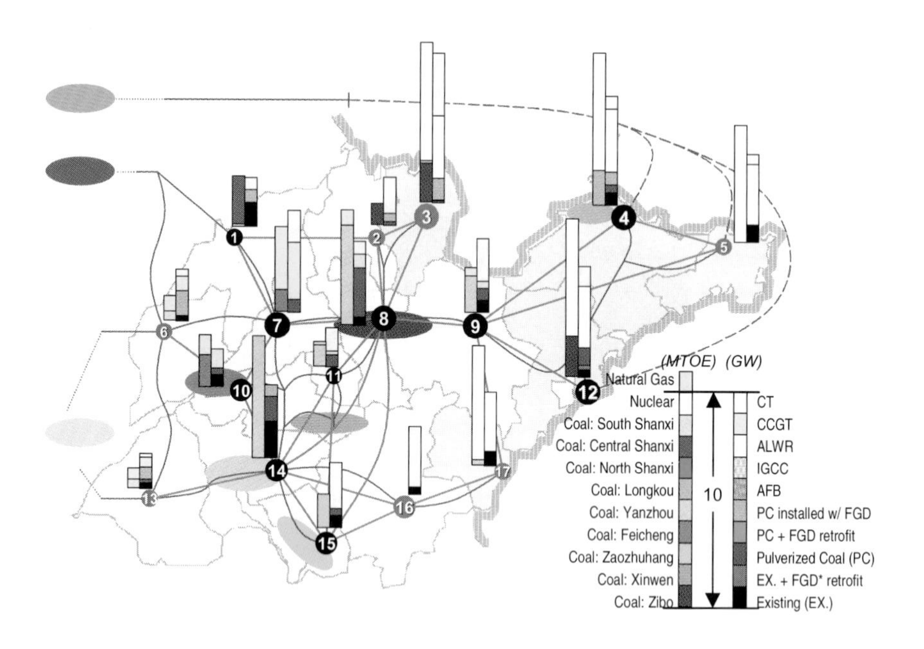

*CT: Combustion Turbine, CCGT: Combined Cycle Gas Turbine, AFB: Advanced Light Water Reactor, IGCC: Integrated Gasification Combined Cycle, AFB: Atmospheric Fluidized-Bed, FGD: Flue Gas Desulfurization

Figure 7.40 Plant location and local fuel consumption in 2020 in the CLSC case

3.2.3. Costs

The total cumulative cost over the period 2005-2024 and its specification for each simulated case, normalized by the total cumulative cost of the Business-As-Usual (BAU) case, are presented in Fig 7.43. The figure shows that the total cost increases as the regulation on the emission become more stringent: The total cost in the BAU case is naturally less than in the other cases. However, it must be noted that the costs of electricity production calculated in this analysis do not include external costs such as damage costs to the environment and health. They should be included for rational economic assessments.

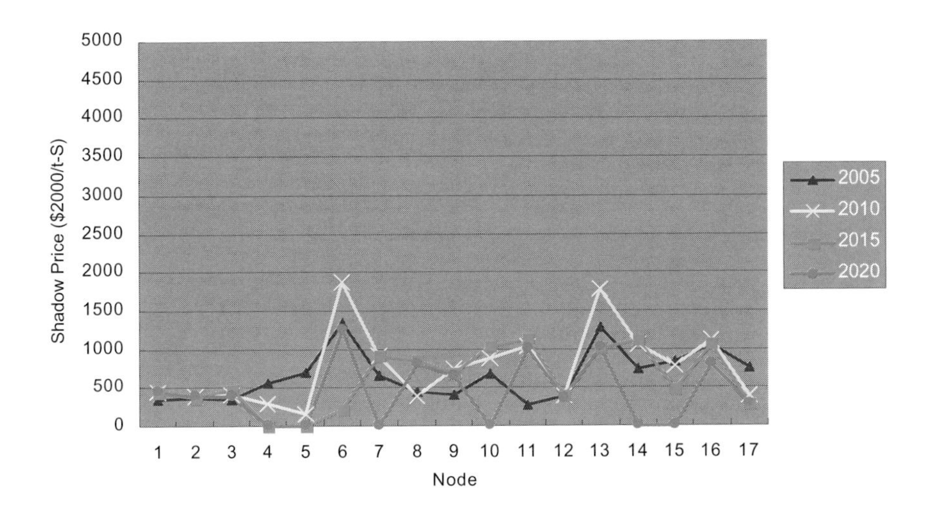

Figure 7.41 Theoretical sulfur tax in the CLSC case

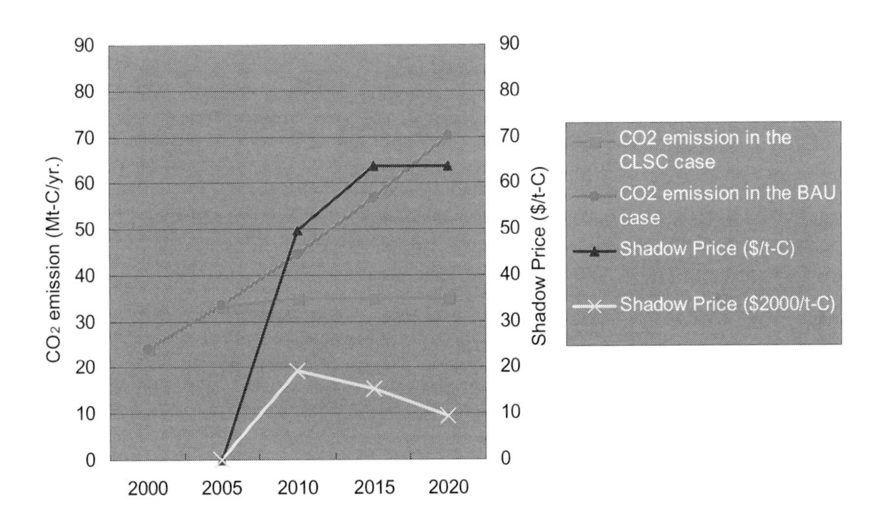

Figure 7.42 Theoretical carbon tax in the CLSC case

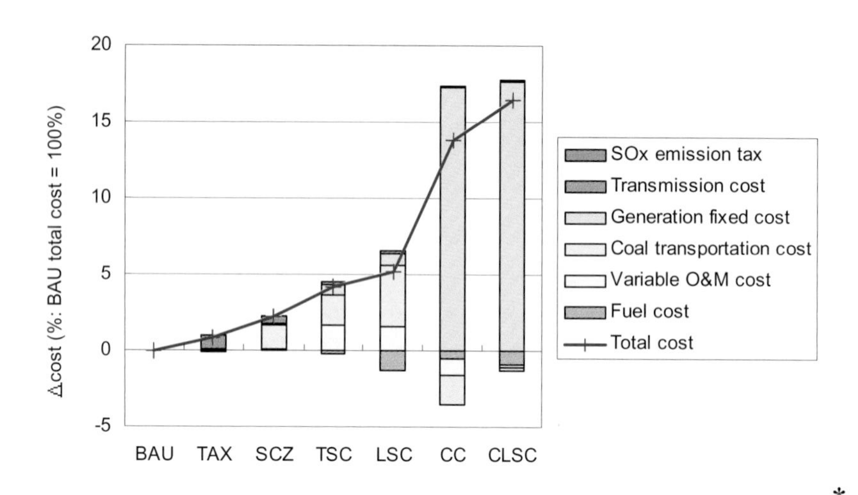

*

BAU: Business As Usual case, SCZ: SOx Control Zones case, TSC: Total SO_x Control case, LSC: Local SO_x Control case, CC: CO_2 Control case, CLSC: CO_2 and SO_x Control case

Figure 7. 43 Total cost

4. CONCLUDING REMARKS

The results of these cases modeled with ETM are summarized in Table 7.14. In the Business-As-Usual (BAU) case, with no environmental constraints, the least-cost generating option is the pulverized coal-fired unit without a Flue Gas Desulfurization (FGD) system. The sulfur content of the coal used in this case is relatively high; provincial SOx emissions in 2020 reach a level over three times those in 2000 in proportion to demand growth. Similarly, 2020 CO_2 emissions in the BAU case almost triple those of 2000. These results would not be changed significantly by imposing a sulfur tax on the SOx emission which is now in practice on a trial basis in the SOx Control Zones (TAX case).

Even in the SOx Control Zones (SCZ) case, where the installation of desulfurization technologies is required when using high-sulfur coal in addition to the taxation in the SOx Control Zones, total SOx emission of the province would be reduced to a lesser extent. In the control zones, though the utilization of the desulfurization technology is scarcely promoted, the SOx emissions are reduced through utilizing lower-sulfur coal and decreasing generation as compared with the BAU case. However, SOx emissions in the cities outside of the control zones are much increased due to increased generation without SO_x reduction measures.

The results of the quantitative emission control cases indicate the least-cost options to control emissions at the desired level. In the Total SOx Control (TSC) case, where total SOx emission in the province is kept below the 2000 level, pulverized coal-fired units with FGD systems attain an increased share of electricity generation. Also, the sulfur content of coal used at the coal-fired plants without a

FGD system is reduced. The SOx reduction options implemented in the Local SOx Control (LSC) case, in which the SOx emission of each city is kept under the 2000 level, are basically the same as the TSC case. However, the cost of SOx control is increased because the locating flexibility of the emission source is constrained. The results of these SOx control cases indicate that controlling the SOx emission does not contribute to reduce the CO_2 emissions, but actually increase it slightly because net thermal efficiency is lowered by the FGD system. The measure adopted to reduce CO_2 emissions is the utilization of nuclear power and natural gas (the CO_2 Control (CC) case and the CO2 and Local SOx Control (CLSC) case). Hence, controlling CO_2 emissions also contributes to reducing SOx emissions. For the CC case, in which CO_2 emissions are kept below 50 percent of 2020 emissions in the BAU case, total SOx emission in the province increase is suppressed significantly below the BAU case. In the CLSC case, for the same reason, SOx reduction measures are no longer as substantial as in the LSC case. However, it costs much more to control CO2 emission itself than to control SOx emissions.

ETM analysis of the seven scenarios leads to several conclusions:

- Pulverized coal-fired plants without a flue gas desulfurization (FGD) system using domestic coal would be the most economical generating option for the Shandong power system if there are no constraints on SO_x and CO_2.
- The current trial sulfur tax rate, 0.2yuan/kg-SO_2 ($50/t-S), is insufficient to control SO_x emissions.
- The current trial policy in the Acid Rain Control Zones and SO_2 Control Zones contains a loophole that allows the transfer of high-sulfur coal use to a non-control zone.
- The utilization of FGD technology when burning high-sulfur coal or shifting from high-sulfur to low-sulfur coal are two economically viable and effective measures in reducing SO_x emissions. However, controls for SO_x emissions do not contribute to the control of CO_2 emissions.
- The utilization of both nuclear and natural gas are required for CO_2 emission control; they also contribute to sulfur emission control.

In the ETM, clean coal technologies such as integrated gasification combined cycle and atmospheric fluidized bed plants are also considered as electricity-generating options, but they do not contribute to optimal solutions in any of the cases. Although they have an advantage over conventional coal-fired technology in thermal efficiency and low SO_x emission rates, thus contributing to the reduction of both SO_x and CO_2 emission, they are too expensive in terms of both the capital and operating costs assumed in the model analysis. The cost reduction of these technologies would help reduce not only SOx emissions, but also CO2 emissions.

The utilization of natural gas in a Combined Cycle Gas Turbine (CCGT) also appears to not be economical for SO_x emission control due to the assumed high price of natural gas. Also, this analysis showed that for CO_2 emission control CCGT was less economical than the nuclear option. It should be noted, however, that projecting

Table 7.14 Results' summary of simulated cases with ETM.

Simulation case*				BAU	TAX	SCZ	TSC	LSC	CC	CLSC
Average Cost of energy ($/MWh$_e$, nominal)**				28.4	28.6	28.9	29.5	29.8	32.6	33.1
SO_2 emission in 2020	SO_x Control Zones		Mt-S (% of 2000)	1.08 (280)	1.02 (265)	0.58 (150)	0.35 (91)	0.39 (100)	0.52 (134)	0.37 (95)
	Outside of the Zones		Mt-S (% of 2000)	0.31 (814)	0.33 (870)	0.49 (1263)	0.08 (195)	0.04 (100)	0.10 (265)	0.03 (72)
	Total		Mt-S (% of 2000)	1.39 (328)	1.36 (320)	1.06 (251)	0.42 (100)	0.42 (100)	0.62 (146)	0.39 (93)
CO_2 emission in 2020	Total		Mt-C (% of 2000)	70.3 (292)	70.3 (292)	70.3 (292)	71.2 (295)	70.9 (294)	35.2 (146)	35.2 (146)
Share of electricity generation (%)**	by power generation technologies	Conventional coal-fired***	w/o FGD	95	94	94	61	66	68	59
			w/ FGD	5	6	6	39	34	0	10
		CCGT***		0	0	0	0	0	7	7
		ALWR***		0	0	0	0	0	24	24
	by zones	SO_x Control Zones		80	79	70	73	79	74	74
		Outside of the Zones		20	21	30	27	21	26	26
Average sulfur content of used coal (%)**	Conventional coal-fired***		w/o FGD	1.22	1.19	0.99	0.81	0.76	1.17	0.89
			w/ FGD	0.79	1.04	1.04	1.33	1.25	0.91	1.28
Average SO_2 emission coefficient (g-SO_2/kWhe)**				10.8	10.5	8.8	5.1	5.1	7.4	5.0
Average CO_2 emission coefficient (g-CO_2/kWhe)**				1010	1010	1011	1019	1016	718	718

* BAU: Business As Usual, SCZ: SO_x Control Zones, TSC: Total SO_x Control, LSC: Local SO_x Control, CC: CO_2 Control, CLSC: CO_2 and SO_x Control

** Average values and share of electricity generation are taken throughout the planning periods (2000-2024),

*** Conventional coal-fired includes existing plants and pulverized coal-fired plants, CCGT: Combined Cycle Gas Turbine, ALWR: Advanced Light Water Reactor

the natural gas price is inherently difficult because it depends on multiple factors, and different assumptions for natural gas prices would produce different results. Because CCGT is already one of the world's most economical and clean generating technologies, the stable and low cost supply of natural gas is expected.

Regardless of objective conditions, environmental policy makers planning Shandong's electric power system should understand that an emission source might change as power transmission schemes develop. The SO_x Control Zones case demonstrates this.

The model analyses presented in this chapter have suggested some possible futures for the Shandong electric power system under the specified conditions. The results, of course, depend on the assumed conditions. The possibilities described here represent only a fraction of all the possible futures. The results of the model analyses should not be taken as predictions but be used as the organized knowledge with which we can choose our future.

REFERENCES

Aratame, Basics of Power System Technology Calculations, Denki-shoin

CETP Database(2001)

Chandler, W., Yuan G., Logan, J. and Zhou, D. (1998), China's Electric Power Options: An Analysis of Economic and Environmental Costs, Battelle Pacific Northwest Laboratory Report PNWD-2433 (June, 1998).

Chugoku Furyoku (1997), Japan, State Statistical Bureau, People's Republic of China and Soken.Inc.

Handbook of Energy & Economic Statistics in Japan (2001), Japan, The Energy Data and Modeling Center.

Imanaka, T. and Yamaji, K.(1999) A Study on Locational Configuration of Distributed Generators Under Uncertain Demand Growth, Japan, *The Transactions of The Institute of Electrical Engineers of Japan*, Vol.119-B, No.10.

Kaigai-denryoku (1997) , July.

SEPCO – Shandong Electric Power Group Corporation. (2000), SEPCO Annual Report 1999.

Xue, B. and Eliasson, B. (2000) China Shandong Province Energy and Emissions: Status, Statistics and Prospects, 1999, ABB Environmental Affairs Brochure, Stockholm, Sweden.

CHAPTER 8

LIFE CYCLE ASSESSMENT

ROBERTO DONES, XIN ZHOU AND CHUNXIU TIAN

1. INTRODUCTION

1.1 Full energy chain analysis

Energy systems analysts will underestimate the environmental burden of each unit of electricity if they only consider the effects of the power plant that produces it. They must take into consideration the impacts of all the industrial activities required to produce energy, from mining of the ore to conversion to electricity. The assessment of an entire technology system should include the management of waste streams as well as the energy and materials invested. For renewables, the burdens from manufacturing of all components of the power units should be compared to the burdens from the process chains for fossil and nuclear energy carriers.

The economic effects of an energy chain are reflected in the costs and ultimately in the price of the electricity sold to final customers. The impacts of energy chains, like the effects of pollution on ecosystems, human heath, and the built environment, may also negatively affect the economy through paths not included in the current price of electricity. The estimation and inclusion of these external or societal costs reveals the total costs of electricity, which is the main aim of CETP.

While the destructive impacts of energy production on the equilibrium of natural processes and the quality of human life in relation to the natural environment may not be quantifiable in monetary terms, it may still be measured with specific indicators. The difficulty of representing the negative effects of human activities in terms of cost can make consideration of these aspects less attractive to decision-makers; compared to quantifiable impacts, these effects may appear less immediate. Multi-Criteria Decision Analysis (MCDA) uses qualitative scales to include these more elusive impacts in policy making in way that complements rankings based on total costs.

The application of Life Cycle Assessment (LCA) methodology to energy systems yields a comprehensive inventory of quantifiable burdens to the environment by studying all stages of energy chains over the entire lifetime of all component facilities, thereby placing all technological options on the same level. This element of the methodology is often referred to as Life Cycle Inventory assessment (LCI) (ISO, 1998). The result of this accounting process should ideally be included in appropriate impact assessments to give a quantitative estimation of each chain's overall effects. Simplified Life Cycle Impact Assessment (LCIA) methods are being developed within the framework of LCA (ISO, 2000).

For CETP, the scope of the LCA task has been restricted to the estimation of energy chain-specific environmental inventories. A sufficiently complete database could be achieved in practical terms only for main air pollutant species and solid wastes, but scattered pieces of information were available for other burdens. No methods of LCIA have been used because: a) another major CETP task already deals with the estimation of the effects of air pollution and its associated external costs, using Environmental Impact Assessment (EIA) methods; b) the LCIA schemes in the best case only roughly estimate potential harm, and would contribute no additional insights; and c) considering the limited set of environmental inventories here collected, the use of a LCIA method, which in principle should unify different kinds of burdens under one or a few indicators, would have been superfluous.

Besides the separated inventories of individual chemical species, this LCA study also uses two partial aggregations to illustrate important effects more synthetically. The first is the total warming potential of greenhouse gas (GHG) emissions, combining CO_2, CH_4 and N_2O over the 100-year time horizon recommended by the Intergovernmental Panel on Climate Change (IPCC, 2001). The second is the health effects of SO_2, NO_x and particulates in terms of Years of Life Lost (YOLL), i.e. a measure of premature deaths due to pollution, using the factors determined by the IEA task for the year 2020.

1.2 Methodology and main assumptions

The LCA study has been performed at various levels by:

1. Modeling current coal energy chains associated with grid-connected power plants operating in Shandong today, in order to define a reference and estimate its total burdens and comparing the results with European chains.
2. Modeling future energy chains, which have been identified within CETP as the most suitable candidates for the time frame of 20 years considered for the Shandong power supply. These include: the coal chains associated with subcritical pulverized coal (PC) with or without flue gas desulfurization (FGD); atmospheric fluidized bed combustion (AFBC); and integrated gasification combined cycle (IGCC) power plants; the chains for gas combined cycle (GCC) power plants, for both cases of pipelined gas and liquefied natural gas (LNG); the nuclear chain, with light water reactors (LWR); and wind power, in spite of its limited potential in Shandong.
3. Carrying out sensitivity analyses on key parameters in order to quantify the effects of data uncertainties.
4. Comparing selected environmental burdens across the different energy chains.
5. Comparing selected chain-related burdens in the year 2020 for the scenarios that have been chosen from the Electricity Sector Simulation

(ESS) for the MCDA task; though these results may not reflect all the burdens over 20 years, because most of the innovative technologies as well as the non-fossil plants will be installed in the second decade, they are a meaningful measure of the trends under different policies. However, due to the ESS-defined yearly differences among single contributions from different technologies to the calculated supply mixes, variations of the burdens' ranking may occur for different years; this will not be addressed here.

The initial ambition was to build a database of environmental burdens from current and future energy chains associated with Shandong power plants that would be as complete as possible, using the background experience and tools of the Swiss LCA studies on energy systems for Western European conditions in the mid 1990's (Frischknecht et al., 1996) and in 2020-2030 (Dones et al., 1996). However, it was impossible to achieve this goal given the limitation of resources, time, and contacts with potential data providers. Taking into account the nature of the three main future energy chains of interest for Shandong (which exhibit remarkably different environmental burdens, thus allowing a relatively comfortable margin for uncertainties), first priority was given to quantify the total chain-related emissions of the major airborne pollutants (SO_2, NO_x, particulates, greenhouse gases), solid wastes, and land use, in order to integrate the results into other CETP tasks. Many other types of data have also been collected when available, but not systematically. Hence, the present study may be considered a streamlined LCA investigation of energy chains specific to Shandong power plants. While fulfilling the mission of this task within CETP, the results of this research also form the basis for studying other processes, different scenarios, other Chinese provinces and China in general. While striving to reduce arbitrary assumptions to the bare minimum, the task researchers used direct extrapolation from the Swiss LCA studies when appropriate, to cover missing direct information, e.g. for the nuclear chain.

The limited availability of data also required the authors to use representative values or values from one single year (1998, 1999, and around 2020) instead of information stretching over the lifetime of the technical systems, as is proper for an ideal LCA study. However, considering the uncertainties in the historical data and the intrinsically open possibilities for the future, this procedure should not introduce significant variations in the burdens, especially in relative terms across the fuel chains considered. Conservative assumptions have been used whenever possible.

Because prior LCA studies have determined that their contribution to total environmental burdens is minor, the authors did not analyze the construction and decommissioning phases for the various facilities. However, inventories of a few construction materials have been included, adapted to Chinese conditions, i.e. lower manufacturing efficiency and higher specific emissions associated with the electricity supply mix compared to Western Europe.

Inventory data have been collected from many sources. The Shandong Electric Power Research Institute (SEPRI) in Jinan contributed direct information on selected burdens of the power produced by the Shandong Electric Power Corporation (SEPCO) in 1998-1999. This data set constitutes the basis for the LCA of the current grid-connected situation in Shandong; it has been used as the reference set. The Policy Research Center for Environment and Economics (PRCEE) in Beijing contributed threefold: first, by surveying selected major coal mines in Shandong and Shanxi with the support of the China Coal Information Institute (CCII) in Beijing; second, by direct contacts with Chinese organizations involved in the energy sector; and third, by searching Chinese statistics. The analysis of future chains associated with selected scenarios for Shandong is based primarily on the past Swiss LCA studies on energy chains and on data from the international technical literature and, to a lesser degree, from the press. Data on future trends for specific Chinese energy-related industrial activities included in the considered chains were available in only a very few cases.

The power production supplying the grid in Shandong is predominantly based on pulverized coal (PC) plants, most of them operated by the state-owned utility SEPCO; they provide around 86 percent of total electricity. Because collection of data about current units began in 1999, 1998 was chosen as the reference year. Information on the sulfur content in the coal burned in 1999 was also requested, because the 1998 governmental directive for Acid Rain and Sulfur Dioxide Control Zones mandated that power-plant coal must be under one percent sulfur content if plants were to avoid the imposition of installing FDG before year 2010.

The three SEPCO power plants still using oil in 1998 have since been converted to burn coal. Given that there is no data on their operation following this changeover, they are not included in the analysis. In any case, oil will probably not be used in future base load power plants, and it is not considered as an option for future supply in Shandong. SEPCO power stations are often made up of several PC units of differing capacities. For modeling purposes, the coal power units have been subdivided into six classes on the basis of the installed capacity: <100 MW (herewith often denominated "small units"), 100 MW, 125 MW, 210 MW, 300 MW, and 600 MW. The last two units constitute the standard size of base-load PC plants manufactured in China today. The lignite mine-mouth power plant in Longkou has been considered separately.

Though different technologies for controlling SO_2 and particulates may be used by individual units at the same power plant, all the units share the same coal source and quality. The authors averaged technology characteristics for the different size classes, but each average unit burned a different plant-specific coal. Small units not owned by SEPCO supply about 15 percent of the total electricity to the provincial grid. Emissions from these facilities have been extrapolated from similar SEPCO plants.

As mentioned above, the future coal power plant technologies considered were selected in collaboration with other teams within CETP. Technologies such as supercritical PC and pressurized fluidized bed combustion power plant (PFBC) were not considered for the next 20 years, due to current lack of domestic technology and high costs. Several FGD options are suitable candidates for Chinese conditions, in particular the wet process, the plants using seawater, and the simplified retrofit plants using sorbent injection. For the purposes of this LCA study, only the wet type with 90 percent efficiency was considered as representative for this technology.

The current coal chain has been described in detail, particularly in terms of the origin of the coal supplied to major SEPCO stations. Preliminary analysis identified mining as the potentially most important environmental loader in the upstream chain. A survey was performed on three major coal bureaus in Shandong as well as three in Shanxi to directly collect information on energy uses, self-production of electricity, and emissions to the environment. Though incomplete, the data were used to construct a model of the underground mining and processing of steam coal in these two provinces. Transport services have been described by extrapolating results of LCA studies performed for Europe, due to the lack of sufficiently detailed information on the Chinese case. However, considering the relatively small contribution of transport to most of the burdens and efficiency improvements expected in the future, these modeling assumptions should be sufficient for the scope of the study.

Though not domestically manufactured, the candidate technology for natural-gas base-load power plants is GCC, because of its very high net efficiency. Some units are already installed in Guangdong, fuelled with liquefied natural gas (LNG) shipped from Southeast Asia. Four origins of the gas and two process chains have been taken into account: pipelining gas from offshore fields in Shandong, from South Siberia (Irkutztk), from Xinjiang; and liquefying the gas extracted in Indonesia and transported as LNG to a hypothetical terminal in Qingdao. The best available technology for pumping the gas, the lowest leak rates for the pipelines, and the energy requirements for state-of-the-art LNG liquefaction have been considered, though these are not likely to be installed until the second decade. When data were lacking, conservative assumptions have been made, for example, all extracted natural gas, independent of origin, is considered sour, maximizing the total SO_2 emitted from the chain. However, in spite of this extreme assumption, the sulfur emission rate from the gas chain still remains far lower than that from the coal chains.

Chinese policy for nuclear energy currently favors Light Water Reactor (LWR) technology, though other plant types are being commissioned (e.g., CANDU). Considering the current slow-down of nuclear power programs in China, long construction periods of several years for such plants, and uncertainties for the future, this study conservatively models a standard PWR (Pressurized Water Reactor) of technology available today operating with the best current parameters, though construction of advanced power plants cannot be excluded. Based on the scanty information available, the reference nuclear chain chosen was an open cycle without reprocessing. This chain produces conservatively high non-radioactive emissions for comparison with other energy chains. Though there is no published evidence of natural uranium in China sufficient to feed a large nuclear capacity, this study takes into account the expectations of Chinese nuclear authorities that 50 percent of the uranium to be used in the generation of nuclear power will be domestically extracted through in-situ leaching (ISL), and the rest made available from the spot-market. Centrifugal enrichment is assumed to cover future Chinese requirements entirely. It is assumed that the spent fuel will simply be deposited in deep geological strata after encapsulation. These assumptions are not to be taken as policy indicators for Chinese conditions (reprocessing is considered and currently tested); they apply only to this study's preliminary estimate of some burdens associated with the local nuclear chain.

Because it is the most promising—though potentially rather limited—renewable electricity production technology in Shandong, a typical wind turbine of medium capacity has been modeled to put this option into perspective. However, wind power is not explicitly considered in the CETP scenarios because its contribution to the total electricity supply integrated over 20 years could only be very marginal; wind power cannot be considered a significant substitute for more polluting technologies. Hydropower is not a suitable option for Shandong,[1] either locally for geographical reasons or as a distant source located in Central China, because existing plans call for the transportation of hydro electricity to other provinces in the southeast.

The electricity generation mix most likely to occur in Shandong in the next 10 to 15 years will not be dramatically different from what it is today. It will continue to be based mainly on PC technology, eventually with FGD. Advanced coal technologies are likely to be implemented relatively late and to a limited extent in Shandong, and will only mildly influence the mix by the end of the time period assumed for this analysis. The contributions of nuclear, natural gas, and wind will be relatively small. Hence, one simplified mix for modeling future electricity supplies to energy chains has been defined. It is conservatively based on coal technologies only, thus somewhat maximizing the emissions from combustion for future scenarios. This electricity supply mix is composed of 50 percent from current power plants and 50 percent from yet-to-be-constructed 600 MW PC's with FGD. However, the current plants will contribute more than half of future total emissions this mix will produce. It will be shown that changing the composition of the supply mix does not greatly affect the ranking of fuel chains. The mix used to model electricity supply from the Shanxi provincial grid to local mines is arbitrarily assumed to be composed of 300 MW PC units in 1999 and 600 MW PC units with FGD in the future. All supplies are supposed to be delivered at medium voltage.

Additionally, on the basis of former LCA studies for Western Europe, indirect contributions to the burdens from material uses throughout the entire gas and coal chains are estimated to be very small, less than one percent even taking into account the lower manufacturing efficiency and more polluting energy mix in China. The uncertainties for several factors such as losses during transportation and mining, or large spontaneous fires in coal mine areas are much greater than the variations in the contributions of different electricity mixes to total pollution or in detailed account of material use for construction of facilities.

These considerations support the choice of using a static approach for the LCA modeling of future energy chains. This has been achieved by means of several simplifying assumptions, like the modeling of services such as transport means and boilers using the results of previous LCA studies extrapolated to Chinese conditions. Two base cases were considered, using data from the years 1998-1999 and from the year 2020. The level of future electricity demand is not important for LCA because all figures are normalized by the unit of electricity.

In this study the term "full energy chain analysis" is sometimes substituted for "LCA", though it is actually a subset of LCA. In contexts where the complexity of the interrelations of a chain related to a specific power plant technology with other industrial sectors like material manufacturing is more explicit, the term energy system has been employed.

1.3 Integration of LCA with other CETP Tasks

The results of LCA give an environmental ranking for single technologies and mixes of technologies. However, LCA is only a part of the environmental package produced by CETP; EIA is also conducted. The 3×12 scenarios selected within the ESS task for the MCDA exercise are used here for the scenario analysis. Furthermore, a few criteria used in the MCDA task have been defined based on the full energy chain approach. These are:[2] a) global warming, measured by the total emission of GHG per unit of electric service;[3] b) public health impact from air pollution, using YOLL per unit of electric service as indicator; c) resource consumption, measured in percent of mass consumption of non-renewable energy resources related to proven resources world-wide at current costs; d) wastes, a criterion made of two sub-criteria: total mass and confinement time required for the critical wastes; and, e) total land use per unit of electric service.[4]

For the preparation of the input matrix for MCDA, LCA provided the GHG emissions from the upstream part of all energy chains as well as the indirect emissions from material and energy uses (if meaningful) for the power plants if not accounted for explicitly or implicitly in the ESS task, to avoid double counting.[5] As an example of the implicit assumptions, all electricity requirements supplied from the Shandong grid to mining and transport were assumed to be included in the total demand of electricity considered by ESS.

The results obtained for the scenario analysis by calculating the selected burdens with LCA data for the year 2020 only may differ from the results obtained by ESS using LCA inputs. On the one hand, the ranking of different scenarios using integral burdens over 20 years may somewhat differ from the ranking for a single year. On the other hand, variations in the operational data for single plants or technologies over the years, as implemented within ESS, may determine some changes in the ranking. Moreover, the assumptions for the shares of Shanxi vs. Shandong coal differ in the two analyses. Furthermore, for scenarios including demand side management ESS recalculated the burdens by dividing them by the full demand level. Here, the burdens have been normalized by the electricity actually supplied. In conclusion, though the LCA results for the electricity supplied in scenarios for the year 2020 represent future conditions, they may not be entirely representative of integral emissions over two decades.

The total annual YOLL per scenario provided to MCDA have been calculated by EIA on the basis of energy chain-related emissions using LCA and ESS data. The resource consumption calculated for MCDA was limited to the direct use of fuel in the power plant (ESS data). Inclusion of upstream chains would not have changed the scenario ranking; to include them would have mainly penalized the natural gas chain, thus only lowering further the ranking of any mix including it.[6] In this task, the authors estimated the total mass of the wastes as well as all land uses.

LCA results were not fully integrated within the ESS, EEM, and ETM tasks. For ESS, systematic inclusion of the burdens from the upstream chains was not considered necessary because the most important results for utilities (costs and air emissions from the operation of power plants) are produced directly by the simulation. LCA results for air pollution from fossil chains are dominated by direct emissions from power plants. Hence, in most cases the inclusion of LCA results for entire chains would have only amplified the differences among scenarios. Nevertheless, the inclusion of LCA does change the ranking of a few scenarios for burdens other than air emissions (e.g., solid waste). The effects of an LCA assessment of demand side management may also influence the results, but the scope of this project does not include any way to assess them.

Besides the electricity sector analyzed for macro-regions in China, including Shandong, EEM modeled China's other energy sectors for which a consistent application of LCA would have required an extended approach. Both aspects were well beyond the scope of the present task. In ETM, optimal routes are calculated taking into account several single mines, for which a consistent application of LCA would have required data that were unavailable or limited. The results of the latter two optimization methodologies may have been influenced by the inclusion of burdens from full energy chains, especially for particle emissions, solid wastes, land use, and GHG produced when there is a high rate of spontaneous combustion of coal or coal waste. The LCA results used for the 2020 scenarios of the MCDA exercise provide an initial insight into the effects of including full chains in energy-economy or transport models.

2. THE COAL CHAIN

2.1 General

China's most important domestic energy resource is coal. Although Shandong Province is relatively rich in coal, there is not enough to meet the rapid rate of development anticipated in the next two decades. Currently in Shandong, coal is practically the only fuel used for electricity production in grid-connected power plants, especially those that are state owned. This reflects the impact of the national governmental directive issued in mid-1980's to convert as many oil power plants as possible to coal. The objective was to divert oil to the petrochemical industry to foster the national development of light industry and road transport (Wang, 1999). This changeover occurred fairly late in Shandong compared to other provinces, probably due to the relatively easy access to domestic oil. (Shandong possesses the largest active Chinese oil fields, those of Shengli.)

The dominance of coal power plants will continue in the near future, which establishes a firm basis for realistic scenarios. As soon as this fact was recognized and the technical literature explored, it became clear that important data were unavailable. This made an accurate description of local conditions supplying coal to Shandong power stations rather difficult. The authors devoted considerable effort collecting data from mine operators, to combine with data received from the Shandong state utility.

2.2 Status of the coal chain associated with Shandong power plants

The coal energy chain can be divided into three main steps: mining, transport, and power plant use. In this study, the authors have established a reference case by assessing the status of the current chain for the supply of coal to Shandong power stations. Transmission of electricity will be treated in the section describing the modeling.

2.2.1 Mining in Shandong and Shanxi: survey of major coal bureaus

On the basis of the origin of the coal supplied to specific major power stations in Shandong, the six most important supplier coal bureaus were determined. They are: Yanzhou, Xinwen, and Zibo in Shandong Province (SD); and, Yangquan, Jincheng, and Xishan in Shanxi Province (SX). An on-field survey in these bureaus was performed in May-June 2001 by PRCEE in collaboration with CCII.

A coal bureau (CB) is a state organization or a company which includes a group of coal mines (CM) scattered on the same field or in different fields. Almost every CM in the CB's investigated has one or more of its own coal preparation plants and transportation system. A CM may produce its own power; derive it from another CM in the same CB; or draw power from the provincial grid. When several self-supplied power plants are operating in one CB, they may be grouped into one managing company, such as the Yanzhou Coal Bureau Electricity Generation Company.

A properly designed questionnaire was worked out, focusing on the main production characteristic, electricity generated on-site, and environmental burdens. The six CB's did not provide sufficient data to compare the full range of items included in the questionnaire. Direct interviews with operators of coal preparation plants and power plants for self-supply could be carried out in only a few cases. The CB's are very sensitive to information concerning environmental burdens, and not prepared to make them public. Hence, the collected information often lacks internal consistency, preventing reliable conclusions on the size of pollution. This makes it difficult to define a coherent set of parameters to describe mining for steam coal in Shandong and Shanxi. A solution to this problem is described in the subsequent section on modeling.

Since the characteristics of coal seams and coalfields differ greatly among CB's and even among CM's within the same CB, the mining technologies and the quality of run-of-mine (ROM) coal may also greatly vary. This may result in substantial differences in coal preparation techniques and variety and quality of final products. Average values and ranges are summarized here. Complete information from the survey is available in the Technical Report of the LCA task (Dones et al., 2003).

Underground mining

A variety of coal mining technologies are used by different CB's, including fully mechanized, conventionally mechanized, and blast technology. Fully mechanized mining takes place at the long-wall working face for cutting, loading, transporting and hydraulic supporting. The mechanization rate in the surveyed CB's spans between 63 percent and 100 percent.

Electricity is the only form of energy used in underground mining. Energy consumption per tonne of ROM coal at the shaft mouth differs from one CB to another. Major energy-consuming underground activities include mining operations, raw coal transportation, water pumping, and air ventilation. The differences in electricity consumption per tonne of ROM coal at the shaft mouth of various CM's reflect different technologies, mechanization rates, transport distances, equipment, and working conditions. A higher rate of mechanization generally requires more electricity consumption per tonne. However, energy intensity also depends on the efficiency of machinery and production. The roof-carving mining method is the least electricity intensive.

Generally, electricity consumption in a CB is calculated on the basis of total electricity consumption per tonne of raw coal. Only in some cases can raw coal be assumed to be steam coal, the product counted for LCA modeling. In the surveyed CB's, electricity consumption varies from the rather low value of about 14 kWh per tonne of raw coal in Jincheng (SX) (12 kWh for underground mining only) to 71 kWh in Zibo (SD). For comparison, from (Hinrichs et al., 1999) the average electricity requirements for coal production per tonne of hard coal equivalent (with LHV of 29.3 MJ_{th}/kg) in the USA is 38.2 kWh, in South Africa 20.3 kWh, in Germany 76.2 kWh, in Poland 92.6 kWh, and in Russia nearly 124 kWh. However, considering that the ratio of underground to open-pit mining varies greatly among these countries,[7] a direct comparison is not possible. The Swiss LCA study (Frischknecht et al., 1996) assumed an average consumption of 85 kWh per tonne of produced hard coal from Australian, North American, South African, and UCTE[8] underground mining, including coal preparation.[9]

The share of electricity uses for strictly mining-related activities in a CB is on the order of 60 percent. About 75 percent of this figure is for underground mining (excluding underground transportation), 14 percent for coal preparation (preliminary and wet preparation), seven percent for total transportation needs (including underground and over ground transportation), and four percent for other mining-related activities. These data are summarized in Table 8.1, showing also non-mining industrial activities (such as a cement factory or tertiary industry), and household's needs. The most energy intensive activity is clearly underground mining, consuming about 45 percent of the total electricity uses in a CB.

Shandong and Shanxi coal mining faces growing shortages of water, and severe potential impacts from pollutants like phenols, cyanide, ammonia, sulfides, Cr^{6+}, As, Cd, Pb, Hg, Zn, and Cu when water is discharged from mines into freshwater bodies. Mining water is usually collected through a drainage system and treated in above-ground plants for recycling in activities such as electricity generation and coal preparation. Only small quantities of wastewater are released into water bodies after treatment, complying with national standards for wastewater discharge. In several cases, such as Yankuang, the recycling rate can be up to 100 percent.

Table 8.1 Average composition of electricity consumption in selected Shandong and Shanxi coal bureaus in 1998.

Total electricity consumption (%)		Electricity consumption of mining-related activities (sum=100%)			
Non-mining & household	Mining	Underground mining [a]	Coal preparation	Transportation	Other mining activities
40	60	75	15	6	4

[a] Excluding underground transportation.

Dust from the working face is not emitted to the open air, but may still cause serious health problems to miners if not abated underground. Safety measures include spraying water onto the working face. Dust emission during underground transportation is also controlled by spraying water. Data on radioactive emissions from underground mining were not available, though on average the radioactivity of Chinese coal is much higher than the average worldwide (UNSCEAR, 1993; Pan et al., 1999; Guo et al., 1996).

Coal preparation

ROM coal is typically ill-suited for the market, except in the Yanzhou CB where ROM coal is relatively high quality in terms of sulfur content (0.6-0.8 percent), ash content, size, and amount of debris. This ROM coal requires preliminary mechanical preparation to become raw coal. This process usually consists of crushing and screening and discarding large pieces of debris (usually greater than 50 mm). A typical flowchart for mechanical coal preparation is shown in Figure 8.1.

Part of the raw coal is further processed, usually by jigging and/or dense medium separation, to produce clean coal, including coking coal and steam coal. In the surveyed CB's, at the end of the 1990's the share of raw coal going through further preparation ranged from 30 percent to 76 percent. This is quite high compared to the average national level in 1998 of 18 percent and the average level for state-owned major coal mines of 43 percent (CCIY, 1999). In the surveyed CB's the coal recovery rate—that processed into clean coal products—varies from 59 percent (1:1.7) to 74 percent (1:1.3) depending on the technology used, the quality of raw coal, and the quality demanded of final products. The average for state-owned major coal mines in 1998 was 69 percent (1:1.4) (CCIY, 1999). Different coal preparation technologies are used to produce a variety of clean and graded coal products to meet the quality demands of consumers. Coal preparation in the six CB's surveyed is mainly designed to produce coking coal rather than steam coal.

While raw coal may be directly supplied to small power plants, steam coal for large power stations is usually miscellaneous coal, i.e. a mixture of raw coal, clean coal, and middlings. The middlings are a by-product of coal preparation with an ash content falling somewhere between the amount contained in debris and in clean coal. The minimum quality requirements for miscellaneous coal for large power plants should include ash content of 28 percent to 30 percent, sulfur content of one percent, and LHV of 21-25 MJ_{th}/kg. The mixing ratio to obtain these minimum characteristics is driven by cost optimization. Prepared, unblended steam coal—

"Premium steam coal", is usually exported, mainly to Japan and South Korea,[10] and is not available for domestic power stations.

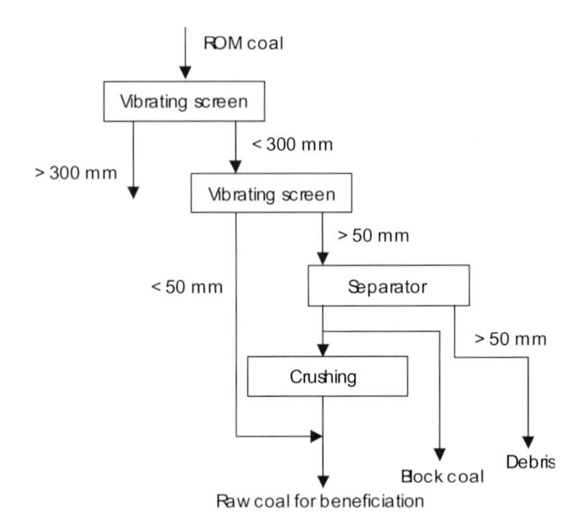

Figure 8.1 Flowchart of typical preliminary coal preparation.

Wet processes are currently used for coal preparation in the six surveyed CB's. The technologies include jigging, dense medium separation, and flotation. Each of these can be applied as independent processes or combined into a single process. In typical jigging, the middlings and the light part are first separated by a jig. The light part is screened to produce the final product, clean coal. The slurry is then dewatered through centrifuge, thickening, and press filter. These pieces of equipment are usually stacked: the upper layer is the jig, the middle layer the centrifuge, and the lower layer the press filter. All water used for coal preparation is recycled mining wastewater from treatment plants. The wastewater from coal preparation processes is often fully recycled. To convert one tonne of raw coal to coking coal requires 0.2-0.4 m^3 water. Only 0.02-0.05 m^3 is needed to convert one tonne of raw coal into steam coal. With wet methods, the level of dust emission is rather low.

Electricity consumption per tonne of prepared raw coal has been obtained by dividing the total electricity annually consumed in the coal preparation plant by the total raw coal processed. The values range from six to nine kWh per tonne of raw coal throughput. However, it is difficult to attribute specific electricity consumption to the unit mass of a single prepared product (coking coal or steam coal or middlings) because they are usually separated in a single process.

Electricity generation at mines for self-supply
Current policy for the construction and operation of power plants for electricity self-supply in Chinese coal mines includes the following elements:

- The installed capacity of each unit should be greater than 500 kW and the unit should not exceed its service life.
- The plants should use debris (also named gangue) and slurry as their major fuel. The LHV of the blended fuel must be ≤ 12.55 MJ_{th}/kg. Since 1997, Chinese policy has been to shut down highly polluting and inefficient small power units. Small plants operated by industries are also included in the shutdown list. The small power plants within coal mines which burn residues are included in a different category, to comply with the policy pursuing comprehensive utilization of resources. This is why small power plants at mines can continue to operate and the construction of new ones is approved. These plants should not be fueled by high quality coal.
- Plants for which construction has been approved after 1 March 1998 should use fluidized bed boilers.
- Plants using fuel with a sulfur content greater than one percent must adopt desulfurization technology after 1 March 1998, along with large power plants in sulfur control zones.
- Emissions should comply with specific standards.[11]

As stated above, the unit capacity and total capacity of power plants at mines for self-supply of electricity are usually relatively small. The most diffused plant sizes currently operating in Shandong and Shanxi are six MW and 12 MW. Only very few units in the surveyed CB's in Shanxi have higher capacities, specifically one with 35 MW and one with 54 MW. Total capacity installed per CM ranges from 1.1 MW (Zhongshun in Zibo CB) to 70 MW (2×35MW, for mine No. 3 in Yangquan CB), up to a maximum within a CB of 78 MW in Shandong (Yankuang Group) and 106 MW in Shanxi (Yangquan CB).

Old units are of the bubbling bed type, while newer units include PC and FBC. On average about 13 percent (12.3 percent to 16 percent) of the electricity produced is for auxiliary power. The coal burned in these power plants is mostly the residue from coal processing, sometimes mixed with coal of higher LHV. A few plants, operating or planned, burn coalbed methane; these are described in the section below on methane emissions.

The average gross efficiency declared by mine operators is approximately 30 percent (range 26 percent to 33 percent). However, these values appear too high considering the size, the age of some units, and the fuel that is supposed to be burned according to the regulations. The equivalent standard coal consumption rate, again declared at mines, was 474-915 grams coal equivalent (gce) per kWh generated. However, these values do not appear to be fully consistent. Standard coal is defined as having LHV equal to 29.3 MJ_{th}/kg. The data from SEPRI for provincial grid-connected small units of capacity below 65 MW (e.g., Liaocheng, Nanding, Jining) are: gross efficiency of 31 percent to 34 percent and consumption rate of coal falling between 360 gce/kW and 400 gce/kWh, which is inversely (and consistently) proportional to the efficiency. Hence, the efficiency corresponding to the maximum consumption of coal declared by mine operators would be 13 percent to 14 percent. This order of magnitude is more credible and comparable to efficiencies provided in the literature for old and small units in China. These discrepancies could not be entirely resolved for this study.

Not all CM's have power plants to supply themselves with electricity. For instance, the Yanzhou CB with eight mines has six power plants, each serving a major mine, but the Xinwen CB with thirteen mines has only two power plants. Some power plants produce enough electricity for an entire coal mine district (CMD),[12] including peak load demand. Surplus power is either dispatched to the provincial grid or to another nearby CM within the same CB. Coal bureaus that are not self-sufficient import electricity from the provincial grid, especially for peak load regulation. A coal mine with its own electricity generation plant usually also manages its the distribution network within its district. This network is connected with the provincial grid. Transmission lines within a CMD are not very long. The transmission is usually at medium (35 kV) or low voltage (6 kV or below). The losses from transmission are around eight percent. The transmission lines from the provincial grid to a CM are usually at medium voltage, with transmission losses estimated by mine operators at about four to five percent

Power plants are equipped with dust control devices. However, their efficiency ranges from a really poor 50 percent up to 99 percent, depending on the technology used. Water scrubbers usually have low efficiency while electrostatic precipitator have the highest efficiency. No scrubbers have been installed to date. In FBC plants, SO_2 emissions are controlled by the addition of lime-gypsum, but the used technology is not very efficient.[13] The height of power plant chimneys within the six surveyed CB's was reported to be 100 m for most unit; stacks for three plants in the Yangquan CB are said to be 80 m, 120 m, and 150 m high.[14]

Within CB's, industrial boilers provide heat exclusively to non-mining industrial activities, such as an affiliated cement factory, and to households for heating and warm water in winter. Co-generating units are also contributing to district heating. Other energy sources such as diesel and gasoline are used almost exclusively for transportation, but their amounts were not available.

Coal transportation overview

Transportation can have different characteristics for different CB's and even within a bureau. Conveyor belts or, alternatively, mine hoists followed by conveyor belts are used to transport ROM coal from the working face to underground or surface coal storage in almost all CM's. The width of belts, the speed of conveyor transportation, available power, all affect the volume of ROM coal transported per working hour. None of the six surveyed CB's provided details; the data combined the electricity consumption rates for mining and underground transportation.

Once above ground, the ROM coal is transported by conveyor belt either to open coal storage or directly to the coal preparation plant. In most large mines, nearly all ROM coal is prepared to different degrees. Prepared coal is transported by conveyor belt to the top of silos. The transportation of gangue and slurry from the washing plant to waste piles is also by conveyor belt. Only if the quality of ROM coal is acceptable to specific consumers, such as local small power plants or industrial boilers, it is sold directly to them and delivered by open trucks. Mine operators estimate that the average loss during truck transportation is around five percent, though power plant operators (SEPRI) presume lower losses of two to three percent.

The total volume of all coal products (coking coal, steam coal, and mixed coal) transported from CM's by truck is much lower than the volume transported by train. The trains are owned by either mine operators or the state if the national railway crosses the CDM. When a national railway is not available, mine-owned trains may transport coal to shunting yards where it is transshipped to state-owned freight trains to its final destination. Some coal is transported long distance to harbors for maritime transportation. The distance to be traveled between the mine and the connection with the national railway varies from several kilometers to 1200 kilometers. All mine-owned trains in the six CB's are pulled by old steam locomotives. In some cases, the middlings are transported directly to a large power plant by mine-owned railway. For example, middlings are transported from the Yanzhou CB to the Zouxian power station, 10 km away. The loss during long transportation by train declared by mine operators is around three percent. Again, power plant operators (SEPRI) estimate a lower average loss, two percent.

Methane emission

Most of the CM's in Shandong are low methane mines. For this reason, it is not economical to collect the methane, which is entirely released to the air through the ventilation system as a safety measure. Coalbed methane (CBM) emission ranges from less than one m^3 per tonne of ROM coal, as in Yanzhou CB, to a few m^3 per tonne, for example 3.6 m^3 per tonne from the Huafeng CM in the Xinwen CB.

Shanxi mines are relatively gassy. The reserves[15] of CBM in Shanxi account for 11 percent of the total national reserve (Su et al., 1999). CBM generation in Yangquan mines ranges between seven m^3 and 40 m^3 per tonne of ROM coal, with an average of 23 m^3/tonne of raw coal. Of the total generation, about 26 percent is drained while the remaining 74 percent is vented. An average 40 percent of captured methane is utilized as town gas.[16] CBM generation from various mines within the Jincheng CB ranges from 10 m^3/tonne of ROM coal to 20 m^3/tonne, with the average around 15 m^3/tonne of raw coal. Of the CBM liberated, around 11 percent is captured. An average of 59 percent of the captured methane is currently utilized as town gas and in two power plants to generate electricity for mine uses. One plant with 2×120 kW units is installed in the Panzhuang CM, the other of 4×400 kW in Sihe. One of the major constrains for the further expansion of the use of CBM has been the lack of town gas pipeline networks.

Spontaneous combustion of coal and coal waste

All surveyed CB's claimed that there have been no major underground fires in recent years. When fires start underground, the operation of the mine is stopped and the drift affected is isolated to prevent the spreading of the fire. The prevention of underground fires is a high safety priority. Preventive measures include spraying water on the working face and on coal stored underground. As small illegal mines have been shut down, sales of state-owned major CM's have been steadily growing to satisfy the demand. One side effect has been a reduction of the volume of coal stockpiles either underground or above-ground, which has also reduced fire risk. No information was received on fire events in abandoned coal mines. The spontaneous combustion of waste heaps was said to be a rare event in these six CB's. Waste piles usually contain small quantities of combustibles, and do not easily self-ignite; any waste containing a significant amount of coal is burned in local power plants.

Summary results from the survey

The task researchers merged the data from technical and statistical Chinese literature with the information gathered during the PRCEE survey. Tables 8.2 and 8.3 show the electricity consumption in the year 1998 in selected Shandong and Shanxi CB's, respectively. The values are normalized per unit of raw coal (RC). Furthermore, though the electricity consumption data for mining in large mines may be considered reliable, no complete statistical or direct information is available on electricity uses for the operations following extraction, e.g. transport of the ore, preliminary coal preparation, and washing. On the basis of the conclusions from the survey, the total electricity use per tonne for coal produced has been set equal to the electricity demand in underground mining plus 25 percent for all other electricity use. (See Tables 8.10 and 8.11.)

The CB's included in Table 8.2 produce 63 percent of the total coal mined in Shandong in 1998. The CB's included in Table 8.3 cover barely 21 percent of total coal production in Shanxi (40 percent of ministerial and province-affiliated state-owned mines) (NCIA, 1999). Of the total coal production from the above CB's in SX, nearly 93 percent is from major mines. For Shandong, the electricity use data cover 92 percent of total coal production from the listed CB's. Thus the electricity consumption data, available only for these major mines, appears to be a statistically meaningful set.

Table 8.4 shows data for the year 1998 on electricity self-generation and associated airborne pollutants for the production of raw coal (RC) in Shandong. Table 8.5 shows the corresponding figures for Shanxi. It is (intrinsically) impossible to differentiate which electricity is used for what activity within a CB. However, the amount of self-supply is in general less than the total used for mining activities. Therefore it can be reasonably assumed that self-generated electricity is used entirely for mine operation and is supplemented by electricity from the grid to cover the total needs, if necessary.

For solid wastes, the amounts of 239 kg and 134 kg per tonne of product coal are estimated for SD and SX, respectively, averaging the data from the survey. For the land use, considering only total waste piles in China, an upper value of nearly 0.01 m^2 additionally affected per tonne of product coal can be estimated (PRCEE). Moreover, the average area disturbed by subsidence from Shandong major mines is nearly 0.13 m^2/t(RC). This study proposes 0.1 m^2 per tonne of product coal for both SD and SX, to somehow take into account all factors.

Comparing the statistical data for all bureaus with the average calculated from the data obtained with the survey demonstrates that the survey can be assumed as quite representative for the environmental burdens associated with coal supply from large mines. However, data remain highly incomplete and discrepancies could not be solved.

Table 8.2 Electricity consumption for the production of raw coal in Shandong in 1998; from statistics (NCIA, 1999) and own survey in Zibo, Yankuang, and Xinwen coal bureaus.

Coal bureau	Electricity self-generation (GWh)	Total electricity use [a]				Electricity use for raw coal mining					
		Use (GWh)	Fraction electricity self-generated	Coal production all mines (MtRC)	Electricity consumption rate (kWh/tRC)	Use (GWh)	Fraction of total	Fraction electricity self-generated	Coal production major mines (MtRC)	Fraction production major mines to total	Consumption rate in major mines (kWh/tRC)
Zibo	170	417	0.41	5.14	81	253	0.61	0.67	3.59	0.70	71
Xinwen	158	652	0.24	11.73	56	399	0.61	0.40	11.17	0.95	36
Zaozhuang	109	585	0.19	8.07	73	140	0.24	0.78	5.64	0.70	25
Feicheng	-	277		5.78	48	206	0.74		5.78	1.00	36
Yankuang [b]	322	621	0.52	21.79	29	333	0.54	0.97	21.79	1.00	15
Fangzi	-	58		0.29	202	34	0.57		0.29	1.00	116
Longkou	-	89		3.43	26	55	0.62		3.43	1.00	16
Linyin	3	66	0.05	0.63	105	35	0.53	0.09	0.58	0.92	60
Total	762	2765	0.28	56.86 [c]		1455	0.53	0.52	52.27 [d]	0.92	
Average					49						28 [e]
Total survey	650	1690	0.38	38.66		986	0.58	0.66	36.55 [f]		
Av. survey					44						27 [g]

[a] Electricity used by all industrial (mining and non mining) activities of the enterprise, excluding residential uses. All mines, major as well as minor, are included.
[b] Former Yanzhou.
[c] 63 percent of total coal production in Shandong
[d] 92 percent of total coal production in the listed CB's.
[e] Adding 25 percent to approximately take into account other electricity uses besides those for underground mining, 37 kWh/tRC is calculated.
[f] 95 percent of the total extracted in the surveyed Bureau, which makes a good statistics.
[g] Adding 25 percent to approximately take into account other electricity uses besides those for underground mining, 36 kWh/tRC is calculated.

Table 8.3 Electricity consumption for the production of raw coal in Shanxi in 1998; from statistics (NCIA, 1999) and own survey in Yangquan, Jincheng, and Xishan coal bureaus.

Coal bureau	Electricity self-generation (GWh)	Total electricity use [a]				Electricity use for raw coal mining					
		Use (GWh)	Fraction electricity self-generated	Coal production all mines (MtRC)	Electricity consumption rate (kWh/tRC)	Use (GWh)	Fraction of total	Fraction electricity self-generated	Coal production major mines (MtRC)	Fraction production major mines to total	Consumption rate in major mines (kWh/tRC)
Yangquan	508	705	0.72	12.30	57	397	0.56	1.28	11.58	0.94	34
Xishan	74	654	0.11	16.12	41	295	0.43	0.25	14.49	0.90	20
Fenxi		222	-	4.86	46	91	0.41	0.00	4.13	0.85	22
Luan		340	-	12.19	28	161	0.47	0.00	11.71	0.96	14
Xuangang		95	-	1.40	68	57	0.60	0.00	1.4	1.00	41
Jincheng	220 [b]	252	0.87	10.81	23	131	0.52	1.68	9.46	0.88	14
Huozhou		188	-	5.98	31	111	0.59	0.00	5.98	1.00	19
Total	802	2516	0.33	63.66 [c]		1242	0.51	0.65	58.75 [d]	0.92	
Average					39						21 [e]
Total survey	802	1611	0.50	39.23		823	0.51	0.73 [f]	35.53	0.91	
Av. survey					41						23 [g]

[a] Electricity consumed by all industrial (mining and non mining) activities of the enterprise, excluding residential uses. All mines, major as well as minor, are included.

[b] NCIA (1999) reports barely 51.8 GWh generated in year 1998. The value from the survey has been adopted, being more consistent with the declared installed capacity of 54 MW.

[c] 21 percent of total coal production in Shanxi (307.19 Mt) or 40 percent of Ministerial and Provincial affiliated State-owned mines (159.64 Mt) (NCIA, 1999). Of this value, nearly 93 percent are considered for the electricity consumption for raw coal mining.

[d] 92 percent of coal production from all mines included here.

[e] Adding 25 percent to approximately take into account other electricity uses besides those for underground mining, 28 kWh/tRC is calculated.

[f] Corrected to account of the overproduction of Yangquan and Jincheng; their fractions are set to one.

[g] Adding 25 percent to approximately take into account other electricity uses besides those for underground mining, 31 kWh/tRC is calculated.

Table 8.4 Electricity self-generation and associated airborne pollutants for selected coal bureaus in Shandong in 1998; from statistics (CEOCI, 2000) and own survey.

Coal bureau	Installed capacity (MW)	Generation (gross) (GWh)	SO_2 (g/kWh generated)	TSP (g/kWh generated)	Coal consumption per kWh generated			Fuel used	
					Standard (gce/kWh)	Actual * (g/kWh)	Type	Amount (kt)	LHV * (MJ_{th}/kg)
Zibo	24 [a]	170	25 [b]	20 [c]	446 [d]	1483	Debris	253	8.8
Zaozhuang	18	109	25 [e]	20 [f]	622	1875	Debris	205	9.7
Xinwen	12 [g] 12 [g]	85 [h] 73	10 [i] 25 [e]	20 [j] 20 [k]	549	1610	Debris	254	10.0
Yankuang	18 36 [n]	123 200	11 4 [o]	8.2 [l] 2 [p]	547 499	758 1318	Debris Slurry	93 [m] 263	21.1 11.1
Linyi	3	3	25 [e]	20 [f]	921	2603	Debris	8	10.4
Total	123	763						1076	
Average (g/kWh generated) (g/kWh at busbar) [r]			16 18	13 15	525 [q]	1410	Debris		11.8

[*] Calculated. [a] The 1.1 MW unit at Zhongshun mine is not accounted for. No data on emissions were provided. [b] No clue about emissions from the survey. This figure has been estimated in analogy with the small units in Nanding which use coal from Zibo. [c] Only data available from the survey: 390 mg/Nm^3 in Zhongshun and 188 mg/Nm^3 in Lingzi. This value assumed equal to non-SEPCO grid-connected small power plants. [d] Zibo reported for the three plants values in sharp contrast: 856, 637, and 915 gce/kWh. This discrepancy could not be solved. [e] Arbitrary, based on the value assumed (SEPRI) for small power plants in SD (13 g/kWh) and the ratio of LHV (factor of about 2). [f] Assumed equal to the average emitted by small power plants in SD. [g] Above: 2×6 MW FBC power plant in Huafeng CM; below: 2×6 MW in Xiezhuang CM. [h] Value from the survey. [i] The CB provided no data. This figure is roughly estimated considering the given sulfur content (1.56 percent), the average LHV (10.3 MJ_{th}/kg), the gross plant efficiency (34 percent, which seems too high), and the declared (realistic) efficiency for sulfur removal of 54%, using for analogy the emission rate from the Laiwu power plant. [j] ESP efficiency of 99 percent, as declared by the CB. However, the emission rate given as <2500 mg/Nm^3 is greater than the limit of 1000 mg/Nm^3 permitted by the law GB-13'223-96 for ash content of >40 percent and use of ESP. The declared TSP emission would be within the legal limit of 3300 mg/Nm^3 if the particulate control were not ESP. The number in the table is a guess (same rate as the highest from small power plants without ESP in SD). [k] Assumed equal to the average emitted by small plants in SD. [l] Number from the survey. It appears too low compared to small grid-connected units: possibly 20 g/kWh. [m] This declared amount, perhaps only debris, is totally inconsistent with the calculated LVH. Most probably, middlings or raw coal are also added. [n] This should correspond to the FBC plants of Xing Longzhuang (2×6 MW), Bao Diang (2×6 MW), and Dong Tan (3×12 MW) mines (information from the survey). However, there is evident contradiction in the installed capacity, unless the two units of the Dong Tan Mine were not operational in 1998. [o] Calculated using data given. The low emission can be explained by the good quality of coal in this CB. [p] Calculated using data given. However, it appears too low even compared with the Nan Tun Mine power plant emission of 8.2 g/kWh. [q] Average weighted with the electricity generation. [r] Calculated assuming auxiliary power of 13 percent.

Table 8.5 Electricity self-generation and associated airborne pollutants for selected coal bureaus in Shanxi in 1998; from statistics (NCIA, 1999) and own survey.

Coal bureau	Installed capacity (MW)	Generation (gross) (GWh)	SO_2 (g/kWh generated)	TSP (g/kWh generated)	Coal consumption per kWh generated			Fuel used	
					Standard (gce/kWh)	Actual * (g/kWh)	Type	Amount (kt)	LHV * (MJ_{th}/kg)
Yangquan	106 [a]	508	41-80 [b]	50 [c]	727 [d]	1150	Debris	584 [e]	10.8
Xishan	12	74 [f]	25 [g]	9.5 [g]	572 [f]	1568	Debris	116	10.7
Jincheng [h]	54	222	5 [i]	5 [j]	642	-	Mix [k]	na	-
Total	172	804							
Average (g/kWh generated)			30-54	33					
(g/kWh at busbar) [l]			35-62	38					

[*] Calculated. [a] From the survey, these should be the bubbling bed power plants #1 (3×6 MW) and #2 (3×6 MW), and the PC plant #3 (2×35 MW). [b] Interval worked out from the (contradictory) information gathered during the survey on the annual coal consumption, assuming a realistic LHV. [c] No clue from the survey. However, all plants declare an efficiency of removal of 90 percent, which looks appropriate for cyclones and venturis. Therefore, also considering the expected high ash content of mixtures used for fuel (debris have 50 percent ash content) and the very low electric efficiency of the bubbling bed plants, an extremely high value per unit of electricity for the TSP is expected. The value is deduced by analogy with small units in the Shiheng power station (SD). [d] The standard coal equivalent provided by the CB during the survey were 642, 647, and 637 gce/kWh for the three power plants (see note a). [e] For the blend of debris and low quality coal, 1000 kt were declared during the survey. Using the ratio given for debris to low quality fuel of 1:1, about 500 kt would be calculated for each species, closer to the value in the table. If the fuel consumption given at the survey is true, the statistics may not report the entire consumed mass of fuel blend. With another interpretation, the value in the table may relate to raw coal rather than to the actual fuel (with 50 percent ash). Hence, no conclusion can be drawn because of lack of exact definitions of the fuel parameters and how the information is reported into statistics. [f] From the survey, the reported capacity is the same but the generated electricity is about 60 GWh and the consumption rate of coal equivalent 512 gce/kWh. [g] Emission rate declared by the CB during the survey. However, this value could not be checked against other data. [h] Here reported are the values from the survey, which seem allegedly more realistic than the dramatically different values from the statistics: 12 MW installed, nearly 52 GWh produced in 1998, 578 gce/kWh, debris as fuel, and an unrealistic value of 11.1 kt of fuel used. [i] No data were available from statistics and survey on SO_2 and TSP. This number is deduced taking into account the very low average sulfur content in the coal products, 0.37 percent for raw coal. [j] Only information is that the plant is equipped with 99 percent efficiency ESP, hence the expected order of magnitude of TSP emission is of a few grams per kWh. [k] Mix of debris, slurry, and slug. [l] Calculated assuming an average electricity self-consumption at plant of 13 percent.

2.2.2 Transport of coal from mines to power plants

An overview of the coal consumption and sources for 17 SEPCO-owned major coal power stations in Shandong for the year 1998 is provided in Table 8.6. More detailed data have been collected by PRCEE from the Shandong Lu Neng Fuel Company and SEPRI, including plant-by-plant distances.

Table 8.6 Coal consumption and sources for 17 selected major coal power stations in Shandong in 1998 (SEPRI).

Power station	Installed capacity (MW)	Units and capacity(MW)	Annual coal use (kt/a)	Sources (Province) (kt/a)			
				Shandong	Shanxi	Henan	Inner Mongolia
Zouxian	2400	4×300, 2×600	4990	4990			
Laiwu	375	3×125	875	875			
Dezhou	1200	4×300	2826		2826		
Heze	250	2×125	581		581		
Shiheng	1335	4×300 2×25+35+50	3068	2301	613.6	153.4	
Yantai	200	2×100, 2×50	750	487.5	262.5		
Qingdao	711	2×300, 3×12+3×25	1654	661.6	827	165.4	
Liaocheng	300	2×100, 2×50	674		674		
Jining	300	2×100, 2×50	702	702			
Longkou	1000	4×200, 2×100	2490	1618.5	871.5		
Weihai	850	2×300, 2×125	1070		856		214
Huangdao	670	2×210, 2×125	1755		1755		
Huangtai	925	2×300, 2×100, 35+40+50	2250	900	1080	270	
Nanding	274	2×65,50+2×17+ 2×6+4×12	890	890			
Linyi	362	2×125, 2×50+12	694	69.4	555.2	69.4	
Shiliquan	1225	2×300, 5×125	2652	2338.6	513.4		
Weifang	600	2×300	1362	408.6	953.4		
Total	12977		29483	16242.2	12368.6	658.2	214
fraction %				55%	42%	2%	1%

The current status of coal supply routes to Shandong is illustrated in Figure 8.2. The distance covered by train from Shanxi mines vary between a minimum of nearly 300 km from Jincheng to Heze power plant, to a maximum of about 1260 km from Luan to Huangdao. However coal can also be transported by train from Datong (about 742 km) or Shenhua Trading Co. (1100 km) circumventing Beijing from south to the port of Qinhuangdao (Hebei), and then by ship to Weihai (222 nm) and Longkou (264 nm).[17] From Henan province, the minimum distance is about 500 km to Linyi, and the maximum 1120 km to Qingdao. In Shandong, power plants are often located only a few kilometers to a few dozen kilometers from local mines, and the coal transported by train or by truck (Shiliquan, Shiheng, Zouxian, Laiwu,

Jining, Nanding). The maximum distance traveled by train within the province is from Zibo to Qingdao (283 km). For comparison, the average distance coal travels in China is nearly 370 km, as estimated by PRCEE on the basis of statistical transport distances, coal shares of total transported commodities, and annual fuel consumption. About 18 percent of all the coal supplied to the 17 major power stations given in Table 8.6 is transported by trucks from nearby mines.

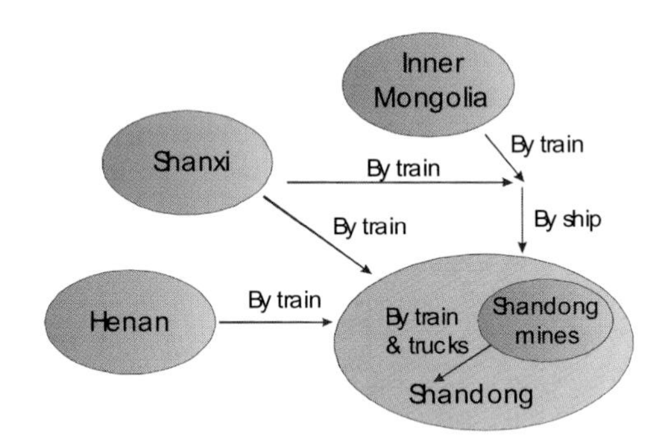

Figure 8.2 Main origins and transport routes of coal to Shandong.

2.2.3 Power plants

Power generation in Shandong is dominated by coal. In this section, the status of large oil plants in 1998 has been also included to complete the picture of grid-connected electricity production in Shandong by that year, the reference for this study. These oil plants were converted to coal in 1999-2000. The existing grid-connected power plants in Shandong can be broadly categorized into two groups: centrally scheduled and non-centrally scheduled plants. The entire grid and the scheduling for dispatching are managed by the state utility SEPCO.

In early 2000, SEPRI provided detailed data from 1998 on 20 major thermal power stations owned by SEPCO. Fuel type, feed rate, and average characteristics for 1998 as well as sulfur content in 1999 are given in Table 8.7. This set for 20 plants has been complemented with key information about the remaining scheduled (less than 10 percent of total energy) and nonscheduled plants. Although the many small plants constituting the nonscheduled park are minor contributors to total electricity generation, they contribute significantly to the environmental damages, especially for their emission of particulates (TSP). However, the importance of small power plants will decrease in the future due to the current national policy of closing highly polluting small units before end of 2003.

Table 8.8 shows an overview of the installed capacity, electricity production, and total air emissions of grid-connected power plants in Shandong in 1998 (SEPRI). The yearly emissions of NO_x and TSP have been extrapolated by SEPRI from a few random measurements (criteria not given). Emissions of SO_2 have been calculated stoichiometrically from the average sulfur content of the coal, subtracting small amounts (five percent) retained in the abatement of fly ashes. Emissions of CO_2 are roughly estimated using IPCC recommended factors for sub-bituminous coal.[18]

Table 8.7 Fuel type and characteristics for grid-connected power plants in Shandong in 1998, and sulfur content in 1999 (SEPRI).

Power plant	Fuel type [a]	Use rate g/kWh generated	Fuel characteristics in 1998			1999
			LHV (MJ_{th}/kg)	Ash (%)	S (%)	S (%)
Zouxian	bituminous	413	22.56	21.5	0.80	0.75
Laiwu	bituminous	474	21.55	27.9	2.58	1.20
Huaneng Dezhou	meager	453	21.42	27.3	1.23	1.01
Heze	anthracite	428	23.75	24.0	0.50	0.59
Shiheng	bituminous	448	22.90	23.9	1.48	0.99
Yantai	meager & bituminous	502	22.02	26.8	1.62	1.39
Qingdao	meager	466	22.19	26.0	2.31	1.52
Jining	meager	489	22.60	21.1	0.61	0.55
Liaocheng	meager	431	24.45	23.9	0.42	0.33
Longkou	lignite [b]	545	20.47 [a]	16.6	0.52	0.46
Huaneng Weihai	bituminous	439	23.52	13.2	0.75	0.67
Huangtai	meager	445	23.42	23.8	1.48	0.82
Huangdao	meager	473	21.47	28.8	1.34	0.98
Nanding	meager	557	21.88	28.6	2.00	1.50
Linyi	meager	447	24.58	19.9	0.44	0.81
Shiliquan	bituminous	442	22.47	22.7	1.18	0.99
Weifang	meager	426	22.68	26.8	1.24	1.16
Baiyanghe	oil	300	39.37	-	1.58	0.80
Xindian	residual oil	266	39.60	-	1.87	0.97
Zhanhua	oil	280	40.50	-	0.97	0.97

[a] The first 17 plants listed are coal fired, the last three oil fired.

[b] Lignite is extracted on site, and makes 65 percent of the fuel burned. The rest 35 percent is transported from Datong (SX); hence, it is most probably hard coal. This would explain the relatively high LHV, well above the range characterizing lignite.

Table 8.9 shows the forecast for total installed capacity, power production, and air emissions of grid-connected power plants in Shandong in 2003, assuming the coal has the same sulfur content it had in 1999 (information received from SEPRI at end of year 2000). The table shows a meaningful reduction compared to 1998. The table also includes SO_2 emission reductions expected from the installation of one FGD for the two 300 MW units in Huangtai (Jinan). Chinese authorities set a deadline of 2003 for the shut down of inefficient small power plants, unless refurbished or converted to cogenerating units. By weighting the sulfur content in the fuel by the power generated, the average in 1998 for the SEPCO major coal stations of 1.2 percent, dropped to approximately 0.9 percent in 1999 after the implementation of directives to reduce SO_2 below one percent or install FGD.

Table 8.8 Total installed capacity, power production, and emissions to air of grid-connected power plants in Shandong in 1998 (SEPRI).

Grid-connected power stations	Capacity (MW)	Generation (TWh)	SO_2 (kt/a)	NO_x (kt/a)	TSP (kt/a)	CO_2 (kt/a)
Centrally scheduled plants						
Zouxian	2400	12.1	71.39	45.4	5.33	10907
Huaneng Dezhou	1200	6.24	62.67	25.66	1.64	6218
Shiheng	1335	6.84	76.85	27.61	18.09	6751
Shiliquan [a]	1225	6	57.75	26.1	14.55	6334
Lougkou	1000	4.57	22	21.2	3.4	4565
Huangtai [b]	925	5.06	52.6	20.43	3.63	4949
Huangdao	670	3.71	33	15.9	5.15	3861
Qingdao [b,c]	711	3.44	63	14.9	6.13	3638
Huaneng Weihai [d]	850	2.51	11.38	8.51	2.22	2260
Weifang	600	3.2	29.93	12.37	3.76	2996
Laiwu	375	1.85	29.32	7.9	8.18	1923
Jining	300	1.44	7.02	6.37	13.68	1539
Liaocheng	300	1.57	4.8	5.8	1.21	1482
Heze	250	1.36	4.65	5.17	5.93	1383
Yantai	200	1.49	20.6	5.05	1.66	1650
Nanding [b]	274	1.6	32.07	8.08	11.8	1957
Linyi	362	1.55	5.19	6.3	8.54	1526
Xindian (oil)	600	2.04	19.29	6.9	0	1526
Zhanhua (oil)	250	1.2	6.3	4.2	0	1044
Baiyanghe (oil)	150	0.48	4.57	1.81	0	189
Sum of plants above	13977	68.25	614.4	275.7	114.9	66698
Other scheduled plants [e]						
Wusuotun	55	0.28	3.5	2.0	5.4	377
Zhangdian [b]	50	0.12	1.5	0.8	2.3	162
Tengzhou	30+7.5	0.18	2.2	1.3	3.5	243
Liyan	50+50	0.55	6.9	3.9	10.6	742
Shengli	400	2.57	19	8.5	1.9	2064
Sum of 5 plants above	642.5	3.7	33.1	16.5	23.7	3588
Non-centrally scheduled (coal-fired)	2726.5	12	156	88.2	242.3	16899
Total Shandong	17346	84	803.5	380.4	380.9	87185

[a] CO_2, SO_2 and NO_x annual emissions are about 27 percent greater and TSP 52 percent greater than communicated previously by SEPRI and used in the EIA task.

[b] Huangtai, Qingdao, and Nanding include cogenerating units. Zhangdian is a cogenerating plant.

[c] Emissions are slightly lower than communicated previously by SEPRI and used in the EIA task.

[d] SO_2, NO_x and CO_2 annual emissions are 25 to 29 percent greater and TSP 3.3 times greater than communicated previously by SEPRI and used in the EIA task.

[e] Assumption for small plants: dust removal efficiency is 85 percent, except for the Shengli plant equipped with ESP; coal quality is the same of the average for the 20 major plants.

Table 8.9 Estimated total installed capacity, power production, and emissions to air of grid-connected power plants in Shandong in year 2003, assuming sulfur content of coal as in 1999 and installation of one FGD for the two 300 MW units in Huangtai (SEPRI).

Grid-connected power stations	Capacity (MW)	Generation (TWh)	SO_2 (kt/a)	NO_x (kt/a)	TSP (kt/a)	CO_2 (kt/a)
Centrally scheduled plants						
Zouxian	2400	12.1	71.39	45.4	5.33	10907
Huaneng Dezhou	1200	6.24	50.2	25.66	1.64	6218
Shiheng	1200	6.11	43.1	23.25	2.89	5687
Shiliquan	1225	6	48.1	26.1	14.55	6334
Lougkou	1000	4.57	22	21.2	3.4	4565
Huangtai+FGD [a,b]	800	4.27	22.1 (8.9) [b]	16.24	2.9	3937
Huangdao	670	3.71	24.4	15.9	5.15	3861
Qingdao [a]	600	2.88	31.9	11.74	1.43	2822
Huaneng Weihai	850	2.51	11.38	8.51	2.22	2260
Weifang	600	3.2	29.93	12.37	3.76	2996
Laiwu	375	1.85	13.5	7.9	8.18	1923
Jining	200	0.87	3.84	3.74	5.08	904
Liaocheng	300	1.57	4.8	5.8	1.21	1482
Heze	250	1.36	4.65	5.17	5.93	1383
Yantai	100	0.82	9.4	2.57	0.7	858
Nanding [a]	130	0.75	10	3.35	0.8	811
Linyi	350	1.45	8.4	5.75	4.79	1392
Xindian [c]	--	--	--	--	--	--
Zhanhua [c]	--	--	--	--	--	--
Baiyanghe [c]	--	--	--	--	--	--
Sum of plants above [c]	12250	60.26	409 (396) [b]	240.7	70.0	58340
Newly installed plants 1998-2003	6940	40.7	283.0	150.6	18.3	36630
Other sched. plants [d]						
Wusuotun	55	0.28	3.5	2.0	5.4	377
Zhangdian [a]	50	0.12	1.5	0.8	2.3	162
Tengzhou	30	0.14	1.7	1.0	2.7	184
Liyan	50+50	0.55	6.9	3.9	10.6	742
Shengli	400	2.57	19	8.5	1.9	2064
Sum of 5 plants above	635	3.66	32.6	16.2	22.9	3529
Non-centrally sched. (coal-fired)	1363	6.1	78	44	121	8450
Total Shandong [b]	21200	111	803 (789) [b]	451	232	106949

[a] Huangtai, Qingdao, and Nanding include cogenerating units. Zhangdian is a cogenerating plant.
[b] Within brackets the values assuming one FGD for the two 300 MW units in Huangtai (Jinan).
[c] Plants were being converted to coal at the end of the 1990's. Final characteristics not known in year 2000 when this table has been compiled.
[d] Assumption for small plants: dust removal efficiency is 85 percent, except for the Shengli plant equipped with ESP; coal quality is the same of the average for the 20 major plants.

In 1998, SEPCO cogenerating units made a total of 552 MW installed with net production of 2.63 TWh. In the same year, non-centrally scheduled cogenerating plants produced 6.62 TWh net electricity; they were about 56 percent (1515 MW) of the electric capacity of the non-centrally scheduled park. Cogeneration has not been modeled here. The non-centrally scheduled plants are rather small. Stacks for most SEPCO units <50MW are 80 m or 100 m high (SEPRI). Without other information, the same height can be assumed for non-centrally scheduled plants, although from visual estimation of stacks for plants in Chinese towns it may be even shorter.

2.3 Modeling the coal chain for current and future supply to the Shandong grid

2.3.1 General

Figure 8.3 shows the simplified model developed for the coal chain associated with current coal power plants in Shandong and with the future coal-based scenario assumed for the calculations. Mining, transport, power plants, transmission lines, and generation mixes are described in the following sections.

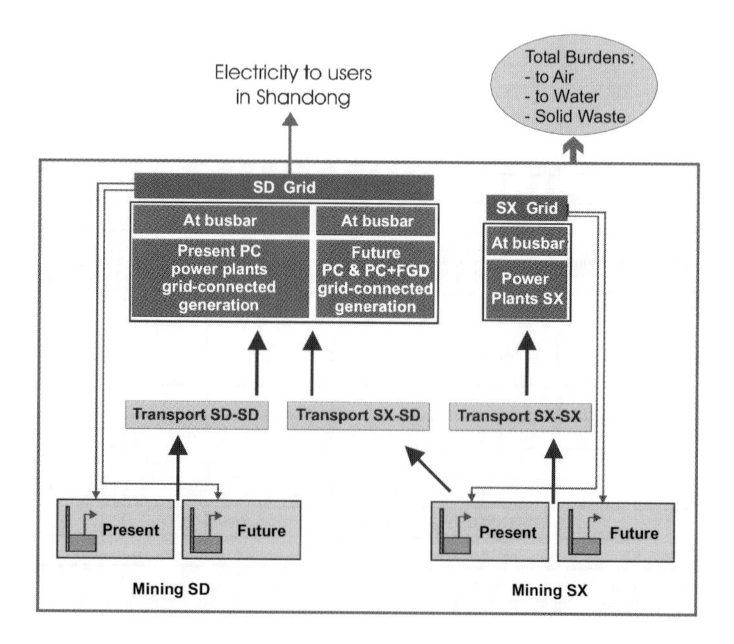

Figure 8.3: Simplified model of the coal chain associated with current coal power plants in Shandong and with the future scenario assumed for LCA calculations.

2.3.2 Mining

The task researchers adopted a simplified approach because of the difficulty of collecting homogeneous data on all mines involved in supplying the Shandong power plants. Estimates were made for four cases selected to represent current average conditions and most probable cases in the future. These cases are illustrated in Figure 8.4, together with transport. Shanxi province, the major exporter of steam coal to Shandong, represents all exporting provinces.

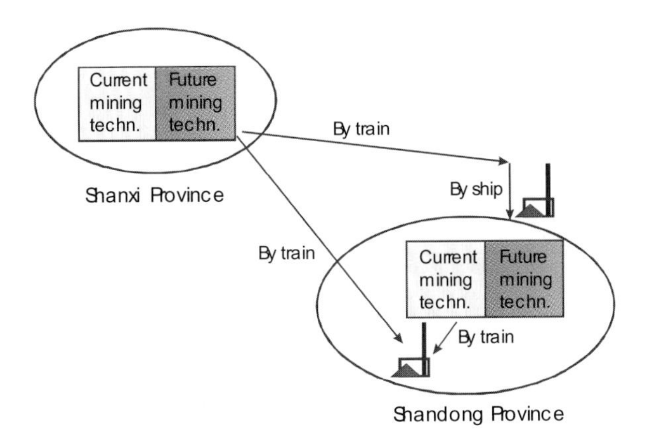

Figure 8.4 Simplified model of current and future coal supply from Shandong and Shanxi.

In practical terms the important differences between present and future mining conditions are changes in the intensity of electricity use and in the (likely) reduction of air emissions per unit product coal. Tables 8.10 and 8.11 summarize information on material and electricity requirements taken from statistics and the survey, and the assumptions about future conditions. The airborne emissions from local power plants are given in Tables 8.4 and 8.5. The methane releases from current mines were discussed previously. In the calculation, the values from the survey are assumed. The value of 16.9 kg of methane per tonne of raw coal as reported for year 1990 by (Rui et al., 1994) has been adopted to represent the average coalbed methane emissions from underground mining in China. For future conditions, all emissions of methane are assumed to drop by one third.

The CB's contacted during the survey reported no major fires in their mines. This information has been adopted in modeling the coal chains associated with Shandong (and SX) coal power plants. However, no final statement can be made on this issue, because no independent way of controlling this information and quantifying the extension of possible fires was available.[19] The literature on uncontrolled fires reports huge quantities of coal lost every year in China. Recently, the Chinese media have been reporting on the seriousness of this issue (Shi, 2000). Several cooperative international projects are monitoring and helping extinguish large fires that occur especially in arid regions (Vekerdy et al., 1999; Sauer, 2002). As reported in (Walker, 1999), apparently the most complete survey on coal fires worldwide, the available estimates for China range from 10 Mt/a to 200 Mt/a. The author considers the upper figure more credible given the presumed loss of 140 Mt/a of coal in Xinjiang alone.[20]

Table 8.10 Material and electricity requirements for coal mining in Shandong; from statistics and own survey.

	Wood (kg/tRC)	Explosive (kg/tRC)	Steel (kg/tRC)	Electricity consumption (kWh/tRC)	Share of electricity self-generation to total consumed (%)
Statistics	1.70	0.17	0.83	28 (37) [49] [a]	52 (39) [28] [a]
Survey [b]	0.73	0.23	0.74	27 (36) [44] [a]	66 (49) [38] [a]
	(kg/tSC)	(kg/tSC)	(kg/tSC)	(kWh/tSC)	(%)
Survey	0.78	0.26	0.80	28 (35) [c]	(50) [d]
Future [e]	0.78 [d]	0.26 [d]	0.80 [d]	38	(60) [d]

RC = raw coal; SC = product steam coal.

[a] Outside parentheses: exclusively electricity for underground mining in major mines. Inside parentheses: 25 percent added to the previous value, in order to include operations following underground mining. Inside square brackets: all electricity uses within coal bureaus.

[b] Value weighted with the production of raw coal in the surveyed mines only.

[c] Outside parentheses: rate for steam coal only. Inside parentheses: rate for all coal products.

[d] Assumed (no information available for future conditions).

[e] Starting from data from the survey as reference. Assuming all coal is cleaned.

Table 8.11 Material and electricity requirements for coal mining in Shanxi; from statistics and own survey.

	Wood (kg/tRC)	Explosive (kg/tRC)	Steel (kg/tRC)	Electricity consumption (kWh/tRC)	Share of electricity self-generation to total consumed (%)
Statistics	1.78	0.05	0.76	21 (30) [39] [a]	65 (46) [33] [a]
Survey [b]	1.84	0.05	0.87	23 (33) [41] [a]	73 (63) [50] [a]
	(kg/tSC)	(kg/tSC)	(kg/tSC)	(kWh/tSC)	(%)
Survey	2.41	0.06	1.10	37 (34) [c]	(60) [d]
Future [e]	2.41 [d]	0.06 [d]	1.10 [d]	48	(75) [d]

RC = raw coal; SC = product steam coal.

[a] Outside parentheses: exclusively electricity for underground mining in major mines. Inside parentheses: 25 percent added to the previous value, in order to include operations following underground mining. Inside square brackets: all electricity uses within coal bureaus.

[b] Value weighted with the production of raw coal in the surveyed mines only.

[c] Outside parentheses: rate for steam coal only. Inside parentheses: rate for all coal products.

[d] Assumed (no information available for future conditions).

[e] Starting from data from the survey as reference. Assuming all coal is cleaned.

These values would mean an additional emission of greenhouse gases from a few percent up to a few tens of percent of what is already produced by the controlled combustion of the coal produced. In estimating environmental burdens from the entire coal chain for the production of electricity, man-made fires should be differentiated from naturally occurring fires. Walker (1999) mentions Zhang (1998) who reports on historical records of the phenomenon, which has increased spectacularly in the past 40 years in proportion to the enormous expansion of mining activities. Considering that the interval of uncertainty for the data is very large, no conclusive statement can be made for energy chain analysis. However, because the subject is important, a value for the GHG produced by coal fires has been arbitrarily assumed to be equal to 15 percent of the complete combustion of the coal used in China.[21] If an average 2.2 kg(CO_2) are produced per kg of coal burned, the above assumption would add approximately 0.33 kg(CO_2-equiv.) per kg of coal to the chain. Moreover, assuming that the average coal of the northern regions of China, which are apparently more exposed to the risk of fire, has one percent sulfur, the burning of extra 15 percent of coal would produce an additional 3 g(SO_2)/kg coal.[22] A significant, but arbitrary, reduction of 30 percent of the emissions stemming from coal fires has been credited for the future, though the closure of small inefficient mines may increase the risk of fires in abandoned mines.

2.3.3 Transport

As illustrated above, the large power stations in Shandong include units of different capacity. The steam coal supplied to any one station is a blend from different origins that is shared among all units. In the modeling of power plants by capacity classes, the information on the actual routes for supply is preserved in this analysis. The average for each class of power plants, though artificially constructed, still gives an accurate picture of average current conditions in Shandong.

Transport routes and the shares of coal from different origins in the future are assumed to remain the same as in 1998. By 2020, the share of Shandong/Shanxi coal which will supply newly built plants has been assumed to be 40/60. Within CETP, the ESS as well as the ETM tasks work with different scenarios and different proportions over 20-30 years, due to the detailed treatment of transport and the consideration of the quality of the coal that these simulation methods allow. However, for the MCDA scenarios the LCA assumptions for the imported coal have been changed to comply with those of the ESS task. A sensitivity analysis has been performed in order to calculate the range of variation that can be expected for different origins of the coal supply, though no specific mine has been taken into account.

Modeling transport of coal by train to Shandong power plants presents two uncertainties: the shares of steam, diesel, and electric trains, and the unknown environmental burdens associated with these transport means. According to Zhang and Folmer (1996) in 1990, 29 percent of the total gross converted ton-kilometers was still transported in China by steam locomotives; only 19 percent was transported by electric locomotives; and the rest by diesel engines. The same authors reported that mainly medium-sized gasoline-fuelled trucks are used for road freight transport; these consume 20 to 30 percent more fuel than foreign trucks of similar types. Ships consume approximately 15 percent more than in industrialized countries. However, due to the dramatically growing transport sector and the introduction of newly constructed trucks, general performance may have improved, though it can be argued that the newest trucks may not be used for coal transportation. A multiplication factor of 1.3 for trucks and 1.15 for ships has been used to calculate the environmental burdens for these transportation means in the Swiss study of current European energy systems. It is assumed that in the near future average Chinese trucks and ships will reach today's European standards.

No information could be gathered about the type and efficiency of Chinese freight trains. Therefore, an approximate approach for the estimation of the emissions associated with railway transport has been taken. Emissions have been arbitrarily assumed to be three times greater than those calculated for the average European freight transport by electric trains in the Swiss LCA study of UCTE energy systems in mid 1990's. On the one hand, this should include the effect of the different electricity mixes and the average efficiencies of the power plants for the two cases. On the other hand, it should somehow cover the increased emissions of the diesel (or steam) locomotives compared to electric ones. For future train transport, the multiplication factor has been reduced to two, crediting a greatly increased diffusion of electric trains but recognizing that the electricity is still mostly produced by coal. Important for the calculation of total land use is also the factor assumed for the land occupied by railways. Lacking a specifically Chinese study, the land use estimated for the European railway has been used, i.e. 0.00014 m^2/tkm.

For the non-SEPCO grid-connected small power plants, due to the lack of direct information a simplified assumption has been made hypothesizing that coal supplied from local mines travels an average distance of 20 km by truck.

According to fuel management regulation of state power companies, total transportation losses (which include also loading-unloading) should not be higher than 1.2 percent for train transportation, and 1.5 percent for combined land and sea transportation (SEPRI). Truck losses can be estimated at roughly 2.5 percent.

SEPRI reported that the reduction in the average sulfur content of coal supplied to SEPCO power plants in 1999 has been achieved mostly by blending the coal at the supplying mines, though their shares remained the same as in 1998. This may be explained by the wide range in quality of the coal from different mines within each district.

2.3.4 Power plants

The classes used to model the currently operating PC power plants are: two classes with <100 MW for scheduled and unscheduled plants; 100 MW; 125 MW; 210 MW; 300 MW; and 600 MW. The last two units constitute the standard size of base load for PC plants manufactured in China today. The characteristics for each class are averaged over the units belonging to 17 major SEPCO power stations. The lignite mine-mouth power plant in Longkou, including 2×100 MW and 4×200 MW units, and burning a blend of the local lignite and Shanxi coal, has been separately considered. Future plants modeled are: 600 MW PC with FGD, AFBC, and IGCC.

Figure 8.5 shows the average fractions for auxiliary power in 1998 for the modeled classes of current power plants. The average fraction over all plants is 6.5 percent (4195 GWh self used over 64530 GWh produced). For the small grid-connected non-SEPCO power plants it has been assumed to be approximately 12 percent.

In previous LCA studies on current and future energy systems, the contribution of energy and materials needed for plant construction to the total environmental burdens of a hard coal chain has been estimated to be on the order of 0.5 percent. Because this is such a small figure, a search for China-specific figures on plant construction was given low priority. It is here assumed that the material use for the

300 MW unit installed in Shandong is on the order of the values assumed in Frischknecht et al. (1996) for current hard coal plants in Western Europe. The material use for other unit sizes has been roughly scaled down from the above reference. Although the emissions associated with material manufacturing have been estimated to be three times greater than assessed for European conditions,[23] their influence on the total burdens associated with the average electricity at busbar today in Shandong is negligible, an expected consequence of the high direct emissions from the operation of power plants.

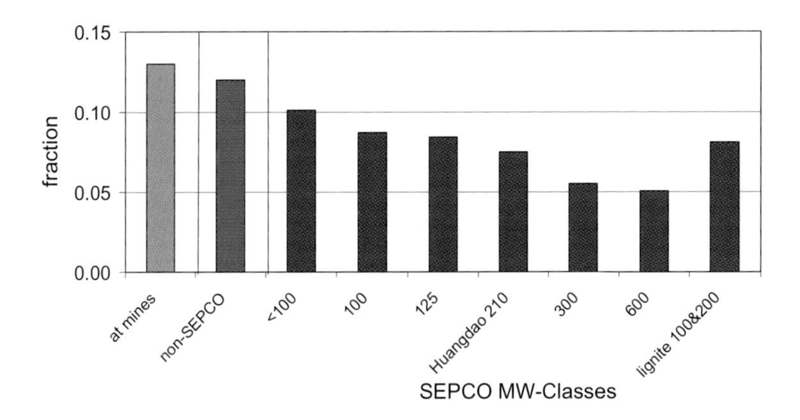

Figure 8.5 Auxiliary power consumed as a fraction of gross generation by grid-connected coal power plats in Shandong (1998) and small mine-mouth plants.

The air emission rates and solid waste production for the modeled classes of power plants are given in Table 8.12. The emission of each single unit has been weighted by its energy output. Also included in the table are the emission rates assumed for future technologies. As expected, the emissions of NO_x generally decrease as the capacity of the power plants increase, because of the increased net efficiency. A notable exception is the 210 MW class, including only two units in the Huangdao plant.

The emission rates of SO_2 directly produced by power plants depend on three factors: the sulfur content in the fuel, the removal rate from the exhaust, and the net efficiency of the plant. Hence, the emissions calculated for each class should not be taken to be strictly related to the characteristics of the class itself. Table 8.12 shows two sets of values for the SO_2 emissions, for 1998 and 1999. The decrease achieved in 1999 on average values is remarkable. This is the year in which the directive to reduce the sulfur content of coal to one percent or less, unless a plant would be retrofitted with an FGD, was first implemented. This is the only piece of information received from SEPRI about year 1999. The values for the non-SEPCO small plants have been guessed (SEPRI).

Future units of the PC with FGD, AFBC, and IGCC technologies are assumed to burn coal of the same quality used in 1999 in the Zouxian plant, 60 percent supplied by Shanxi and 40 percent by Shandong mines. They will remove sulfur with respective efficiencies of 90 percent, 95 percent and 99 percent.

The emission rate of particulates depends primarily on the removal efficiency of the specific control system. All 300 MW and 600 MW units are equipped with electrostatic precipitators (ESP), with an efficiency 99 percent. Also both 210 MW units in Huangdao have ESP, but of lower efficiency. The 125 MW and 100 MW units have venturi or ESP. Smaller units may have multi-tube cyclone or venturi particle removal systems that are not very efficient but that do not meaningfully affect the already low net efficiency of the plant. Some stations may have ESP in common for several units including small ones.

Table 8.12 Emission rates from current power plant classes in Shandong and assumed for coal technologies used in LCA future scenarios.

Installed capacity class (MW)	Emissions to air (g/kWh at busbar)					Solid waste production (kg/kWh)
	SO_2 in 1998	SO_2 in 1999	NO_x	TSP	CO_2	
(non SEPCO)<100 [a]	14.82	14.82	8.38	23.03	1605	0.182
<100	15.46	11.66	5.38	9.09	1357	0.155
100	8.55	6.41	4.30	2.35	1114	0.121
125	9.51	7.24	4.43	3.20	1090	0.115
210 [b]	8.78	6.42	4.64	1.79	1125 [b]	0.148
300	10.08	7.76	4.10	0.58	1002	0.109
600	6.01	5.64	3.79	0.48	918	0.088
Lignite 100/200 [c]	5.25	4.64	5.06	0.81	1090 [c]	0.099
600 & FGD [d]	0.96 [e]	0.73	3.83	0.16 [f]	927	0.115 [g]
AFBC [h]	0.47 [e]	0.36	1.42	0.46	940	0.128 [i]
IGCC [h]	0.08 [e]	0.06	0.40	0.39	735	0.077

[a] Assumed values (SEPRI).
[b] Two out of four units in the Huangdao power plant. Their gross efficiency is about comparable with the average efficiency of the 125 MW class.
[c] Longkou, fueled with 65 percent on-site lignite. Possible under-estimation of CO_2 by about 8 percent.
[d] Gross efficiency conservatively assumed equal to subcritical PC's today in China.
[e] Although clean coal technologies are suited for high sulfur coal, here the average coal used in year 1998 has been adopted. However, the assumed FGD efficiency is 90 percent, while it can reach 95 percent, though at the expenses of the total plant efficiency. AFBC has been credited 95 percent removal efficiency, IGCC 99 percent (Oskarsson et al., 1997).
[f] Assuming that the FGD further reduces the emission of particulates.
[g] Arbitrarily assuming an increase of 30 percent of total waste due to FGD compared with a plant of same capacity and same fuel but without FGD.
[h] After (Oskarsson et al., 1997).
[i] Assumed 1.5 greater than the waste from a PC (without FGD) of equal capacity.

The solid waste production rate for a PC unit without FGD depends on the ash content of the fuel and the net electric efficiency. It may be argued that changing the coal blend for most of the plants between 1998 and 1999 may have changed the average ash content of the mixes and hence the solid waste produced. However, this information was not available. The operation of a wet FGD produces an amount of gypsum two to three times greater than the mass of solid waste produced by the plant. However, because a great part of the gypsum will most probably be recycled, it is arbitrarily assumed here that total waste due to FGD will increase by 30 percent compared with a unit of same capacity and same fuel but without FGD.

Most likely, FGD's will be installed in large units burning high sulfur coal (one to three percent) for which the wet scrubber is the preferred technology. This study models a 600 MW unit equipped with 90 percent efficient wet FGD, assuming 1998 average coal characteristics. The extent of application of FGD's in existing as well as newly built PC's is highly uncertain. This is because on the one hand current Chinese policy permits the operation of PC without FGD if its fuel is less than one percent sulfur, a threshold that might be lowered to 0.75 percent before the end of this decade. On the other hand, coal-washing rates at mines may increase, which would expand the available stock of low sulfur (and lower ash) steam coal. Because an LCA assessment of future systems is already affected by several other uncertainties, no attempt has been made to model other options for FGD (e.g., sea water or injection scrubbers).

The land use per unit of energy for all SEPCO-owned units has been assumed to be equal, because the various units in one station share the same ground. Besides, it is not known whether the size of the stock of coal, which occupies a large part of the total affected area, is based on the same requirements for all stations. On the basis of values found in the literature for large coal power stations and used in the Swiss LCA study,[24] and of visual inspection at the Huangtai power plant, a land use of 300 m^2/MW installed has been assumed, which gives approximately 1.2 m^2/GWh. Individual small units are assumed to have a higher land use of 1.5 m^2/GWh, due to lower efficiency. Large units with FGD are arbitrarily assumed to occupy 30 percent more land due to the FGD itself and the additional waste produced.

The dose contribution of radioactive emissions from power plant operation to the health effects of the coal chain depends in large part on whether particulate and sulfur control technologies are in place, which reduce the airborne fraction of the radioactive emissions, and to a lesser degree on plant efficiency. Radioactive emissions from Chinese coal plants may have substantial impacts. UNSCEAR (1993) reports that due to the significantly high radioactivity of average Chinese coal, the collective dose from the operation of PC's in China can be 10-100 times higher than from the operation of PC's elsewhere. Specific data for Shandong plants were lacking, and the values from the literature are incomplete (Pan et al, 1999). However, including the effects of radioactivity would have made worse the already poor performance of the coal chains. Hence, this missing information should not affect the conclusions of the study in terms of ranking technologies and scenarios.

2.3.5 Transmission lines

SEPRI provided some information about high and medium voltage transmission lines as well as medium to high capacity transformers for the provincial Shandong grid in 1998. This was not sufficient to allow a quantification of the environmental burdens associated with these transmission lines, and no information was available for these types of transformers. No information existed on the low voltage grid, which is not managed by SEPCO. This is possibly a serious limitation, because relatively high material consumption and hence high burdens may be expected.

SEPRI claims about six percent transmission losses in the high/medium voltage grid operated by SEPCO. According to the estimation of some coal mine operators approached during the PRCEE survey, the transmission losses of the Shandong provincial grid are six to eight percent for high voltage, eight to ten percent for medium voltage, and ten to twelve percent for low voltage (6 kV) transmission segments. Total network losses would be over 22 percent. This evaluation, though not supported by further evidence, goes along with information reported in the Chinese press about high losses that in the recent past were sometimes around 40 percent of electricity produced (China Daily, 12/6/2000).

For the losses, mine operators' higher estimates for the medium and low voltage networks have been adopted for the reference years 1998-1999, whereas the lowest were used for future conditions up to 2020. The losses reported by SEPRI were assumed for the high voltage transmission lines only, constant over the considered time frame.

Considering that the aim of CETP is to estimate the total burdens associated with electricity services, a preliminary guess of the burdens from materials use in the network has been attempted based on engineering judgment. The materials for the high, medium, and low voltage Shandong networks today were extrapolated from material use rates estimated in the existing European LCA studies, by multiplying the corresponding set with the arbitrary factors 1.1, 1.25, and 1.5, respectively. However, future material use for Shandong networks is assumed to be equal to today's European values.

Only the direct emission of N_2O, a strong greenhouse gas, has been considered to occur during the operation of the high voltage lines. This gas is produced due to the corona effect at an estimated rate of 0.018 g/kWh, which equals about five g(CO_2-equiv.)/kWh. Because no information could be obtained, the emission of the even stronger greenhouse gas SF_6 from transformers has not been accounted for. This gas contributes approximately two g(CO_2-equiv.)/kWh for Western European conditions.

2.3.6 Supply mixes

Figure 8.6 shows the shares of the different classes of power plants for the 1998 Shandong grid mixes as well as the future mix assumed for this study. The generation mix is also given because these figures are regularly reported in Chinese statistics.

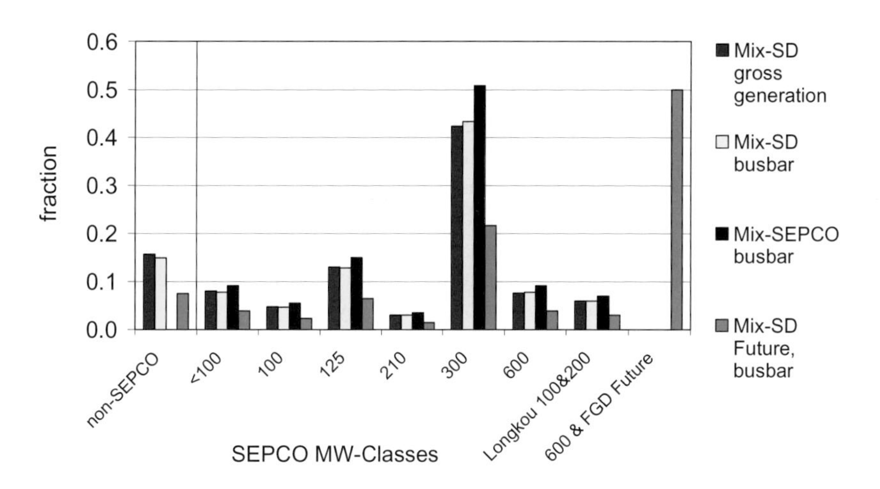

Figure 8.6 Shares of electricity mix by power plant classes for 1998 and the future.

All non-SEPCO units are <100 MW each, i.e. they are eligible to be shut down before year 2003 to comply with the current state policy. Their 1998 share of the total supply was a substantial 15 percent. We can expect that their contribution to pollution for this year was high. No official information on the actual rate of closure for those units in 1998-2002 has been at hand. However, SEPRI's guess is that about half of the non-centrally scheduled (coal-fired) plants in Shandong will be closed, while the others may be refurbished (e.g., as cogeneration plants) and kept running. Some plants will be left open as exceptions to the general rule (see Table 8.8 above).

For the LCA calculations concerning future energy chains, a likely "future" Shandong scenario or mix of technologies has been defined; it is composed of 50 percent current (1999) power plants, with the same relative share of PC classes as in the present SEPCO mix, and 50 percent 600 MW PC's with FGD.

2.4 Results

In this section, the results obtained for the coal chains associated with the reference classes of power plants in Shandong are illustrated with examples of selected environmental burdens, chosen on the basis of their importance and the completeness of the information throughout the chain. These are the greenhouse gases, SO_2, NO_x, TSP, total solid wastes, and land use. Emissions to water are not considered, because the limited and scattered data available do not allow a discussion of the full chain. Contributions from the various parts of the chain are discussed separately when they are relevant.

Considering different possible uses of LCA results, various conditions are considered, namely the electricity generated, the electricity at busbar of the power plants (i.e. the net electricity supplied to networks, to allow comparison among different energy carriers), and the electricity at various voltage levels. The results will illustrate the effects of the relatively high network losses for different electricity services.

To make the discussion as complete as possible, the effects of spontaneous fires in coalfields and mine wastes are included in the coal chain assumed here to represent contemporary average Chinese conditions. To put the results of this study into perspective, the air releases will be compared with those from chains associated with Western European plants analyzed in the Swiss LCA studies.

In Figure 8.7 the average coal requirements for the 1998 classes of power plants and the 600 MW PC with FGD, normalized by the electricity generated and at busbar, are compared with total requirements, which include the effect of energy use throughout the energy chain and losses during transportation. The average direct coal consumption rate for the Shandong generation mix in 1998 is 483 g(coal)/kWh, which corresponds to 523 g(coal)/kWh at the busbar; the full-chain approach gives 543 g(coal)/kWh at the busbar. Hence, the full chain analysis increases coal requirements by about four percent.

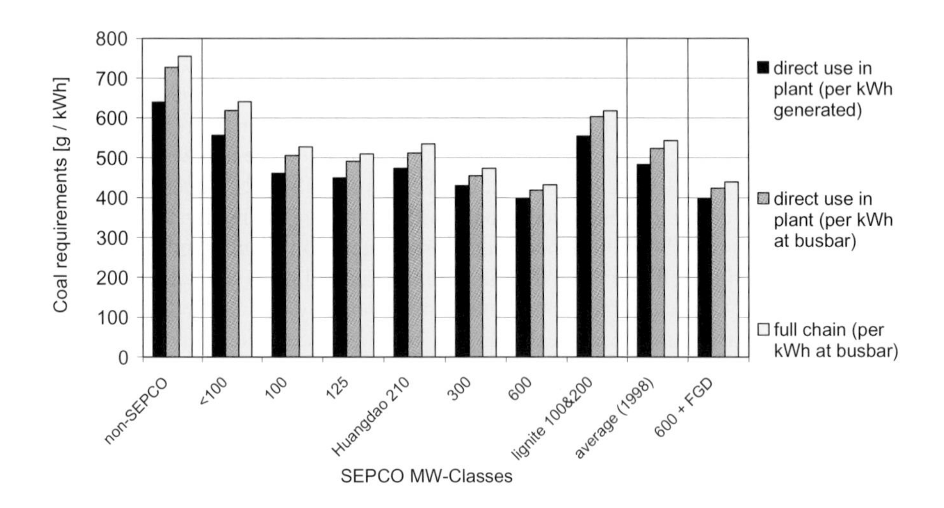

Figure 8.7 Coal requirements by class of Shandong PC power plants (1998 and future).

Table 8.13 shows selected environmental burdens per unit of electricity at busbar for the chains associated with power plant classes in Shandong for the years 1998-1999. Table 8.14 shows the same burden rates calculated for 2020.

Table 8.13 Selected environmental burdens calculated for the chains associated with coal power plant classes at the end of the 1990's.

Installed capacity class (MW)	Emissions to air (g/kWh at busbar)					Solid waste production (kg/kWh)
	SO_2 in 1998	SO_2 in 1999	NO_x	TSP	GHG [a]	
(non SEPCO)<100	15.24	15.21	8.65	23.40	1680	0.370
<100	16.07	12.24	5.77	9.97	1529	0.290
100	9.11	6.95	4.67	3.24	1284	0.225
125	10.05	7.76	4.78	4.02	1256	0.214
210 [b]	9.66	7.28	5.27	3.44	1428	0.227
300	10.54	8.20	4.39	1.25	1141	0.203
600	6.24	5.85	3.93	0.69	956	0.195
Lignite 100/200 [c]	5.80	5.17	5.40	1.55	1238	0.228

[a] Given in g(CO_2-equiv.)/kWh.
[b] Two out of four units in the Huangdao power plant.
[c] Longkou power plant, using a blend with 65 percent on-site lignite.

Table 8.14 Selected LCA-based environmental burdens for the chains associated with coal power plant classes, partly used for the calculation of future scenarios in year 2020.

Installed capacity class (MW)	Emissions to air (g/kWh at busbar)					Solid waste production (kg/kWh)
	SO_2 in 1998	SO_2 in 1999	NO_x	TSP	GHG [a]	
(non-SEPCO)<100 [b]	15.19	15.18	8.70	23.22	1689	0.376
<100 [b]	15.98	12.17	5.76	9.49	1500	0.303
100 [b]	9.02	6.88	4.65	2.74	1250	0.239
125 [b]	9.98	7.70	4.78	3.57	1225	0.228
210 [b,c]	9.51	7.15	5.18	2.50	1348	0.254
300 [b]	10.48	8.15	4.39	0.87	1114	0.214
600 [b,d]	6.46	6.08	4.14	0.91	1048	0.185
Lignite 100/200 [b,e]	5.74	5.13	5.46	1.24	1224	0.243
600 & FGD [b,f]	1.41	1.19	4.18	0.59	1059	0.213
AFBC	0.90	0.79	1.75	0.86	1063	0.220
IGCC	0.44	0.42	0.68	0.73	839	0.155

[a] Given in g(CO_2-equiv.)/kWh.
[b] Used for the calculation of the LCA future mix for Shandong.
[c] Two out of four units in the Huangdao power plant.
[d] Assuming 60 percent SX coal. In 1998 only coal from SD was used in the Zouxian plant.
[e] Longkou power plant, using a blend with 65 percent on-site lignite.
[f] Gross efficiency equal to current maximum for PC; 60 percent of coal from SX.

In Table 8.15, the current SEPCO and Shandong provincial grid mixes are compared with the assumed future mix. For illustration purposes only, the results for an average Chinese chain (using the assumptions described above) associated with two typical PC plant types, a 300 MW unit without FGD (current) and a 600 MW unit with FGD (future), are also shown.

Table 8.15 Selected total environmental burdens associated with Shandong grid mixes and with two standard PC units using Chinese average conditions for the upstream chain.

Supply Mix or Unit (MW)	Emissions to air (g/kWh at busbar)					Solid waste production (kg/kWh)
	SO_2 in 1998	SO_2 in 1999	NO_x	TSP	GHG [a]	
SEPCO Mix 1998	10.14	7.97	4.65	2.61	1201	0.225
Shandong Mix 1998	10.90	9.05	5.24	5.70	1272	0.247
Future SD Mix	6.13	5.10	4.72	2.98	1159	0.231
300 [b] [China]	11.54	9.22	4.42	0.93	1286 [c]	0.206
600&FGD [China] [c]	2.31	2.08	4.12	0.48	1190 [c]	0.205

[a] Given in g(CO_2-equiv.)/kWh.
[b] Without FGD.
[c] Assuming that spontaneous combustion in China contributes an additional 15 percent of GHG emissions beyond that produced by controlled combustion of coal.

In Table 8.16 the selected total environmental burdens associated with the current SEPCO mix, with the Shandong provincial grid mix, and with the assumed future mix, are given for three different voltage levels. However, as already stated above, the network losses considered are only meant to be a first guess, because no precise information for today nor any trend for the future was available. However, it should not be difficult to recalculate the burdens for different conditions should more accurate figures be obtained.

Figure 8.8 reports the GHG emission rates for the chains associated with different Shandong coal power plant classes at the end of the 1990's. It also shows the GHG emission rate for the UCTE mix in the first half of the 1990's as calculated in the Swiss LCA study of current European energy systems. This European mix includes nearly 48 percent fossil power generation (10.5 percent lignite, 17.5 percent hard coal, 9.7 percent oil, and 9.8 percent industrial and natural gases), 37 percent nuclear, 14.3 percent hydro, and 1.2 percent other. Shandong mixes exhibit much higher values than the UCTE average, because coal is practically the only fuel used. In the next two decades, the GHG emission rate for the generation mix in the countries already belonging to UCTE in 1995 should not change much.[25] A value of 500 g(CO_2-equiv.)/kWh at the busbar can be used as an approximation for purposes of comparison.

Table 8.16 Selected total environmental burdens associated with the current SEPCO and Shandong provincial grid mixes, and with the assumed LCA future mix, given for three different voltage levels.

Supply Mix	Emissions to air					Solid waste
	SO_2 in 1998	SO_2 in 1999	NO_x	TSP	GHG [a]	Production[b]
	(g/kWh at high voltage)					(kg/kWh)
SEPCO Mix 1998	10.76	8.46	4.93	2.78	1280	0.239
Shandong Mix 1998	11.67	9.69	5.61	6.11	1368	0.265
Future SD Mix	6.51	5.42	5.01	3.16	1235	0.245
	(g/kWh at medium voltage)					(kg/kWh)
SEPCO Mix 1998	11.73	9.22	5.37	3.03	1395	0.260
Shandong Mix 1998	13.02	10.83	6.24	6.79	1525	0.294
Future SD Mix	7.15	5.96	5.47	3.46	1351	0.267
	(g/kWh at low voltage)					(kg/kWh)
SEPCO Mix 1998	13.37	10.51	6.13	3.45	1591	0.297
Shandong Mix 1998	15.42	12.92	7.19	7.84	1784	0.340
Future SD Mix	8.32	7.00	6.12	3.90	1530	0.300

[a] Given in g(CO_2-equiv.)/kWh.
[b] Same voltage level as for air emissions on the same row.

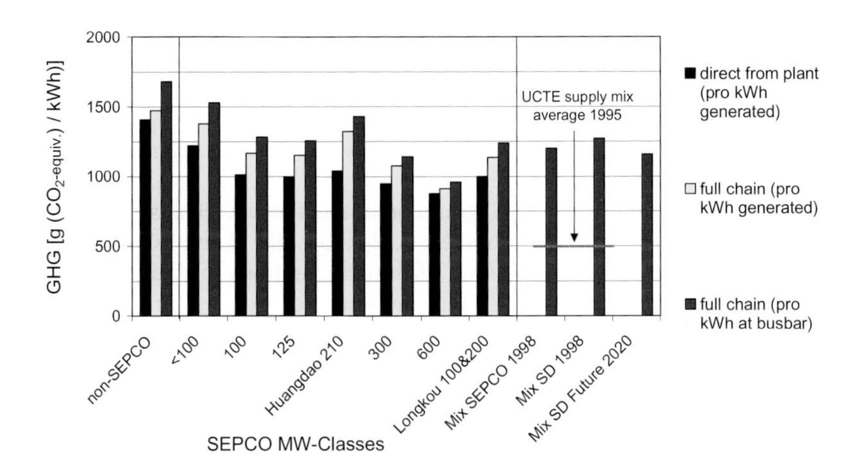

Figure 8.8 Greenhouse gas emission rates from chains associated with Shandong coal power plants and mixes (1998 and 2020).

The contributions from the upstream chains, explicitly illustrated in Figure 8.9, differ among the classes depending on the routes of the coal supply. The amount of methane from mining depends on the share of SX coal in the blend. For this reason it is highest for Huangdao (8 g/kWh at busbar) to which only SX coal is supplied, and negligible for small non-SEPCO plants. The picture also shows the preliminary average values calculated for SX and Chinese coal chains, assuming their power plant mixes contribute the same emissions as the SD mix does. The upstream part of the chain associated with Chinese PC's includes the country average emission rate of coalbed methane and the (large) value for spontaneous combustion in coal mines and waste heaps (section 2.3.2). The upstream part of the chain associated with the PC mix for SX does not include any coalfield fires, along with what has been reported by the operators of the surveyed CB's, though such events are recorded for arid areas in north of the province (Bökemeier and Elleringmann, 2002). These LCA-based GHG emissions for SD, SX, and China grid-connected PC mixes should be regarded as providing intervals of possible values, rather than precise estimations.

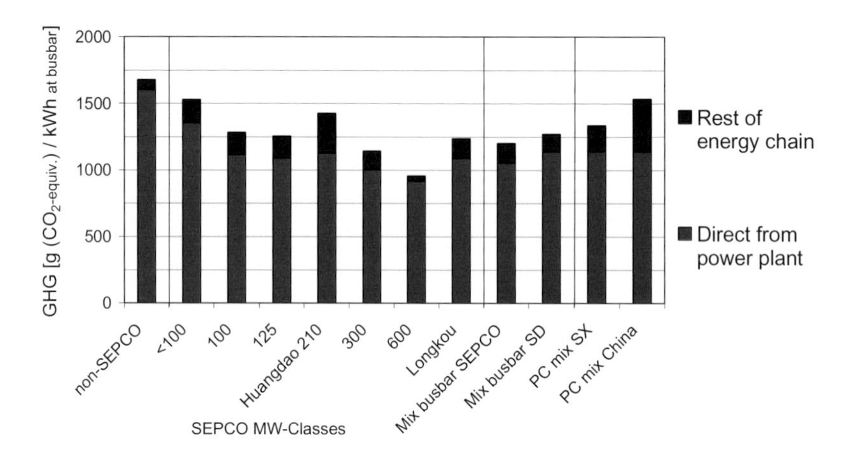

Figure 8.9 Greenhouse gas emission rates from chains associated with Shandong coal power plant classes and mixes in 1998: contributions from power plants and upstream chains.

Figure 8.10 shows the GHG emission rates for the chains associated with different coal power plants used for the calculation of the future LCA mix as well as other scenarios. Besides consideration of future conditions for the mines as well as for transport as described above, a change has been considered for the 600 MW unit class compared to the situation in 1998-1999: In the future, 60 percent of the coal supply will come from Shanxi and 40 percent from Shandong. This distribution also pertains to the 600 MW PC with FGD. The major reduction calculated for IGCC is entirely the result of net plant efficiency. The AFBC emits slightly more GHG than the 600 MW PC because the effect of a somewhat higher net efficiency is more than offset by the substantial increase of N_2O (this species alone accounts for about 70 g(CO_2-equiv.)/kWh), due to the relatively lower temperature of combustion.

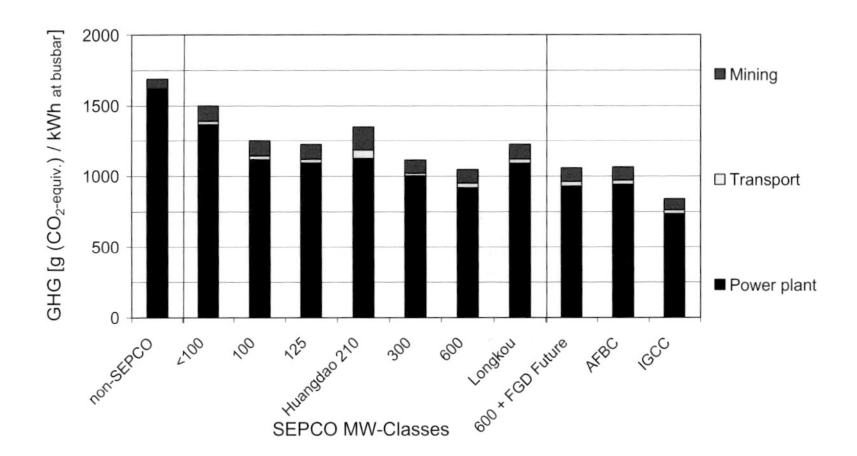

Figure 8.10 Greenhouse gas emission rates from chains associated with Shandong coal power plants in 1998 and technologies used for the calculation of future scenarios.

Figure 8.11 shows the 1998 SO_2 emission rates for different Shandong coal power plant classes and mixes. The average for the Shandong mixes is 10-11 g/kWh at the busbar. The picture also shows the average (3.8 g/kWh at the busbar) and the min-max range (0.9-8.5 g/kWh at the busbar) of the LCA values for SO_2 emissions for Western European hard coal power plants as estimated in the Swiss LCA study for countries' averages in the first half of the 1990's (today the values are somehow lower). The average SO_2 emission rates estimated in Dones et al. (1996) for coal plants in 2010-2030, again for Western Europe, should decrease to 0.7 g/kWh at the busbar as advanced coal technologies are installed and aging PC's shut down.

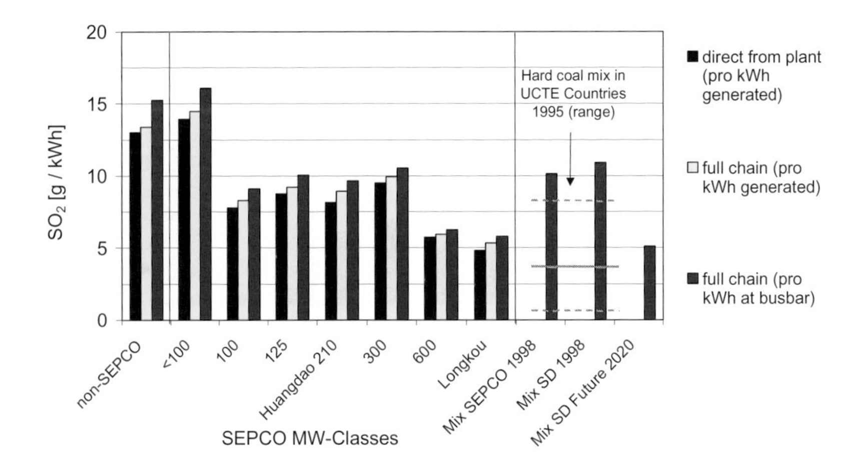

Figure 8.11 SO_2 emission rates from chains associated with Shandong coal power plants (1998) and mixes (1998 and 2020).

Figure 8.12 shows the SO_2 emission rates in 1999. With the use of fuel with lower sulfur content in 1999, the direct emission rates from SEPCO plants decreased six to 25 percent compared to 1998; the values of the two Shandong mixes reached the level of the highest European value in 1995 (average in the countries of former Yugoslavia). However, these values were still more than twice the average UCTE and nearly ten times higher than the minimum (Austria, where all plants have FGD).

Figure 8.13 summarizes the results for SO_2 emission rates for the individual steps of the chains associated with different coal power plants in 1999 and advanced technologies used for the calculation of the future LCA mix as well as scenarios.

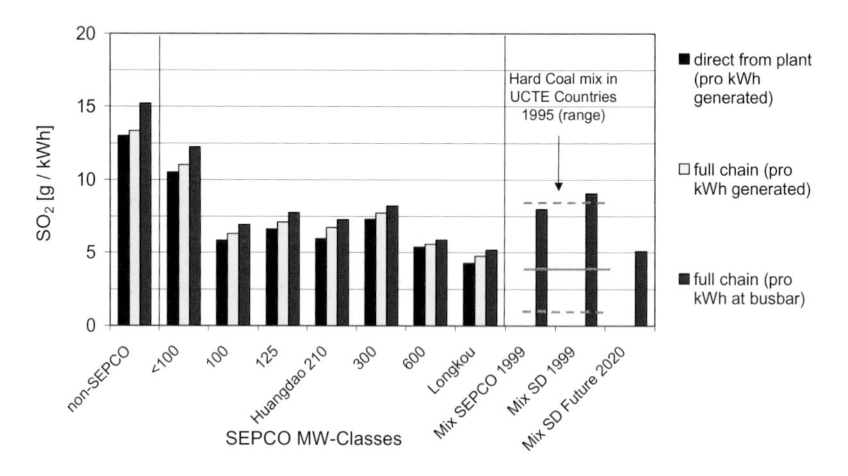

Figure 8.12 SO_2 emission rates from chains associated with Shandong coal power plants (1999) and mixes (1999 and 2020)

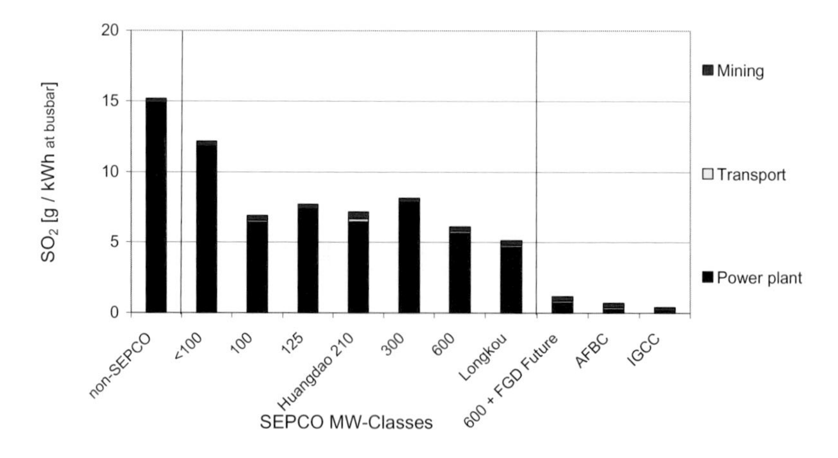

Figure 8.13 SO_2 emission rates from chains associated with Shandong coal power plants in 1998 and technologies used for the calculation of future scenarios.

The contributions of mining and transport to the total for the various power plant classes in 1999 range from three percent (non-SEPCO small plants, assumed to be supplied by truck from nearby mines) up to 12 percent for the Huangdao 200 MW units (all coal supply coming from the distant Shanxi mines). Due to high retention levels, direct emissions from the power plants using FGD's, fluidized bed, or gasification are relatively small, thus making mining and transport meaningful contributors in relative terms to the total calculated for these specific chains.

Figures 8.14 and 8.15 report the results for NO_x emission rates.

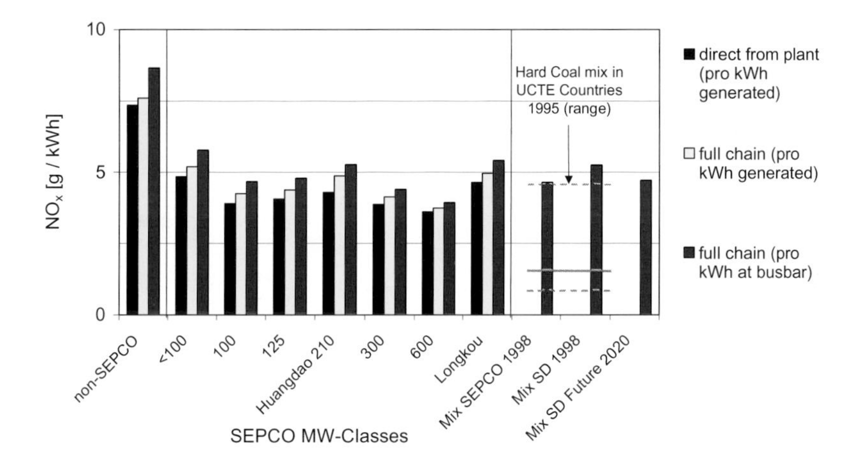

Figure 8.14 NO_x emission rates from chains associated with Shandong coal power plants and mixes (1998 and 2020).

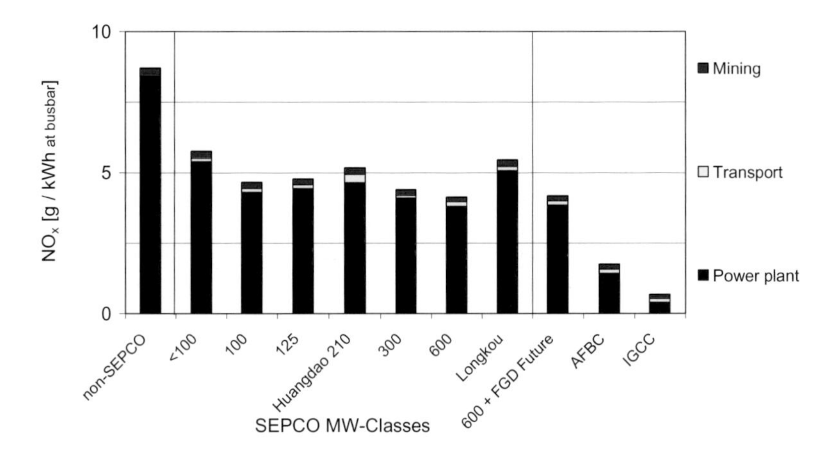

Figure 8.15 NO_x emission rates from chains associated with Shandong coal power plants in 1998 and technologies used for the calculation of future scenarios.

Figure 8.14 also gives the minimum, average, and maximum chain-related NO_x emissions (0.9 g/kWh, 1.6 g/kWh, and 4.6 g/kWh, respectively) calculated in the Swiss LCA study for Western European countries' averages for hard coal power plants in the first half of the 1990's. Direct emissions from power plants are nearly proportional to plant efficiency. Plants with AFBC and IGCC technologies show dramatically reduced emissions. The upstream chain contributes only a few percentage points to the total NO_x, as illustrated in Figure 8.15. Transport contributes about half the total associated with the upstream chain, when part of the coal is transported from Shanxi.

Figure 8.16 shows the particle emission rates. Because small plants use relatively inefficient particle removal systems, their emission rates are remarkably high, up to 20 times greater than for large plants using ESP. The picture also gives the minimum, average, and maximum particle emissions (0.8 g/kWh, 1.2 g/kWh, and 2.5 g/kWh, respectively) calculated for chains associated with European plants.

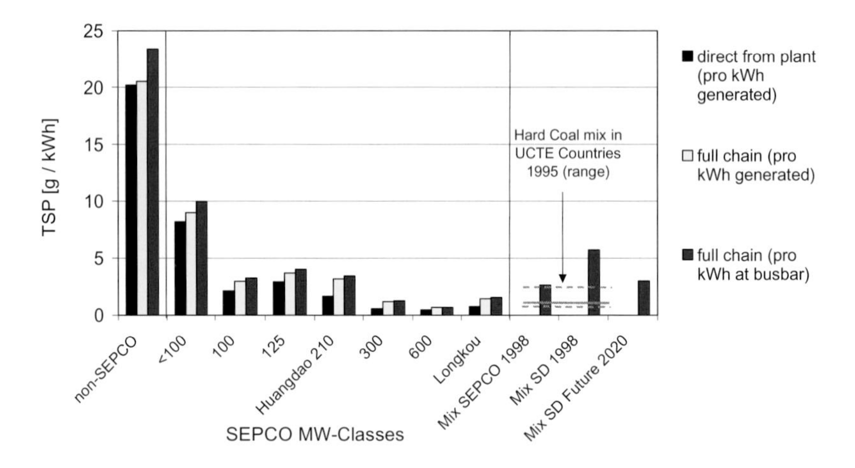

Figure 8.16 TSP emission rates from chains associated with Shandong coal power plants and mixes (1998 and 2020).

As can be seen from Figure 8.17, the amount of particulates from the upstream chain becomes meaningful in relative terms to total only for large units and long-distance transport of coal, with contributions from small mine-mouth power plants. The upstream-chain part is only about one percent for small non-SEPCO units because of the extremely large direct emissions from the power plant's stack and short transport distances, but nearly 55 percent for the 300 MW units. The size distribution of TSP from upstream processes (not available for this study) may differ from the finer particulate emissions directly from power plants, thus causing different health effects. However, this issue has not been addressed here.

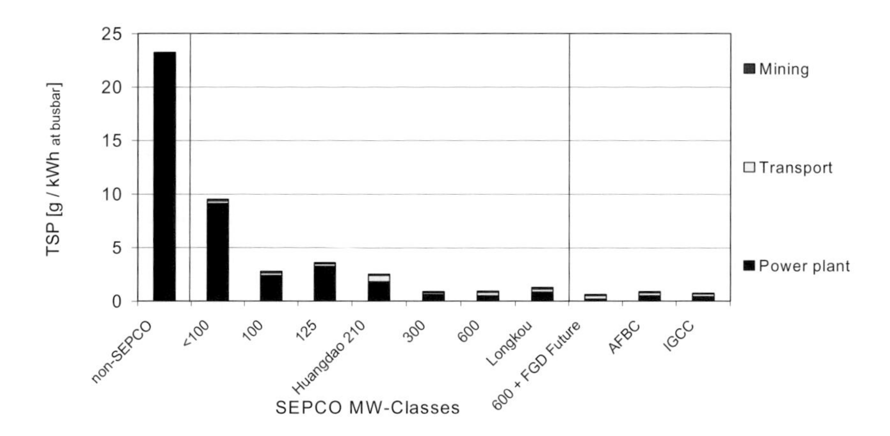

Figure 8.17 TSP emission rates from chains associated with Shandong coal power plants in 1998 and technologies used for the calculation of future scenarios.

Figure 8.18 compares the solid waste production rates. The minimum, average, and maximum total solid waste production averages (0.184 kg/kWh, 0.237 kg/kWh, and 0.302 kg/kWh, respectively) for Western Europe are also given. The solid waste amounts associated with the electricity from the Shandong mixes are comparable to average European hard coal energy chains, for different reasons. The higher wastes produced by washing the coal burned in Europe is offset by lower ash production at the power plant. European plants have a higher average efficiency but their transport requirements for coal imported from other continents is very high.

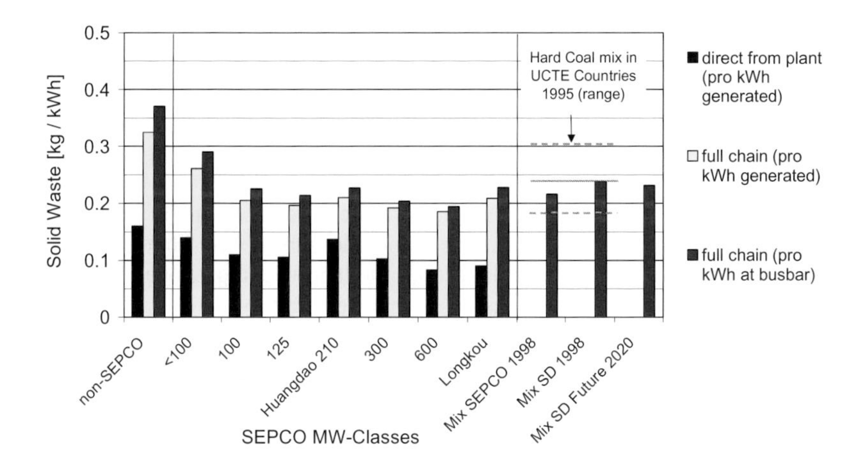

Figure 8.18 Solid waste production rates from chains associated with Shandong coal power plants and mixes (1998 and 2020).

Three main factors dominate the amount of the waste produced: the ash content of the coal; the efficiency of the plant; and, the solid wastes left at mines per tonne of coal output. Normally, the upstream part of the chain contributes about half to the total, as can be seen in Figure 8.19. Chains including greater share of coal from SD also exhibit a higher amount of waste from mining. Transport by train may contribute a few percentage points to the total. The assumption for the recycling rate of the wastes controls the ranking. In particular, the gypsum from FGD is assumed to be largely utilized by the building industry (up to 90 percent). Conversely, the different composition of the waste from AFBC makes it difficult to recycle, and must be disposed of as landfill. The lower waste production rate for the IGCC chain is an effect of the relatively high net efficiency of the plant.

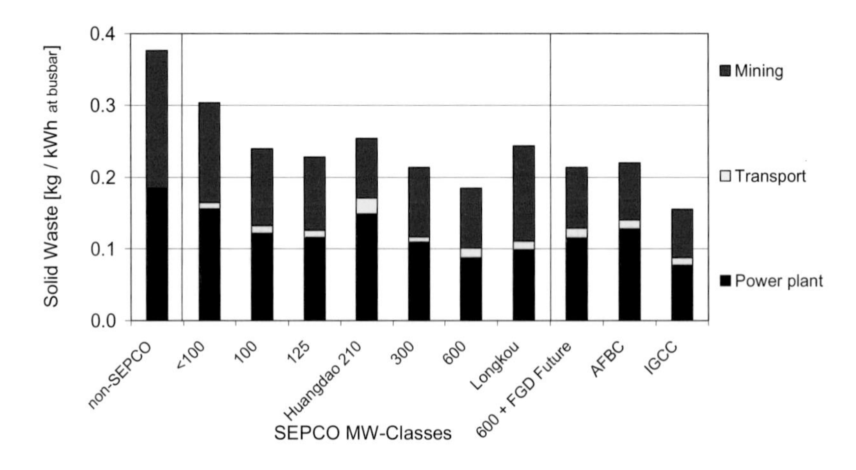

Figure 8.19 Solid waste production rates from chains associated with Shandong coal power plants in 1998 and technologies used for the calculation of future scenarios.

Figure 8.20 shows land use for the chains under consideration here. The differences are determined by the efficiency of the plant and the transport requirements. However, the values should not be compared too strictly, because they depend strongly on two key assumptions: first, the burdens at mines are assumed to be equal for all provinces, and somehow include subsidence (though this is not strictly a land use but a disturbance); second, the land use impacts per tonne of coal by kilometer transported by train has been imported from previous LCA studies for European conditions.

The land for transport should not meaningfully change from the present to the future. The area for mining and subsidence is not easy to reduce. Figure 8.20 does not show any effect of landfill disposal of the solid waste produced at the power plant. There is insufficient information on the current recycling rate of bottom ash in Shandong and the disposal strategy (including reclamation) of what is left.

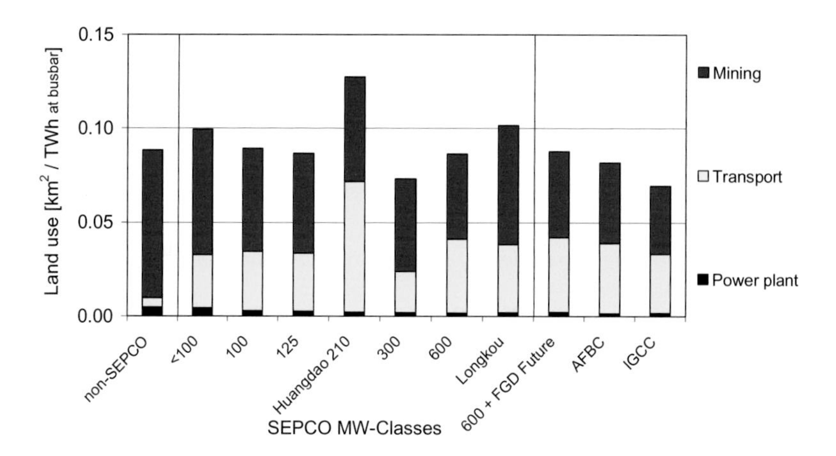

Figure 8.20 Land use rates from chains associated with Shandong coal power plants and mixes (1998 and 2020).

3. THE NATURAL GAS CHAIN

3.1 General

Natural gas will probably contribute more and more to the Chinese effort to diversify fuel use. When locally available, domestic gas will increasingly be used if competitive production costs are established. If neither gas, coal, nor oil resources are easily available, gas has higher chances to be imported (Girdis et al., 2000). However, the need to improve air quality in highly populated areas may promote the use of the gas either via pipeline (as for the Beijing area) or as Liquefied Natural Gas (LNG) when appropriate coastal terminals are available.

Shandong province benefits from considerable coal reserves, and hosts the largest oilfield in China, Shengli. While the production of coal in Shandong and the import of coal from nearby provinces have been steadily increasing, raw oil production in Shengli has stagnated since the end of the 1980's and has been steadily decreasing since 1992 (Wang, 1999). Therefore, while it could be argued that the conditions for a large gas market are not ideal, the forecast steady increase in the demand for energy in Shandong may call for growing use of natural gas in the province. Furthermore, Qingdao on Shandong's southeast coast has been identified as a potential LNG port along with Shanghai and Fuzhou in East China, and Guangzhou in the South (IEA, 2000).

The specific use of gas will influence the amount of pollution avoided. Gas should have the highest priority in the residential sector, where coal combustion is inefficient and produces high sulfur emissions, and in industry (e.g., fertilizer production). However, most of the gas will probably be consumed in the power sector, which appears to be the most realistic economic driving force in creating a gas market for other end users (Logan and Luo, 1999). Furthermore, while equally or less equally developing provinces nearby may compete with Shandong for gas pipelined from the north or west, the presence of pipelines facilitates the diffusion of gas. However, for this LCA study none of the above factors has been taken into account; pipelines and the LNG chain are assumed to supply power generation only. This somehow increases the environmental burdens.

3.2 Status and possible expansion of the natural gas chain in China

Numerous accounts of resources and reserves in China exist in the literature. As reported in Girdis et al. (2000), official Chinese sources estimate possible domestic gas resources at 51.7 trillion cubic meters (TCM), though proven reserves in 1996 were only 1.7 TCM. However, MN (2001) mentions a lower estimate of 38 TCM, including 29.9 TCM onshore and 8.1 TCM offshore (AGOC, 2000). According to Logan and Luo (1999), proven natural gas reserves in China range from 1.2 to 5.3 TCM, whereas according to BP (2001) at the end 2000 they were 1.43 TCM, less than one percent of the world total of 150 TCM. A similar value can be found in DOE/EIA (2001a), 1.36 TCM for China's proven reserves as of January 2001.

Most of the gas is onshore, in the Sichuan basin, in the Ordos basin,[26] and in the western provinces of Gansu, Ningxia, Qinghai, and Xinjiang. As reported in DOE/EIA (2001a), during 2000 China announced the discovery of a large natural gas field, estimated to contain more than 200 billion cubic meters (BCM), located in the northern part of the promising Tarim Basin in Xinjiang. Another announcement concerned the find of a massive field of approximately 400 BCM in the Ordos Basin.[27] Four pipelines are already transporting gas from the Changqing field in Shaanxi to Beijing (853 km), Tianjin, Xi'an and Yinchuan. In addition, there have been recent discoveries offshore in East and South China, among them a field in the Bohai Sea off the coasts of Shandong in July 1999 (Logan and Luo, 1999).

Currently recovered volumes of coalbed methane (CBM) in China amount to only about 0.5 BCM, but potential reserves are estimated at 30-35 TCM (Logan and Luo, 1999; Logan and Zhang, 1998). Besides the usefulness of methane as an energy resource, its extraction before mining improves mine productivity and safety, and decreases the amount of man-made GHG emissions by turning methane into carbon dioxide. In perspective, enhanced CBM recovery from deep coal seams, as well as oil recovery, could be achievable with the (costly) injection of CO_2 produced by power plants, thus further limiting CO_2 emissions per unit of electricity.

In 1997, China produced approximately 21.4 BCM of gas, more than 7.5 BCM of it in Sichuan. Total production in China is expected to be more than 30 BCM by 2005, according to official sources. More than 8000 km pipelines existed in China in 1997 (Girdis et al., 2000). A more recent source claimed a national total of 12000 km, with networks existing in Sichuan province and Chongqing, reflecting very rapid development (AGOC, 2000).

Figure 8.21 summarizes the most likely routes for supplying natural gas to Shandong.

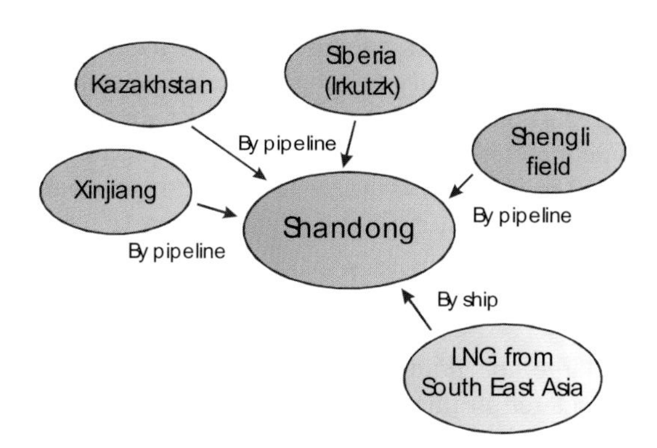

Figure 8.21 Possible future routes for supply of natural gas to Shandong.

Girdis et al. (2000) and Ivanov (1999) have reported the identification of two macro-regions that are potential exporters of pipelined gas to the economically fast growing and densely populated eastern coastal provinces of China. These are first, the Sakhalin Islands, Sakha Republic or Yakutia, Kovykta (Irkutskaya Oblast) in East Siberia, and the Krasnoyarsk region in northeast Asia and second, Turkmenistan, Uzbekistan, and Kazakhstan in central Asia, 6000 km from the east coast of China. These central Asian countries are trying to open up new markets to their east to reduce their reliance on Russian pipelines to export to Western countries. This study considers gas piped from the Irkutsk basin to represent supplies from North Asia, and the remote, and thus potentially expensive, gas piped from the Tarim basin in Xinjiang to represent the second region.

In February 2001 a Russian gas company and China National Petroleum Corp. signed an agreement to jointly develop the Kovykta gas field, 350 km north-northeast of Irkutsk (Girdis et al., 2000). The $6 billion project will deliver an initial flow of 12 BCM per year to Beijing, approximately 3000 km away. Other sources report higher investment costs, nearly $7 billion according to Logan and Zhang (1998) for a supply of gas up to 30 BCM per year. Though BP has identified several obstacles hindering the possible extension of the pipeline beyond Beijing (Girdis et al., 2000), South Siberia still remains the most attractive source of piped gas to Shandong, estimated for this study to be 4000 km away. Similar gas flows of 20-30 BCM/year would be necessary from the Central Asian region to keep costs at a competitive level. In 2000 the government announced a decision to allocate to the power sector 60 percent of the 12 BCM of gas output from a planned pipeline of approximately 4,200 km between Xinjiang and Shanghai. Its construction by PetroChina Co. began in the same year (MN, 2000; PD, 2001).

Worldwide LNG demand is expected to grow significantly over the next two decades. In 1995, the nominal liquefaction capacity of 20 plants amounted to 103 BCM/year. This is expected to grow by an additional 84 to 126 BCM/year by 2010. Geographically, China's coastal provinces have access to many potential sources of supply, including Australia, Indonesia, Malaysia, and Qatar (Girdis et al., 2000). It is forecasted that in year 2010 China may demand as much as 5.6 BCM/year, to compare to 98 BCM/year for Japan (Girdis et al., 2000; after

Cedigaz, 1996). LNG is an easier solution to implement than pipeline projects because of its modularity, flexibility of supply, lower investment requirements, and simpler contracts. LNG contracts are bilateral, unlike the multilateral agreements required when a pipeline crosses one or more countries between the producing and consuming countries (Girdis et al., 2000). A few modern gas combined cycle (GCC) plants burning LNG already exist in Guangdong province.

3.3 Modeling the natural gas chain for future electricity supply to Shandong

Considering the large uncertainties and the many source options for gas, a few cases have been considered in this study to build an average model to describe the use of the GCC and gas turbine (GT) technologies for base loads and peak loads, respectively. The guiding principle for modeling the total environmental inventories of the natural gas chain has been to estimate the orders of magnitude of a few selected environmental burdens rather than to make a detailed assessment. The accurateness of the description of plausible gas chains for Shandong is constrained by the lack of direct information. Generic data have been taken from the international literature. When it was unavailable, speculative but conservative assumptions (in the direction of maximizing the burdens) have been made by extrapolating figures from studies of other circumstances.

Figure 8.22 shows the chains defined for this LCA study; these correspond to the four routes supposedly most suitable for bringing gas to Shandong. These selected routes do not cover all possibilities, but represent typical distances that gas must travel to reach the province. Piping from the three basins in Shandong (offshore), South Siberia (Irkutsk), and Xinjiang, plus the option of using LNG shipped to the port of Qingdao from Indonesian fields, similarly to gas supply to Japan, were considered realistic options.

A technical solution suggested by one CETP stakeholder and considered by the ESS task, but not included in the MCDA scenarios, posits the construction of power plants in domestic gas fields that would supply electricity at long distance. To simulate such possibility, a chain with GCC located in Xinjiang and including a dedicated high voltage direct current (DC) transmission line is also discussed in a sensitivity analysis in a separate section.[28] This alternative is called "gas by wire" for short. The relevant chain is also included in Figure 8.22 for completeness. Factors essential for making this option possible, including the availability of cooling and service water on the site of the presumably large power plants, and the suitability of the dedicated line that should cross many other developing provinces also needing electric energy, are not addressed here.

The steps of a generic gas chain cause relatively low direct emissions of SO_2 and TSP and it is assumed that the manufacturing of the materials requires more energy than in Europe. We therefore calculate that the material use in pipelines and in the power plant contributes nearly 30 percent to total SO_2 and about 85 percent to total TSP emitted by the entire chain. Transport requirements by truck and train have been addressed using approximate values extrapolated from the Swiss LCA study for European conditions; these contribute negligibly to total emissions.

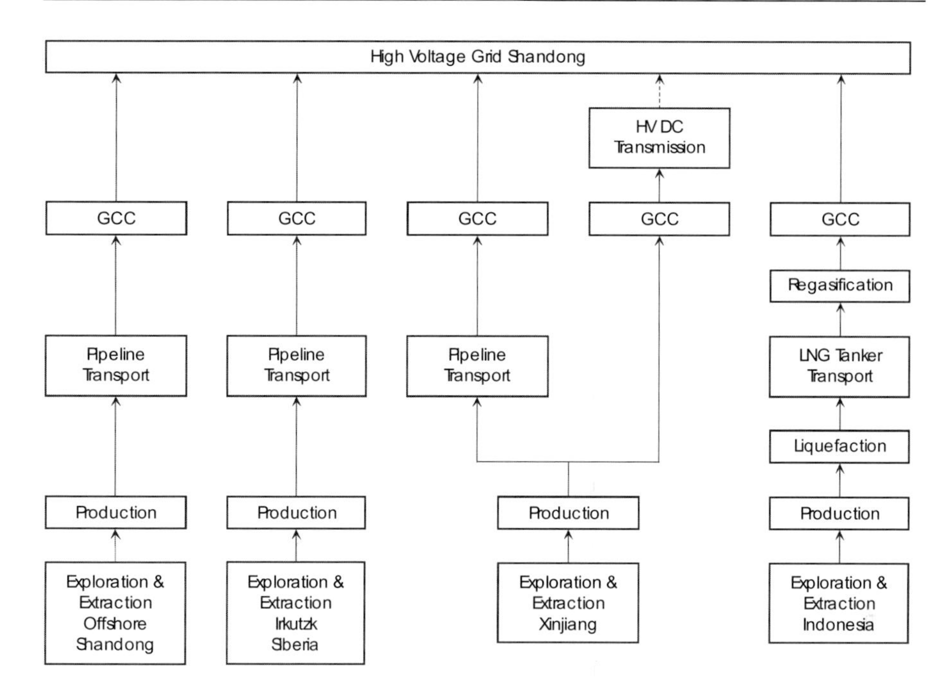

Figure 8.22 Natural gas energy chains modeled for future electricity supply to Shandong.

This study includes basically only energy self-supply burning part of the extracted gas. The few relatively small electricity uses in some stages of the gas chain have only a negligible influence on total emissions, except for TSP. However, even for this species the differences between emissions from gas and coal chains are so great that the influence of electricity uses can be ignored.

3.3.1 Exploration and extraction

Because information about location-specific values is not available, energy as well as emissions and solid waste have been assumed to be the same for all four cases. Using the Swiss LCA study for European gas power plants, the energy requirements are evaluated at approximately 0.028 kWh_{th}/m^3 of gas extracted. The direct methane emissions from leakage have been assumed to be 0.4 percent of the gas extracted. The direct particulate emissions from the operation approaches 10^{-4} g/m^3. Solid waste is estimated at about 14 g/m^3, land use barely $3 \cdot 10^{-7}$ m^2/m^3.

3.3.2 Production

The percent of sour gas is unknown for the regions considered. It has been assumed, very conservatively, that all gas requires treatment to extract the sulfur.[29] Despite this factor, the gas chain exhibits the smallest total emissions of SO_2 of all other energy chains. The energy requirements are assumed to be 0.065 kWh_{th}/m^3 of gas extracted. Methane losses are assumed at 0.2 percent of the extracted gas, the average of the values estimated for the Dutch and Russian production of natural gas;

this equals about 1.44 g/m^3 of gas produced. This and the gas released from extraction make up 0.6 percent of the total extracted gas, value adapted from EPA (1998). DOE/EIA (2001b) reports that a total 0.5 percent of the natural gas production in the USA was vented and flared in the period 1967-2000. Emissions of pollutants that occur from flaring include 10 g(CO_2)/m^3, 0.006 g(NO_x)/m^3, and 0.017 g(SO_2)/m^3. Land use is estimated at about $5 \cdot 10^{-6}$ m^2/m^3.

3.3.3 Pipeline transport

Gas turbines are used in pumping stations to transport gas long distances in high-pressure pipelines. The related gas use as well as the total gas leakage have been assumed to be proportional to the distance traveled. The approximate distances considered for gas piped from south Siberia and Xinjiang are 4000 km and 6000 km, respectively.

According to Gazprom/EPA (1996) and RFSHEM (1997) as reported in Popov (2001), the average current Russian transmission segment requires approximately 10 percent of throughput for its own energy needs. However, a potential for important improvement exists, for example substituting 25 percent efficiency old compressors with 36 percent efficiency modern ones, increasing the maintenance level, and replacing leaking components with low-leaking or non-leaking ones. These improvements have been credited for hypothetical future projects transporting gas to Shandong. From the literature, values in the range 1.4 percent to three percent per 1000 km can be found for gas burned for pumping (Frischknecht et al., 1996). The assumed rate for the gas used in GT at pumping stations is 1.5%/1000 km, assuming that in the future China will use the best technology available today.

The total mass of steel used for the pipe over its lifetime of 20 years has been divided by the total gas assumed to be transported. This includes 600 BCM from the South Siberian reservoir and 400 BCM from Xinjang (if this region fulfills the potential claimed by Chinese sources). To calculate land use for a partially above-ground pipeline a rate of 2000 m^2/km has been assumed, normalizing it by the total transported volumes of the gas for the two cases. The pipes are assumed to be left in place, and their steel considered solid waste. This is a conservative estimate.

3.3.4 Gas leaks for piped gas

No information could be gathered about losses and energy requirements for pumping natural gas in modern Chinese transmission pipelines. It was necessary to use the international literature to form an estimate of these two key factors. The focus has been put on the Siberian gas industry.

Popov (2001) has compared four studies of the amount of leaks from Russian pipelines. Two are governmental reports, RFSHEM (1997) and ICRFCC (1998), which present estimates for the entire Russian natural gas sector. Popov considers these accounts to be very uncertain because they are based on generic emission factors and on very aggregated sectoral statistics. Gazprom and USEPA conducted the third study on methane emissions from compressor stations in the Saratov and Moscow regions (Gazprom/EPA, 1996). However, a few technological emissions such as compressor exhaust and start-up and shutdown emissions were not included. The fourth and most recent study (Dedikov et al., 1999) was jointly performed by

Gazprom and Ruhrgas on all segments of the gas chain controlled by Gazprom. This does not include distribution inside Russia. The study made extensive measurements in 1996-1997 in North-Western Siberia. Researchers found that the methane losses from gas production and processing are relatively small, 0.06 percent of the total output. The study divides these emissions into leaks (0.02 percent) and intentional emissions caused by venting, depressurization, and repair work on wells (0.04 percent), but does not clarify how these values were extrapolated to the entire basin or to total production in Russia. The study assumed that 70 percent of the intentional methane releases are flared, but the resulting CO_2 emissions were not included in the total emissions. Leaks during transmission are on the order of 0.9 percent, including around 0.2 percent from venting during maintenance and leaks from valves, and about 0.7 percent from compressor stations. The measurements showed that a few major leaks account for most of the total methane releases. The estimate of one percent total (8,200 m^3/km/year, with 50 percent uncertainty) is consistent with preliminary estimates indicating maximum emissions at 1.5-1.8%/year (Grizenko, 1997).[30] Summarizing, the range of results from the four studies falls between nearly one percent (Dedikov et al., 1999) and less than three percent (from governmental reports) of natural gas production. A more accurate comparison is not possible, due to insufficiently reported details.

After long distance transport at the level of 70 bar above atmospheric pressure, the pressure is reduced to less than one bar in the regional distribution grid supplying power plants. The bulk of the losses from the gas sector are presumably occurring from local distribution networks supplying industrial and residential sectors at a pressure around 0.1 bar. Average leakage from modern West European high-pressure pipelines is about 0.02 percent for distances on the order of 1000 km, whereas in the low-pressure grid leakage may be up to 0.9 percent. Average values for older networks or for networks with insufficient maintenance may be higher, which could explain the difference of one to two percent in the Russian estimates discussed by Popov (2001). The distribution segment has not been explicitly modeled here, but included in the transport module.

Values estimated by Western industry for methane emissions from the entire natural gas supply chain (production, transmission, distribution, and use) range 0.1 to 0.5 percent of the gas consumed, according to IGU (1997) as reported by Dedikov et al. (1999). However, the estimate of EPA (1998) for US conditions is 0.4 percent for extraction and 0.2 percent for processing, as mentioned above, giving a total one order of magnitude higher than claimed by Dedikov et al. (1999) for Russia. Additionally, EPA estimated 0.6 percent losses from transmission and storage, and 0.4 percent from distribution, for a total of 1.6 percent for the sector (EPA, 1998).

In this study, the leakage rate during pipeline transport has been assumed to be one percent for the gas from Siberia, or 0.25%/1000 km. Application of the same rate to the gas from Xinjang gives 1.5 percent total leakage. The Shandong offshore fields are assumed to be 150 km far from the plant and the gas leak rate to be conservatively higher than for long-distance transport (0.2 percent of production).

3.3.5 Liquefaction to LNG

For the purpose of scenario analysis this study assumes that in support of policies for the expansion of gas usage of the order of several GW of GCC, countries producing gas will build large base-load liquefaction plants. These facilities will produce several thousand tonnes of LNG per day, mainly for transport by tanker.

A complete description of the existing processes for natural gas liquefaction can be found in Finn et al. (1999). The refrigerant and the type of refrigeration characterize these processes. The refrigerant may be part of the gas feed (open-cycle process) or a separate continuously recirculating fluid (closed-cycle process). Three main types of refrigeration cycle are used: cascade, mixed-refrigerant, and expander cycle. This study assumes that the mixed-refrigerant cycle (MRC) with a propane pre-cooling process will represent the average method the liquefaction industry will use to producing LNG for shipment to East China after 2010. The energy intensity required to produce LNG depends on the feed and ambient conditions, and on compressor efficiency. MRC requires a specific power of about 0.38 kWh per kg of LNG produced, a factor 1.15 greater than for a cascade cycle (Finn et al., 1999). The most widely diffused compressors in base-load liquefaction are centrifugal. Here it is assumed that these compressors are driven by gas turbines, which in real applications may be rated up to 80 MW, with an efficiency of 36 percent. The assumed energy requirement for liquefaction is approximately seven percent of the gas produced. This value is consistent with the current energy requirements in the range of eight to nine percent (EIA, 1997), when higher efficiencies are postulated for the future. No information was available about land use, which has been arbitrarily assumed to be three times higher than that for regasification because of the space needs of compressors and coolers.

3.3.6 LNG shipment

LNG tankers are very costly because of their double hulls and special lining (EIA, 1997). A small fraction of the cargo is used to keep the rest refrigerated by evaporation (boil-off). Around 0.15 to 0.25 percent of the cargo per day is used to power the ship, which will travel about 480 nautical miles. Assuming that the route from the liquefaction plant in Indonesia to the port of Qingdao is approximately equal to the route to Japan (about 2,400 nautical miles), the total self-consumption of the cargo can been estimated at 1.34 percent of the initial load (EIA, 1997). Were the LNG shipped from the Persian Gulf, the distance traveled and the associated gas losses would nearly treble. The calculation of the emissions from powering the tanker has been made assuming stoichiometric combustion of about 26 g/m^3 of natural gas-equivalent delivered, which with a gas to liquid ratio of 610 means 0.43 g/m^3 of LNG delivered. The emissions of air pollutants are derived in proportion to the emission rates of a standard gas turbine.

No information was available for methane losses during shipping of LNG. However, it is believed that it is a small fraction of the vaporized gas, arbitrarily assumed here at two percent.[31] Losses during loading and unloading of the liquefied gas at terminals could not be estimated; it is probably only a fraction of a percentage point. In considering the full recycling of materials, solid wastes from LNG tanker transportation have not been tallied. The land use associated with specialized harbors has also not been accounted for, but it is most likely very small considering their long lifetime.

3.3.7 Regasification

The regasification plants are located at LNG terminals, which also include storage tanks. According to EIA (1997) regasification utilizes 2.5 percent of the LNG delivered, which corresponds to 0.36 kWh_{th} per kg of gas produced. It is arbitrarily assumed that two percent of the gas necessary for regasification is leaked as methane. A land use of $5 \cdot 10^{-7}$ m^2/kg has been estimated on the basis of information contained in CMS (2001) referring to a US facility. Most of the material used for plant components is assumed to be recycled, hence not accounted for as solid waste.

3.3.8 Gas Combined Cycle power plant

Only the GCC technology has been assumed here because its very attractive characteristics make it the ideal candidate for a base-load gas power plant. Its net efficiency has been conservatively assumed at 55 percent, though values in the range 55 to 57 percent are reached today. The gas consumption rate is 0.1 m^3/kWh_{th} or 0.182 m^3/kWh at the busbar. Because of missing data on the composition of the gas, a LHV of approximately 50 MJ_{th}/kg or 36 MJ_{th}/Nm^3 is assumed for gas from all sources. Furthermore, the LHV of LNG may differ from the value of the original extracted gas because cooling to reach temperatures below the liquefaction point of about -162°C causes the separation of components such as oxygen, carbon dioxide, and sulfur compounds. However, again because there is no data on the composition of the gas to be liquefied, the same LHV has been assumed for the resulting LNG.

The specific emission of CO_2, assuming stoichiometric oxidation of the carbon in the molecule of methane, has been calculated at 360 g(CO_2)/kWh at the busbar. Except for NO_x with 0.196 g/kWh, the amount of other pollutant species emitted are relatively small: 0.04 g(CH_4)/kWh; 0.007 g(N_2O)/kWh; 0.003 g(SO_2)/kWh; and, 0.002 g(TSP)/kWh. Considering the compactness of the plant and its very high efficiency, the assumed land use rate is rather small, about $5 \cdot 10^{-7}$ m^2/kWh. The solid wastes from the dismantling of the power plant are assumed to be negligible because most of the material can be recycled.

3.3.9 Gas turbines and boilers used in the gas as well as in other energy chains

All turbines and boilers used throughout the gas chain have been assumed to have the same emission rates per unit of thermal energy considered for GCC.

3.4 Results

Table 8.17 shows selected results for chains associated with base-load GCC and the four routes of gas supply. In addition, the table shows the results of a sort of average chain, defined arbitrarily to include 20 percent gas from offshore Shandong (SD), 50 percent from South Siberia (SIB), ten percent from Xinjiang (XJ), and 20 percent LNG from Indonesia. This average chain will be used to compare gas with coal and nuclear. The table is completed by the results for gas turbines used for peak load management, supplied by natural gas coming half from offshore Shandong and half from South Siberia. However, a direct comparison between GCC and turbines is not very meaningful because the two technologies provide different services. Apart from the NO_x emission rate, which is high for the intrinsic characteristics of combustion for a GT, emissions depend on the net efficiency assumed for the GT (36 percent) and the chosen supply routes.

Table 8.17 Selected environmental burdens from energy chains associated with different natural gas supply routes to future base-load GCC plants in Shandong.

Burden		SD	SIB	XJ	LNG	Average for GCC	Gas turbine
GHG	g(CO_2-equiv.)/kWh	390	444	472	435	434	639
SO_2	g/kWh	0.042	0.047	0.052	0.049	0.047	0.058
NO_x	g/kWh	0.224	0.308	0.356	0.374	0.309	2.049
TSP	g/kWh	0.006	0.011	0.017	0.006	0.009	0.009
Solid wastes	g/kWh	2.8	3.2	3.6	3.3	3.2	4.5
Land use	km^2/kWh	0.0023	0.0052	0.0089	0.0029	0.0045	0.0072

[a] kWh at busbar

Figures 8.23 through 8.26 show selected results for the four origins of the natural gas. The variation percent for GHG relative to the minimum is 21 percent. Total emissions of CO_2 vary by 11 percent whereas methane can change by a factor of three. Considering that the GHG emission factor for the operation of GCC is fixed for all chains, the differences are roughly proportional to the distance from the source of the gas for pipeline transport. The calculated contribution of methane to total GHG emission reaches 14 percent in the chain for Xinjiang gas. The LNG chain exhibits lower GHG emission than the chain importing the gas into Shandong via pipelines.

These GHG emissions are here compared with the LCA estimates for current and future chains associated with European natural gas power plants. The GHG calculated in Frischknecht et al. (1996) for chains associated with the best performing (42 percent efficiency) single-cycle gas power plants burning only natural gas in the mid-1990's are in the range 540-660 g(CO_2-equiv.)/kWh, depending on the origin of the gas. The upper value reflects a large share of gas from Russia (up to 71 percent for Austria), including a two percent leakage of methane. Dones et al. (1996) calculate a total GHG of 391 g(CO_2-equiv.)/kWh assuming 60 percent net efficiency for GCC installed in Switzerland by the 2020's, and one third of the gas imported from Russia with relevant leakage reduced to one percent.

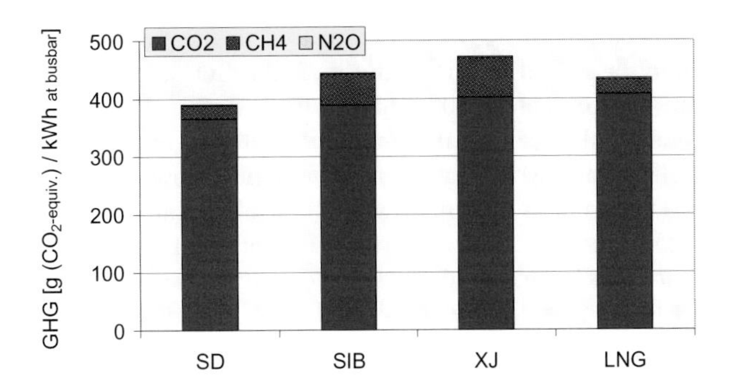

Figure 8.23 Greenhouse gas emission rates by main species per kWh at the busbar of GCC power plants from the natural gas chain for different origins of the gas.

As illustrated in Figure 8.24, emissions of SO_2 vary nearly 25 percent and NO_x nearly 66 percent, whereas TSP values change within a factor of three. These differences originate solely from the turbines utilized for powering the pump stations in pipelines and, in the case of the LNG chain, from the energy uses for liquefaction, transport and regasification. In case of pipelines, these differences are directly proportional to the pipeline length.

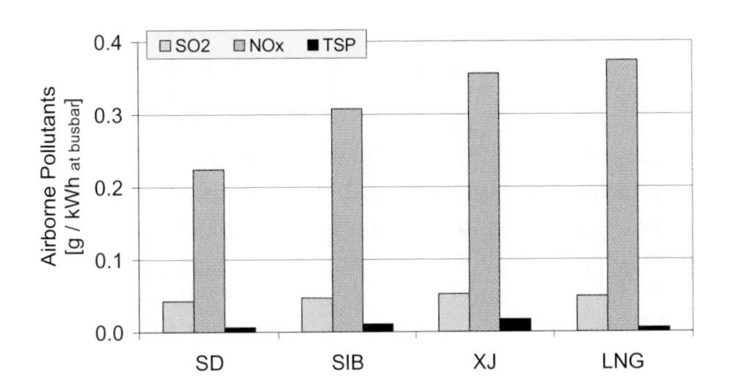

Figure 8.24 Emission rates of selected airborne pollutants per kWh at the busbar of GCC power plants from the natural gas chain for different origins of the gas.

However, it must be emphasized that unlike the quite reliable figures for the differences in NO_x based on average combustion characteristics and efficiency of turbines, the figures for SO_2 emissions depend directly on the assumed sulfur content of the extracted gas, and total SO_2 and TSP on the assumed emission factors for the GT. Considering that the key input values maximize the burdens, for example by using current rather than speculative future technologies and including sweetening for all gas, smaller values for SO_2 and TSP emissions can be expected.

The maximum SO_2 emission rate per unit of kWh from the gas chain is at least ten times lower, and TSP at least 30 times lower than the corresponding emission from the clean coal chain. However, compared to the future coal-based electricity mix in Shandong assumed in this study for 2020, SO_2 and TSP emissions from the gas chains are approximately 100 and 170 times lower, respectively. Furthermore, these emissions would occur at the production sites in remote areas and distributed along the pipelines, and would therefore have a much lower impact on human health than emissions from plants nearby or within densely populated areas.

Figure 8.25 shows the rate of production of total solid waste. The variation among the four cases is nearly 30 percent. Total solid waste from the gas chains is at least 65 times lower than for the coal-based electricity mix in Shandong assumed in this study for 2020.

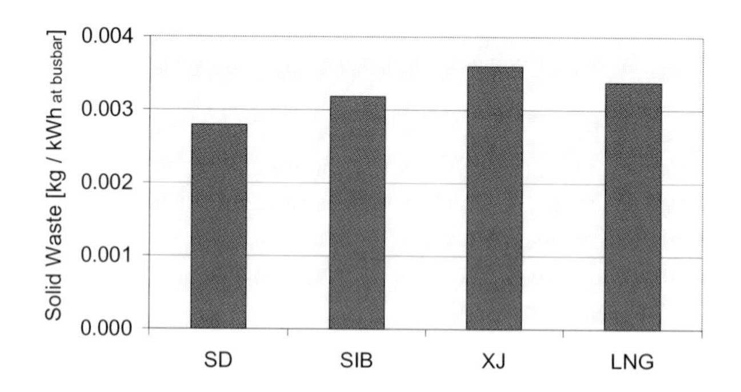

Figure 8.25 Solid waste production per kWh at the busbar of GCC power plants from the natural gas chain for different origins of the gas

Figure 8.26 shows the land use for the four routes. The highest value, for XJ, is nearly 3.5 times greater than the lowest, for SD, but the values depend on assumptions regarding pipelines; underground pipes would give no surface disturbance during operation. However, despite of the extreme-case assumption made for the XJ case, land use is still ten times less than that for the coal-based electricity mix assumed in this study for 2020.

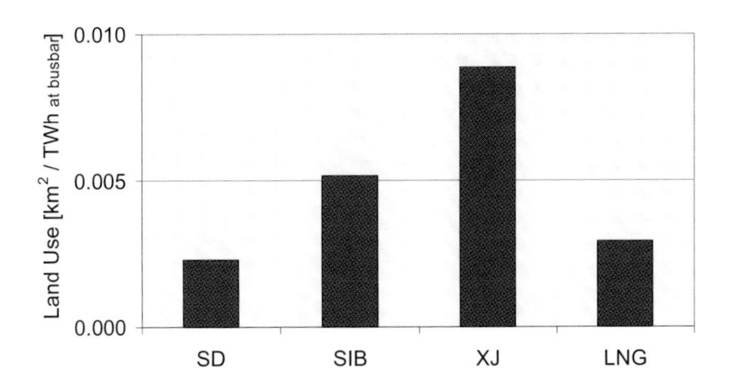

Figure 8.26 Land use per TWh at the busbar of GCC power plants from the natural gas chain for different origins of the gas.

Figures 8.27 through 8.29 compare contributions to total air emission rates per kWh at the busbar of future GCC plants installed in Shandong from the component steps in natural gas chains. Each figure includes synoptically four graphs illustrating the contributions for the chains of gas from offshore Shandong, South Siberia, Xinjiang, and LNG from Indonesia. It can be clearly seen that the length of pipeline transport increases the proportion of the contribution of this step to total emissions, with the Xinjiang chain contributing the most. The relative increase is dramatic for GHG emissions, particularly methane.

Liquefaction is the most energy-intensive activity in the upstream chain for LNG. Liquefaction, transport by LNG tanker, and regasification together produce about 43 g(CO_2-equiv.) per kWh at the busbar, less than the GHG emission emitted during transport from South Siberia (nearly 56 g(CO_2-equiv.) per kWh) or from Xinjinag (about 82 g(CO_2-equiv.) per kWh). The GHG emitted from exploration and extraction is mostly methane. The contributions from nitrous oxide are marginal—2 g(CO_2-equiv.)/kWh mainly from power plant operation.

Figure 8.29 shows that the largest emission of SO_2 occurs during production, when sulfur is separated from the gas. For the LNG chain, liquefaction, transport by LNG tanker, and regasification together contribute about 40 percent of total NO_x emissions. The bulk of solid waste is produced during exploration and extraction of the gas, though very long pipelines may contribute ten to 15 percent of the total if they are left on site at the end of their lifetime. No information on solid waste from the handling stages of LNG was available, but they presumably generate only small amounts. The share of land used by pipelines increases with increasing distance, up to 75 percent of the total for the chain originating in Xinjiang. The rest is shared about equally by the power plant and the extraction/production stages. In case of LNG, about 65 percent of land use occurs in the upstream stages.

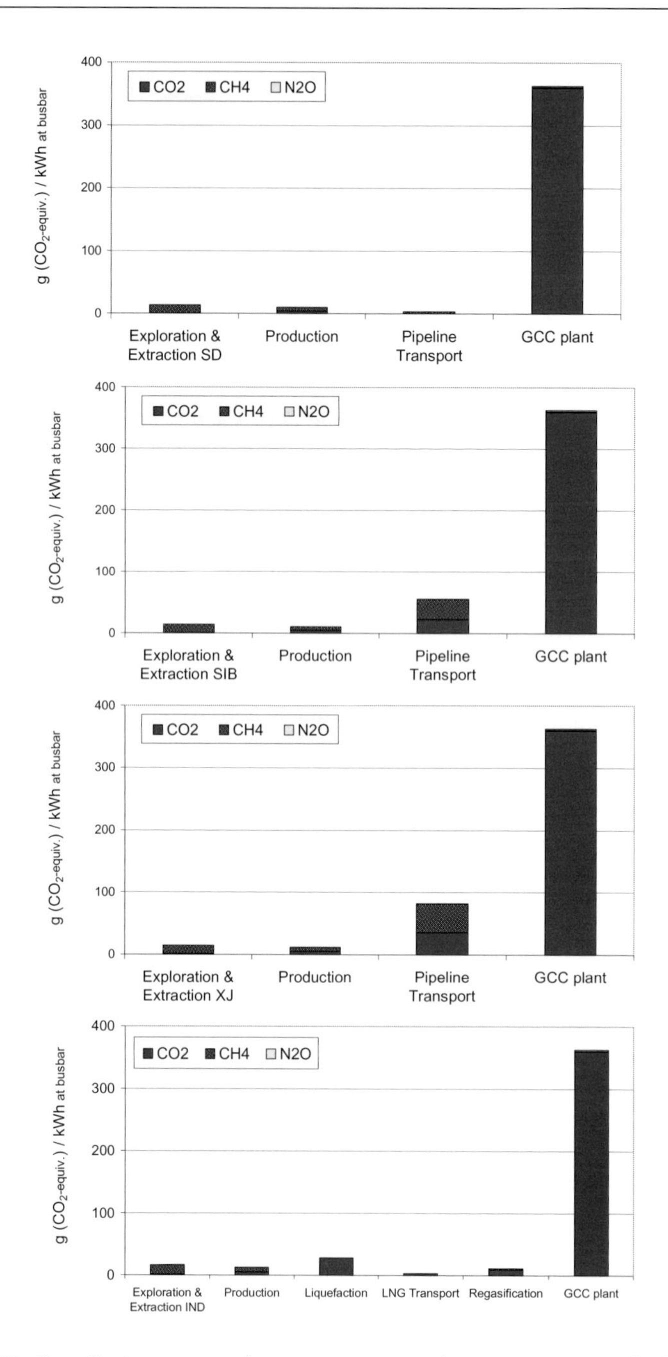

Figure 8.27 Contributions to greenhouse gas emissions by main species per kWh at the busbar of GCC from steps of the natural gas chain (starting from the top: gas from Shandong, South Siberia, Xinjiang, and LNG from Indonesia).

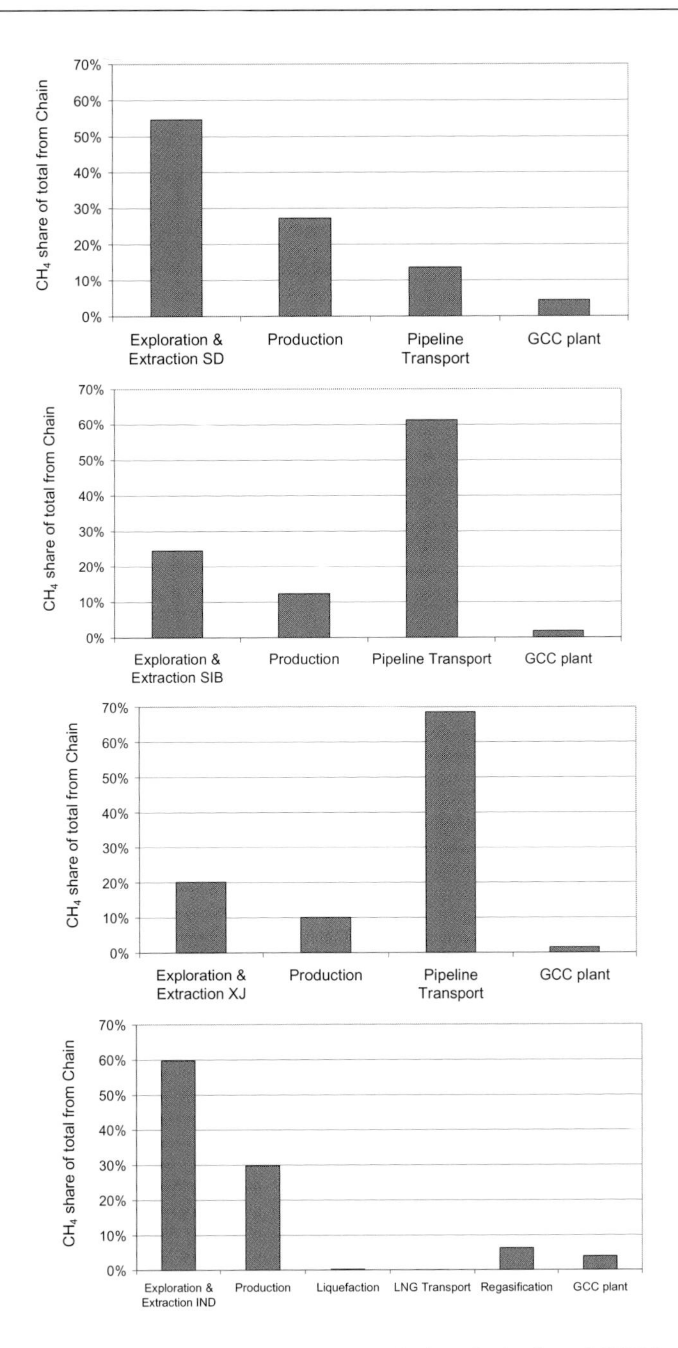

Figure 8.28 Contributions to methane emissions per kWh at the busbar of GCC from steps of the natural gas chain (starting from the top: gas from Shandong, South Siberia, Xinjiang, and LNG from Indonesia).

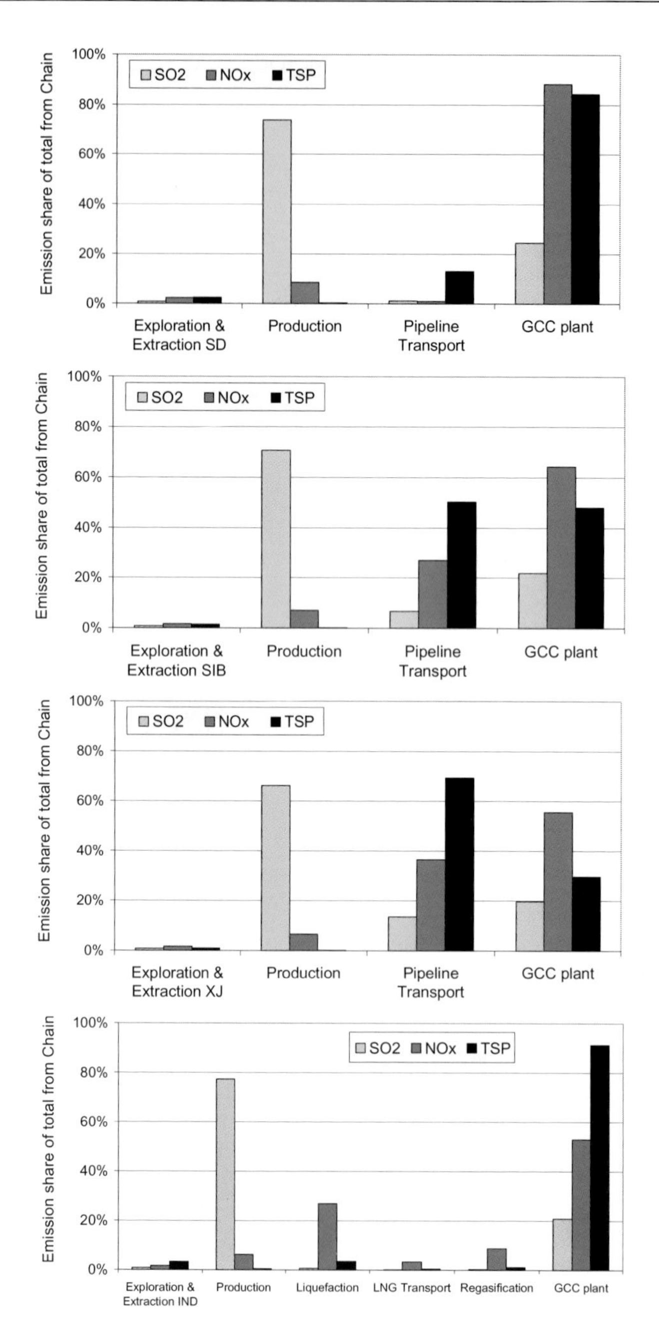

Figure 8.29 Contributions to selected airborne emissions per kWh at the busbar of GCC from steps of the natural gas chain (starting from the top: gas from Shandong, South Siberia, Xinjiang, and LNG from Indonesia).

4. THE NUCLEAR CHAIN

4.1 General

Considering the relative current and expected contributions from coal and nuclear generation to the Shandong electricity supply, and the relative impacts per GWh of these two energy chains, this task has placed a lower priority on finding original LCA data for the nuclear energy chain in China. In particular, direct connections with Chinese nuclear authorities, industry, and research were not featured in this project. However, a literature search on the present Chinese nuclear energy chain has been attempted to establish a reference point on which the characteristics of a future chain might be based. The rather scanty technical information that exists is insufficient for this purpose. Nevertheless, a generic chain could be reasonably defined for this streamlined study by adapting European and American data.

4.2 Status of the nuclear chain in China

The total civilian nuclear capacity installed in China in 1998 was 2268 MW (gross). Net nuclear electricity production was 14.15 TWh in 1999, 15.96 TWh in 2000, and 16.68 TWh in 2001, contributing slightly above one percent to China's overall electricity production. The average load factor for Chinese nuclear power plants (NPP) has increased from 74.6 percent in 1999 (NEW, 7-8/2000) to 84.2 percent in 2000 and 87.9 percent in 2001 (NucNet, 2002).

In 2001 three commercial NPP's were operational. One is a domestically designed, constructed, and operated 300 MW PWR unit in Zhejiang Qinshan (Phase I). It was first connected to the grid in December 1991, and reached full power in July 1992. Two 984 MW Framatome PWR units were installed in Guangdong Daya Bay; Unit 1 began commercial operation in February 1994, and Unit 2 in May 1994. Qinshan Phase II includes two domestically designed and built 650 MW PWRs; the first reactor was connected to grid in February 2002, the second is scheduled for connection at the end of 2002. Six more units are under construction at three different sites, totaling 5544 MW. Qinshan Phase III comprises two 728 MW units of the AECL CANDU-6 PHWR type; they are scheduled to begin operation in 2003. In Guangdong Ling Ao, two 984 MW Framatome PWR units are scheduled to enter service in 2002 and 2003, respectively. At the third site, Jiangsu Lianyungang Tianwan (north of Shanghai), the first of two improved units originally of the Russian-designed 1060 MW VVER 1000/428 NPP-91-type, will be completed in 2004 (NEW, 7-8/2000).

Several commercial projects have been proposed (NEW, 3-4/2000). Only one, with 2×1000 MW units to be build in Haiyang, concerns Shandong. The Guangdong nuclear power group has proposed the construction of six 1000 MW units in Yangjiang county; their construction could begin in 2002 with first power production scheduled for 2009 and completion of last units six years later. Further proposed plants include units for Qinshan phase IV; 2×1000 MW in Zhejiang

Sanmen; and, 2×1000 MW in Hui An (Fujian). Current plans foresee an installed capacity in China between 40 and 50 GW by the year 2020.

According to Qian (1999) China's nuclear chain will depend on its own uranium supply, minimizing dependence on foreign trade. However, Chinese authorities have not provided data on domestic uranium production levels sufficiently detailed to determine whether this is a realistic goal. The Uranium Institute (UI)[32] (Kidd, 1999) estimated that cumulative production amounted to about 3000 tonnes of uranium, mostly for military purposes, of which 500-1500 tU were exported in the late 1980's, while NEA/IAEA (2002) reports a cumulative production of 7435 tU between 1990 and 2000. The reserves are estimated by the UI at around 67000 tU, with current annual production believed to be 500 tU (Kidd, 1999), while NEA/IAEA (2002) reports a production of 700 tU for both 1999 and 2000, known resources at 73000 tU, and undiscovered conventional resources at 1.77 MtU.[33]

As described above, China should have at least nine GW of nuclear power available before year 2010. The UI has posited three scenarios for Chinese uranium requirements for power generation (Kidd, 1999). In the first, at the high end, two units a year from 2005 on beyond the ones already planned would produce 30 GW by 2020, requiring 3000 tU by 2010 and 5500 tU by 2020. The low-end scenario, in which installed nuclear capacity remains constant between 2010 and 2020 at the level achieved in 2010 (about 10 GW), would require 1700 tU per annum after 2010. The UI also developed a reference scenario with output and requirements falling between these two. It is believed that virtually no deposits discovered up to now could be mined economically by Western standards, considering current low market prices for yellowcake, as China has only scattered, small, and low-grade seams (Kidd, 1999). Chinese sources assert the increasing utilization of less expensive chemical mining techniques (Lu, 1999). However, international experts argue that it is hard to imagine that China would reach a domestic production of 1000 tU per annum. The UI concludes that China can hardly remain self-sufficient in uranium supplies unless its nuclear capacity does not grow beyond the existing and planned capacity (Kidd, 1999).

Technical information on uranium mining and milling as well as other industrial activities in the Chinese nuclear chain is scanty. Conventional mining has been done for a few decades at Hengyang (this mine is now in stand-by) with a nominal capacity of 500-1000 tU/a, at Fuzhou with 300 tU/a nominal, and Chongyi with 120 t/a nominal (NEA/IAEA, 2002). However, several additional, scattered and small Chinese uranium deposits exist, with relatively low-grade ore that are thus expensive to exploit by conventional methods (Wang and Dai, 1993). Since the early 1980's the focus of the exploration for uranium resources has shifted from the granitic, carbonaceous-siliceous-pelitic and volcanic deposits in the south to the sandstone deposits in northern and northwest China (e.g. the reserve base in the Yili Basin, Xinjiang, accounts for 9000 tU). These resources can be exploited using chemical extraction. The corresponding reserves are only about 25 percent of total (NEA/IAEA, 2002). Chemical extraction techniques in China use in-situ leaching (ISL), including underground stope leaching following blasting, and pile leaching (heap leaching following acid curing). New production centers already established include the Yining (Yili Basin, Xinjiang) ISL facility (200 tU/a nominal capacity), the Lantian (Shanxi) heap-leaching facility (100 tU/a nominal, 2000 tU total), and the Benxi (Liaoning) mine, which uses heap leaching following acid curing (120 tU/a nominal). These three sites have increased production from 300 tU in 1998 to 320 tU in 2000. Five small uranium mines have been decommissioned, and eight mine-mill complexes are in various stages of decommissioning (NEA/IAEA, 2002).

Very little information could be gathered on uranium conversion. The Jiunquan (Subei, Gansu) Atomic Energy Complex converts enriched UF_6 to UF_4 for further processing into metal (CEIP, 1998).

Information about uranium enrichment at Chinese facilities is incomplete and contradictory. However, the trend is similar to what has happened elsewhere: use of the existing but obsolete and highly energy intensive gaseous diffusion technology, with gradual substitution by centrifuge enrichment (Lenders, 2001); the laser technique is being studied (Qian, 1999). One source (CEIP, 1998) mentions the following large installations: a gaseous diffusion plant in Heping (Sichuan), operational since 1974 (FAS, 2000); a gaseous diffusion plant in Lanzhou (Gansu); a new diffusion plant reported to be under construction in Lanzhou, for export of low enriched uranium; and a Russian-supplied centrifuge plant under construction, with a capacity of 200000 SWU/year in Chengdu (Sichuan).[34] Another source that explicitly refers to the Chinese military program (Zheng, 1996) mentions three enrichment installations: the Lanzhou plant and two in Helanshan (Ningxia Hui).

According to FAS (2000), Lanzhou was the site of the first gaseous diffusion enrichment plant, which began operation in 1957, and probably provided the highly enriched uranium for the first military nuclear test in October 1964. The plant appears to be connected to thermal coal and hydroelectric power stations. An intergovernmental agreement was signed in December 1992 with Russia for nuclear projects in China. The first phase of construction by Russia of a gas centrifuge enrichment plant with a capacity of 200000 SWU/year in Chengdu (Sichuan) was completed in 1996. An intergovernmental protocol within the above-mentioned agreement, signed in December 1996, envisaged two additional gas centrifuge plants in Lanzhou, to be commissioned in the mid 2000's, complying with IAEA safeguards (FAS, 2000). Some sources assert that a third centrifuge facility is under construction in Hanzhong (Shaanxi) (FAS, 1997), to be operational in the early 2000's, also subject to IAEA safeguards. In contradiction with the above, Qian (1999) mentions that the first centrifugal plant introduced from Russia has been in operation since April 1999 at Hanzhong.

According to (DOE/EIA, 2000), CNEIC has delivered uranium concentrate and enrichment services to US utilities since 1997. In particular, in 1999 China provided 145000 SWU, i.e. 1.4 percent of total US purchases of enriched uranium.

In 1987, the first production line of fuel assemblies for PWR was put into operation in Yibin (Sichuan) to support the initial core and the refueling of the 300 MW reactor Qinshan I. Fuel assemblies have also been exported to Pakistan. In 1991, the factory (CNEIC) introduced Fragema technology from France. As a result of this co-operation, the fuel fabrication plant has been manufacturing all reloads for the Daya Bay units since 1994. It complies with internationally advanced manufacturing standards, and is being back-fitted to produce high burn-up fuel.

Preliminary work has started in order to produce fuel also for PHWR and VVER (Watteau et al., 2000). In addition to Yibin, another source (CEIP, 1998) mentions the Nuclear Fuel Component Plant in Baotou (Mongolia).

A decision on pursuing the reprocessing of commercial spent fuel, which would close the nuclear fuel cycle, was made in the mid 1980's (Qian, 1999). A pilot reprocessing plant has been under construction at Lanzhou since 1994, and is scheduled to begin hot-operation in the 2000's with a nominal capacity of 100 kg heavy metal (HM) per day (Qian, 1999; NIR, 2000; FAS, 2000). A pilot-scale reprocessing plant using the Belgian PAMELA-type technology is being built at the Sichuan Nuclear Fuel Plant (NIR, 2000).

In 1992, SEPA issued "The Policy and Principles on Disposal of Low- and Intermediate-level Radioactive Wastes" with priorities defined as follows: radioactive wastes for temporary storage shall be solidified as early as possible; in principle, long-term storage facilities for liquid waste at NPP's will not be approved; temporary storage of low- and intermediate-level radioactive liquid as well as solid wastes from NPP operation is currently limited to the duration of five years; regional repositories for disposal of low and intermediate-level radwastes shall be built (Pan and Qu, 1999). Radwaste conditioning includes vitrification of high-level waste (HLW) and solidification of intermediate- as well as low-level wastes (ILW, LLW) using cement or bitumen matrixes. A vitrification facility (possibly a pilot plant) began cold testing in 1996.

It is reported (NIR, 2000) that four regional sites have been proposed for near-surface disposal of solid conditioned LLW. Two sites have been selected to date, one in Lanzhou, the other adjacent to the Qinshan NPP in the south. Sites for repositories in Northwest and South China were approved by SEPA in 1994 and 1995.[35] The northwest repository began operation in 1999, while the southern repository in Beilong (Guangxi) should have started in year 2000 (Pan and Qu, 1999).

A long-term research and development program on deep underground disposal for HLW has been underway since 1986, including survey of geological formations and experiments with nuclide migration (Pan and Qu, 1999). Since 1989, work has focused on potential sites in the Beishan area (Gansu), one of 21 potential candidates originally considered (NIR, 2000). The initial goal of confirming the site as the final repository by 2010 has turned out to be unrealistic. The new target for repository construction is now 2050. According to (NIR, 2000), a Central Spent Fuel Storage Facility is under construction in Lanzhou, with a design similar to the Swedish CLAB facility.

4.3 Modeling the nuclear chain for Shandong electricity supply scenarios

A clear picture could not be firmly established for the locations, installed capacities, and throughputs for all nuclear chain facilities, as summarized above. Moreover, in the open international literature there is a total lack of data on material and energy requirements, as well as radioactive and non-radioactive emissions for operational Chinese nuclear facilities, which hinders the establishment of even a simplified LCA study. However, the aim of CETP was to analyze electricity supply scenarios in the

Shandong province for 20 years to come, not to study individual energy chains in great detail. Furthermore, if any nuclear reactor ever does supply electricity to the Shandong grid it will most realistically occur in the second decade. By then, taking into account the opening of international markets and beneficial effects of international cooperation, the best-practice technical standards will be introduced throughout the entire chain. This has already been demonstrated in the construction of the fuel fabrication plant in Yibin in partnership with Fragema; the cooperation with Framatome for the Daya Bay NPP; with Westinghouse for refurbishment work at Qinshan; and the collaboration with Russian companies in building gas centrifuge enrichment plants.

The analysis of future nuclear scenarios for Shandong has been accomplished by adapting models and data already developed for different contexts. Therefore, the results obtained here for the future Chinese nuclear energy chain should not be accepted uncritically, nor used in other comparisons, but applied only in the context of scenario analysis. This original LCA study effectively provides an initial evaluation of the relative differences between nuclear and other energy chains. In the following, the modeled energy chain will refer to China rather than Shandong, because the various activities composing the chain occur throughout the country (only power plants would be sited in Shandong, according to the available information), and because natural uranium is internationally traded.

In order to define the average future nuclear energy chain for China and estimate the selected environmental burdens, a few pivotal elements must be defined. The most important elements from the point of view of LCA are: the type of power plant and its operation; the mining methods and their share of the total; the process used for enrichment and the electricity source for its supply; and whether the spent fuel is reprocessed, i.e. whether the chain is an open or closed cycle.

Considering Chinese priorities, the most probable reactor candidate is of the LWR type (Qian, 1999), either of Western or Russian technology. From the point of view of environmental inventories and on the basis of the results of the Swiss LCA studies, no substantial differences are expected between the burdens incurred by PWR's and BWR's, either for the current or more advanced models. Therefore, the analysis has been made assuming a modern high performance PWR.

Although China's declared policy seems to pursue the reprocessing of spent fuel, there is no information available stating whether the uranium and plutonium extracted are or would be recycled as MOX fuel in reactors. Considering also the missing information on reprocessing technology in China, an open cycle has been assumed here as basis. In principle, an analysis of open cycles should point higher burdens, thus making the conclusions of the comparison with other energy chains more defensible.

Figure 8.30 shows a flow chart of the assumed future Chinese simplified nuclear chain for LWR's.

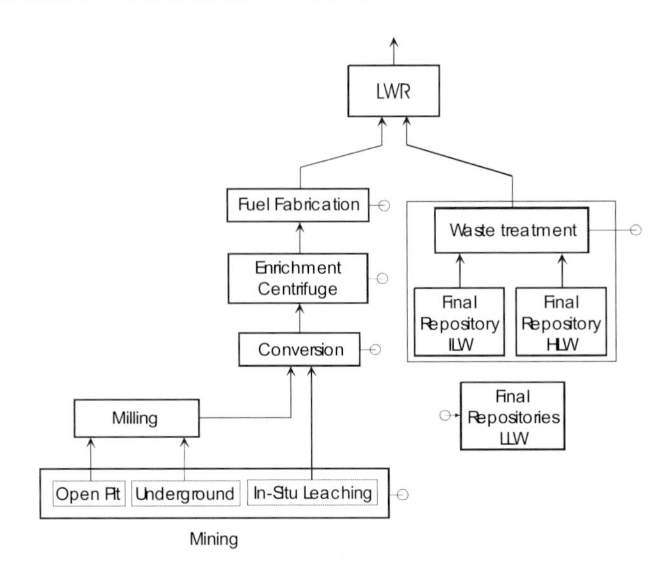

Figure 8.30 Model of the nuclear open cycle assumed as reference for China.

In the figure, the upstream part of the chain is depicted on the left hand side; the downstream part is on the right hand side. Both parts contribute environmental burdens to the electricity finally produced at the power plant (hence the direction of the arrows in the picture).

Three different mining processes, two mechanical (open pit, underground) and one chemical, have been individually modeled. The reason lies not only in the considerable differences of environmental burdens, but also in the fact that mechanical mining needs to be followed by milling the ore to concentrate the uranium into yellowcake, whereas with the ISL technique the uranium is directly obtained by chemical separation. Enrichment of the fissile isotope U-235 is assumed to be obtained exclusively by centrifuge. Sensitivity analyses have been performed on this type of enrichment process as well as on the electricity supply mix. The results are discussed in the section on sensitivities.

For the downstream processes, the waste treatment assumed in this base nuclear chain involves the encapsulation of HLW, i.e. the unprocessed spent fuel elements, and the conditioning of ILW including operational waste and that from decommissioning the power plant. Because the Swiss LCA study shows that the conditioning plant contributes very low burdens relative to other steps of the chain, it has not been explicitly modeled in this study. The final repositories of ILW and HLW have been lumped into one module. The depositories for LLW are considered in a rather simplified form, because of their minimal material and energy requirements relative to the whole chain.

Because important information about radioactive emissions to air and water as well as about the production of radioactive wastes from the existing Chinese nuclear installations is missing, they will not be addressed in this study. However, because of the increasing level of international industrial cooperation, and Chinese desire to comply with IAEA standards, these emissions should be of the order of magnitude assessed for Western nuclear chains in the future. Several studies comparing different energy sources have determined that the external costs from a nuclear chain for European LWR's mid of the 1990's (including short-term and long-term radioactive emissions as well as non-radioactive emissions) are about one order of magnitude lower, per unit of electricity supplied, than the external costs from European modern coal systems (Rabl and Spadaro, 2001; ExternE, 1999). The difference should remain high with future technologies for European conditions (Hirschberg et al., 2000). Furthermore, average Chinese coal contains substantially higher radioactivity than coal used to fire European power plants, to the extent that the collective dose from PC operation and ash recycling may be much higher than the collective dose from the nuclear chain. In the preliminary assessment performed by Pan et al. (1999), this difference amounted to a factor of about 23. However, this factor disregards long-term exposure to uranium mill tailings and radioactive coal mine emissions (radon, etc.). Probably, the two chains may have comparable effects from total radioactivity. Therefore, while this LCA study cannot pretend to be complete from the point of view of environmental burdens, it can be argued that it provides an adequate first ranking of energy chains. Nevertheless, it is highly recommended that follow-up studies on Chinese energy chains should also address radioactive emissions to air and water.

The transport requirements have been included based on the travel distances estimated for chains associated with European LWR's. Several reasons justify this simplified assumption. First, the very high energy density of the uranium fuel means that relatively low masses must be transported per unit of produced electricity, especially when compared with the huge transport requirements for coal. Second, distances traveled by nuclear fuel in china may not differ from the long range transport from North American nuclear facilities in the upstream part of the chain (mining/milling, conversion, enrichment), as well as the long distances traveled within Europe between enrichment plants and fuel fabrication plants, and between LWR's and reprocessing plants (in France and UK), between these and the intermediate depository and after to the final repositories. A third factor is that intensive development of nuclear power in China will require access to the international market of natural uranium. The results obtained with and without transport requirements defined as sketched above are two percent for GHG, three percent for NO_x (a little higher than GHG due to emissions from trucks), but below one percent for SO_2 and TSP. Any error introduced by not knowing the specific transport distances between Chinese nuclear facilities should not affect the results.

The electricity supply for all needs throughout the nuclear chain has been assumed to follow the simplified LCA future scenario, at medium voltage level. This possibly introduces an overestimation of the emissions of pollutants associated with electricity use, because the average future electricity mix in China will also include a meaningful share of hydropower and a possibly growing nuclear capacity. Also other countries providing nuclear services may have less polluting power technologies, and will most probably rely less on conventional coal units than China.

4.3.1 Mining

Modeling mining has multiple difficulties. On the one hand, the characteristics of conventional surface and underground mining from the point of view of the environmental burdens per unit of product uranium may dramatically change with the geology and climate of the location and the ore grade of the seams. On the other hand, the reclamation practices for mines (and mills) may greatly influence the long-term effects from radioactive releases, in particular of gaseous Rn-222, acids, and heavy metals in leachates to water bodies. Chemical mining by ISL may affect groundwater but significantly reduces radioactive solid waste per unit mass of uranium product. Therefore, it is impossible to define one standard mining activity and hence highly speculative—indeed, indefensible—to build one LCA module for current uranium mining for supply to Chinese reactors. In addition, it is impossible to neatly define the shares of the three mining types today and in the future. The only information gathered concerned the Chinese policy of increasing ISL, with the target of up to 50 percent of the future domestic production of natural uranium.

In this LCA study the environmental burdens of open pit and underground mining as modeled in (Frischknecht et al., 1996) were used, without any extrapolation to the Chinese context nor to possible future supply scenarios for uranium, because of complete lack of any basis for justifying it. These data reflect worldwide averages and have been derived from the literature.

Values for open pit mining used as main inputs to the present analysis, given per kg of natural uranium, are: 16 kWh_{th} of diesel consumed for powering the mine; 56 g of particulate matter emitted from mining operations excluding combustion; and land use of 0.2 m^2. Corresponding values for underground mining, given again per kg of natural uranium, are: 83 kWh_{th} of diesel fuel; 40 g of particulate matter; and land use of 0.007 m^2. For both mining techniques, the solid wastes are assumed to be used to back fill the pit, and the mine area to be reclaimed. Hence, the bulk of the radioactive solid wastes from mining are considered to be concentrated at mills.

It seems that no LCA study exists yet on ISL. Moreover, only a few attempts at describing the practice and some of the effects of chemical mining on the environment around the world have been made (Mudd, 2000a, Mudd, 2000b). Even if it were possible, no unified average picture exists yet. Investigations on ISL are recommended for future work. At any rate, considering the importance of ISL for Chinese conditions, a simplified module has been included. Its input data are based on engineering judgment on the order of magnitude of specific requirements, relative to other mining and milling methods. Arbitrary assumptions for ISL mining for the present analysis are: one tenth of the energy intensity of underground mining and milling; no particulate matters emitted from the process; land use double that for underground mining; and LLW produced is on the order of five percent of radwaste from conventional milling.

The shares assumed for the three extraction methods are based on a rough extrapolation from the single piece of information gathered on Chinese policy goals for uranium mining: 50 percent of ISL and the rest divided between open pit and underground mining according to the distribution of these techniques worldwide today, i.e. 23 percent and 27 percent, respectively.[36] As explained before, use of domestic uranium alone is highly unrealistic because of the limitation of Chinese resources, the scattered distribution of small seams, and the current availability of low-priced uranium from the international spot market.

4.3.2 Milling

Milling generates some important burdens that characterize the nuclear chain, in particular the tailings (LLW) and the long-term airborne emissions of Rn-222 from them. These aspects have been addressed in the Swiss LCA study, and partially used here. Main input values to the chain calculation are: nearly 140 kWh_{th} of gas/diesel per kg product natural uranium in yellowcake; land use of 0.1 m^2 per kg natural uranium; tailings are 0.25 m^3 per kg natural uranium, which represents the largest contribution to total radioactive waste volume for the chain, though much less potentially harmful than ILW and HLW. Milling operations generate meaningful particle emissions, estimated at 220 grams per kg of natural uranium in yellowcake.

4.3.3 Conversion

Although no information could be collected for the Chinese nuclear chain about this step, in which the yellowcake is converted into UF_6, it has been included in the chain for completeness. The main input values for the base chain, given per kg product natural U in UF_6, are: 10.3 kWh electricity; 198 kWh_{th} of diesel/gas; land use of $5.5 \cdot 10^{-4}$ m^2; and LLW of $6.1 \cdot 10^{-4}$ m^3.

4.3.4 Enrichment

Centrifuge appears to be the technology most likely to be employed for enrichment. This process is progressively gaining market share worldwide over diffusion.[37] It entails much lower energy intensity; lower costs; flexibility in obtaining different degrees of enrichment for different batches thus meeting the specifications of the customers; and smaller plant dimensions. Additionally, the diffusion plants worldwide are aging, having been built in the 1950's-1970's.

In this study, the electricity intensity has been defined rather conservatively at 100 kWh/SWU in order to cover possible differences between the Russian technology used in China and the Urenco plants in Europe. Due to lack of information on the Chinese plant(s), the value used here is double the consumption intensity of modern Urenco plants (Urenco, 1995). Other key factors assumed, all given per unit of SWU, are: 1.62 kg natural uranium in UF_6 from conversion; 130 kWh_{th} of diesel oil for boilers; land use of 0.05 m^2; LLW of $2.6 \cdot 10^{-3}$ m^3; and about 1 kg of non-radioactive solid waste.

Chlorofluorocarbons (CFC) and Hydrofluorocarbons (HFC), gas with relatively high global warming potential (GWP) used in enrichment processes as cooling fluids, have also been included in the assessment, despite the lack of information about Chinese and Russian technology. Conservatively, the emissions have been assumed to be double the upper value for the Urenco plants in the mid-1990's, giving about 26 g(CO_2-equiv.)/SWU. However, the stratospheric ozone-builder gases are banned by the Montreal Protocol and they must be substituted by less harmful ones, though these may still have high GWP. These changes have not been credited, because contribution of CFCs and HFCs to the total GHG calculated for the base nuclear chain is of the order of only one percent.

4.3.5 Fuel fabrication

To obtain one kg uranium enriched 3.7 percent into the fissile isotope U-235, 4.91 SWU are necessary. Other main input factors per kg of enriched uranium in fuel assemblies are: 22.2 kWh electricity; nearly 28 kWh_{th} of diesel oil; land use of $4 \cdot 10^{-3}$ m^2; LLW of $2.7 \cdot 10^{-4}$ m^3; and about 2.1 kg of non radioactive solid waste.

4.3.6 Nuclear power plant: PWR

The LWR has been assumed as the reference technology for the nuclear chain. However, the need to model a major nuclear policy up to the year 2020 could involve consideration of conventional (Western PWR and BWR, or Russian VVER) and advanced LWR's (for example AP600, AP1000,[38] APWR, EPR, ABWR, and ESBWR). This would require the modeling of several reactor types. Therefore, it could be argued that modeling only one type would not be sufficient for scenario analysis. However this study has only included one reactor type, because the conservative and defensible results are sufficient to make a correct comparison of the relative differences between the major fuels, useful for decision-making.

Because any reactor built in Shandong around the end of the 2010's would probably be a conventional LWR, a currently operating Western PWR of 1000 MW capacity with best operational performance has been adopted as reference. This reactor is fuelled with 3.7 percent enriched uranium, has relatively high average fuel burn-up of 43 MWd_{th} per kg enriched uranium, net efficiency just above 31 percent, and a corresponding requirement of nearly 3.2 kg enriched uranium per GWh. This choice leads to a general overestimation of the burdens associated with material and energy uses throughout the nuclear chain, because advanced LWR's will have higher efficiency (about 33 percent net), higher fuel burn-up (some models up to 55-60 MWd_{th} per kg enriched uranium), lower fuel requirements (below three kg of enriched uranium per GWh), and longer designed operational life (60 years vs. 30-40 today). However, a reduction of the burdens would just further improve an environmental performance already much better than that of coal chains.[39] Hence, the choice of a conventional PWR matches the goal of this study.

Main materials used in the construction of the nuclear power plant (as well other installations throughout the chain) have been included in the analysis. However, they contribute marginally to total burdens, e.g. only three percent to total GHG. Also the electric energy of non-nuclear origin and the fuels invested for the construction of the plant, as well as the fuel for testing the emergency diesels have been accounted for, making the bulk of the indirect emissions associated with the power plant (these correspond, for example, to one third of the total GHG from the nuclear chain).

During its operation, the power plant produces HLW for approximately 0.0038 m^3/GWh. This number has been generated by hypothesizing a simple casing or cylindrical cask containing four spent fuel assemblies, to be disposed of directly in this form in deep geological strata. The volume includes the cask. Considering the entire lifetime of the plant, the total produced ILW are estimated at about 0.0314 m^3/GWh, including wastes from decommissioning.

4.3.7 Spent fuel disposal in final repositories

The material and energy used for encapsulating the spent fuel assemblies have not been included, for lack of information. However, energy use and emission rates from this activity should be relatively small, considering the relative simplicity of the operation and the relatively low waste volumes. The energy rates that should be used for the construction and operation of the formerly proposed (Nagra, 1985) Swiss final repository for HLW in deep geological granite strata (geosphere) and of the ILW repository of a different configuration (horizontal caverns in a mountain), have been evaluated separately in Frischknecht et al. (1996). Because there is no information about the Chinese concept and preliminary design for nuclear repositories, the Swiss LCA inventories have been directly imported into this study, but merged in one module.

The digging of each repository generates about one Mm^3 of solid waste, which gives approximately 19000 kg per m^3 of ILW or HLW. In conclusion, this simple adoption for the module of nuclear repositories does not pretend to be an assessment of the Chinese case, but only an initial guess suggesting the relative contribution of the repository step to total emissions from the nuclear chain.

4.3.8 Low level waste depository

This module is extremely simplified, containing only the assumed requirement of energy and land use of 2 m^2 per m^3 of LLW.

4.4 Results

Table 8.18 shows selected results, including the total burdens calculated for the assumed nuclear chain and the part from material uses. For comparison, the value estimated in Dones et al. (1996) for the total GHG for future open cycle with reprocessing, associated with advanced reactors in Switzerland, is about six g(CO_2-equiv.)/kWh, whereas if enrichment supplies to the Swiss utilities were only from centrifuge plants, the total for the same chain today would be about seven g(CO_2-equiv.)/kWh. The total GHG calculated for the UCTE and Swiss nuclear chains in mid 1990's, using actual average enrichment supplies (a mix of diffusion and centrifuge from various origins) and average fuel burn-ups, were about 16 g(CO_2-equiv.)/kWh and 18 g(CO_2-equiv.)/kWh, respectively.

Figures 8.31 through 8.36 show for the selected environmental burdens the contributions calculated for each step of the chain normalized to the electricity produced at the power plant.[40] The electricity uses for the construction of the power plant and for the operation of the centrifuge facility originate more than 70 percent of total GHG, as can be seen in Figure 8.31.

Table 8.18 Results for selected environmental burdens for the assumed nuclear energy chain.

Burden		Total	Contribution (%) from materials
GHG	g(CO_2-equiv.)/ kWh at busbar	8.7	3%
SO_2	g/kWh at busbar	0.080	64%
NO_x		0.053	10%
TSP		0.033	37%
Non-radioactive solid waste	kg/kWh at busbar	0.0034	6%
Radioactive solid waste*		0.0069	0%
Total solid waste		0.0103	2%
Land use	km^2/TWh at busbar	0.0128	9%

* Total mass of low, intermediate, and high radioactive wastes, whose breakdown is in the table below:

LLW	m^3/kWh at busbar	$3.40 \cdot 10^{-6}$
ILW		$3.14 \cdot 10^{-8}$
HLW		$3.80 \cdot 10^{-9}$

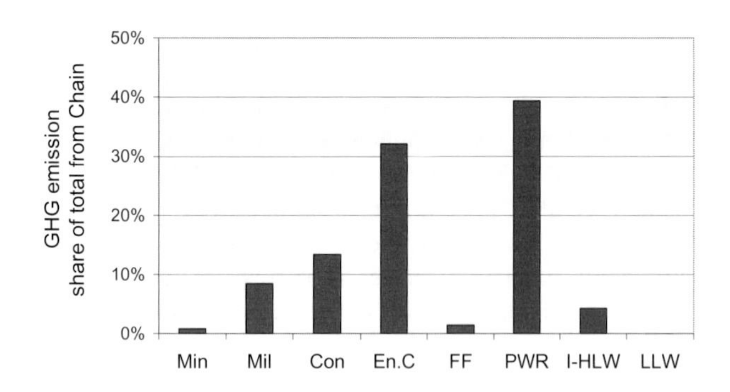

Figure 8.31 Greenhouse gas emissions from individual steps of the open nuclear cycle: LCA-based shares of total from chain.

Mining shows a minor contribution in spite of the conservative assumption that the electricity is supplied by diesel generators, which are less efficient compared to large power plants. Also the fuel fabrication plant and the depositories contribute marginally. The electricity for constructing the power plant is also the major cause of the large calculated values for airborne pollutants from PWR's (Figure 8.32). The contribution to total NO_x emissions from materials is somewhat lower than the corresponding share for other airborne pollutants, because NO_x is directly emitted from different processes in the chain, in addition to the direct emissions from fuel combustion. Mills emit particulates directly; adding the particulates associated with energy use, the total emission from milling contribute nearly 15 percent of the total from the chain. The direct emissions from mechanical mining are not meaningful contributors.[41] Particulate from other steps of the chain originate from energy uses.

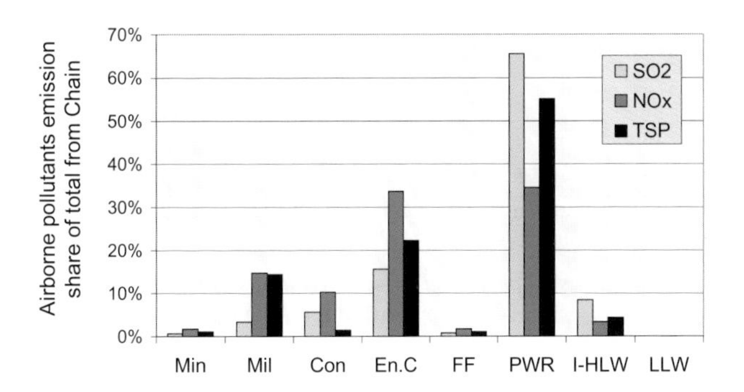

Figure 8.32 Emissions of selected airborne pollutants from individual steps of the open nuclear cycle: LCA-based shares of total from chain.

The total mass of solid waste as well as its two component classes—non-radioactive and radioactive, the latter lumping the volumes of ILW, HLW, and HLW—are shown in Figures 8.33 through 8.35. The power plant contributes the highest share of non-radioactive wastes in the course of decommissioning it. Also the I/HLW repositories contribute substantially to the total due to the excavated rock. Waste from enrichment originates indirectly from electricity use. The predominant contributor to the total volume of radioactive waste is the LLW in tailing ponds consequent to milling; the total mass of ILW and HLW, all generated by the operation and decommissioning of the power plant, is much smaller (Table 8.18). Naturally, the radiotoxicity of wastes should be taken into account as well as their volume. This is not discussed here because LCA studies actual emissions to the environment and not potential hazards, which are prevented by the conditioning of the radioactive wastes. Moreover, the lack of specific data on Chinese practices and prospective developments hinders any comparison. In relative terms, milling exhibits the greatest production rate (per kWh) of total solid waste. However, its absolute value can substantially change depending on the ore grade. The harm from mill tailings can dramatically change depending on the location, meteorology, and reclamation practice. However, local effects are left to specific assessments, which are beyond the scope of this project. Further work shall investigate specific Chinese conditions and consider the uncertainties in the future trade of uranium products.

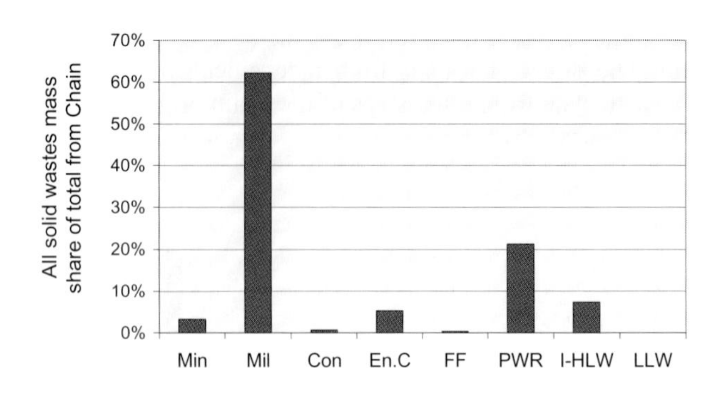

Figure 8.33 Solid wastes produced from individual steps of the open nuclear cycle: LCA-based shares of total from chain.

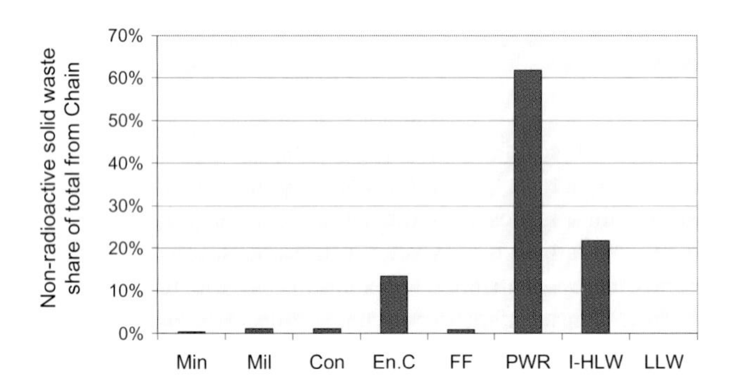

Figure 8.34 Non-radioactive solid wastes produced from individual steps of the open nuclear cycle: LCA-based shares of total from chain.

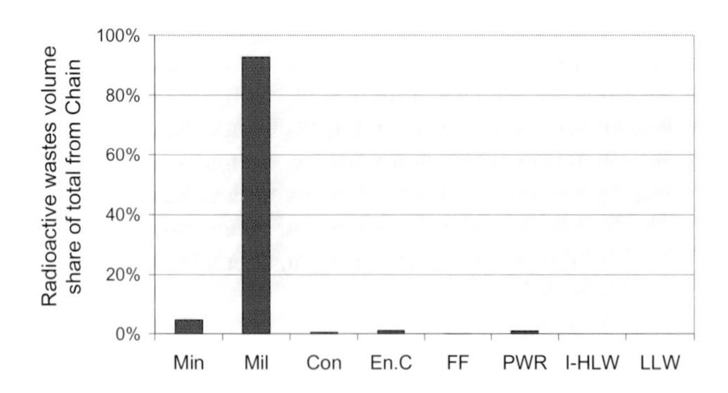

Figure 8.35 Radioactive solid wastes (LLW, ILW, and HLW) produced from individual steps of the open nuclear cycle: LCA-based shares of total from chain.

Figure 8.36 shows the distribution of land use burdens within the nuclear chain. The mill tailing pond is the dominant factor, in spite of the high share of ISL mining assumed. More than 85 percent of the total land use calculated for the chain comes from the power plant, the enrichment facility, and the mill. The LLW depository also contributes a few percent to total.

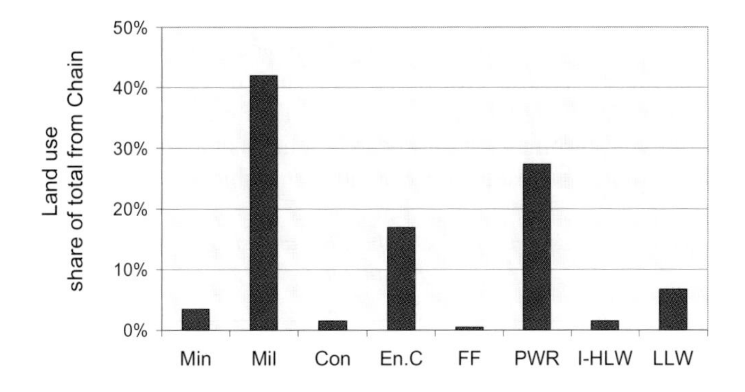

Figure 8.36 Land use from individual steps of the open nuclear cycle: LCA-based shares of total from chain.

5. WIND POWER

5.1 Potential in Shandong

One wind farm already exists in Shandong, in the island of Changdao off the northern coast of the peninsula, with a total capacity of 9×600 kW. At the end of 2000, there was not yet any plan for the construction of other units in the province.

The total theoretical wind potential at 10 m height in China has been estimated at 3226 GW, with only 253 GW practically achievable. Of these, barely four GW may be found in Shandong province (CACETC, 2000). Most of the total practical potential in China lies in Inner Mongolia, with nearly 62 GW, and in other northern and western provinces, while eastern coastal provinces each have lower total potential than Shandong. Using an optimistic capacity factor of 0.2 and exploiting the full potential yields an upper limit of seven TWh/year, which is two to three percent of the demand levels expected for 2020.

However, wind power has not been explicitly included in the ESS task scenarios. A specified fixed contribution of wind power to all scenarios would not change their relative rankings. Even large investments in this renewable energy technology would provide a minimal share of total demand. Nevertheless, considering that wind power is apparently one of the renewable technologies most likely to receive further attention in Shandong as well as in other provinces in China,[42] a simplified inventory of a wind turbine has been included in this study for estimating the order of magnitude of the associated burdens for comparison with other electricity conversion systems.

5.2 Modeling a wind turbine for future electric supply to Shandong

The environmental inventory for an onshore 800 kW wind turbine established in the Swiss LCA study on current European energy systems has been approximately extrapolated to Chinese conditions taking into account the differences between UCTE and Chinese average electricity and fuel mixes, and the lower energy efficiency for material manufacturing in China. Total GHG and NO_x emission rates associated with wind turbines for European conditions have been multiplied by two; SO_2 and TSP by four. The results are illustrated in Table 8.19.

Table 8.19 Preliminary LCA-based selected burdens associated with an 800 kW wind turbine in Shandong.

Burden	Unit	
GHG	g(CO_2-equiv.)/kWh at busbar	19.5
SO_2	g/kWh at busbar	0.22
NO_x		0.04
TSP		0.16
Waste	kg/kWh at busbar	0.004
Land use	m^2/GWh at busbar	0.86

6. COMPARISON OF CURRENT AND FUTURE ENERGY CHAINS

Figures 8.37 through 8.43 show comparisons of selected environmental burdens, per unit of kWh at the busbar, from the full energy chains associated with the power plant mix of the Shandong provincial grid in 1998, the mix of SEPCO-owned plants for the same year (which is a subset of the previous mix), the "current mix" i.e. the mix of PC technologies already installed today that are assumed to be still running in year 2020, the 600 MW PC representative of the best Chinese technology today, the 600 MW PC with FGD, a mixture of AFBC and IGCC (50 percent each) named "Clean Coal", the GCC fuelled with natural gas, the open nuclear chain with LWR, and wind turbines.[43] The emissions are divided into two parts according to their origin, the first directly stemming from power plants during their operation, the second from the rest of the corresponding energy chain.

The total GHG emissions shown in Figure 8.37 are an aggregation of different species as CO_2 equivalents, using GWP for a 100-year horizon (IPCC, 2001). The technologies analyzed can be roughly divided into three groups: i) the technologies and mixes based on coal, which exhibit the highest emissions in the range of approximately 950-1270 g(CO_2-equiv.)/kWh at the busbar, with the rest of the chain contributing 11 to 12 percent of the total; ii) the natural gas chain with an average of nearly 435 g(CO_2-equiv.)/kWh at the busbar with the rest of the chain making up

16 percent of the total; and, iii) the nuclear and wind technologies, with the lowest emissions of nearly 10 g(CO_2-equiv.)/kWh and 20 g(CO_2-equiv.)/kWh, respectively, both stemming from the upstream and downstream energy chain and the construction of the power plant.

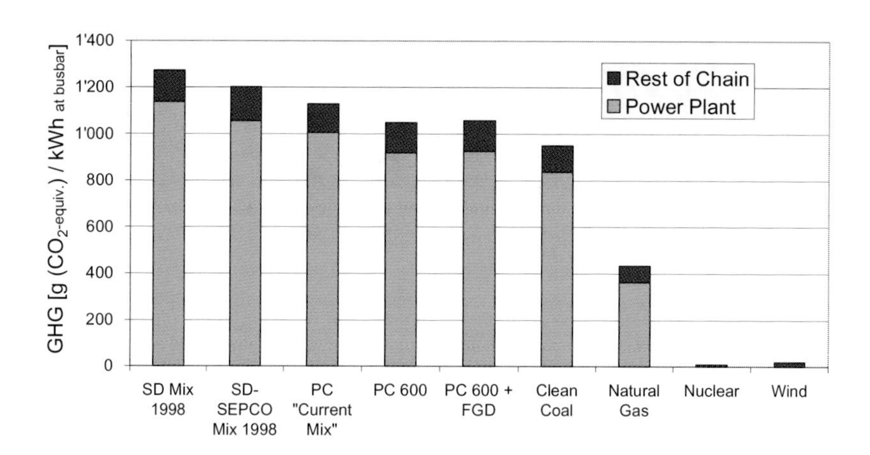

Figure 8.37 Greenhouse gas emission rates from chains associated with Shandong current and future power plants and mixes: contributions from power plants and upstream chains.

Figure 8.38 shows a comparison of SO_2 emission rates. The analyzed technologies, either single or in mixes, can be roughly divided into three groups, but these are different from those discussed for GHG: i) the coal technologies and coal mixes based on current coal power plants without active sulfur removal systems, which exhibit the highest emissions of SO_2 in the range of approximately 6-11 g/kWh at the busbar, depending on the sulfur content in the input coal and the average plant efficiency, with the rest of the chain contributing four to seven percent of the total; ii) the coal technologies with active sulfur removal (FGD, FBC, gasification), with LCA emissions 0.7-1.4 g/kWh at the busbar, for sulfur content in the coal reflecting approximately average conditions in 1998 and 30 to 60 percent contributions from the chain; and, iii) natural gas CC, nuclear, and wind energy chains, with 0.05 g/kWh, 0.08 g/kWh, and around 0.2 g/kWh, respectively.

The NO_x emission rates are given in Figure 8.39. Again, the chains associated with conventional coal technologies or mixes exhibit roughly one order of magnitude higher emission rates than the natural gas chain, and two orders of magnitude higher than nuclear or wind chains. While the bulk of the emissions from the coal systems with PC originate at power plants, over 60 percent of the total from the gas chain stems from upstream activities.

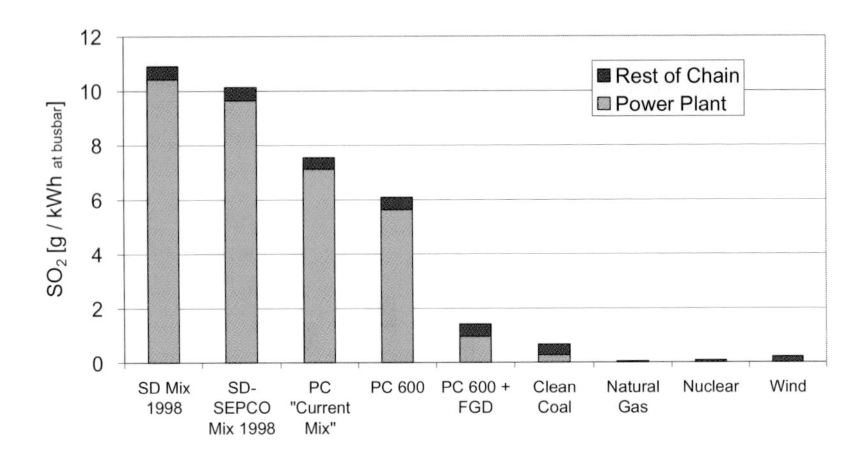

Figure 8.38 SO_2 emission rates from chains associated with Shandong current and future power plants and mixes: contributions from power plants and upstream chains.

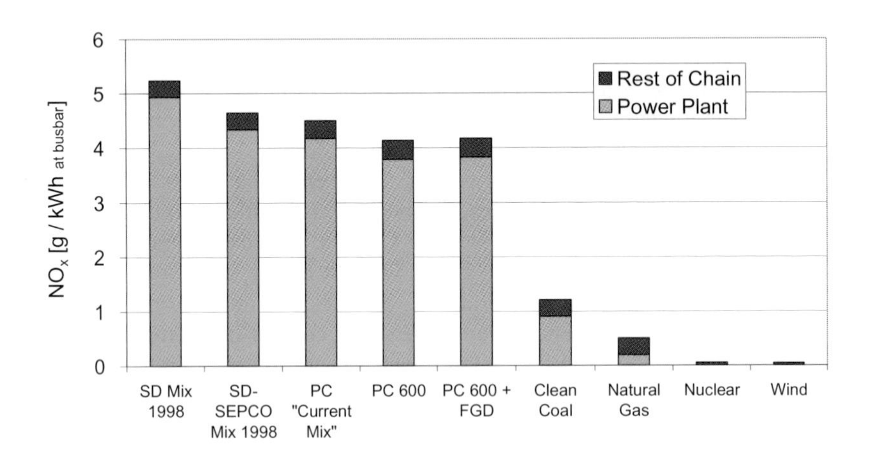

Figure 8.39 NO_x emission rates from chains associated with Shandong current and future power plants and mixes: contributions from power plants and upstream chains.

The particulate emission rates are illustrated in Figure 8.40. The level strongly depends on the particulate control technology at coal power plants, which is extremely poor for the small-capacity PC units. This is reflected in the very high emissions rates calculated for the 1998 mixes. The emissions at coal mines are credited they will reduce, but not below a certain limit. Those associated with transport are proportional to the distance involved, which partly explains the differences in the values for the rest of the coal chain in the picture. All coal technologies and chains exhibit particle emission at least two orders of magnitude higher than the total emission from the gas or nuclear chain. The emission from wind power depends on the indirect (through energy uses) releases from material manufacturing. Considering the rough extrapolation made here for the total emissions associated with wind power, the related values shown in these pictures should be taken only as orders of magnitude.

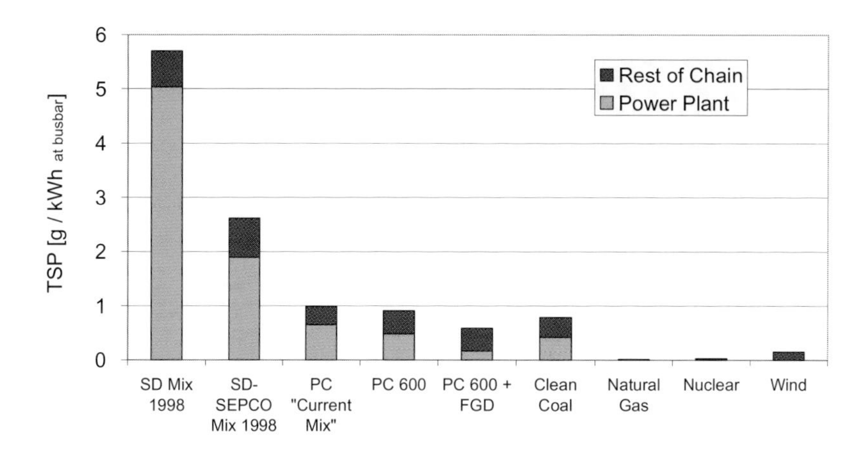

Figure 8.40 TSP emission rates from chains associated with Shandong current and future power plants and mixes: contributions from power plants and upstream chains.

Figure 8.41 is derived from an approximate calculation on the basis of EIA task results, assuming a certain background pollution, unspecified locations (in Shandong) for the various emission sources within a chain, and the expected change of population density over 20 years. The figure compares mortality effects in terms of YOLL calculated for SO_2, NO_x and particulate emission rates. As for the GHG shown in Figure 8.37 above, the YOLL represents aggregated effects of airborne pollutants, but goes beyond the simple inventories, which strictly pertain in an LCI study. The factors used here are: 420 YOLL/kt(SO_2), 270 YOLL/kt(NO_x), and 300 YOLL/kt(TSP). These values are approximate, because they implicitly assume that the direct and indirect emissions from any step of the chain are occurring in areas with common characteristics, and thus do not reflect differences in distribution of point sources and population. The EIA chapter in this volume (Chapter 9) and the associated technical report (Hirschberg et al., 2003) detail the scientific basis and modeling characteristics for determining the YOLL factors for the species. Health effects from repositories of conditioned radioactive solid wastes were not considered in this analysis, but results of previous studies have shown that they are of very minor importance compared to effects from emissions from other steps of the chain.

With the introduction of a highly efficient particle removal technology (ESP), the installation of plants with higher net efficiency, the use of or refurbishment with scrubbers for PC's, and the introduction of clean coal technologies, the health effects associated with the unit electricity supplied by coal chains to the grid fall steeply by a factor of about ten. However, the environmental performances of natural gas, nuclear, and wind technologies still remain substantially higher, yielding only seven to 20 percent of the impacts calculated for the clean coal mix.

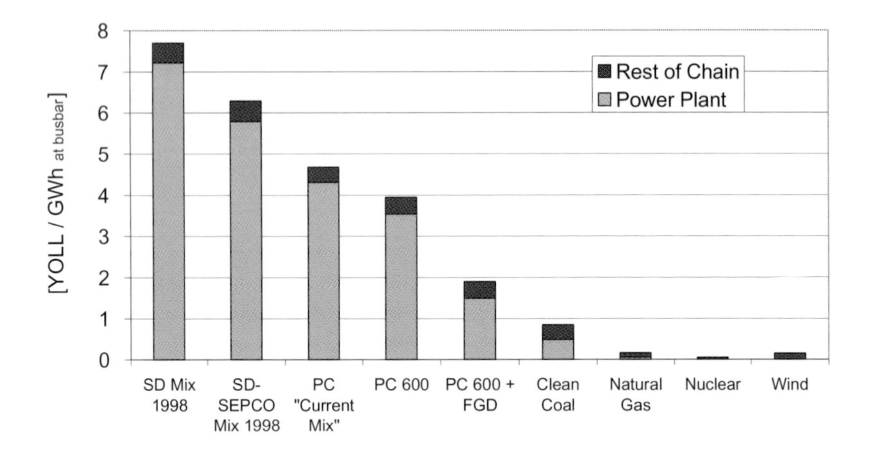

Figure 8.41 Mortality from air pollution per unit of electricity at busbar from chains associated with Shandong current and future power plants and mixes: contributions from power plants and upstream chains.

Figure 8.42 shows the solid waste production rates. Although the actual rates for the various coal systems may differ from those assumed in this analysis, depending on the amount of recycling of wastes produced during power plant operation, it is apparent that in general this energy chain presents the largest waste production.

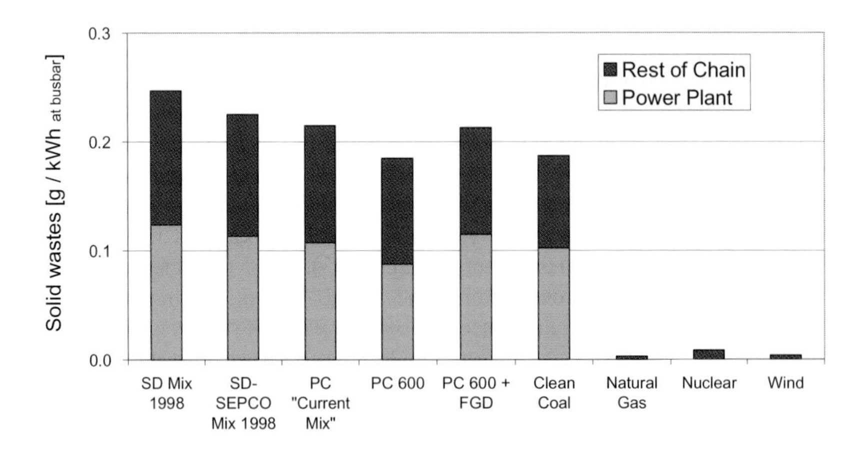

Figure 8.42 Solid waste production from chains associated with Shandong current and future power plants and mixes: contributions from power plants and upstream chains.

Of the 0.01 kg(solid waste)/kWh calculated for the nuclear chain, less than one percent is conditioned ILW and HLW but two thirds is total LLW. This study does not proceed beyond the assessment of the quantities of main waste categories into analysis of the effects of waste disposal on the environment and humans.[44] However, the issue of the long confinement time of highly radioactive or toxic/carcinogenic substances is qualitatively addressed in the MCDA task.

Figure 8.43 shows the land use. The estimates do not include landfill of solid wastes from the coal plants (bottom ash, gypsum), which may be substantial considering the amount of waste generated, and may dramatically change depending on the degree of recycling. However, it is clear that such comparison would only underscore the already large differences amongst the various chains. In addition, the area disturbed by coal mining also includes the surfaces affected by subsidence phenomena specifically estimated for Shandong. Coal chains have land use from nearly 0.07 km^2/TWh at the busbar for IGCC up to nearly 0.09 km^2/TWh for PC's. Estimates of the land use for the nuclear and the gas chains are significantly lower, 0.012 km^2/TWh[45] and 0.006 km^2/TWh at the busbar, respectively.

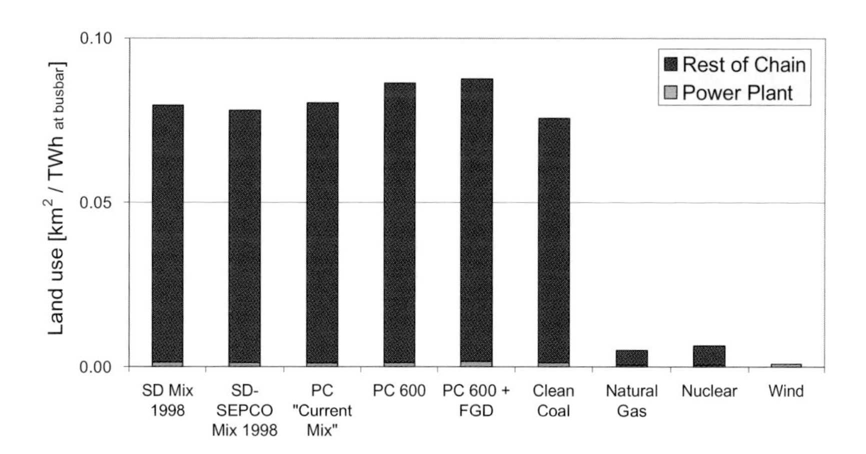

Figure 8.43 Land use from chains associated with Shandong current and future power plants and mixes: contributions from power plants and upstream chains.

The value assumed for land occupation for coal mining corresponds to about half of that calculated for the rest of the chain, the other half being due to transport, as can be seen in Figure 8.20. Unless the assumptions for transport in this study strongly overestimate the relevant land use in China, the ranking among the different energy chain is not determined by the assumptions for coal mining.

Land use for the nuclear chain may increase by roughly ten percent if the share of conventional mining followed by milling, and hence the area occupied by the waste ponds, reflects the worldwide average, instead of the increased share for ISL assumed here to reflect likely future Chinese conditions. Hence, the ranking seems rather robust. It should be emphasized that this study does not address the type of land affected and its transformation, nor the duration of the land disturbance by industrial activities. Surfaces affected within or near inhabited areas as they are in several coal districts are treated with the same weight as gas fields and coal and uranium mines in remote areas.

The above pictures represent average conditions. Possible variation in key factors for the various energy chains are addressed case by case in the following section on sensitivities. However, the ranking does not change in relative terms, even under extreme conditions such as the use of obsolete and uneconomical diffusion enrichment in the nuclear chain. The values for wind turbines extrapolated for this study are believed to lie near the upper limit. However, considering the continuous improvement of turbine technology and the reduction of energy intensity for material manufacturing in China, their environmental burdens can only decrease, thus preserving the environmental ranking of wind power compared to coal technologies.

In conclusion, the environmental performance of all coal systems is the worst amongst the considered energy chains. However, with the introduction of improved or advanced coal technologies, the harm to human health corresponding to the unit of electric service may be reduced meaningfully. The bulk of future electricity production in Shandong most likely will be still based on coal, considering the geopolitical conditions of the province. The use of cleaner coal technologies complemented by natural gas and nuclear, and supplemented by some renewables like wind,[46] may allow the increase of electricity production to match the increasing demand without further increasing the environmental effects compared to today.

However, though for the time being a second priority issue in China, GHG is likely to increase with the demand unless a strong nuclear policy is pursued. Issues such as solid waste may be addressed by increasing the level of recycling, following the examples of other countries in the world, and in the case of FGD residues by coupling the installation of these facilities with the creation of a market for gypsum (if not radioactive). Further research on the utilization of residues from various types of FBC and the choice of a suitable process for those from IGCC should reach maturity before major construction programs are undertaken to meet the energy demand. Reutilization of coal mine debris to prevent subsidence as well as conditioning and reclamation of waste heaps should be pursued to decrease pollution of water bodies and to regain or protect land.

7. SENSITIVITY ANALYSIS FOR THE ENERGY CHAINS

Two sensitivity analyses have been performed, one varying the generation mix for the electricity required by the three fuel chains analyzed, the other testing the extreme values for key parameters. The complete results are given in the technical report of the task. In this volume only the conclusions are illustrated in order to frame the uncertainties of the simplified analysis.

7.1 Coal chain sensitivities

7.1.1. Electricity supply mix to the Shandong grid

The coal energy chain is loaded by indirect burdens derived from electricity mixes powering railway transport and mines. As shown before, the two grids in SD and SX have been modeled individually, though it is planned to connect the different Chinese grids before the end of the 2010's. This approach is sufficient for the aim of the study. It is highly likely that in SX the predominant fuel will remain coal, considering the great abundance of this resource in the province. For Shandong, CETP has assessed numerous scenarios based on different mixes of fuels. However, the LCA assessment of coal chains has been performed using one future supply strategy conservatively based on coal only and on current plant technologies (partially including FGD). The question is whether the results for these reference coal chains can be applied in calculating the LCA burdens of other supply strategies considered in the MCDA exercise.

A sensitivity analysis has been performed first using the reference coal chains for calculating LCA emissions for a few MCDA scenarios in 2020, selected in order to encompass all represented strategies, then using these results to recalculate the coal chain specific burdens. The goal of this sensitivity was to demonstrate that it is not necessary to dynamically model all feedbacks from the electricity mix scenario by scenario. The conclusion of the calculations is that the LCA burdens for future coal energy chains associated with Shandong power plants are in general not sensitive to variations in the electricity mix used to meet electricity requirements within the chain itself. The differences calculated for the selected burdens here discussed remain on the order of one percent or below. The greatest difference is calculated for the total emission of SO_2 but still below 1.5 percent. The above conclusion has been obtained using snapshots of the supply situation in 1998/1999 and 2020. Moreover, this conclusion is strengthen by the fact that changes in the supply mix for the next twenty years should occur gradually whatsoever energy policy will be actually implemented, thus fully justifying the use of a mix of current and future coal power plants for modeling electricity supply to Shandong and eventually to coal chains for other provinces in China. Hence, the focus of any follow-up LCA study on Chinese coal chains should rather be on a more accurate quantification of the environmental burdens from small mine-mouth power plants and coal transport.

7.1.2 Origin of the coal

Though unrealistic, two opposite cases have been defined in order to estimate the influence of the origin of the coal alone on the total burdens calculated for the chain associated with Shandong power plants. In one case only domestic coal is burned in all plants, in the other only coal from Shanxi. Present as well as future conditions were analyzed. For each single coal technology or mix, the emission rates of pollutants from the associated full energy chain increase when SX coal is used instead of SD. Due to the assumptions made for the future reduction of emissions associated with transport by train and the increase of methane recovery rate at SX large mines, the use of SD coal decreases GHG emissions from 21 percent for the actual mix to 13 percent for the future. The SO_2 differences obtained range narrowly between six and nine percent for the current technologies without FGD or for mixed technologies, because of the relatively high emission rates from conventional power plants, that overshadow the contributions from the rest of the chain. However, for plants with FGD or clean coal technologies, switching from SD to SX coal may increase the chain-related SO_2 emission rate by 50 percent and even 100 percent in relative terms. The variation in NO_x emissions remains relatively small for mixes and PC's, but reaches 20 percent for clean coal technologies because of the reduced emission rate at the plants. The most dramatic changes are observed for TSP. Though for the considered mixes the variation remains around 20 percent, for technologies with high rate of particle removal (plants with ESP, FGD, and intrinsically IGCC) this difference may be greater than a factor of two. Conversely, the total amount of solid waste for SD mixes decreases by around 10 percent when SX coal is imported. The calculated total land use increases by about a factor of two when plants are fueled by SX coal, due to the substantial increase in transport requirements.

Therefore, it can be concluded that the assumed ratio of coal imported vs. coal produced from domestic SD mines is one important factor for the comprehensive estimation of scenarios burdens. For this reason, when analyzing the MCDA scenarios the shares of SD and SX coal assumed in the ESS task were used. For future studies or policy applications, it is recommended that either different coal supply scenarios be analyzed for their environmental effects, or that coal origin be considered as an uncertainty factor whose environmental burdens can be quantified using the information from this analysis.

7.1.3 Power supply to mines

The survey performed at the main SD and SX coal bureaus has not resulted in a fully consistent description of the performance of local power plants; information about trends for future development was also inadequate. The cautious attitude of mine operators may be due to their hesitancy in spreading information about the ecological consequences of mining, or due to their positions on how different energy policies may affect them. Sensitivity analyses have been performed on the origin of the electricity used in mining and on the emission rates of main pollutants stemming from small mine-mouth power plants.

For the origin of the power supplied, two extreme cases were defined: electricity only from provincial grids vs. electricity only from units at mines. For the sensitivity on emissions, only the supply of electricity from power plants at mines is considered in order to maximize the emission rates per unit of coal produced, and the average case is compared with the maximum emission rates.

Electricity self-production at mines vs. supply from grids. In general, the use of electricity from the grid results in lower burdens for the whole energy chain. This is because the electricity supplied to grids is generated by plants with higher net efficiency than the small plants at mines, which more than compensate transmission losses. Large, grid-connected plants also have highly efficient particle removal equipment. However, the calculated differences for LCA emission rates from SD current and future mixes remain on the order of one percent for GHG; five percent for SO_2; two percent for NO_x, and one percent for TSP, because of the relatively low energy intensity of mine operation. These percentages may increase somewhat in the modeling of future individual coal technologies, because of the higher relative influence of the upstream chain to the total. Solid waste and land use also change by only a few percent. However, from the point of view of rational use of available resources, burning sludge, slurry and middlings at mines certainly has the advantage of saving approximately one to three percent of steam coal per unit of electricity delivered to customers. Correspondingly, the transport burdens for the chain are also a little bit lower when residuals are burned at mines instead of importing electricity from grids. Other environmental factors not considered here, like the use of freshwater or water pollutants, may have provided a more complete picture. However, insufficient information prevents a solid discussion of a wider spectrum of environmental burdens; follow-up studies may remedy this situation.

From another point of view, mine operators may pay more when purchasing electricity from the grid, especially in a liberalized market. Differences in the cost of coal dependent on the electricity supplied to mines was not explicitly included in CETP, therefore no final statements can be attempted on this issue. To compare these two electricity production modes to support specific decisions, it would be necessary to set up a dedicated comprehensive multi-region optimization model taking into account environmental, economic, and risk factors. Naturally, the preceding considerations are not valid for large mine-mouth plants, especially of the AFBC type or PC with FGD, which present environmental as well as economic advantages over equivalent plants located far from mines, while depending on the distance from customers and network losses.

Electricity self-production at mines: average vs. maximum mine energy uses and airborne emissions. Two cases have been considered to estimate the effects of key factors concerning the electricity self-production at mines on chain-related burdens: the electricity intensity per unit of coal produced and the TSP and SO_2 emission rates. Effects of average reference values for these parameters have been compared with the effects of relatively large values. In both cases mines are assumed to use self-generated electricity exclusively.

The difference of the GHG emissions calculated in the two cases for the various mixes and classes of coal units varies on the order of a few percent. Because of the very high emission rates assumed, based on the high sulfur content of the mining residues burned as fuel as well as on the very low efficiency of current small power plants at mines, the consequence for chain-related SO_2 is a difference of about 14 percent. Again, when sulfur releases are drastically reduced at the power plant, the effect on the emissions at mines becomes more important in relative terms: the difference in the chain-based SO_2 emission rates calculated for the two cases of this sensitivity is above 50 percent for the PC with FGD and up to 80 percent for clean coal technologies. The NO_x emissions have not been modified for this sensitivity, because the mine bureaus were unable to provide any information. However, it is probable that NO_x emission rates from small power plants at mines are higher that emission rates from large base load units. The variation of TSP between the two cases must be explained case by case, depending on the efficiency of the particle removal technology used in the existing PC units. For the current electricity mix, the effect is similar to the SO_2 emission rate explained above. For the future mix, the difference is minimal because a major reduction of the TSP emission rate at mines has been credited. For current individual technologies, the variation in the LCA-based TSP emission rate under the worst conditions at mine power plants grows with the increasing efficiency of the base load plant unit and of its ESP, up to a few dozen percent. If the hypothesized reduction of TSP emissions from mine-mouth plants in the future does not occur, the variation between the two sensitivity cases for the chains associated with future power plant technologies would reach a factor of two to three. The total chain-related solid waste increases only a few percent if energy intensity at mines increases. However, because of missing information from the bureaus, no conclusion can be made about the real effects of power plant operation at mines on this burden.

In conclusion, uncertainties can substantially affect the relative contributions of mine-based power plants to chain-related pollutant emissions for future technologies. Nevertheless, the absolute values of the emission rates from coal chains should decrease with times as large base load power plants supplying the provincial grid improve. In the reference case analyzed in this study, a meaningful reduction of environmental burdens from small plants at mines has been credited based on the implementation of a mild environmental policy. If this does not occur, and if the energy intensity for producing coal is either substantially higher than the average declared today, or will increase in the future, mine-based power plant emissions could become increasingly important from the ecological and health viewpoints, in relative terms to total emission.

7.2 Natural gas chain sensitivity

7.2.1 Gas by wire from Xinjiang

This study estimates the environmental burdens from a simplified natural gas chain considering hypothetical base load GCC installed in Xinjiang and supplying electricity to the Shandong grid by means of a dedicated high voltage DC transmission line. This option, named "gas-by-wire", was once mentioned by CETP stakeholders, though it appears not realistic. This chain performs better than chains involving long-distance transport of gas or LNG, scoring 417 g(CO_2-equiv.)/kWh at the Shandong high-voltage grid. Naturally, the leakages of methane and the energy uses for gas transportation are avoided. Due to the assumed production of 0.018 g(N_2O)/kWh for corona effect at the transmission lines,[47] this species provides nearly two percent of total GHG compared to the lower contribution (< 0.5%) in other gas chains. The net GHG emission per unit of electricity service from the hypothetical GCC in Xinjiang is higher than for a similar unit installed in Shandong: due to the transmission losses, assumed to be 1%/1000 km, approximately six percent more electricity must be produced by the plant to deliver the same amount of electricity to the high voltage Shandong grid. The DC lines contribute to total emissions also indirectly, for their material use. However, the values are minimal, though it must be warned that the material use here considered corresponds to the high voltage alternate current grid as modeled in the Swiss LCA studies.

7.3 Nuclear chain sensitivities

7.3.1 Uranium enrichment process and electricity supply mix

Uranium enrichment and the origin of the electricity supplied to it are crucial to evaluate the non-radioactive burdens associated with the full energy chain for LWR power plants. For the 20-year time frame considered here, the centrifuge is the enrichment technology that will most likely be used.[48] However, considering the incomplete information available, it cannot be definitively stated that China is not currently using diffusion to enrich uranium for power production. Therefore, a comparison of the two processes is meaningful.

Centrifuge is assumed to require 100 kWh/SWU and diffusion 3000 kWh/SWU. Both values are rather conservative, because the efficiency of the facilities has been improving worldwide.[49] Considering that the diffusion technology most probably available in China must have been developed in the late 1950's, and that it has been used for military purposes, and therefore without economic constraints for the optimization of energy requirements to run the process, the value assumed here for electricity intensity is the greatest found in the literature for US conditions in the late 1970's. However, the French Eurodif plant works with about 2300 kWh/SWU. This facility is supplied directly by the nuclear power plants on the same site. However, the Chinese existing diffusion plant(s) are most likely supplied by coal power plants, less likely by hydropower, or by a mix of the two. For simplicity, in this analysis all enrichment units are assumed to be fed by the SD mixes already used for the sensitivity on coal chains (section 7.1.1).

If centrifuge is utilized, the calculated ratio of total electricity needs in the chain to total electricity production at power plant is 0.4 percent, whereas with diffusion this ratio goes up to nearly five percent.[50] The centrifuge plant contributes 39 percent to total electricity uses per kWh produced at the NPP, for diffusion technology the contribution dramatically increases to 95 percent. As an effect of these differences, very large variations are calculated for the environmental indicators. The indirect contributions from the coal chain through electricity uses are dramatic for the chain with diffusion enrichment. The largest increment is calculated for GHG, greater by a factor of eight than in the chain with centrifuge technology. Large increases are also calculated for airborne pollutants. Due to the constant mass of radioactive waste production rates from both chains,[51] the total amount of solid wastes increases by a factor of only about two. The land use changes less dramatically than other burdens, only by a few tenths of a percent.

In general, for each burden the total rate calculated for the chain with centrifuge is less sensitive to the electricity supply mix that that for the chain with diffusion enrichment. Moreover, for all selected burdens except land use, the current SD mix produces the highest burdens. Logically, when analyzing scenarios entailing a substantial nuclear capacity, it would be more consistent to model the electricity uses using a supply mix that includes it. However, when utilizing a mix based exclusively on coal PC, the results thus maximized not only represent a scenario closer to the nuclear option in its initial stage of development, when it contributes little to the mixes, as of today, but also some of the uncertainties that should be covered. The results for the GHG emission rates for the chains with centrifuge enrichment are all below ten g(CO_2-equiv.)/kWh at the busbar, with minimum values around seven g(CO_2-equiv.)/kWh for the scenarios with mixed fuels. The highest calculated GHG emission from the nuclear chain using diffusion technology approaches 80 g(CO_2-equiv.)/kWh; the minimum is about 45 g(CO_2-equiv.)/kWh. Calculated SO_2 emission rates for the chain with centrifuge are in the range 0.06 to 0.1 g/kWh at the busbar; when diffusion is utilized, the range is 0.1-0.6 g/kWh. Emission rates of NO_x for the chain with centrifuge range from 0.04 to nearly 0.06 g/kWh at the busbar. When diffusion is utilized, NO_x emission rates vary in the range 0.13 to 0.34 g/kWh. The results for TSP exhibit the strongest relative variation among the selected burdens. The ratio between TSP values for the chain with diffusion vs. the chain with centrifuge for an equal supply scenario varies between a minimum of two and a maximum of 7.5.

The LCA-based non-radioactive burdens from the nuclear chain clearly depend to a large degree on the enrichment process involved. Nevertheless, the differences between the two cases tend to decrease in the future, as the environmental performance of all the electricity mixes improves. The calculated air emission rates, and hence the associated health effects, from the considered nuclear chains remain somewhat overestimated, but are still well below the rates calculated for the coal chains. Comparing the nuclear chains with the natural gas chain reveals substantial differences for the SO_2, TSP, solid wastes, and land use burdens, with gas being distinctly less polluting. Conversely, the difference for GHG (factor five to 70) and NO_x (factor one to eight) are unfavorable to gas. Even if the worst environmentally performing nuclear chain were used to calculate the MCDA scenario burdens, the conclusions for scenario analysis would not change.

7.3.2 Open chain with reprocessing

An open cycle with reprocessing, i.e. an open chain with separation and conditioning (vitrification) of solid wastes from spent fuel, but without recycling the extracted plutonium in the reactor, has been assessed. The assumed alternative chain is illustrated in Figure 8.44 for the case of enrichment with centrifuge. A fully closed cycle, with utilization into LWR's of the plutonium and possibly also uranium separated from the spent fuel, should give comparable or lower environmental burdens than open cycle(s).

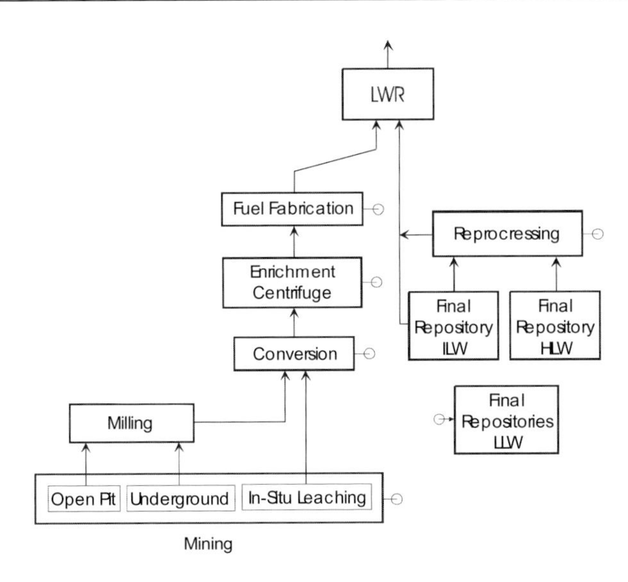

Figure 8.44 Model of nuclear open cycle with reprocessing.

Again, due to lack of definitive information about reprocessing in China, the modeling already performed for the Swiss LCA study on current energy systems has been imported into this analysis. This assessment combined characteristics of the two large European commercial plants in the UK (THORP in Sellafield) and in France (La Hague). Naturally, the contribution of reprocessing to the selected total burdens is the same in absolute values, but different in relative terms, for the two chains with centrifuge and diffusion enrichment. For instance, the GHG increases 18 percent in the chain with centrifuge but only about two percent in the chain with diffusion. The total volume of radioactive wastes changes only insignificantly because basically only the total volumes of high and intermediate level radioactive waste should change by reprocessing the fuel, which adds only a minor contribution to the total radioactive volume. Naturally, volume alone is by no means a final measure of potential harm. All solid waste classes are combined in one because of the definition of corresponding sub-criterion selected for the MCDA task. The completely different conditions that HLW and ILW impose are reflected in a second sub-criterion expressing how long the critical wastes for each energy chain must be confined.

The results of the sensitivity analysis of reprocessing show that its inclusion into the chain does not change the differences between the burdens calculated for nuclear and the two fossil chains.

7.3.3 Uranium mining exclusively with domestic in-situ leaching

A simplified sensitivity analysis has been performed assuming only domestic ISL uranium mining, though this is probably an unrealistic option for covering future demand. The results for most of the selected pollutants do not substantially change. Nonetheless, for the intrinsic characteristics of ISL, the total radioactive solid waste, which means only LLW in this case, and subsequently the total solid waste decrease considerably. The same applies to the total land use, which decreases to less than half of the value for the reference chain. It must be stressed again that other ISL-related impacts, like possible radioactive releases into groundwater, may dramatically increase in the chain.

8. ANALYSIS OF SCENARIOS SELECTED FOR THE MCDA TASK

As a last step of this assessment, the total environmental burdens for the 36 scenarios chosen for the MCDA task from the far more numerous set of ESS are compared for the year 2020 using exclusively LCA results. Calculations for cumulative results over 20 years have not been performed. The reason is that considering that most of the innovative fossil technologies as well as the nuclear plants would most likely be installed in the second decade, the comparison of the burdens for year 2020 is a meaningful measure of the trends the different policies, represented by the twelve selected strategies, would establish. Additionally, a dynamic approach was not considered appropriate because the limitation of information increases the level of approximation, which would make the major effort required by a more complex analysis unrewarding.

For the MCDA calculations, LCA provides the indicators of environmental burdens from the coal and natural gas upstream chains, while the characteristics of the power plants are those chosen by ESS. Conversely, for the nuclear chain the total burdens are those calculated by the LCA task. Therefore, the LCA rankings of the scenarios may differ somewhat from those calculated by ESS for the MCDA tasks. This should not be seen as a weakness, but as a necessary effect of the application of different approaches to system analysis, each internally consistent, whose results may concur to support policy making.

8.1 Description of scenarios and assumptions

Table 8.20 gives the LCA-based SO_2, NO_x, and TSP emission rates for the chains associated with the six technology classes used in the scenario analysis. These species determine the mortality in terms of Years of Life Lost (YOLL), also shown in Table 8.20, as an indicator of health impacts, calculated assuming the factors estimated by the EIA task for 2020.

Wind power is not explicitly used in the ESS scenarios for Shandong. However, if wind turbines were included, besides the fact that their substitution potential is very small (about three percent of the lowest level of electricity demand in the most optimistic case), the order of magnitude of the emissions per unit of electricity is about comparable with nuclear and gas. It can be implicitly assumed that a marginal contribution from wind would not visibly change the total calculated burdens, and hence the ranking of the scenarios

Table 8.20 LCA-based emission rates of SO_2, NO_x and TSP used for the assessment of the MCDA scenarios for year 2020 (kWh at busbar).

Power plant mix or type [a]		SO_2 (g/kWh)	NO_x (g/kWh)	TSP (g/kWh)	Mortality (YOLL/GWh)
PC "Current Mix" [b]	Direct PP	7.11	4.18	0.65	4.31
	LCA PP	7.12	4.19	0.66	4.32
	Total chain	7.54	4.50	0.99	4.68
PC 600 MW [b]	Direct PP	5.64	3.79	0.48	3.54
	LCA PP	5.64	3.79	0.49	3.54
	Total chain	6.08	4.14	0.91	3.94
PC 600MW & FGD[c]	Direct PP	0.96	3.83	0.16	1.48
	LCA PP	0.96	3.83	0.17	1.49
	Total chain	1.41	4.18	0.59	1.90
"Clean Coal" [c]	Direct PP	0.28	0.91	0.42	0.49
	LCA PP	0.30	0.97	0.45	0.52
	Total chain	0.67	1.21	0.79	0.85
Natural Gas CC	Direct PP	0.003	0.196	0.002	0.05
	LCA PP	0.010	0.198	0.005	0.06
	Total chain	0.047	0.309	0.009	0.11
Nuclear [d]	Direct PP	--	--	--	0.00
	LCA PP	0.05	0.02	0.02	0.03
	Total chain	0.08	0.05	0.03	0.06

[a] All figures rounded to the second or third decimal to show differences. "Direct PP" means releases during operation of the power plant. "LCA PP" includes emissions associated with the materials used for construction; in case of nuclear power plant, it also includes the emissions from the testing of emergency diesel generators and the energy used for construction of the plant.
[b] Without FGD. Coal characteristics for the corresponding SD units in 1999.
[c] Arbitrary mix of 50 percent each AFBC and IGCC. Coal characteristics derived from the average for SD 300 MW units in 1998 (average sulfur content in the coal was nearly 1.2 percent).
[d] Effects of radioactive airborne emissions is not included.

Total GHG as well as its main component species are included in Table 8.21. The material use influences the GHG only marginally, in the worst case of PC "current mix" being on the order of a fraction of a percent. Carbon dioxide is the main contributor, with methane from gas pipelines and coal mining as well as N_2O from AFBC's influencing the total by several percent.

Figure 8.45 shows the contributions of the analyzed technology classes to the electricity supplied in 2020 for the ESS scenarios selected for the MCDA task. The 36 scenarios are grouped under three futures, which depend on uncertainties in the three areas of demand growth, cost of coal, and cost of natural gas. Each future is identified by a three-character suffix: FIB means moderate demand growth and base cost for fossil fuels; FAF means moderate demand growth with a high cost for coal and a low cost for natural gas; SUB represents high demand growth, low cost for coal, and base cost for natural gas. A strategy is labeled by the first two clusters of letters to the left. For accurate definitions, the reader should refer to the ESS chapter in this volume (Chapter 6). In the bar charts, for each group of 12 strategies, the six on the left-hand side use only fossil fuels. The remaining six scenarios include gas as well as nuclear. Clean coal technologies are present in the LONLAM, TONLAG, and TONPAS scenarios.

Table 8.21 LCA-based emission rates of greenhouse gases used for the assessment of the MCDA scenarios for year 2020 (kWh at busbar).

Power plant mix or type [a]		GHG (g(CO_2-equiv.) /kWh)	CO_2 (g/kWh)	CH_4 (g/kWh)	N_2O (g/kWh)
PC "Current Mix"	Direct PP	1006	1006	na	na
	LCA PP	1008	1007	0.007	$1.8 \cdot 10^{-5}$
	Total chain	1129	1065	2.74	$1.0 \cdot 10^{-3}$
PC 600 MW	Direct PP	919	919	na	na
	LCA PP	919	919	0.005	$1.3 \cdot 10^{-5}$
	Total chain	1048	983	2.83	$1.3 \cdot 10^{-3}$
PC 600 MW & FGD	Direct PP	928	928	na	na
	LCA PP	929	929	0.005	$1.4 \cdot 10^{-5}$
	Total chain	1059	993	2.86	$1.4 \cdot 10^{-3}$
"Clean Coal" [b]	Direct PP	837	802	na	na
	LCA PP	838	803	0.004	$1.2 \cdot 10^{-1}$
	Total chain	951	859	2.47	$1.2 \cdot 10^{-1}$
Natural Gas CC [c]	Direct PP	363	360	0.040	$7.0 \cdot 10^{-3}$
	LCA PP	363	360	0.043	$7.0 \cdot 10^{-3}$
	Total chain	434	390	1.85	$7.5 \cdot 10^{-3}$
Nuclear	Direct PP	--	--	--	--
	LCA PP	3	3	0.017	$8.7 \cdot 10^{-5}$
	Total chain	9	8	0.03	$4.6 \cdot 10^{-4}$

na = not available.

[a] Rounded values. Some figures are given with three digits in order to appreciate the differences. "Direct PP" identifies the releases during operation of the power plant. "LCA PP" includes emissions associated with the materials used for construction; in case of nuclear power plant, it also includes the emissions from the testing of emergency diesel generators and the energy used for construction of the plant.

[b] Arbitrary mix of 50 percent each AFBC and IGCC.

[c] Averaged considering different origins of the gas.

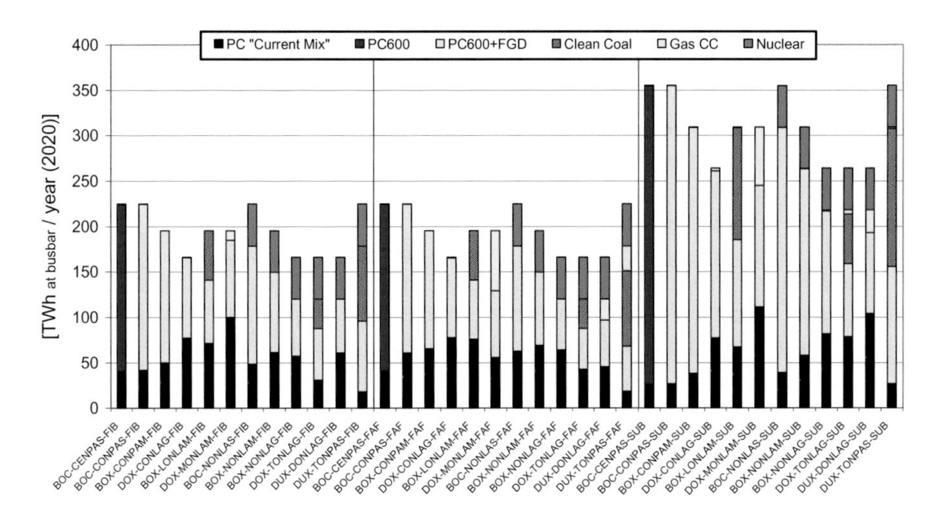

Figure 8.45 Electricity supply shares in 2020 by generation technology for the MCDA scenarios.

For each future, three supply levels are defined, depending on the existence or not of Demand Side Management (DSM).[52] However, no LCA analysis of DSM has been attempted for this project. The variations that this analysis would introduce are difficult to estimate. Possible efficiency variations of supply technologies, as modeled in ESS, have also not been considered. However, it is estimated that these would have changed the results by only few percent, probably remaining within the uncertainty range for all the LCA calculations.

In a scenario name, the first group of three letters describes the possible actions hypothesized for power plants existing in 1999. From left to right the key is: retiring inefficient, and thus uneconomical, units (D, if not B); retrofitting with FGD (U, if not O); and, fueling with coal prepared with simple dry methods (X, if not C). These conditions have been implemented in this LCA task in a simplified way. The "current mix" group has the same composition of current PC's in all scenarios, with exception of the DUX strategies.

Considering that the average efficiency of retrofit FGD systems is in the range 50 to 80 percent (Fukusawa, 1997) and that not all existing plants would be retrofitted, a reduction of 50 percent of the SO_2 emitted by the power plants of the "current mix" has been assumed for the DUX strategies. The decrease in the net efficiency of the plant due to the operation of the scrubber, less than one percent, which would increase the emission rates of e.g. CO_2, has not been included because of other sources of uncertainties. Also no credit has been given to the possible reduction of TSP emission rates due to scrubbing.

No differentiation has been made here between scenarios with and without dry coal preparation as defined within the ESS task. For the simple methods envisaged by ESS, i.e. jigging and screening already implemented today for steam coal at large mines, the additional electricity use is rather small. In the context of scenario analysis, four kWh per tonne of steam coal produced have been added to current mining energy requirements in order to maintain consistency with ESS assumptions of partial cleaning of the coal. However, the energy requirements associated with more intense coal processing for future conditions have been included to some extent in the analysis of coal mining for future chains (see Tables 8.10 and 8.11).

Another approximation made here concerned the efficiency of the scrubbers for newly built power plants, assumed constant at 90 percent whereas in ESS cheaper seawater scrubbers with 80 percent removal efficiency were assumed for coastal locations in addition to wet/dry FGD's with 90 percent efficiency.

Table 8.22 presents the coal requirements for the four coal technologies considered. Direct needs are compared with total LCA requirements calculated to be about four percent higher due to the electricity requirements of mining and transport.

Table 8.22 Coal requirements for the coal technology mixes and classes used in the estimation of MCDA scenarios for year 2020.

Coal power plant mix or type	Coal requirements (g/kWh at busbar)	
	Direct to power plant	LCA calculated
PC "Current Mix"	467	486
PC 600 MW	418	435
PC 600 MW & FGD	422	440
"Clean Coal"	365	380

Figure 8.46 shows the annual coal supply from within and outside Shandong, assumed for the selected strategies. The imported coal is assumed to come from Shanxi. The scenarios assume a range 41-152 Mt total coal use in 2020, depending on the total electricity demand, the share of coal systems, and the technology used for coal power plants (hence their efficiency).

The shares of imported (SX) coal to the total supply vary in the range of 59 to 94 percent (average nearly 80 percent). The very high imports in some of the scenarios appear unrealistic[53] and differ from the assumptions made for the LCA calculations for the coal chains previously described (60 percent coal from Shanxi in 2020). However, to be as consistent as possible with the assumptions made in the ESS task, the LCA calculations shown here consider the given shares. A significant consequence is the increase of GHG due to longer average transport distance (1000 km assumed from Shanxi, against 150 km within Shandong) and higher methane emission from Shanxi mines.

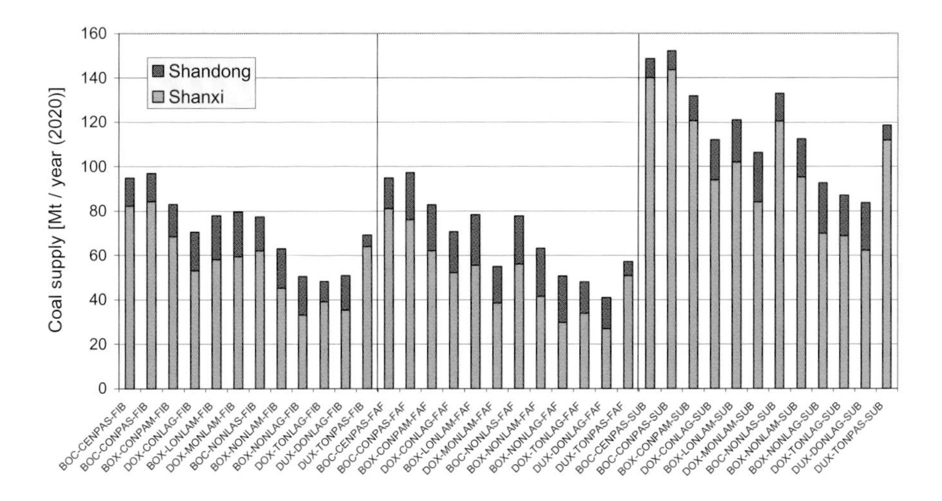

Figure 8.46 Annual coal supply in 2020 from within and outside Shandong, assumed in the ESS scenarios selected for the MCDA Task.

8.2 Results

The GHG emission rates as well as their breakdown into the contributions from the different classes of technologies are illustrated in Figure 8.47. The coal chains dominate in all scenarios. Only MONLAM-FAF has a relatively high contribution of GHG from gas, 17 percent of the total for 34 percent of the annual electricity supplied by GCC. Obviously, substituting coal with nuclear drastically reduces GHG. Because of the higher shares of coal in all SUB strategies with mixed fuels, their average GHG rate is somewhat higher than for the corresponding strategies for futures with lower demand growth.

All Shandong scenarios display GHG emission rate in the range 700 to 1120 g(CO_2-equiv.)/kWh at the busbar, always higher than the approximately 500 g(CO_2-equiv.)/kWh for the UCTE mix. The PC-dominated strategies show GHG emission rate at around 1100 g(CO_2-equiv.)/kWh. Although COMPAM and CONLAG are characterized by lower electricity produced than CENPAS and CONPAS within the same future group (see Figure 8.45), the GHG emissions are practically the same.[54] The explanation lies in the simplifying assumptions for the "current mix", constant for all scenarios. When clean coal technologies are contributing to the coal generation mix (as in LONLAM, TONLAG, and TONPAS strategies), the GHG emission rate is reduced by a few percent. When GCC's are installed (as in the MONLAM, DONLAG, and TONPAS), the fossil fuel switch from coal to gas causes a sizeable decrease of GHG (the range is 850 to 980 g(CO_2-equiv.)/kWh).[55] When nuclear is also contributing to the generation mix, the GHG emission rate is further reduced, down to 700-900 g(CO_2-equiv.)/kWh.

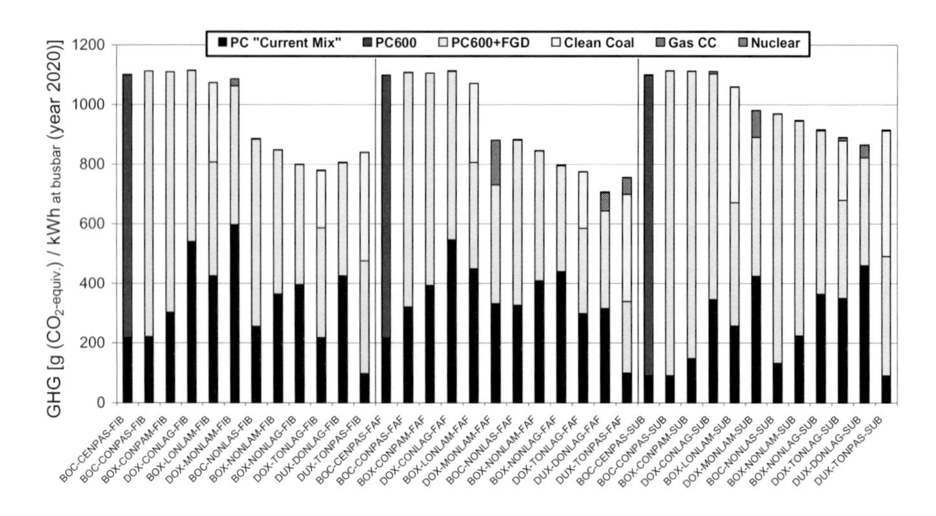

Figure 8.47 Greenhouse gas emission rates (full chain) by generation technology for the MCDA scenarios for year 2020.

Figure 8.48 shows the LCA-calculated contribution of the three main GHG species to total emission rates. The non-CO_2 species contribute six to ten percent of the total. Methane, predominantly from SX mines, is the second contributor with six to nine percent. Nitrous oxide contribution is barely visible, slightly less than two percent of the total for scenarios with AFBC's.

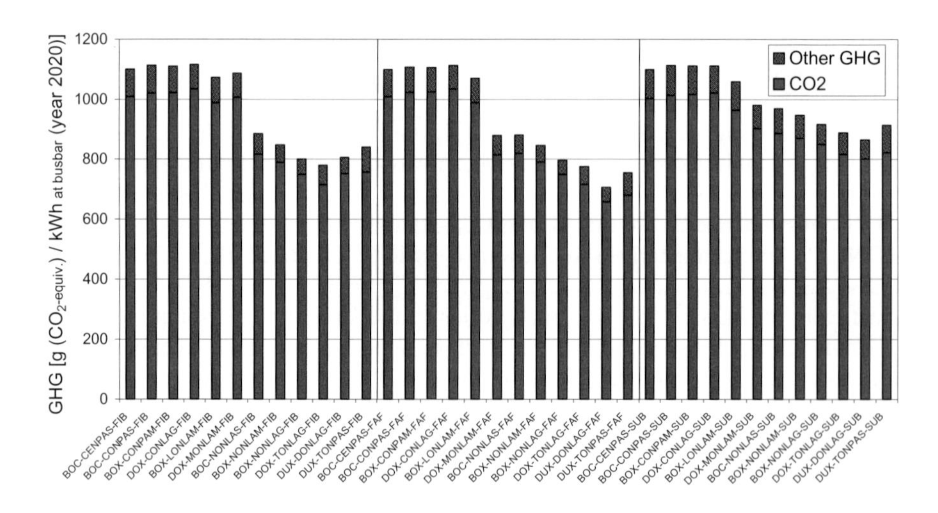

Figure 8.48 Greenhouse gas emission rates by main species for the MCDA scenarios for year 2020.

The scenario with the highest calculated non-CO_2 contribution is CONPAS-SUB with 99 g(CO_2-equiv.)/kWh at the busbar, because it has the highest electricity demand combined with the highest installed capacity of units with FGD and the highest import of coal from SX. The consideration of GHG species other than CO_2 does not substantially change the ranking of the 12 strategies within the three selected futures.

Figure 8.49 shows the LCA-calculated annual GHG emissions in million tonnes of CO_2-equivalent in 2020. The contributions from the operation of power plants are given separately from the rest of the energy chains. These add rather homogeneously in all scenarios in the range of 13 to 17 percent, depending mainly on the amount of coal transported from SX, the amount of gas transported long-distance, and the nuclear capacity. Within each future group, the 12 strategies exhibit similar rank, with a few exceptions due to changing shares of specific technologies for a few strategies. With an average GHG total emission for the Shandong mix in 1998 of 1270 g(CO_2-equiv.)/kWh at the busbar and 78.5 TWh supplied, the total GHG emission was about 100 million tonnes of CO_2-equivalent. Therefore, the results in Figure 8.49 show that in the case of moderate electricity demand growth, high DSM, and a large share of non-coal technologies the total annual GHG for Shandong electricity increases just a few tenths of a percent compared with the current annual value, whereas with high demand and conventional coal only, the annual GHG may increase up to a factor of four.

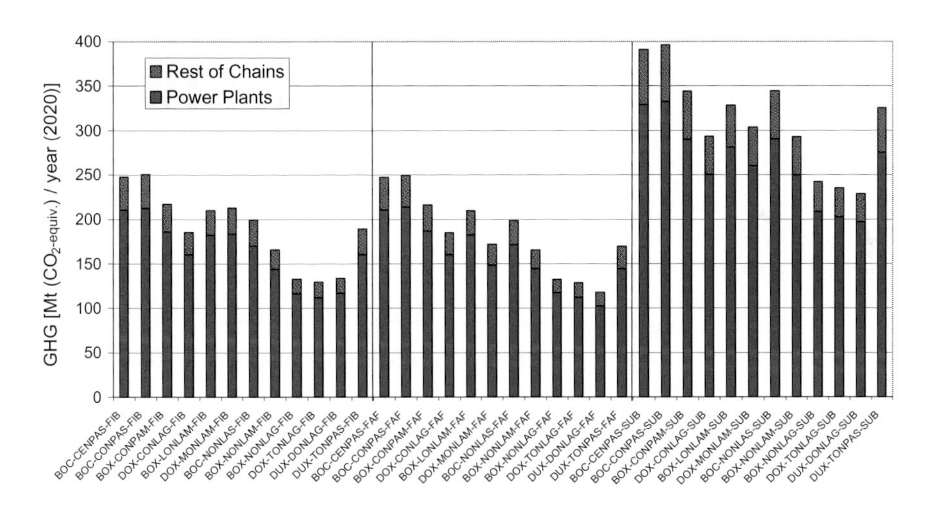

Figure 8.49 Annual GHG emissions from power plants and the rest of energy chains for the MCDA scenarios in year 2020.

Figure 8.50 shows the breakdown of the full-chain contributions to the SO_2 emission rates from the six technologies. Emissions from the coal systems dominate, even in strategies including extensive implementation of sulfur emission control in newly built or refurbished coal plants and important shares of natural gas and nuclear. The upstream part of the coal chain contributes eight to 46 percent; the higher values are obtained for the TONPAS strategy with the cleanest coal power plants. The shares of "current mix" plants and of modern PC without FGD (CENPAS) are of course the most important factors. Changing the key assumptions for these two classes would directly influence the average sulfur dioxide emission rates. For example, if the coal burned in these plants contained less sulfur than in 1999 or if they were equipped with simplified FGD's, the emission rate calculated for these scenarios would substantially decrease. The lowest emission rates are calculated for the TONPAS strategy, which combines the lowest contributions from the "current mix" with the use of scrubbers in new PC's as well as refurbishing old plants, employment of AFBC/IGCC, and nuclear power. For comparison, full-chain SO_2 from future UCTE coal plant mixes is estimated between 0.7 g/kWh and 1 g/kWh at the busbar.

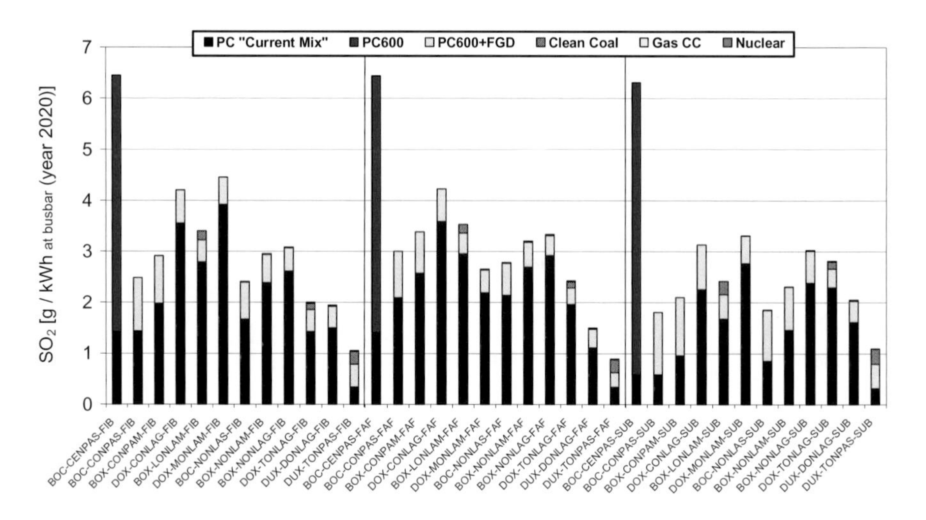

Figure 8.50 SO_2 emission rates (full chain) by generation technology for the MCDA scenarios for year 2020.

Figure 8.51 shows the LCA-calculated annual SO_2 emissions in kilotonnes per year in 2020, separating the power plants from the rest of the chains. Across the future groups the ranking of the twelve strategies changes due to differences in the share of specific supply technologies in corresponding strategies. For example, MONLAM-FIB and MONLAM-SUB are manifestly second in ranking within their respective FIB and SUB future groups, while their symmetric scenario MONLAM-FAF is below the other scenarios based on only coal within the FAF future group, because of the high share of gas displacing coal. From the definition of scenarios (see Figure 8.45), the share of current PC's is the lowest in MONLAM-FAF (43 percent of total coal vs. 54 percent and 46 percent for FIB and SUB, respectively), and at the same time the percent contribution of gas is the highest among the three scenarios MONLAM (34 percent of total supply of electricity at busbar vs. five percent and 21 percent for FIB and SUB, respectively).

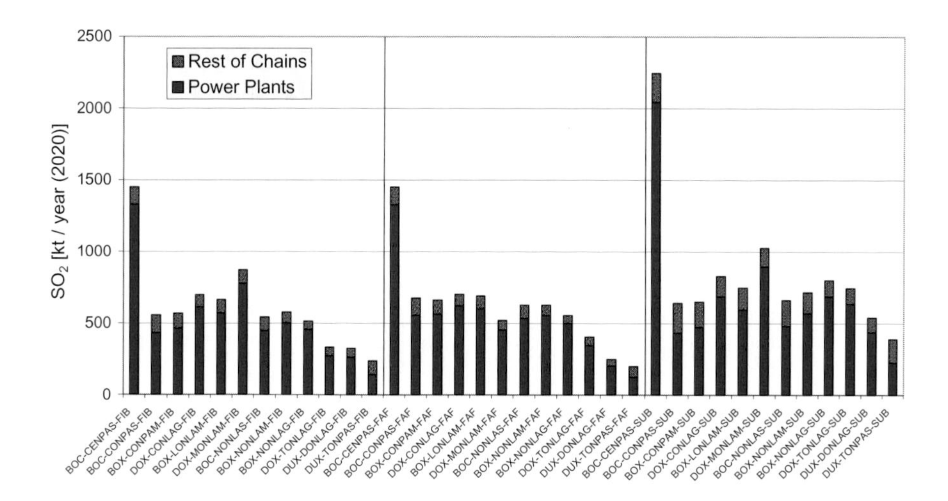

Figure 8.51 Annual SO_2 emissions from power plants and the rest of energy chains for the MCDA scenarios in year 2020.

The clear message from the analysis is that in the case of implementation of no environmental policy as in CENPAS, where only PC's without FGD are installed, the total sulfur emissions may increase dramatically in proportion to the demand growth. Obviously, the construction of new plants with scrubbers reduces this dangerous level dramatically. In Shandong, the annual levels from grid-connected PC's in 1998 and 1999 can be calculated using the average SO_2 emission of nearly 11 g(SO_2)/kWh at the busbar and nine g(SO_2)/kWh, respectively, for an energy supply of approximately 78.5 TWh and 81.3 TWh at the busbar, respectively. The resulting annual SO_2 emissions are calculated at about 864 kilotonnes for 1998 and 732 kilotonnes for 1999.[56] This would mean that in the next 20 years the sulfur emissions may be significantly decreased or at a minimum stabilized with environmentally oriented policies, though demand may increase by a factor of about two or more.

In general, unless specific interventions are imposed locally, the highest priority today for reducing sulfur emissions should be on coal power plants. However, with the dramatic reduction of SO_2 emission rates from power plants, and increasing demand for coal, the relative importance of emissions related to mining operation and transport will increase, calling for increased use of FBC technology or refurbishment of existing small power plants with FGD's for powering coal mines.

Figure 8.52 shows the NO_x emission rates. The emissions from coal systems are overwhelming even in scenarios with substantial gas and nuclear capacity. Figure 8.53 shows the annual NO_x emission in year 2020.

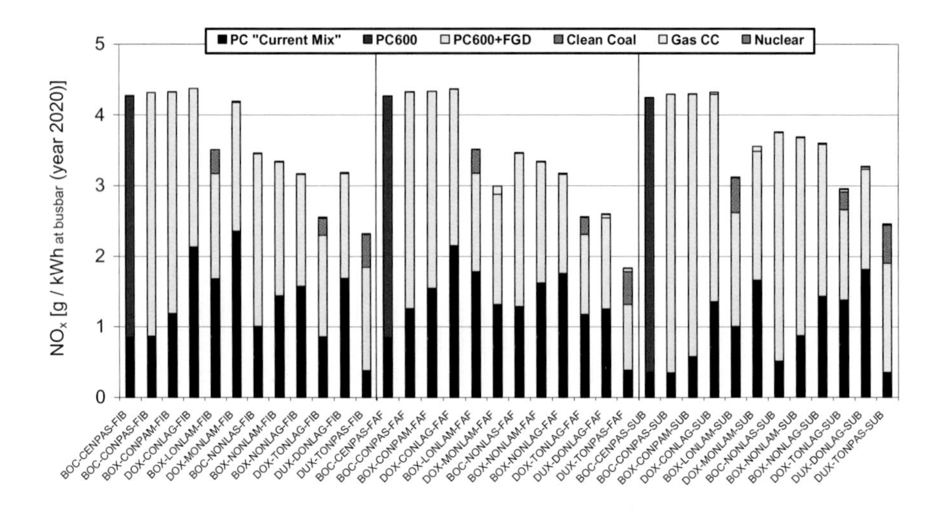

Figure 8.52 NO_x emission rates (full chain) by generation technology for the MCDA scenarios for year 2020.

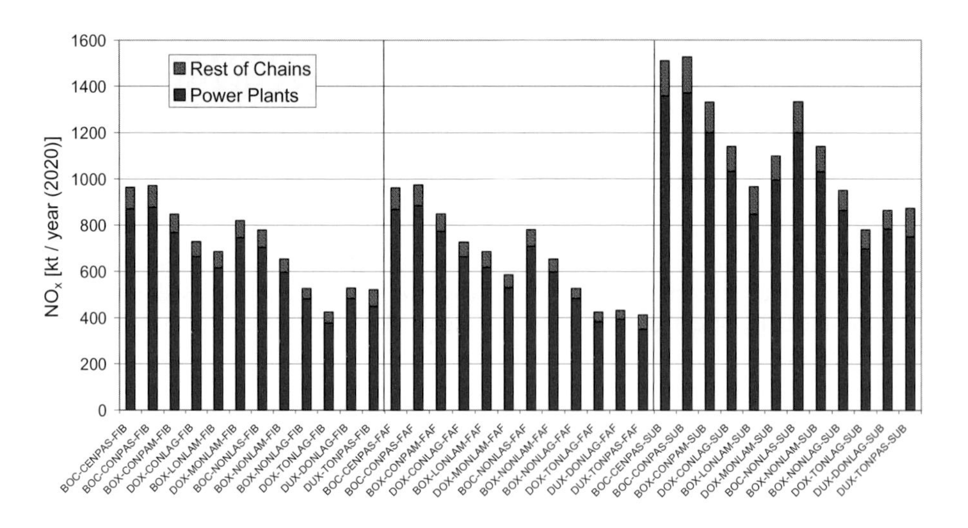

Figure 8.53 Annual NO_x emissions from power plants and the rest of energy chains for the MCDA scenarios in year 2020.

The releases not originating at power plants contribute eight to 15 percent. The shares of specific types of coal power plants are the key factors. The TONPAS strategy exhibits the greatest relative contribution from upstream chains because it has the highest generation from clean coal power plants and the lowest generation from current PC's. The total NO_x emissions in 1998 are calculated at about 408 kt.

This would mean that environmental policies oriented towards reducing SO_2 but not directly at nitrogen oxides would stabilize but hardly reduce emissions of this species from the current level.

Figure 8.54 illustrates the contributions of the different classes of technologies to the emission rates of particulates. It is apparent that practically only the coal chains contribute.

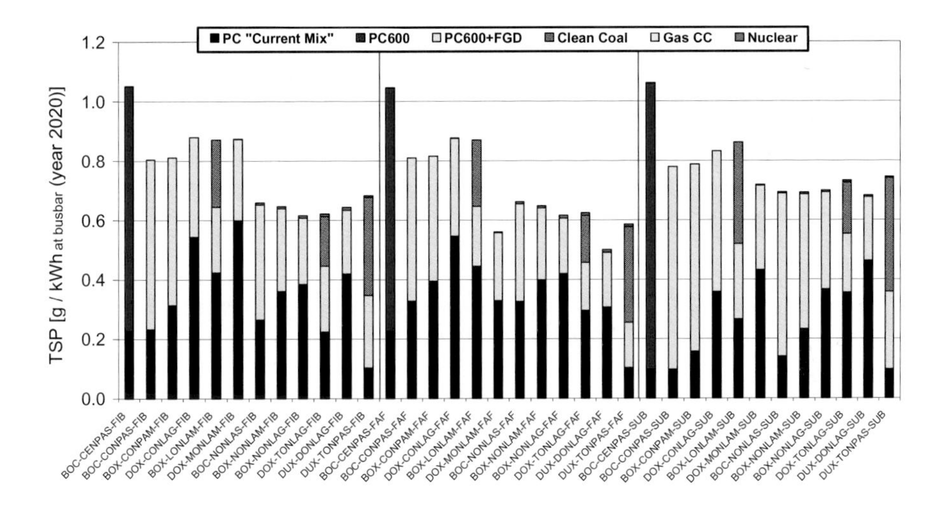

Figure 8.54 TSP emission rates (full chain) by generation technology for the MCDA scenarios for year 2020.

Figure 8.55 shows the LCA-calculated annual TSP emissions in kilotonnes for the assumed scenarios in 2020. The share of specific coal power plant technologies is again the key factor in explaining the differences. In 1998 the total TSP emission corresponding to power production in Shandong was approximately 448 kilotonnes, of which 66 percent were directly emitted from non-SEPCO small units. Hence, the results show that the implementation of any of the twelve strategies reduces the annual TSP emission, when also reduction measures are taken at mine-mouth small power plants. The CENPAS scenarios show again the worst performance.

When FGD's are installed in all new PC's, as in CONPAS, the emissions directly from PC's decrease substantially. However, while it is easy to explain the differences for the above mentioned two strategies, it is difficult to compare other strategies within one future group in detail because the total energy supplied at busbar may differ (see Figure 8.45). It is worth mentioning that excluding the CENPAS scenarios, the others show results in a relatively narrower range than for other emission species, due to the higher relative contributions from coal mining and transport.

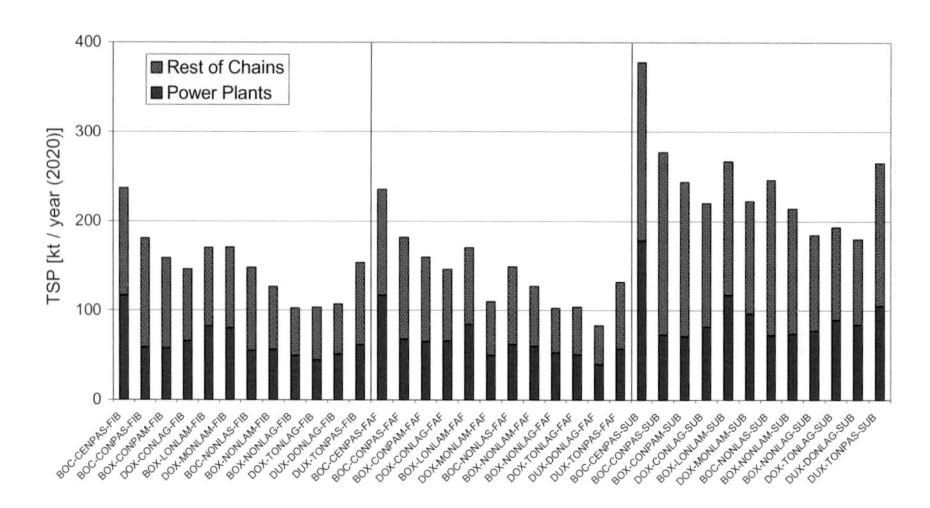

Figure 8.55 Annual TSP emissions from power plants and the rest of energy chains for the MCDA scenarios in year 2020.

Figure 8.56 shows the contributions from the energy chains associated with the six classes of power plants to solid waste production rates. The contributions from the natural gas and nuclear chains are negligible. However, the characteristic production of highly radioactive wastes in the nuclear chain requires engineering solutions for their conditioning and long-term confinement,[57] though the volumes are relatively very small.

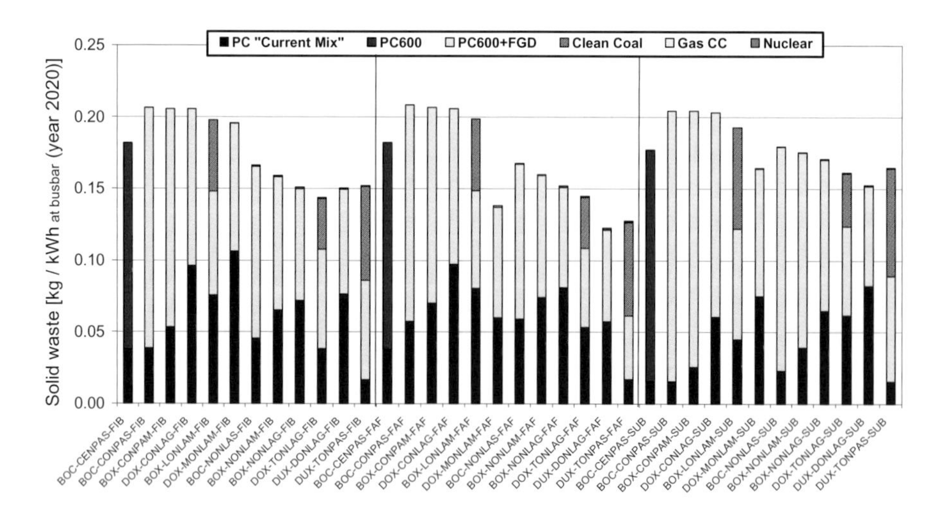

Figure 8.56 Solid waste production rates (full chain) by generation technology for the MCDA scenarios for year 2020.

The rest of the chains contribute 43 to 50 percent of the total. Differently from the airborne emissions analyzed above, the CENPAS strategy exhibits lower solid waste than other "coal only" strategies (CONPAS, CONPAM, CONLAG, and LONLAM) that are additionally burdened by the un-recycled waste from FGD's[58] and clean coal technologies, but also because it has a very high share of Shanxi coal.

Figure 8.57 shows the calculated annual wastes in 2020. The forecasted increase in the share of SX coal will contribute to the decrease of the total wastes for the corresponding coal chain. Current (1998) PC total wastes production rates have been estimated at about 0.25 kg/kWh. This adds up to an annual production of about 18.8 million tonnes, of which 4.3 million tonnes (or 23 percent, for nearly 15 percent of electricity supplied) are produced by small, non-SEPCO units. The conclusion of the analysis is that it will be extremely difficult to keep the annual generation of waste at the current level unless a high recycling is pursued, fuels alternative to coal are used, and strong DSM implemented.

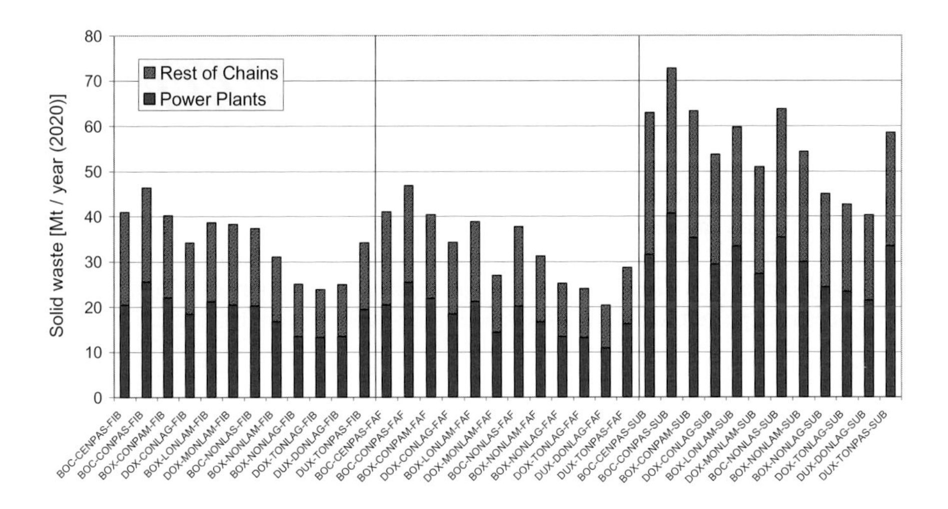

Figure 8.57 Annual total solid waste production from power plants and the rest of energy chains for the MCDA scenarios in year 2020.

The total land use is dominated by the coal chains, as can be seen from Figure 8.58. Though the contribution of nuclear to total land use is of the order of a few percent only, the land used for depository of LLW from mining and milling may remain contaminated by natural radioactive isotopes, which requires costly operations of reclamation and control of leachates for many decades. Moreover, while ISL infrastructure occupies only a small amount of land, the possible contamination of groundwater is an issue of environmental concern that could not be addressed here.

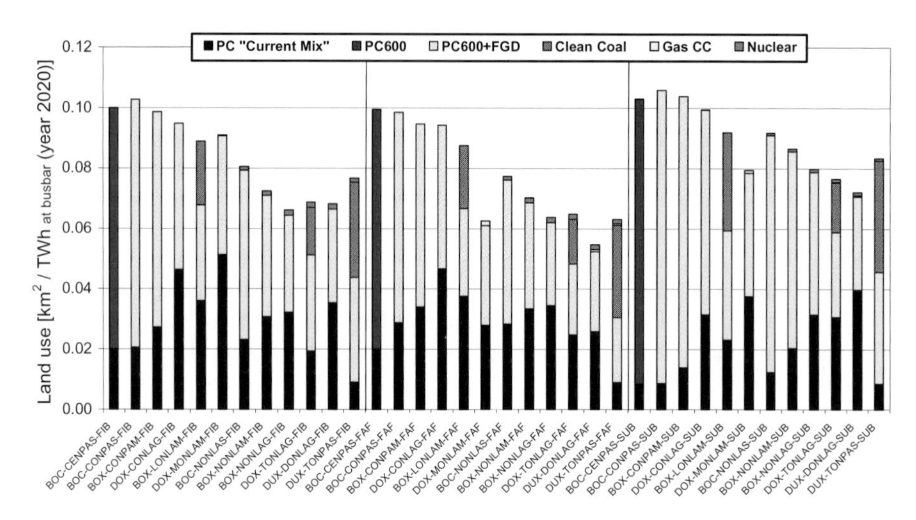

Figure 8.58 Land use rates (full chain) by generation technology for the MCDA scenarios for year 2020.

Figure 8.59 shows the annual total land use. Reflecting the results for coal chains, power plants cause only one to two percent of the total burden. The total land use calculated for electricity supply to Shandong in 1998 is about 6 km^2/TWh at the busbar. If the several assumptions of this study are realistic, it will be impossible to keep the land use at the current annual rate because, in order to cope with the increased demand of fuel, a substantial increase in the import of coal from other provinces is expected, with the associated higher burdens from transport.

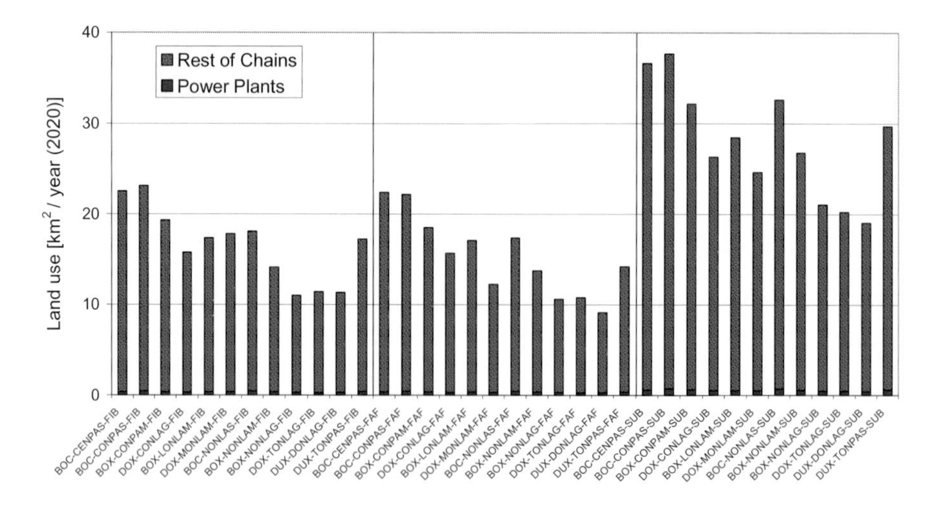

Figure 8.59 Annual land use from power plants and the rest of energy chains for the MCDA scenarios in year 2020.

8.3 Health impacts calculated with LCA airborne emissions

The life-cycle inventories of the airborne species SO_2, NO_x and TSP are used in the EIA task. Using the year 2020 mortality factors for the unit mass of each pollutant species given in the EIA chapter (also adopted for the chain comparison in Section 6 above), and the emission inventories given in Table 8.20, the LCA-based health effects in YOLL per GWh for the technology classes considered in this scenario analysis are calculated as shown in Figure 8.60.

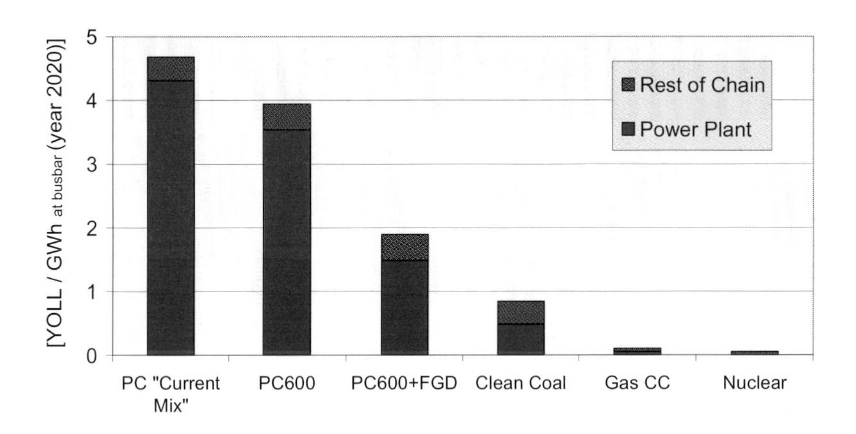

Figure 8.60 LCA-based health effects for technologies considered in the LCA analysis of the MCDA scenarios in year 2020 (excluding GHG).

For coal technology, a decisive reduction of health damage can be obtained only through the installation of scrubbers or the use of advanced technology. Considering the very low values for the natural gas or nuclear chains, substitution of current PC's with these technologies dramatically reduces the effects of annual air pollution on human heath per unit of electricity service. Figure 8.61 shows the YOLL calculated for the scenarios. As for the emission inventories of the single species and burdens reported before, the results may differ somewhat from the ones calculated by EIA and ESS, though the upstream input for fuel-chain-related emissions is consistent. However, the advantage of the LCA approach is that the understanding of the ranking of scenarios is more straightforward.

One firm conclusion is that the health effects originate almost entirely from the coal systems; quite obviously the performance of a mix improves as conventional PC technology is replaced by PC with scrubbers, clean coal, natural gas, and nuclear. For most of the scenarios, SO_2 is the major cause of mortality, up to 65 percent. For all scenarios TSP exhibits the smallest share to total, between seven and 17 percent. The emissions of NO_x cause greater health effects than SO_2 only in a few strategies including widespread implementation of sulfur control technologies.

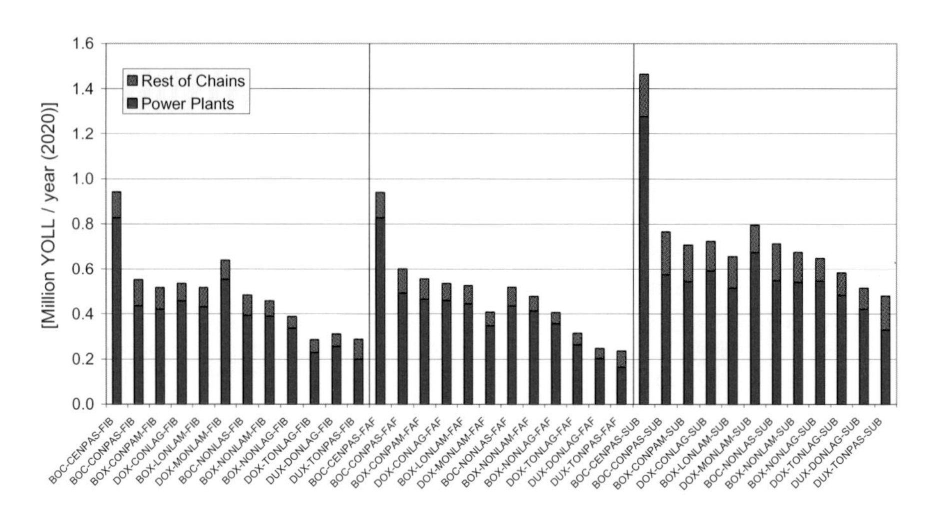

Figure 8.61 LCA-based annual health effects from power plants and the rest of energy chains for the MCDA scenarios in year 2020.

9. CONCLUSIONS

This LCA study provides a first sound picture of the serious environmental impacts from the energy chains associated with coal power plants in Shandong, which are supplied by major mines. The burdens from future chains are speculative, but based on realistic assumptions and extrapolated from results of assessments on similar technologies for other conditions in Western European countries. In addition to the ranking among selected environmental inventories for coal, natural gas, nuclear, and wind technologies, the study gives a quantitative evaluation of the burden levels from the various stages of individual chains.

Reasonably complete data could be collected for only a few important environmental burdens. The air emission species considered are: SO_2, NO_x, particulates and the greenhouse gases CO_2, CH_4 and N_2O. Only limited information was gathered on solid waste from the operation of power plants and coal mines. However, they are sufficient to delineate a first picture of the amount of wastes from the full chain. Waterborne emissions as well as radioactive emissions to air and water from Chinese nuclear and coal facilities could not be included. Even though a survey was carried out in order to gain direct information from Shandong and Shanxi coal mines, information on operating characteristics and pollutant emissions from small power plants at mine sites remains limited. The application of upper-range estimates for emissions and energy intensity in mining operation significantly changes the relative contribution of the mining step to total emissions from a chain.

The analysis shows that the coal chains relevant to Shandong and to China in general generate environmental burdens substantially higher than those from chains associated with Western European hard coal power plants. While the power plants produce most of the pollution, emissions from coal mining can be considerable, for example up to 25 percent of total GHG when high rates of spontaneous combustion mine fires are assumed across China. The environmental performance of coal chains is generally the worst amongst the energy chains considered for all selected burdens. Although the introduction of improved or new coal technologies may meaningfully reduce the harm to human health from major airborne pollutants related to the unit of supplied electricity, the effects of the coal chain will remain still higher than those from the same species directly emitted or associated with the natural gas or the nuclear chains. Therefore, improved coal technologies should be complemented by gas and/or nuclear generation to meet the growing demand for electricity and to control emissions.

The robustness of the results has been tested by varying key parameters in sensitivity analyses; an accurate uncertainty analysis is not possible. It is estimated that the range of total burdens from current coal chains shall stay within ten percent of the calculated averages. One key factor is the origin of the coal, which may influence the calculation of total burdens by about ten percent, because of conditions at the mines and the burdens that are proportional to transport distances. Hence, the assumption about the share of coal imported into the province is an important factor in estimating the burdens for different scenarios. In the following, the results on selected burdens estimated for the assessed energy chains and for future scenarios in 2020 are summarized and compared with Western European figures.

The greenhouse gas emissions assessed for the Shandong grid mix and SEPCO mix at the busbar in 1999 are approximately 1270 g(CO_2-equiv.)/kWh and 1200 g(CO_2-equiv.)/kWh, respectively, about 12 percent of each originating from the upstream chain. To put these figures into a country-wide perspective, if the average power plant sizes and coal quality were the same for China as for Shandong, and if average country-wide conditions were assumed for mining and transportation, a preliminary value calculated would be about 1540 g(CO_2-equiv.)/kWh at the busbar, of which about 25 percent comes from the upstream chain. However, if the reference average coal power plant for China is a 300 MW PC, a value of nearly 1400 g(CO_2-equiv.)/kWh at the busbar would result. Using the low end of the estimated range for coalfield fires would further bring down these chain-specific emissions by about seven percent.

For comparison, the average full chain GHG emissions rate from hard coal technologies in UCTE in 1995 has been calculated in the Swiss LCA study on Western European energy systems to approach 1080 g(CO_2-equiv.)/kWh at the busbar, with a range of approximately 980-1260 g(CO_2-equiv.)/kWh for UCTE country-specific hard coal mixes, depending on the average plant efficiency. Based almost entirely on coal, the current (1999) Shandong mix at busbar emits nearly 2.5 times more chain-related GHG per unit electricity than the UCTE mix 1995,[59] whereas the SEPCO mix, based on more efficient units, emits 2.3 times more. The values for the MCDA scenarios in 2020 are in the range 1060-1120 g(CO_2-equiv.)/kWh at the busbar for the case of PC power plants, and in the approximate range 700-980 g(CO_2-equiv.)/kWh at the busbar for mixes including clean coal technologies, nuclear and/or natural gas. For comparison, the expected values for the UCTE mix around 2010 are on the order of 500 g(CO_2-equiv.)/kWh at the busbar, which might further decrease if policies supporting the expansion of renewables are combined with increased use of natural gas and there is no major substitution of fossil for nuclear capacity.

Underground coal mining contributes about 80 percent to the upstream emissions of GHG for chains associated with current Shandong power plants. This value includes only major mines, but a similar conclusion could be expected when minor mines are also included, because their contributions to future total supply to power plants are small, and their energy intensity is lower due to the high share of manual work (Horii, 2001). The important contribution from coal mining in the Chinese case can be compared with the contribution of about five percent from coal mining to total GHG from chains associated with Western European coal power plants. This relatively low percentage is mainly due to the partial recovery of methane from underground mines in the countries supplying the coal to Western Europe, and the comparatively low rate of major coalfield or mine fires in the USA and Australia, or even the near absence of such fires in Europe and South Africa. Hence, there is a strong potential for reducing the GHG emissions from the Chinese coal chain by intervening at the mine fields, in addition to the reduction to be obtained by increasing the efficiency of power plants. Improvements in this sense have been credited for the future coal chains addressed in this study.

The transportation of coal to Shandong power plants contributes only minimally to total GHG, especially when local coal is used, but it may reach about five percent of the total when the coal is transported long distance. However, because these values are affected by an uncertainty for the type and efficiency of the locomotives used, they should be considered only as an evaluation of the order of magnitude.

With the average sulfur content in 1998 of 1.2 percent, weighted with the electricity production of seventeen SEPCO main coal power stations (range 0.42-2.58 percent), the average SO_2 emission rates for the SEPCO and Shandong grid mixes were in the range of 10-11 g/kWh at the busbar. The upper value corresponds to about three times the average LCA-based SO_2 emission for the UCTE mix 1995 (nearly 3 g/kWh), but one order of magnitude higher than the lowest average emission rate estimated for a UCTE country in that year (range 0.1-6.7 g/kWh),[60] due to the systematic use of scrubbers in coal plants. With the mandated use of fuel with lower sulfur content in 1999, for an average of about 0.91 percent (range 0.33-1.52 percent), the values of the two Shandong mixes became of the same order of the highest average value estimated for the coal mix in a UCTE country in 1995 (8.5 g/kWh at the busbar), but still more than the double the average UCTE.

For the Shandong mixes the upstream chain contributes about five percent of total SO_2 emissions. This percentage about doubles if all coal comes from Shanxi, due to the combined effect of higher transport distance and the relatively high sulfur emission rate per unit mass of coal produced there. For future technologies aimed at controlling sulfur emissions from power plants, the contribution from the upstream chain would increase in relative terms: 40 percent for 600 MW PC with FGD, 60 percent for AFBC, and nearly 90 percent for IGCC.

The average UCTE coal plant mixes around 2020 should have low LCA-based SO_2 emissions of 0.7 g/kWh at the busbar, as clean coal technologies (mostly PFBC and supercritical PC with FGD) are progressively installed in place of the aging subcritical PC's. The values estimated for the MCDA scenarios for Shandong in 2020 are above six g/kWh at the busbar for the case of only PC's that do not have scrubbers but burn coal with relatively low sulfur content. This means a reduction of 30 percent from the rate calculated for the SD mix in 1999, but considering the much higher total demand for electricity this would originate higher annual sulfur emissions than today. For scenarios based only on coal but with FGD's installed in all plants built after 2000 (as assumed by the ESS task) the SO_2 emissions are calculated in the range of 2-4 g/kWh at the busbar. The range of specific SO_2 emissions for supply mixes, which include clean coal, nuclear, and/or natural gas technologies is 1-3 g/kWh.

The NO_x emissions associated with the Shandong grid mix and SEPCO mix at busbar in 1998 are approximately 5.3 g/kWh and 4.7 g/kWh, respectively. The upstream chain contributes eight percent. For AFBC and IGCC the share from the upstream chain increases in relative terms to 20 percent and 40 percent, respectively. The values for NO_x emissions for the current Shandong coal chains are around a factor of three greater than the average estimated for the UCTE hard coal chains in 1995 (the country-specific range was 0.9-4.6 g/kWh). However, the factor would rise to around six if the Shandong mixes were compared with the 1995 UCTE electricity production mix which exhibited a NO_x average emission of 0.84 g/kWh at the busbar (the country-specific production mixes averaged 0.04-2.84 g/kWh). The NO_x emission rates estimated for the 2020 MCDA scenarios, in which the coal share does not go below 60 percent of total supply, are in the range 1.8-4.4 g/kWh at the busbar, thus remaining well above the average for UCTE in 1995.

The total particulate emission rates associated with the Shandong grid mix and SEPCO mix at busbar in 1998 are about 5.7 g/kWh and 2.6 g/kWh, respectively. The upstream chains contribute 12 and 28 percent respectively, with mining and transport supplying about half each. The total emission rates are 2.3 to 5 times greater than the average estimated for the UCTE hard coal mix in 1995: the average was below 1.2 g/kWh, whereas the UCTE country-specific range was 0.8 to 2.5 g/kWh, which may have decreased in more recent years. However, the differences are dramatic when comparing the Shandong grid mix and SEPCO mix with the 1995 UCTE electricity production mix, which exhibited an emission of particulate of barely 0.15 g/kWh at the busbar (the country-specific production mixes averaged 0.02-0.84 g/kWh). The rates estimated for the MCDA scenarios for 2020 are in the range 0.5-1.1 g/kWh at the busbar, thus aligning these mixes with UCTE standards in the 1990's.

The production rates of total solid waste associated with the Shandong grid mix and SEPCO mix at busbar in 1998 were about 0.24 kg/kWh and 0.22 kg/kWh, respectively, of which the upstream chains contribute about 48 percent. The average calculated for the UCTE mix of hard coal power plants in 1995 was also nearly 0.24 kg/kWh (range 0.18-0.30 kg/kWh), but the coal was assumed to be entirely washed. A different perspective is given by comparing the above factors for Shandong mixes with the 1995 UCTE electricity production mix which had nearly 0.07 kg of solid wastes per kWh at the busbar (the country-specific production mixes averaged 0.02-0.31 kg/kWh). The total production rates of solid wastes estimated for the 2020 MCDA scenarios decrease to 0.12-0.21 g/kWh at the busbar.

The total land use estimated for the various coal chains is in the range 70-130 m^2/GWh. The key factors are the distance that the coal must be transported by train from the mine to the power plant, the size of the mine waste heaps, and the area affected by subsidence phenomena. The power plants themselves do not contribute meaningfully to the total.

From this assessment, the Shandong current supply mixes based almost exclusively on coal plants compare poorly with Western European country mixes. However, the environmental performance related to the electricity supply services in 2020, using coal alone or coal mixed with gas and/or nuclear technologies, should generally improve, though indicators like GHG, SO_2 and NO_x may remain relatively high. However, integral emissions over a year may increase compared to today, as is expected to be the case for GHG and NO_x.

For the analysis of the natural gas chain, four hypothetically representative routes have been considered, three with gas piped from Shandong (offshore), South Siberia (Irkutsk), and Xinjiang, and one in which LNG is transported from South East Asia (Indonesia) to the potential terminal in Qingdao. The basic data for modeling the future natural gas chain associated with GCC's in Shandong have been derived from representative international literature and existing European LCA studies. The pumping energy and leakage rate for pipeline transport, as well as the energy requirements for the LNG refrigeration chain have been estimated on the basis of current best practice.

The range calculated for the GHG emissions for piped gas is 390 to 470 g(CO_2-equiv.)/kWh at the busbar, where the rate increases with the distance traveled. In the worst case—gas transported for 6000 km from gas fields in Xinjiang—the contribution of the upstream chain is 23 percent, whereas for the nearest domestic potential fields it is only seven percent. For LNG, the total has been estimated at 435 g(CO_2-equiv.)/kWh at the busbar, of which 17 percent from the upstream chain. The total GHG emission for the gas transported from South Siberia is comparable to that calculated for the LNG chain. It can be concluded that the full chain must be assessed in order to achieve a correct picture of total GHG emissions, though in the context of national accounting the emissions taking place outside the political borders of a country may need to be ruled out.

Other selected airborne emissions for the four natural gas routes are in the ranges of 0.04-0.05 g(SO_2)/kWh; 0.22-0.37 g(NO_x)/kWh; and, 0.006-0.017 g(TSP)/kWh. The highest values for SO_2 and TSP arise from the chain for Xinjiang gas; for NO_x the highest emissions come from the LNG chain. However, the results for SO_2 should be taken with prudence, because assumptions maximizing the quantity of sulfur in the extracted gas have been arbitrarily made in the absence of specific information. Solid wastes are produced in very small quantities in the range of 2.8-3.6 g/kWh. Total land use depends strongly on the assumption made about pipelines: here the range for the three land routes is approximately two to nine m^2/GWh, whereas for LNG it has been calculated to be three m^2/GWh.

For the nuclear chain based on LWR, an open cycle without reprocessing of the spent fuel has been conservatively considered, in order to maximize the non-radioactive emissions. The key assumption for the estimation of these burdens is on the uranium enrichment process. Centrifuge technology has been assumed for all enrichment services, replacing the aging and much less energy efficient diffusion plants. However, even with the extreme assumption of only diffusion enrichment powered by coal-burning plants, the nuclear chain would still have non-radioactive air emissions one order of magnitude lower than the chain associated with clean coal technologies. For uranium mining, it has been necessary to estimate specific Chinese conditions in spite of poor information, assuming that 50 percent of uranium will be obtained by chemical extraction (ISL). The spent fuel is assumed to be simply deposited deep underground after encapsulation.

Though it has been impossible to inventory radioactive emissions from Chinese facilities, data from other LCA studies could have been imported for use in scenario calculations. The results of heath impact studies of the full nuclear chain for European conditions show that the radioactive emissions contribute minimally to total effects. However, these figures have not been included to avoid excessive speculation. Certainly, this missing element as well as others listed in the "Outlook" section (below) for all chains deserve further study in order to support more complete and conclusive comparisons of the total environmental burdens from energy conversion under Chinese conditions. In terms of airborne emission rates, GHG scores nine g(CO_2-equiv.)/kWh; total SO_2 emission is 0.08 g/kWh; NO_x 0.05 g/kWh; and particulates 0.03 g/kWh. The non-radioactive solid wastes are approximately 3.4 g/kWh; the low-level radioactive waste is less than seven g/kWh; the intermediate-level radioactive wastes, including parts of the decommissioned power plant, are around 0.07 g/kWh; and the high-level radioactive wastes below 0.008 g/kWh. The total mass of solid wastes is therefore about 10 g/kWh. The land use has been calculated at about 6 m^2/GWh.

Though not explicitly included in the MCDA scenarios, the estimation of total emissions associated with an 800 kW wind turbine for allegedly optimal wind conditions in Shandong has been performed. The estimate was obtained by extrapolating to approximate Chinese conditions the LCA results obtained using average European conditions for the manufacturing of materials. The calculated emission rates of air pollutants are: GHG of nearly 20 g(CO_2-equiv.)/kWh; 0.22 g(SO_2)/kWh; 0.04 g(NO_x)/kWh; and 0.16 g(TSP)/kWh. The solid wastes are just four g/kWh, because most of the materials except the foundations can be recycled. The land use has been estimated at about 0.9 m^2/GWh, assuming that half of the installed capacity will be offshore.

Energy policy decision-makers need to know whether full chain environmental analysis yields a different ranking of options for electric power production than an analysis using only the direct burdens from power plants. The assessment of the total burdens calculated for the year 2020 strategies selected for the MCDA task shows how such inclusion of upstream and, for nuclear, downstream chains change the ranking of these options, especially for particulate emissions, production of solid wastes, and land use. Conversely, the ranking of scenarios for GHG, SO_2, and NO_x is not influenced greatly by the inclusion of upstream chains, due to the large airborne emissions from coal power plants, especially those heavily polluting units operating today that may still be running without major refurbishment in the future.

The LCA results calculated for scenarios in 2020 may not reflect the integral of the burdens over the next 20 years, because most of the innovative technologies as well as the non-fossil plants will be installed in the second decade. However, they are a meaningful measure of the policy trends they represent. However, due to yearly discrepancies among the individual contributions of different technologies to the calculated supply mixes, as implemented within the ESS task, variations in the burden ranking of different scenarios may occur for different years.

This study quantifies the potential for reductions of total emissions, which can assist energy policy decision-makers effectively for establishing a priority list for technology choices. The current Chinese policy of shutting down very polluting and inefficient small units is certainly highly beneficial. This study further suggests that a first priority must be to increase the net efficiency of coal power plants, extending the use of highly efficient particulate control technologies, introducing scrubbers, and at later stages possibly promoting NO_x control technologies.

However, increasing attention should also be paid to pollutant emissions at mines, especially particulates and SO_2 released from small power plants used at the mine-mouth for self-supply of electricity. The shut-down of these obsolete power plants should be coupled with the installation of more efficient units that may combine the burning of residuals from mining and washing—making full use of on-site resources—with pollution control. However, this choice should be evaluated economically and ecologically against other solutions. A rational option, if physical conditions (e.g., sufficient water resources) as well economic conditions (price of coal, price of electricity from the grid, privatization of energy enterprises) permit, appears to be installation of large mine-mouth plants. In this case, mining residuals cannot fuel them. These units would not only generate electricity to cover the entire needs of the mines, but also dispatch the excess electricity to grids. Lower electricity costs at mines may also facilitate investments in coal-washing technology. However, a comparison between different options for power supply to mines was beyond the scope of this analysis and therefore no conclusion can be drawn here.

Furthermore, increased recovery of methane in Shanxi's gassy mines should be pursued for its twofold effect: rational use of resources and decrease of the coal-mining contribution to greenhouse gas emissions. The same arguments support a strong policy to reduce spontaneous combustion of coalfields and coal waste stockpiles. Use of scrubbers, AFBC, or IGCC technology for base load power plants should also encourage the utilization of local Shandong coal with relatively high sulfur content. Such a policy would have the dual effect of reducing the amount of low-sulfur coal transported from the outer provinces, and freeing the supply of low-sulfur coal for applications for which the installation of expensive and energy-consuming emission control devices may not be as economical as it is for large power plants.

Recycling of solid wastes from coal power plants should be pursued to the greatest extent possible, while recognizing the limitations that may arise from the composition of the waste (for advanced technologies like AFBC and IGCC) and the radioactivity content in the ashes. The allocation of gypsum from FGD's to the building industry should be pursued already in the planning stage. Solid wastes from coal mining and coal washing with valuable energetic content are already being utilized for on-site energy conversion, as mentioned above. Other uses of mining waste lie in the construction sector and in the prevention of subsidence by refilling shafts and tunnels. Conditioning and reclamation of waste heaps should be implemented as a regular practice to decrease pollution of water bodies, to regain or protect land, and to prevent spontaneous fires.

For the natural gas chain, if the gas is piped, the pumping efficiency and the control of leaks should be maximized to achieve the highest net efficiency of the entire system and to reduce greenhouse gas emissions. Efficiency improvements should also be attained throughout the LNG chain. Natural gas could be used mostly for applications other than power production, for example for space heating or fertilizer production as mentioned in several studies for China. This allocation would concentrate the burning of highly polluting fuels such as coal in large facilities that can be equipped with relatively expensive pollution removal technology. This policy choice is a matter of economics and the rational use of resources that should be assessed on a case-by-case basis. Suitable LCA studies would help by putting various alternative uses of the gas into perspective. This has not been attempted here because the project has been focused on the comparison of electricity production systems only.

Compared to coal, the nuclear chain exhibits much lower inventories of the airborne species discussed in this study. The centrifuge process for the isotopic enrichment of uranium is preferable to diffusion from both environmental and economic perspectives. Though not quantified here, the possible effect to groundwater by in-situ leaching of uranium must be prevented. The radioactive emissions to air and water from the facilities in the nuclear chain should comply with international standards. Radioactive solid wastes should be conditioned and eventually disposed of in suitable repositories, their hazard minimized to the greatest extent possible. Under these conditions, the health effects from total radioactive emissions are assumed to be minor.

Though not fully modeled in this study, certainly one of the strongest recommendations is to promote the reduction of the still large losses in the electric networks at all voltage levels, in order to minimize the costs and the environmental burdens per unit of electric service. This is already one of the current goals for energy policy in China.

Considering the objective limitations of the data and the focus on chains associated with power plants installed or to be built in Shandong province, the results of this analysis should be considered with appropriate caution when used in a different context. Possible consequences of environmental factors not fully addressed here because of missing information or quantified on the basis of engineering judgment only, should not be ignored or underestimated at any stage in the decision making process. Nonetheless, the large differences assessed for the selected environmental burdens among coal, natural gas, nuclear, and wind energy systems are such that the main conclusions of the LCA as well as other CETP tasks using LCA results can be considered sufficiently robust.

10. OUTLOOK

The LCA work performed for CETP on Chinese energy chains has been pioneering. It has established a framework for the assessment of environmental inventories of energy systems for the first time at such a scale. The aim was at providing Chinese decision-makers with technical information to optimally allocate energy-related resources with minimum negative consequences for the environment. A network has been established for collaboration between the universities and research institutions of AGS with equivalent Chinese organizations, the Shandong state utility, state and provincial environmental regulators, and between these and other Chinese groups and industrial companies. A simplified and flexible LCA analytical tool has been developed that could be easily used for extended input sets, for other case studies beside Shandong's, and/or for other countries beside China, in line with CETP goal of developing a methodology for integrated assessment that could be applied to different conditions. However, to continue this research, these international as well as Chinese networks should be maintained and extended in order to ameliorate the quality and completeness of the data.

Specific research projects should be assigned to study environmental burdens that could not be analyzed here, and the results used to perform more advanced LCA studies on Chinese energy chains. Heating and transport systems should be addressed in addition to electricity systems. The most important burdens yet to be assessed are: use of freshwater for all industrial activities considered; leachates from spoil heaps at coal mines; the full set of airborne emission species from coal power plants, especially from mine-mouth small power plants; information on releases to groundwater from in-situ leaching from uranium mining in China; radioactive releases from nuclear facilities in China; radioactive isotopes content of coal and its release through the power plant stack. Should future LCA studies also include hydropower, the amounts of construction materials for dams and methane releases from Chinese reservoirs would also be necessary. If biomass were considered as a suitable option for some Chinese regions, specific information would be required.

Should more accurate and extended data be available, the use of a software tool which would fully represent all feedbacks within the energy systems and from other industrial sectors of concern, as implemented in the Swiss LCA studies (database *ecoinvent*), would be conceptually easy to implement, though highly demanding in terms of resources to be invested. This tool would also allow uncertainty analysis, which is missing here though uncertainties are somewhat encompassed by means of sensitivity analyses.

The integration of the LCA methodology with other analysis tasks is an improvement over previous energy system studies. Total (direct plus indirect) inventories for major environmental burdens were produced by combining ESS generation by technology and LCA burden factors per GWh, in order to provide inputs to the EIA as well as MCDA analyses. However, full integration of LCA with the EES, EEM, and ETM tasks was not pursued. ESS generates results on costs and air emissions from operation of power plants of immediate interest for utilities, and consideration of full energy chains would further complicate the understanding of results. The inclusion of emissions from upstream chains would have only amplified the differences in most cases, though for other burdens the effect would have been more pronounced. EEM analyzes complete Chinese energy sectors, and integration of full energy chains would have required a much extended application of that approach to other systems besides conversion to electricity. LCA figures were not explicitly included in ETM because they were not sufficiently detailed. The results of the latter two optimization methodologies may have been somewhat influenced by inclusion of the full energy chain burdens, especially when caps are assumed for the total emissions from sectors taken separately, and for burdens like TSP emission, solid wastes, and land use. The results obtained in this LCA study can serve as a first exploration of the effects of the inclusion of full energy chains in energy-economy or transport models. The partial or full integration of LCA into EEM and ETM modeling tasks may occur in follow-up studies for Shandong or in applying the CETP methodology to other provinces of China or other countries, if they are useful in accomplishing the goals of these studies.

REFERENCES

AGOC (2000). *China sees natural gas as a supplement to the national energy supply*. Alexander's Gas and Oil Connections, *5*, 2 (7 February 2000). Retrieved from: http://www.gasandoil.com/goc/news/nts00621.htm

Bökemeier, R., and Elleringmann, S. (2002). Höllenfahrt durch China. *GEO Magazin*, *9* (100-124).

BP (2001). *Statistical review of world energy*. London: British Petroleum.

CACETC 2000: *China wind power – Study Report*. Wind Power Expert Team, 28 February 2000. Retrieved from: http://www.climatetech.net/pdf/china_wind.pdf

Cameco (2002). *Enrichment Backgrounder*. Cameco news, Updates 2002. Retrieved from: http://www.cameco.com/media_gateway/news_releases/2002/2002-july-22backgrounder.php

CCIY (1999). *China Coal Industry Yearbook*. Beijing: China Coal Industry Publishing House.

CCIY (1998). *China Coal Industry Yearbook*. Beijing: China Coal Industry Publishing House.

Cedigaz (1996). *Natural Gas in the World: 1996 Survey*. Paris: Cedigaz.

CEIP (1998). China: Nuclear weapons systems. In: W.J. Rodney et al., *Tracking Nuclear Proliferation 1998*. China Endowment for International Peace. Retrieved from: http://www.ceip.org/programs/npp/nppchnct.htm

CEOCI (2000). *Status of State-owned Major Coal Mines and Coal Dressing Plants 1999*. Beijing: Center for Economic Operation of Coal Industries.

CESY (1999). *China Energy Statistical Yearbook*. Beijing: China Statistical Publishing House.

CESY (1998). *China Energy Statistical Yearbook*. Beijing: China Statistical Publishing House.

China Daily 12/6/2000. *Power generation progresses steadily*. Beijing.

CMS (2001). *CMS Energy – Trunkline LNG*. Retrieved from: http://www.cmspanhandlecompanies.com/documents/ CMS_Brochure_030102.pdf

CORINAIR (1996). *Joint EMEP/CORINAIR atmospheric emission inventory guidebook*. McInnes G. (Ed.), Vol.I and II. Copenhagen, Denmark: European Environment Agency.

Dedikov, J.V., Akopova, G., Gladkaja, N., Piotrovskij, A., Markellov, V., Kaesler, H., Ramm, A.,

Müller von Blumencron, A., and Lelieveld, J. (1999). Estimating methane releases from natural gas production and transmission in Russia. In: *Atmospheric Environment*, *33,* 3291-3299.

DOE/EIA (2001a). *International Energy Outlook 2001*. US Department of Energy, Energy Information Administration, Office of Integrated Analysis and Forecasting, DOE/EIA-0484(2001). Washington DC. Retrieved from: http://www.eia.doe.gov/oiaf/ieo/index.html

DOE/EIA (2001b). *Historical Natural Gas Annual 1930 Through 2000*. US Department of Energy, Energy Information Administration, DOE/EIA-E-0110(00). Washington DC. Retrieved from: http://www.eia.doe.gov/pub/oil_gas/natural_gas/data_publications/historical_natural_gas_annual/current/pdf/table_05.pdf

DOE/EIA (2000). *Uranium Industry Annual 1999*. US Department of Energy, Energy Information Administration, Office of Coal, Nuclear, Electric and Alternate Fuels, DOE/EIA-0478(99). Washington DC.

Dones, R., Gantner, U., Hirschberg, S., Doka, G., and Knoepfel, I. (1996). *Environmental Inventories for Future Electricity Supply Systems for Switzerland*. PSI Report No.96-07. Villigen, Switzerland.

Dones, R., Zhou, X., and Tian, C. (2003). *Environmental inventories for current and future energy chains associated with electricity supply in Shandong Province, China – Technical report of the LCA Task for the CETP Project*. Villigen, Switzerland: PSI.

EIA (1997). Worldwide Natural Gas Supply and Demand and the Outlook for Global LNG Trade. In: *Natural Gas Monthly*, August 1997, US Energy Information Administration. Retrieved from: http://tonto.eia.doe.gov/FTPROOT/features/world_ng.pdf

EPA (1998). *Inventory of U.S. Greenhouse Gas Emissions and Sinks: 1990-1996*. Office of Policy, Planning and Evaluation, EPA 236-R-98-006. Washington DC: US Environmental Protection Agency. [Quoted in (Popov, 2001)]

ExternE (1999). *ExterneE – Externalities of Energy – Vol. 10: National Implementation*. European Commission, Directorate-General XII, Science Research and Development (EUR 18528). Luxembourg: Office for Official Publications of the European Communities. Available at: http://ExternE.jrc.es/publica.html.

FAS (2000). Washington DC: Federation of American Scientists. Retrieved from: http://fas.org/nuke/guide/china/facility/uranium-mines.htm

FAS (1997). Washington DC: Federation of American Scientists. Retrieved from: http://fas.org/nuke/guide/china/facility/chengdu.htm

Finn, A.J., Johnson, G.L., and Tomlison, T.R. (1999). Developments in natural gas liquefaction. In: *Hydrocarbon Processing*, Gulf Publishing Company, *78*, 4. Retrieved from: http://www.hydrocarbonprocessing.com/

Frischknecht, R. (Ed.), Bollens, U., Bosshart, S., Ciot, M., Ciseri, L., Doka, G., Dones, R., Gantner, U., Hischier. R., and Martin, A. (1996). *Ökoinventare von Energiesystemen — Grundlagen für den ökologischen Vergleich von Energiesystemen und den Einbezug von Energiesystemen in Ökobilanzen für die Schweiz – 3rd Ed*. Zurich: ETHZ/PSI.

Fukusawa, K. (1997). *Low cost, retrofit FGD systems*. IEA Coal Research, IEAPER/34. London.

Gazprom/EPA (1996). *Methane Leak Measurements at Selected Natural Gas Pipelines Compressor Stations in Russia* (Draft). Moscow. [Quoted in (Popov, 2001).]

Girdis, D., Tavoulareas, S., and Tomkins, R. (2000). *Liquefied Natural Gas in China*. World Bank Discussion Paper No.414. Washington DC.

Grizenko, A.N., Akopova, G., and Gladkaja N. (1997). *Russian gas industry and the problem of greenhouse gas emissions*. Presented in Aberdeen, available from Gazprom (April 1997). [Quoted in (Dedikov et al., 1999).]

Guo, Y., Wu, J., Gou, Q., Shu, X., and Xiue, Y. (1996). Radioactive gas measurement in underground coal mines. In Y. Guo and T.S. Golosinski (Eds.), *Mining science and Technology*, Proceedings of the '96 International Symposium on Mining Science and technology, Xuzhou, Jiangsu, China, 16-18 October 1996 (503-506). Rotterdam/Brookfield: Balkema.

Hinrichs, W., Atmaca, T., Neumann, W., and Thormann, A. (1999). Stoffmengenflüsse und Energiebedarf bei der Gewinnung ausgewählter mineralischer Rohstoffe – Teilstudie Steinkohle. Geologisches Jahrbuch – Sonderheft, Bundesamt für Geowissenschaften und Rohstoffe und Staatlichen Geologischen Diensten in der Bundesrepublik Deutschland. Hannover.

Hirschberg, S., Dones, R., and Gantner, U. (2000). Use of External Cost Assessment and Multi-Criteria Decision Analysis for Comparative Evaluation of Options for Electricity Supply. In: S. Kondo and K. Furuta (Eds.), Proceedings of the 5th International Conference on Probabilistic Safety Assessment and Management PSAM-5, Osaka, 27 November-1 December (pp. 289-296). Tokyo: Universal Academy Press, Inc.

Hirschberg, S., Heck, T., Gantner, U., Lu, Y. Spadaro, J. V., Krewitt, W., Trukenmüller, A. and Zhao, Y. (2003). Environmental impact and external cost assessment in the China Energy Technology Program. Villigen, Switzerland: PSI.

Horii, N. (2001). Coal Industry: Development of small coal mines in market transition and its externality. In: N. Horii and S. Gu (Eds.), Transformation of China's energy industries in market transition and its prospects (pp. 23-62). Chiba, Japan: Institute of Developing Economies, Japan External Trade Organization.

ICRFCC (1998). Second National Communication of the Russian Federation Under the United Nations Framework Convention on Climate Change. Interagency Commission of the Russian Federation on Climate Change. Moscow (in Russian). [Quoted in (Popov, 2001).]

IEA (2000). China's Worldwide Quest for Energy Security. Paris: OECD/IEA. Available at: http://cdnet.stic.gov.tw/ebooks/OECD/54.pdf

IGU (1997). Task force gas, the environment: report on environmental care in the gas business. In: Proceedings of the 20th World Gas Conference, Copenhagen, 10-13 June 1997 (pp. 63-66). International Gas Union. [Quoted in (Dedikov et al., 1999).]

IPCC (2001). Climate Change 2001: The Scientific Basis. J.T. Houghton, L.G. Meira Filho, B.A. Callander, N. Harris, A. Kattenberg, and K. Maskell. Intergovernmental Panel on Climate Change. Cambridge, UK: Cambridge University Press.

IPCC (1995). IPCC guidelines for national greenhouse gas inventories. Volume 3. Greenhouse gas inventory reference manual. Geneva, Switzerland: World Meteorological Organization.

ISO (2000). Environmental management – Life cycle assessment – Life cycle impact assessment. International Standard, ISO 14042:2000(E).

ISO (1998). Environmental management – Life cycle assessment – Goal and scope definition and inventory analysis. International Standard, ISO 14041:1998(E).

ITC (2001). International Institute for Aerospace Survey and Earth Sciences, update March 2001. Enschede, the Netherlands. Retrieved from: http://www.itc.nl/~prakash/coalfire/distribution_china.html

Ivanov, V. (1999). Energy Mega-Projects Will Change Northeast Asia. First draft presented at the Conference on Economic Cooperation in Northeast Asia, Ulaanbaatar, Mongolia, 9 June 1999. Retrieved from: http://www.tradp.org/textonly/ivanov.htm

Kidd, S. (1999). Forecasters probing: How will China fuel its growing nuclear power program? Nuclear Europe Worldscan, 11-12/1999.

Lenders, M. (2001). Uranium enrichment by gaseous centrifuge. Deutsches Atomforum Annual Meeting on Nuclear Technology, Dresden, 16 May 2001. Retrieved from: http://www.urenco.com/pdf/atomforum_May_2001.pdf

Liu, C., Li, S., Qiao, Q., Wang, J., and Pan, Z. (1998). Management of spontaneous combustion in coal mine waste tips of China. Water, Air, and Soil Pollution, 103 (441-444). The Netherlands: Kluwer Academic Publisher.

Logan, J., and Luo, D. (1999). Natural Gas and China's Environment, IEA-China Natural Gas Industry Conference, Beijing, 9-10 November 1999. Retrieved from: http://www.pnl.gov/china/pubs.htm

Logan, J., and Zhang, J. (1998). Powering non-nuclear growth in China with natural gas and renewable energy technologies. China Environment Series 2, Summer 1998. Retrieved from: http://ecsp.si.edu/ecsplib.nsf/6b5e482eec6e8a27852565d1000e1a4c/54b8301f5e346cf08525664b00716a08?OpenDocument

Lu, Y. (1999). China's uranium mining, Nuclear Europe Worldscan, 11-12/1999.

Marland, G., Boden, T., and Andres, R.J. (1995). Carbon dioxide emissions from fossil fuel burning: emission coefficients and the global contribution of Eastern European countries. Quarterly Journal of the Hungarian Meteorological Service, 99, 3-4 (157-170). [Quoted in (IEA, 1997).]

MN (2001). China to Fill Oil Gap With Natural Gas. Muzi News (6 February 2001). Beijing. Retrieved from: http://news.muzi.com/ll/english/1044829.shtml

Mohrhauer, H. (1995). Entwicklung bei der Uran-Anreicherung. atw, 40(8/9).

Mudd, G.M. (2001a). Critical review of acid in situ leach uranium mining: 1. USA and Australia. Environmental Geology, 41 (390-403).

Mudd G.M. (2001b). Critical review of acid in situ leach uranium mining: 2. Soviet Block and Asia. Environmental Geology, 41 (404-416).

Nagra (1985). Radioaktive Abfälle: Eigenschaften und Zuteilung auf die Endlager-Typen. NGB 85-02. Baden, Switzerland: Nagra.

NCIA (1999). Statistical Annual Report of Coal Industry 1998 (Aug. 1999). Beijing: National Coal Industry Agency.

NEA/IAEA (2002). Uranium 2001: Resources, Production and Demand. Paris: OECD.

NEW 7-8/2000. Nuclear Europe Worldscan 7-8/2000.

NEW 3-4/2000. Nuclear Europe Worldscan 3-4/2000.

NIR (2000). The virtual Repository – China. (January 2000). Nuclear Inforing. Retrieved from: http://cobweb.quantisci.co.uk/VRepository/chin.htm

NucNet (2002). NucNet News, 140/02. Retrieved from http://www.world-nuclear.org

Oskarsson, K., Berglund, A., Deling, R., Snellman, U., Stenbäck, O., and Fritz, J.J. (1997). A Planner's Guide for Selecting Clean-Coal Technologies for Power Plants. World Bank Technical Paper No.387. Washington DC.

Pan, Z. and Qu, Z. (1999). Radioactive waste management in China. Nuclear Europe Worldscan, 11-12/1999.

PD (2000). Rich Natural Gas Reserves Found in North China. People's Daily (2 December 2000). Retrieved from: http://english.peopledaily.com.cn/200012/02/eng20001202_56746.html

Popov, I. (2001). Estimating Methane Emissions From the Russian Natural Gas Sector. Pacific Northwest National Laboratory, Advanced International Studies Unit, PNNL-13462. Washington DC. Retrieved from: http://www.pnl.gov/aisu/pubs/Gazprom.PDF

Qian, F. (1999). Introduction to China's strategy on nuclear fuel development and cycle, Nuclear Europe Worldscan, 11-12/1999.

Rabchuk, V.I., Ilkevich, N.I., and Kononov, Y.D. (1991). A study of methane leakage in the Soviet natural gas supply system, Report to Batelle Pacific Northwest Laboratory. Irkutsk, USSR: Siberian Energy Institute. [Quoted in (Dedikov et al., 1999).]

Rabl, A., and Spadaro, J. (2002). The ExternE Project: Methodology, Objectives, and Limitations. In: Externalities and Energy Policy: The Life Cycle Analysis Approach, Workshop Proceedings, Paris 15 16 November 2001 (47-61). Paris: OECD. Retrieved from: http://www.nea.fr/html/ndd/reports/nea3676-externalities.pdf

RFSHEM (1997). Russian Federation Climate Change Country Study. Volume 1. Inventory of Technogenic GHG Emissions. Final report, Moscow: Russian Federal Service for Hydrometeorology and Environmental Monitoring. [Quoted in (Popov, 2001).]

Rui S. et al. (1994). Coal Industry: Sustainable Development and the Environment. Beijing: Coal Industry Publishing House.

Sauer, H. D. (2002). Verbrennung am falschen Ort. Neue Zürcher Zeitung 15/5/2002, Zurich.

Shi, W. (2000). Personal communication. Beijing.

Sinton, J.E., and Fridley, D.G. (2000). What goes up: recent trends in China's energy consumption. Energy Policy 28, 671-687.

Smith, I.M. (1997). GHG emission factors for coal - the complete fuel cycle. IEACR/98. London: IEA Coal Research.

Su, S., et al. (1999). Environmental Protection of Shanxi Coal Industry, China Environmental Science Press, (1999).

UNSCEAR (1993). 1993 Report to the General Assembly, with scientific annexes. United Nations sales publication E.94.IX.2. New York.

Urenco (1995). The Uranium Enrichment Plant Almelo, Information booklet. Almelo, the Netherlands.

Vekerdy, Z., Wang, F., Zhang, J.M., and Prakash, A. (1999). Requirements for the integration of remote sensing and field data in a GIS for the management of fire fighting in coalfields. In: Proceedings of 2nd International Symposium on Operationalization of Remote Sensing, Enschede, The Netherlands, 16-20 August 1999. Retrieved from: http://www.itc.nl/personal/vekerdy/pdf/ORS_99_Vekerdy.pdf

Walker, S. (1999). Uncontrolled fires in coal and coal wastes. London: IEA Coal Research.

Wang, H.H. (1999). China's Oil Industry and Market. The Netherlands: Elsevier.

Wang, J., and Dai, Y.N. (1993). In-situ leaching of uranium in China. In: Uranium in situ leaching, IAEA TECDOC-720 (pp. 129-132). Vienna: IAEA.

Watteau, M., Chen, B., and Zhang, S. (2000), Fuel technology transfer to China: ten years and still growing. Nuclear Europe Worldscan, 1-2/2000.

Wise (2002). World Nuclear Fuel Facilities. Update 18 November 2002. Retrieved from: http://www.antenna.nl/wise/uranium/efac.html#ENRG

WNA (2002). World Uranium Mining – July 2002. World Nuclear Association. Retrieved from: http://www.world-nuclear.org/info/inf23.htm

Yang, Q. (1999). The next-generation reactor AC600/1000. Nuclear Europe Worldscan, 11-12/1999.

Zhang, X. (1998). Coal fires in Northwest China. Detection, monitoring and prediction using remote sensing data. PhD Thesis, ITC publication No.58, International Institute for Aerospace Survey and Earth Sciences. Enschede, the Netherlands. [(Quoted in Walker 1999).]

Zhang, Z.X., and Folmer, H. (1996). The Chinese Energy System: Implications for Future Carbon Dioxide Emissions in China. The Journal of Energy and Development, 21(1) 1-44.

Zheng, Y. (1996). China's Nuclear Arsenal. National University of Singapore. Retrieved from http://www.kimsoft.com/korea/ch-war.htm

Zittel, W. (1997). Untersuchung zum Kenntnisstand über Methanemissionen beim Export von Erdgas aus Russland nach Deutschland. Ottobrunn, Germany: Ludwig-Bölkow-Systemtechnik Gm.

NOTES

[1] Shandong has 30 small hydropower plants with nominal capacity in the range 0.5-5.4 MW, for a total of nearly 47 MW (SEPRI). One 4×250 MW pumping storage plant for peak load management is under construction in the south of Shandong; it should start operation in 2005.

[2] Water pollution has not been considered because in several environmental assessments for Europe it was found that the health effects of water emissions associated with energy systems were of secondary importance to those of airborne pollutants. However, these studies have not included the consequences of chemical extraction of uranium, because this technique has gained importance worldwide only in the past decade. Furthermore, only a few pieces of information were available to the authors about releases to water from the coal chains in Shandong and Shanxi provinces. Therefore, no statement can be made in this study about the significance of water pollution for the current situation in China; this will be pursued in follow-up studies.

[3] China is not committed to the Kyoto Protocol limitation of emissions of GHG in the short run. Emission limits are based on national inventories thus excluding any releases that occur outside political boundaries.

[4] This criterion was originally defined to exclude the land affected by solid wastes, to prevent a possible double-count. The same argument would apply to the fuel transport burden criterion. For the results of the LCA task to be as complete as possible, the land use includes waste disposal and transport. The LCA results are also transferred to the corresponding MCDA criterion. This does not constitute double counting for the MCDA exercise for two reasons. First, the waste criterion expresses the risks associated with all of the solid waste, and not just the surface use by waste treatment and disposal. Second, the transport burden criterion reflects the fact that the transport of all energy carriers will increase from year 2000 to year 2020, in response to the growing demand for energy, thus competing more and more with transport of other commodities and passenger traffic on limited roads and railways; the amount of land used per unit of produced electricity, however, will not increase in response to these changes.

[5] The main reason for this choice was to allow these factors to accommodate the detailed assumptions made for the operation of single power plants or technologies within ESS, which may change with different scenarios. Average factors for classes of plants or single technologies are considered here instead.

[6] This is due to the exploitation rate of world-wide natural gas resources, which is increasing more rapidly than the rate for coal.

[7] About 97 percent of the coal extracted in China is mined underground.

[8] Union for the Co-ordination of Transmission of Electricity. Members in year 2002 are: Austria, Belgium, Bosnia-Herzegovina, Czech Republic, Croatia, Denmark (associated member), France, Germany, Greece, Hungary, Italy, Luxembourg, Macedonia, the Netherlands, Poland, Portugal, Slovak Republic, Slovenia, Spain, Switzerland, and the Federal Republic of Yugoslavia. The CENTREL countries Czech Republic, Hungary, Poland, and Slovak Republic have officially joined UCTE in 2001.

[9] Hence a high solid waste rate of 540 kg per tonne of produced coal.

[10] Chinese coal export in 1998 included roughly 75 percent steam coal and 25 percent coke (Sinton and Fridley, 2000).

[11] These standards were not provided.

[12] Definition not provided. Presumably this means a group of mines located in one geographical area and belonging to one CB.

[13] The removal efficiency was not reported. It is presumably on the order of 50 percent.

[14] Very small units of only a few MW may have shorter stacks.

[15] Unspecified whether proven or total.

[16] It is unclear whether the percentages given by the Yangquan CB refer to the total volume of gas released or to the methane content. One possible interpretation is that the drained gas is comprised of 40 percent methane, which represents 26 percent of the total methane generated by mining operations in the Bureau.

[17] The information on the miles from Qinhuangdao to Weihai and Longkou comes from the China's maritime transportation map; however, straight distances taken from a regular map of China give about 170 nm and 145 nm, respectively.

[18] For sub-bituminous coal a value of about 96 g(CO_2)/MJth is suggested in (IPCC, 1995; Smith, 1997; Marland et al., 1995). With a LHV of about 22.5 MJth/kg, average for Shandong coal, nearly 2.2 kg(CO_2)/kg coal are produced. This value was used by SEPRI and is adopted here. The interval given by (CORINAIR, 1996) for sub-bituminous coal used in Europe (17.4-23.9 MJth/kg) is 90.9-115.1 g(CO_2)/MJth. Nevertheless, a certain difference with respect to international averages may be expected for the CO_2 emission per unit of mass of Chinese coal, because of the generally higher content of ashes. This would slightly increase the emission rate, due to the heat capacity of the ash. However, recalculating the emissions based on fuel compositions would have been a rather difficult exercise, because of the variability of the coal batches delivered to the power stations. The only exceptions considered by SEPRI were the emission rate of 2.38 kg(CO_2)/kg coal (anthracite) from the Heze power plant, but still complying with IPCC, and the value for the Longkou power plant. For the latter, using a blend of lignite and sub-bituminous coal, the CO_2 emission was calculated assuming 50 percent total carbon content in the fuel, which gives about 1.8 kg(CO_2)/kg coal. However, this value may have been underestimated. In general, without other information on the composition of the coal used, an uncertainty of a few percent can be expected for the calculated values of CO_2 emissions.

[19] Incidentally, maps of China roughly indicating the major areas affected by spontaneous combustion fires in coalfields and mines include, besides the heavily affected Xinjiang, Ningxia, Inner Mongolia, and North-Eastern provinces, also areas in Shanxi (ITC, 2001) and even in Shandong (Walker, 1999). Sauer (2002) mentions that the arid regions around the bend of the Yellow River are reported to be affected by such events. Liu et al. (1998) mention that the first case of extinguishing spontaneous combustion fires in coal waste heaps occurred in 1981 at the Fenxi CB in Shanxi. Bökemeier and Elleringmann (2002) witnessed and documented such events in the north of the Shanxi province as well as in Xinjiang and Northern Mongolia.

[20] Chinese sources tend for the lower range. Sauer (2002) reports a loss of 20 Mt/year but without quoting the original source.

[21] The uncertainties include the amount of coal loss and the methane formed because of the low temperature of combustion. Sauer (2002) mentions that the GHG produced by these spontaneous combustion events is the double of what would be produced if combustion were complete.

[22] The conditions of spontaneous combustion determine the release of various other gases and the production of tar and benzene that may affect the groundwater (Sauer, 2002). Selected air pollutants developed by the combustion of one tonne of typical coal mine waste in China are given in (Liu et al., 1998). Unfortunately, neither the composition of this waste nor the amount burned in one year in China are given in the reference.

[23] This is an arbitrary assumption designed to encompass all uncertainties on the energy intensities of material manufacturing in China and the difference in composition of the energy mixes.

[24] Values between 100 m^2/MWe and 3'700 m^2/MWe, where the upper values include the area occupied by bottom ashes. When ashes are recycled, the upper value is estimated to be about 600 m^2/MWe.

[25] This reflects the effects of countervailing tendencies including the slow but steady growth of the electricity demand; a higher share of fossil fuels; switch from oil to natural gas; slow decrease of the nuclear and hydro shares; implementation of new coal and gas technologies; a relatively small use of new renewables; and, possibly, further reduction of methane release rates from gas pipelines and underground coal mines.

[26] The Ordos Basin, sprawling across Shaanxi, Shanxi, and Gansu provinces, and the Ningxia Hui and Inner Mongolia Autonomous Regions covers an area of 400000 km^2 (PD, 2000).

[27] Located in Ih Ju League, Inner Mongolia, it covers an area of about 5000 km^2.

[28] No reference in the literature has been found on this option.

[29] The amount of sour gas extracted was about 14 percent of the total used in UCTE countries in mid-1990's.

[30] Zittel (1997) surveyed the previous estimates and speculations on leakage from Russian natural gas production and transport systems. The minimum value found was 1.2 percent (range given in Gazprom (1996) was 1.2-1.4 percent from production and transmission) and the maximum 9.35 percent (Rabchuk et al., 1991). The interval given for "total emissions" was 4-9.35 percent. Zittel suggested, assuming current worse conditions, a maximum loss of 1.8 percent.

[31] About 0.02 percent is assumed to be released from gas turbines.

[32] Now World Nuclear Association.

[33] No cost range is given. No differentiation between estimated additional resources and speculative resources is given. Total world-wide undiscovered conventional resources of uranium reported by countries is 9.94 MtU (NEA/IAEA, 2002).

[34] One centrifugal enrichment plant with 200000 SWU/year capacity covers about 13.5 TWh, nearly all the nuclear electricity production in China in 1999. This value is calculated assuming: 3.4 percent average enrichment of the fuel; 4.33 SWU per kg of enriched uranium; and, fuel burn-up of 40 MWd_{th} per kg of enriched uranium. The previous in turn means approximately 3.4 kg of enriched uranium per GWh.

[35] The radioactive level of the waste is not reported in the source.

[36] Roughly based on the available shares for uranium production worldwide. WNA (2002) reports for 2001 a production by chemical extraction of 31 percent (ISL 16 percent, by-product 15 percent), open pit 29 percent, and underground 40 percent. NEA/IAEA (2002) reports for the period 1998-2000 a steady increase of chemical extraction from 21 to 29 percent, a steady decrease of open pit from 39 to 28 percent, and a fluctuating production by underground mining in the range 36 to 43 percent.

[37] Around the mid-1990's, of the enriched uranium available on the spot market 83 percent came from diffusion and 17 percent from centrifuge services (Mohrhauer, 1995). Using the total installed capacity worldwide in 2001, and excluding the Paducah diffusion plant in the USA, closed since May 2001, the installed capacity of centrifuge approaches 50% of the total (Wise, 2002). Excluding down-blending of Highly Enriched Uranium (HEU) from dismantled Soviet nuclear weapons, centrifuge enrichment services cover about 53 percent of the total worldwide supply of 35 MSWU (Lenders, 2001; Cameco, 2002).

[38] The Nuclear Power Institute of China (NPIC) has developed the conceptual design of the AC600 based on the Westinghouse AP600; the AC600 can be scaled up to 1000 MW (Yang, 1999).

[39] Coal power chains in China may pose a relatively high risk from radioactive emissions. Pan et al. (1999) report from different Chinese studies that collective doses from airborne radon and fly-ash due to PC operation are one to two orders of magnitude higher than from the NPP in Daya Bay. Boiler ash from radioactive coal also retains non-volatile radioactive isotopes. When recycled for brick or concrete fabrication, the ashes may significantly add to the collective dose, more than trebling the dose from power plant operation. However, the results reported in Pan et al. (1999) are not complete (no doses from mining of the coal are given; no estimation of long term emissions from uranium mill tailings), and no specific values for Shandong and Shanxi coal were available. For these reasons, radioactive emissions could not be included in this study.

[40] Legend for the pictures: Min = Mining; Mil = Milling and tailings; Con = Conversion; En.C = Enrichment Centrifuge; FF = Fuel Fabrication; PWR = Pressurized Water Reactor; I-HLW = Intermediate and Highly Radioactive Waste Repository; LLW = Low Level Waste Depository.

[41] Open pit mining, which produces substantial amount of dust, comprises only 23 percent of all mining.

[42] Apart from the 4×250 MW pumping storage plant under construction, Shandong has no hydro potential. The huge potential for hydropower in the central south provinces is being partially exploited.

[43] Wind is not explicitly included in the MCDA scenarios.

[44] This also applies to classes of non-radioactive wastes that were ignored here for lack of detailed information. Also the health effects of the recycling of radioactive coal ash has not been addressed here.

[45] If only ISL mining were used, the calculated land use for the nuclear chain would diminish to 0.003 km^2/TWh.

[46] Transmission of electricity from hydropower plants in South-Central China to Shandong seems highly unlikely considering distances and distribution of potential customers in East and South China. Though this argument may also be applied to piped natural gas, the availability of major ports for LNG terminals and the demand for gas for applications other than power production may be the driving forces behind importation of this fuel, despite geopolitical obstacles.

[47] This is a common assumption for the high voltage grid also.

[48] Research on the Atomic Vapor Laser Isotope Separation (AVLIS) has been suspended in the USA in 1999 and will end in France in 2003 (Lenders, 2001). AVLIS would use about the same electricity as centrifuge (Dones et al., 1996), 40 kWh/SWU according to (Lenders, 2001).

[49] Urenco has achieved electricity consumption of 50 kWh/SWU (Lenders, 2001).

[50] Because mining and milling electricity requirements are assumed to be fully covered by diesel plants, they are not included in these figures, which consider electricity from grids only.

[51] Equal mass production rate of LLW is assumed for both enrichment processes. They contribute marginally to the total LLW calculated for the chain.

[52] The last letter in the central string of the name identifies this characteristic: S = no DSM; M = 10 percent demand reduction for DSM; and, G = 20 percent reduction.

[53] The BOC-CENPAS-SUB and BOC-CONPAS-SUB would imply the transport of 140 or more million tonnes of coal from SX, which is more than ten times the amount imported in 1998 for the SEPCO plants (approximately 13 Mt, or 40 percent of the total supply to SEPCO units).

[54] No discount has been applied here for DSM, i.e. total annual emissions are divided by the electricity actually supplied, not by the level without DSM.

[55] The contribution of gas systems to total supply may change with the future (see ESS Chapter).

[56] These values may differ somewhat from those reported by other CETP tasks. The values are determined in part by assumptions that must be made within each task in order to assure internal consistency. Furthermore, it is important to remember that each task is also affected by uncertainties and approximations. The most important thing, in meeting CETP's goals for informing energy policy, is to have the results of the different research tasks converge into a single picture, in which small differences become insignificant.

[57] As mentioned in Sections 1.3 and 7.3.2, this aspect has been included in the MCDA task by combining the sub-criterion "Confinement Time of Critical (hazardous) Wastes" with "Amount of Wastes", to produce the criterion "Wastes".

[58] The differences shown here may change depending on how much of the gypsum produced by wet FGD's is recycled. It is assumed here that the FGD increases the direct solid waste produced by a PC by a factor 1.3.

[59] Mainly made with nearly 48 percent fossil, 37 percent nuclear, and 14 percent hydropower. The CENTREL countries, full UCTE members since 2001, have a higher share of fossil which adds about 2.5 percent to total fossil share in UCTE in 2001.

[60] This does not include the CENTREL countries. One country, Greece, exhibited 16 g/kWh, because it has a large share (over 70 percent) of low quality lignite coal plants with no scrubbers and low average plant efficiency.

CHAPTER 9

ENVIRONMENTAL IMPACT AND EXTERNAL COST ASSESSMENT

STEFAN HIRSCHBERG, THOMAS HECK, URS GANTNER, YONGQI LU, JOSEPH V. SPADARO, WOLFRAM KREWITT, ALFRED TRUKENMÜLLER AND YIHONG ZHAO

1. INTRODUCTION

Aside from tremendous benefits the conversion of energy can be harmful to the human health and environment. Assessment of public health effects associated with air pollution caused by various means of electricity generation was one of the central goals of CETP. Currently, China is experiencing rapid economic growth, and this trend is expected to continue. But the damage to air, soil and water quality backfires on the rate of growth. Economic growth is accompanied by increasing electricity demand, with coal as the dominant energy source. This development implies that, along with the dominant health effects, sulfur deposition (and the resulting acidification) stand out as the primary environmental issues to be addressed. Examples of the impacts include effects such as chronic bronchitis, reduction in lifetime expectancy, or adverse effects on the environment, such as loss of crops.

The awareness about environmental issues has been steadily increasing in China. In the last decade efforts have been undertaken to develop a national strategy for dealing with the problem. These efforts involve emission control requirements for major air pollutants, including establishment of SO_2 and acid control zones. The evaluation of the efficiency of these measures, possible enhancements and implications of failure to implement the policy in the future have been subject to a number of analyses with varying scope and level of detail.

This chapter presents the state-of-the-art CETP study of health and environmental impacts of air pollution in China and addresses the corresponding consequences of selecting various supply options for the future. The analyses performed produced many results and insights relevant for the whole of China. However, the detailed modeling of the future electricity supply scenarios was focused on the Shandong Province. For the purpose of this work it is essential to note that this province is among the most industrialized and energy-intensive regions in the country. At present, the electricity produced in Shandong is almost entirely based on coal; there are no significant hydro resources, and the local environmental conditions are not highly favorable for wind, solar photovoltaic or biomass electricity production.

In the present work, the assessed impacts were also used as the basis for monetary evaluation, this also providing an estimate of corresponding welfare losses. The estimated impacts are considered robust and, if needed, can be used as the basis for decision-making independently of the monetary values.

The analyses summarized in this chapter utilize some of the results of Life Cycle Assessment (LCA), covered in Chapter 8, and selected scenarios modeled in Electric Sector Simulation (ESS) Task (Chapter 6). One of the health impact indicators estimated here, i.e. reduction in lifetime expectancy, served as an input to the Multi-criteria Decision Analysis (Chapter 11). The full account of the work carried out within the present task is provided in the technical report (Hirschberg et al., 2003b). While this chapter addresses the impacts associated with normal operation, the accidents are covered in Chapter 10 and in more detail in the corresponding technical report (Hirschberg et al., 2003a).

2. OBJECTIVES, SCOPE AND SUB-TASKS

2.1 Objectives

The main goals for this task were defined as follows:

- To estimate health and environmental impacts of outdoor air pollution in China, with main emphasis on the electricity sector in the Shandong Province;
- To assess the corresponding damage costs thus enabling the estimation of the *"true"* costs of electricity generation, reflecting the extent of impacts on health and environment;
- To carry out a comparative evaluation of alternative supply options and pollution control measures that could be implemented in the future, based on the input from relevant CETP tasks and other sources.

2.2 Scope

A number of factors need to be considered when defining the analysis scope.

2.2.1 Time horizon

The reference evaluations were carried out for the current situation. Scenario analyses use the time horizon of 20-30 years from now, consistent with energy and electricity modeling tasks within CETP. Among the time-related aspects that need to be reflected in the analysis are: energy and electricity demand development, evolution of energy and electricity mixes, population growth, economic growth and prospective technological advancements.

2.2.2 Geographical boundaries

Air pollutants can be dispersed over very long distances and the associated impacts need to be accounted for. The boundaries used in the assessments are model-dependent but the general principle has been to account for all contributions as long as they are significant. In other words, the dispersion of pollutants from the source is pursued up to the distance at which their contributions to the damage become negligible. Boundaries with regard to considered sources of pollution are addressed under the next item.

2.2.3 Air pollution sources

Electric power sector in Shandong was treated in this research in the most detailed manner. The second priority level was the power sector in China, followed by other air pollution sources in China. The scope of the models used covers also to some extent air pollution sources in neighbor countries but no major efforts were invested in qualifying and updating the extensive data specific for these countries and used in the models. The robustness of the results of the present work corresponds to the priorities outlined here. However, the treatment of transboundary pollution originating from China can be considered as quite reliable. Scenarios for the future were analyzed only for the power sector in Shandong.

Apart from power plants whole energy chains were considered when estimating the final damages and their costs. This is in accordance with the overall approach used in CETP and with the general sustainability principles.

2.2.4 Air pollutants and other emissions

The following emissions were considered:

- **Sulfur dioxide (SO_2).** SO_2 is held responsible for a variety of environmental damages, particularly on human health, ecosystems, crops and building materials. As a precursor of sulfates, sulfur dioxide is indirectly a major contributor to long-term ("chronic") mortality and several morbidity effects. Also evidence of the acute health effects associated with SO_2 is now available. SO_2 can cause direct economic damages by reducing crop yields. Acidification resulting from SO_2 emissions is also hazardous to natural ecosystems.
- **Nitrogen oxides (NO_x).** Currently, the direct effects of NO_x emissions are not being assessed though positive associations between NO_x and daily mortality or respiratory hospital admissions in several European cities have been reported. NO_x is a precursor to nitrates (secondary particulates). The view is supported that the apparent NO_x effects may be due to particulates, or at least, are highly dependent on background particulate levels.
- **Particulates.** Ambient air pollution particulates are a complex mixture, varying in size and in composition. Many epidemiological studies found evidence of adverse acute health effects of particulate air pollution. There is also strong epidemiological evidence of chronic health effects. Because particulate air pollution is a complex mixture rather than a single substance, there is a lot of diversity in how particulate air pollution is characterized in various epidemiological studies. Internationally, there are many studies

showing acute health effects of particulates expressed as PM_{10} (inhalable particulates), or total suspended particulates (TSP). In Europe, a number of studies refer to black smoke (BS). Some studies, mostly from North America, show the effect of finer fractions such as $PM_{2.5}$ or sulfates. There is some evidence, that these fine fractions are associated with greater risks than PM_{10}. It may also be that the toxicity of particulates is greater according to their acidity, and less according to their solubility. For modeling, we distinguish between "primary particulates" which are emitted directly from the emission sources and the "secondary particulates", i.e. sulfates and nitrates, which are formed in the atmosphere from SO_2 and NO_x emissions. Secondary particulates are assumed to cause similar health impacts as primary particulates, but because of some epidemiological evidence sulfate aerosols are assumed to be more harmful as primary PM_{10} and nitrates.

- **Greenhouse Gases (GHGs).** GHG emissions can potentially lead to global climate change and are considered in this work as important attributes of the analyzed scenarios. The Life Cycle Assessment (LCA) Task generated detailed and reliable GHG inventories. Damages associated with GHG emissions are estimated here but are considered as much less robust than the impacts of air pollutant emissions.
- **Radiation.** Impacts on human health and ecosystems can occur when radioactive substances are set free into the environment. During normal operation of nuclear power plants, the emissions of radioactivity are at very low, not harmful levels. The associated damages are negligible from the practical point of view. Risks related to hypothetical severe nuclear accidents that might lead to large releases of radioactivity are addressed in detail in Chapter 10.

2.2.5 Impacts not considered

First, this work addresses exclusively impacts of outdoor air pollution. Indoor pollution in China is a major issue but its evaluation was beyond the goals and scope of CETP. Among outdoor air pollutants the impacts of ozone were not addressed though there exists epidemiological evidence on acute health effects caused by ozone. There was no model available that could be applied for China and in relative terms the ozone-specific health effects are for the Chinese conditions considered small in comparison with those due to the air pollutants mentioned above.

Water pollution is a threat particularly where water resources are scarce. Fresh water supply is a serious issue for many parts of China and the problem is expected to become even more acute in the future. Fresh water scarcity in Shandong has been taken into account in the establishment of Electric Sector Simulation (ESS) scenarios insofar as clean coal technologies demanding very high fresh water consumption have not been considered (Chapter 6). Water pollution has been seen as a minor problem for electricity generation in the European impact studies (European Commission, 1999). The assessment of water pollution was restricted basically to acidification. A major reason for this simplification was the assumption that the control of water emissions in Europe is already rather strict so that no severe damages on water resources may be expected. This assumption may not hold if the conditions are less restrictive. Detailed dispersion models including water and soil are still under development in the European framework. It may be worthwhile to follow these developments and consider in the future the potential implications for China.

Finally, impacts such as noise and visual amenity were not analyzed in the present work. In relative terms and considering the technologies of interest for Shandong in particular, these impacts are of low significance compared to the dominant ones.

2.2.6 Analytical implications of the selected scope

In order to meet the objectives within the above summarized scope, two analytical approaches were employed in this research:

- "Impact pathway approach" for the estimation of environmental damage costs associated with the various energy chains relevant for electricity generation in the Shandong province and with air pollution damages in China in general. The approach and its applications are described in detail in Section 3.
- Simulation-based analysis of acidification for selected regional scenarios. This approach and its applications are described in detail in Section 4.

2.3 Sub-tasks

The following tasks were implemented in this activity:

Initial screening analysis. This sub-task defined the energy chains to be addressed in the study and identified priority impact pathways.

Energy chain, technology, receptor and environment characterization. This includes technology description and magnitude of fuel chain activities, type and magnitude of emissions and other residual burdens, and data on receptors and environment.

Initial application of a simplified approach to damage assessment. This sub-task generated preliminary, rough results for damages associated with electricity generation.

Adaptation of the impact pathway approach model to the Chinese conditions. Implementation of the modifications of the software used for this purpose required an extensive data input that has been assembled during a relatively long period of time.

Estimation of environmental damage costs of air pollution, based on the detailed model. This includes estimation of changes in pollutant concentrations, calculation of expected impacts and translating impacts into economic damages.

Quantification of prioritized damages from energy chain stages other than power plants. The prioritization sub-task (sub-task 1 above) identified impact pathways beyond air pollution from power plants that under the Chinese condition needed to be analyzed with regard to the associated damages. This sub-task was carried out in close co-operation with the LCA Task within CETP. The associated impacts were addressed at a lower level of detail.

Evaluation of impacts and damage cost for the reference case and for future scenarios. The reference case corresponds to the current situation and covers China and all pollutant emission sources. Selected future scenarios for electricity supply in Shandong, originating from the ESS Task within CETP, were analyzed.

Scenario-based assessment of acidification. This sub-task, based on the RAINS-Asia model, generated SO_2-depositions in terms of annual loads and excesses of critical loads (as a measure of damages to ecosystems), and provided control costs for emissions (total and marginal costs to achieve specified emission targets in a given region).

Uncertainty and sensitivity analysis. Major sources of uncertainty were identified and their impacts on the results were evaluated. Sensitivity analyses also included application of the simplified approach as a mean to examine some of the assumptions and provide a basis for improving the simplified model and thus enable resource-efficient future applications.

Synthesis of results. The results from the various sub-tasks were integrated and conclusions were drawn.

3. EXTERNAL COST ASSESSMENT

3.1 Concept and basic approaches to estimation

By *"externalities"* we understand economic consequences of an activity (such as electricity production) that accrue to society, but are not explicitly accounted for in the decision making of activity participants. In the literature externalities have been alternatively called side effects, spillover effects, secondary effects and external economies/diseconomies. In economic terms detrimental consequences are called *"external costs"*; positive consequences are called *"external benefits"*.

Two fundamental types of externalities can be distinguished: environmental and non-environmental. Examples of environmental externalities include impacts on public and occupational health (mortality, morbidity), impacts on agriculture and forests, biodiversity effects, aquatic impacts (ground water, surface water), impacts on materials (such as buildings, cultural objects), and global impacts (greenhouse effect). The costs associated with damages caused by air pollution are called environmental external costs since they are normally not included in the price of electricity. Among non-environmental externalities are public infrastructure, security of supply of strategic goods, and government actions (such as R&D expenditures).

Externalities arise due to the imperfections and/or non-existence of markets. For instance, there is no market for clean air and water. In a system of perfect competitive markets and in the absence of externalities, prices constitute the instrument for efficient resource allocation both on the production and consumption sides of the economy. Externalities are the source of misallocation of resources thus generating welfare losses to the society.

From the economic point of view complete elimination of externalities is neither practical nor desirable. For example, there is an optimum level of environmental degradation and this is not at zero level. From the point of view of the society as a whole the industry should reduce its discharge of pollution to the point where the sum of the cost of pollution and cost of pollution control is a minimum. In other words, beyond some point the cost of reducing pollution exceeds the benefits.

The impact of every industrial activity in a society includes, on one side, external costs like pollution, and also external benefits like improvement of the standard of living, employment or economic development. From society's point of view the price of a product should reflect all the involved external costs and benefits. When this is the case the costs and benefits are considered as *"internalized"*.

Most attention has been directed towards internalization of external costs. With respect to external benefits the major ones are usually considered as already internalized by the existing market processes. There is, however, no general consensus about this point of view.

Knowledge of external costs is useful when evaluating abatement measures and setting environmental standards. Comparing external costs with pollution control costs is a powerful tool for rational decision-making. By adding external costs to the internal ones the principle of "true costs" is approached. This can lead to more efficient production, abatement measures, new technologies and processes, as well as changes of behavior and attitudes.

Four basic approaches to the estimation of environmental external costs can be distinguished (see e.g. Krupnick et al. (1995)):

- *"Top-down"* approach (pioneered by Hohmeyer (1988)). For example, in order to estimate damages from fossil fuels three steps are executed: (1) Identification of other studies' estimates of the total health costs attributed to air pollution; (2) Estimation of the fraction of the total emissions that originates from electric power generation from fossil fuels; (3) Multiplication of the total health costs according to (1) by the fraction according to (2). The method is obviously quite rough, relies on previous estimates of total damages, does not account for the different steps of energy chains and has no capability to account for site-specific effects.
- *"Pollution control costs"* as a surrogate for damages, e.g. Bernow et al. (1990). The rationale for this simple and thus attractive approach are the difficulties and uncertainties associated with the estimation of damage costs. The method is based on two not generally valid (actually frequently flawed) assumptions: (1) The marginal costs of abating emissions equal the marginal damages; (2) Environmental regulation is economically efficient. The assumptions are arbitrary since the regulators cannot decide on optimal policies without knowing the damage costs. Nevertheless, the approach may be useful in the context of analyzing external costs of Greenhouse Gas (GHG) emissions, a global problem characterized by overwhelming

uncertainties when attempting to estimate the *"true"* damage costs. Critical comments on the use of control costs are provided in Pearce (1995).

- *Limited "Bottom-up"* approaches, e.g. Ottinger et al (1990); Pearce et al. (1992). Five steps were included in the Ottinger study (also referred to as Pace study; the study by Pearce is similar in spirit): (1) Estimate of emissions; (2) Estimate of dispersal of pollutants; (3) Determination of the population, flora and fauna exposed to the pollutants; (4) Estimation of the impacts from exposure to the pollutants; (5) Calculation of the monetary cost of the impact. When calculating damages the method relies heavily on the estimates from previous studies; thus, no data are collected on the primary level. The account for different steps of energy chains is more extensive than in *"top-down"* studies but still limited. Finally, the site-specific effects are not considered.
- *Full scope "Bottom-up"* approach originally stemming from the EC/US study on external costs of energy chains (European Commission, 1995; ORNL&RfF, 1994). The *"impact pathway"* approach (also called *"damage function"* approach) was employed in these studies. This methodology covers all relevant stages in each energy chain (extraction, fuel processing, transport, power generation, waste management and storage) and accounts for site-specific effects. Details are provided in the following sections. The approach has been applied to a variety of technologies at specific sites in Europe (European Commission, 1999; Hirschberg et al., 2000). The full scope "*bottom-up"* approach is today considered as the state-of-the-art method and has been implemented in this research.

3.2 Impact pathway approach and the EcoSense model

The present study uses the impact pathway approach supported by the EcoSense software. EcoSense is an integrated impact assessment model that has been developed by Stuttgart University within the ExternE Project (European Commission, 1999) in order to facilitate a standardized application of the impact pathway approach. Basically, the EcoSense model processes the following steps:

- Setup of emission data
- Modeling of atmospheric transport and chemical conversion
- Assessment of impacts to receptors (e.g. humans or crops)
- Monetary valuation

The multi-source version of the model, earlier developed and used for external cost assessment in Western Europe, was extensively modified for application to Chinese conditions. A China-specific "Reference Environment Database", involving receptor data, meteorological data and emission data, was developed and implemented in co-operation between PSI, Stuttgart University, partially using inputs from Chinese organizations.

Figure 9.1 shows the structure of the EcoSense China/Asia model.

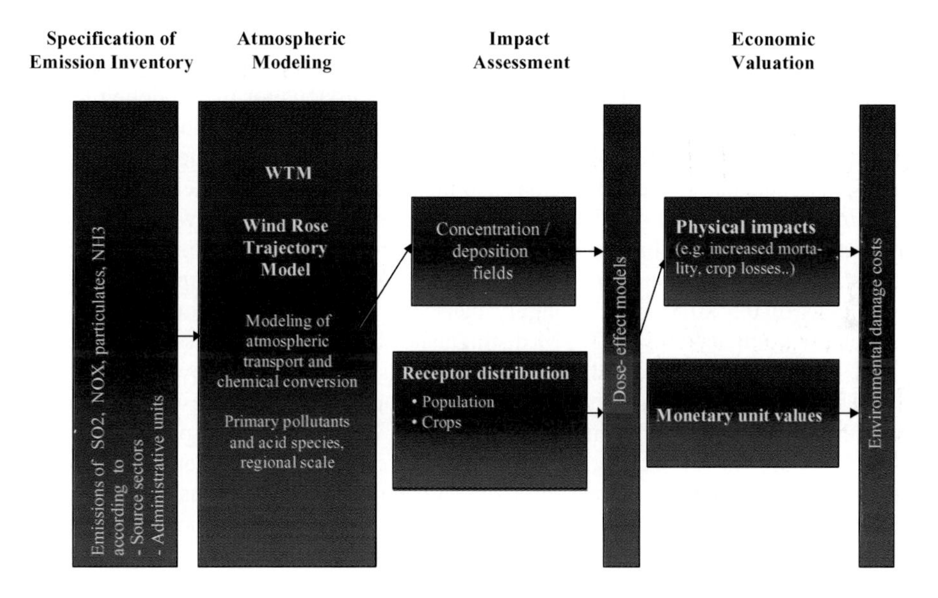

Figure 9.1 The structure of EcoSense China/Asia.

EcoSense China/Asia allows calculations to be carried out of fuel cycle externalities according to the impact pathway approach, and supports the assessment of damage resulting from the operation of a single point source (e.g. a power plant) at a given location, taking into account pre-defined background conditions. Furthermore, the effects of changes in emissions from specific sectors (e.g. energy, industry, transport, households, etc.) can be assessed on a continental, national, provincial or community level.

In the following description is provided of the various elements of the methodology. For more details we refer to the technical report (Hirschberg et al., 2003b) and to the original literature. The emphasis here is on new developments and on the implementation of the EcoSense China/Asia version.

3.2.1 Emissions and atmospheric modeling

The emissions from a source are transported by wind through the atmosphere and mix up with emissions from other sources. Some of the pollutants take part in chemical reactions when they interact with other components in the atmosphere. To simulate these atmospheric transport and conversion processes, EcoSense uses the Windrose Trajectory Model (WTM) (Trukenmüller and Friedrich, 1995). WTM is a user-configurable Lagrangian trajectory model based on the windrose approach of the Harwell Trajectory Model developed by Derwent et al. (1988). (Generally, a "Lagrangian" model follows an air parcel, which moves continuously with the local wind. By contrast, a so-called "Eulerian" model considers fixed cells and simulates how species enter and leave each cell through the walls (Seinfeld and Pandis 1998)).

An atmospheric transport model like the WTM of EcoSense, which includes chemical reactions, has to take into account the whole background of relevant species, even if only a single emission source (like a power plant) is investigated. Thus, EcoSense contains a complete database of emissions for the modeling region. In practice, the spatial resolution is limited. Discrete areas (polygons representing administrative units or cells of a regular grid) are used to specify the location of the emissions. The background emissions and the receptor data in EcoSense are given in administrative units (for China: country, provinces, counties), which are more transparent for users than grid cells. For the atmospheric modeling, all data are transformed to a regular grid. EcoSense China/Asia uses a geographical grid with grid cells of 0.5°x0.5° which roughly means 50x50km^2. Concerning the height at which emissions take place WTM distinguishes between high emission sources (at least 100 m above ground or higher) and low emission sources (below 100 m).

The WTM model calculates concentrations and depositions resulting from the given emissions. The concentrations refer to a well-mixed atmospheric boundary layer and are assumed to be constant for different heights of receptors above the ground. This is a simple but appropriate approach for the scales considered here, i.e. for the characterization of long-term medium to long-range dispersion of pollutants. For the impact assessment of a specific emission set (e.g. a single power plant or the whole power sector of a province), two runs are needed, one with the background emissions excluding the specific emission set and one with the background emissions plus the specific emission set. The difference in concentrations resulting from these two runs represents the marginal (incremental) change resulting from the specific emission set.

Due to the non-linearities in the model, the changes of concentrations depend on the input emissions in a non-trivial way.

3.2.2 Exposure-response models

The dose-effect models recommended in ExternE (European Commission, 2000) are also used in the EcoSense China/Asia.

For the basic health effect functions, a linear relation between marginal changes of concentrations and health impacts was assumed. The slopes of the exposure-response functions used for impact assessment of mortality and morbidity are summarized in Table 9.1 and the endpoints are commented below.

For the relevant pollutants it is assumed that there is a threshold at the individual level, in the sense that most people are not realistically at risk of severe acute health effects at current background levels of air pollution. There is however no good evidence of a threshold at the population level; i.e. it appears that, for a large population even at low background concentrations, some vulnerable people may be exposed part of the time to concentrations which do have an adverse effect. The World Health Organization (WHO, 1999) adopted a 'no threshold' position for particulates and ozone. ExternE recommended to quantify health effects from particulates, SO_2, and ozone on a 'no-threshold' basis.

Table 9.1 Quantification of human health impacts. The exposure response slope, f_{er}, has units of [cases/(yr-person-µg/m^3)] for morbidity, and [YOLL/(yr-person-µg/m^3)] for mortality (adopted from European Commission (2000)).

Receptor	Impact Category	Reference	Pollutant	f_{er}
ASTHMATICS				
Adults	Bronchodilator usage	Dusseldorp *et al*, 1995	PM_{10}, Nitrates, Sulfates	0.163 0.163 0.272
	Cough	Dusseldorp *et al*, 1995	PM_{10}, Nitrates, Sulfates	0.168 0.168 0.280
	Lower respiratory symptoms (wheeze)	Dusseldorp *et al*, 1995	PM_{10}, Nitrates, Sulfates	0.061 0.061 0.101
Children	Bronchodilator usage	Roemer *et al*, 1993	PM_{10}, Nitrates, Sulfates	0.078 0.078 0.129
	Cough	Pope and Dockery, 1992	PM_{10}, Nitrates, Sulfates	0.133 0.133 0.223
	Lower respiratory symptoms (wheeze)	Roemer *et al*, 1993	PM_{10}, Nitrates, Sulfates	0.103 0.103 0.172
ELDERLY 65+				
	Congestive heart failure	Schwartz and Morris, 1995	PM_{10}, Nitrates, Sulfates	1.85E-5 1.85E-5 3.09E-5
CHILDREN				
	Chronic cough	Dockery *et al*, 1989	PM_{10}, Nitrates, Sulfates	2.07E-3 2.07E-3 3.46E-3
ADULTS				
	Restricted activity days	Ostro, 1987	PM_{10}, Nitrates, Sulfates	0.025 0.025 0.042
	Chronic bronchitis	Abbey *et al*, 1995	PM_{10}, Nitrates, Sulfates	2.45E-5 2.45E-5 3.90E-5

Receptor	Impact Category	Reference	Pollutant	f_{er}
ENTIRE POPULATION				
	Respiratory hospital admissions (RHA)	Dab *et al*, 1996	PM_{10},	2.07E-6
			Nitrates,	2.07E-6
			Sulfates	3.46E-6
		Ponce de Leon, 1996	SO_2	2.04E-6
	Cerebrovascular hospital admissions	Wordley *et al*, 1997	PM_{10},	5.04E-6
			Nitrates,	5.04E-6
			Sulfates	8.42E-6
	Acute mortality (YOLL)[a]	Anderson *et al*, 1996, Touloumi *et al*, 1996	SO_2	5.4E-6
	Chronic mortality (YOLL)[a]	Pope *et al*, 1995	PM_{10},	1.57E-4
			Nitrates,	1.57E-4
			$PM_{2.5}$,	2.60E-4
			Sulfates	2.60E-4

[a] Converted to YOLL from increase of risk rates given in European Commission (2000).

Mortality effects from air pollution. Acute mortality effects are associated with correlations between short-term (order of days) changes in concentrations of air pollutants and short-term changes in mortality. Chronic mortality refers to long-term (order of years) effects of air pollution. Because of its importance for the final results, more space is devoted here to the assessment of chronic mortality.

The measure used to quantify mortality effects is the reduction of life expectancy expressed as *"Years of Life Lost"* (YOLL). The individual YOLL are summed up to total YOLL for the whole affected population. Mortality studies usually measure changes in mortality rates of the population. Non-trivial extra steps are needed to convert mortality rates into YOLL.

In a large study by American Cancer Society (ACS) Pope et al. (1995) followed more than 500'000 adult individuals in 151 US cities over a period of about 8 years (1982-1989). An association was found between mortality and air pollution after adjusting for age, sex, race, current cigarette smokers, former cigarette smokers, pipe/cigar smokers, exposure to passive smoking, occupational exposure, education, body-mass index (BMI), and alcohol use. The adjusted mortality rate was about 17% higher for a difference of 24.5 $\mu g/m^3$ $PM_{2.5}$ between the highest and lowest polluted areas. For sulfates, an increase of mortality of about 15% was associated with an increase of 19.9 $\mu g/m^3$ SO_4. The strongest associations were observed for cardiopulmonary disease and lung cancer, with insignificant associations with death due to other causes. Results from the ACS study were used in ExternE to derive a dose-effect model. The ACS study is still ongoing. Recently, the results for a 16-years follow-up period (1982-1998) have been published (Pope et al,. 2002) confirming again the correlation between ambient fine particulate concentrations and mortality.

Chronic mortality effects have also been studied for the Chinese city Shenyang (Xu, 1998). Mortality rates were compared for three classes of areas: areas with low, middle and high concentrations of SO_2 and TSP. The time period was one year (1992). The systematic increase of age-adjusted mortality rate with increasing pollution was confirmed. Exposure-response relationships were most striking for children and elder people. The data were analyzed for two confounding factors, commuters between e.g. low and highly polluted areas and persons who have lived elsewhere before. The influence of cigarette smoking was not addressed.

Because of the long follow up period, the large sample size and the extensive consideration of possibly confounding effects like cigarette smoking, the ACS study (Pope et al., 1995, Pope et al., 2002) stands out as the most reliable basis for the assessment of chronic mortality effects from air pollution. Thus, the corresponding exposure-response functions have been used also in EcoSense China/Asia. Nevertheless, the transfer of the results derived for US conditions to other regions is an issue that has been intensively discussed in the ExternE studies (European Commission, 1999; European Commission, 2000). For detailed discussion of the increase of chronic mortality and of the derivation of YOLL from the mortality risk ratio we refer to (Hirschberg et al., 2003b). Here we limit ourselves to stating the main points.

The chronic mortality functions of the US study (Pope et al., 1995) were scaled down in European Commission (2000) by a factor of two when they were transferred to Europe. The reason was the observation that the exposure-response functions for acute mortality showed about twice as high a slope in US studies as in the European ones. But later re-analyses of the US data by NMMAPS (National Morbidity, Mortality, and Air Pollution Study) corrected the US functions for acute mortality downwards so that it is unclear at present whether the chronic YOLL functions still have to be rescaled and how much (Hurley, 2002). A further down-scaling factor 1.5 was assumed in European Commission (2000) to take into account historic exposure given that the exposure to air pollution may have been higher at the time before the period investigated in the US studies. In total the YOLL were scaled down by a factor of about 3. A further minor correction originates from the extension of the increased mortality risk to persons of age below 30. The function, which has finally been used in EcoSense, is 0.00026 YOLL per person per year per $\mu g/m^3$ $PM_{2.5}$ or sulfates.

The question arises how the function for chronic mortality should be transferred to China. In the Shenyang study (Xu, 1998), a relative risk of 1.7 was found for a TSP concentration difference of 157 $\mu g/m^3$ between the most polluted (518 $\mu g/m^3$; mortality 697/100'000) and the least polluted areas (361 $\mu g/m^3$; mortality 409/100'000). In Xu (1998) the composition of TSP is not given so that the essential $PM_{2.5}$ or PM_{10} functions cannot be derived directly. A Chinese study of TSP composition (Ning Da-Tong et al., 1996) found a relatively high $PM_{2.5}$ fraction of about 50% in the total TSP in Shenyang. The TSP composition was analyzed for years 1986/1987 whereas the chronic mortality study refers to the year 1992. If we assume that the composition holds for the period of the health study as well, the difference in TSP concentration of 157 $\mu g/m^3$ would correspond roughly to 80 $\mu g/m^3$ $PM_{2.5}$. With a Cox proportional hazard model, this yields an increased mortality of 0.67% per $\mu g/m^3$ $PM_{2.5}$, which is almost the same as the original value in Pope et al. (1995). This is an indication against down-scaling of the original US function, different from the assumption in the European study. However, there is some uncertainty about this point. Ning Da-Tong et al. (1996) report only 230 $\mu g/m^3$ TSP for Shenyang in winter, and even less in summer. Xu (1998) cites measured values of 610-1550 μg/m^3 TSP for five days in January in the years 1982-1984 and states that it is not clear where the discrepancies in TSP measurement come from.

We adhere to the careful estimates for the chronic YOLL functions assumed in the later ExternE (European Commission, 2000) study. Nevertheless, the Chinese study gives some indication that the mortality effects could be closer to the original US (Pope, 1995) function than to the down-scaled ExternE function. If this could be confirmed, the real YOLL in China would be higher than the values presented in our study (roughly by a factor of two: factor of three from ExternE scaling and factor of 0.75 from dynamic YOLL model for China (see uncertainty analysis)). As will be showed later, this would strengthen our final conclusions.

The associations between day-to-day changes in the concentration of fine particles, SO_2, and ozone and changes in daily death rate are well established. As the 'chronic' mortality impacts from fine particles include acute effects, we do not quantify acute mortality effects from fine particles separately, to avoid double counting. Exposure-response functions used for the quantification of acute mortality impacts from SO_2 are derived from results of the APHEA study (Anderson et al., 1996; Touloumi et al., 1996); effects are considered as additive. The corresponding change in mortality risk is 0.072% per (μg/m^3) SO_2. Like in the case of chronic mortality, the percent change in death rate is converted to a loss of life expectancy (Years of Life Lost) in the exposed population. The loss is assumed to be about 0.75 YOLL per acute death (European Commission, 1999). Considering a base line mortality of about 10^{-2}, the resulting function is about 5.4E-6 YOLL/capita/year/(μg/m^3 SO_2). For EcoSense China/Asia, the European YOLL function was used without further investigating the influence of the differences in the Chinese population structure. The error from this simplification is finally very small because of the minor contribution of acute YOLL to the total YOLL.

Morbidity effects from air pollution. All morbidity effects accounted for in this study are included in Table 9.1. Here, only the two dominant ones are commented on.

Chronic bronchitis. Chronic bronchitis accounts for a significant part of the morbidity damage costs of air pollution. The assessment of chronic bronchitis impacts from fine particles is based on the analysis of a longitudinal study of Seventh Day Adventists in the USA (Abbey et al., 1995). In the regression analysis, development of new cases of definite symptoms of chronic bronchitis was significantly related to the long term ambient PM_{10} concentrations. Adjustment was made for years smoked, years lived with a smoker, years worked with a smoker, age, gender and education.

EcoSense China/Asia assumes 2.45E-5 cases/(μg/m^3) for PM_{10} per adult (aged 27+ years) according to the recommendations in (European Commission 2000). In an air pollution study for Shanghai and five other cities, Lvovsky et al. (2000) referred basically to the same longitudinal study (Abbey et al., 1993). Lvovsky et al. (2000) used for PM_{10} the lower boundary 3.06E-5 cases/(μg/m3) of a 95% confidence interval (3.06E-3 to 9.18E-5), which they extracted from Ostro (1994).

Clear Water Blue Skies study of the World Bank (World Bank, 1997) assumes 6.1E-5 cases/(μg/m3 PM_{10}) per person.

Restricted activity days. Ostro (1987) used morbidity data collected from the total population sampled in the HIS (US Health Interview Survey conducted annually by the National Center for Health Statistics) for the years 1976 to 1981. Both Working Days Lost and Restricted Activity Days (RAD) were studied, in separate analyses, for each of these six years. Although Working Days Lost appeared to be related to fine particles, there were major year-by-year differences in the estimated coefficients. Results for RADs, which are based on about 12'000 subjects per year, showed a more consistent relationship with fine particles; these results are recommended in ExternE for impact assessment.

Effects on crop yields. To estimate the effects of air pollution on crop yields, Chinese exposure-response functions have been integrated into EcoSense China/Asia. According to SEPA (1998), the approximate relation between crop production and SO_2 concentration is:

$$Y(c) = A + B \cdot c$$

where

Y:	Relative yield
A:	Constant
B:	Constant
c:	SO_2-concentration in mg/m^3.

The constants depend on the crop type as shown in the table 9.2.

Table 9.2 Crop-specific constants for estimating the dependence between crop production and pollution.

Crop Type	A	B
Rice	26.01	-2.85
Rape	31.12	-15.81
Radish	105.57	-56.97
Tomato[a]	92.70	-34.67
Wheat	23.52	-6.33
Barley	34.11	-12.22
Bean	43.69	-30.14
Cotton	30.60	-7.70
Soya Bean	40.82	-11.75

[a] function used for impacts on vegetables in general

For a change of concentration Δc, the impact on crop yield (i.e. the change ΔP of the annual crop production P) is calculated from

$$\Delta P = P * \left(\frac{Y(c + \Delta c)}{Y(c)} - 1 \right).$$

For the other crops for which no Chinese exposure-response functions were available, functions from the European ExternE study (European Commission, 1999) were used. For the assessment of effects from SO_2 on crops ExternE recommends an adapted function based on the one suggested by Baker et al. (1986). The function assumes that yield will increase with SO_2 from 0 to 6.8 ppb, and decline thereafter. The function is used to quantify changes in crop yield for potato and sugar beet.

3.2.3 Monetary valuation

Monetization is a convenient method for aggregating health impacts and environmental burdens having different physical units into a single damage estimate or indicator. Moreover, an economic assessment is advisable when comparing the benefits and costs of abatement measures, technological choices or policy regulations. Without economic evaluation, one risks making decisions that may lead to substantial welfare losses or improper allocation of resources.

To obtain the damage costs, one multiplies the number of impacts (for example, the number of asthma attacks) by the cost per case (US$ per asthma attack). For health impacts, the unit costs include the cost of illness, wage and productivity losses, which are market based factors, as well as non-market costs that take into account an individual's willingness-to-pay (WTP) to avoid the risk of pain and suffering. Economists have developed several techniques for valuing non-market goods. In recent years, contingent valuation has become the method of choice, which obtains WTP estimates by asking individuals how much money they are willing to pay to achieve a benefit. For mortality impacts, one needs to determine the Value of a Life Year Lost (VLYL), which in turn is based on the so called Value of Statistical Life (VSL), the amount of money that society is willing to pay to avoid an anonymous premature death. The median values for VSL and VLYL (for valuing long-term mortality impacts) in most industrialized countries (European Commission, 1999) are, respectively, 3.1 million and 110'000 $US\$_{2000}$ (undiscounted).

Non-market goods are difficult to value economically. This is especially true for estimating the VSL. It involves an ethical choice, and now there is emerging consensus in industrialized countries that one should base it on individual preferences (WTP) rather than the so called human capital approach, which is solely based on an individual's wage losses, plus interest earnings, due to accidental death. The human capital approach takes the point of view of the social value of a person, which often is inconsistent with an individual's own risk aversion attitudes.

The distribution of VSL results of individual preferences from studies carried out in industrialized countries tends to have a lognormal shape (Ives et al., 1993). The spread is so large that the ratio of the upper and lower bounds corresponding to one geometric standard deviation is greater than one order of magnitude. Published studies for North America indicate that a value in the range of 1 to 5 million US$ is a reasonable choice for the value of VSL, with 3.1 million US$ being the statistical geometric mean of this range.

The unit costs recommended by ExternE for monetizing different health impacts, including short- and long-term mortality (cost per YOLL), were adopted as the starting point in the present research. However, for countries outside of Europe (EU15), the European unit damage costs are multiplied by the ratio of Purchasing Power Parity GNP (PPPGNP) values for the new location and Europe ($PPPGNP_{EU15}$ = 17'764 $US\$_{1994}$) as explained in Markandya (1997). This adjustment is needed to account for income differences between countries, and consequently, in the willingness to pay regarding valuation of health impacts by individuals in different parts of the world. The PPPGNP ratio may be raised to an exponent, the elasticity factor, which reflects an individual's commitment to spend a larger share of disposal income to protect against adverse health impacts. Typical elasticity factors lie in the range 0.3 and 1 (Markandya, 1997). Markandya has recommended an elasticity factor of one, which has been consistently assumed in the present study.

The unit cost values for crops are based on market values. As crop damages are relatively small, they are estimated simply on the basis of quantity multiplied by constant price, without consideration of induced effects (compensatory producer behavior).

Table 9.3 shows the monetary unit costs assumed in EcoSense China/Asia for China.

3.2.4 Uncertainties

The uncertainties in the calculation of the environmental burdens are difficult to assess due to gaps in present working knowledge; they are also relatively large by comparison to the estimates themselves. In the assessment of global warming impacts, for instance, the 95% confidence interval (CI) covers an amazing four orders of magnitude.

On a positive note, the environmental impact assessment field is rapidly evolving and uncertainties in atmospheric dispersion models and Exposure-Response (ER) functions should become smaller in the coming years as new information and data are collected. Uncertainties associated with future scenarios and policy and ethical decisions, on the other hand, will undoubtedly be more difficult to control or eliminate.

ER functions belong to the most uncertain elements in Impact Pathways Analysis. In the current epidemiological literature, results on uncertainty ranges reflect only the statistically significant deviations in the data and neglect the systematic errors, biases and confounders. Specification of likely or appropriate overall confidence intervals is based on expert judgment and is aggravated by incomplete epidemiological knowledge. Nevertheless, despite the large uncertainties in the results, it is still possible to reach meaningful conclusions on important policy issues, technological choices or ranking of abatement measures (Rabl et al., 1998).

Table 9.3 Monetary unit costs used for China.

Impact	Monetary value US$$_{2000}$	
Health effects		
Year of life lost	15'710	per YOLL
Chronic bronchitis	25'400	per case
Cerebrovascular hospital admission	2510	per case
Respiratory hospital admission	650	per case
Congestive heart failure	490	per case
Chronic cough in children	36	per case
Restricted activity day	17	per case
Cough	7	per case
Bronchodilator usage	6	per case
Lower respiratory symptom	1	per case
Crops		
Barley – yield loss	75	per ton
(Oats – yield loss) [a]	70	per ton
Potato – yield loss	120	per ton
Rice – yield loss	225	per ton
(Rye – yield loss) [a]	190	per ton
Sugar beet – yield loss	60	per ton
(Sunflower seed – yield loss) [a]	325	per ton
(Tobacco – yield loss) [a]	50'000	per ton
Wheat – yield loss	125	per ton
Cotton – yield loss	1400	per ton
Soya bean – yield loss	300	per ton
Vegetables – yield loss	135	per ton
Rape – yield loss	110	per ton
Corn, Maize – yield loss	175	per ton

[a] Damages not calculated in EcoSense China/Asia because of lacking crop data or lacking exposure-response functions.

It is important to distinguish between policy decisions that are binary (ex., choice between nuclear or coal fired power plant) or continuous (ex., what limit to set for the SO_2 emissions from a power plant). For binary decisions, the situation is sometimes simple because the uncertainty, even if very large, has no effect if it does not change the ranking. For continuous choices the effect of uncertainty can also be surprisingly small because near an optimum the total social cost varies only slowly as individual cost components are varied.

The impact pathway analysis is essentially multiplicative, when summed over all receptor sites to obtain the total damage caused by a pollutant. According to the central limit theorem of statistics, the "natural" distribution for multiplicative processes is a lognormal function, the same way that the normal (Gaussian) distribution is the natural distribution for additive processes. Although the lognormal distribution becomes rigorous only in the limit of infinitely many variables, in practice it can be a good approximation even for a few factors, provided the distributions with the largest spread are not too far from lognormal. For many environmental impacts, the lognormal model for the result seems quite relevant because the distributions of the individual factors are not far from lognormality (Rabl and Spadaro, 1999).

If the uncertainty of each of the steps of the pathway analysis (i=1...n) is characterized by a geometric standard deviation σ_{gi}, the uncertainty of the final result σ_g is given by Rabl and Spadaro (1999):

$$[ln(\sigma_g)]^2 = [ln(\sigma_{g1})]^2 + [ln(\sigma_{g2})]^2 + ... + [ln(\sigma_{gn})]^2 \cdot$$

It involves a quadratic sum of the terms of the pathway steps, and it is dominated by the largest errors. The central limit theorem implies that the distribution of the result is likely to be approximately lognormal. As illustration, Table 9.4 shows the uncertainty analysis for several types of impacts, including chronic and acute mortality and hospital admissions due to particulates (Rabl and Spadaro, 1999). Chronic mortality is especially important because it accounts for nearly three-quarters of the total damage cost attributable to particles. The uncertainty for acute mortality is very large because of lack of information about the life span reduction. However, in the case of particles, it can be shown that if the life span reduction for acute mortality is less than 1 year, acute effects contribute less than 1% of the total YOLL impact (Rabl, 1998).

The values of σ_g for damage cost estimates for cancers due to exposure to heavy metals and radionuclides lie in the range of 6 to 10 (Rabl and Spadaro, 1999).

In section 3.4.4 the uncertainty parameters specific for the present study will be discussed and the associated results obtained for China will be presented.

Table 9.4 Uncertainty of damage cost estimates for health impacts due to particles (Rabl and Spadaro, 1999).

	Lognormality	**Chronic Mortality**	**Acute Mortality**	**Hospitalization**
Emission data (including TSP→PM_x)	probably yes	1.2	1.2	1.2
Dispersion	yes	2	2	2
Regression of ER function	no	1.3	1.3	1.3
Transfer of ER function (composition)	?	2	2	2
Cost per day	?			1.2
Duration	probably yes			1.2
YOLL (years of life lost)	probably yes	1.5	4	
VSL (value of statistical life)	Yes	2	2	
Value of YOLL, given VSL	?	1.3	1.3	
Latency (including discount rate)	probably yes	1.4		
Total σ_g [a]		**4.0**	**6.6**	**2.9**
Without VSL [b]		3.2	5.6	2.9

[a] Global or overall geometric standard deviation based on the equation above.

[b] Excludes uncertainty due to valuation of life, because it is of political rather than scientific nature.

3.3 Technical details of EcoSense China/Asia implementation

This section summarizes the emission inventory, atmospheric modeling adaptations and receptor data implemented in EcoSense for the purpose of the present study.

3.3.1 Emission database and modeling areas

Figure 9.2 shows the modeling areas used in EcoSense China/Asia. The atmospheric model WTM runs in the air quality modeling domain. For each grid cell of this domain, emission data and meteorological data are implemented. The impacts on receptors are assessed on the smaller impact assessment domain, centered around Shandong and the other highly populated areas of China. The air quality modeling domain is always larger than the impact assessment domain in order to have well-defined boundary conditions for the impact assessment (the receptors are affected also by pollutants coming from outside). The spatial resolution corresponds approximately to 50 km x 50 km, resulting in 20'083 grid cells (more than 50 million km^2) in the air quality modeling area, and in 10'692 grid cells (more than 26 million km^2) in the impact assessment area.

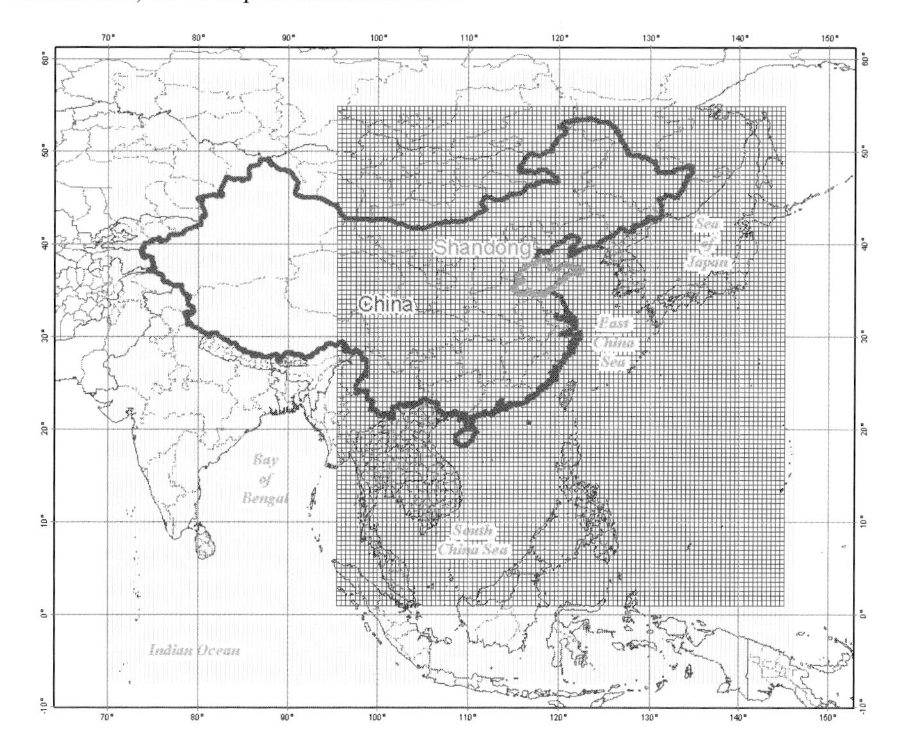

Figure 9.2 Modeling areas for EcoSense China/Asia: the larger area is the air quality modeling domain for the Windrose Trajectory Model and the smaller, red area is the impact assessment modeling domain.

Emission scenarios can be created in two ways in EcoSense. The emission scenario manager (Figure 9.3) of the multi-source module shows the full emission database and allows modifications on administrative and sectoral level. This is appropriate for analysis of complete emission sectors (e.g. the contribution of the transport sector of a country can be assessed by setting the emissions from this sector to zero and relating the results to those obtained for the full background calculation). The module makes it also easy to create specific emission reduction scenarios e.g. 90% reduction of SO_2 from all power plants of a region. The resolution of the administrative units (county level in China) is comparable to the resolution of the grid so that in principle the multi-source module can also be used to model a single emission source. Additionally, a single-source module exists enabling the user to define the characteristics of a single power plant. The single-source module adds the emissions from the power plant directly to the background emissions in the grid cell that corresponds to the plant's longitude/latitude. The subsequent calculation is then the same for the single-source and the multi-source scenarios. In EcoSense Europe, the local impacts have also been calculated with higher spatial resolution in the vicinity of the power plant, using the Industrial Source Complex Model (ISC), a Gaussian plume model developed by the US-EPA (Brode and Wang, 1992). Generally, the differences due to the higher spatial resolution of the receptor data and the more detailed meteorological data were not very significant. At the same time assembling the full set of local data for China was not feasible. Therefore, EcoSense China/Asia uses only WTM for atmospheric modeling. The implications of this simplification have been addressed in sensitivity analysis (section 3.4.4).

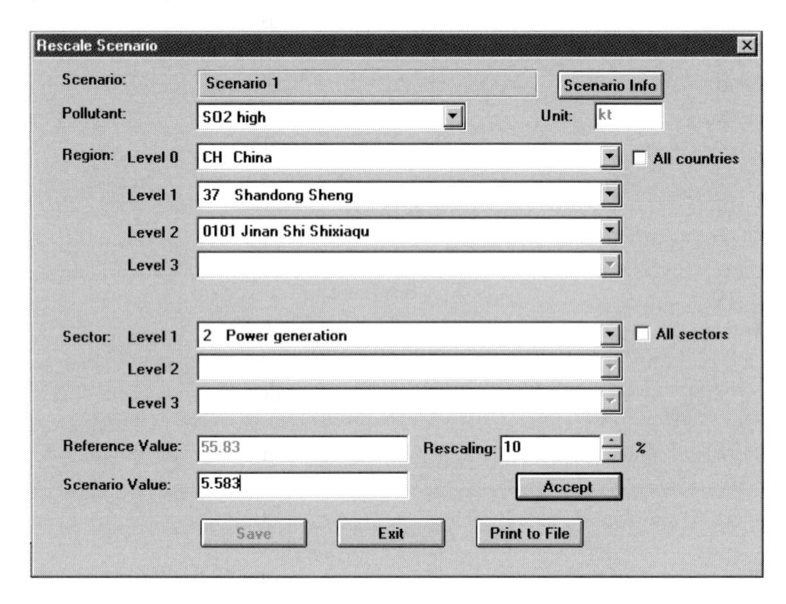

Figure 9.3 EcoSense emission scenario manager. The user can view the emission database and create scenarios.

3.3.2 Administrative units

To create the geographical database for EcoSense China/Asia, Geographical Information System (GIS) data for China and the surrounding countries were needed. The boundaries of the administrative units of China are available in GIS ARC/INFO format on county level. For the other countries, only the boundaries of the countries were available. For the purpose of the project, this resolution is sufficient. Consequently, the emission scenario manager and the reference environment database in EcoSense work on three administrative levels for China (country level, province level and county level) and on one level for all other countries. The geographical grid (0.5 degree x 0.5 degree), necessary for the atmospheric modelling, was created and intersected with the administrative units using the GIS ARC/INFO.

In total there are 32 provinces and 2404 counties; in addition, Taiwan is divided into 21 counties. The following data sources were used in this context:

- CIESIN (Center for International Earth Science Information Network), http://sedac.ciesin.org/china/ (CIESIN, 2000)
- ArcView GIS Version 3.1, data set "country", ESRI, CD, 1998 (ESRI, 1998).

3.3.3 Emissions

The emission database in EcoSense China/Asia used the EDGAR inventory for year 1990 as the starting point (EDGAR, 2000). For CETP, the emission database has been updated to the reference year 1998 as far as feasible.

The global EDGAR database provides emission data on a 1° x 1° grid for SO_2, NO_x and NH_3. The multi-source version of EcoSense needs emission data on administrative unit level. The emission data from EDGAR were transformed from the 1° x 1° grid to the administrative units and integrated into the EcoSense database. Consistency checks of EDGAR were made and the results were found satisfactory.

The emissions from China have been updated to the year 1998 in EcoSense using different references (China Statistical Yearbook 1998; Hao et.al., 2002; Sun and Wang, 1997; Streets and Waldhoff, 2000). With the exception of Shandong's power sector for which the emissions from the twenty largest power plants have been included in detail, the emissions have been updated on province or country level, i.e. generally the relative spatial distribution was adopted from the 1990 EDGAR database.

In face of the CETP focus, the most detailed update was made for the power sector of the Shandong province (Table 9.5). About 30 % of the total SO_2 and NO_x emissions originate from small plants, not included in the table but accounted for in the overall inventory. Particulate emissions from the transport sector were not included in the inventory. Particulate emissions from power plants are assumed as PM_{10} for impact assessment, according to the recommendations in ExternE (European Commission, 1999).

Table 9.6 provides the emission inventories for the current situation in China and the Shandong Province. The specific contributions from the power sector are also given.

Table 9.5 Characteristics of the 20 largest power plants in Shandong, year 1998. Source: Shandong Electric Power Corporation (SEPCO).

Power plant name	Capa-city (MW)	LHV (kJ/kg)	Sulfur content (%)	Ash content (%)	Hours of opera-tion	Annual generation 1998 (GWh/yr)	NO_x emi. (kt/yr)	SO_2 emi. (kt/yr)	CO_2 emi. (kt/yr)	Parti-culate emi. (kt/yr)
Zouxian	2400	22560	0.80	21.5	5040	12098	45.31	71.42	10969	5.29
Laiwu	375	21548	2.58	27.9	4922	1845	7.90	29.32	1923	8.18
Weifang	600	22675	1.24	26.8	5328	3197	12.37	29.93	2996	3.76
Huaneng Dezhou	1200	21421	1.23	27.3	5203	6243	25.66	62.67	6218	1.64
Heze	250	23750	0.50	24.0	5436	1359	5.17	4.65	1279	5.93
Shiheng	1335	22900	1.48	23.9	5124	6841	27.61	76.92	6746	18.11
Yantai	300	22024	1.62	26.8	4977	1493	5.05	20.60	1650	1.66
Qingdao	711	22193	2.31	26.0	4994	3551	11.66	50.38	2822	1.43
Liaocheng	300	24450	0.42	23.9	5217	1565	5.78	4.81	1482	1.21
Jining	300	22600	0.61	21.1	4790	1437	6.37	7.02	1539	13.68
Longkou	1000	20471	0.52	16.6	4567	4567	21.21	22.00	5022	3.40
Huaneng Weihai	850	23520	0.75	13.2	2871	2440	8.60	12.00	2354	2.24
Huangdao	670	21466	1.34	28.8	5534	3708	15.93	33.03	3861	5.15
Huangtai	925	23421	1.48	23.8	5469	5059	20.43	52.57	4950	3.63
Nanding	274	21878	2.00	28.6	5828	1597	8.08	32.07	1957	11.80
Linyi	362	24575	0.44	19.9	4287	1552	6.30	5.19	1526	8.54
Shiliquan	1225	22470	1.18	22.7	5313	6508	20.59	45.58	4998	9.55
Xindian	600	39601	1.87	Oil	3408	2045	6.90	19.29	1683	n.a.
Baiyanghe	150	39374	1.58	Oil	3213	482	1.81	4.57	447	n.a.
Zhanhuat	250	40503	0.97	Oil	4780	1195	4.20	6.30	1044	n.a.
Total/Av. for the 20 Plants	14077		1.21			68782	266.93	590.32	65466	105.20
Total, all SEPCO Plants	17430					84300				

n.a.=not available

Table 9.6 Emission inventory for the reference year 1998.

	SO_x in kt/year	NO_x in kt/year	PM in kt/year	NH_3 in kt/year
China				
all sectors	22'500	11'184	27'740	11'499
power sector	6970	4234	1659	n.a.
Shandong				
all sectors	2152	755	1278	985
power sector	839	378	131	n.a.

3.3.4 Atmospheric modeling

At the heart of Derwent's model (Derwent et al., 1988) is a hypothetical air parcel that travels along a straight trajectory and carries air pollutants towards a receptor. The vertical extension of the air parcel is determined by the "effective" mixing height h. The travel of the parcel extends over a period of four days, about the three- to fourfold atmospheric residence time of the model species. Hence the initial pollutant concentrations in the air parcel contribute only few percent to the final concentrations at the receptor, so that initial concentrations can be neglected and set to zero. To account for the effect of emissions all around the receptor, trajectories from 24 equally spaced directions are considered. The travel speed of the air parcel and the frequency of trajectories on the long-term average are determined by the boundary layer windrose, why we call this type of model a Windrose Trajectory Model (WTM). In practice, only one trajectory per direction is computed, and the resulting concentrations are weighted with the frequency of winds coming from this direction. The travel speed determines (i) the distance the air parcel can cover within four days and (ii) how long it takes the air parcel to cross one square of the emission grid, i. e. how much emissions the parcel receives from each emission grid square along the trajectory. In EcoSense China, the whole procedure is repeated for 6'079 receptor grid squares.

Model species include primary particles, oxidized sulfur or SO_x (SO_2, H_2SO_4, $(NH_4)_2SO_4$) , oxidized nitrogen or NO_y (NO, NO_2, HNO_3, NH_4NO_3, non-specific nitrate aerosol), and reduced nitrogen or NH_x (NH_3, NH_4NO_3, $(NH_4)_2SO_4$); their interaction in the chemical scheme is illustrated in Figure 9.4.

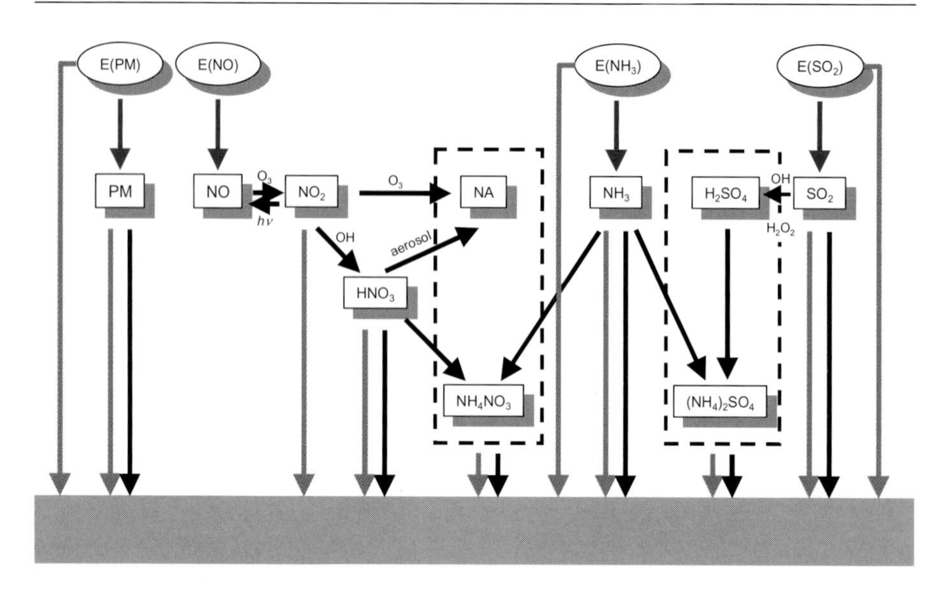

Figure 9.4 Coupled life cycles of the air pollutants (presentation modified from Barrett et al. (1995)). PM is short for particulate matter, NA for non-specific nitrate aerosol. Total nitrate aerosol (left dashed box) includes NA and NH_4NO_3, sulfate aerosol (right dashed box) includes H_2SO_4 and $(NH_4)_2SO_4$. Emission is indicated by red, chemical conversion by black, dry deposition by green and wet deposition by blue arrows.

In the following we give a brief overview on the implementation of the atmospheric model in the China/Asia version of EcoSense. Only most important parameters and modifications are addressed. For more details we refer to the technical report.

Unlike Derwent's model that for the entire model domain employs a precipitation rate of 1000 mm a^{-1} and for most receptor points the same windrose, the EcoSense China WTM uses geographically variable precipitation rates and receptor-specific windroses.

Precipitation.

- Sources:
  1) The Global Precipitation Climatology Centre (GPCC, 1998).
  2) Global Precipitation Climatology Project (GPCP) Version 2 Combination (Susskind et al., 1997, Huffman and Bolvin, 2000).
- Data set method: Conventionally measured data from raingauge networks over land only (1); gridded analysis based on gauge measurements where available and satellite estimates of rainfall elsewhere, e.g. over the sea (2).
- Time period: 1961-90 (1), 1979-99 (2).
- Resolution: 2.5 degrees x 2.5 degrees longitude/latitude.

GPCC data have been used over land and GPCP Version 2 data over the sea. Figure 9.5 shows the distribution of precipitation in the area of interest.

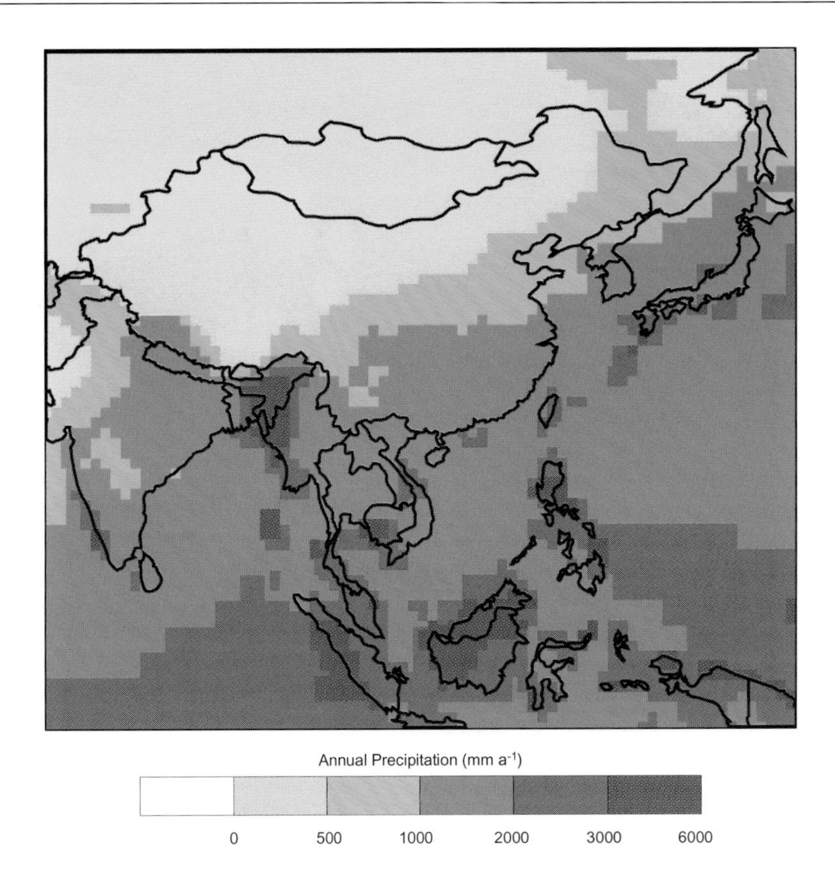

Figure 9.5 Mean annual precipitation. Data sources are GPCC (1998) over land (1961–90) and GPCP (Susskind et al. 1997; Huffman and Bolvin 2000) over the sea (1979–99).

Wind speed.

- Sources: United States of America National Centers for Environmental Prediction (NCEP) and National Center for Atmospheric Research (NCAR) (Kalnay et al., 1996; Kistler et al., 2001).
- Time period: 1989-1998.
- Resolution: 2.5 degrees x 2.5 degrees longitude/latitude.

Receptor-specific wind speeds used in the EcoSense China WTM are shown in Figure 9.6 together with exemplary frequency distributions of wind direction at three sites.

Recalculated mixing height. Mixing height, that is the depth of the atmospheric boundary layer, depends on wind shear and solar heating of the ground. Due to the diurnal cycle of solar radiation, mixing height also experiences a strong diurnal variation.

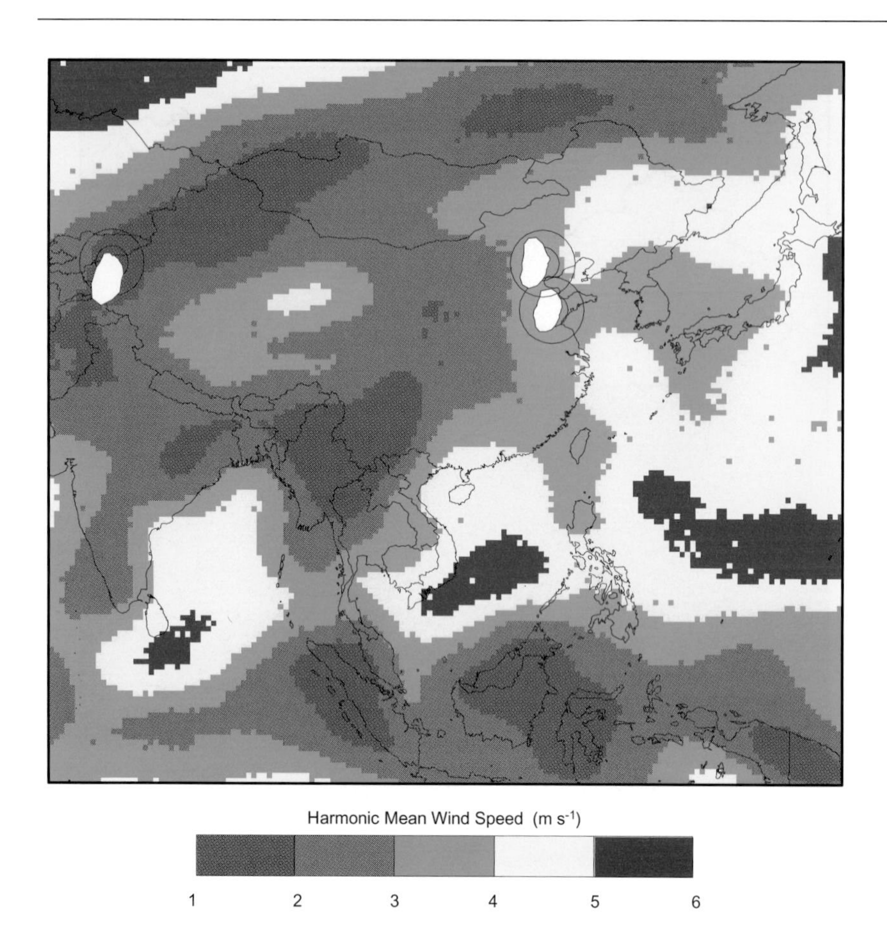

Figure 9.6 Receptor-specific wind speed data used in the EcoSense China WTM. Harmonic mean 1989–1998 at the 925 hPa pressure level derived from the NCEP/NCAR 50-Year Reanalysis (Kalnay et al., 1996; Kistler et al., 2001). Site-specific frequency distributions of wind direction are employed at each grid square – here three exemplary frequency distributions are presented as white shapes for Beijing, Jinan, and Kashi. Below such a shape, the inner circle indicates the frequency for equal distribution of all directions, the outer circle twice this frequency.

The following analysis steps were involved in the estimation of the mixing height for the Shandong Province:

- Calculation of 6-hourly time series of mixing height (from 1990 meteorology from U.S. NCEP Reanalysis project) at each grid cell.
- Calculation of mixing height statistics (harmonic mean of 0h, 6h 12h, 24h, 48h and 96h moving maximum) at each grid cell.
- Exponential fit to statistics at each grid cell.
- Determination of "effective mixing height" based on an average pollutant life time of 31.5 hours (mean of SO_x and NO_y life time).

- Population weighted harmonic mean over all grid cells.

Figure 9.7 shows mixing height in the modeling area. The result is 620 m for China, compared to 800 m for Europe according to Derwent (1988).

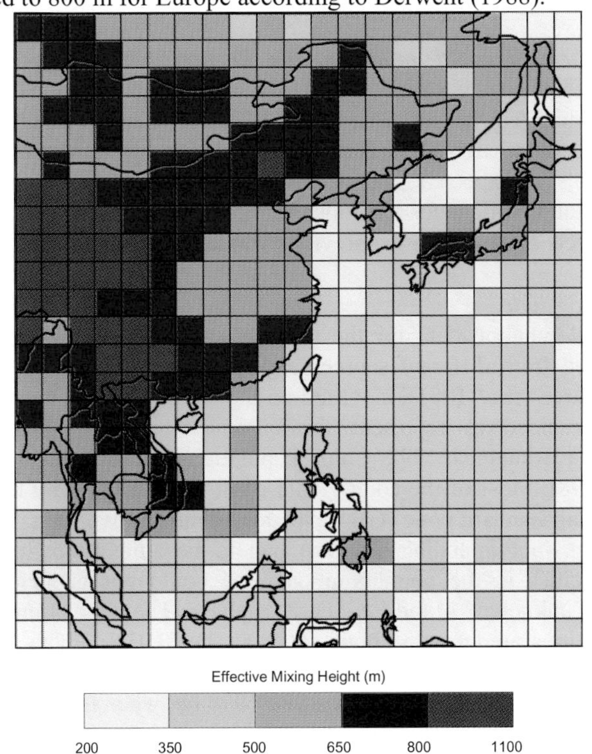

Figure 9.7 Effective mixing height.

Background concentrations. Ozone and OH radical concentrations were increased based on Logan (1999) and Spivakovsky et al. (2000), respectively.

Sulfur chemistry. Modifications were based on Carmichael et al. (2001):

- SO_2 -> SO_4 conversion rate 7.15E-6 per second (old value was 2.80E-6 per second).
- SO_2 dry deposition velocity 0.2 cm per second (old value was 0.5 cm per second).
- This leads to new "local deposition" parameters for SO_2 (ratio of emissions which is deposited in local grid cell): low sources: 3% (earlier 8%); high sources: 1% (earlier 2%).
- Sulfate dry deposition velocity 0.2 cm per second (old value was 0.1 cm per second).

- Nitrates and particulates are treated like sulfates. This leads to a new "local deposition" parameters for particulates: low sources: 3% (was 1.6%); high sources: 1% (was 0.4%).

Overall consequences of modifications in atmospheric modeling. The most remarkable changes are connected to sulfur chemistry: less SO_2 is lost to dry deposition while conversion to sulfate is faster, so more sulfate aerosols are expected. On the other hand, sulfate aerosols exhibit faster dry deposition, so less sulfate aerosols are expected. The two effects more or less cancel each other.

Nitrates and particulates have also higher deposition velocities, so they should have less effects compared to the previous version.

3.3.5 Receptor data

Population. Population data for the Chinese counties and for the other countries were collected from different sources and integrated into the EcoSense database. The main source was CIESIN as based on the 1990 census (CIESIN, 2000), which has been extended to represent the current situation using various sources (Encyclopedia Britannica, 1998; China Population Statistics Yearbook, 1998; ESRI, 1996 and 1998). Most of the county data could be identified by the Chinese “Guo Biao” National Standard code (GB 1990) for administrative units. About 150 of the 2404 counties (not including Taiwan) had to be adjusted by their names. Few counties or cities in the population database could not be identified. They are indicated as “unknown” in the EcoSense database. The population of this unknown counties or cities amounts to about 360’000 (i.e. 0.03% of the total population). Most of the unidentified counties are located in the far West of China. The population database covers data from 1990-1998 but when it comes to details it may not be fully consistent. Migration and growth have had impacts resulting in some changes in population structure during this period, not reflected in the database used. Nevertheless, the contribution of such inconsistencies to the total uncertainties is small (see uncertainty section). Figure 9.8 shows the distribution of the population in the EcoSense modeling grid.

Crop production data. Overall crop production data and the corresponding data by province, as implemented in EcoSense China/Asia Version 1.0 for reference year 1998, are shown in Tables 9.7 and 9.8 respectively.

3.4 Estimates of external costs for China and Shandong

This section presents first a short survey of results obtained in earlier studies. This is followed by results of this study, including: reference impacts and external costs for China and Shandong, corresponding estimates for future electricity supply scenarios for Shandong, and uncertainty and sensitivity analysis.

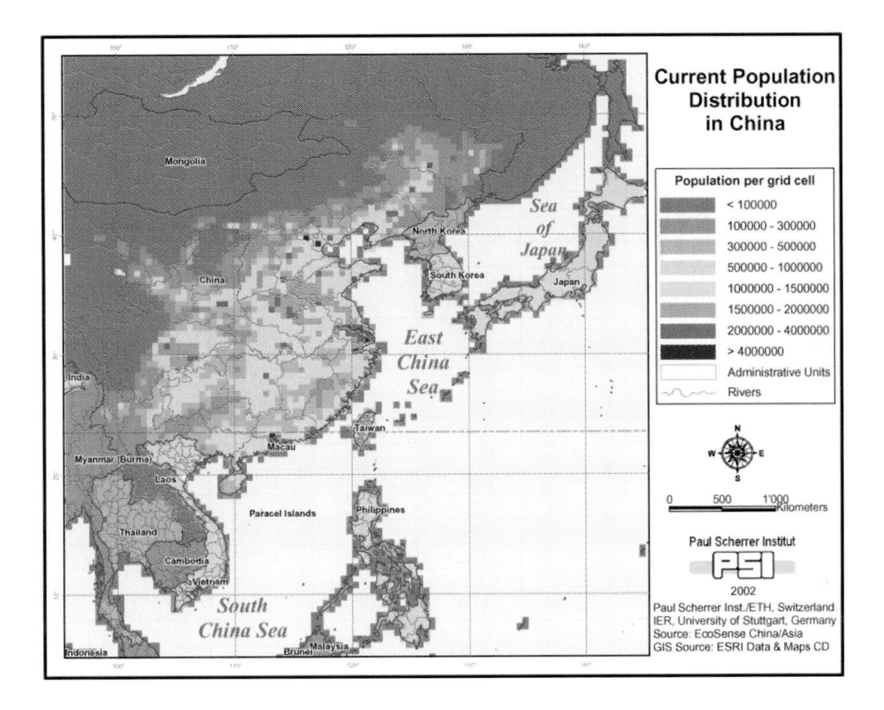

Figure 9.8 Population distribution used in EcoSense China/Asia (grid cell size 0.5°x0.5°).

Table 9.7 Overall crop production data in EcoSense China/Asia.

Agricultural Product	**China and Taiwan in kt/year**	**Rest of Modelling Area in kt/year**
Wheat	123'289	105'416
Barley	3700	12'668
Sugarbeet	78'898	17'720
Potato	43'039	59'569
Tobacco (*)	4253	n.a.
Rye (*)	700	4840
Oats (*)	700	4412
Rice	200'734	285'661
Soya bean	18'755	n.a.
Cotton	181'633	n.a.
Vegetables	37'264'971	n.a.
Rape	8687	n.a.
Corn	104'310	n.a.

(*) no exposure response function available

Table 9.8 Crop production data by Province and for Taiwan as implemented in EcoSense China/Asia.

in 10'000t	Wheat	Soya Bean	Cotton	Vegetables	Rice	Corn	Rape
China	***12200.5***	***1853.6***	***18166.3***	***3739325***	***19527.2***	***10222.1***	***850.4***
Beijing	96.4	2.8	176	4280	15.8	118.9	0
Tianjing	77.4	4.6	148.7	7605	51.1	65.5	0
Hebei	1330.7	68	787.7	192380	102.4	1009.5	3.5
Shanxi	348.6	37.8	299	70000	3.8	350.1	1.5
Inner-Mongolia	307.9	118.7	184.5	12	70.6	677.9	8.9
Liaoning	56.5	37.2	662.9	19439	385.7	668.3	0.1
Ji Ling	13	73.7	416.3	0	376.2	1260.3	0
He-Long-Jing	328.4	588.7	591.5	0	860.9	1165.9	2.9
Shanghai	34.3	2.1	175.1	8704	165.9	4.8	10.4
Jiangsu	1064.7	92.1	1054.6	428964	1931.2	243.7	11.5
Zhejiang	72.1	34.6	637.7	57901	1238	14.1	46
Anhui	941.2	100.4	708.8	260000	1290.2	229.1	139.7
Fujian	17.2	25.3	744.3	0	739.3	8.9	2
Jiangxi	10.7	39	836.1	156222	1636	8.4	68.1
Shandong	2241.3	85.8	1430.6	410000	112.1	1106	4.4
Henan	2372.4	107.3	1098	660100	342.9	807.7	42
Hubei	446.8	54.3	1139.4	425000	1818.5	161	147.5
Hunan	29.1	56	958.5	210775	2495.8	76.5	104.8
Guangdong	5.6	23.1	1585.6	0	1582.5	51.1	1.1
Guangxi	4.8	40.6	897.1	1176	1269.3	165.5	12.3
Hainan	0	2.3	186.1	0	168.3	4.2	0
Sichuang	611.1	75.8	1721.7	82199	1664.6	580.7	104.8
Guizhou	107.9	33.4	580.5	854	462	263.6	53
Yun-nan	165	80	427.5	518	530.2	363.2	12.8
Tibet	28.3	1.5	0	0	0.5	1.4	3.3
Shaanxi	562.7	26.7	342.2	50365	93.4	271.4	25.8
Ganshu	328.3	9.4	175.1	12828	4.8	166.9	17.3
Qinghai	78.3	13.2	23	0	0	0	17.8
Ningxia	82.2	6.9	31.5	3	59.9	83.2	0
Xinjiang	437.6	12.3	146.3	680000	55.3	294.3	8.9
Taiwan	128.4	21.9			546.4	208.9	18.3

Source: China Statistical Yearbook '98

3.4.1 Earlier health and environmental impact cost studies for China

Numerous studies were carried out with the goal to estimate damage costs associated with pollution in China. The vast majority of such studies were carried out by Chinese researchers or with support from them. Table 9.9, though certainly not complete, summarizes in chronological order the results of major studies; the table largely builds on a comparison of the research on the issue of pollution-related losses (CAES, 1999).

Table 9.9 Estimated losses due to pollution in China according to previous studies[1].

Source	Assessment method	Range of estimation	Ref. Year	Loss, Billion Yuan[2]/ year	Share of GDP, %
Miao Fan-ju, Xiu Wen (1985)[3]	Market valuation	Air pollution loss in Beijing	1983	0.496	3.6
Xiu Shou-bo (1986)[3]	Subgroup assessment (sum over pollution of air, water, waste...)	National air pollution loss	1980	4.4	1.2
	Integral assessment (prediction for year 2000)		2000	20.0	1.2
Liu Wen et al. (1989)[3]	Benefit transfer	National environ. loss	1980	44.4	10.44
	Valuation by experts	Thereof,			
		Water pollution		29.4	6.9
	Market valuation	Air pollution		8.5	2.0
		Solid wastes		4.05	0.95
		Noise		2.42	0.55
		National ecology damage	1980	26.5	6.23
		National environ. loss	2000	96.8	5.5
Guo Xiao-ming and Zhang Hui-qing (1990)	Market valuation	National environ. loss	1981-1985	38.156	6.74
	Opportunity cost				
	Project cost	Thereof,			
	Human capital	Water pollution		15.662	2.77
		Air pollution		12.400	2.19
		Solid wastes		0.574	0.10
		Pesticides		9.520	1.68
		National ecology damage	1981-1985	49.752	8.90
Xiu Fang et al. (1992)[3]	Human capital Medicine cost	Health effect of enterprises' pollution in selected regions (village and town)	1989	0.0093	-
Qiu Geping (1992)[3]		National environ. loss	about 1990	95.0	6.75
		Thereof,			
		Water pollution		40.0	
		Air pollution		30.0	
		Solid wastes and Pesticides		25.0	
Ge Ji-qi (1992)[3]	Market valuation	Waste water in Tai Lake	1985-1988	1.008	3.12
Wang Xiao-jing, Chen Guo-jie (1993)[3]	Travel cost method	Usage value of Three Gorges (expenditures on travels)	1990	0.834	-
	Contingent Valuation Method (CVM)	Entity value	1990	2.634	
Li Jin-Chang (1994)[3]	Integral assessment	National environmental, ecological and resource loss	1994	200.0	20.0
Jin Jian-ming et al. (1994)	Market valuation Recovery cost	National ecological loss	1985	83.1	19.37
		Thereof,			
		Shichuang		10.2	
		Shandong		2.9	
		Ningxia		0.15	
Guo Shi-qing et al. (1994)[3]	Market valuation	Agricultural loss from environ. pollution	1988	12.5	-
Wu Gang, Zhang Jing-yang (1994)[3]	Market valuation	Acid rain's effect on forest in Chongqing	1993	1.307 Y/hectar /year	-
Sima Jian (1996)[3]	Reference summary (based on review of existing studies) Extrapolation	National environ. loss National ecological loss	1991-1995	133.0 90.5	-

Source	Assessment method	Range of estimation	Ref. Year	Loss, Billion Yuan[2]/ year	Share of GDP, %
World Bank (1997)	Human capital Willingness to pay	Air, indoor air, lead, water, acid rain	1995 1995	24.228 G$ 53.589 G$ Billion US$	3.5 7.7
Xu Songling (1997)	Direct loss, indirect loss and remediation cost	Ecological loss: Thereof, Forest Pasture Agriculture land Catchment	1993	110 38 25 36 11	4.0
Xia Guang[3] (1998)	Market valuation Opportunity cost	National environ. loss Thereof, Water pollution Air pollution Land occupation of solid waste	1992	98.61 35.60 57.89 5.12	4.04
Smil Vaclaw (1998)[3]	Benefit transfer Human capital	National environ. loss	1988	40.0	3.0
Chen Shijie (1999)	Human capital method	Hangzhou city, air pollution	1999	0.78	3.99 (GDP in the mentioned city only)
Zhou Yuexian and Li Hong, (1999)	Human capital method	Luo Yang city, air pollution	1999	0.1504	-
Chongqing Medical University (2000)	Human capital Willingness to pay	Chongqing city, air pollution	2000	1.05 3.27	7.3 22.9 (GDP in the mentioned city only)
Lvovsky et al. (2000)	Willingness to pay	Shanghai city, air pollution of fossil fuel Health Non-health	2000	 0.730 G$ 0.096 G$ Billion US$	-

[1] The numbers in the table are cited as they appear in the survey (CAES, 1999). Some of them seem to be not internally consistent.
[2] Costs in Billion Yuan, except for studies which provided results directly in US $; 1 US $ approximately corresponds to 8 Yuan.
[3] Reference in (CAES, 1999), not included in the reference list for this chapter.

The studies compiled in the table have different scopes, highly varying level of detail and employ a wide spectrum of methods for the estimate of impacts and the associated costs. To our knowledge no previous study carried out a comprehensive simulation of the dispersion of pollutants and their interactions with the receptors.

Of particular importance for the monetary damage estimates is the choice of the method for transferring the physical impacts into monetary values. The most common approach used in China is the Human Capital (HC) method. It calculates the value of life as the present value of net foregone earnings (given death). The cost of illness is composed of medical costs, costs of patient work time losses and costs of the family member's work time losses due to bed caring. The cost of premature death is estimated based on the net GNP (value added) per capita. The Human Capital method leads to much lower prices of risk reduction than those resulting from use of the market approach (WTP; see Section 3.2.3).

The "Clear Water and Blue Skies" study (World Bank, 1997) used both the WTP and HC methods, thus presenting two sets of results. The HC approach has been subject to extensive criticism. For example, ECON (2000) states that the principal reason for rejecting the HC approach is that it does not express the price of risk reduction. ("There is no logical reason why an individual should equate the price he is willing to pay for risk reduction to his remaining lifetime earnings. The two are simply different concepts.") The HC approach does not address the welfare of people who do not have an income (e.g. retired), who also exhibit willingness to pay for the prevention of disease or premature death. In the present work the WTP approach was adopted; however, the robustness of the cost-benefit analysis was examined by employing the HC approach in a sensitivity calculation (see Section 3.4.4).

Practically, all earlier studies show high environmental losses both in absolute terms as well as in relation to GNP, with the tendency towards higher numbers in more recent studies. The monetary valuation of ecological damages, which include such effects as deforestation and depletion of natural resources, is subject to still much higher uncertainties than estimation of pollution-related losses. The latest published studies establish that losses caused by air pollution dominate over the corresponding damages associated with water pollution though the latter remain to be of high concern.

The present study focused on detailed modeling and estimation of air pollution impacts, with particular emphasis on the electricity sector.

3.4.2 Current impacts and external costs

Overall damage estimates. Year 1998 is used here as reference for the current situation in China since the Shandong Electric Power Corporation (SEPCO) supplied detailed major pollutant emission data for all larger power plants in Shandong for this year, and the corresponding data available for later years, were not complete. It should be noted that, as a result of partial implementation of new environmental policy, manifested by wider use of coal with lower sulfur content, the emissions of SO_2 from large power plants in the Shandong Province decreased in 1999. At the same time, the electricity production continued to grow. The present total SO_2 emissions from the Shandong power sector are judged to be close to those in 1998. Consequently, the 1998 scenario is considered representative for the current situation.

Figure 9.9 shows the mortality effects in terms of millions of years of life lost (YOLL) per year caused by outdoor air pollution in China in general, and in the Shandong Province in particular. The contribution from the power sector in these two areas is also highlighted. Chronic effects strongly dominate mortality. Roughly, each premature death in Europe due to air pollutants corresponds to about 11 YOLL for chronic fatalities, and to 0.5 to 0.75 YOLL for acute deaths. Starting from 9.1 million YOLL per year (Figure 9.9), this means that close to one million people die prematurely in China every year as the result of outdoor air pollution. The damage costs associated with health effects including mortality are illustrated in Figure 9.10. According to the present study mortality costs constitute about 72% and morbidity costs about 27% of the total estimated damage; damage to crops is small compared with health effects. The total loss corresponds to 6-7% of PPP GNP for China. It should be emphasized that, due to the nature of the method used, the estimated total damage cost expressed as a share of GNP is independent of the assumed GNP per capita (i.e. whether PPP or standard GNP is used).

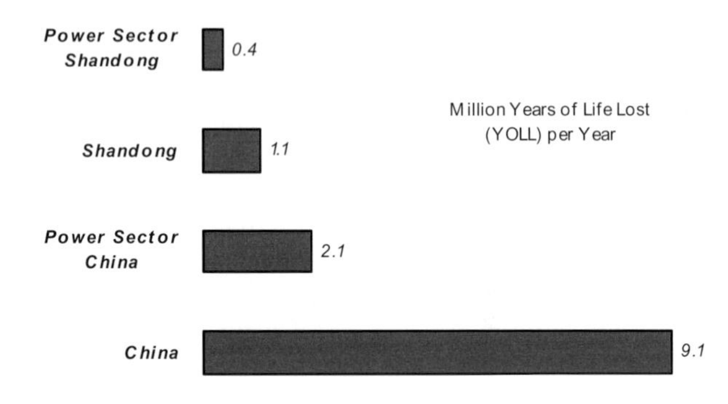

Figure 9.9 Mortality due to outdoor air pollution for the reference year 1998; estimates based on multi-source analyses.

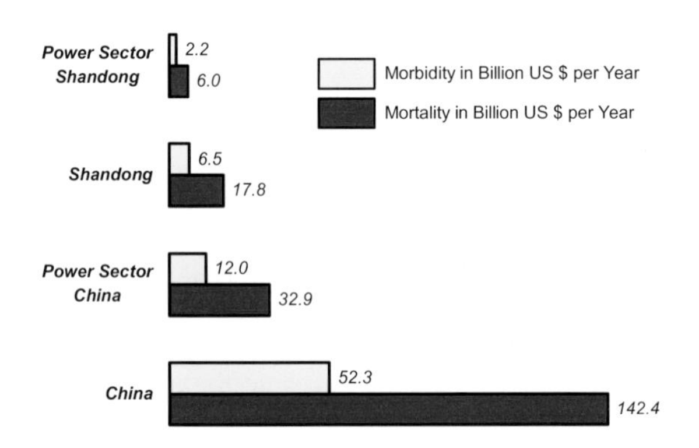

Figure 9.10 Damage costs due to outdoor air pollution for the reference year 1998; estimates based on multi-source analyses.

Distribution of damages. The spatial distributions of mortality due to air-pollutant emissions from all sectors in China and from China's power sector are shown in Figures 9.11 and 9.12.

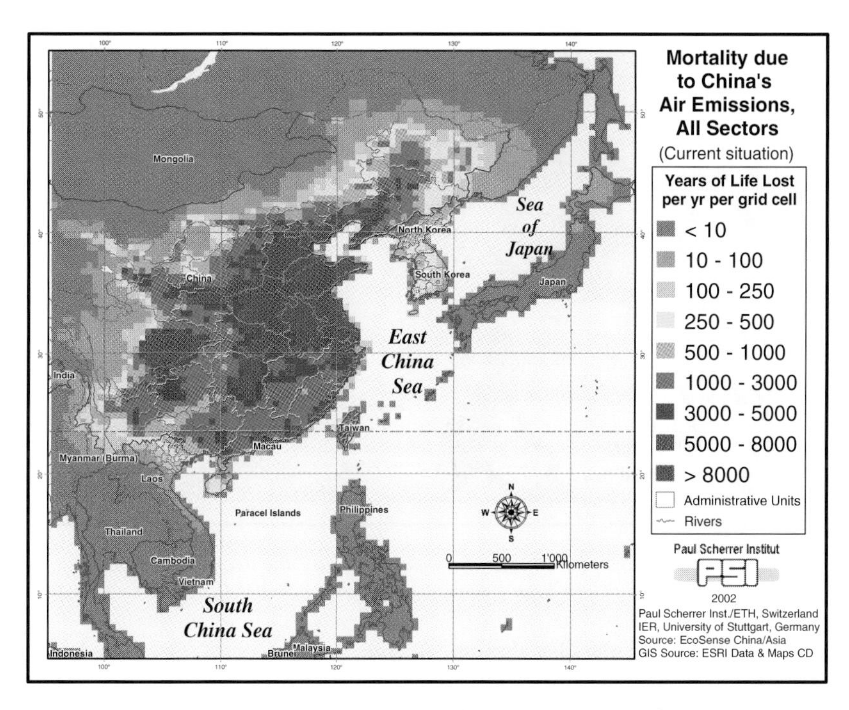

Figure 9.11 Multi-source-based distribution of mortality due to current emissions of air pollutants from all sectors in China (grid cell size 0.5°x0.5°).

Figures 9.13 and 9.14 illustrate the corresponding spatial distributions for the Shandong Province. The major part of the damage to health is caused by secondary particulates formed by chemical transformation of SO_2 and NO_x into sulfates and nitrates, respectively. In relative terms sulfates are dominant and SO_2 is thus indirectly the major contributor, followed by nitrates and primary particulates. Mortality effects directly caused by SO_2 are relatively small. Generally, there is a large spatial variation of mortality, with the highest damage occurring in areas with intensive industrial activities and high density of population.

The relative contribution to external costs of morbidity from specific effects is shown in Figure 9.15, based on calculations for the Shandong power sector. The pattern is similar for other cases. Chronic bronchitis and restricted activity days dominate the morbidity costs.

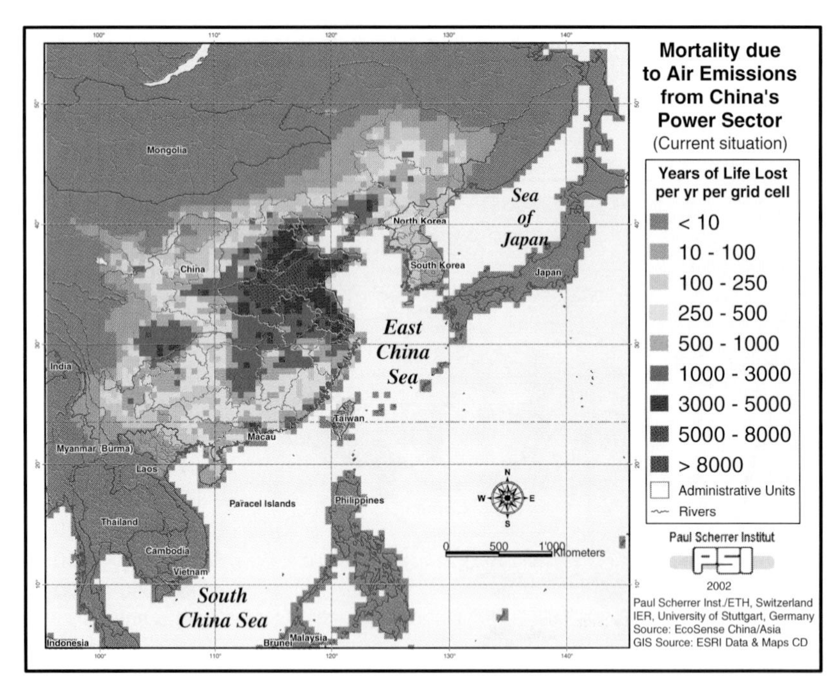

Figure 9.12 Multi-source-based distribution of mortality due to current emissions of air pollutants from China's power sector (grid cell size 0.5°x0.5°).

Figure 9.16 shows the distribution of crop damage costs, associated with air pollution originating from Shandong power plants. About 8% of China's cotton is produced in Shandong. Because of the relatively high price of cotton it dominates the monetary damage to crops. The losses based on mass are dominated by wheat (31% of the total).

Damage dependence on plant location and technology. EcoSense single-source calculations using the same reference power plant were carried out for all 21 locations in Shandong, including all sites of larger power plants currently operating in Shandong. The reference plant used for the comparison is based on conventional coal power plant technology, as implemented in the currently operating 860 MWe Huangtai plant in Jinan, Shandong. Jinan is the largest city in Shandong, and is located inland. For the reference power plant, technical parameters in table 9.10 were used, reflecting essentially the actual performance of the Huangtai plant.

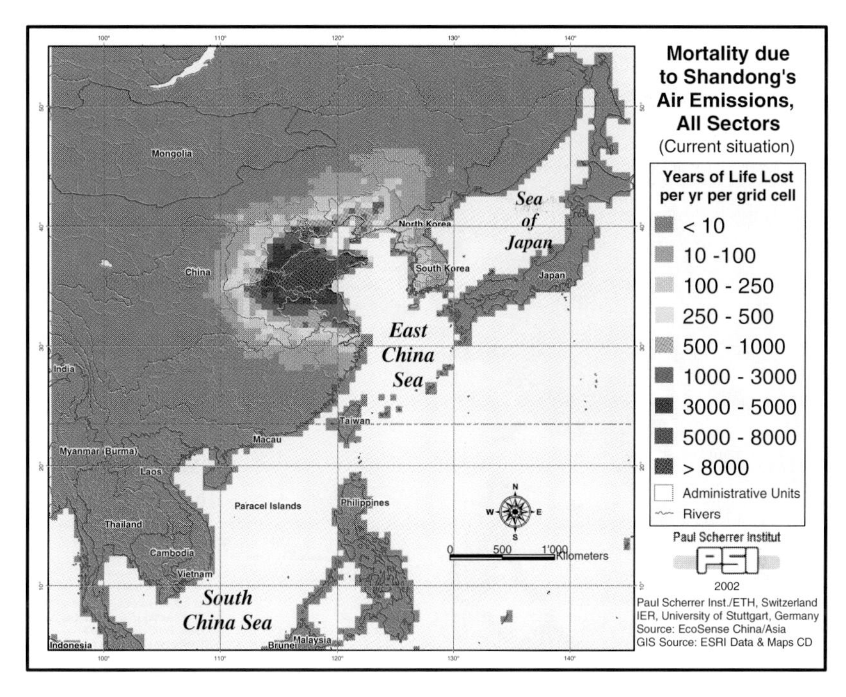

Figure 9.13 Multi-source-based distribution of mortality due to current emissions of air pollutants from all sectors in Shandong (grid cell size 0.5°x0.5°).

Table 9.11 shows in detail the dependence of health impacts on the 21 locations of the power plant for the reference year 1998. The most dominant impacts due to sulfates differ up to a factor of about four between the different locations. The values for direct PM and nitrates show even higher variations up to a factor of six to seven. Direct impacts due to SO_2, which was not converted to sulfates, vary by a factor of about five.

Some examples of the impact of location and technology on mortality resulting from the emission of air pollutants are shown in Figure 9.17. The results for the previously mentioned Jinan site are compared to Heze and Weihai, which are located in the south-western part of the province and in the northern part of the Shandong peninsula, respectively. Furthermore, the impact of retro-fitting the Huangtai plant with a Flue Gas Desulfurization (FGD) unit having 95% SO_2 removal efficiency, and contributing to a further reduction of PM emissions beyond the already implemented abatement measures, was investigated. Finally, the impact to health and environment associated with the installation of the advanced, gas-fired, combined-cycle plant (ABB technology; now Alstom), was also estimated. The gas technology used here is characterized by very high efficiency and extremely low pollutant emissions.

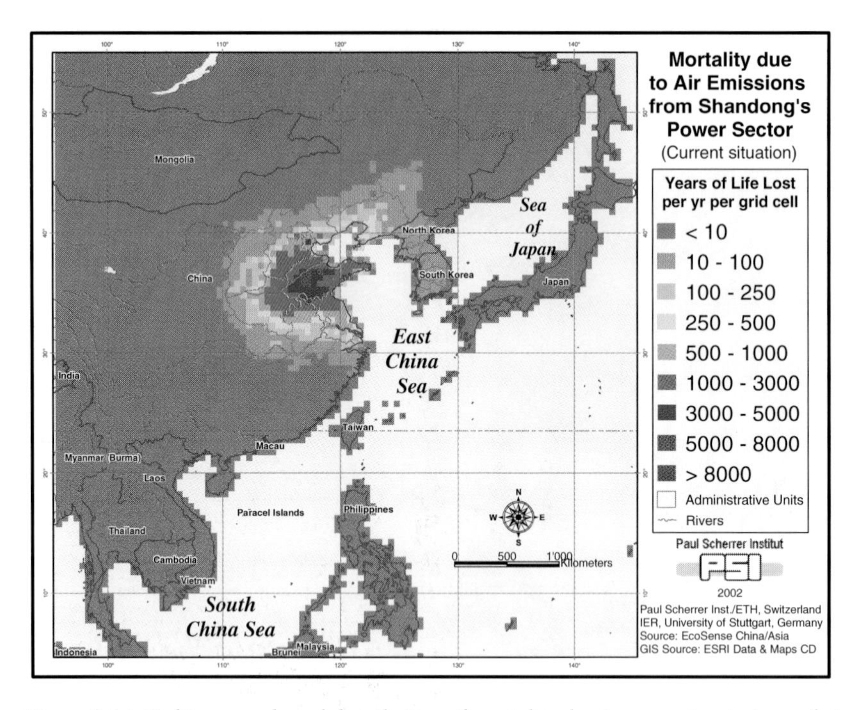

Figure 9.14 Multi-source-based distribution of mortality due to current emissions of air pollutants from Shandong's power sector (grid cell size 0.5°x0.5°).

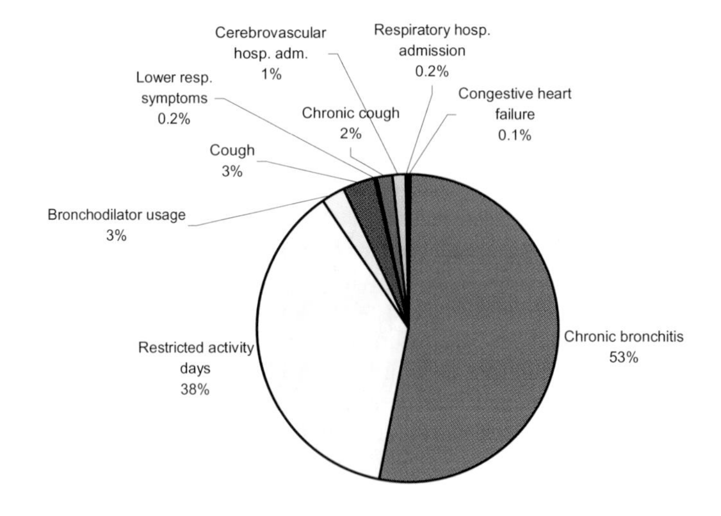

Figure 9.15 Distribution of external costs of morbidity effects due to air emissions from Shandong power sector.

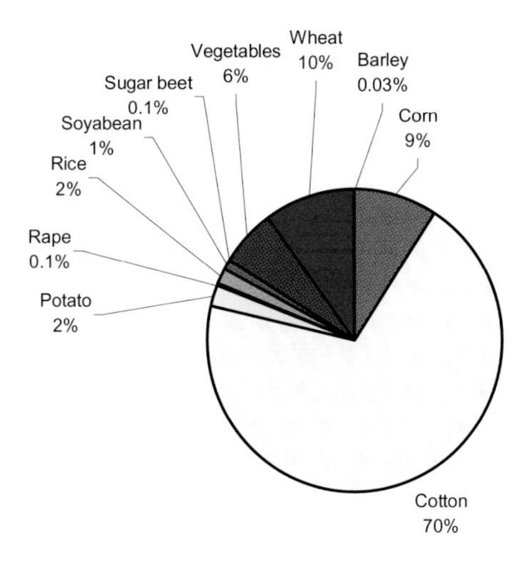

Figure 9.16 Distribution of external costs of crop yield losses due to air emissions from Shandong power sector.

Table 9.10 Parameters characterizing the reference coal power plant.

Characteristic	**Performance**
Capacity	925 MW
Electricity sent out	860 MW
Full load hours per year	5469 h
SO_2 Emissions	3800 mg/Nm^3
NO_x Emissions	1500 mg/Nm^3
PM Emissions	500 mg/Nm^3
Flue gas volume stream	2’500’000 Nm^3/h
Annual SO_2 Emissions	52’000 tons/a
Annual NO_x Emissions	20’500 tons/a
Annual PM Emissions	6800 tons/a
Stack height	220 m

Table 9.11 Health impacts per year for reference power plant in Shandong 1998.

Location	Jinan	Dezhou	Liao cheng	Shiheng	Laiwu	Qingdao	Huang dao
Chronic mortality – Years of Life Lost							
PM_{10}	1.8E+03	1.7E+03	2.0E+03	1.9E+03	1.6E+03	1.1E+03	6.6E+02
Sulfates	1.9E+04	1.9E+04	2.2E+04	1.9E+04	1.7E+04	1.1E+04	1.0E+04
Nitrates	3.8E+03	4.2E+03	5.3E+03	3.2E+03	3.3E+03	2.0E+03	2.0E+03
Acute mortality – Years of Life Lost							
SO_2	5.0E+02	4.7E+02	5.5E+02	5.3E+02	4.8E+02	3.2E+02	2.3E+02
Respiratory hospital admissions							
PM_{10}	2.3E+01	2.2E+01	2.6E+01	2.4E+01	2.1E+01	1.4E+01	8.7E+00
Sulfates	2.5E+02	2.5E+02	2.9E+02	2.5E+02	2.3E+02	1.4E+02	1.4E+02
Nitrates	5.0E+01	5.5E+01	7.0E+01	4.2E+01	4.3E+01	2.6E+01	2.7E+01
SO_2	1.9E+02	1.8E+02	2.1E+02	2.0E+02	1.8E+02	1.2E+02	8.7E+01
Cerebrovascular hospital admissions							
PM_{10}	5.7E+01	5.4E+01	6.4E+01	6.0E+01	5.2E+01	3.4E+01	2.1E+01
Sulfates	6.0E+02	6.1E+02	7.0E+02	6.1E+02	5.5E+02	3.4E+02	3.3E+02
Nitrates	1.2E+02	1.3E+02	1.7E+02	1.0E+02	1.1E+02	6.4E+01	6.6E+01
Congestive heart failure							
PM_{10}	2.9E+01	2.8E+01	3.3E+01	3.1E+01	2.7E+01	1.7E+01	1.1E+01
Sulfates	3.1E+02	3.1E+02	3.6E+02	3.1E+02	2.8E+02	1.8E+02	1.7E+02
Nitrates	6.2E+01	6.9E+01	8.7E+01	5.3E+01	5.4E+01	3.3E+01	3.4E+01
Cases of chronic bronchitis							
PM_{10}	2.2E+02	2.1E+02	2.5E+02	2.3E+02	2.0E+02	1.3E+02	8.2E+01
Sulfates	2.2E+03	2.3E+03	2.6E+03	2.3E+03	2.0E+03	1.3E+03	1.2E+03
Nitrates	4.7E+02	5.2E+02	6.6E+02	4.0E+02	4.1E+02	2.5E+02	2.6E+02
Episodes of chronic cough							
PM_{10}	4.7E+03	4.5E+03	5.3E+03	4.9E+03	4.3E+03	2.8E+03	1.7E+03
Sulfates	4.9E+04	5.0E+04	5.8E+04	5.0E+04	4.5E+04	2.8E+04	2.7E+04
Nitrates	9.9E+03	1.1E+04	1.4E+04	8.4E+03	8.7E+03	5.3E+03	5.4E+03
Cases of bronchodilator usage asthma –adults							
PM_{10}	3.7E+04	3.5E+04	4.1E+04	3.8E+04	3.4E+04	2.2E+04	1.4E+04
Sulfates	3.9E+05	3.9E+05	4.5E+05	3.9E+05	3.5E+05	2.2E+05	2.1E+05
Nitrates	7.8E+04	8.6E+04	1.1E+05	6.6E+04	6.8E+04	4.2E+04	4.2E+04
Cases of bronchodilator usage asthma –children							
PM_{10}	7.4E+03	7.0E+03	8.3E+03	7.7E+03	6.7E+03	4.4E+03	2.7E+03
Sulfates	7.7E+04	7.8E+04	9.0E+04	7.8E+04	7.1E+04	4.4E+04	4.3E+04
Nitrates	1.6E+04	1.7E+04	2.2E+04	1.3E+04	1.4E+04	8.3E+03	8.5E+03
Cough days in asthmatics –adults							
PM_{10}	3.8E+04	3.6E+04	4.2E+04	4.0E+04	3.5E+04	2.2E+04	1.4E+04
Sulfates	4.0E+05	4.0E+05	4.6E+05	4.0E+05	3.6E+05	2.3E+05	2.2E+05
Nitrates	8.0E+04	8.8E+04	1.1E+05	6.8E+04	7.0E+04	4.3E+04	4.4E+04
Cough days in asthmatics –children							
PM_{10}	1.3E+04	1.2E+04	1.4E+04	1.3E+04	1.2E+04	7.5E+03	4.7E+03
Sulfates	1.3E+05	1.4E+05	1.6E+05	1.4E+05	1.2E+05	7.6E+04	7.3E+04
Nitrates	2.7E+04	3.0E+04	3.8E+04	2.3E+04	2.3E+04	1.4E+04	1.5E+04
Days with lower respiratory symptoms asthma –adults							
PM_{10}	1.4E+04	1.3E+04	1.5E+04	1.4E+04	1.3E+04	8.1E+03	5.1E+03
Sulfates	1.4E+05	1.5E+05	1.7E+05	1.5E+05	1.3E+05	8.2E+04	7.9E+04
Nitrates	2.9E+04	3.2E+04	4.1E+04	2.5E+04	2.5E+04	1.5E+04	1.6E+04
Days with lower respiratory symptoms asthma –children							
PM_{10}	9.8E+03	9.3E+03	1.1E+04	1.0E+04	9.0E+03	5.8E+03	3.6E+03
Sulfates	1.0E+05	1.0E+05	1.2E+05	1.0E+05	9.4E+04	5.8E+04	5.7E+04
Nitrates	2.1E+04	2.3E+04	2.9E+04	1.8E+04	1.8E+04	1.1E+04	1.1E+04
Restricted activity days							
PM_{10}	2.3E+05	2.2E+05	2.5E+05	2.4E+05	2.1E+05	1.3E+05	8.4E+04
Sulfates	2.4E+06	2.4E+06	2.8E+06	2.4E+06	2.2E+06	1.4E+06	1.3E+06
Nitrates	4.8E+05	5.3E+05	6.7E+05	4.1E+05	4.2E+05	2.6E+05	2.6E+05

Table 9.11 Health impacts per year for reference power plant in Shandong 1998 (continued).

Location	Rizhao	Wanggezhuang	Jimo	Zhanhua	Nanding	Xindian	Weifang
Chronic mortality – Years of Life Lost							
PM_{10}	1.2E+03	9.5E+02	1.1E+03	1.0E+03	1.5E+03	1.5E+03	1.2E+03
Sulfates	1.4E+04	1.2E+04	1.1E+04	1.4E+04	1.5E+04	1.5E+04	1.3E+04
Nitrates	3.3E+03	2.7E+03	2.0E+03	3.3E+03	3.4E+03	3.4E+03	3.0E+03
Acute mortality – Years of Life Lost							
SO_2	3.8E+02	3.1E+02	3.2E+02	3.0E+02	4.3E+02	4.3E+02	3.6E+02
Respiratory hospital admissions							
PM_{10}	1.5E+01	1.3E+01	1.4E+01	1.4E+01	2.0E+01	2.0E+01	1.6E+01
Sulfates	1.9E+02	1.6E+02	1.4E+02	1.8E+02	2.1E+02	2.1E+02	1.8E+02
Nitrates	4.4E+01	3.5E+01	2.6E+01	4.4E+01	4.5E+01	4.5E+01	3.9E+01
SO_2	1.4E+02	1.2E+02	1.2E+02	1.1E+02	1.7E+02	1.7E+02	1.4E+02
Cerebrovascular hospital admissions							
PM_{10}	3.7E+01	3.0E+01	3.4E+01	3.3E+01	4.8E+01	4.8E+01	3.9E+01
Sulfates	4.6E+02	3.9E+02	3.4E+02	4.5E+02	5.0E+02	5.0E+02	4.3E+02
Nitrates	1.1E+02	8.6E+01	6.4E+01	1.1E+02	1.1E+02	1.1E+02	9.5E+01
Congestive heart failure							
PM_{10}	1.9E+01	1.6E+01	1.7E+01	1.7E+01	2.5E+01	2.5E+01	2.0E+01
Sulfates	2.3E+02	2.0E+02	1.8E+02	2.3E+02	2.6E+02	2.6E+02	2.2E+02
Nitrates	5.5E+01	4.4E+01	3.3E+01	5.5E+01	5.6E+01	5.6E+01	4.9E+01
Cases of chronic bronchitis							
PM_{10}	1.5E+02	1.2E+02	1.3E+02	1.3E+02	1.9E+02	1.9E+02	1.5E+02
Sulfates	1.7E+03	1.5E+03	1.3E+03	1.7E+03	1.8E+03	1.8E+03	1.6E+03
Nitrates	4.2E+02	3.4E+02	2.5E+02	4.2E+02	4.2E+02	4.2E+02	3.7E+02
Episodes of chronic cough							
PM_{10}	3.1E+03	2.5E+03	2.8E+03	2.7E+03	4.0E+03	4.0E+03	3.2E+03
Sulfates	3.7E+04	3.2E+04	2.8E+04	3.7E+04	4.1E+04	4.1E+04	3.5E+04
Nitrates	8.8E+03	7.1E+03	5.3E+03	8.8E+03	8.9E+03	8.9E+03	7.8E+03
Cases of bronchodilator usage asthma –adults							
PM_{10}	2.4E+04	2.0E+04	2.2E+04	2.1E+04	3.1E+04	3.1E+04	2.5E+04
Sulfates	2.9E+05	2.5E+05	2.2E+05	2.9E+05	3.2E+05	3.2E+05	2.8E+05
Nitrates	6.9E+04	5.6E+04	4.2E+04	6.9E+04	7.0E+04	7.0E+04	6.1E+04
Cases of bronchodilator usage asthma –children							
PM_{10}	4.8E+03	3.9E+03	4.4E+03	4.3E+03	6.2E+03	6.2E+03	5.1E+03
Sulfates	5.9E+04	5.1E+04	4.4E+04	5.7E+04	6.4E+04	6.4E+04	5.5E+04
Nitrates	1.4E+04	1.1E+04	8.3E+03	1.4E+04	1.4E+04	1.4E+04	1.2E+04
Cough days in asthmatics -adults							
PM_{10}	2.5E+04	2.0E+04	2.2E+04	2.2E+04	3.2E+04	3.2E+04	2.6E+04
Sulfates	3.0E+05	2.6E+05	2.3E+05	2.9E+05	3.3E+05	3.3E+05	2.8E+05
Nitrates	7.1E+04	5.7E+04	4.3E+04	7.1E+04	7.2E+04	7.2E+04	6.3E+04
Cough days in asthmatics -children							
PM_{10}	8.3E+03	6.8E+03	7.5E+03	7.3E+03	1.1E+04	1.1E+04	8.8E+03
Sulfates	1.0E+05	8.7E+04	7.6E+04	9.9E+04	1.1E+05	1.1E+05	9.5E+04
Nitrates	2.4E+04	1.9E+04	1.4E+04	2.4E+04	2.4E+04	2.4E+04	2.1E+04
Days with lower respiratory symptoms asthma -adults							
PM_{10}	8.9E+03	7.3E+03	8.1E+03	7.9E+03	1.2E+04	1.2E+04	9.5E+03
Sulfates	1.1E+05	9.4E+04	8.2E+04	1.1E+05	1.2E+05	1.2E+05	1.0E+05
Nitrates	2.6E+04	2.1E+04	1.5E+04	2.6E+04	2.6E+04	2.6E+04	2.3E+04
Days with lower respiratory symptoms asthma -children							
PM_{10}	6.4E+03	5.2E+03	5.8E+03	5.6E+03	8.3E+03	8.3E+03	6.8E+03
Sulfates	7.8E+04	6.7E+04	5.8E+04	7.6E+04	8.5E+04	8.5E+04	7.3E+04
Nitrates	1.8E+04	1.5E+04	1.1E+04	1.8E+04	1.9E+04	1.9E+04	1.6E+04
Restricted activity days							
PM_{10}	1.5E+05	1.2E+05	1.3E+05	1.3E+05	1.9E+05	1.9E+05	1.6E+05
Sulfates	1.8E+06	1.6E+06	1.4E+06	1.8E+06	2.0E+06	2.0E+06	1.7E+06
Nitrates	4.2E+05	3.4E+05	2.6E+05	4.3E+05	4.3E+05	4.3E+05	3.8E+05

Table 9.11 Health impacts per year for reference power plant in Shandong 1998 (continued).

Location	Yantai	Weihai	Longkou	Heze	Shiliquan	Jining	Zouxian
Chronic mortality – Years of Life Lost							
PM_{10}	5.6E+02	3.0E+02	7.1E+02	2.1E+03	1.7E+03	1.9E+03	1.9E+03
Sulfates	7.8E+03	5.8E+03	9.9E+03	2.4E+04	1.8E+04	2.0E+04	2.0E+04
Nitrates	1.4E+03	8.9E+02	2.0E+03	5.8E+03	3.2E+03	3.3E+03	3.3E+03
Acute mortality – Years of Life Lost							
SO_2	1.8E+02	1.1E+02	2.2E+02	6.0E+02	5.4E+02	5.7E+02	5.7E+02
Respiratory hospital admissions							
PM_{10}	7.4E+00	4.0E+00	9.4E+00	2.8E+01	2.3E+01	2.5E+01	2.5E+01
Sulfates	1.0E+02	7.7E+01	1.3E+02	3.1E+02	2.4E+02	2.7E+02	2.7E+02
Nitrates	1.9E+01	1.2E+01	2.6E+01	7.7E+01	4.2E+01	4.3E+01	4.3E+01
SO_2	6.8E+01	4.3E+01	8.4E+01	2.3E+02	2.1E+02	2.2E+02	2.2E+02
Cerebrovascular hospital admissions							
PM_{10}	1.8E+01	9.6E+00	2.3E+01	6.8E+01	5.5E+01	6.1E+01	6.1E+01
Sulfates	2.5E+02	1.9E+02	3.2E+02	7.6E+02	5.8E+02	6.5E+02	6.5E+02
Nitrates	4.5E+01	2.9E+01	6.3E+01	1.9E+02	1.0E+02	1.1E+02	1.1E+02
Congestive heart failure							
PM_{10}	9.2E+00	4.9E+00	1.2E+01	3.5E+01	2.8E+01	3.2E+01	3.2E+01
Sulfates	1.3E+02	9.7E+01	1.6E+02	3.9E+02	3.0E+02	3.3E+02	3.3E+02
Nitrates	2.3E+01	1.5E+01	3.2E+01	9.6E+01	5.3E+01	5.4E+01	5.4E+01
Cases of chronic bronchitis							
PM_{10}	7.0E+01	3.7E+01	8.9E+01	2.6E+02	2.1E+02	2.4E+02	2.4E+02
Sulfates	9.3E+02	7.0E+02	1.2E+03	2.8E+03	2.2E+03	2.4E+03	2.4E+03
Nitrates	1.8E+02	1.1E+02	2.5E+02	7.3E+02	4.0E+02	4.1E+02	4.1E+02
Episodes of chronic cough							
PM_{10}	1.5E+03	7.9E+02	1.9E+03	5.5E+03	4.5E+03	5.0E+03	5.0E+03
Sulfates	2.1E+04	1.5E+04	2.6E+04	6.3E+04	4.8E+04	5.3E+04	5.3E+04
Nitrates	3.7E+03	2.4E+03	5.2E+03	1.5E+04	8.5E+03	8.7E+03	8.7E+03
Cases of bronchodilator usage asthma –adults							
PM_{10}	1.2E+04	6.2E+03	1.5E+04	4.4E+04	3.6E+04	4.0E+04	4.0E+04
Sulfates	1.6E+05	1.2E+05	2.1E+05	4.9E+05	3.8E+05	4.2E+05	4.2E+05
Nitrates	2.9E+04	1.8E+04	4.1E+04	1.2E+05	6.7E+04	6.8E+04	6.8E+04
Cases of bronchodilator usage asthma –children							
PM_{10}	2.3E+03	1.2E+03	2.9E+03	8.7E+03	7.1E+03	7.9E+03	7.9E+03
Sulfates	3.2E+04	2.4E+04	4.1E+04	9.8E+04	7.5E+04	8.3E+04	8.3E+04
Nitrates	5.9E+03	3.7E+03	8.2E+03	2.4E+04	1.3E+04	1.4E+04	1.4E+04
Cough days in asthmatics -adults							
PM_{10}	1.2E+04	6.4E+03	1.5E+04	4.5E+04	3.7E+04	4.1E+04	4.1E+04
Sulfates	1.7E+05	1.2E+05	2.1E+05	5.1E+05	3.9E+05	4.3E+05	4.3E+05
Nitrates	3.0E+04	1.9E+04	4.2E+04	1.2E+05	6.9E+04	7.0E+04	7.0E+04
Cough days in asthmatics -children							
PM_{10}	4.0E+03	2.1E+03	5.1E+03	1.5E+04	1.2E+04	1.4E+04	1.4E+04
Sulfates	5.6E+04	4.2E+04	7.1E+04	1.7E+05	1.3E+05	1.4E+05	1.4E+05
Nitrates	1.0E+04	6.4E+03	1.4E+04	4.2E+04	2.3E+04	2.3E+04	2.3E+04
Days with lower respiratory symptoms asthma –adults							
PM_{10}	4.3E+03	2.3E+03	5.5E+03	1.6E+04	1.3E+04	1.5E+04	1.5E+04
Sulfates	6.0E+04	4.5E+04	7.7E+04	1.8E+05	1.4E+05	1.5E+05	1.5E+05
Nitrates	1.1E+04	6.9E+03	1.5E+04	4.5E+04	2.5E+04	2.5E+04	2.5E+04
Days with lower respiratory symptoms asthma –children							
PM_{10}	3.1E+03	1.7E+03	3.9E+03	1.2E+04	9.5E+03	1.1E+04	1.1E+04
Sulfates	4.3E+04	3.2E+04	5.5E+04	1.3E+05	1.0E+05	1.1E+05	1.1E+05
Nitrates	7.8E+03	4.9E+03	1.1E+04	3.2E+04	1.8E+04	1.8E+04	1.8E+04
Restricted activity days							
PM_{10}	7.1E+04	3.8E+04	9.0E+04	2.7E+05	2.2E+05	2.4E+05	2.4E+05
Sulfates	1.0E+06	7.5E+05	1.3E+06	3.0E+06	2.3E+06	2.6E+06	2.6E+06
Nitrates	1.8E+05	1.1E+05	2.5E+05	7.4E+05	4.1E+05	4.2E+05	4.2E+05

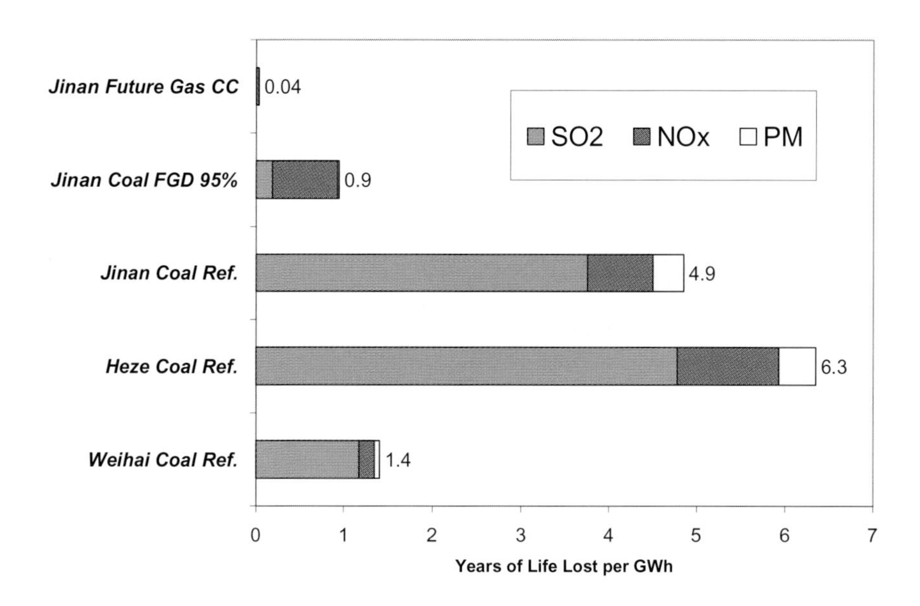

Figure 9.17 Mortality dependence on plant location and technology. For comparison, the reference coal power plant has been used for all locations.

The installation of a FGD unit at the Huangtai plant would reduce damages caused by pollutants by more than a factor of five. The damage caused by air pollutants emitted by the advanced, combined-cycle gas plant is practically negligible. As expected, there is a strong damage dependence on plant location and technology used. The variation in the results obtained for the different sites can be explained by the differences in regional population density (in particular, the share of land area differs strongly from site to site), by differences in ammonia (NH_3) background concentrations in affected areas (Figure 9.18), and by differences in meteorological conditions (e.g. region-specific dominant wind directions, Figure 9.19). The reason for the importance of ammonia is that it reacts with SO_2 in the atmosphere to form sulfates that in turn dominate the health effects. SO_2 is emitted predominantly from power plants and industry whereas ammonia mainly originates from agriculture. Thus the spatial distributions of power plants and industrial sites on the one hand and agricultural production on the other hand have to be considered. The maps of SO_2 and NH_3 concentration distributions show that the highest concentrations for each one of the species strongly overlap. These areas are also densely populated. We can expect therefore a high ammonium sulfate production and high health damages from ammonium sulfates in China.

The distribution of mortality associated with two identical reference coal plants located in Jinan and Weihai is illustrated in Figures 9.20 and 9.21, respectively. The relative differences in the level of damage are clearly visible. These figures can also be compared with the distribution of mortality due to air pollutant emissions from the whole power sector in Shandong (Figure 9.14).

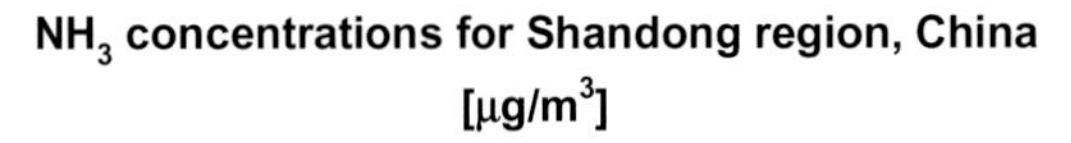

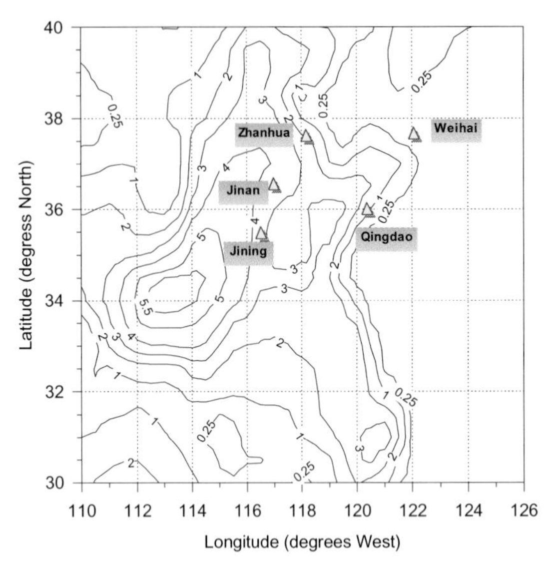

Figure 9.18 NH_3 concentration contours across the Shandong province.

Figure 9.19 Windrose data (wind direction distribution in %) for selected sites in Shandong (mean wind speed approximately 4 m/s).

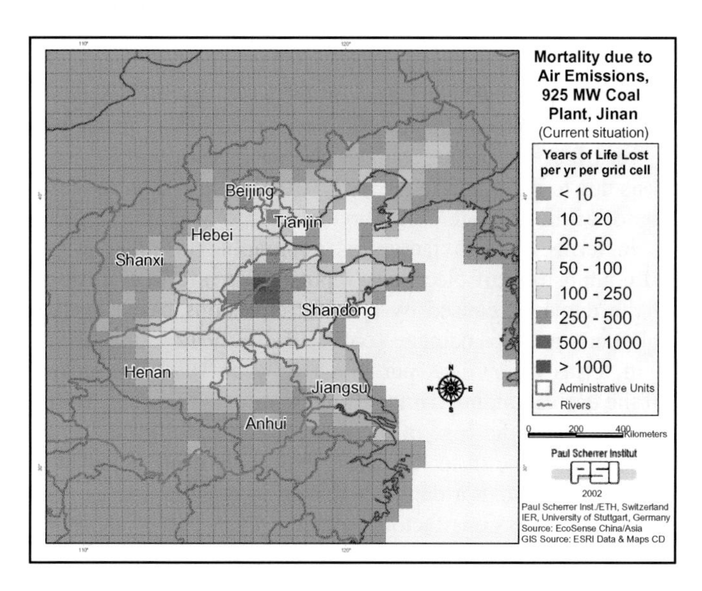

Figure 9.20 Single-source-based distribution of mortality due to current emissions of air pollutants from reference power plant in Jinan (grid cell size 0.5°x0.5°).

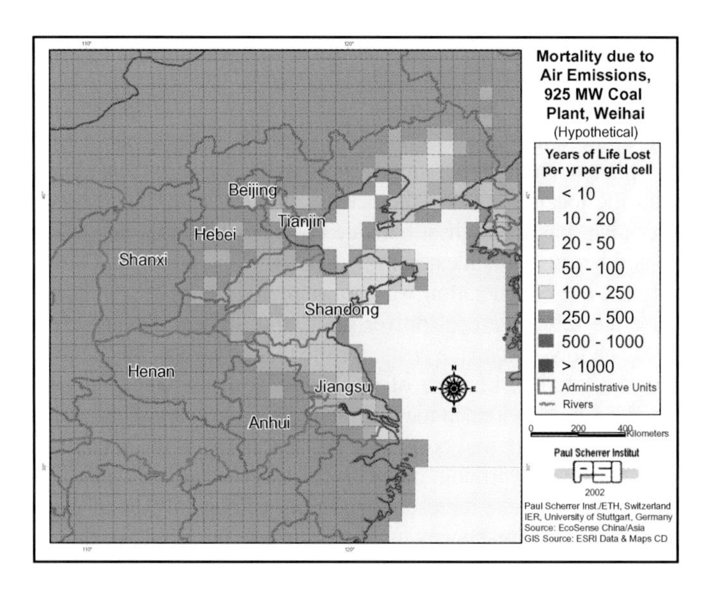

Figure 9.21 Single-source-based distribution of mortality due to current emissions of air pollutants from reference power plant in Weihai (grid cell size 0.5°x0.5°).

Integration of LCA and EIA. Until now only the damages caused by the emissions from power plants have been considered. In order to account for the damages associated with whole energy chains also contributions from stages other than the power plant need to be considered. Based on the LCA inventories established in the corresponding CETP task (see Chapter 8), it was possible to quantify the dominant contributions from the rest of the chains. The coal chain was subject to the most detailed analysis due to its importance. To the extent allowed by the resolution of the available inventories the damage estimation takes into account the specific structure of the chains in terms of such factors as geographic location of various stages, transports and composition of electricity inputs. This information was used for the estimation of the damages caused by specific emissions based on the normalized (per ton) pollutant-specific damage costs. Table 9.12 summarizes the chain contributions in terms of YOLLs per GWh in year 1998 and provides also the comparison of the corresponding damages due to direct emissions from power plants (see also Figure 9.17). In the case of the nuclear chain, for radiation the maximum YOLL estimates for Europe from Krewitt (1996) were used, adjusted by the higher population density of China. For damages related to other emissions from the rest of the nuclear chain, the emissions factors from the LCA part (Chapter 8) were used. Radioactive emissions from the coal chain have been neglected. There is some evidence that collective doses per GWh due to radioactive emissions related to coal-fired power chains in China could be comparable or even higher than those related to average UNSCEAR nuclear power chains (Pan, 1999).

For the current coal plants (conventional coal, no FGD) contributions from the rest of the chain are of the order of 10% of the damages due to power plant emissions. Given FGD implementation, the rest of the chain becomes a much more significant contributor but the overall damages are reduced to moderate levels. For gas and nuclear the rest of the chain is actually dominant, but the overall estimates are very low.

To show the influence of the location, a reference coal power plant has been used in the figures 9.17, 9.20 and 9.21. By contrast, Figure 9.22 shows the external costs per kWh of power plants actually operating in various locations in Shandong. The values depend now on the location and on the actual emission data of the power plant. The Nanding power plant has the highest damage costs per kWh because its emissions per kWh are high and because it is located in the center of the province. (Nanding includes cogeneration units, but cogeneration was not modeled here.) If only the locations are compared using the reference coal power plant, the highest damage costs have been found for Heze. The lowest damage costs, for the real as well as for the reference power plant, are caused at Waihei because of the location on the coast, far away from the highly populated agglomerations in the central and western parts of Shandong.

The estimate of total external costs also takes account of the much more uncertain damages caused by global warming, based on the FUND 2.0 model (Toll and Downing, 2000). The estimates used here represent maximum values. In view of the limitations of the approach used, and the enormous uncertainties involved in the estimation of economic consequences related to global warming, their validity may be questionable. Thus, the global warming component of the external costs is in relative terms much less robust than the corresponding damage costs due to major air pollutants.

Table 9.12 Normalized mortality effects caused by direct emissions from power plants and rest of the chain. All results are for the current situation with regard to the population density but FGD backfits and energy carriers considered as candidates for the future (natural gas and nuclear) are included.

Energy Carrier/ Technology	Contribution from Power Plant[1] [YOLL/GWh]	Contribution from Rest of the Chain [YOLL/GWh]
Coal/Conventional Ref. Plant[2]	4.9	0.27
Coal/Conventional low Sulfur[2]	2.7	0.26
Coal/Conventional plus FGD (removal efficiency: 95% SO_2)[2]	0.9	0.28
Coal/Conventional plus Dry FGD (removal efficiency: 80% SO_2, 50% NO_x)[2]	1.2	0.26
Coal/Conventional plus Wet FGD (removal efficiency: 90% SO_2, 50% NO_x)[2]	0.8	0.26
Coal AFBC[2]	0.4	0.25
Coal IGCC[2]	0.2	0.21
Natural Gas/Combined Cycle	0.04	0.06
Nuclear/Conventional	0.002[3]	0.08[4]

[1]For all fossil plants reference location in Jinan has been used.
[2]Current coal mix: 60% from Shandong, 40% from Shanxi.
[3]Radioactive emissions, upper range.
[4]Includes non-radioactive and (upper range) radioactive emissions.

"True" costs of electricity generation. One of the central goals of CETP was to estimate the *"true"* cost of electricity generation in China and in the Shandong province in particular. The *"true"* costs are composed of the internal generation costs and the external environmental ones. Based on the assessed impacts and monetary unit costs provided in Table 9.3, external costs can now be assessed. The internal costs used here originate from the ESS Task (Chapter 6). The capacity factor was assumed to be the same for all technologies (54%).

Figures 9.23, 9.24 and 9.25 show the *"true"* costs of electricity generation, with specific contributions by various components. The costs are provided for a variety of coal technologies in few selected locations in Shandong, and for natural gas in Jinan and for nuclear power located on the coast. It should be noted that severe accident contributions (see Chapter 10) to the damage costs have low significance in comparison with external costs due to normal operation. All results provided in the figures include the contributions from the rest of the chain. For comparison the average external costs for Shandong's power sector are shown.

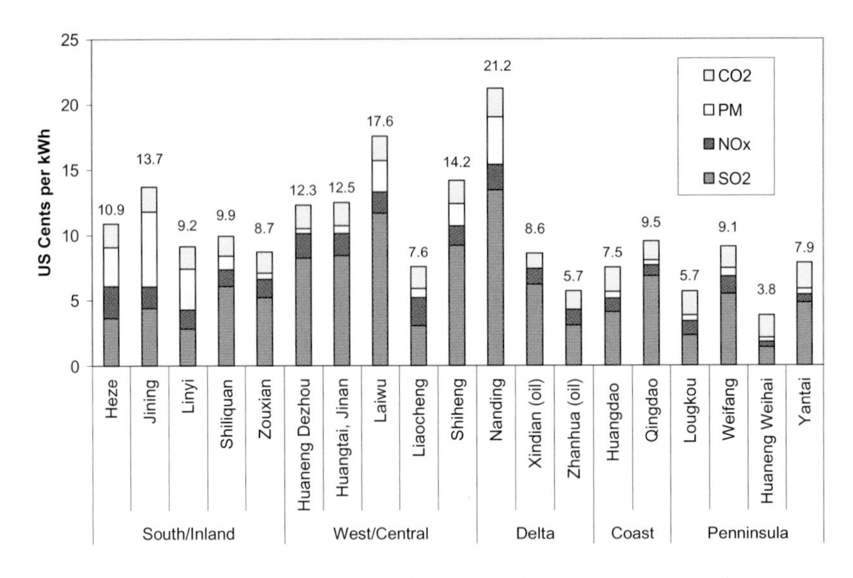

Figure 9.22 External costs per kWh electricity of power plants currently operating in Shandong. Values include LCA contributions except for oil power plants. Cogeneration is not considered.

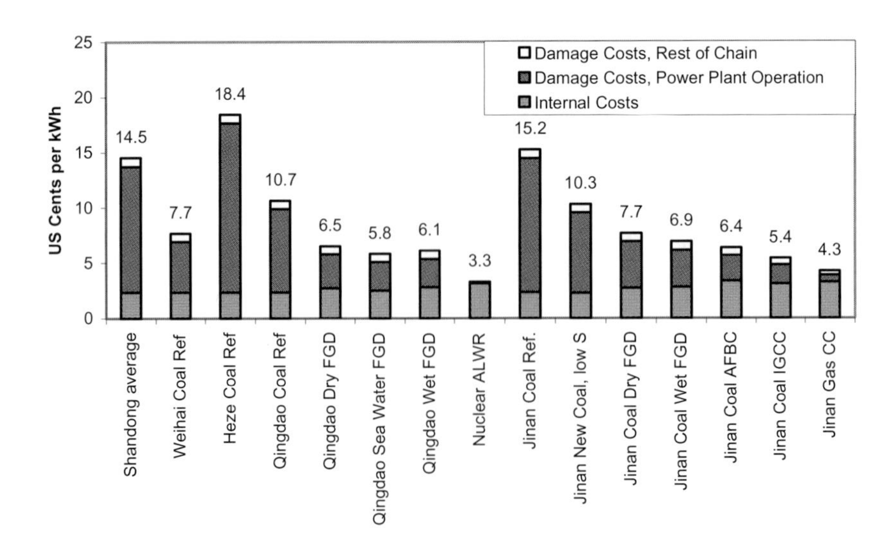

Figure 9.23 "True" costs of electricity generation by various means in Shandong; contributions to external costs from power plants and rest of the chain are shown.

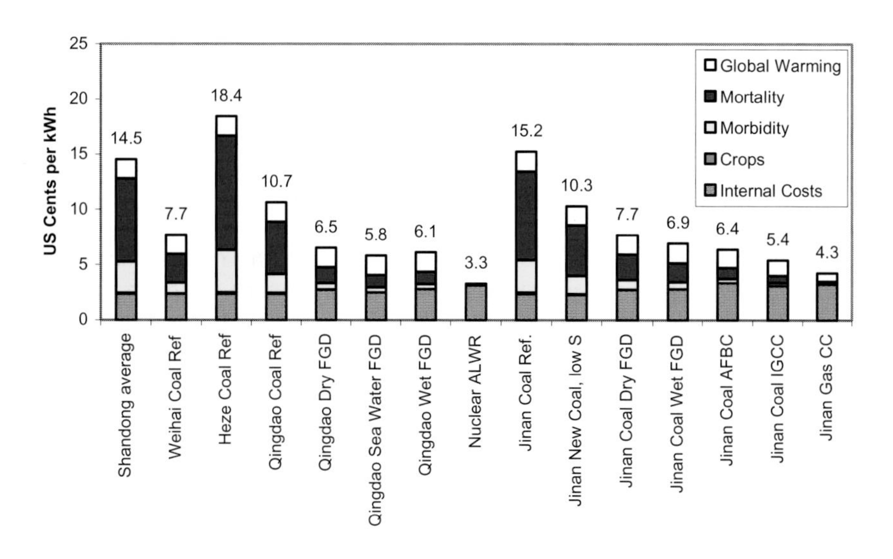

Figure 9.24 "True" costs of electricity generation by various means in Shandong; contributions to external costs from mortality, morbidity, crops and global warming are shown.

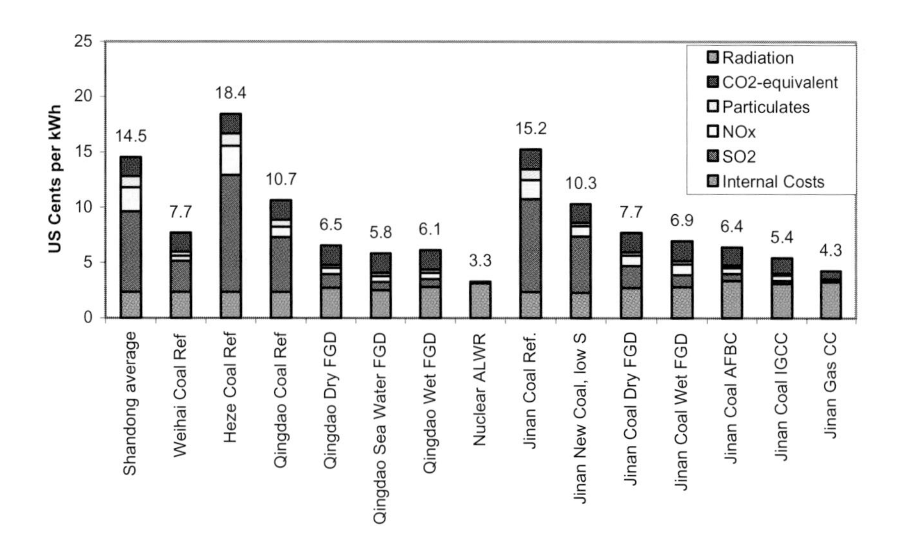

Figure 9.25 "True" costs of electricity generation by various means in Shandong; contributions to external costs from major pollutants and global warming are shown.

The results shown in the figures reflect the high impacts of currently operating plants as well as location dependencies, illustrated earlier in Figure 9.17. The resulting monetary damages stemming from pollutant emissions from the Shandong power sector correspond on average to 10.5 US cents per kWh for the full chain (without Global Warming), with 9.9 US cents per kWh for the power plant operation alone; the figures for the total China power sector are 3.9 and 5.1 US cents per kWh (power plant emissions only, without and with Global Warming, respectively). The average results for China are much lower than those for Shandong due to three reasons: extensive use of hydro in China (practically no hydro in Shandong), lower average damages per ton of pollutant emitted, lower average emission factors in China than in Shandong according to official Chinese statistics. The external costs clearly exceed generation costs. The considered more environmentally friendly options for the future exhibit low external costs, which in spite of higher internal costs results in total *("true")* costs being lower than for the present technologies. As can be seen for the current systems, and under Chinese conditions, the contribution from global warming is in any case relatively small compared to the damage caused by air pollution.

Parameterization of impacts and damage costs.. The previously shown (Table 9.11) quantification of impacts for the various locations in Shandong, allows for the parameterization of damage costs, i.e. they can be given per ton of pollutant emitted, thus providing a convenient basis for simple and quick estimation of damage costs associated with a wide range of technologies. The only inputs required are technology-specific emissions of the major pollutants, which may be combined with location-specific normalized damages to assess the resulting damage costs. This approach has been implemented in the simulation available on the CETP DVD (see Chapter 12), allowing the users flexible explorations of the findings of the present work. Thus, the user can choose anyone of the current sites in Shandong, the electricity generation technology with its characteristics (power level, load factor, efficiency, abatement systems performance), and estimate the associated impacts and external costs within few seconds.

Figures 9.26 and 9.27 show the mortality and monetary damages per ton of pollutant, respectively. Results are provided here for few selected locations in Shandong, and for Shandong and China (average).

In Figure 9.28 the relative distribution of all calculated external costs per emitted ton of each major pollutant is shown for a reference power plant located in Jinan, demonstrating on a single plant level the relative dominance of mortality, followed by significant morbidity contributions. Attention needs to be paid to the logarithmic scale used in the diagram.

In spite of the (PPP) GNP-based reduction of the unit damage costs by a factor of seven in comparison with Western Europe, the dominant SO_2-related cost damage is comparable to estimates for typical European conditions. The mortality (expressed in YOLL per ton SO_2) estimated in the present study for China is in fact a factor of seven higher than the average for the European Union (Krewitt et al., 2001). This is due to the much higher population density in the most affected regions of China, and to high ammonia concentrations in the vicinity of population centers (as mentioned before ammonia plays an important role in the transformations of primary pollutants into sulfates and nitrates). Compared to Germany (Krewitt et al., 1997), the NH_3 to SO_2 ratio in China is relatively high in the grid cells with high SO_2 emissions. This means that there is enough ammonia for each additional ton of SO_2 to cause a maximum impact.

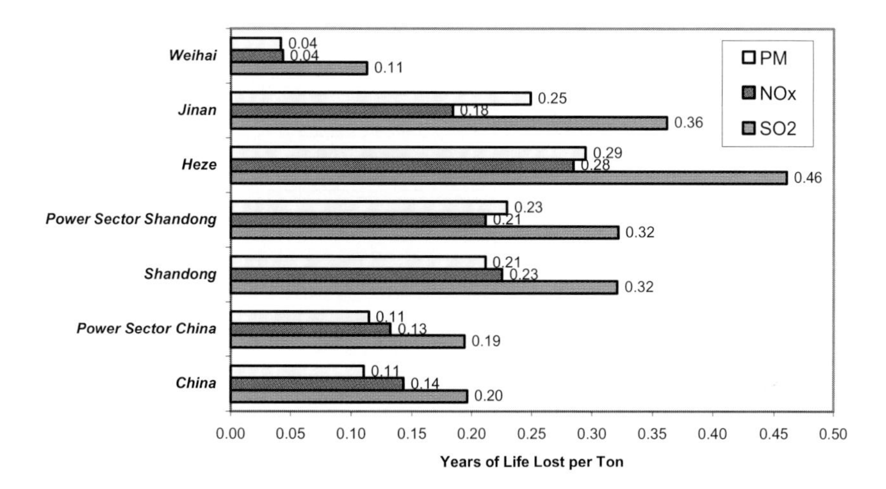

Figure 9.26 Mortality per ton of pollutant.

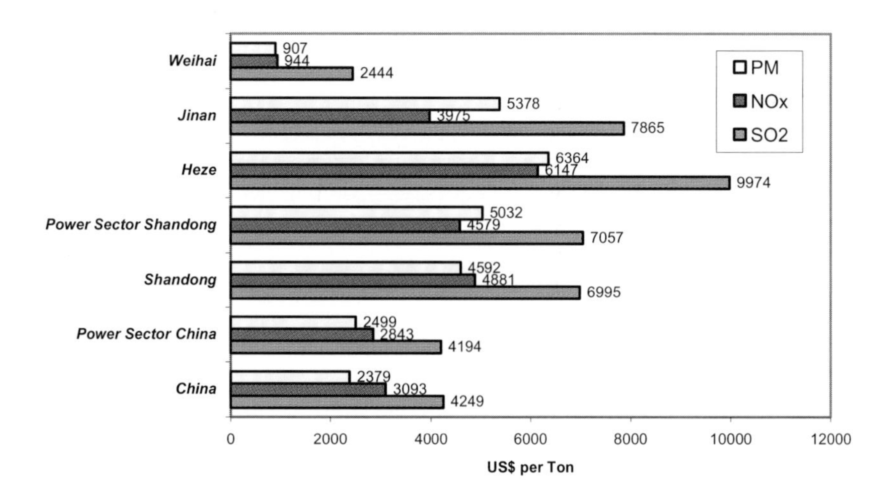

Figure 9.27 Monetary damages per ton of pollutant.

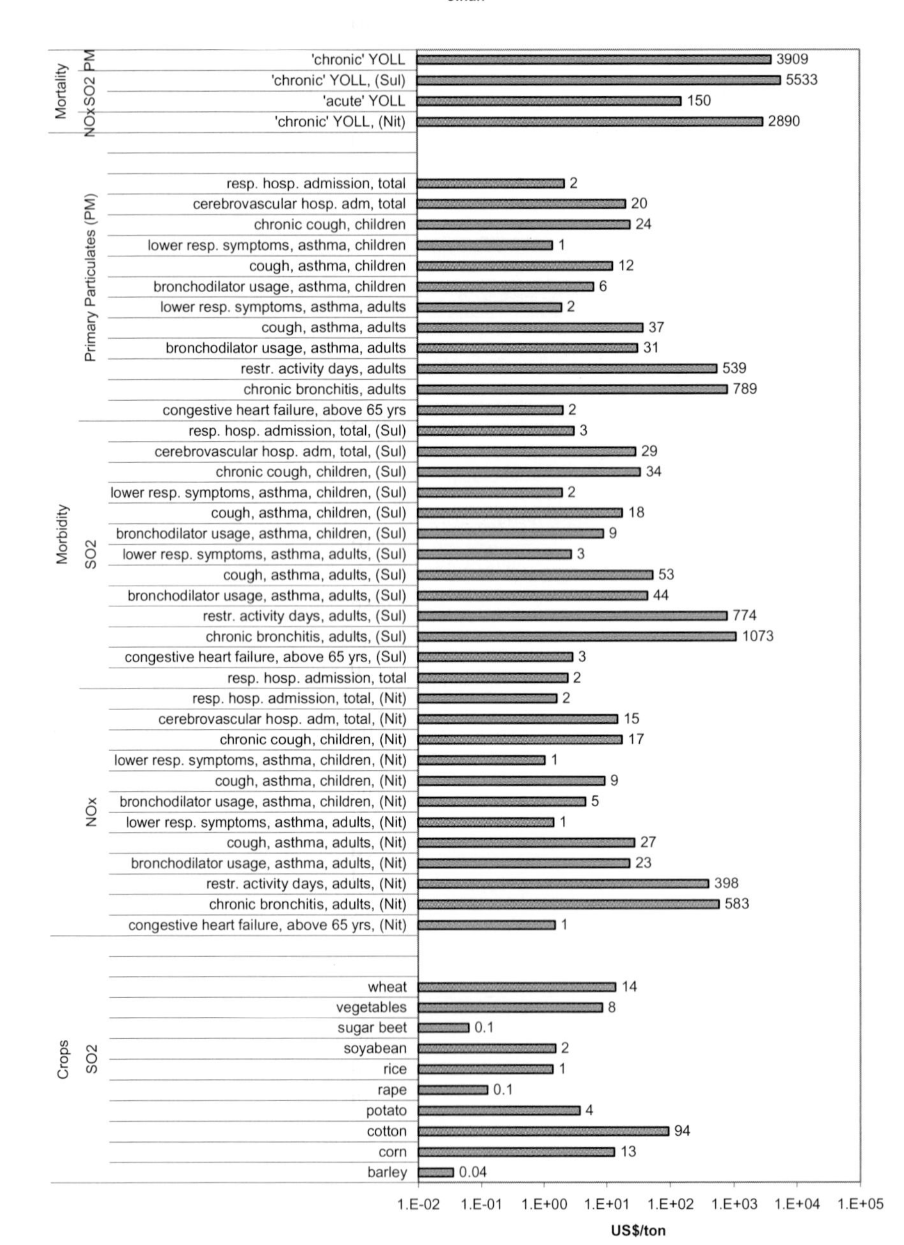

Figure 9.28 Damages per ton of pollutant emitted from reference power plant in Jinan.

3.4.3 External costs of future electricity supply scenarios for Shandong

Electricity supply scenarios for the years 2000 to 2024 were analyzed for Shandong Province. The scenarios of interest were selected from 18'144 scenarios modeled in the Electric Sector Simulation (ESS) Task (Chapter 6). At the core of the ESS methodology is a model that simulates electrical system dispatch, and is based on the marginal cost of generation from individual units. Thousands of different scenarios can be designed, generated, modeled, analyzed and presented, along with such attributes as costs, emissions, risks, and use of resources. Scenarios are created by combining multi-option strategies with future uncertainties. Strategy options include improvement of existing power generation (retirement of old units, FGD retro-fits, coal treatment), technologies for new generation (conventional coal, IGCC, nuclear, natural gas), and demand-side management (load management and end-use efficiency). Uncertainties considered in the CETP ESS implementations include demand growth and fuel prices.

External costs were estimated by PSI for more than 100 selected scenarios. These calculations were based on the annual emission trajectories for Shandong province, which was divided into five regions for this purpose. New capacity was distributed to these regions, with the choice of technology depending upon the scenario. Where site specific information or technology data was lacking, distribution was assumed to be proportional to current regional generation. Damage per ton of pollutant by region and year were combined with projected emissions to determine the total future damage. Combining scenario-specific emissions with damage costs per emitted ton is based on the assumption of linear exposure-response functions without thresholds. As shown before, the impact to health is dominated by sulfates, which are formed from SO_2 and NH_3 emissions through chemical reactions in the atmosphere. The YOLLs per ton of SO_2 in China are high compared with corresponding values in other regions of the world (see Krewitt et al. (2001)). An emitted ton of SO_2 has a high impact in China, due to the high population density and the high NH_3 emission background (mainly from agriculture and Nature). A high impact per ton of SO_2 can also be expected in the future, because of the growing population and the expanding agricultural production. The formation of nitrates from NO_x depends, in a more complicated way, on the prevailing atmospheric conditions. Thus, the extrapolation into the future of environment damage due to NO_x is more uncertain.

The results from 36 selected scenarios are shown here: these scenarios form the set chosen for the Multi-Criteria Decision Analysis (MCDA), described in Chapter 11. Health impacts formed inputs to the analysis, along with economic and LCA-based indicators. Table 9.13 defines the composition of the 12 strategies compared here. The set chosen is representative, as it includes the bounding cases as well as a number of potentially attractive environmentally friendly options (relative to the current situation). Table 9.14 defines the three futures combined with the 12 strategies to form the 36 selected scenarios. It should be noted that the 'moderate' growth in electricity demand, postulated for two of the futures, actually averages out at about 5.1% per year, meaning that Shandong electricity demand will more than triple over the next 20 years.

Figures 29 - 34 show the estimated mortality and external costs for the selected scenarios for the year 2020. The external costs include apart from mortality also morbidity, damages to crops and the highly uncertain global warming contributions. The contributions of the rest of the chain to the total damage costs vary between about 9 and 28%. The lower range applies to the "dirtiest" scenarios while the upper range is characteristic for the "cleanest" ones. In the latter case the power plant damage contributions have been reduced so much that the rest of the chain plays in relative terms a more significant role. Some technological improvements were credited also for the rest of the chain but for the (most important) coal chain these changes are compensated by the increase in population density and structural changes (more coal from Shanxi); these two factors tend to increase the impacts so the overall effect is that contribution from the rest of the coal chain remains about constant over the next 20 years.

Table 9.13 Characteristics of the MCDA scenario set (from Electric Sector Simulation Task).

	Existing Generation			*New Generation*					*Demand-Side*	
Multi-Option Strategy	Retire Units	FGD Retrofits	Prepared Coal*	Conv. Coal w/o FGD	Conv. Coal with FGD	IGCC (2012)	Nuclear (2010)	Nat.Gas (2015)	Load Mgt.	End-Use Effic.
BOC-CENPAS				•						
BOC-CONPAS					•					
BOX-CONPAM			•		•					10%
DOX-CONLAG	•		•		•				•	20%
BOX-LONLAM			•		•	•			•	10%
DOX-MONLAM	•		•		•			•	•	10%
BOC-NONLAS					•		•		•	
BOX-NONLAM			•		•		•		•	10%
BOX-NONLAG			•		•		•		•	20%
DOX-TONLAG	•		•		•	•	•	•	•	20%
DUX-DONLAG	•	•	•		•		•	•	•	20%
DUX-TONPAS	•	•	•		•	•	•	•		

** For "Existing" Units only*

The merits of clean coal technologies in general, and FGD in particular, of fuel diversification (i.e. the use of natural gas and/or nuclear), and of demand-side management, are evident for all considered futures. It should be noted, however, that the potential for expansion of natural gas as a fuel in the electricity sector in Shandong is quite limited. Furthermore, the feasibility of extensive demand-side management, and its costs, were not studied in detail within the CETP. Some of the low-pollutant-emission scenarios (e.g. DUX-TONPAS, DUX-TONLAG) also have the advantage of emitting much lower amounts of greenhouse gases (GHGs) than scenarios that do not rely on demand-side management and/or nuclear energy. Clean coal technology as such cannot significantly help in curbing GHG emissions, which will continue to increase as they are linked to the demand growth. Thus, supply-side strategies that reduce the projected emissions of primary pollutants (in the first place SO_2), while limiting the increase in GHG emissions, are of high interest.

Table 9.14 Key assumptions about future developments for the selected scenarios (from Electric Sector Simulation Task).

FIB	F	Moderate Economic Growth (5.11%/yr-Shandong Grid GWh)
	I	Business as Usual Coal (Status Quo mining technology and transportation costs)
	B	Base Gas (¥26/GJ, "Reference" natural gas costs from ERI/PNL & Pew Center Reports)
FAF	F	Moderate Economic Growth (5.11%/yr-Shandong Grid GWh)
	A	Aggravated Coal (Periodic Transportation Cost Increases reflecting transport bottlenecks)
	F	Low Gas (¥15/GJ, Lower inital cost uncertainty)
SUB	S	Strong Economy (7.66%/yr-Shandong Grid GWh)
	U	Productive Coal (Modernization of mining reduces cost/improves quality)
	B	Base Gas (¥26/GJ, "Reference" natural gas costs from ERI/PNL & Pew Center Reports)

Figure 9.35 shows results from a cost-benefit analysis for two selected electricity supply strategies for Shandong Province in the year 2020 in relation to a strategy based on the use of coal without scrubbers (FGD). The two cleaner scenarios shown are: (1) conventional coal with FGD; and (2) improvements of existing generation—retirement of aging units, retrofits (including FGD), coal treatment—together with clean/advanced coal technologies, nuclear and some natural gas. Since the relative difference in internal costs is quite small, the total (internal plus external) costs of much more environmentally friendly strategies are clearly lower than the full costs of seemingly cheap, “dirty”, non-sustainable strategies.

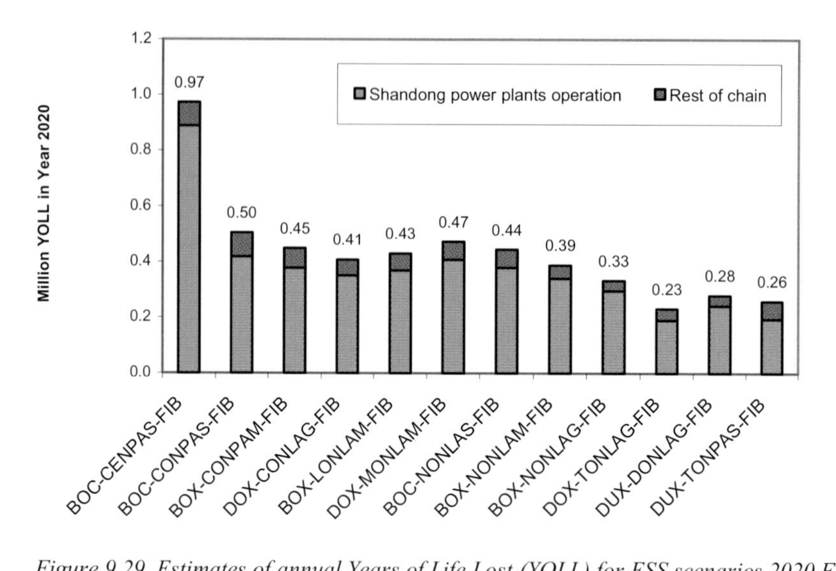

Figure 9.29 Estimates of annual Years of Life Lost (YOLL) for ESS scenarios 2020 FIB, including LCA contributions.

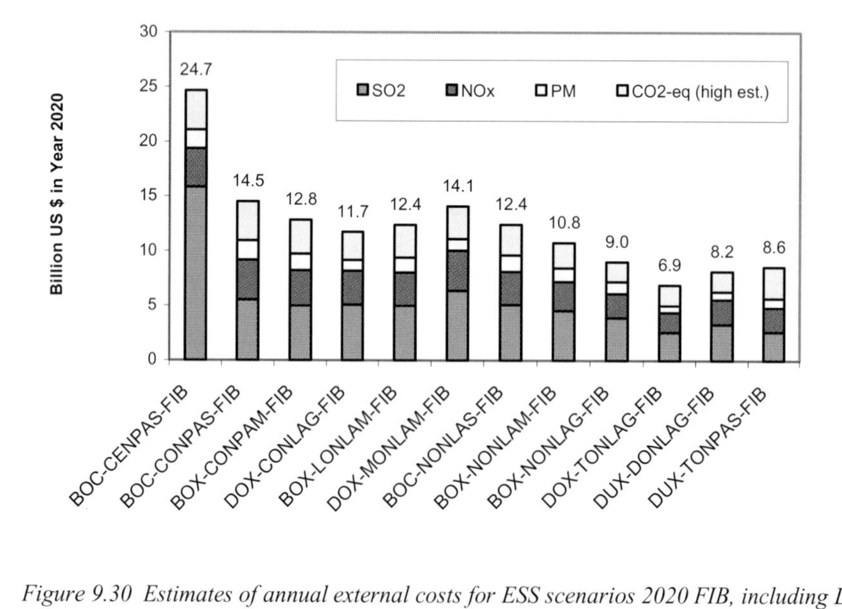

Figure 9.30 Estimates of annual external costs for ESS scenarios 2020 FIB, including LCA contributions.

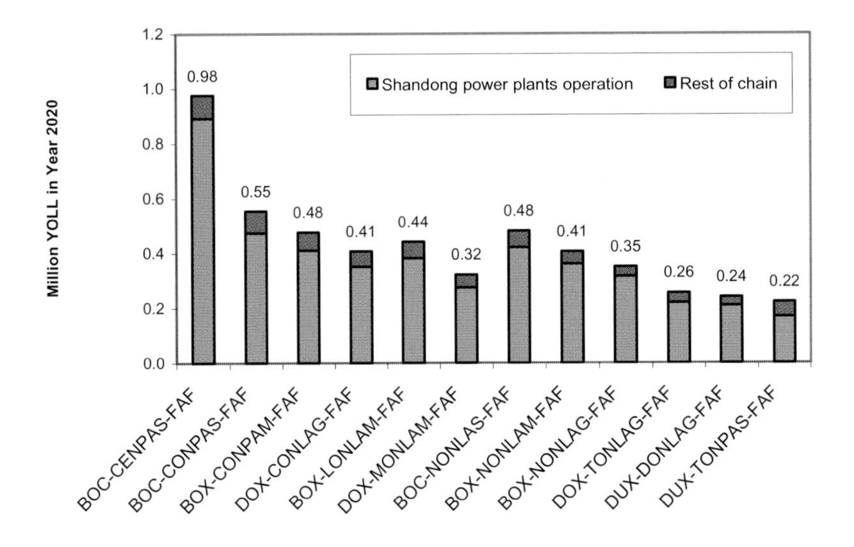

Figure 9.31 Estimates of annual Years of Life Lost (YOLL) for ESS scenarios 2020 FAF, including LCA contributions.

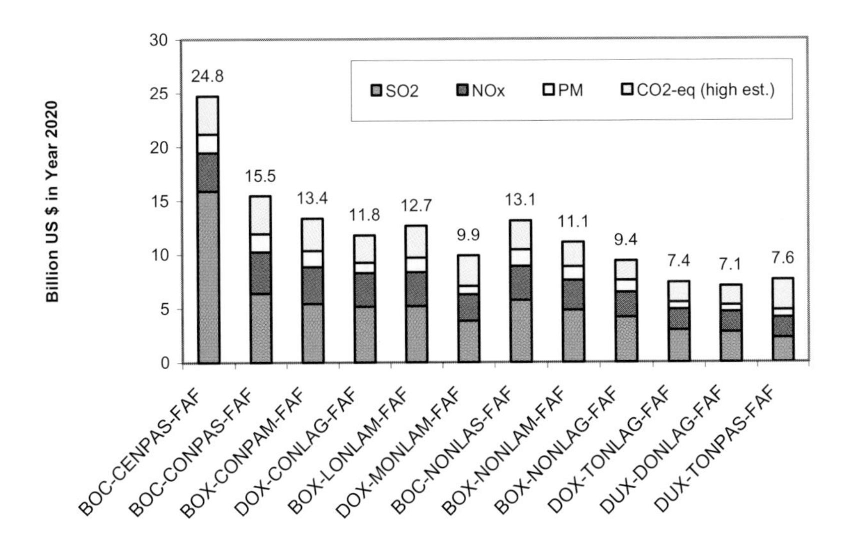

Figure 9.32 Estimates of annual external costs for ESS scenarios 2020 FAF, including LCA contributions.

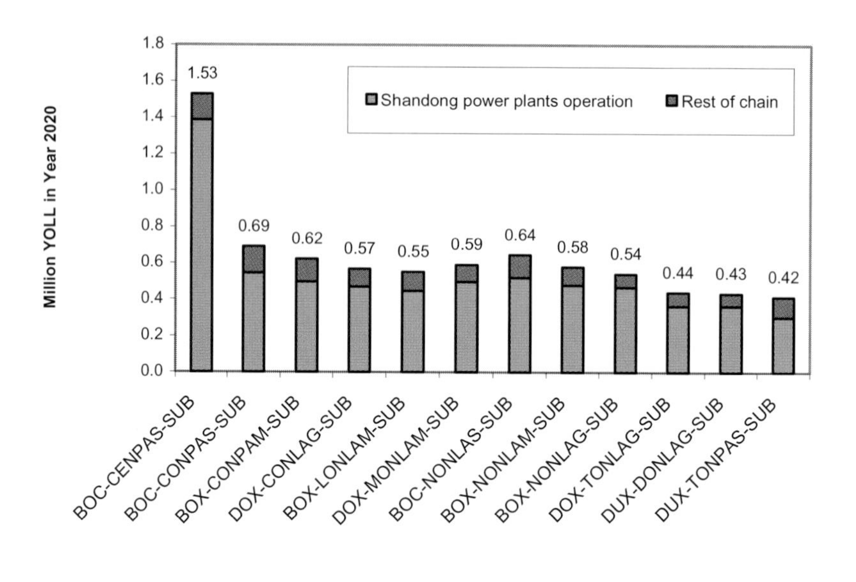

Figure 9.33 Estimates of annual Years of Life Lost (YOLL) for ESS scenarios 2020 SUB, including LCA contributions.

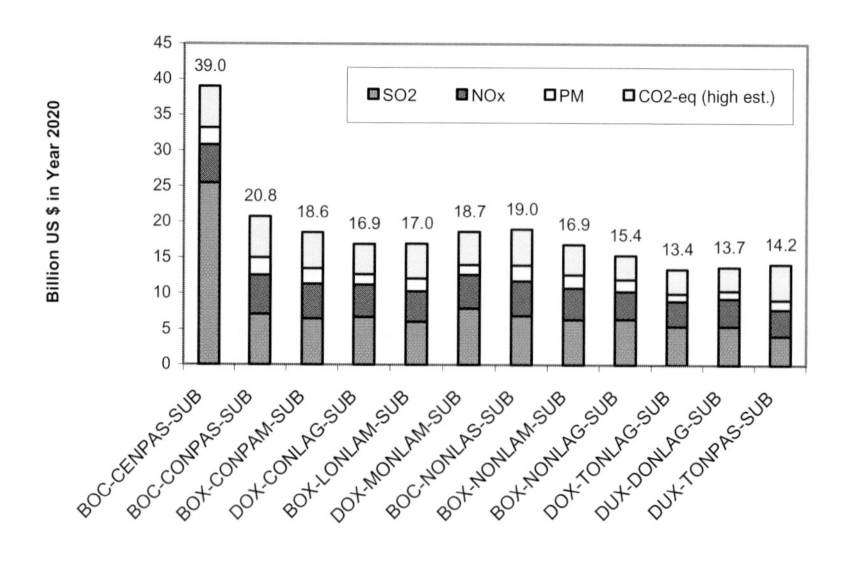

Figure 9.34 Estimates of annual external costs for ESS scenarios 2020 FAF, including LCA contributions.

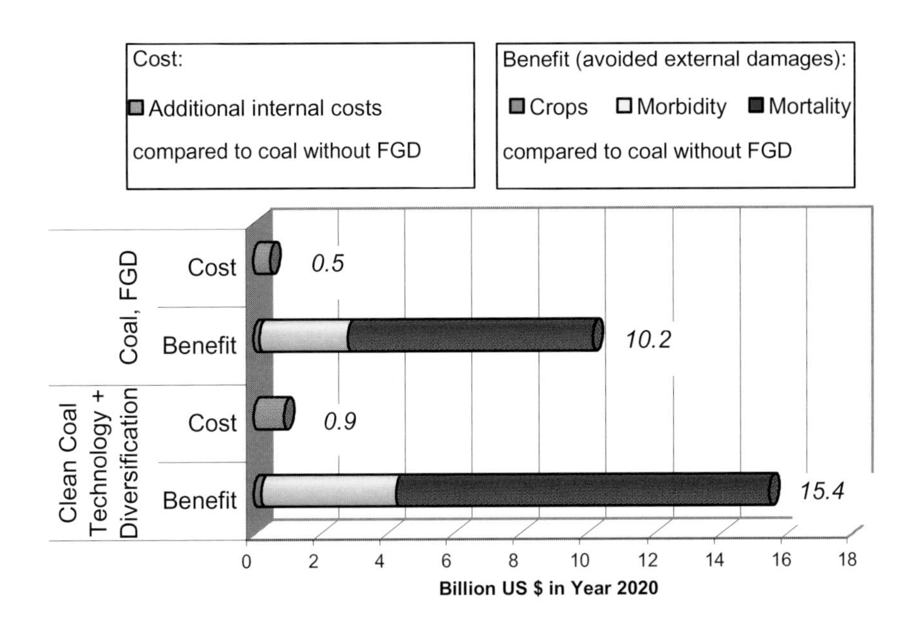

Figure 9.35 Cost-benefit analysis of "clean" versus "dirty" strategies (FIB future), including LCA contributions. CO_2 damage costs are not considered though their inclusion would further strengthen the benefits of cleaner strategies.

3.4.4 Uncertainty and sensitivity analysis

A single aggregate estimate of external costs depends on a large number of input data and on a series of models applied consecutively. The estimation of uncertainties in environmental impact assessment is a difficult task even for the European conditions where good quality data are relatively abundant. The situation is still more demanding in the present study because the uncertainties of various Chinese-specific input data are not well known and generally the access to relevant data was not straight-forward.

Here we provide a summary of the analysis of uncertainties; a more detailed account can be found in the technical report (Hirschberg et al., 2003b). A strict uncertainty analysis is not possible at the moment and lies beyond the scope of the present study. Instead, we discuss uncertainties of the EcoSense China/Asia model based on the experience with the European model and arrive at plausible quantitative estimates whenever possible. Since the EcoSense China/Asia model emerged from the adaptation of the European EcoSense model, we refer here to the uncertainty discussion in the methodology part of the ExternE report (European Commission, 1999). The focus in the quantification of uncertainties associated with air pollution damages assessed in CETP is mainly on the "natural science" part of the assessment, i.e. concerns emissions, atmospheric modeling, and exposure-response functions.

The monetary valuation may be regarded as unproblematic for goods like crop yields since real market prices form the estimation basis. However, for non-market goods like risk of mortality, the valuation raises its own problems. The uncertainty of the monetary valuation of YOLL under the WTP assumption was considered also in the ExternE study and showed to contribute only with an additional geometric standard deviation of about two. But the problem is more the monetary valuation method itself. Thus, sensitivity analyses were performed in the present work, demonstrating that the conclusions from the cost-benefit considerations are robust also when the WTP approach is substituted by the human capital method (which results in much lower damage costs but whose basis is highly questionable).

We refer to section 3.2.4 for the basic discussion of the approach to uncertainty estimation. For the uncertainty analysis of the impact assessment, two basic measures are important (for details see e.g. Bean (2001), European Commission (1999), Rabl and Spadaro (1999)):

- Geometric mean value (or geometric expected value) μ_g
- Geometric standard deviation σ_g.

For a lognormal distribution, multiplicative confidence intervals around the geometric mean can be derived easily from the geometric standard deviation. The 68% confidence interval is approximately $[\mu_g / \sigma_g, \mu_g * \sigma_g]$, and the 95% confidence interval is approximately $[\mu_g / (\sigma_g)^2, \mu_g * (\sigma_g)^2]$ (European Commission, 1999; Rabl and Spadaro, 1999).

In section 3.2.4 the formula for estimating the total standard deviation of the product based on the uncertainties in each step of the pathway analysis, has been provided.

Uncertainties of emission data. For the European emission inventory used in ExternE, the assumed uncertainties of PM_{10}, SO_2 and NO_x were about +/-10 to 20 %. For the uncertainty calculation, a general geometric standard deviation $\sigma_g = 1.2$ was assumed for emissions (European Commission, 1999).

As elaborated in section 3.3.3 the EcoSense China emission database uses a variety of sources; EDGAR (2000) was employed as the starting point, subject to an update to the current conditions. EDGAR gives some estimates for uncertainties in overall global and regional emission data.

For SO_2 and NO_x emission data from all sources, a general uncertainty of +/-50% is quoted (EDGAR, 2000). Current SO_x emission data for China have been estimated to be accurate within a range of about 10% (Streets and Waldhoff, 2000). The difference between Streets and Waldhoff (2000) and EDGAR (2000) for NO_x emissions in China in 1990 is about 15%, which is remarkably good. In principle, SO_2 emissions can be estimated relatively accurately because they depend mainly on the sulfur content of the fuel if no abatement technology is applied. Nevertheless, a source of uncertainty for SO_2 estimates in China is the large variation of sulfur content of Chinese coal. In general, NO_x emission factors are more uncertain than SO_2 emission factors because the formation of NO_x depends significantly on the details of the combustion process (e.g. temperature).

Uncertainties of NH_3 emission factors for Europe typically lie between 20% and 50% (Suutari et al. 2001). For China the spatial NH_3 emission distribution from EDGAR was updated using data from Sun and Wang (1997). In their paper, standard emission factors from literature were assumed for ammonia emissions from livestock and industrial activities that together cover the major part of all ammonia emissions. For fertilizer application, specific emission factors adjusted to the situation in China were applied. Given that the inventories of livestock and industrial production are reliable, the uncertainties of NH_3 emissions in China should not differ too much from the uncertainties assumed in the European projects.

The uncertainties of particulate emission inventories for China are problematic. Remarks in Xu (1998) indicate that there are significant inconsistencies in concentration measurements of TSP. Furthermore the distribution of particle sizes (i.e. PM_{10} or $PM_{2.5}$) is often unclear. Nevertheless, the uncertainties in the background emission data for particulates are in general not relevant for our conclusions concerning single emission sources because the particulates are not reactive in the atmosphere, i.e. the results do not depend on the background concentrations (different from the case of other major pollutants). The only results directly affected by the uncertainties of the PM background emission database are our estimates of the non-dominant total damages due to primary particulates from all emission sources for China and Shandong, which thus may be considered more uncertain than the other estimates.

It is likely that the update reduced the uncertainty of overall emission values. But the uncertainties of the spatial distribution due to the uncertainties in the original EDGAR data and due to the transformation from the EDGAR 1°x1° grid to the administrative units and to the 0.5°x0.5° grid used in EcoSense China remains.

For the uncertainty of the Chinese emission data implemented in EcoSense China, we generally assume a geometric standard deviation σ_g = 1.5, which is supported by the uncertainty estimate of the EDGAR database.

Atmospheric modeling. For the European atmospheric modeling, a geometric standard deviation σ_g = 2 was assumed (European Commission, 1999). In the original ExternE study, this estimate refers to dispersion modeling without chemical reactions only, i.e. to non-reactive particulates. Nevertheless, the uncertainties related to the other pollutants including chemical reactions have been ranked into the same uncertainty class in ExternE (with a question mark for nitrates).

For EcoSense China, wind data were adopted from the NCEP/NCAR Reanalysis Project and adjusted to the EcoSense model. The NCEP/NCAR Reanalysis group gives some estimates of uncertainties in the wind data, based on comparisons of differences between monthly means from different NCEP models and from other operational systems (Kalnay et.al., 1996). The differences are mostly of the order of 1 m/s. In the atmospheric modeling area used in EcoSense, the average wind speed is about 3.8 m/s. Thus, the uncertainty of the wind data is of the order of 20-30%.

Some aspects of the atmospheric model were further advanced in the China version of EcoSense. For example, the mixing height and the photolysis rates were calculated much more precisely than in the European case. The reaction rate coefficients for the chemistry were updated according to the most recent publications (though the differences to the old values are small). Long time series were used for the calculation of the wind and precipitation data. The wind data was derived from 6-hourly global wind fields covering a period of ten years (1989-1998). This makes the annual average values used in the model quite robust.

Because there are no better quantitative estimates for the contribution of the chemical modeling to the uncertainty, we compared the output of two models, EcoSense China/Asia and RAINS Asia (Regional Air Pollution Information and Simulation for Asia) model (Amann and Dhoondia, 2001), developed independently by IIASA. RAINS uses linear transfer matrices constructed from results of a long-range atmospheric transport model, while each EcoSense run includes a full-scope non-linear Lagrangian dispersion model calculation. We refer to Section 4 for more details on the RAINS-Asia model and its application within CETP.

Figure 9.36 shows SO_2 concentrations from EcoSense China versus SO_2 concentrations from RAINS-Asia for the grid cells in the overlapping area of both models. The values from EcoSense were transformed from the 0.5° x 0.5° grid to the 1° x 1° grid used by RAINS. The figure shows that for single grid cells the values from both models can significantly differ, which is not very surprising.

For 68% of the relevant grid cells with significant concentrations, the two models yield the same average concentrations within a multiplicative range of about 2.1. If this is taken as an indication for a lognormal 68% confidence interval and the uncertainties of emissions are accounted for using a geometric standard deviation of 1.5 (see above), the resulting geometric standard deviation of the atmospheric modeling would be about 1.9. 95% of the significant cells match within a factor 5.4. This is formally consistent with a geometric standard deviation of about 2.1 for the atmospheric modeling.

In summary, the uncertainty of the atmospheric modeling in EcoSense China/Asia seems to be of the same order as in the European model. Thus we assume $\sigma_g = 2$ for the atmospheric modeling of particulates and sulfates, which are most essential for the final results. For the complex nitrogen chemistry, the uncertainty might be higher.

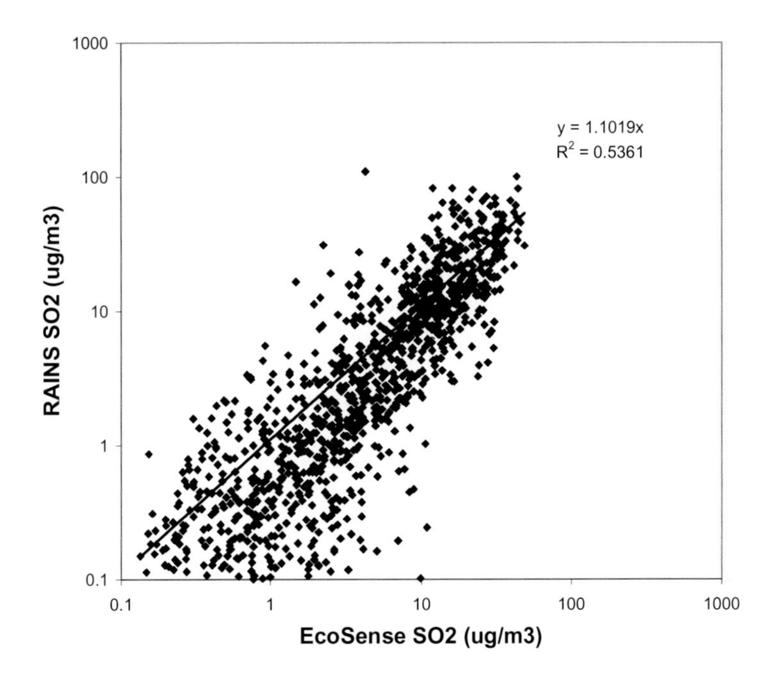

Figure 9.36 Comparison of SO_2 concentrations from EcoSense China/Asia and from RAINS-Asia (reference year 1998).

Exposure-response functions and receptor data.

(a) Exposure-response functions for human health.

Based on Pope et al. (1995) the geometric standard deviation for sulfates and $PM_{2.5}$ was assessed in ExternE at 1.3 (European Commission, 1999; Rabl and Spadaro, 1999). The mortality rates have to be converted into YOLL. This step introduces additional uncertainties because it is not clear in detail which parts of the population are affected, how the repair mechanisms in the body may compensate a pollution pulse later, and how long the latency is.

For the conversion of relative risk of mortality into YOLL and for latency, geometric standard deviation of 1.5 and 1.4, respectively, was assumed in European Commission (1999). Using the results based on different conversion methods as an indicator for the uncertainty, we conclude that σ_g for YOLL conversion should be somewhat higher for China (1.8) than for EU15 (1.5). For latency, the factor σ_g = 1.4 is considered to be valid also for China

In ExternE (European Commission, 1999), the uncertainty of the transfer of exposure-response function from US to Europe was assumed to correspond to σ_g = 2. In face of the discussions during the various phases of ExternE and the relatively large scaling factor used in European Commission (2000), we judge the transfer uncertainty to be higher. On the other hand, the Shenyang study (Xu, 1998), showed the order of magnitude consistent with the original US data. We assume here a slightly increased factor $\sigma_g = 2.3$.

For the Chinese case, there is another source of uncertainty, which was not explicitly covered in the European study. This is the uncertainty in the population data. Along with the population growth, a significant migration from rural to urban areas changes the spatial distribution. The trend is expected to continue in the future. It is likely that our database fairly well represents the current rural population distribution but underestimates the population in urban areas. To account for the uncertainty in population data, a geometric standard deviation of 1.1 was included in the uncertainty estimate. Nevertheless, the uncertainty in population data is small compared to the other uncertainties and has marginal impact on the final uncertainty estimate.

Table 9.15 summarizes the uncertainties related to the dominant end-point "chronic mortality".

Table 9.15 Estimated uncertainties in terms of geometric standard deviations for end-point "chronic mortality".

	Chronic Mortality ExternE Europe (European Commission, 1999)	**Chronic Mortality EIA China (this study)**
Emission data	1.2	1.5
Atmospheric modeling WTM	2	2
Exposure-response function, original study	1.3	1.3
Transfer of exposure-response function to other region	2	2.3
YOLL calculation from mortality	1.5	1.8
Latency	1.4	1.4
Population data	-	1.1
Total (without monetary valuation)	3.2	3.9

In total, the geometric standard deviation of the chronic YOLL values is estimated at 3.9 for the "natural science" part in the Chinese case (the valuation problem is treated separately in the sensitivity analysis below). While this estimate cannot be regarded as precise, it indicates that the 68% confidence interval for dominant chronic mortality lies within a factor of four in relation to the estimated mean value. Referring to the discussion in 3.2.2 of the transfer of the results of Pope's study (1995) to the Chinese conditions in view of Xu's results (1998), it is possible that the YOLL estimates of the present study may underestimate the real YOLL by up to a factor of two.

The uncertainties of acute YOLL are higher than the uncertainties of chronic YOLL, but according to European Commission (1999) lie in the same uncertainty range. However, this is of secondary importance since acute mortality has low significance for the overall results.

Table 9.16 summarizes the estimated uncertainties of morbidity damage costs (European Commission, 1999). The uncertainty estimates referred to concern: SO_2, sulfates, PM_{10} and $PM_{2.5}$. The impacts from nitrates are generally more uncertain.

Table 9.16 Uncertainty estimates of morbidity end-points. Uncertainty rating A corresponds to σ_g = 2.5 to 4; uncertainty rating B corresponds to σ_g = 4 to 6 (European Commission, 1999).

Receptor	Impact Category	Reference	Uncertainty rating
ASTHMATICS			
Adults	Bronchodilator usage	Dusseldorp *et al* , 1995	B
	Cough	Dusseldorp *et al* , 1995	A
	Lower respiratory symptoms (wheeze)	Dusseldorp *et al* , 1995	A
Children	Bronchodilator usage	Roemer *et al* , 1993	B
	Cough	Pope and Dockery, 1992	A
	Lower respiratory symptoms (wheeze)	Roemer *et al* , 1993	A
ELDERLY 65+			
	Congestive heart failure	Schwartz and Morris, 1995	B
CHILDREN			
	Chronic cough	Dockery *et al* , 1989	B
ADULTS			
	Restricted activity days	Ostro, 1987	B
	Chronic bronchitis	Abbey *et al* , 1995	A
ENTIRE POPULATION			
	Respiratory hospital admissions (RHA)	Dab *et al* , 1996 Ponce de Leon, 1996	A
	Cerebrovascular hospital admissions	Wordley *et al* , 1997	B

As pointed out in the methodology section, we use a comparably low estimate for cases of chronic bronchitis that account for the major part of morbidity damage costs. Several other studies applied higher values for this crucial exposure-response function (World Bank, 1997; European Commission, 1999; Lvovski, 2000; ECON, 2000).

(b) Exposure-response functions for crops.
The EcoSense database contains a standard set of exposure-response functions for crops from ExternE (European Commission, 1999). For the China version of EcoSense some functions were replaced and some new functions were added, using Chinese source (SEPA, 1998). This concerns: rice, rape radish, tomato, wheat,

barley, bean, cotton and soya bean. The uncertainty of damages for crops due to SO_2 (including uncertainty of prices) was estimated in ExternE at $\sigma_g = 2.5$ to 4 (European Commission, 1999). However, the influence on the final result is small since the estimated crop damages are rather insignificant compared to health damages.

Sensitivity due to unknown locations of emission sources in the LCA chain. The results for the whole chain may vary depending on the exact location of up-stream stages of the chain (such as coal mine) that significantly contribute to the overall damages. This issue was explored based on multi- and single-source EcoSense runs. The latter allowed investigation of the influence of mine location on the damage contribution from the full chain; in this context both mines in Shandong and Shanxi province were investigated. For all analyzed cases the influence of the location uncertainty has been shown to be practically negligible.

Sensitivity due to changes in background emissions. Particularly the ammonia background emissions have strong influence on the impacts. If there are high SO_2 emissions in a low NH_3 background, the formation of ammonium sulfate will be limited by NH_3. In this situation an increase of NH_3 will lead directly to an (almost linear) increase of ammonium sulfate. An additional ton of SO_2 will not increase the impacts from ammonium sulfate further because the reaction is already saturated. On the other hand, if the NH_3 emissions are high but SO_2 emissions are relatively low, the SO_2 is the limiting factor for the sulfate production. In this case, an additional ton of SO_2 will cause maximum impacts.

As mentioned before the NH_3 to SO_2 ratio in China is relatively high in the grid cells with high SO_2 emissions. If instead the NH_3 emissions in the relevant Chinese areas would be significantly lower, the impacts would be lower because then several grid cells would fall below the saturation line for the ammonia sulfate reaction.

For sensitivity analysis, the ammonia background emissions were changed in the total modeling area of EcoSense China/Asia. Figure 9.37 shows damage costs and mortality results for the power sector of Shandong, calculated for different NH_3 emission backgrounds. When the ammonia background emissions are doubled, the changes of the final results are very small (about 5% only). On the other hand, a reduction of the NH_3 background by 50% has much more influence on the results. This shows the saturation effect in the formation of ammonia sulfates and ammonia nitrates. Due to the high agricultural production, the current NH_3 emissions are already so high that a further increase of ammonia emissions has not much influence on the sulfate (nitrate) formation per ton SO_2 (per ton NO_x, resp.). Consequently, a further increase of NH_3 emissions would not significantly change the damages per ton SO_2 or NO_x per person or per crop yield unit. Nevertheless, the total mortality and morbidity damages per ton SO_2 or NO_x increase with the growing population. The same is true for crop damages in case of a further increase of the agricultural production.

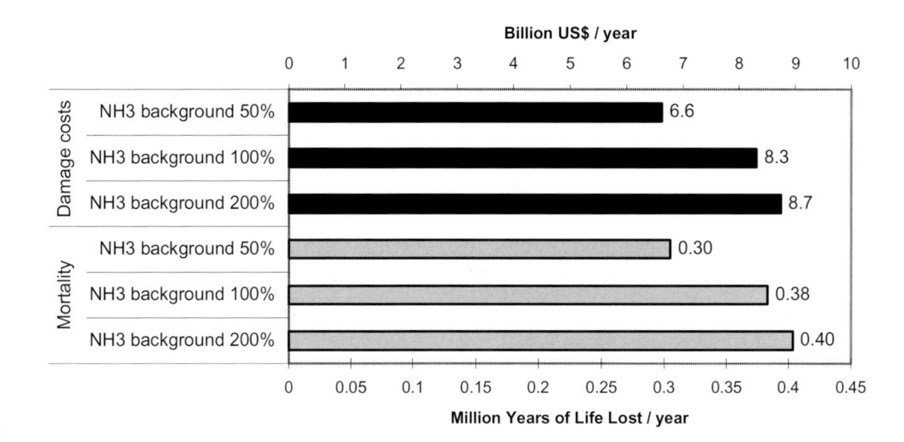

Figure 9.37 Sensitivity of and mortality and damage costs induced by Shandong power sector to changes of the NH_3 background emissions; reference year 1998.

In the analysis of future scenarios the crucial assumption was made that ammonia emissions will stay constant in the analyzed period while the increase in population density was accounted for. Significant reduction of NH_3 emissions in China is considered to be rather unlikely. In the European reference scenario for 2010 (Amann et al., 1998), a 12% decrease is assumed for NH_3, while SO_2 is to be reduced by 70%, compared to year 1990.

Sensitivity to alternative approaches to monetary valuation. For monetizing impacts, two methods are commonly discussed, the Willingness-to-pay (WTP) approach and the Human Capital approach.

- Willingness-to-pay (WTP) approach: The valuation is based on an individual's willingness to pay to avoid a negative impact or to reduce the risk that it occurs. Similarly one can employ the willingness to accept (WTA) payment as a compensation for an occurring impact.
- Human Capital (HC) approach: Morbidity and mortality are valued as lost productivity plus explicit expenditures (such as costs of medical care).

WTP and HC lead to very different values, particularly for mortality. ExternE (European Commission, 1999) and ECON (2000) used only the WTP approach, discarding the Human Capital approach as being inappropriate. We refer here to the criticism of the human capital approach summarized in section 3.4.1. The "Clear Water Blue Skies" study (World Bank, 1997) and the EHS study for Hong Kong (Yee, 1998) applied both the WTP as well as HC method.

If the estimates of specific damage unit costs based on WTP are available for one country but need to be used for another one with unknown WTP, the question arises what approach to use for such an adjustment (e.g. transfer of US values to China). In the present study transfer by ratio of Purchase Power Parity (PPP) GNP per capita was employed, following Markandya (1997) and recommendation by ExternE

(European Commission, 1999). By contrast, the "Clear Water Blue Skies" study (World Bank, 1997), used standard GNP per capita for scaling purpose. The latter method leads to results lower by about a factor of four.

In the following the numerical consequences of the use of the various approaches to monetization are demonstrated with regard to mortality and morbidity, and finally also for the cost-benefit analysis carried out on the supply scenario level.

(a) Mortality.
All approaches mentioned above refer to the Value of Statistical Life (VSL). In the case of EcoSense China European VSL was used as the reference value. Apart from the differences in transfer methods, the various studies use different reference years, which leads to further discrepancies (due to changes in GNP per capita). According to our EcoSense simulation, the air pollution from the power sector in Shandong affects mainly receptors in the province itself. The GNP per capita in Shandong is roughly twice the GNP per capita assumed in "Clear Water Blue Skies" for China. Therefore, if the GNP-weighted VSL from Clear Water Blue Skies is to be used, it has to be scaled by a factor 2 for the application concerning the Shandong Province.

The table below gives an overview on the different values of statistical life.

Table 9.17 Value of Statistical Life (VSL) according to various studies.

Study	VSL (US$)
ExternE Europe	3'000'000
EcoSense China PPP GNP per capita	430'000
WTP Clear Water Blue Skies GNP per capita China	60'000
WTP Clear Water Blue Skies GNP per capita Shandong	120'000
WTP ECON 2000	100'000
WTP EHS Hong Kong transfer based on PPP GNP per capita to China	114'286
WTP EHS Hong Kong original	800'000

The Human Capital approach for VSL refers to the annual wages per capita (World Bank,1997; Yee, 1998). Therefore the GNP per capita is used directly for the human capital value of life years lost (VLYL), which is needed for the monetization of the years of life lost (YOLL) calculated in EcoSense (Table 9.18). The discrepancy between the two methods is quite overwhelming.

Table 9.18 Valuation of YOLL - Value of Life Year Lost (VLYL).

Study	VLYL (US $)
EcoSense China PPP GNP per capita	15'710
Human Capital - GNP/capita Shandong	1'000

(b) Morbidity.
For the purpose of sensitivity analysis, the Human Capital and WTP values from Clear Water Blue Skies (World Bank, 1997) were applied where it was possible, i.e. for the endpoints compatible with EcoSense modeling. The results are summarized

in Table 9.19, demonstrating again large discrepancies, though in most cases less dramatic than in the case of VSL and VLYL.

Table 9.19 Valuation of morbidity impacts.

Impact Category	**Willingness-To-Pay**			**Human Capital**	
	EcoSense China/Asia	WTP Clear Water Blue Skies: original	WTP Clear Water Blue Skies: GNP-corr.	Human Capital Clear Water Blue Skies: original	Human Capital Clear Water Blue Skies: GNP-corr.
	US$/case	US$/case	US$/case	US$/case	US$/case
Congestive heart failure	490	-	490	-	490
Chronic bronchitis	25'400	6'100	12'200	900	1'800
Restricted activity days	17	1.8	3.6	1.8	3.6
Bronchodilator usage	6	-	6	-	6
Cough	7	-	7	-	7
Lower resp. symptoms	1	-	1	-	1
Chronic cough	36	-	36	-	36
Cerebrovascular hospital admission	2'510	(810)[1]	1'620	(810)[1]	1'620
Respiratory hospital admission	650	210	420	210	420

[1] Clear Water Blue Skies shows no value for cerebrovascular hospital admission; therefore the EcoSense value was scaled using the ratio from respiratory hospital admission.

(c) Application of different monetary valuations to ESS 2020 scenarios.
The two following diagrams shows cost-benefit analyses for two ESS scenarios ("Coal with FGD" and "Clean Coal Technology + Diversification") under two different valuations used in the past studies. For both cases the differences relative to the ESS scenario "Coal without FGD" are shown. We refer to Figure 9.34 for the reference EcoSense China valuation. Two additional sensitivity cases not shown here, were analyzed – one based on the Shandong-adjusted WTP scheme employed in "Clear Water Blue Sky" study (World Bank, 1997), and one based on the adjusted WTP scheme used in the Hong Kong study (Yee, 1998).

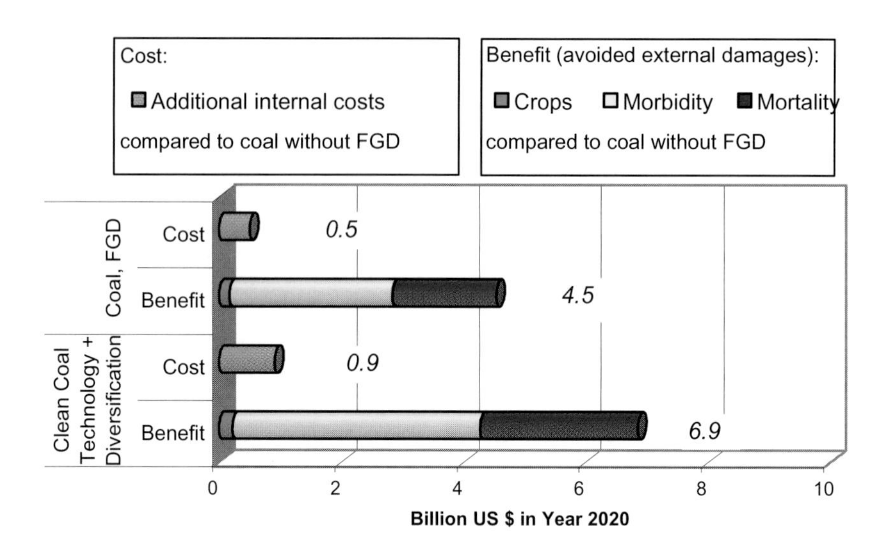

Figure 9.38 Cost-benefit analysis of "clean" versus "dirty" strategies (FIB future), including LCA contributions, based on WTP-scheme according to ECON (2000).

The valuation according to ECON (2000) reduces the benefits (avoided damages) in comparison with the EcoSense evaluation of the present study roughly by a factor of two; in the case of use of the human capital approach the reduction roughly corresponds to a factor of nine. The two additional sensitivity cases not shown here are rather close to the ECON-case.

The damages avoided by implementation of cleaner technologies exceed in all analyzed sensitivity cases the increase in internal costs, thus demonstrating the robustness of the findings. It should be noted that we abstained in this study from adjusting the WTP-based valuation by the projected future increase in GNP per capita. Given an annual growth in China in the range of 5-7% in the next 20 years, this would lead to even higher estimates of damages by a factor of about 2.5 to 3.5.

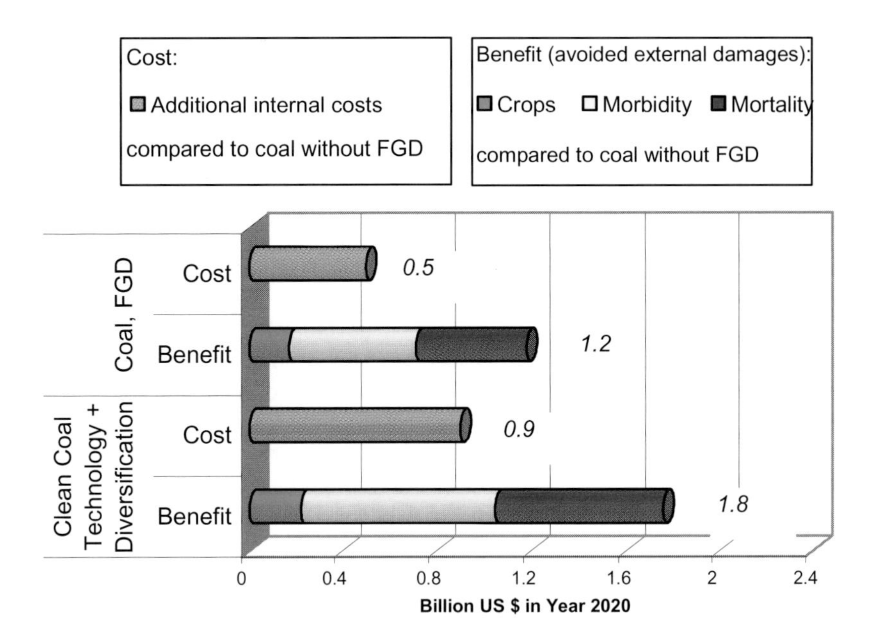

Figure 9.39 Cost-benefit analysis of "clean" versus "dirty" strategies (FIB future), based on WTP-scheme according to Human Capital approach.

3.4.5 Application of simplified approach to impact and external cost assessment

As evident from the preceding sections the application of a detailed external cost assessment is a rather complex undertaking. First, it requires access to a code having capabilities similar to EcoSense China. Second, the list of input data is exhaustive, but more importantly, all values must be known a priori to carrying out the assessment. Such a detailed and extensive compilation of data requirements is generally not likely to be readily available, especially in many developing countries, which stand to greatly benefit from including economic and environmental accounting in future energy planning strategies.

In the early phase of CETP, a simplified approach for the estimation of health impacts was applied. The reasons for employing this approach in CETP were the following: (a) Uncertainties with regard to the prospects for assembling the full set of data with adequate quality for the full scope assessment; (b) Need to obtain rather quickly preliminary results to guide the priorities. In view of the successful implementation and application of EcoSense China the main driving factors for CETP interest in the simplified approach changed somewhat in the course of the project. As a spin-off effect of the detailed EcoSense-based modeling within CETP it was possible to "calibrate" the simplified approach, thus assuring that its future uses have a sound basis. Furthermore, the estimates generated by the simplified approach may be seen as a sanity check of the results achieved using the much more detailed approach.

The earliest attempt to develop a simplified model for damage estimation was undertaken by Curtiss and Rabl (1996). Further developments were made by Spadaro (1999), who continued this work with support of the International Atomic Energy Agency (IAEA).

The simplified approach contains a suite of simple models, collectively named the Uniform World Models (UWM), which rely on a limited number of key parameters for assessing the physical impacts and damage costs to human health. Although not discussed here, simplified approaches for estimating the impacts to crops and building materials have also been developed (Spadaro, 1999). Impact estimates are calculated assuming an elevated, fixed point source with a constant emission rate. The damage assessment function, expressed in integral form, is solved analytically using various simplifying assumptions concerning the receptor distribution across the impact domain, the near-source (local) meteorological statistics and the source parameters; especially important is the stack height. The accuracy of the estimates depends on the data specified by the user and gradually improves with the amount and kind of available input data. In the simplest approximation, only the receptor density within a circle of radius 500 to 1000-km centered at the source is all that is required to predict the impact. Compared to a detailed assessment, the simplified damage results are accurate to an order of magnitude in the worse case scenario (simplest approximation for an emission source in the vicinity of a large urban center), but quite often discrepancies are less than a factor of two.

The Simple Uniform World Model (SUWM). The starting point in the derivation of the *Simple* Uniform World Model relation is the impact assessment formula which assumes a continuous receptor distribution:

$$I = \int_{area} \rho(r)\ F_{er}(r, C(Q))\ dA$$

ρ is the receptor density, F_{er} is the exposure response function, C is the incremental change in ambient air concentration at the earth's surface due to an emission rate Q, A is the impact area and r is the source-receptor position vector. The origin of the coordinate system is fixed at the location of the pollution source. Since the concentration increase C varies linearly with Q, the equation above is equally valid for steady state or transient emission sources. Impacts (I) are integrated over all receptors at risk. Usually, the impact range extends from a few hundred to several thousand kilometers from the emission source, depending on the pollutant's atmospheric residence time (Spadaro, 1999).

To find an "approximate" solution to the impact equation, several simplifications are adopted:

1. The receptors at risk are uniformly distributed across the computational domain:

$$\rho(r) = \rho_{uni} = constant$$

2. The exposure response function may be written in the following simplified form:

$$F_{er}(r, C(Q)) = f_{er}(r)\, C(r, Q) \quad [for\ all\ values\ of\ C]$$

The ER function slope f_{er} is independent of airborne concentration. In other words, the ER function is assumed to be *linear* over the entire range of relevant ambient concentration values. Furthermore, one implicitly assumes that the ER function has *no threshold* limit, meaning that there is no minimum concentration value below which the pollutant impact is zero. If a threshold value does indeed exist, it is assumed that the background ambient level everywhere in the computational area is already above this limit.

3. The ER slope is constant for all values of **r**:

$$f_{er}(r) = f_{er,uni} = constant$$

4. The concentration C(**r**,Q) is proportional to the pollutant's atmospheric removal rate:

$$M(r) = k(r)\, C(r, Q)$$

M(**r**) is the pollutant removal flux along the earth's surface and k(**r**) is a proportionality constant. M(**r**) has units of mass per unit time per unit surface area (horizontal), k(**r**) has units of velocity and, from here on, will be referred to as the pollutant's *depletion velocity* in air.

5. The depletion velocity is constant throughout the impact domain:

$$k(r) = k_{uni} = constant$$

This implies that the atmospheric dispersion parameters are independent of the geographical site of the emission source.

Substituting the modelling assumption equations into the impact equation the impact formulas for airborne primary and secondary pollutants can be obtained:

Primary pollutants

$$I_{suwm} = \frac{\rho_{uni,p}\, f_{er,uni,p}\, Q}{k_p}$$

Secondary pollutants

$$I_{suwm} = \frac{\rho_{uni,s}\, f_{er,uni,s}\, Q}{k_{s,eff}}; \quad k_{s,eff} = \frac{k_s \bullet k_p}{k_{p \to s}}.$$

Q is the emission rate of the primary species (mass per unit time). The subscripts p and s refer to primary and secondary pollutants, respectively. k_s is the depletion velocity of the secondary pollutant. The chemical transformation of the primary pollutant into secondary species is characterized by the transformation velocity $k_{p \to s}$.

The depletion velocities for primary and secondary airborne substances are estimated using a simplified mathematical model for predicting the marginal changes in the regional concentration values as a function of the downwind distance from the source. Measured data or computer simulations are used to evaluate any unknown coefficients.

The depletion velocities of several common air pollutants for selected sites and areas around the world are summarized in Table 9.20. The values for Chinese sites were derived based on the EcoSense China results. The larger the value of k, the quicker the pollutant disappears from the atmosphere because of dry deposition, precipitation and/or by chemical transformation into a secondary compound.

Table 9.20 Depletion velocities for selected sites and regions around the world.

Southeast Asia: **China**

Site characteristics	**PM_{10} (cm/s)**	**SO_2 (cm/s)**	**NO_x (cm/s)**	**Sulfates (cm/s)**	**Nitrates (cm/s)**
BEIJING [40.0°N, 116.3°W] → inland site ⇒ mean wind speed: 4.1 m/s precipitation: 60 cm/yr	0.64	0.84	1.44	1.77	0.82
JINAN [36.7°N, 117.0°W] → inland site ⇒ mean wind speed: 4.0 m/s precipitation: 70 cm/yr	0.53	0.49	0.73	0.90	1.00
JINING [35.4°N, 116.6°W] → inland site ⇒ mean wind speed: 3.8 m/s precipitation: 74 cm/yr	0.73	0.63	0.71	0.76	0.72
KUNMING [22°N, 101.2°W] → inland site ⇒ mean wind speed: 1.6 m/s precipitation: 166 cm/yr	1.83	1.16	0.90	2.26	0.81
LANZHOU [36.7°N, 103°W] → inland site ⇒ mean wind speed: 3.0 m/s precipitation: 37 cm/yr	0.99	1.06	2.35	2.06	0.67
QINGDAO [36.1°N, 120.3°W] → coastal site ⇒ mean wind speed: 4.2 m/s precipitation: 75 cm/yr	0.93	0.84	0.65	0.97	1.17
SHANGHAI [32.2°N, 118.2°W] → inland site ⇒ mean wind speed: 4.1 m/s precipitation: 99 cm/yr	0.96	1.06	1.40	2.27	1.02

Site characteristics	PM_{10} (cm/s)	SO_2 (cm/s)	NO_x (cm/s)	Sulfates (cm/s)	Nitrates (cm/s)
YANTAI [37.5°N, 121.4°W] → coastal site ⇒ mean wind speed: 4.3 m/s precipitation: 73 cm/yr	0.84	0.74	0.65	1.00	1.04
SHANDONG PROVINCE	0.74	0.66	0.96	0.90	0.97

Southeast Asia: **Thailand**

PHICHIT [16.3°N, 100.3°W] → inland site ⇒ mean wind speed: 3.0 m/s precipitation: 125 cm/yr	0.65	0.87	1.53	1.82	0.76

Europe

WESTERN EUROPE	0.67	0.73	1.47	1.73	0.71
POLAND	0.86	0.89	1.05	1.98	1.29

North America: **USA**

ALBANY, NY [42.7°N, 73.8°E] → inland site	1.00	-	-	-	-

South America: **Argentina**

BUENOS AIRES [35°S, 60°E] → inland site ⇒ mean wind speed: 1.9 m/s precipitation: 92 cm/yr	2.19	2.08	0.40	3.63	1.59

South America: **Brazil**

NORTH AMAZONAS [0°S, 65°E;] → inland site ⇒ mean wind speed: 6.9 m/s precipitation: 270 cm/yr	2.86	1.38	2.26	4.76	3.00
BELO HORIZONTE [20°S, 44.4°E] → inland site ⇒ mean wind speed: 2.5 m/s precipitation: 147 cm/yr	1.26	0.84	1.49	3.11	1.33

South America: **Paraguay**

Site characteristics	PM_{10} (cm/s)	SO_2 (cm/s)	NO_x (cm/s)	Sulfates (cm/s)	Nitrates (cm/s)
CAPITAN P. LAGERENZA [20°S, 60°E] → inland site ⇒ mean wind speed: 5.1 m/s precipitation: 103 cm/yr	1.13	1.05	2.13	3.13	1.04

For comparison with the estimates obtained for China the health impacts and damage costs for various pollutants emitted in Western Europe are summarized in Table 9.21. Results are normalized per 1000 tons of precursor emissions. A uniform population density of 80 persons/km^2 is assumed, which is the Western European population averaged over both land and water surfaces. Depletion velocities are taken from Table 9.20. In addition to the mean values, the last two columns in the table provide multipliers for taking into account variations due to source location and stack height (Spadaro, 1999).

The Robust Uniform World Model (RUWM). The Robust Uniform World Model improves on the simple version by including in the impact analysis the local receptor distribution, local meteorological data and the "effective" stack height of the emission source (defined as the actual stack height plus plume rise). Three major improvements to SUWM are introduced in RUWM:

1. The impact domain is separated into local and regional parts.
2. The effective stack height appears "explicitly" in the impact assessment function.
3. Local meteorological data and space-dependent receptor data are utilized.

Since in the present work only SUWM was utilized for details we refer to the technical report.

The AirPacts Program. The Simple and Robust versions of the Uniform World Models are implemented in the AirPacts software (Spadaro, 1999). The AirPacts suite of programs enables the user to obtain a preliminary estimate of the physical impacts and associated damage costs to human health, agricultural crops and man-made materials due to atmospheric emissions of the following pollutants: particulate matter (PM), sulfur dioxide (SO_2), nitrogen oxides (NO_x), carbon monoxide (CO), and secondary species such as nitrate and sulfate aerosols.

The AirPacts software is a 32-bit application written in Microsoft Visual Basic 6.0 for the Windows platform. The program has been tested on Windows 95, 98 and Windows NT4.0 operating systems. Contact information: J.Spadaro@iaea.org.

The most important limitation of the SUWM is that local details are not important in the analysis. The AirPacts package contains modules which, if required and the more detailed data are available, enable the user to carry out more sophisticated and detailed analyses. In Table 9.22 the input data requirements for each model are summarized.

Table 9.21 Health impacts (number of cases) and aggregate damage costs by type of pollutant for Western European conditions.

Pollutant	Asthma [5] (*adults & children*)	Chronic cough (*children*)	Restricted activity days (*adults*)	Respiratory hospital admissions	Years of Life Lost (*YOLL*)	Millions of $US\$_{2000}$	Site multiplier (rural ↔ urban)	Stack height multiplier (225 m ↔ 0 m)
Particles (PM_{10}) [1]	4940	190	7200	0.78	59	8.22	0.5 ↔ 6	0.6 ↔ 15
SO_2				0.71	1.9	0.33	0.5 ↔ 6	0.6 ↔ 15
Sulfates	3200	120	4680	0.51	38	5.27	0.7 ↔ 1.4	≈ 1
Nitrates	4670	175	6970	0.74	56	7.76	0.7 ↔ 1.4	≈ 1
Ozone [2] (via NO_2)			8300		5	1.19	?	?
Ozone [2] (via VOC)			5000		3	0.74	?	?
CO						8.4×10^{-5}	0.5 ↔ 6	0.6 ↔ 15
As (cancer) [3]					1136	170	0.5 ↔ 6	0.6 ↔ 15
Cd (cancer) [3]					139	20.8	0.5 ↔ 6	0.6 ↔ 15
Cr - VI (cancer) [3]					932	140	0.5 ↔ 6	0.6 ↔ 15
Ni (cancer) [3]					19	2.85	0.5 ↔ 6	0.6 ↔ 15
Dioxin (cancer) [4]					1.2×10^{8}	1.85×10^{7}	0.7 ↔ 1.4	≈ 1

Values obtained using the *Simple* Uniform World Model for an emission rate of 1000 tons per year, depletion velocities from Table 9.20 and population density of 80 pers/km^2. Receptor groups: 76% adults, 24% children, 13% over 65 years and 3.5% asthmatics.

ρ_{uni} = 80 (persons/km^2) × Receptor group (receptor/person) = [receptors/km^2]. ER function has units of annual cases per **receptor** per μg/m^3.

Notes:

[1] For $PM_{2.5}$ estimates, multiply values by 1.67 (ExternE, 1999).

[2] Estimates based on the work by Rabl and Eyre (1998); YOLLs are estimated for acute mortality only.

[3] 1 cancer equals 10 YOLL and the unit cost per cancer incidence is 1.5 million US$ (Rabl, Spadaro and McGavran, 1998).

[4] Estimates account for both inhalation and ingestion pathways (Rabl, Spadaro and McGavran, 1998).

[5] Asthma impacts include incidences of bronchodilator usage, coughing and wheezing.

Table 9.22 Input data requirements for the simplified models included in the AirPacts software (Spadaro, 1999). Input data can be specified for the entire impact domain (1000 x 1000 km area about the source) or for any portion(s) thereof.

PARAMETER	SUWM	RUWM		QUERI			URBAN
		Est. #1	Est. #2	Est. #1	Est. #2	Est. #3	
Local characteristics							
• Urban or rural location		✓	✓	✓	✓	✓	Applies to urban sites only ✓
• Receptor density		✓	✓	‡	✓		
• Receptor data (5 by 5 km^2)		†	†			✓	
Regional characteristics • Receptor density	✓	✓	✓	✓	✓	✓	✓
Local weather data							
• Mean wind speed			✓				✓
• Mean ambient temperature			✓				✓
• Pasquill class distribution			✓				✓
• Detailed hourly data			§			✓	§
Stack data							
• Height			✓		✓	✓	✓
• Exit diameter			✓			✓	✓
• Exhaust gas temperature			✓	‡	‡	✓	✓
• Exhaust gas velocity			✓	‡	‡	✓	✓
• Pollutant emissions	✓	✓	✓	✓	✓	✓	✓
• Pollutant depletion velocity	✓	✓	✓	✓	✓	✓	✓
Other							
• ER functions	✓	✓	✓	✓	✓	✓	✓

✓ mandatory input data
† can be substituted for the local receptor density
§ can be substituted for mean weather statistics
‡ if known an improved impact estimate will be calculated

Detailed vs. Simplified analyses: A comparison of results. How accurate is the impact predicted by the Uniform World Models? Comparisons with results calculated using a detailed impact assessment approach (EcoSense), are tabulated in

Table 9.23 and also shown graphically in Figures 9.40 – 9.43 for single point sources located at selected sites around the world, including China. In the detailed assessments, the local impacts are estimated using concentration data from long-range atmospheric transport models. That is, no dedicated model for local impact analysis (ISC) has been employed in the various studies to improve the local assessment beyond the WTM-analysis.

Background data and comparison of impact results for the Shandong province in China are given in Table 9.24. In this context we also refer to Figures 9.18 and 9.19. Both primary and secondary pollutants are considered. No detailed local analysis has been performed, as is the case for the results in Table 9.23.

In view of the comparisons in Table 9.23 (Figures 9.40 – 9.43), the Simple Uniform World Model results appear to be reasonably accurate with deviations not much greater than ±50%. The level of achieved accuracy is quite remarkable considering the functional simplicity of the relationships in impact equations for primary and secondary pollutants. The impact, in fact, is the product of three terms: (1) the exposure response function for a given health impact; (2) the exposed population at risk (adults, children, etc.) averaged over land and water surfaces for a circular area with radius of approximately 1000 km; and (3) the pollutant's mean atmospheric concentration across the range of the domain (The pollutant's mean local concentration (< 50 km) is typically two orders of magnitude higher that the average regional concentration (from 50 km and up to 1500 km from the source.):

$$C_{mean} = C_{uni} = \frac{Q}{\pi R_{Impact}^2 k}$$

The difference between the results of the *Simple* Uniform World Model and the detailed assessment for a particular location arises because the SUWM modelling assumptions are not completely satisfied. This is quite understandable, as in real life the world is not uniform. The discrepancy between the simplified and detailed models is usually largest for primary pollutants such as particles, SO_2 and NO_x when these are emitted into the air in the proximity of large urbanized areas. Under these conditions, the SUWM **under**-estimates the health impact by as much as a factor of six for typical stack heights (≈100 m). For the secondary species like sulfate and nitrate aerosols, which in the China/Shandong case strongly dominate the results, deviations are typically much smaller, i.e. less than ±35%. Impact estimates for secondary pollutants are much less sensitive to local conditions (stack height, population distribution and weather statistics, particularly wind direction and wind speed), unless the precursor pollutant is very reactive (HCl, for example). The SUWM, usually, **over**-estimates the damage costs when sources are located near or along the coast because the dispersed plume moves over water part of the time, in which case the real health impacts are zero. Spatial variations of SO_2, NO_x and NH_3 also play a significant role. Based on the experience from European studies, use of the RUWM improves the agreement with EcoSense results. We abstain here from further elaborating this statement since RUWM was not used for China and refer to the technical report for a more detailed discussion.

Results for the Shandong province in China are given in Table 9.24 (for background data see also Figures 9.18 and 9.19). For the Chinese analyses, the range of the impact domain for inland locations R_{Impact} is 650 km, which is considerably smaller than the appropriate value for European calculations.

Table 9.23 Health cost comparisons from air pollution for selected sites around the world using the detailed Impact Pathways and SUWM approaches – data are plotted in Figures 9.40 – 9.43.

Southeast Asia: **China** (PPPGNP = 2510 $)

Site characteristics [latitude, longitude in degrees; receptor density in pers/km^2; impact domain size in km]	**Public health damage cost** (US$_{2000}$ per kg) **SUWM**[(A)] **assessment**	**Detailed assessment**	**Deviation** SUWM / detailed analysis
BEIJING [40.0°N, 116.3°W; 203; 650]	PM_{10} = 3.09	2.43	1.27
→ inland site (166 pers/km^2, <50 km)	SO_2 = 0.10	0.09	1.18
	Sulfates = 1.85	1.59	1.16
	Nitrates = 2.42	2.06	1.18
JINAN [36.7°N, 117.0°W; 323; 650]	PM_{10} = 5.93	5.08	1.17
→ inland site (797 pers/km^2, <50 km)	SO_2 = 0.28	0.24	1.16
	Sulfates = 5.78	6.96	0.83
	Nitrates = 3.14	3.58	0.88
JINING [35.4°N, 116.6°W; 354; 650]	PM_{10} = 4.72	5.46	0.86
→ inland site (654 pers/km^2, <50 km)	SO_2 = 0.24	0.27	0.89
	Sulfates = 7.51	7.49	1.00
	Nitrates = 4.79	3.13	1.53
KUNMING [22°N, 101.2°W; 102; 1000]	PM_{10} = 0.53	0.55	0.98
→ inland site (22 pers/km^2, <50 km)	SO_2 = 0.036	0.028	1.30
	Sulfates = 0.72	0.70	1.04
	Nitrates = 1.23	0.93	1.33
LANZHOU [36.7°N, 103°W; 87; 1000]	PM_{10} = 0.86	0.89	0.97
→ inland site (78 pers/km^2, <50 km)	SO_2 = 0.035	0.034	1.04
	Sulfates = 0.68	0.69	0.99
	Nitrates = 1.26	0.96	1.31
QINGDAO [36.1°N, 120.3°W; 213; 1000]	PM_{10} = 2.23	2.99	0.75
→ coastal site (537 pers/km^2, <50 km)	SO_2 = 0.11	0.15	0.72
	Sulfates = 3.54	3.84	0.92
	Nitrates = 1.77	1.89	0.94
SHANGHAI [32.2°N, 118.2°W; 337; 650]	PM_{10} = 3.44	4.05	0.85
→ inland site (633 pers/km^2, <50 km)	SO_2 = 0.13	0.16	0.83
	Sulfates = 2.39	3.16	0.76
	Nitrates = 3.22	4.05	0.80
YANTAI [37.5°N, 121.4°W; 195; 1000]	PM_{10} = 2.26	1.56	1.45
→ coastal site (51 pers/km^2, <50 km)	SO_2 = 0.11	0.08	1.39
	Sulfates = 3.14	2.69	1.17
	Nitrates = 1.82	1.30	1.40

Southeast Asia: **Thailand** (PPPGNP = 6970 $)

Site characteristics [latitude, longitude in degrees; receptor density in pers/km^2; impact domain size in km]	**Public health damage cost** (US$_{2000}$ per kg)		**Deviation** SUWM detailed analysis
	SUWM[A] **assessment**	**Detailed assessment**	
PHICHIT [16.3°N, 100.3°W; 78; 500] → inland site (168 pers/km^2, <50 km)	PM_{10} = 3.23	3.86	0.84
	SO_2 = 0.13	0.13	0.83
	Sulfates = 1.92	2.75	0.70
	Nitrates = 2.79	3.70	0.75

NB, Detailed results from EcoSense China/Asia, ver. 1.0 (this work). Local impact assessment based on Windrose Trajectory Model (WTM) concentration data; WTM is a simplified implementation of the more complex Harwell Trajectory Dispersion Model (HTM) developed in the UK in the mid eighties (Derwent and Nodop, 1986).

Europe: **EU-15** (PPPGNP = 17'764 $)

Site characteristics	SUWM assessment	Detailed assessment	Deviation
CENTRAL EUROPE [80 pers/km^2, mean over land and water, excluding Scandinavian countries] → 32 sites are available for comparisons (up to 1200 pers/km^2, <50 km)	PM_{10} = 8.22	7.58[C] [3.06-30.6]	1.08
	SO_2 = 0.33		
	Sulfates = 5.27	4.97[C] [3.05-8.20]	1.06
	Nitrates = 7.76	5.33[C] [2.46-10.5]	1.43
FINLAND [5--15 pers/km^2, depending on site] → 3 sites are available for comparisons	PM_{10} = 0.89[B]	0.94[C] [0.72-1.40]	0.94
	SO_2 = 0.035[B]		
	Sulfates = 0.57[B]	0.72[C] [0.53-0.83]	0.79
	Nitrates = 0.84[B]	0.62[C] [0.47-0.76]	1.33
IRELAND [20--30 pers/km^2, depending on site] → 2 sites are available for comparisons	PM_{10} = 2.51[B]	2.09[C] [1.50-2.90]	1.20
	SO_2 = 0.098[B]		
	Sulfates = 1.61[B]	2.04[C] [1.46-2.78]	0.80
	Nitrates = 2.37[B]	1.54[C] [1.46-1.61]	1.51
SPAIN [45--65 pers/km^2, depending on site] → 12 sites are available for comparisons	PM_{10} = 5.56[B]	3.74[C] [2.37-10.9]	1.48
	SO_2 = 0.23[B]		
	Sulfates = 3.56[B]	3.44[C] [2.22-5.03]	1.03
	Nitrates = 5.25[B]	3.79[C] [1.53-6.40]	1.36

Europe: **Poland** (PPPGNP = 5480 $)

Site characteristics [latitude, longitude in degrees; receptor density in pers/km^2; impact domain size in km]	Public health damage cost ($US\$_{2000}$ per kg) SUWM[A] assessment	 Detailed assessment	Deviation SUWM / detailed analysis
[47-63 pers/km^2, depending on site] → 10 sites are available for comparisons	PM_{10} = 1.84[B]	1.93[C] [0.78-21.2]	0.95
	SO_2 = 0.057[B]	0.065[C] [0.03-0.10]	0.88
	Sulfates = 0.97[B]	1.01[C] [0.71-1.30]	0.96
	Nitrates = 0.89[B]	0.92[C] [0.72-1.22]	0.97

NB, Detailed results from EcoSense Europe, version 2.0 (Krewitt et al., 1995). local impact assessment based on WTM concentration data. In the comparisons, the ExternE 1999 (European Commission, 1999) health damage costs have been revised to reflect the recent changes in ERFs and unit damage costs recommended in the ExternE 2000 literature (European Commission, 2000). ExternE 2000 health impact estimates are approximately half the values calculated by ExternE 1999 (European Commission, 1999), except for SO_2 damages, which are 17% higher.

North America: **USA** (PPPGNP = 25'880 $)

ALBANY, NEW YORK [42.7°N, 73.8°E; 45; 1000] → inland site (225 pers/km^2, <30 km and 56 pers/km^2,<80km)	PM_{10} = 4.52 PM_{10} = 2.80[D]	2.88	1.56 0.97[D]

NB, Detailed results based on concentration data calculated by the EXMOD model (ESEERCO, 1995); for comparisons, the ExternE 2000 ERFs and unit costs have been assumed (European Commission, 2000).

South America: **Argentina** (PPPGNP = 8720 $)

Site characteristics [latitude, longitude in degrees; receptor density in pers/km^2; impact domain size in km]	Public health damage cost ($US\$_{2000}$ per kg) SUWM[A] assessment	 Detailed assessment	Deviation SUWM / detailed analysis
BUENOS AIRES [35°S, 60°E; 10.5; 600] → inland site approximately 100 km from the capital (14 pers/km^2, <50 km)	PM_{10} = 0.16 SO_2 = 0.0074 Sulfates = 0.16 Nitrates = 0.22	0.22 0.0099 0.12 0.18	0.73 0.77 1.29 1.21

South America: **Brazil** (PPPGNP = 5400 $)

Site characteristics [latitude, longitude in degrees; receptor density in pers/km^2; impact domain size in km]	Public health damage cost ($US\$_{2000}$ per kg) SUWM[A] assessment	Detailed assessment	Deviation SUWM / detailed analys
NORTH AMAZONAS (near Venezuela) [0°S, 65°E; 6.9; 600] → inland site (0.3 pers/km^2, <50 km and 2.6 pers/km^2, <125 km)	PM_{10} = 0.051	0.045	1.12
	SO_2 = 0.0026	0.0023	1.12
	Sulfates = 0.050	0.035	1.42
	Nitrates = 0.048	0.036	1.35
BELO HORIZONTE [20°S, 44.4°E; 57.5; 600] → inland site (398 pers/km^2, <50 km)	PM_{10} = 0.85[E]	0.82[E]	1.04
	SO_2 = 0.056[E]	0.054[E]	1.04
	Sulfates = 0.64[E]	0.69[E]	0.91
	Nitrates = 0.91[E]	1.11[E]	0.80

South America: **Paraguay** (PPPGNP = 3550 $)

Site characteristics	SUWM[A] assessment	Detailed assessment	Deviation
CAPITAN PABLO LAGERENZA [20°S, 60°E; 7.9; 600] → inland site (16.8 pers/km^2, <50 km)	PM_{10} = 0.096	0.12	0.79
	SO_2 = 0.0045	0.0057	0.80
	Sulfates = 0.058	0.081	0.72
	Nitrates = 0.11	0.13	0.81

NB, Detailed results from EcoSense Latin America, version 1.0 (Krewitt et al., 2001); local impact assessment based on WTM concentration data.

Notes for Table 9.23

[A] Health costs for the *Simple* Uniform World Model approach are calculated using the relationship:

$$\text{SUWM} = \frac{(\text{pollutant emission rate}) \times (\text{receptor density})}{(\text{depletion velocity})} \times \sum_{\text{health impact j}} \text{ERF}_j \times \$_{\text{case, j}}$$

where, ERF_j and $\$_{case,\ j}$ are, respectively, the exposure response function and unit damage cost for health impact j. For sulfate and nitrate damage estimates, the pollutant emission rates used in the UWM formula are those for SO_2 and NO_x, respectively. For countries outside of Europe (EU-15), the European unit damage costs are scaled by the ratio of PPPGNP values for the new location and Europe as recommended by Markandya (1997). PPPGNP is the purchasing power parity gross national product per capita.

The *Simple* Uniform World Model assumes the pollutant emission rate is constant throughout the year (*otherwise, the mean annual emission rate should be used in the formula*), the receptor density is uniform across the impact domain (*the population is averaged over land and water surfaces*) and the pollutant's depletion velocity is independent of the source-receptor distance (*the atmospheric dispersion characteristics and the pollutant removal mechanisms [dry and wet deposition rates and chemical transformation] are independent of location*).

(B) Geometric mean of damage cost for stated receptor density range.
(C) Geometric mean of stated damage cost interval; the ratio of geometric means for the SUWM and Detailed assessment is used for deviation calculations.
(D) Damage cost from QUERI model (Estimate #1); QUERI is a simplified impact assessment model (one of four computer tools included in the AirPacts software (Spadaro, 1999), which improves on the SUWM estimate using *correction* factors to account for differences in the local and regional population densities.
(E) Includes *only* the damage cost assessment beyond 50 km from the emission source; a local analysis is required to calculate the total impact.

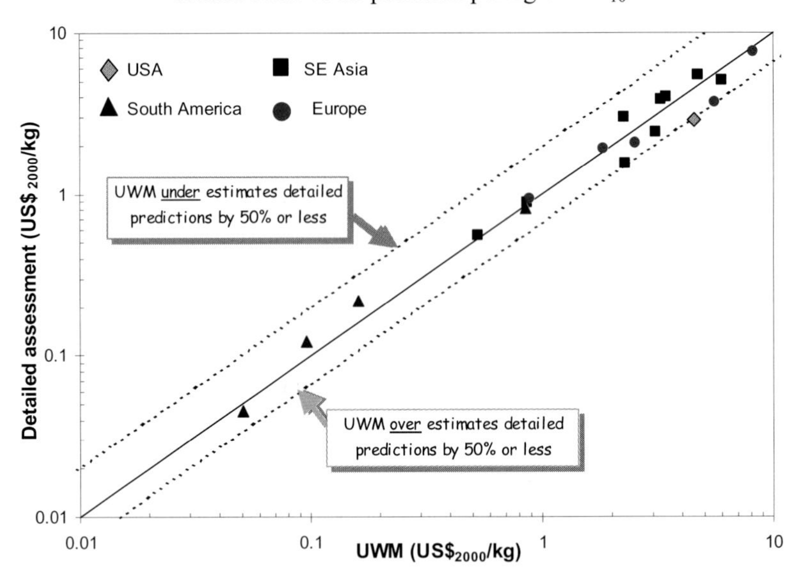

Figure 9.40 Detailed vs. Simple *Uniform World Model impact assessment results for PM_{10} (data from Table 9.23).*

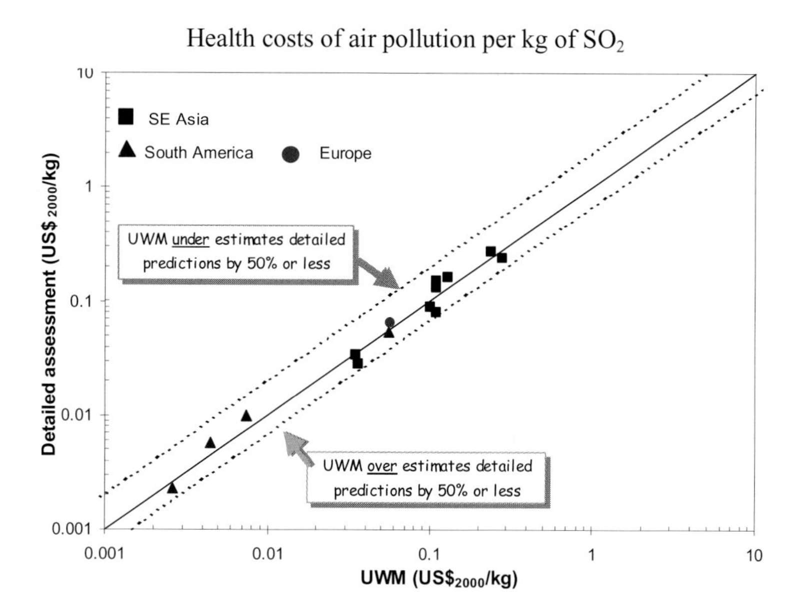

Figure 9.41 Detailed vs. Simple *Uniform World Model impact assessment results for SO_2 (data from Table 9.23).*

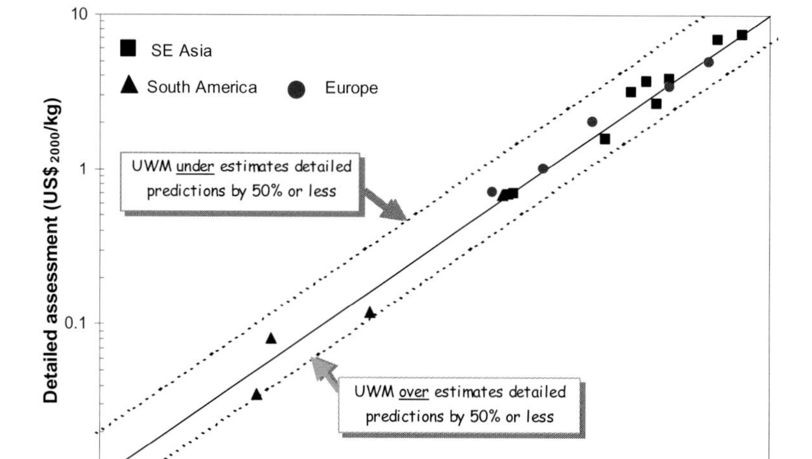

Figure 9.42 Detailed vs. Simple *Uniform World Model impact assessment results for sulfates (data from Table 9.23).*

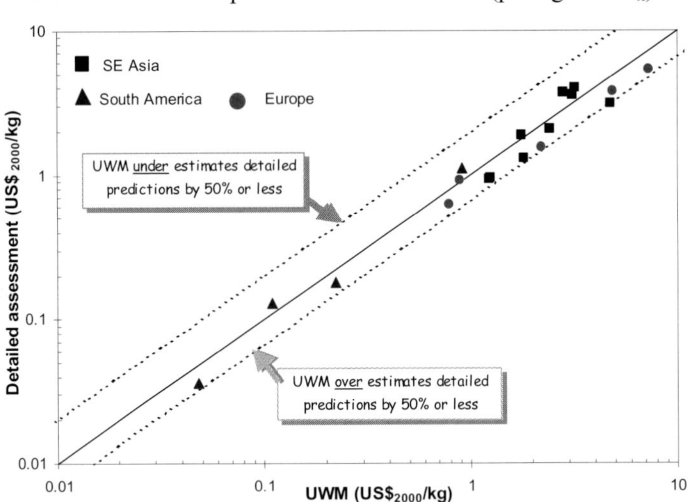

Figure 9.43 Detailed vs. Simple *Uniform World Model impact assessment results for nitrates (data from Table 9.23).*

This impact range reflects the significantly higher regional receptor density in China, which is nearly four times higher than the corresponding value for Europe. The pollutants' depletion velocities for Shandong have been listed in Table 9.20 and the ER functions in Table 9.1.

As seen in Table 9.24, the discrepancy between EcoSense China/Asia and SUWM impact results is less than ±40% for all reference sites not located along the south- and north-east coasts of Shandong. Typically, the SUWM **over**-estimates up to 65% the impacts for coastal and peninsula sites, which is entirely consistent with results obtained in ExternE for the London and Barcelona locations in Europe. The only notable exception occurs when the source is located near Weihai at the tip of the northeastern coast of Shandong. For this site, SUWM results are *higher* than EcoSense predictions by a factor of two to three. The wider margin of error can be explained by noting that the windrose distribution at the Weihai location is highly "seaward" skewed (see Figure 9.19). Therefore, most of the time the pollution plume is dispersed over the sea, where no health impacts are incurred. The assumption that the windrose distribution is uniform in the derivation of the SUWM impact equation is clearly not satisfied here; hence, it is no surprise that the simplified approach gives poor results. Incidentally, the ratio of inland to seaward wind occurrence is roughly one-third.

It is worth mentioning here that the use of "empirical" depletion velocities (i.e., depletion velocities obtained by equating SUWM and EcoSense results) as surrogates for "actual" depletion velocities (those calculated according to basic physics and chemistry principles) is inappropriate when the regional receptor density is high (several hundred persons per km^2) or when the emission source is located along the coast. For the Weihai case study, for instance, the depletion velocity is not higher, rather the SUWM damage result is lower because the plume is transported over an uninhabited area (the sea) most of the time. The depletion velocity only characterizes the air concentration, whereas the impact is the product of concentration and the density of receptors at risk (which in turn depends on R_{Impact}).

To summarize, the most influential factors contributing to the deviation in the results calculated by the Uniform World Models and a detailed impact assessment include the following: (i) type of pollutant (primary vs. secondary species); (ii) local conditions (population distribution, meteorological data and stack parameters); (iii) prevalent wind direction (receptors upstream or downstream of emission source); (iv) precipitation rates and concentrations of precursor and atmospheric species participating in chemical interactions; (v) source location (rural vs. urban surroundings and inland vs. coastal sites or islands) and (vi) ratio of local (less than 50 km from source) to regional (beyond 50 km from source) population densities.

At the regional level, non-linear effects in chemistry occur when the precursor pollutants cannot be transformed into secondary species because of insufficient quantities of ammonia (NH_3) and/or hydroxyl radicals (OH) present in the atmosphere. Atmospheric saturation of these species may occur when the aggregate emissions from all sources distributed across the impact domain is very large. Typically, this means on an annual basis hundreds of thousands of tons of a pollutant being emitted into the air. When such a situation arises, the primary pollutant cannot be "effectively" depleted from the air by chemical conversion; and consequently, its impact to public health increases and that of its secondary species decreases, with both effects varying non-linearly. In this case, the depletion velocity k is not constant as assumed in the derivation of the Simple Uniform World Model, and the anticipated impact from primary (secondary) species will certainly be *under-* (*over-*) estimated as compared to values calculated with detailed atmospheric transport models that take into consideration the consequences of chemical reactions in the air.

Non-linear chemistry effects for pollutants emitted from the entire German power sector have been reported by Krewitt et al. (1999), who have argued that differences in the spatial variations in SO_2, NO_x and NH_3 levels have contributed to the lack of formation of sulfate and nitrate aerosols as compared to atmospheric levels present when individual power stations are considered (marginal emissions).

From a comparison of impact estimates by the *Simple* UWM and the single- and multi-source versions of the EcoSense model for Europe (European Commission, 1999) it is clear that the SUWM is sufficiently adequate to predict the impacts (and damage costs) for both individual and aggregate emissions of PM_{10} and SO_2 via sulfates, with deviations less than ±35% for particulates and ±20% for sulfates. For nitrates, the SUWM predictions are in-line with single-source EcoSense results (deviations less than 14%), but ***over***-estimate the EcoSense multi-source estimates by a factor between 3 and 4. The non-linear chemistry effect is much more pronounced for nitrates than it is for sulfates because the latter are more reactive in the air and would tend to consume the existing stock of atmospheric ammonia before it can be scavenged by NO_x (Lee and Watkins, 1998).

Table 9.24 EcoSense vs. SUWM: Comparison of mortality impact results for selected sites in Shandong province, China.

Site	Latitude (deg)	Longitude (deg)	Population (pers/km^2)	EcoSense vs. SUWM Impact (YOLL) PM (Eco)	PM (SUWM)	Δ (SUWM/Eco)	SO_2 (Eco)	SO_2 (SUWM)	Δ (SUWM/Eco)	Sulf (Eco)	Sulf (SUWM)	Δ (SUWM/Eco)	Nitr (Eco)	Nitr (SUWM)	Δ (SUWM/Eco)
Weifang (Delta)	36.71	122.10	314	179	211	1.18	6.9	8.0	1.16	250	291	1.16	144	162	1.12
Zhanhua (Delta)	37.70	118.13	308	150	207	1.38	5.7	7.9	1.38	262	286	1.09	162	159	0.98
Jinan (West/Central inland)	36.66	116.98	323	260	217	0.83	9.6	8.3	0.86	356	299	0.84	183	166	0.91
Laiwu (West/Central inland)	36.19	117.66	329	238	221	0.93	9.2	8.4	0.92	323	305	0.94	160	169	1.06
Shiheng (West/Central inland)	36.20	116.50	334	272	225	0.83	10.2	8.6	0.84	360	310	0.86	155	172	1.11
Jining (South inland)	35.40	116.59	354	279	238	0.85	11.0	9.1	0.83	383	328	0.86	160	182	1.14
Heze (South inland)	35.23	115.45	359	308	241	0.78	11.5	9.2	0.79	454	333	0.73	283	185	0.65
Shiliquan (South inland)	34.80	117.53	357	252	240	0.95	10.4	9.1	0.88	346	331	0.96	157	184	1.17
Qingdao (Coast)	36.10	120.33	213	153	143	0.94	6.0	5.4	0.90	196	197	1.01	97	110	1.13
Rizhao (Coast)	35.42	119.45	220	170	148	0.87	7.2	5.6	0.78	268	204	0.76	162	113	0.70
Huangdao (Coast)	35.95	120.23	212	96	142	1.49	4.3	5.4	1.26	191	197	1.03	99	109	1.10
Weihai (Peninsula)	37.50	122.10	197	40	132	3.34	1.8	5.0	2.85	94	183	1.95	38	101	2.64
Yantai (Peninsula)	37.53	121.36	195	80	131	1.64	3.2	5.0	1.54	138	181	1.31	66	100	1.51
Longkou (Peninsula)	37.63	120.30	205	103	138	1.34	4.1	5.2	1.27	182	190	1.04	95	106	1.12

(1) YOLL impact is for SO_2 acute mortality (ERF = 5.34E-6 YOLL/pers/μg/m^3) and chronic mortality for the remaining pollutants (ERF = 1.57E-4 YOLL/pers/μg/m^3 for particulates and nitrates; 2.6E-4 for sulfates).
(2) Pollutant emission rate is the same for all pollutants, namely 1000 tons per year for PM_{10}, SO_2 and NO_x.
(3) Mean depletion velocities (cm/sec) for the Shandong region: 0.741 for PM_{10}, 0.662 for SO_2, 0.962 for NO_x, 0.903 for sulfates (Sulf) and 0.967 for nitrates (Nitr). These are the geometric means of calculated values for selected sites across the province.
(4) Receptor densities are calculated assuming a domain size of radius 650 km for inland locations and 1000 km for coastal and peninsula sites. The origin of the coordinate system is centered at the emission source.
(5) EcoSense (Eco) results are from China version of EcoSense, single source model (ver. 1.0), according to this work.

Concluding remarks on simplified approach. The main assumptions of the Uniform World Models include: (i) linearity of the exposure response function, with no threshold value or background concentration above such a limit, if one does exist, (ii) uniformity of the atmospheric dispersion parameters, and (iii) steady state conditions. Mass conservation allows the integral in the damage function expression to be solved in closed form. The Uniform World Model is exact when the receptor density and source distributions are uniform throughout the impact domain.

The "Simple" model requires five input parameters: the pollutant emission rate, the slope of the exposure response function, the depletion velocity, the average receptor density across the impact domain (averaged over land and water for a circle having a radius of 500 to 1000 km) and the economic value. Estimates are not site specific. The "Robust" version improves on the results for primary pollutants by taking into account local receptor density, stack parameters and weather conditions near the source, thereby allowing for site variability and dependency on stack height.

Despite the simplifications evoked in the Uniform World Models, the impact estimates are reasonably close to the values that are obtained following the detailed Impact Pathways procedure.

4. ACIDIFICATION IN CHINA AND SHANDONG

The state-of-the-art assessment of external costs, though ambitious in terms of its scope and level of detail, does not provide a complete coverage of the potentially important consequences, as pointed out in the review by Krewitt (2002). Thus, estimations of the impact to the ecosystem (apart from agricultural losses) remain extremely uncertain. Some assessments of the costs of forest decline due to acid rain have been carried out in the past, but their validity is highly questionable. Since SO_2 deposition, and the resulting acidification, is one of the most important environmental issues in China, the CETP approach employed the most versatile tool for analyzing acidification at a regional level: the RAINS-ASIA (Regional Air Pollution Information and Simulation for Asia) model (Amann and Dhoondia, 2001). The latest version of the code (RAINS 7.52) became available in 2001, allowing the full implementation of the acidification analysis within CETP. In this section we summarize the CETP work based on RAINS-ASIA. A much more detailed account can be found in the technical report (Hirschberg et al., 2003b).

4.1 The RAINS 7.52 Model of Air Pollution

The main features of RAINS ASIA 2 model include (Amann and Dhoondia, 2001):

- Estimates of present and future (1990-2030) sulfur emissions in Asia for the baseline economic development scenarios
- More than 100 administrative regions and 400 individual large point sources.
- Assessment of the effectiveness of emission control measures.
- Cost estimates of emission control strategies.
- Assessment of the atmospheric dispersion of SO_2 emissions.

- Impacts of emission control measures on regional sulfur deposition and SO_2 concentrations levels.
- Optimization tool to identify least-cost emission control measures.
- On-line help system.

The various sub-models are organized into three modules:

- **The energy-emissions module (EMCO).** EMCO estimates current and future levels of emissions of sulfur dioxide (SO_2) in Asia. These estimates are based on national statistics and projections of economic activity, energy consumption levels, fuel characteristics, etc., taking into account implemented and possible emission control measures. EMCO also estimates costs for the reduction of emissions and provides also "cost curves" that rank the abatement measures applicable in each region and LPS according to their cost-effectiveness. These cost curves are used as input for the optimization module (OPT). EMCO includes 38 administrative regions and 213 individual large point sources (LPSs) in China.
- **The deposition and critical loads assessment module (DEP).** The DEP module provides estimates of acid deposition loads throughout the region being studied as a function of changing emissions. The depositions are in turn compared with maps of environmental sensitivities (the "*critical load*" maps). Critical loads represent the maximum long-term deposition levels which can be tolerated by sensitive ecosystems without damage (Hettellingh et al., 1995). The emission data used as input to DEP are produced by the EMCO module. Resulting deposition fields and critical loads achievement can be used to construct target deposition levels used as input into the optimization described below.
- **The optimization module (OPT).** The OPT module identifies, for a given set of regional target depositions levels (which can be derived for example from critical loads), the cost-minimal allocation of measures to reduce emissions. This optimization takes into account that some emission sources are linked via the atmosphere to sensitive receptors more strongly than others (as reflected in the atmospheric transfer matrices), and that some sources are cheaper to control than others (as captured by the cost curves). Figure 9.44 shows the general flow of information in RAINS-ASIA, with utilization of the optimization model.

4.2 Energy demand and supply

4.2.1 Background

Estimation of the level of acidification requires as an input establishment of one or several energy scenarios. Major effort was invested in the present task to define the reference energy supply scenario for China.

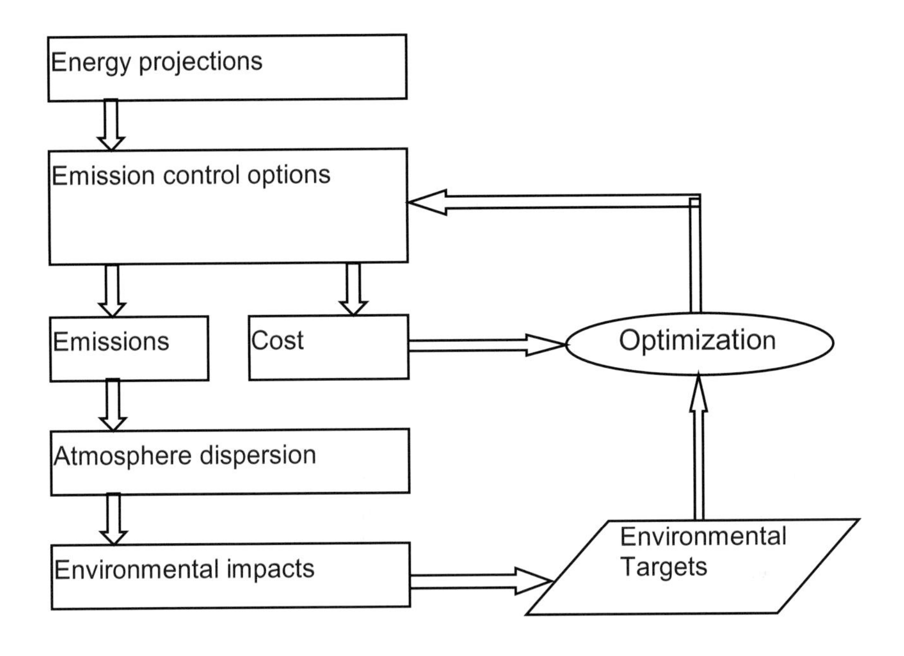

Figure 9.44 Modeling framework in RAINS-ASIA – general flow of information (Amann and Dhoondia, 2001).

The default scenario provided in RAINS-ASIA was not considered relevant for the purpose of this study. It was built on information from the reference year 1995, when energy consumption in China was steadily increasing year-by-year. This trend was decisive for the establishment of the default scenario, as implemented in the newest version of RAINS-ASIA. However, the energy consumption in China reached a peak in 1996 and then, for several reasons started to decline over the following years in spite of the robust economy growth. This unexpected drop in demand follows the sharp decrease in the consumption of coal, as illustrated in Figure 9.45. The total energy consumption in China in year 1999 was practically identical with that from year 1995. Sinton and Fridley (2000) mention a number of possible explanations for the divergent trends in energy and economic growth, such as intentional efforts to curb energy use, unintentional policy impacts, and uncontrollable or unanticipated factors. They consider economic system reforms shutting down inefficient factories and favoring development of more efficient enterprises, consumers switching to higher quality coal, and possibly energy efficiency improvements by normal turnover of capital equipment, as factors having the largest magnitude of effect. What regards electricity consumption it continued to increase after 1996, though at a lower rate than earlier. As a consequence of these developments, previous projections of energy and electricity consumption, not taking into account the recent trends, need to be corrected downwards. This was done in the present work.

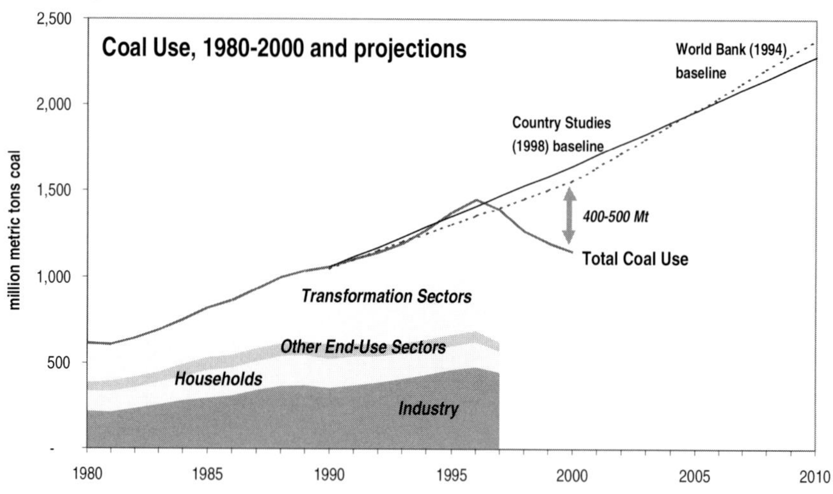

Figure 9.45 Historical development of coal consumption in China, and earlier projections (mainly following Sinton and Fridley (2000), though somewhat modified).

It should be noted that, for the other countries included in the model, the default RAINS-ASIA energy and acidification control scenarios were not modified. Thus, the results of the present study are only valid for China.

4.2.2 Modeling approach

A simulation approach rather than optimization approach for generating regional energy scenarios was employed in this task. The schematic flowchart of this method is displayed in Figure 9.46. The figure highlights the links between the various input parameters and how they are used to compute regional and LPS energy consumption. Throughout the approach, energy supply is assumed to meet energy demand; in other words, the model is demand constrained, not supply constrained. National energy scenarios for each sector are developed at the country level. Fuel consumption is then apportioned to provincial regions using historical regional distributions, and regional socioeconomic projections.

4.2.3 Socio-economic development assumptions

The national population growth in future China was studied in many research projects, and results from most of them are in general convergent. For the purpose of this study it was assumed that the population of China will reach 1.56 billion in year 2030 and that migration of rural population into cities will continue. Thus, the share of urban population will increase in 2030 to about 58% from the current 35%.

Based on the economic and social development target in the blueprint of the Chinese government, the GDP growth rate was here assumed to be as high as 7-8%

during 2000-2010. From then on, stable growth of economy is still anticipated, but the growth is expected to slow down to 6-7% during 2010-2020 and 5-6% during 2020-2030.

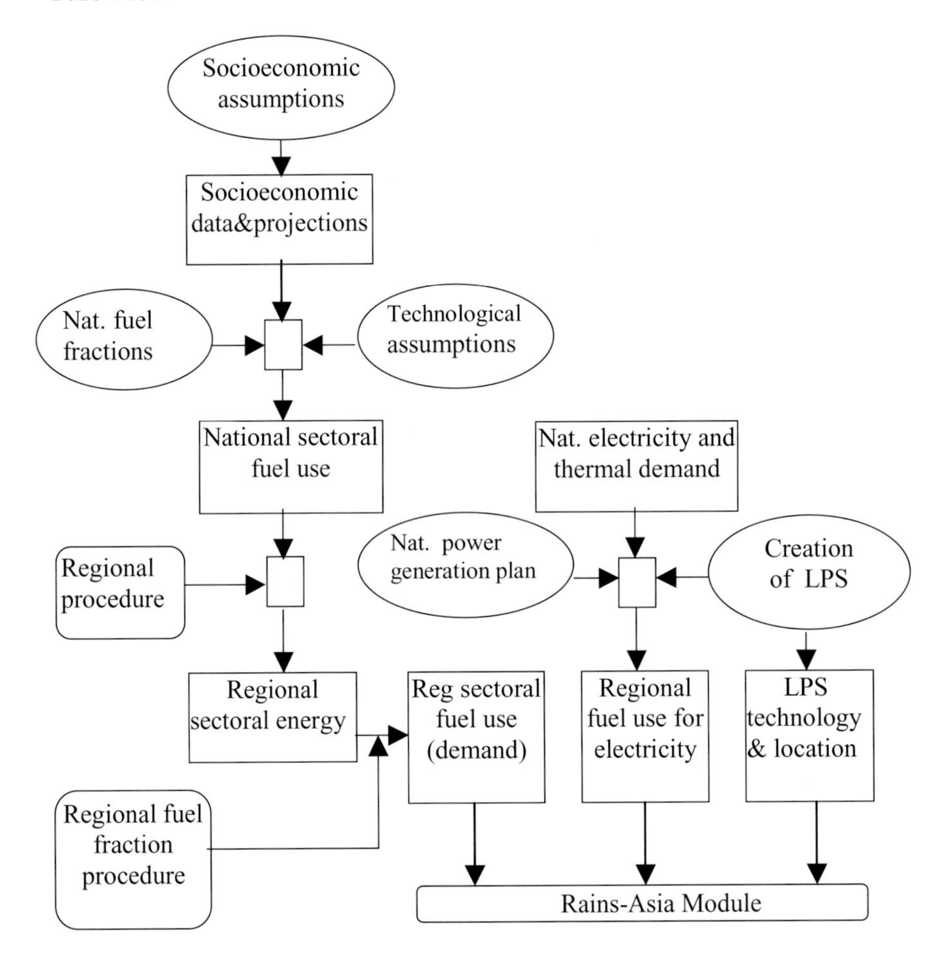

Figure 9.46 Flowchart of energy prediction approach.

4.2.4 Baseline energy scenario

Energy demand analysis. Estimation of energy use was carried out for five end-use sectors and two energy conversion sectors. Sectoral energy consumption was disaggregated into five end-use energy consumption sectors, i.e. residential, service, agricultural, industrial and, transportation. Each end-use energy sector was analyzed in a similar way.

China's economic structure is likely to pursue the same path as experienced in the developed economic entities. The contribution of the service sector to the total economic production will increase, agriculture contribution to GDP will drop and industry GDP share will decrease slightly.

When establishing the reference scenario for China, the trends in the development of energy intensity within the above mentioned sectors were evaluated, and then used for the projections. The energy intensity in China falls presently faster than in most other countries. Between 1990 and 2000 it has been improving at an annual rate of 7.5%, expressed in terms of energy use per GDP, though from an initially very poor level. If China can achieve an annual energy savings rate of 3.5-4.5% over the next three decades, the energy intensity will then decrease to about 0.4 kgce/US$ value added industrial output (from the current of about 1.5), a number still twice as high as the current level in industrialized countries.

Energy supply analysis. One baseline energy supply scenario was developed for China. For this scenario it was assumed that the current efforts in China aiming at increased use of hydro power, natural gas and nuclear power will continue according to the plans.

Here we limit the presentation of the baseline energy supply scenario to the commercial primary energy use and the modal split in power generation. The commercial primary energy use can be deduced from the sum over its use in the five end-use sectors and the two transformation sectors. Figure 9.47 gives the commercial primary energy use by sectors while the corresponding distribution by energy carrier is shown in Figure 9.48.

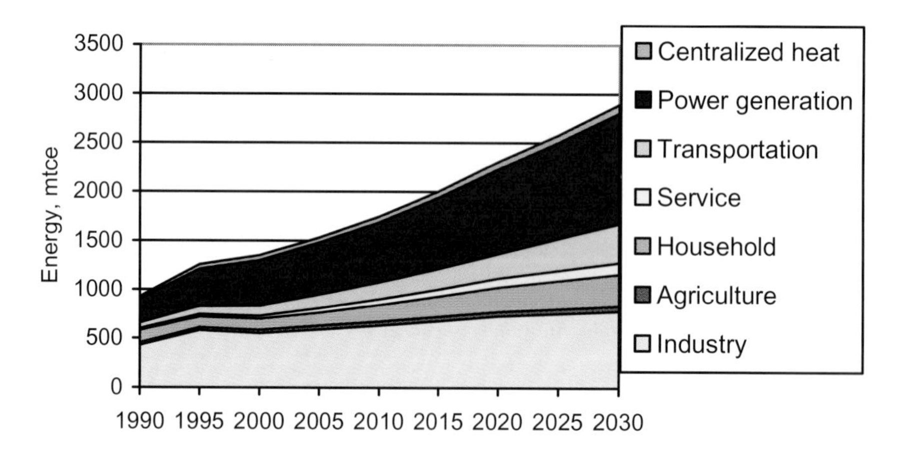

Figure 9.47 Commercial primary energy use by sectors.

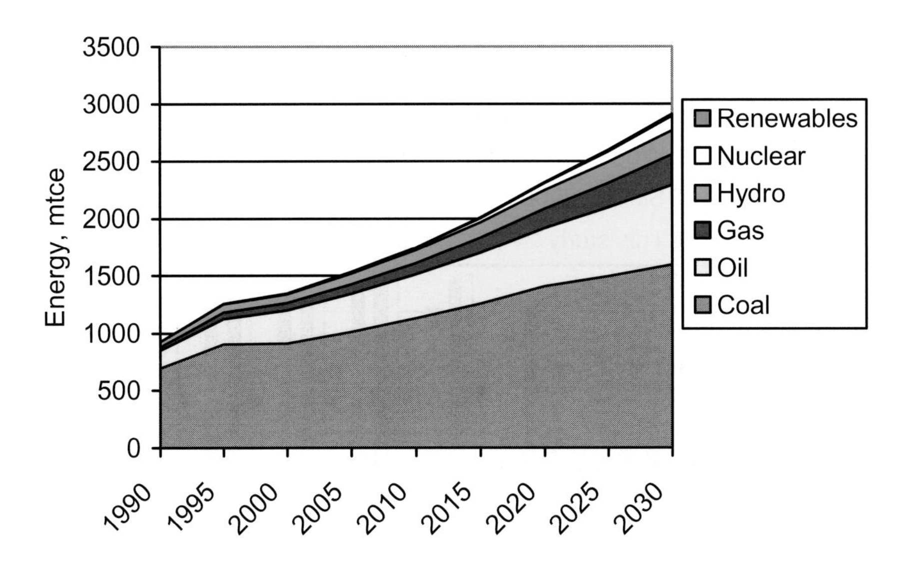

Figure 9.48 Commercial primary energy use by energy carriers.

According to the baseline scenario the total primary energy use by 2030 will be about 2.3 times higher as the current level. However, the explicit consideration of the recent trends leads to a slower growth than in comparison with some other current studies. This is illustrated in Figure 9.49, which compares the present results with statistics until year 2000, RAINS-ASIA default results, the projections by Wu and Chen (2001) and by Sinton and Friedly (2000). The best agreement is with the last mentioned reference.

The primary energy supply structure of the baseline scenario shows improvement, with a decreasing share of coal from currently 68% to 55% over the next three decades; however, the total coal consumption increases about 1.8 times compared to the current level.

The power generation is likely to keep a stable rise in the future. The total generation by 2030 is expected almost to triple compared to the current level. The reference power generation mix in the year 2030 exhibits, in comparison with the current situation, a significantly lower share of coal, somewhat increased share of hydro, a strongly increased share of nuclear, and increased (but in absolute terms small) contributions from natural gas and renewable energies. Natural gas, however, is expected to expand strongly in the context of the total energy use. The baseline evolution of power generation mix in China is shown in Figure 9.50. The projected growth of the total electricity generation in China is somewhat slower than in a number of other recent studies but the differences are mostly smaller in relative sense than in the case of energy.

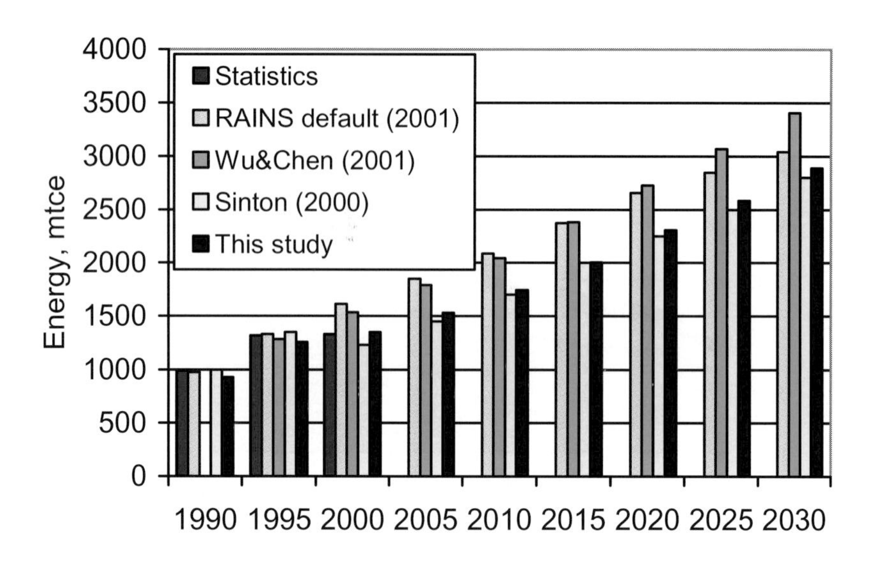

Figure 9.49 Comparison of commercial primary energy use projections for China according to this and other recent studies.

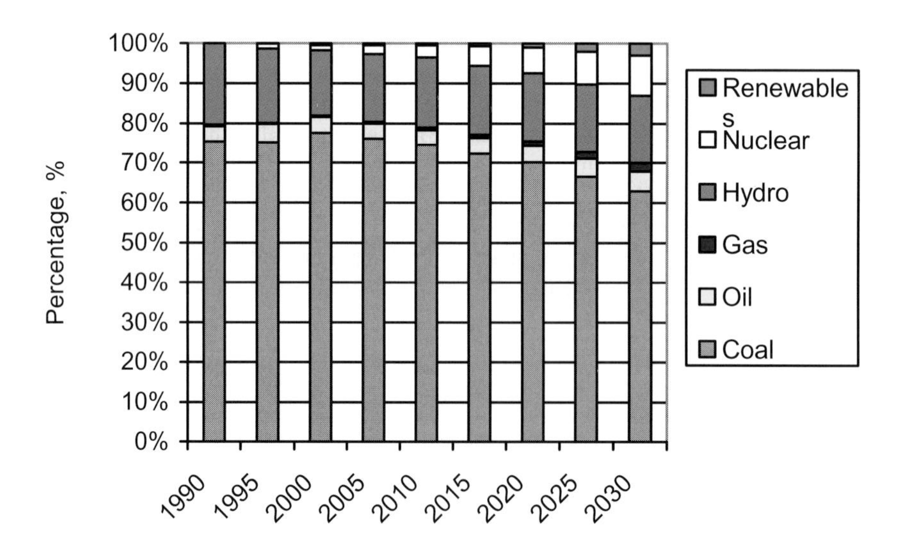

Figure 9.50 Evolution of power generation mix in China.

The prediction of energy demand in the Shandong province is based on the research by the Energy Research Institute (ERI, Beijing), described in detail in Chapter 4. However, in the present task we modified the electricity use by detracting the in-plant consumption. The reference development of the electricity generation in

Shandong is consistent with the moderate growth projections used in the ESS Task within CETP, i.e. it is consistent with the values used in the EcoSense simulations. The RAINS-ASIA default electricity generation projections for Shandong appear to be unrealistically low.

4.3 Analysis of Emission Control Scenarios

The modeling framework for the treatment of emissions and dispersion, as implemented in RAINS-ASIA, is shown in Figure 9.51. This section addresses the emission part while dispersion will be covered in the subsequent one.

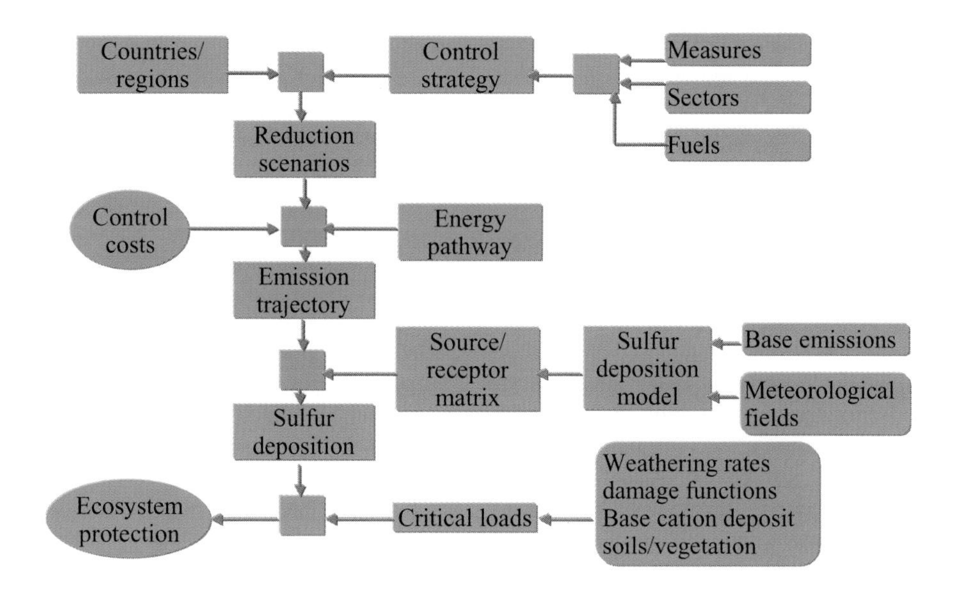

Figure 9.51 Emission and dispersion modeling framework in Rains-Asia.

4.3.1 Technical emission control options

Defining a control strategy means prescribing certain emission control measures for specific emission sources. Reduction strategies can be oriented towards entire regions or countries, towards individual plants (e.g., for any of 213 large point sources in China), to whole economic sectors considered in the database (e.g., for fuel use in the domestic sector, certain industries, etc.), or to specific fuel types (e.g., for high sulfur hard coal).

The following groups of emission control options can be considered in Rains-Asia model (Amann and Dhoondia, 2001):

- energy conservation

- energy mix adjustment
- using low sulfur fuel (naturally occurring low sulfur coal grades, coal washing, heavy fuel oil with low sulfur content, diesel oil with lower sulfur content)
- introducing desulfurization during the combustion process, e.g., by injecting limestone into the furnace or by various types of fluidized bed combustion
- introducing desulfurization of the flue gas after combustion.

The first two emission control options, energy conservation and fuel substitution, relate directly to assumptions developed in the construction of energy scenarios. These options for emissions reductions have been explored in the energy scenario generation procedure and are already reflected in the baseline scenario.

Since it is difficult, and not very useful to describe dozens of commercially available processes without knowing the specific applicability to each individual emission source represented in RAINS-ASIA, the model groups the technologies into categories that describe their major technical and economic features. In practice, for each of the technology categories, one representative process has been selected and introduced into the model.

4.3.2 Low control scenario

The low control scenario is analyzed for the purpose of estimating the emissions and resulting pollution, given no special control efforts in terms of energy efficiency improvements, low sulfur fuel substitution and/or use of technical measures such as desulfurization options. Therefore, this scenario can be regarded as a conservative scenario in terms of these specific sulfur control measures. On the other hand, the baseline energy supply, particularly what concerns electricity, reflects the diversification efforts through increased use of natural gas, hydro and nuclear power. Thus, the low control scenario definitely does not represent the worst case. The future increase of emissions is caused by the increase in energy consumption and lack of the implementation of control measures, which is comparable to the situation in the base year 1990 when practically no controls were applied.

The emissions by fuel in low control scenario are displayed in Figure 9.52. Comparing to 23.2 Mt emissions in 1995, and 24.0 Mt in 2000 the future emissions show a steady increase during the following three decades. Coal is clearly the dominant polluter in China over the analyzed period, having about 93% share of the total emissions. The use of oil roughly contributes about 4% of the total emissions over the time considered.

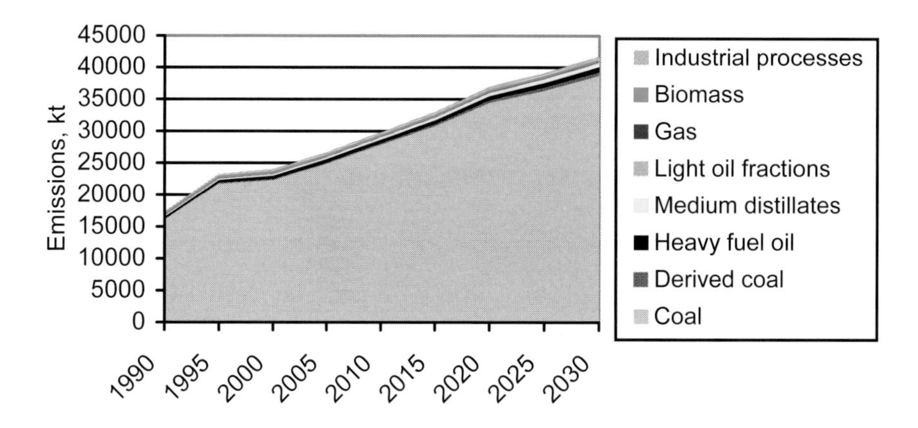

Figure 9.52 SO_2 emissions by fuel in low control scenario.

Figure 9.53 shows the contributions to the total SO_2 emissions by sector. Power generation and industry are the two major sources of SO_2 emissions. For example, in 1995 these two sectors contributed about 42% and 41%, respectively. It should be noted that the historical contribution of power generation to the emissions, as presented here, is significantly higher than the corresponding share according to EcoSense emission database (31% in year 1998; see Table 9.6). We attribute this discrepancy in the first place to differences between EcoSense and RAINS-ASIA with regard to the emission allocation to the power sector and industry, respectively. Specifically, for the Shandong province the relative contribution of the power generation is even significantly higher than the average for China. Given the continued rapid development of power generation, it is likely that its share in the total SO_2 emissions will continue to increase, particularly under the low control scenario.

For the purpose of comparison, Table 9.25 lists the emission predictions from the present and other research efforts. All cases concern "Business-As-Usual" or low control scenarios, except for World Bank (1997) where a range of values is provided with the upper level corresponding to the other studies while the lower range reflects high control strategy. The results cited in UNEP (1996) are identical with upper ranges in World Bank (1997). The results of the current analysis are comparable to those in both reports by the World Bank (1994 and 1997) but are significantly lower than the estimates made by the Chinese researchers in the Eighth Five-year Project (1999). The results in Asia Development Bank (1993) are by far the highest.

The total SO_2 emissions in the Shandong province according to our RAINS-ASIA analysis are for year 2000 identical with the reference emissions of EcoSense for the reference year 1998. The estimates for the next 20 years for the low control scenario show a relatively larger rate of increase for Shandong in comparison with the whole China.

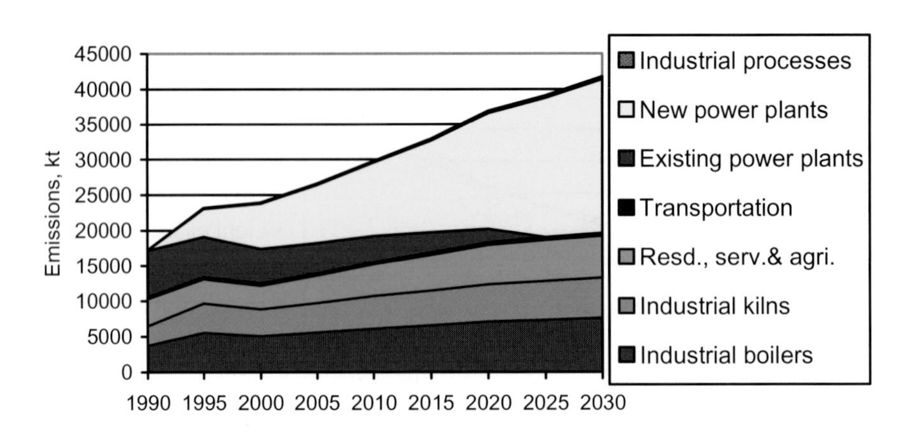

Figure 9.53 SO_2 emissions by sector in low control scenario.

Table 9.25 Evolution of SO_2 emissions in China according to various sources (in Mt).

Study/Year	2000	2010	2020
World Bank (1997)	20.5-23.8	23.5-31.1	22.7-35.5
UNEP (1996)	23.8	31.1	35.5
World Bank (1994)	23.4	28.5	33.1
Asia Development Bank (1993)	27.0	41.0	55.0
8th Five-year Project (1999)	27.3	39.5	42.7
This research	24.0	29.8	36.9

4.3.3 Moderate control scenario

The objective of the moderate control scenario establishment is to describe the development trend of future SO_2 emissions and pollution under the current policy, given that this policy will be fully implemented. Comparing to the low control scenario, this scenario shows how effective the current policy will be, if consequently pursued in the future.

Current policy analysis

A number of measures serve as instruments for the current policy. This includes:

- **SO_2 emission standards.** The emission control requirement for coal end use is embodied in the national standards. There are four kinds of standards used in China, i.e. the air quality standards, emission standards, technological standards, and pollution alert standards. The emission standards are directly associated with the emission control of pollution sources. Currently the control of SO_2 pollution from coal end use in industrial boilers and kilns mainly relies on the control of sulfur content in the coal used. Such requirements of minimum sulfur content can be implicated in the related national standards such as: "Emission standard of air pollutants for coal burning boilers", "Emission standard of air pollutants for industrial kilns", "Emission standard of air pollutants for coking furnaces", "Emission standard of air pollutants for cement plants". For example, for a boiler using coal with sulfur content less or equal than 2%, the maximum SO_2 emission concentration is 1200 mg/m^3; for that with coal sulfur content larger than 2%, the maximum is 1800 mg/m^3. From the standards mentioned above it can be deduced that given no other control efforts the sulfur content of coal used in coal-fired industrial boilers should not exceed 0.9%, while a maximum sulfur content of 1.0% is possible for industrial kilns burning coal.

The standard for pollutant emissions from power plants was formulated in 1996. It provides specific requirements on the plants, whose severity depends on when the plants were constructed. For plants constructed in 1997 or later and located in the national SO_2 and acid rain control zones (see below), additional limits on emission concentrations were also formulated. To match the emission standard, estimations with specific coefficients recommended in the standard indicate that, for example, for a 1200 MW plant in the urban area, sulfur content in coal below 0.9% is required; for a 2400 MW plant in a hilly rural area, a coal with sulfur content below 1% is necessary; and for a 2400 MW plant in a plain rural area, 1.2% sulfur content can be allowed as maximum.

- **Two control zones.** Under the Law on Air Pollution Control of China the SO_2 control zone and acid control zone where the acid deposition (or acid rain) are very serious problems, have been mapped. The SO_2 control zone comprises mainly the cities located in the north of China as the southern cities with air quality problem usually strongly suffer from acid rain and are mapped into the acid rain control zones. The mapped acid rain control zone is mainly located in the south of Yangzi River, and the east of Shi-chuan and Yun-nan provinces, an area of about 800'000 km^2, or 8.4% of the total area of China. SO_2 control zones cover 61 cities and their precinct counties in 14 provinces with an area of about 290'000 km^2, i.e. about 3% of the national area. Thus, the total area of Two Control Zones occupies 11.4% of the national land (Tsinghua University, 1998). Figure 9.54 shows the mapping result for SO_2 and acid rain control.

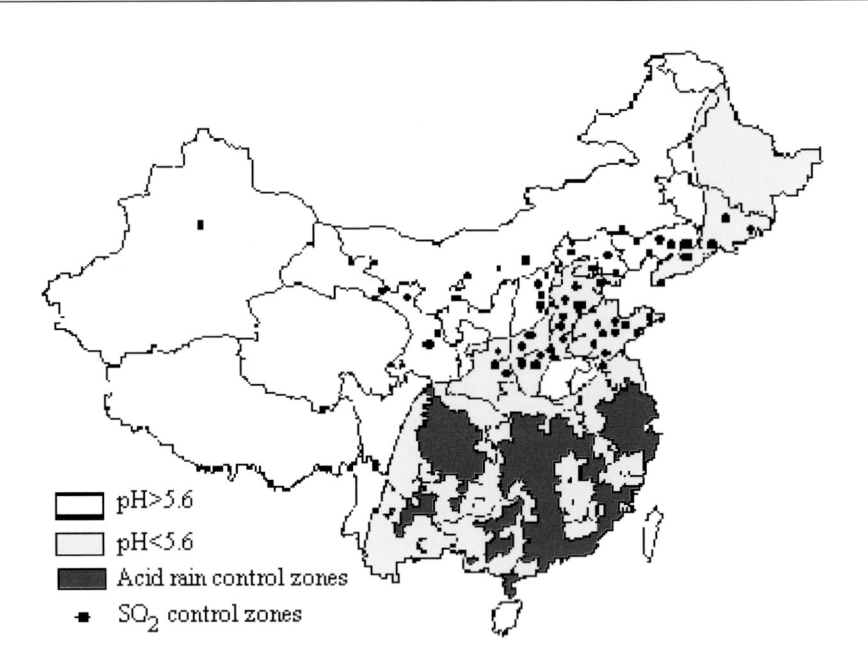

Figure 9.54 Acid rain and SO_2 pollution control zones.

In 1995, the emissions of SO_2 in the two zones, corresponded to about 60% of the total national emissions. Thus, the pollution control in these two zones, could effectively prevent the nationwide deterioration of acid rain pollution. In the two zones, it was required that by the year 2000 the SO_2 emissions are kept at the level of 1995, the industrial sources meet the related national emission standards, and the air quality in most provincial and tourist cities of the two zones meets the national standard. By the year 2010, the control objective in the two zones is to achieve the air quality within the national standard for all cities, and a strong reduction of heavily acid rain-polluted regions with PH $\leq$4.5 in the acid rain control zone.

- **Retirement of small and low efficiency units.** Small units have a larger share in power plant park in China than in other countries. The use of small units results in low efficiency of power generation. Retirement of small and old units is not only beneficial for the modernization of power industry, but also for energy saving and emission control of acid pollutants. The State Power Corporation (SPC) has formulated the retirement plan for small units ($\leq$50 MW) it owns (SPC, 1998). According to this plan all such units should be retired by 2003. However, besides the small units owned by SPC, there are also a lot of small sized units that are affiliated to other owners. To close up these units will be more difficult due to deficiencies of local administrative management and a lack of substituting power supply. In the moderate control scenario it was assumed for the whole country that 50% of all units with capacity 50 MW or less will be retired by 2005, and all units will be phased out by 2010. In the future, also 100 MW units should be progressively substituted according to the policy underway (SEPA, 2002), and we assume that all old units with capacity 100 MW or less will be phased out until 2020.

- **Use of high-efficiency power generation technology.** China as a developing country shows great ambitions to develop clean coal technology for power generation. At the country level, the efforts to improve the generation efficiency focus on two aspects (Jiang Zhesheng, 1995). One is the development of PC units with large capacity and high efficiency. In the next 30 years, conventional PC units will remain to be dominant, but the emphasis will be on the development and promotion of following units: (1) super-critical units with capacity 600 MW in the near term and 1000 MW in the longer term; (2) 300-600 MW anthracite fired units; (3) 300-600 MW air-cooling units; and (4) 300 MW thermal supply units. In parallel, major efforts are made to develop and use IGCC and PFBC-CC technologies. In 2000, the average efficiency was about 38% for new plants. Later on the average generation efficiency for new plants in China is assumed to increase to 40.1% in 2010 and to 42% in 2020. These averages are used in the moderate control scenario calculations for new power plants.
- **FGD in power plants.** FGD research and development have been conducted in China within the past three national five-year plans. However, due to the late introduction of SO_2 emission control laws and standards, its implementation is in an initial stage. According to the Law on Air Pollution Control of China, the newly constructed power plants are required to install FGD or implement other desulfurization measures in case emission standards are not met or local limits on total emissions are exceeded. The exiting plants should meet the standards or emission permits within the deadline set by the State Council or local governments. The recent document of the national 10th five-year plan on environmental protection policy of the State Power Corporation (SPC) highlights the intentions of SPC to promote FGD use in power plants in the next five years (SPC, 2001). During 2000-2005, 20 existing plants with total capacity of 5’700 MW, which currently exceed the emission standard, are supposed to install FGD equipment; 13 power plants located nearby the urban areas, with total capacity of about 3800 MW, plan to install FGD and other abatement technologies to improve the urban air quality, although some of them meet the current emission standard. In addition, among the 18’000 MW coal-fired units to be taken into operation in the following five years, 19 plants of about 6000MW capacity will be using clean coal technology including FGD. If the plans will be implemented, then by 2005 there will be 15’500 MW capacity in China, equipped with appropriate abatement technologies, mainly FGD; this corresponds to about 5.3% of the national total capacity. Most of the affected plants are located in the Two Control Zones. By 2010 all plants in the Two Zones, burning coal with sulfur content $\geq 1\%$, should install FGD; if the sulfur content in the coal being used will not change

significantly, then this means that about 40% of the total generation capacity in the Two Control Zones will need to install FGD by 2010.
Following the moderate control scenario beyond 2010, the emission control will progressively expand into a nationwide scale according to the Regional Acid Rain Control Planning of China (ETARCP, 1999), which means that all power plants burning high sulfur content coal (S≥1%) need to install FGD after 2010. Under this scenario by the year 2020 a nationwide control of acid rain will be implemented meaning that about 40% units that burn coal with sulfur content above 1% will install FGD. The technology will be adopted based on the principle of high efficiency FGD for new units, and middle efficiency FGD for old units. The potential for the use of low sulfur coal as a measure of emission control from large plants is generally limited as the available resources are better used for other applications, where implementation of measures such as FGD is not feasible.
Specifically for Shandong, a province releasing about 10% of national SO_2 emissions and having 10 sub-regions mapped within the SO_2 Control Zone, ambitious plans have been made for FGD implementation at large power plants (Shandong Electricity Management Bureau, 1999). If the current sulfur content in the coal used remains unchanged then FGD implementation level for large plants constructed after 1990 in Shandong should increase from 10% in 2010 to about 50% in 2030.

- **Industrial boilers and kilns.** Industrial boilers and kilns are a very significant source of SO_2 emissions in China, currently sharing about 40% of the total. As most industrial activities are in or close to the cities and have low chimneys (typically 20-60 m), they cause severe local-scale SO_2 pollution. As a large part of industrial boilers and kilns are small it is difficult to control the associated emissions. Furthermore, a considerable amount of coal used has high sulfur content. FGD technology has been partly used for the industrial boilers and kilns in recent years, but there have been many technological problems. Also in view of the costs, FGD use in small boilers and kilns is not encouraged by the government.
The current thermal efficiency of small sized boilers and kilns is 55% and 40%, respectively. The use of low sulfur coal or washed coal can not only control the SO_2 emissions, but also improve the energy conservation. Therefore, low sulfur coal use to substitute the high sulfur coal is the cost-effective measure to control these SO_2 emissions. In the next few years, the east area of China will first promote the use of low sulfur coal. So far in many cities such as Beijing, Shanghai, Liaoning, Jinan, Hangzhou, Nanjing the local governments have already set up the sulfur content cap for coal imported from other regions. However, according to the current regulation only about 5% of all boilers are required to use low sulfur coal or washed coal.

In the moderate control scenario, all industrial boilers using high sulfur coal in the Two Control Zones are assumed to switch to low sulfur coal. By 2010 it is assumed that all industrial boilers in the whole country will have to meet the emissions standard. For industrial kilns the share of low sulfur coal would then expand to 39%, and by 2020 this percentage will hold on but with lower sulfur content coal (0.7%) being used in these kilns.

- **Household sector.** The domestic use of coal includes cooking, hot water, space heating etc. The direct combustion of raw coal in households is the main cause of local air quality problems, and it also wastes much energy due to its low combustion efficiency. The use of cleaner energy, such as low sulfur coal, coal briquette, gas and electricity to replace the raw coal burning is the only effective and manageable measure to reduce the emissions cost-effectively. In the Law on Air Pollution Prevention and Control, it is stated that the State Council and other responsible organizations should put forward measures to improve the fuel consumption mix of cities, develop urban coal gasification, and promote the production and use of coal briquette. SEPA also clearly proclaimed for cities in Two Control Zones to forbid the raw coal burning in households from 1998, and for all large and middle cities to clamp down the raw coal burning by the end of 1999. As a consequence, the local governments are making efforts to enhance the supply of clean energy. In the moderate control scenario the use of low sulfur coal in households is assumed to expand to the national average for the period 2005-2030 of about 39% (24.8% in 1990-2000); for Shandong, where most cities are within the Two Control Zones, much higher shares of low sulfur coal are anticipated (about 54%).

SO_2 emissions in moderate control scenario. Figure 9.55 displays the emissions from each sector under moderate control scenario, which reflects full implementation of the current policy. The effectiveness of this policy is evident as emissions in the future are actually slightly lower than the current level, in spite of the large increase in energy consumption. Power plants and industry dominate the emissions also in the future.

Compared to the low control scenario, the moderate control scenario reflects the effectiveness of current policy that the government intends to carry out. Table 9.26 lists the national SO_2 emission reductions in the moderate control scenario compared to the low control scenario. Among the credited measures FGD in power plants and use of low sulfur coal in industry have the highest emission reduction potential.

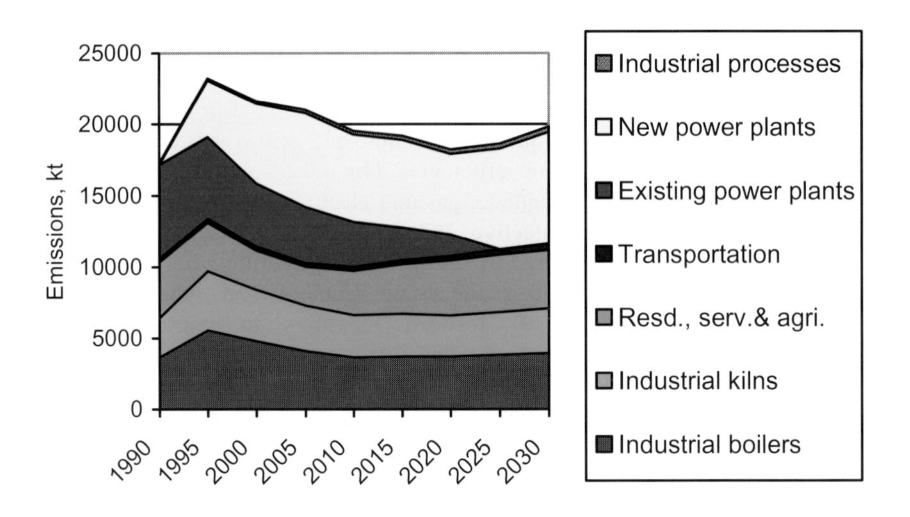

Figure 9.55 SO_2 emissions by sector in moderate control scenario.

Table 9.26 National SO_2 emission reductions in current policy scenario, in kt.

Measures	2000	2005	2010	2015	2020	2025	2030
Energy saving from small unit retirement	241	613	816	867	891	0	0
Energy saving from advanced power technology	0	0	65	173	378	701	871
FGD in power plants	998	1206	3714	6136	9636	11'621	12'682
Low sulfur coal in industry	423	2471	4118	4857	5881	6118	6363
Low sulfur coal in households	675	1330	1553	1735	1870	1988	1962
Total reductions, kt	2337	5620	10'266	13'768	18'656	20'428	21'878

4.3.4 High control scenario

Necessity of enhanced policy scenario setting. In the latest plan of acid rain control, the Chinese government has set up a clear target of acid rain control between year 2000 and 2020 (ETARCP, 1999). The target includes the control of precipitation PH values, depositions and SO_2 emissions. Accordingly, the government sets up the caps of SO_2 emissions for years 2000, 2010 and 2020. Based on the plan, by 2000 the total emissions are to be at most 24.60 Mt, i.e. only 0.90 Mt increase from the level of 1995. By 2010, the emissions should not exceed 20.69 Mt, a reduction of about 3 Mt from 1995. By 2020, the total emissions should be below 16.19 Mt. The current policy as reflected in the moderate control scenario is insufficient to reach this goal. The key factors for an enhanced policy considered here are further improved emission controls in power generation, industry and households.

Enhanced control measures.

- **Power generation.** Since industrial and household boilers have the highest priority for using the limited amounts of good quality coal, enhanced FGD implementation is the most promising measure for power plants. In the enhanced policy from 2005 on, for new units built after 2000 in Two Control Zones and using coal with sulfur content ≥0.7%, installing FGD would be required. The same applies to old units in the Two Zones if the sulfur content is ≥1.0%. From 2010 on, for all units in other regions outside the Two Zones, using coal with ≥1.0% sulfur, would need to install FGD. From 2020 on FGD would be required unless the sulfur content is below 0.7%.

  Inevitably, FGD use will bring about the cost increase of power plants. The cost is different for various FGD technologies. The research shows that installation of wet calcium FGD technology takes about 12-18% of total capital cost in a power plant, while the dry lime stone injection in furnace costs only 3-6% of the total (Hao and Lu, 1999). For the new units the wet FGD is the best technical option, as the SO_2 removal efficiency may be above 95% and results in moderate cost per ton SO_2 removed. However, due to the high capital cost of wet FGD it may be more cost effective for old units to install less efficient but cheaper technologies such as dry FGD. If the units only have short life time left, the best choice may be to shift to the low sulfur coal use in order to meet the standard. In the enhanced policy scenario the share of large power plants in China equipped with FGD increases in year 2030 to about 63% and to an even higher level in Shandong.
- **Industrial boilers and kilns.** As the old boilers and kilns account for about 73% of coal use, the implementation of the new standards is helpful to enhance the control. Substitution by low sulfur coal appears to be the only way in the near future to meet the new standard. We assume that the new standards will get effective first in the Two Control Zones by 2005. It means that all boilers and kilns burning coal with sulfur content ≥0.9% and ≥1%, respectively, will be required to use the low sulfur coal in the Two Zones by 2005.

  By 2010, the use of low sulfur coal in boilers and kilns will not be a sufficient measure. A significant proportion of low sulfur coal is from the coal washing, whose cost levels are comparable to FGD. Simultaneously the capacity of coal washing is limited, for instance the current ratio of coal washing is about 22%. Therefore, in the enhanced control policy, the use of FGD for middle and large boilers and kilns is also credited, so that over 10% of all industrial boilers and kilns is assumed eventually to be equipped with FGD while about 27% will use low sulfur coal. Higher shares are assumed in both cases for Shandong.

- **Households.** Clean coal substitution for the raw coal is still adopted as the main measure in the best control scenario, but with a stricter requirement on sulfur content in coal. Under the enhanced policy the share of low sulfur coal could be about 70% in the whole China already in the mid-term.

SO_2 emissions in high control scenario. The enhanced control results in further emission reductions. Of particular significance are the reductions for power plants. Figure 9.56 compares the predicted emissions with the targets set up by the Chinese government. The enhanced policy leads to meeting the goals though the trajectory is somewhat different.

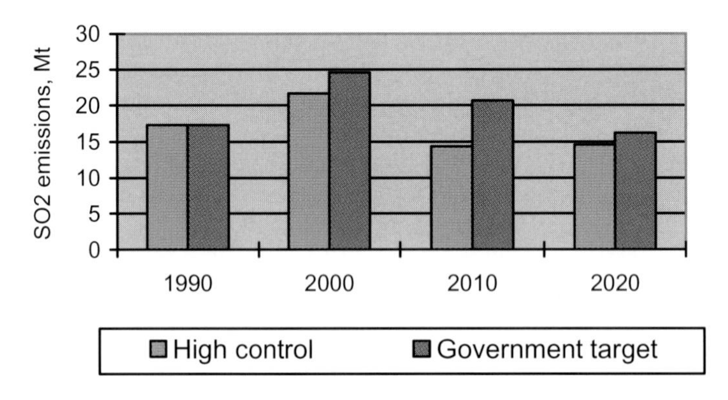

Figure 9.56 SO_2 emissions in high control scenario compared to governmental targets.

SO_2 emissions in Shandong in the enhanced scenario slowly decrease from 1995 to 2005, and then dramatically decrease to around 1155 kt until 2010. After 2010, the emissions keep stable at around 1000 kt until 2030.

4.3.5 Emission scenario comparison

SO_2 emissions resulting from the three alternative control scenarios are shown in Figure 9.57. In relative terms, the "High control" scenario brings larger reductions in SO_2 emissions in comparison to the "Moderate control" option, if Shandong is considered rather than the whole of China.

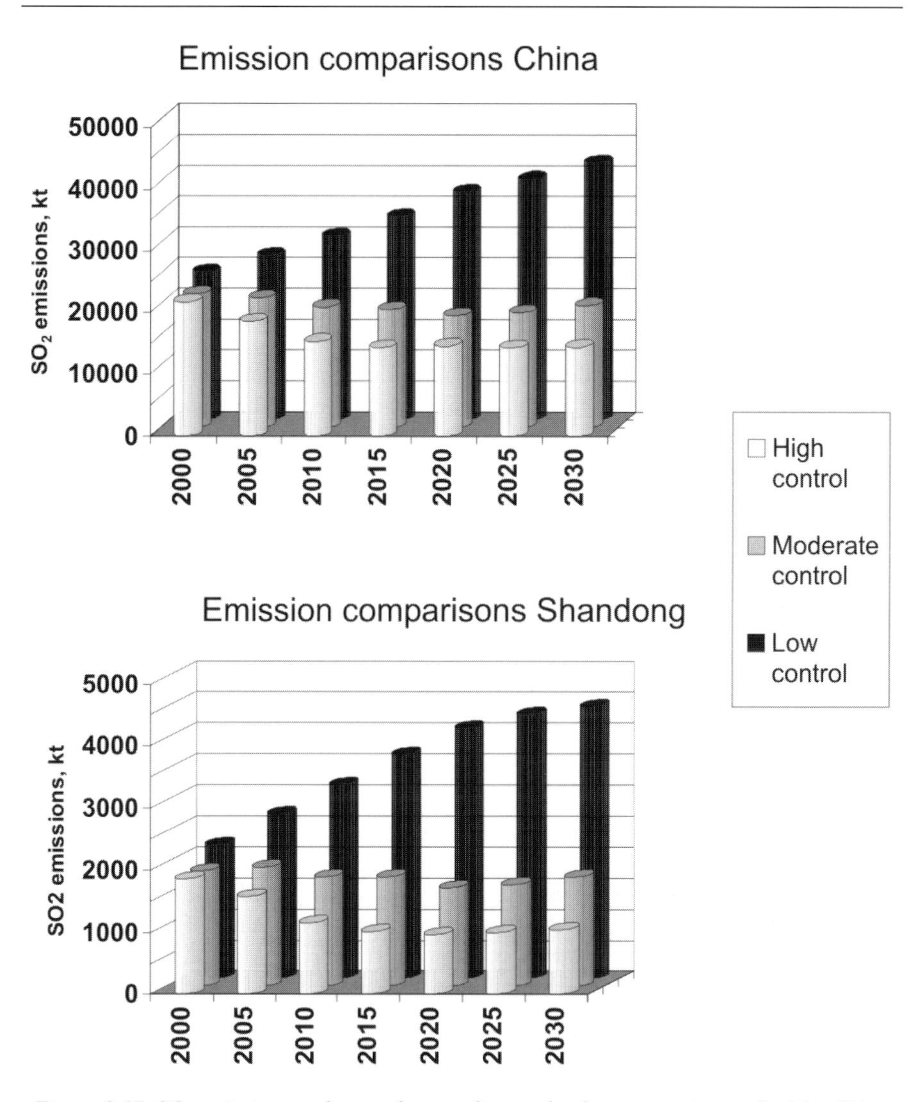

Figure 9.57 SO_2 emissions and control, according to the three scenarios applied for China and Shandong Province.

The difference in control costs increases with time. Compared to the "Low Control" scenario in the year 2020, the estimated additional control costs for the "Moderate control" scenario amount to about 9 billion US$ per year, and to 13 billion US$ per year for the "High control" case, when the whole of China is considered. For Shandong, the corresponding estimates are 1.0 billion and 1.3 billion US$ per year, respectively.

To assess the effectiveness of control scenario, the cost per unit reduction is a widely accepted criterion. The less cost per unit reduction is, the more cost effective the scenario. The analysis shows the cost in the moderate control scenario is about

483 $/t over the period analyzed, while the cost in the high control scenario is about 380 $/t. Therefore, although the total cost is significantly higher in the latter case, it is more cost-effective than the moderate policy scenario. The same applies to Shandong where the unit reduction costs are somewhat lower than the average ones for China.

4.4 Simulation of acid deposition in China

4.4.1 ATMOS model

There are over ten major models in the publications in recent years that have been used to calculate the long range transport of acid gases in the troposphere and the quantities of acids deposited on the surface in East Asia. These models usually consist of: (1) a transport component (or module) to describe the wind speed and direction, the eddy diffusivity and mixing layer height, the temperature, the water vapor, cloud water content, and the radiation intensity of each location as a function of time; (2) a chemical kinetic mechanism to describe the rates of atmospheric reactions, including homogeneous gas-phase, heterogeneous, and liquid phase reactions; and (3) removal modules to describe the dry deposition of material, and the in-cloud and below-cloud removal processes. For the comparison of these models we refer to Carmichael et al. (2000) and to the summary provided in the technical report (Hirschberg et al., 2003b). As a result of the comparison it was concluded that the choice of model parameters is less of a critical result-driving issue in comparison with the spatial and temporal variations in emission inventories and meteorology.

In RAINS, estimates of future deposition of air pollutants are based on transfer matrices for long range transport (Amann and Dhoondia, 2001). Sulfur depositions resulting from the emissions of each country can be calculated. The calculations are based on (linear) source-receptor matrices derived from the ATMOS model of long-range transport of air pollutants in Southeast Asia (compare Arndt et al. (1998)). The Lagrangian model ATMOS is a modified version of the USA National Oceanic Atmospheric Administration, Branching Atmospheric Trajectory (BAT) model. The ATMOS model is a three-dimensional, multiple layer model, and provides a one by one degree resolution of the concentrations and depositions of SO_2 and sulfates. The model separates the vertical dimension into two layers during the day and three layers at night. The day layers are the boundary and upper layers. The night layers are the surface, boundary, and upper layers (Carmichael and Arndt, 1995).

4.4.2 Modeling results for China and Shandong

According to the emission control scenarios, the annual sulfur deposition was calculated. Selected results are presented in following color maps. Figures 9.58-9.61 show the evolution of depositions for the low control scenario. Figures 9.62 and 9.63 display the deposition patterns in year 2020 for moderate and high control scenarios, respectively.

The deposition in China follows closely the spatial distribution and the density of the emissions. The eastern part of China, including North-East China, North China,

East China, South China and South-West China, suffers the serious sulfur deposition at different degrees, while the western part including the West China and North-West China has very low depositions. The highest annual depositions occur in the East China including Zhejiang, Shanghai, Jiangsu and Shandong, and in the South-West China centered at Chongqing and Sichuang.

Consistent with the emission share of LPS, about 17.5% of the current total anthropogenic sulfur emissions in China, the average fraction of sulfur deposition due to LPS amounts to 15.9% of the total. The maximum amount of deposition due to point sources can exceed 80%.

In low control scenario the depositions will rapidly intensify. The heavily deposited area with deposition above 1250 mg/m^2-yr (1 eq/ha/year, the unit used in the figures, corresponds to 1.6 mg/m^2-yr) will expand from some parts of East China and Sichuan basin in 1995 to all parts of East China, South-West China, South-China, North China and east coast of North-East China by 2030. This area will have a share of 40% of the national total by 2030, compared to 10% in 1995. Meanwhile, the center of deposition at Chongqing, Shanghai and Shandong areas will expand by a factor of four.

In the moderate control scenario (Figure 9.62), taking year 2020 as an example, the area with deposition higher than 1250 mg/m^2-yr will be six times smaller than in the low control scenario. Especially after 2010, the growth trend is controlled and the depositions will stabilize. The high deposition area in East China will be two times smaller than in 1995 while the highest deposition center in Sichuan basin will be remarkably improved after 2010.

In the high control scenario, the deposition will continually decrease from 1995 to 2020. The deposition profiles in 2010 and 2020 are similar, with areas having depositions above 1250 mg/m^2-yr almost eliminated, and with a decrease by nearly a factor of two of the area with deposition above 625 mg/m^2-yr, compared to 1995. The high deposition area is in 2020 confined to very limited regions in Chongqing and East China while in the low and moderate control scenarios they constitute 30% and 5%, respectively, of the total area of China.

Shandong local sources contribute 52.3% of the local deposition, of which 35.4% is from the local area sources and the reminder from the local large point sources. 47.7% of the depositions in Shandong are due to emissions from other provinces. Among the external contributors, Hebei, Henan, Jiangsu and Shanxi provinces totally have a share of 25% in Shandong depositions.

However, the impact of dispersion is bilateral. Shandong emission sources also cause the sulfur deposition in the neighboring regions. The range and intensity of such transboundary dispersion from Shandong is shown in Figure 9.64. The regions affected are Hebei, Henan, and Jiansu provinces but also to lower extent the North China, the Middle China, and the neighboring parts of Korea and Japan.

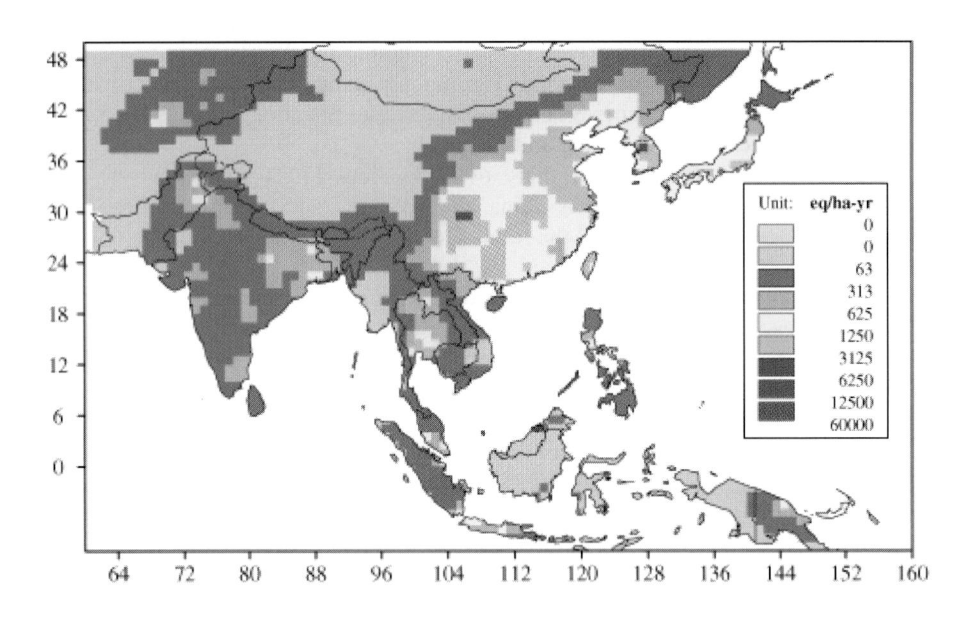

Figure 9.58 Annual sulfur depositions in low control scenario in 2000.

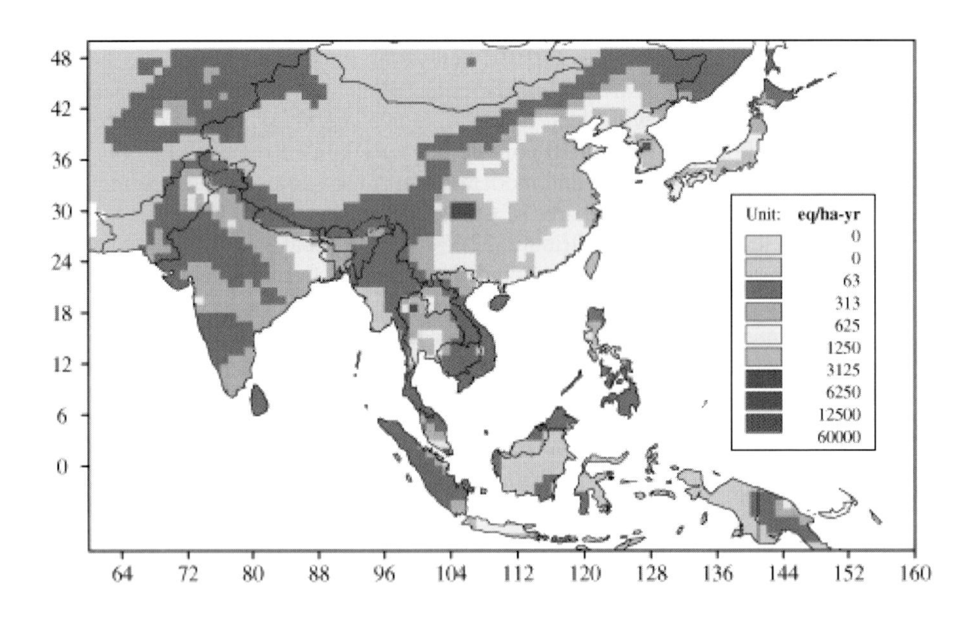

Figure 9.59 Annual sulfur depositions in low control scenario in 2010.

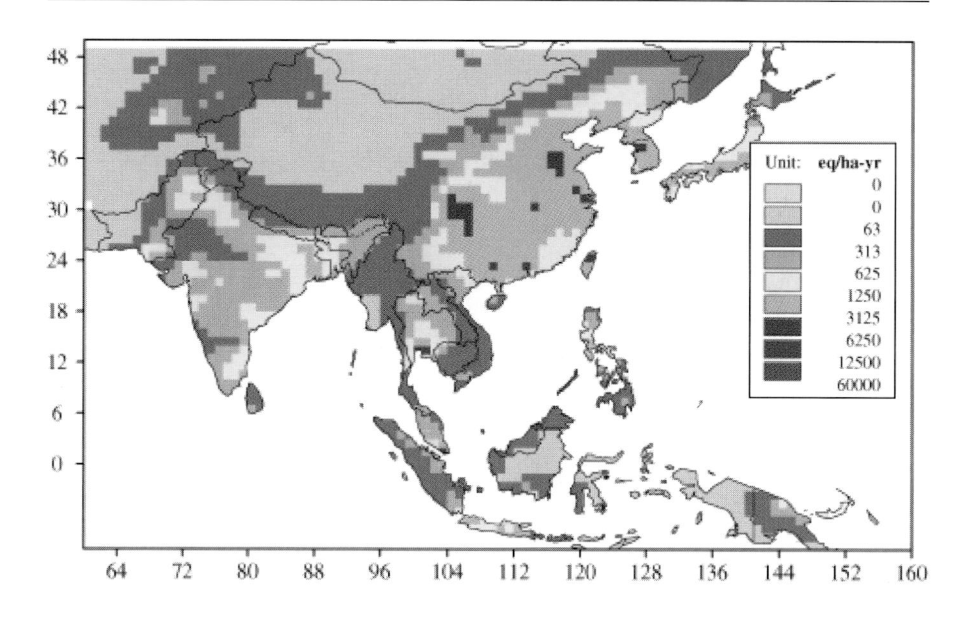

Figure 9.60 Annual sulfur depositions in low control scenario in 2020.

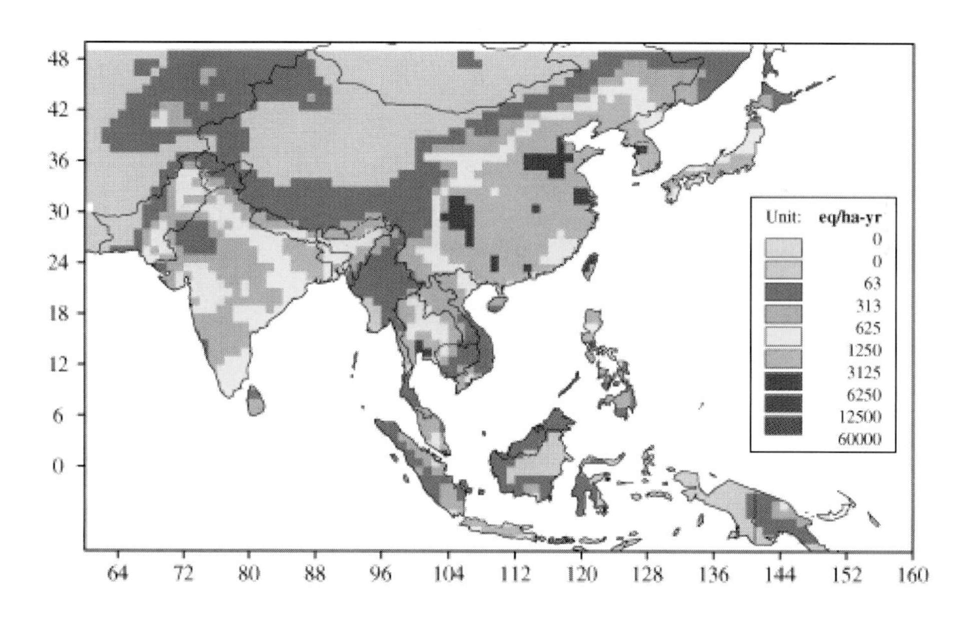

Figure 9.61 Annual sulfur depositions in low control scenario in 2030.

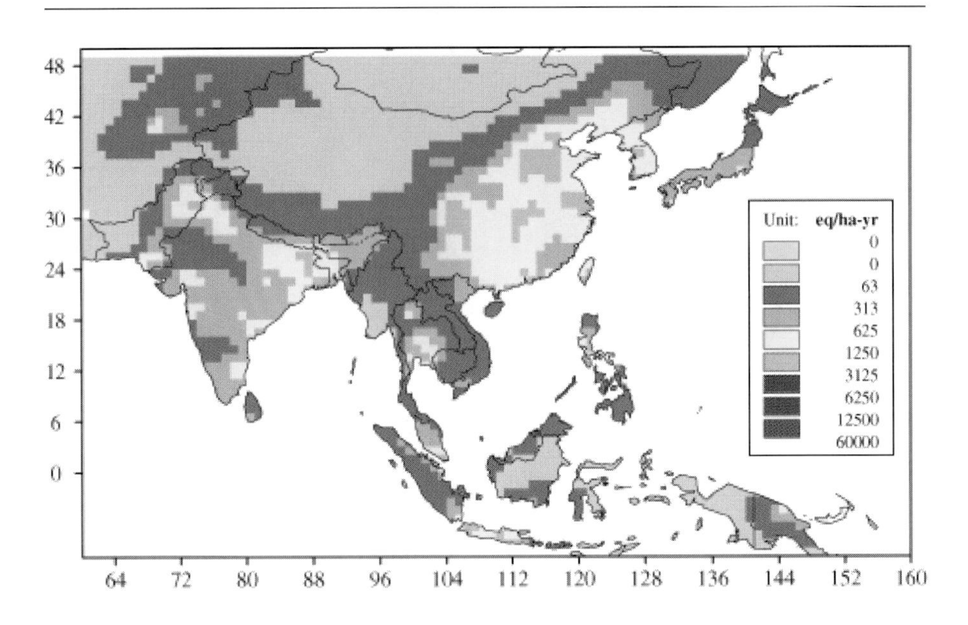

Figure 9.62 Annual sulfur deposition in moderate control scenario in 2020.

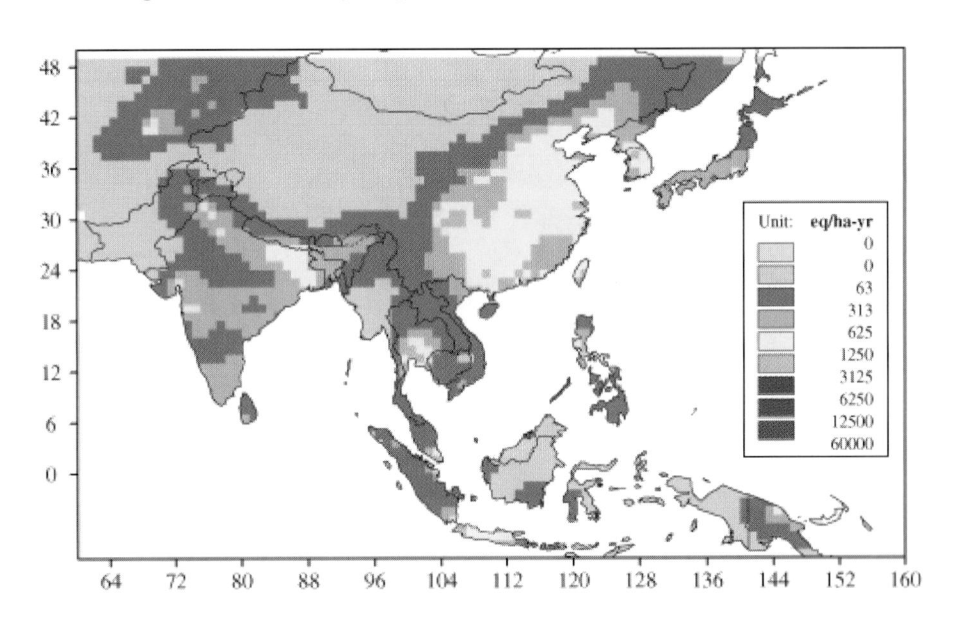

Figure 9.63 Annual sulfur deposition in high control scenario in 2020.

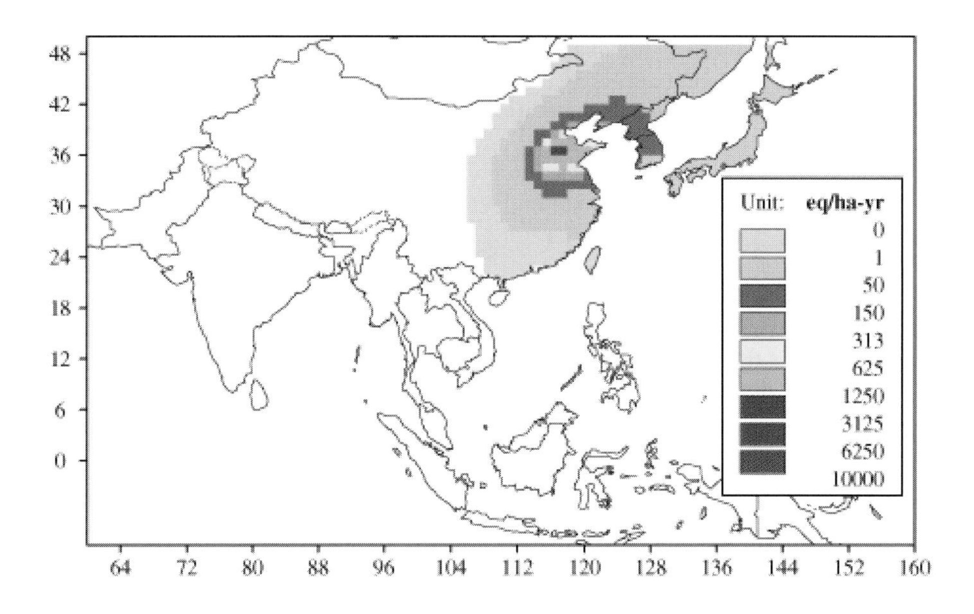

Figure 9.64 The depositions caused by Shandong sources.

4.5 Environmental impact assessment of control scenarios

4.5.1 The Rains-Asia impact assessment approach

The environmental impact of acidic deposition is generally described using the exceedance of the critical loads. Critical loads of acid deposition can be derived either directly from the relationship between atmospheric deposition and effects on "specified sensitive elements" within an ecosystem (ecosystem status) by experimental research, field observation and investigation, or indirectly from critical values for ion concentrations or ion ratios in the ecosystem, based on dose-response relationships between these chemical criteria and the ecosystem status (De Vries, 1993).

In Rains-Asia two methods were adopted to compute the critical loads simultaneously. One method, the relative sensitivity approach, provides a non-parametric quantitative assessment of sensitivity based on climatic factors, factors relating to geology, soil characteristics, vegetation type and land use. The other method, the Steady State Mass Balance Model, is based on a mathematical model to compute the critical loads from the requirement that the supply of acidic compounds should not exceed the ability of a system to buffer the acidity. The objective of applying two methods simultaneously was to (1) combine bio-geochemical considerations by means of the Steady State Mass Balance Model with other ecological aspects using the relative sensitivity approach; (2) assess the reliability of the computed critical loads by comparing the geographical distribution over the Asia region of critical loads to the distributional pattern of classes of relative sensitivity; and (3) ensure availability of an assessment of sensitivity in situations where lack of data would not have allowed computation of critical loads (Hettelingh et al., 1995). This approach was considered to be an improvement over the identification of critical loads and sensitivity classes in Europe where both methods were applied independently of one another (Hettelingh et al., 1992).

4.5.2 Results of critical load mapping

The cover of a 1° x 1° grid cell may contain more than one combination of vegetation classes and biogeochemical characteristics. In mapping critical loads for each grid cell a decision is needed about which ecosystem to represent. As a solution a cumulative distribution function (CDF) of critical loads in grid cell has been computed.

In Rains-Asia model, it has been decided to use a (low) quantile of the CDF $F(x)$ rather then the minimum critical load. Taking the q-th quantile critical load protects a (100 - q)th percentage of the ecosystems. Thus, if a 5-percentile (P5) critical load is chosen, 95% of the ecosystems in each grid will have a low risk of being damaged when acid deposition does not exceed this 5-percentile critical load.

Estimation of low percentile critical load is associated with rather high uncertainties because of the unreliability of the basic data, in general, and the data resolution, in particular. Taking these limitations and current data quality into account, P15 is recommended by the model developer and was chosen as the basis for the following analysis of deposition excess. Figures 9.65 and 9.66 show the P5 and P15 critical loads, respectively.

It can be seen from the figures that the most sensitive (0-500 eq/ha.year) areas are in south-western China such as most of Yun-nan province and parts of Gui-zhou province, the boundary regions of Henan and Hubei provinces, Zhejiang province, Fujian province, south parts of Jiangsu province, Guangdong province, Guanxi province, parts of the Tibetan plateau and in the southern part of the boreal forest that dips into northern China. There are other wide areas of high sensitivity (500-1000 eq/ha.year), from the parts of Henan province, to almost all the other regions in the south of Yangzi River, such as most of Hubei, Hunan, Jiangxi provinces. The high sensitivity regions are largely found on acidic soils and the ecosystem shows a low ability to buffer acidic deposition in the long term. The high critical load areas cover the mainly dry, desert regions; also rich agricultural lands such as in the north-western China and Inter-Mongolia are not overly sensitive to acidic deposition.

In Shandong province, the eastern hill areas are mainly brown soil and partly Albic bleached soil. The middle and south hill areas are brown soil. The north-western plains are Solonchak and damp soil. The north areas are mainly Solonchak soil. In the east, south and middle of the province, the ecosystem is moderately sensitive to the acid deposition, with the critical loads at 500-1000 eq/ha.year, while in the northern and western parts the ecosystem is less sensitive with the critical loads above 1000-2000 eq/ha.year.

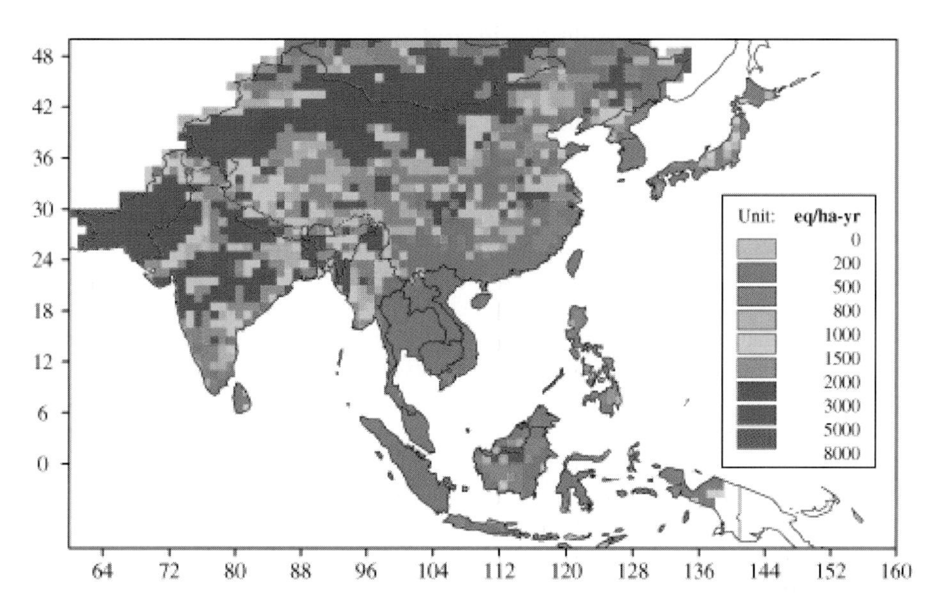

Figure 9.65 5-percentile critical loads.

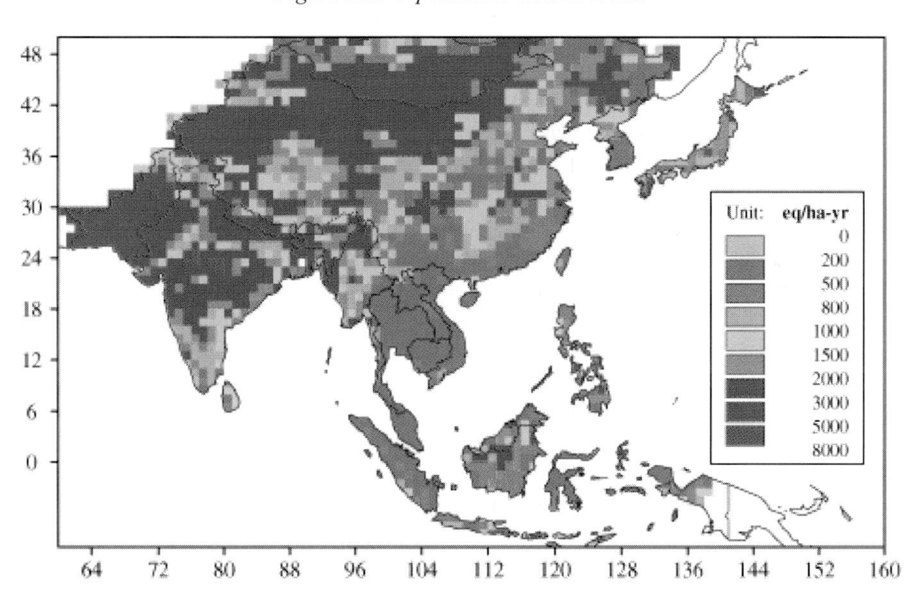

Figure 9.66 15 percentile critical loads.

4.5.3 Scenario analysis of excess acid deposition

Figures 9.67-9.70 show the results in years 2000, 2010, 2020 and 2030, respectively, under the low control scenario. Critical loads are widely exceeded in many regions. While the affected area will extend to the west and the north, the intensity will increase from average 0.5-1 keq/ha/year in 1995 (when 25% of the national ecosystems were at risk) to about 1.5 keq/ha/year by 2030 in affected areas. The south-western (Gui-zhou, Sichuang), the south (Guangxi, Guangdong, Hunan) and east (Shandong, Jiangsu, Shanghai, Zhejiang, Hubei, Jiangxi) of China will suffer a high excess deposition above 2 keq/ha/year. The hot spot further expands in Chongqing city to over 300 km range, and new hot spots will be formed in Guiyang, Shandong, Shanxi, and around Shanghai, spreading to Jiangsu. It should be noted that in this case, the excess deposition will reach unprecedented levels in some regions. Highest excess deposition (up to 5 keq sulfur per hectar per year) is found for some ecosystems in the Sichuan and Chongqing provinces. In accordance with the above national trend, Shandong will develop its excess deposition rapidly. By 2030, almost the whole province will suffer from ecosystem damage due to acid deposition; the southern, the middle and the south-western parts will be most affected with an excess deposition generally above 2 keq/ha/year.

Future excess deposition for the moderate control scenario is displayed in Figure 9.71 for the year 2020. The gains compared to the low control scenario are evident. The exceedance of deposition in Shandong will be 20-30% lower than the current level for the central, eastern and southern regions of the province. However, also the full implementation of the current policy will not eliminate the acidification problem.

Finally, Figure 9.72 displays the results for environmental impact under the high control scenario. In this case by 2010, the excess deposition will shrink to 10% of the national area. Most areas of Anhui, Fujian, and parts of Shanxi, Hubei, Jiangsu, Jiangxi, Zhejiang will reduce sulfur deposition below the critical loads. Meanwhile, the excess deposition intensity will decrease remarkably, from average range 0.5-1 keq/ha/year in 1995 to average 0.2-0.5 keq/ha/year. The hot spot with excess deposition over 2 keq/ha/year will disappear over the whole country, and most areas with acidification problem will be affected in a milder way with the exceedance of critical loads below 1 keq/ha/year. The patterns of excess deposition in 2020 and 2030 stay at about the same level as in 2010 since the emissions stabilize after 2010.

4.6 Optimization of Current Policy Scenario

In the preceding sections moderate control scenario was discussed based on the analysis of current source emission standards and control measures, and an assumption on the extension of the government's current blueprint to the future. Therefore, this scenario, and the resulting sulfur emissions and depositions are considered as reference developments. However, to achieve the same environmental targets, the current policy as simulated in this study, might not be the most cost-effective. The optimization module in Rains-Asia has been employed to identify the cost-effective combination of measures to control emissions and depositions, which would satisfy the goals of the current policy.

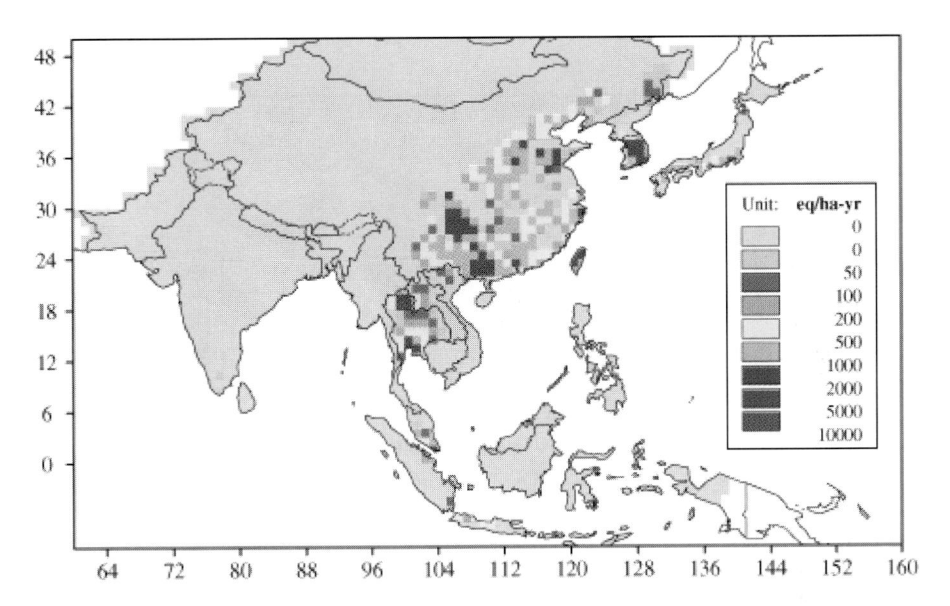

Figure 9.67 Deposition above critical loads in 2000; low control scenario.

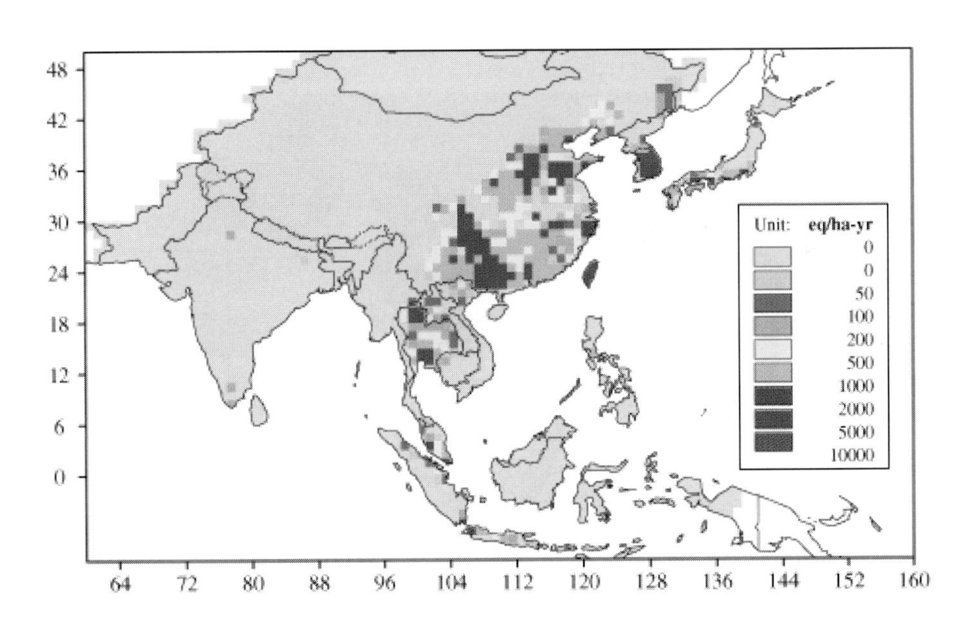

Figure 9.68 Deposition above critical loads in 2010; low control scenario.

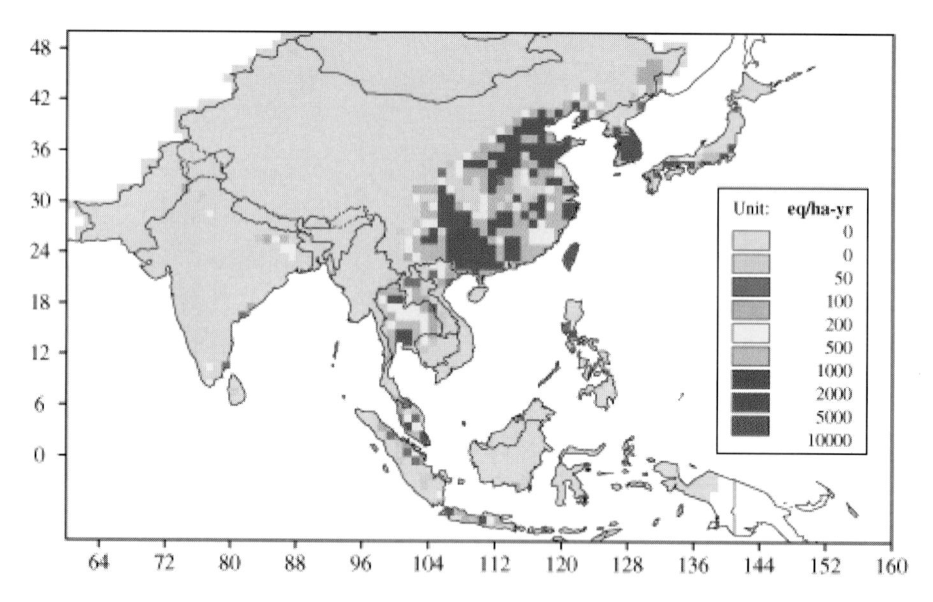

Figure 9.69 Deposition above critical loads in 2020; low control scenario.

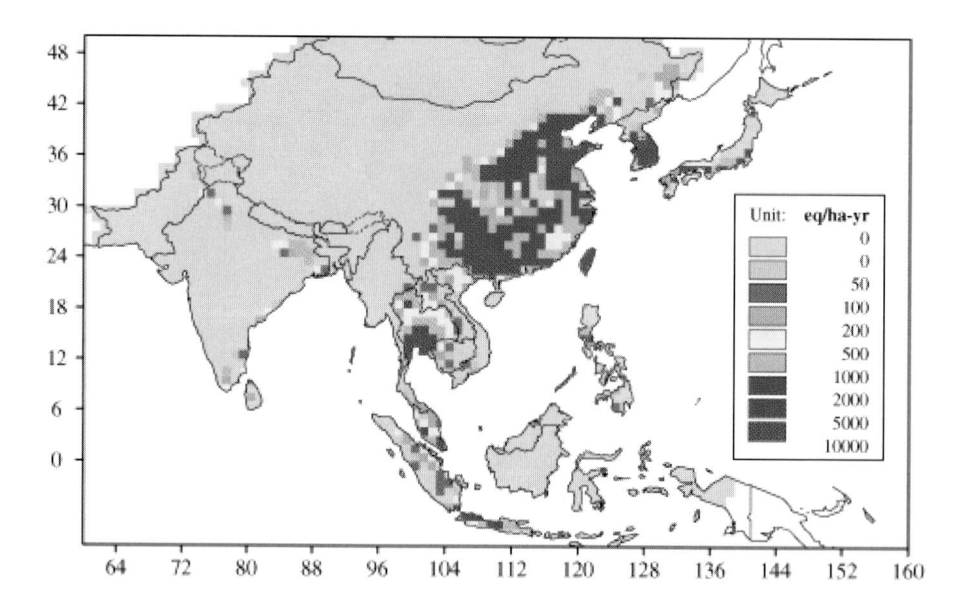

Figure 9.70 Deposition above critical loads in 2030; low control scenario.

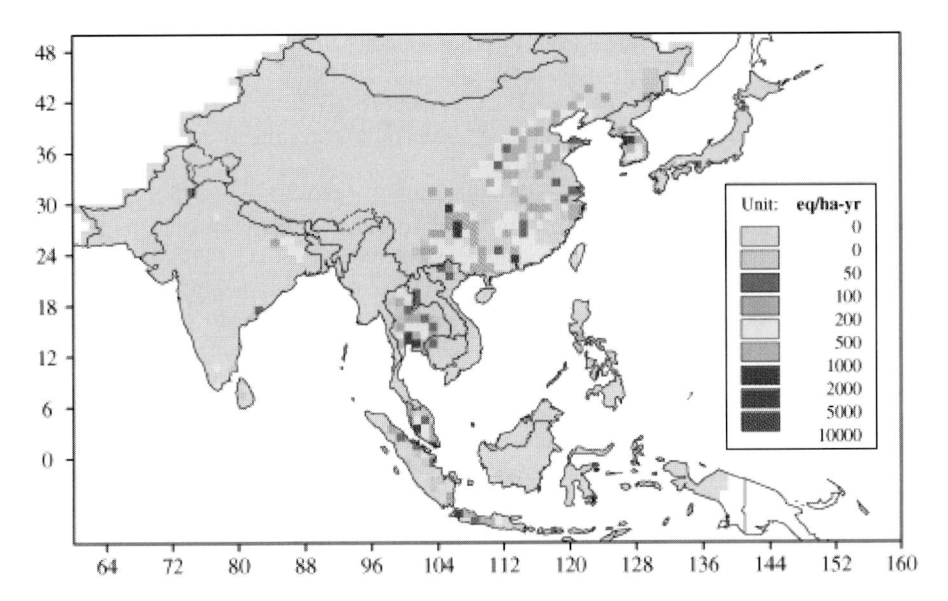

Figure 9.71 Deposition above critical loads in 2020; moderate control scenario.

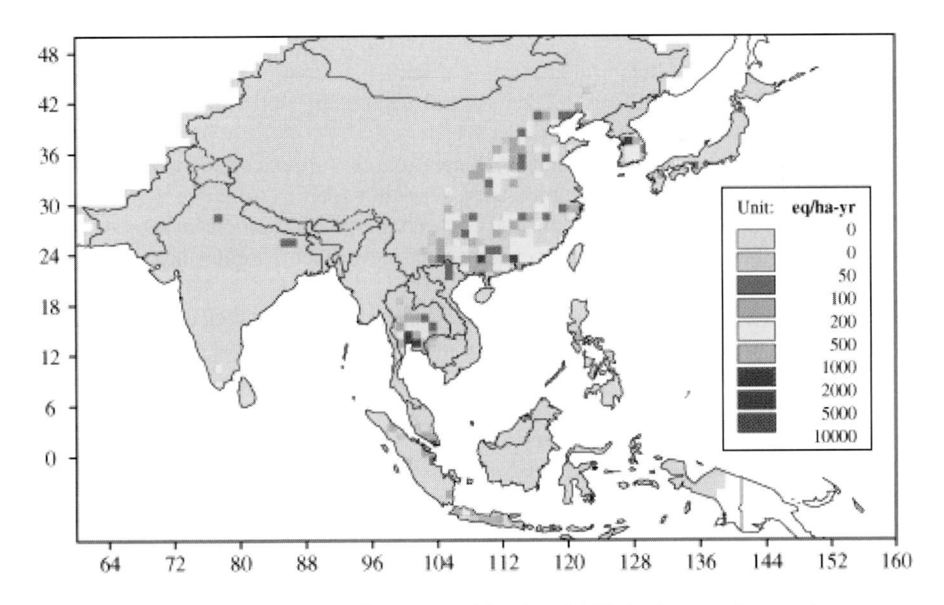

Figure 9.72 Deposition above critical loads in 2020; high control scenario.

4.6.1 Methodology of optimization

The optimization process identifies, for a given set of regional target depositions levels that can be derived from critical loads or other scenarios without optimization, the cost-minimal allocation of measures to reduce emissions. This optimization takes into account that some emission sources are linked via the atmosphere to sensitive receptors more strongly than others (as reflected in the atmospheric transfer matrices), and that some sources are cheaper to control than others.

In the RAINS-Europe model, optimization techniques have been used to identify the cost-minimal allocation of resources in order to achieve environmental targets for acidification, eutrophication and tropospheric ozone (Amann et al., 1999). In particular, for acidification the target was to reduce and ultimately minimize the gap between present deposition of acidifying substances and the ultimate policy goal. Similar to RAINS-Europe (Makowski et al., 1998), a similar optimization feature was implemented in RAINS-ASIA. Since RAINS-ASIA covers only sulfur pollution, the optimization is confined to the cost-effective abatement of sulfur dioxide emissions.

The optimization calculation needs the input of cost curves providing the costs of reducing emissions at the different sources (countries, sub-national regions, or large point sources (LPS), i.e. large power plants) for a selected year. The cost curves are compiled by ranking all available abatement options according to their marginal costs. Consequently, this methodology produces piece-wise linear curves, consisting typically of more than a dozen segments. By this curve, the minimum costs of achieving the emission reductions for each abatement level are identified, using the optimal cost-abatement combination.

Each cost curve is specific to each area source (e. g. provincial area), or each LPS source. In each cost curve, the figures for pollutant removed are incremental, i.e., they show the amount of pollutant abated using a particular control technology above the amount already removed using another control technology previously applied for the same fuel-sector combination.

In the present study China has totally 38 area sources (provinces and large cities) and 213 LPSs. For each source in each prediction year, the specific cost curve is generated in the Rains-Asia model. As an example, the cost curve for a Shandong area source and the cost curve for one LPS in Shandong – Zhou-xian power plant, are displayed in Figures 9.73 and 9.74.

In the figures, the blue curve represents the total costs accumulated, and the red curve represents the marginal cost for each control measure taken. The cost curves begin with the emissions determined by the control strategy adopted in a given emission abatement scenario. In this research, the purpose is to optimally reduce the emissions from the hypothetically unabated Low Control Scenario to achieve the same environmental impact as the current policy scenario. Therefore the cost curve of the low control scenario is adopted as the emission reduction concerns the low control scenario.

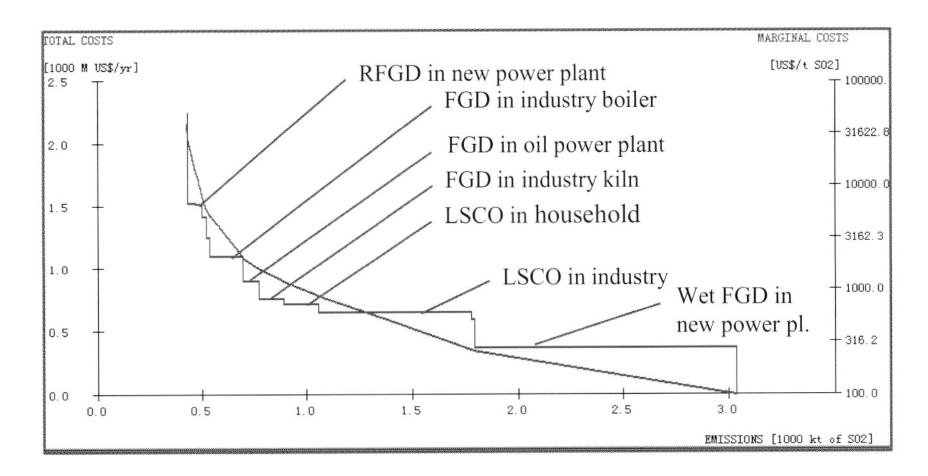

Figure 9.73 Cost curve of Shandong area source in 2030.

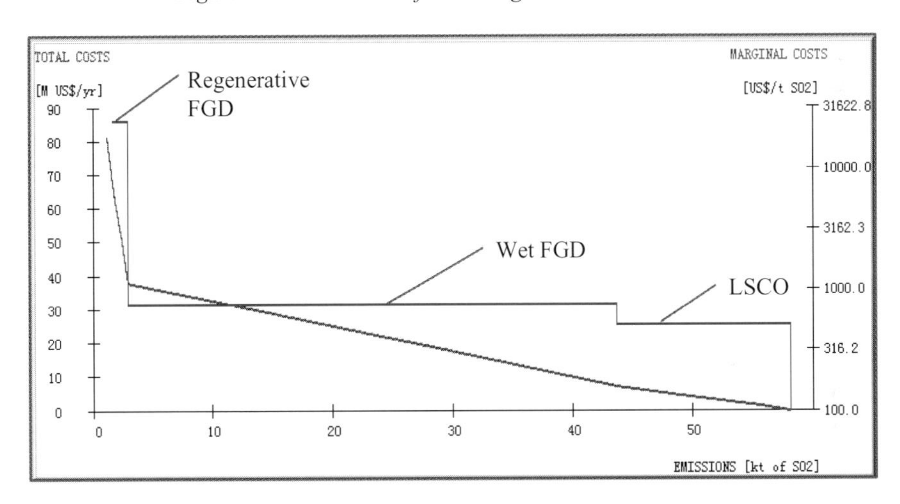

Figure 9.74 Cost curve of Zhouxian power plant in Shandong in 2030.

These two examples are for year 2030. For Shandong province by 2030, the low control scenario shows the total predicted emissions of 4382.69 kt, among which 3034.60 kt are from area sources, and 1348.09 kt from 18 LPS sources represented in RAINS-ASIA. Zhouxian power plant is one of the 18 LPSs in Shandong, and will emit 58.15 kt SO_2 by 2030, if no specific measures will be applied.

Possibilities of further controlling the emissions depend on the type of emission control measures assumed in the control strategy. For the area source shown in Figure 9.73, the least-cost measure is the wet FGD installation in the new coal fired power plants, with the marginal cost of 277 $/t$SO_2$, followed by the low sulfur coal use (LSCO) in industry (600 $/t$SO_2$), LSCO in households (729 $/t$SO_2$), FGD in industrial kilns (821 $/$tSO_2$), FGD in the oil-fired power plants (1211 $/$tSO_2$), FGD in industrial boilers (2088 $/$tSO_2$), and advanced FGD in power plants (6683 $/$tSO_2$). According to the total costs (blue curve), wet FGD in the new coal fired plants and LSCO use in industry are the most effective contributors to controlling the emissions, with the reductions of 1241 kt and 735 kt, respectively.

For the Zhouxian power plant, as displayed in Figure 9.74, three control measures are considered as the potential control technologies. The LSCO use is the most cost-effective with the marginal cost of 501 $/$tSO_2$, and the associated annual reductions can reach 14.42 kt SO_2. If the reductions due to LSCO use are not sufficient, the wet FGD installation has to be considered. The use of wet FGD costs 752 $/$tSO_2$, and can reduce SO_2 emissions by 40.83 kt.

4.6.2 Optimization results

Comparison between optimization scenario and current policy scenario. The optimization calculation was carried out for year 2020. The target is to achieve the same depositions as obtained in the moderate policy scenario (see Figure 9.62).

Deposition targets specify maximum sulfur deposition for selected grid cells in China. Targets are selected for all grid cells in China. Meanwhile, policy constraints can specify the minimum and/or maximum values of emissions in each region/LPS within which the optimization is allowed to modify emissions. In this research, no policy constraints are specified, the model thus takes the whole range of emission reduction options from the cost curves. It means that the optimization varies emissions between the upper and lower bounds specified in the cost curves between maximum and minimum technically feasible reductions. According to the set of target deposition levels specified for all groups of grid cells in China, the optimization identifies the cost-minimal allocation of emission reductions over the whole country, so that sulfur deposition is at all grids at or below the targets.

Figure 9.75 illustrates the comparisons of emissions and costs in the optimization scenario and current policy scenario. The result shows that, to achieve the same depositions as under the current policy scenario, the emissions in the optimization scenario are practically identical, i.e. about 18.2 Mt SO_2. However, the total costs obviously drop, from about 9 billion $/year in the current policy scenario to about 8 billion $/year after optimization, about 10% saving of total yearly costs. The optimization thus identifies the cost-minimal allocation of emission reductions over the whole country and between the different emission sectors, so that sulfur deposition is at all grids at or below the targets.

The comparison between two scenarios for Shandong province is shown in Figure 9.76. After the optimization, the total emissions required to achieve the deposition targets decrease to 1469 kt from 1584 kt SO_2 under current policy scenario, i.e. about 7.3% lower level . By the optimum allocation both over space and source sectors, the total costs in Shandong drop by 9.7%, compared to that in the current policy scenario. This again demonstrates the effectiveness of the optimization process used.

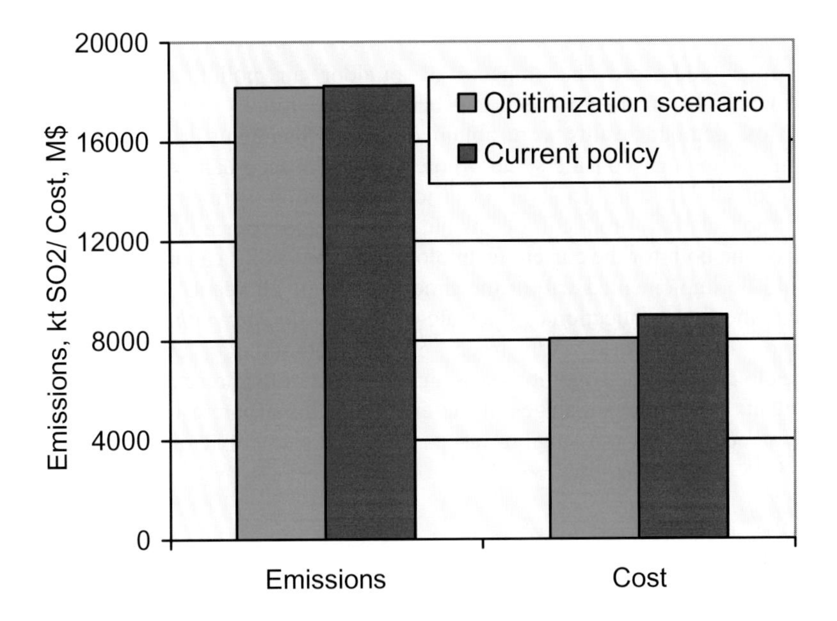

Figure 9.75 Reduction of control costs based on optimization for China (year 2020).

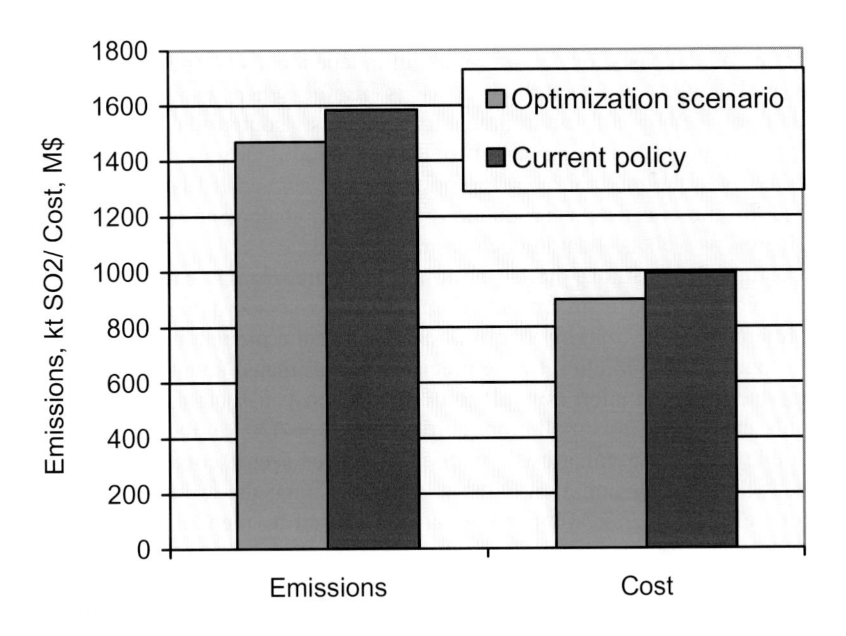

Figure 9.76 Reduction of control costs based on optimization for Shandong (year 2020).

5. CONCLUSIONS

This chapter addressed the impacts of outdoor air pollution on health and environment. The main emphasis has been on the rapidly expanding electricity sector. Most detailed results were obtained for the Shandong Province but a wide spectrum of findings considers the whole China. In accordance with the general objective of CETP *"true"* costs of electricity generation were estimated, reflecting not only the (internal) production costs but also damages to health and environment. This was done both for the current technologies as well as for candidate technologies that could be implemented within the time horizon of 20 years. The present study represents the first application of the state-of-the-art *"impact pathway"* approach to this region of the world. Adjustment of the methodology, existing tools and databases to the Chinese conditions was a central part of the present task. Apart from power plants other relevant stages of the energy chains of interest were considered. Impacts on ecosystems were analyzed separately in a study of acidification.

5.1 Monetized impacts

Examples of impacts estimated in this study include effects such as reduction in lifetime expectancy (expressed in terms of *"Years of Life Lost"*; YOLL), chronic bronchitis, and adverse effects on the environment such as loss of crops. In the present work, the assessed impacts were also used as the basis for monetary evaluation providing an estimate of corresponding welfare losses.

The estimated physical impacts are considered sufficiently robust and, if desired, can be used as the basis for decision-making independently of the monetary values. However, knowledge of external costs is useful when evaluating abatement measures, or setting of environmental standards; comparing external against pollution-control costs is a powerful tool for rational decision-making. By adding external costs to the internal ones, the principle of *"true costs"* is approached. This can lead to more efficient production processes, abatement measures for new technologies, as well as changing behavioral attitudes.

This study demonstrates that air pollution in China strongly backfires on the rate of economic growth. Health impacts dominate the assessed damages. Taking 1998 as the reference year[1], roughly one million Chinese die prematurely each year as a result of outdoor air pollution (more precisely we estimated about 9 Million YOLL per year due to air pollution from all sources). The costs associated with the damage to health caused by this pollution correspond to 6-7% of the GDP, which is comparable to the current growth rates. The power sector contributes about one quarter of the total air pollution damage costs in China; the estimated share of the Shandong electricity generation in the damage caused by the Chinese power sector is in turn about 20%.

Based on the *"Willingness to Pay"* (WTP) approach used in this study, the total mortality costs related to air pollution in China are almost three times higher than the corresponding costs of morbidity. The latter is dominated by chronic bronchitis and restricted activity days. Damages to crops, though significant, are low compared to health impacts.

The major part of the damage to health is caused by secondary particulates formed by chemical transformation of SO_2 and NO_x into sulfates and nitrates, respectively. In relative terms sulfates are dominant and SO_2 is thus indirectly the major contributor, followed by nitrates and primary particulates. The number of YOLL per ton of SO_2 emitted in China is on average almost seven times higher than the average for the European Union; for Shandong the corresponding factor is about 11.

Generally, there is a large spatial variation of health effects, with the highest damage occurring in areas with intensive industrial activities and high density of population. The distribution of impacts is also strongly affected by such factors as ammonia background concentrations in affected areas and by differences in meteorological conditions (e.g. region-specific dominant wind directions). For the current power plant sites in Shandong the difference in chronic mortality (expressed in YOLL per year and assuming same reference emissions of the major pollutants emitted), between the site with highest impacts (Liaocheng) and the site with lowest impacts (Weihai), corresponds to about a factor of four. For the reference plant in Jinan the pollutant emissions at the level of year 1998 cause about 25'000 YOLL per year.

The estimated average *"true"* cost of power generation by means of current power plants (including all relevant stages of energy chains) amounts to 12.8 US cents per kWh generated in Shandong, with the highly uncertain global warming contribution of 1.7 US cents per kWh not included. This cost is dominated (about 80%) by its external component. The contribution from *"rest of the chain"* is typically of the order of 5% for the current plants. For the whole Chinese power sector the assessed average external cost amounts to 3.9 US cents per kWh (only power plant contributions included; damages due to global warming not included); for Shandong the corresponding value is 9.9 US cents per kWh. Thus, the average external cost assessed for China is more than a factor of two lower than for Shandong, which is due to extensive use of hydropower in China, lower average damages per ton of pollutant emitted, and according to the official statistical records lower average emission factors for Chinese coal power plants compared to Shandong's. The estimates for China and Shandong are generally higher than typical for western Europe, depending on high emissions and impacts per unit of pollutant emitted; this is in spite of the (PPP) GNP-based reduction of the unit damage costs by a factor of seven in comparison with western Europe.

The highest external costs per kWh among currently operating plants in Shandong were estimated for the coal power plant in Nanding (21.2 US cents per kWh); lowest for the coal power plant in Weihai (3.8 US cents per kWh); the global warming contribution is included in these estimates.

The options for reducing the impacts in the future include use of coal with low sulfur content, implementation of FGD units, and for the new plants use of advanced clean coal power generation technologies (AFBC and IGCC), and/or diversification of power supply in Shandong by introducing natural gas combined cycle (CC) plants and nuclear energy. Taking the current plant in Jinan and the associated rest of the coal chain as the reference, use of low sulfur coal reduces the mortality (expressed in YOLL per GWh) by a factor of 1.7; implementation of FGD with 95% SO_2 removal efficiency by a factor of 4.4; replacement by AFBC by a factor of 8.0; replacement by IGCC by a factor of about 13; replacement by natural gas CC plant by a factor of about 52 and by nuclear power plant by a factor of 63. The strong dependence between technology and health damages is also reflected in the estimated *"true"* costs which range between 15.2 US cents per kWh for the reference current coal power plant in Jinan down to 3.3 US cents per kWh for the Advanced Light Water Reactor (ALWR). The corresponding estimates for the same power plant but using low sulfur coal is 10.3 US cents per kWh; for the same plant equipped with dry FGD 7.7 US cents per kWh; for AFBC 6.4 US cents per kWh; for IGCC 5.4 US cents per kWh; and finally for natural gas CC 4.3 US cents per kWh. All these estimates are for the Jinan site, with the exception of ALWR; external cost contributions from the rest of the chain are included in all cases.

Total costs estimates, if accepted by decision-makers, are highly attractive as directly comparative aggregated measures of system performance. It has been proposed by some authors that the total system-specific costs can serve as an integrated relative indicator of sustainability since they reflect the economic and environmental efficiency of energy systems. One objection to this proposition is that the social dimension, which plays a central role in the comparative assessment of energy systems, does not come to the surface when the systems ranking is purely based on the total costs. Taking the nuclear power as an example, issues like high level long-lived radioactive wastes or hypothetical severe accidents, contribute marginally to the estimated external costs. At the same time such issues remain controversial and, depending on the socio-political perspectives of those involved, can be of high importance for the decision process. This dimension of the decision problem may be pursued through multi-criteria decision analysis (MCDA), which combines acquired knowledge on technology performance with value-based judgments of the stakeholders. The present task provided one of the indicators used in MCDA as implemented in CETP, i.e. technology-specific reduction of lifetime expectancy (in YOLL per kWh). Thus, the monetized values were not used in this context, as they are in relative terms less robust. First, the step from physical damages to monetized ones is already dependent on value judgments; second, alternative methods for monetization exist and may lead to different results in absolute sense as demonstrated in sensitivity analysis (though the ranking is usually preserved).

The findings on impacts and external costs specific for a variety of technologies are also reflected in the analysis of electricity supply scenarios for the Shandong Province and the years 2000 to 2024. The scenarios of interest were selected from 18'144 scenarios modeled in the Electric Sector Simulation (ESS) task carried out jointly by MIT and ETHZ (see Chapter 6).

The merits of clean coal technologies in terms of the potential to reduce impacts and external costs, of FGD in particular, of fuel diversification (i.e. the use of natural gas and/or nuclear), and of demand-side management, are evident for all considered futures. Due to the low potential for hydropower and for other renewables in Shandong, natural gas and nuclear power represent the dominant diversification option. Future extensive use of natural gas as a fuel in the electricity sector is subject to major uncertainties with regard to the future availability of gas for such uses.

Widespread demand-side management is an attractive option whose feasibility and costs need to be further investigated. Some of the low-pollutant-emission scenarios characterized by substantial diversification (e.g. DUX-TONPAS, DUX-TONLAG) have the advantage of emitting much lower amounts of greenhouse gases (GHGs) than scenarios which do not rely on demand-side management and/or nuclear energy. Clean coal technology as such cannot significantly help in curbing GHG emissions, which will continue to increase as they are linked to the demand growth. Thus, supply-side strategies, which reduce the projected emissions of primary pollutants (in the first place SO_2), while limiting the increase in GHG emissions, are of high interest.

The contributions of the rest of the chain to the total damage costs of the analyzed scenarios vary between about 9 and 28%. The lower range applies to the "dirtiest" scenarios while the upper range is characteristic for the "cleanest" ones. In the latter case the power plant damage contributions have been reduced so much that the rest of the chain plays in relative terms a more significant role.

Typically, the total external costs of "clean" scenarios are a factor of three to four lower than for the most "dirty" ones (depending on whether damages due to global warming are included or not). The total (internal plus external) costs of environmentally friendly strategies based on clean coal technologies and diversification of supply are significantly lower than the seemingly lower costs of "dirty" and non-sustainable strategies using traditional coal technologies. According to the base case evaluation, the benefits in terms of avoided damages are for a strategy using conventional coal technology combined with consequent implementation of FGD 20 times higher than the associated increase in internal costs; in the case of a strategy using clean coal technology and diversification of fuels, the corresponding factor is about 17 (reference year 2020, FIB future).

The results of the complex evaluation of externalities are subject to major uncertainties. Thus, taking as an example chronic mortality which dominates the estimated external costs, the final result is affected by uncertainties in: pollutant emission data, atmospheric modeling, exposure response functions and most importantly their transfer to Chinese conditions, calculation of Years of Life Lost from mortality, latency, and population data. The geometric standard deviation for chronic mortality as estimated for China is 3.9 (without monetary valuation). On the basis of the latest not yet fully published research supporting dose-response functions, we have reason to believe that the mortality effects as estimated in the present research may have been underestimated.

Concerning the monetary valuation, sensitivity analyses were carried out employing alternative approaches, including the Human Capital method. The prevailing opinion of the international scientific community is that the latter approach is inappropriate; nevertheless, it has been employed in the sensitivity analysis since it is widely applied in China. The alternative approaches, particularly the Human Capital method, lead to lower external costs. However, the damages avoided by implementation of cleaner technologies exceed in all analyzed sensitivity cases the increase in internal costs, thus demonstrating the robustness of the findings. Also, we abstained in this study from adjusting the WTP-based valuation by the projected increase in GNP per capita. Given an annual growth in China in the range of 5-7% in the next 20 years, this would lead to even higher estimates of damages by a factor of about 2.5 to 3.5.

Based on the cost-benefit analysis the overall conclusion is that reduction of major air pollutant emissions due to electricity generation in China/Shandong, and of the associated health and environmental damage, is feasible, and economically and socially justified.

5.2 Acidification

The state-of-the-art assessment of external costs, does not provide a complete coverage of the potentially important damages. Thus, estimations of the impacts to the ecosystems (apart from agricultural losses), remain extremely uncertain. Some assessments of the costs of forest decline due to acid rain have been carried out in the past, but their validity is highly questionable. Since SO_2 deposition, and the resulting acidification, is one of the most important environmental issues in China, the CETP approach employed the most versatile tool for analyzing acidification at a regional level: the RAINS-ASIA (Regional Air Pollution Information and Simulation for Asia) model.

Major effort was invested to define the reference energy supply scenario for China. The default scenario provided in RAINS-ASIA was not considered relevant for the purposes of this study. The new version of the model was built on information from the reference year 1995, when energy consumption in China was steadily increasing year-by-year. This trend was decisive for the establishment of the default scenario, as implemented in the newest version of RAINS-ASIA. However, the energy consumption in China reached a peak in 1996 and then, for several reasons, started to decline over the following years. This unexpected drop in demand follows the sharp decrease in the consumption of coal. As a consequence, previous projections of energy consumption, not taking into account the recent developments, need to be corrected downwards. Electricity consumption again increased after 1996, though at a lower rate than before. When establishing the reference scenario for China, the trends in the development of energy intensity were evaluated, and then used for the projections.

The overall projected increase in the use of energy in China between the years 2000 and 2030 corresponds to about a factor two, while the corresponding increase for electricity is about 2.7. The reference power generation mix in the year 2030 exhibits, in comparison with the current situation, a significantly lower share of coal, somewhat increased share of hydro, a strongly increased share of nuclear, and increased (but in absolute terms small) contributions from natural gas and renewable energies. Natural gas, however, is expected to expand strongly in the context of the total energy use. The reference development of the electricity generation in Shandong is consistent with the moderate growth projections used in the ESS task within CETP, i.e. it is consistent with the values used in the EcoSense simulations. The reference scenario involves quite extensive fuel diversification, and thus it definitely does not represent the worst-case development.

Three emission scenarios, i.e. “Low Control”, “Moderate Control” and “High Control” cases, were established and analyzed. “Low Control” corresponds to the reference case. "Moderate Control" reflects current policy goals (assuming that these will be implemented), while "High Control" is the enhancement of these policies. The "Low Control" scenario corresponds to minimal control measures, and does not include implementation of emission standards, limitations of sulfur content in coal, or the use of clean coal technologies. "Moderate Control" includes measures such as national emission standards for power plants, industrial boilers and kilns; the introduction of SO_2 and acid rain control zones; the retirement of small, low-efficiency units; the use of high efficiency power generation technology; and the implementation of FGD at those power plants not currently using coal with relatively low sulfur content. "High Control" means the additional implementation of FGD for boilers, and more radical constraints of sulfur content in coal.

Under the "Low Control" Scenario SO_2 emissions double in the next three decades, with power generation having a 60% share of the total. A heavy regional acid deposition occurs in the Sichuang Basin and the east coastal China, expanding also to the neighboring areas and resulting in doubling of the current depositions in the affected areas. Critical loads calculations demonstrate that the areas with most sensitive ecosystems are unfortunately the ones most heavily exposed to SO_2 depositions. At present, about 25% of the total land area of China is exposed to high acidification risks. Low levels of pollution control will lead to further deterioration with about 40% of the total area exhibiting exceedance of critical loads. Emissions in Shandong province will be doubled in the next three decades, given low control level.

If the current policy as represented by the "Moderate Control" Scenario will be fully implemented, the SO_2 emissions in 2030 will be somewhat lower than in year 2000. Implementation of FGDs, and increased use of low sulfur content coal at power plants, is the most efficient measures for the reduction of emissions in the "Moderate Control" case. Due to the focused control in the two Control Zones the depositions in the most sensitive areas will drop by 2030 to 45% of the current level. Overall, the ecosystems will be protected at a roughly similar level as in 1990, in spite of the rapid economic growth and much increased energy use. Implementation of current policy will keep the emissions in Shandong in year 2030 roughly at the level of year 2000. The exceedance of critical loads Shandong will be, however, 20-30% lower than the current level.

The key factors for an enhanced policy represented by "High Control" Scenario are further improvements of emission controls in power generation, industry and households. Of particular importance are full implementation of FGD for new coal power plants and extension of control zones beyond the two defined by the current policy. As a result, the emissions in year 2030 drop by about one third below the current level and the depositions in the most sensitive areas to about 30%. Thus, high levels of control would reduce areas at risk to about 15% of the total land area. The "High Control" Scenario, with the enhanced control in power plants and industry in Shandong, will achieve a reduction of emissions from 2000 to 2030 by 45%, i.e. in relative terms larger reductions than for the whole of China.

Compared to the "Low Control" scenario in the year 2020, the estimated additional control costs for the "Moderate Control" scenario amount to about 9 billion US$ per year, and to 13 billion US$ per year for the "High Control" case, when the whole of China is considered. For Shandong, the corresponding estimates are 1.0 billion and 1.3 billion US$ per year, respectively.

Optimization of control measures was demonstrated to be feasible by means of cost-minimal allocation of emission reductions. This optimization takes into account that some emission sources are linked via the atmosphere to sensitive receptors more strongly than others, and that some sources are cheaper to control. Consequently, the optimization identifies the minimal-cost allocation of emission reductions over the whole country, and between the different emission sectors, so that sulfur deposition is at or below target levels for all readings. As a result of the optimization, it is possible to achieve the same target deposition for China as under the "Moderate Control" policy, but at 10% lower cost. Roughly the same cost reduction also applies to Shandong when optimization is employed.

It is worth noting that currently 30-40% of acid depositions in Shandong result from emissions in other provinces. Simultaneously, emissions in Shandong have significant impact on other provinces, particularly Jiangsu and Hebei.

5.3 Recommendations on future work

Improvements and extensions of the present work are feasible and recommended. This includes:

- *More detailed assessment of externalities for whole China, including differentiation between the various regions.* The current analysis is detailed only for Shandong. Corresponding implementations for other provinces are rather straightforward but require data collection efforts.
- *External costs of additional technologies.* This concerns technologies that were not explicitly considered in the current externality assessment due to relatively low relevance for Shandong (e.g. hydropower, wind, solar photovoltaic, decentralized cogeneration), or were beyond the scope of CETP (heating systems).
- *Analysis of external costs for the current situation and for energy supply scenarios in specific provinces, regions or in whole China.* Given the extensions covered in the two points above such analyses are feasible. Depending on the scope (electricity only or total energy supply) the externality assessment can be connected with approaches employed in EEM, ESS and ETM tasks of CETP. Generally, the object and scope of evaluation can be adjusted to the specific needs of stakeholders.
- *Extensions and updates of impact and external costs analysis reflecting emerging advancements of state-of-the-art.* This includes new findings on: dose-response functions, monetary valuation of health effects as well as global warming and impacts on ecosystems. Furthermore, the evaluation scope can be extended to include emissions to water and soil. All these aspects are currently addressed in European Union projects with extensive PSI participation. Appropriate transfers of the findings to the Chinese conditions are necessary.
- *Further acidification analysis.* Only one energy supply scenario was analyzed in combination with three emission control strategies. A more

extensive set of scenarios is of interest along with improvements of emission data. Also analyses using the optimization approach should be extended. Trans-boundary effects could be studied in more detail, calling for addressing also the current and prospective developments in neighboring countries. Given extensions of RAINS-ASIA to cover major pollutants other than SO_2, the corresponding analyses could be carried out along with the study of acidification.

REFERENCES

Abbey, D. E., Petersen, F. F., Mills, P. K. and Beeson, W. L. (1993). Long-term ambient concentrations of total suspended particulates, ozone and sulfur dioxide and respiratory symptoms in a non-smoking population. *Archives of Environmental Health*, 48 (1), 33–46.

Abbey, D. E., Lebowitz, M.D., Mills, P. K., Petersen, F. F., Beeson, W. L. and Burchette, R. J. (1995). Long-term ambient concentrations of particulates and oxidants and development of chronic disease in a cohort of nonsmoking California residents. *Inhalation Toxicology*, 7, 19-34.

Amann, M., Bertok, I., Cofala, J., Gyarfas, F., Heyes, C., Klimont, Z., Makowski, M., Schöpp, W. and Syris, S. (1998). *Emission reduction scenarios to control acidification, eutrophication and groundlevel ozone in Europe.* Report prepared for the 22nd Meeting of the UN/ECE Task-Force on Integrated Assessment Modelling. International Institute for Applied Systems Analysis (IIASA), Laxenburg, Austria.

Amann, M., Cofala, J., Heyes, Ch., Klimont, Z., Schöpp, W. (1999). The RAINS model: a tool for assessing regional emission control strategies in Europe. *Pollution Atmospherique*, December 1999, 41-63.

Amann, M. and Dhoondia, J. (2001). *User instruction and manual for RAINS-ASIA*. International Institute for Applied Systems Analysis (IIASA), Laxenburg, Austria.

Anderson, H. R., Ponce de Leon, A, Bland, J. M., Bower, J. S. and Strachan, D. P. (1996). Air pollution and daily mortality in London: 1987-92. *BMJ*, 312, 665-9.

Arndt, R. L., Carmichael, G. R. and Roorda, J. M. (1998). Seasonal source-receptor relationships in Asia. *Atmospheric Environment*, 32, 1397-1406.

Asia Development Bank (1993). *National response strategy for Global Climate Change: People's Republic of China.* Asia Development Bank Report.

Baker, C. K., Colls, J. J., Fullwood, A. E. and Seaton, G. G. R. (1986). Depression of growth and yield in winter barley exposed to sulphur dioxide in the field. *New Phytologist*, 104, 233-241.

Barrett, K., Seland Ø., Foss, A., Mylona, S., Sandnes, H., Styve, H. and Tarrasón, L. (1995). *European transboundary acidifying air pollution. Ten years calculated fields and budges to the end of the first Sulphur Protocol.* EMEP/MSC-W Report 1/95. Meteorological Synthesizing Centre - West, Oslo, Norway.

Bean, M. A. (2001). *Probability: The science of uncertainty.* Brooks/Cole, Pacific Grove, USA.

Bernow, S., Biewald, B. and Marron, D. (1990). *Environmental externalities measurement: quantification, valuation and monetization.* Tellus Institute, Boston, Massachussets, USA.

Brode, R. W. and Wang, J. (1992). *Users' guide for the Industrial Source Complex (ISC2) dispersion models.* Volumes I-III: EPA-450/4-92-008a. EPA-450/4-92-008b. EPA-450/4-92-008c. U.S. Environmental Protection Agency, Research Triangle Park, North Carolina 27711, USA.

CAES (1999). *Handbook of sustainable development index system of urban environment.* Chinese Academy of Environmental Science, Beijing: Chinese Environmental Press.

Carmichael, G. and Arndt, R. L. (1995). RAINS-ASIA: *An assessment model for air pollution in Asia. Chapter 5: Long range transport and deposition of sulfur in Asia.* Report on the World Bank Sponsored Project "Acid Rain and Emission Reductions in Asia", December 1995.

Carmichael, G. R., Calori, G., Hayami, H. and Uno, I. (2000). *Model intercomparison study of long range transport and sulfur deposition in East Asia.* Report draft, 2000.

Carmichael, G. R., Hayami, H., Calori, G., Uno, I., Cho, S. Y., Engardt, M., Kim, S.-B., Ichikawa, Y., Ikeda, Y. Ueda, H. and Amann, M. (2001). Model intercomparison study of long range transport and sulfur deposition in East Asia (MICS-ASIA). *Water, Air, and Soil Pollution*, 130 (1-4), 51-62
.

Chen Shijie (1999). The costs of health effect due to air pollution in Hangzhou. *Chinese Public Health*, 15 (4), 1999, 316-318.

China Statistical Yearbook (1998). *China Statistical Yearbook 1998.* State Statistical Bureau. Beijing, China.

China Population Statistics Yearbook (1998). *China Population Statistics Yearbook 1998.* State Statistical Bureau. Beijing, China.

Chongqingg Medical University (2000). *The health effect of air pollution and its cost in Chongqingg city.* The report of 9th FIY project, 2000.

CIESIN (2000). China Administrative Regions GIS Data: 1:1M, County Level, 1990. Center for International Earth Science Information Network. Retrieved 2000, from http://sedac.ciesin.org/china/.

Curtiss, P. and Rabl, A. (1996). Impacts of air pollution: General relationships and site dependence. *Atmospheric Environment*, 30, 3331-3347.

Dab, W., Quenel, S. M. P., Le Moullec, Y., Le Tertre, A., Thelot, B., Monteil, C., Lameloise, P., Pirard, P., Momas, I., Ferry, R. and Festy, B. (1996). Short term respiratory health effects of ambient air pollution: results of the APHEA project in Paris. *J Epidem Comm Health*, 50 (suppl 1), 42-46.

Derwent, R. G. , Dollard, G. J. and Metcalfe, S. E. (1988). On the nitrogen budget for the United Kingdom and north-west Europe. *Q.J.R. Meteorol. Soc.*, 114, 1127-1152.

Derwent, R. G. and Nodof, K. (1986). Long-range transport and deposition of acidic nitrogen species in North-West Europe. *Nature*, 324, 356-358.

De Vries, W. (1993). Average critical loads for nitrogen and sulfur and its use in acidification abatement policy in the Netherlands, *Water, Air, and Soil Pollution,* 68, 399.

Dockery, D. W., Speizer, F. E., Stram, D. O., Ware, J. H., Spengler, J. D., Ferries, B. G. (1989). Effects of inhalable particles on respiratory health of children. *Am Rev Respir Dis*, 139, 587-594.

Dusseldorp, A., Kruize, H., Brunekreef, B., Hofschreuder, P., de Meer, G. and van Oudvorst, A. B. (1995). Associations of PM_{10} and airborne iron with respiratory health of adults near a steel factory. *Am J Respir Crit Care Med*, 152, 1932-9.

ECON (2000). *An environmental cost model: A rapid assessment method with application to six cities.* Econ-report No.16, 2000.

EDGAR (2000). Emission database for global Atmospheric Research. Version 2.0, 1990. Retrieved from ftp://info.rivm.nl/pub/lae/EDGARV20/.

Eighth Five-year Project (1999). *National plan of regional acid rain control.* Research Report, Beijing, China9.

Encyclopedia Britannica (1998). *CD Britannica 1998.*

ESRI (1996). ArcChina, *Digital map database of China 1:1 000 000. International Version.* CD, National Buraeu of Surveying and Mapping China, ESRI.

ESRI (1998). *ArcView GIS Version 3.1, data set "country".* CD, ESRI.

ESEERCO (1995). *The New York State environmental externalities cost study.* Final Report EP91-50, prepared by Tellus Institute and RCG Hagler, Bailly, Inc., 11 Arlington St., Boston, MA, 02116 USA.

ETARCP (1999). *Regional acid rain control planning of China.* Editing team of acid rain control plan, Beijing, China.

European Commission (1995). *Externalities of Energy (ExternE). Volume 2: Methodology.* European Commission, DG XII, Science, Research and Development, Brussels.

European Commission (1999). *Externalities of fuel cycles. ExternE - Externalities of energy. Volume 7: Methodology 1998 update.* EUR 19083, European Commission, DG XII, Science, Research and Development, JOULE, Brussels.

European Commission (2000). *External costs of energy conversion – improvement of the ExternE methodology and assessment of energy related transport externalities.* Final report prepared for the European Commission, Contract JOS3-CT97-0015. IER, University of Stuttgart, Stuttgart, Germany.

Guo Xiao-ming and Zhang Huiqing (1990). Study on prediction and strategy of China's environment for the year 2000. Beijing: Tsinghua University Press.

GPCC (1998). The global precipitation climatology centre. Information on World Wide Web under http://www.dwd.de/research/gpcc.

Hao Jiming and Lu Yongqi (1999). Technical and economic evaluation of FGD use in coal fired power plants in China. China Environmental Industry, 1999, 5 (3), 45-61.

Hao Jiming, Tian Hezhong and Lu Yongqi (2002). Emission inventories of NO_x from commercial energy consumption in China 1995-1998. *Environmental Science & Technology*, 36 (4), 15 February 2002, 552-560.

Hettelingh, J.-P., Gardner, R. H. and Hordijk, L. (1992). A statistical approach to the regional use of critical loads, *Environmental Pollution*, 77, 177-183.

Hettelingh, J.-P., Chadwick, M. J., Sverdrup, H. and Zhao, D. (Eds.) (1995). *Assessment of environmental effects of acidic deposition.* Chapter 6 of the Report on the World Bank Sponsored Project "RAINS-ASIA: An Assessment Model for Air Pollution in Asia". (http://www.iiasa.ac.at/~rains/asia1/).

Hirschberg, S., Burgherr, P., Spiekerman, G., Cazzoli, E., Vitazek, J. and Cheng, L. (2003a). *Comparative assessment of severe accidents in the Chinese energy sector.* PSI Report. Paul Scherrer Institute, Wuerenlingen and Villigen, Switzerland.

Hirschberg, S., Dones, R., Gantner, U. (2000). Use of external cost assessment and multi-Criteria decision analysis for comparative evaluation of options for electricity supply. *In Proceedings of the 5th International Conference on Probabilistic Safety Assessment and Management*, Osaka. Tokyo: Universal Academy Press. (An updated version of this paper, employing the most recent damage functions was published in: PSI Annual Report 2000 – Annex IV).

Hirschberg, S., Heck, T., Gantner, U., Lu, Y. Spadaro, J. V., Krewitt, W., Trukenmüller, A. and Zhao, Y. (2003b). *Environmental impact and external cost assessment in the China Energy Technology Program.* PSI Report. Paul Scherrer Institute, Wuerenlingen and Villigen, Switzerland.

Hohmeyer, O. (1988).Social costs of energy consumption: External effects of electricity generation in the Federal Republic of Germany. New York: Springer-Verlag.

Huffman, G. J. and Bolvin, D. T. (2000). GPCP Version 2 Combined precipitation data set documentation. SSAI and Laboratory for Atmospheres, NASA Goddard Space Flight Center, Greenbelt, MD, USA. Retrieved 8 December 2000 from ftp://rsd.gsfc.nasa.gov/pub/gpcp-v2/doc/V2_doc.

Hurley, F. (2002). E-Mail communication, 14 June 2002.

Ives, D. P., Kemp, R. V. and Thieme, M. (1993). *The Statistical Value of Life and Safety Investment Research.* Report n°13, February 1993. Environmental Risk Assessment Unit, University of East Anglia, Norwich, England.

Jiang Zhesheng (1995). The idea on future development of clean coal power generation technology in China. *Clean Coal Technology*, 1 (1), 22–27.

Jin Jian-ming and Wang Jun-san (1994). *Ecological damage and its recovery in the typical ecological region.* The final report of the 7th FYP project.

Kalnay, E., Kanamitsu, M., Kistler, R., Collins, W., Deaven, D., Gandin, L., Iredell, M., Saha, S., White, G., Woollen, J., Zhu, Y., Leetmaa, A., Reynolds R., Chelliah, M., Ebisuzaki, W., Higgins, W., Janowiak, J., Mo, K. C., Ropelewski, C. and Wang, J. (1996). The NCEP/NCAR 40–year reanalysis project. *Bulletin of the American Meteorological Society*, 77 (3), 437–471.

Kistler, R., Kalnay, E., Collins, W., Saha, S., White, G., Woollen, J., Chelliah, M., Ebisuzaki, W., Kanamitsu, M., Kousky, V., van den Dool, H., Jenne, R. and Fiorino, M. (2001). The NCEP–NCAR 50–year reanalysis: Monthly means CD–ROM and Documentation. *Bulletin of the American Meteorological Society*. 82 (2), 247–268.

Krewitt, W. (1996). *Quantifizierung und Vergleich der Gesundheitsrisiken verschiedener Stromerzeugungssysteme.* PhD thesis. IER, University of Stuttgart, Germany.

Krewitt, W. (2002). External costs of energy – Do the answers mach the qestions? Looking back at ten years of ExternE. *Energy Policy*, 30, 839–848.

Krewitt, W., Heck, T., Boyd, R. and Eyre, N. (1997). *External costs from electricity generation in Germany and UK.* Final Report 1997, Contract JOS3-CT95-0002, JOS3-CT95-0010. The European Commission, Joule III, ExternE Maintenance / National Implementation / Aggregation.

Krewitt, W., Heck, T., Trukenmüller, A. and Friedrich, R. (1998). Environmental damage costs from fossil electricity generation in Germany and Europe. *Energy Policy*, 27 (3), 173–183.

Krewitt, W., Trukenmüller, A., Bachmann, T. M. and Heck, T. (2001). Country-specific damage factors for air pollutants. A step towards site dependent Life Cycle Impact Assessment. *Int. J. LCA*, 6, 199–210.

Krewitt, W., Trukenmueller, A., Mayerhofer, P. and Friedrich, R. (1995). ECOSENSE - An integrated tool for environmental impact analysis. In H. Kremers and W. Pillmann (Eds.), *Space and Time in Environmental Information Systems.* Umwelt-Informatik aktuell, Band 7. Marburg: Metropolis-Verlag.

Krupnick, A., Burtraw, D. and Palmer, K. (1995). The social benefits of social costing research. Prepared for the European Commission, International Energy Agency and Organization for Economic Cooperation and Development. Workshop on the External Costs of Energy, 30 - 31 January 1995, Brussels, Belgium.

Lee and Watkins (1998): Working paper for ExternE Project.

Logan, J. A. (1999). An analysis of ozonesonde data for the troposphere: Recommendations for testing 3-D models and development of a gridded climatology for tropospheric ozone. *Journal of Geophysical Research – Atmospheres*, 104 (D13) 16115–16149.

Lvovsky, K, Hughes, G., Maddison, D., Ostro, B. and Pearc,e D. (2000). *Environmental costs of fossil fuels: a rapid assessment method with application to six cities.* Environment Department Papers (No.78). The World Bank, October 2000.

Makowski, M., Heyes, Ch. and Schöpp, W. (1998). *The mathematical formulation of the ozone optimization problem in RAINS: A mathematical description of the non-linear optimization problem.* IIASA Technical Note. International Institute for Applied Systems Analysis (IIASA), Laxenburg, Austria.

Markandya, A. (1997). *Monetary valuation issues in extended ExternE.* Working paper prepared for the ExternE Project for the European Commission – DG XII.

Ning Da-Tong, Zhong Liang-XiChung, Yong-Seung (1996). Aerosol size distribution and elemental composition in urban areas of Northern China. *Atmospheric Environment*, 30 (13) 2355-2362.

ORNL&RfF (1994). *Estimating fuel cycle externalities: Analytical methods and issues.* Report No. 2, prepared by Oak Ridge National Laboratory and Resources for the Future. Washington D.C.: McGraw-Hill/Utility Data Institute.

Ostro, B. D. (1987). Air pollution and morbidity revisited: A specification test. *J Environ Econ Manage* 14, 87-98.

Ostro, B. D. (1994). *Estimating the health effects of air pollution: A methodology with application to Jakarta.* Policy Research Working Paper 1301. Policy Research Department, World Bank, Washington, D.C., USA.

Ottinger, R. L., Wooley, D. R., Robinson, N. A., Hodas, D. R. and Babb, S. E. (1990): Environmental costs of electricity. New York: Oceana Publications.

Pan Ziqiang, Chen Zhushou, Zhu Zhiming, and Xiu Binglin (1999). Preliminary research of health and environmental impacts and greenhouse gas emission from coal-fired power and nuclear power chains in China. *International Journal of Global Energy Issues*, 12, 257-270.

Pearce, D., Bann, C. and Georgiou, S. (1992). *The social costs of fuel cycles.* Centre for Social and Economic Research on the Global Environment, University College London, UK.

Pearce, D. (1995). *The development of externality adders in the United Kingdom.* Prepared for the European Commission, International Energy Agency and Organization for Economic Cooperation and Development. Workshop on the External Costs of Energy, 30 - 31 January 1995, Brussels, Belgium.

Ponce de Leon, A., Anderson, H. R., Bland, J. M., Strachan, D. P. and Bower, J. (1996). Effects of air pollution on daily hospital admissions for respiratory disease in London between 1987-88 and 1991-92. *J Epidem Comm Health* 50 (suppl 1) 63-70.

Pope, C. A. III and Dockery, D. W. (1992). Acute health effects of PM10 pollution on symptomatic and asymptomatic children. *Am Rev Respir Dis*, 145, 1123-1126.

Pope, C. A. III, Thun, M. J., Namboodiri, M. M., Dockery, D. W., Evans, J. S., Speizer, F. E. and Heath, C. W. Jr. (1995). Particulate air pollution as predictor of mortality in a prospective study of US adults. *Am J Resp Crit Care Med*, 151, 669-674.

Pope, C.A. III, Burnett, R. T., Thun, M. J., Calle, E. E., Krewski, D, Ito, K. and Thurston, G. D. (2002). Lung cancer, cardiopulmonary mortality, and long-term exposure to fine particulate air pollution. *Journal of the American Medical Association (JAMA)*, 287 (9), 1132-1141.

Rabl, A. (1998). Quantifying the benefits of air pollution control: The interpretation of exposure-response functions for mortality. *Journal of Hazardous Materials*, 61, 91-98

Rabl, A. and Eyre, N. (1998). An estimate of regional and global O_3 damage from precursor NO_x and VOC emissions. *Environment International*, 24, 835-850.

Rabl, A. and J. V. Spadaro (1999). Environmental damages and costs: An analysis of uncertainties. *Environment International*, 25, 29-46.

Rabl, A. Spadaro, J. V. and McGavran, P. D. (1998). Health risks of air pollution from incinerators: A perspective. *Waste Management & Research*, 16, 365-388.

Roemer, W., Hoek, G., Brunekreef, B. (1993). Effect of ambient winter air pollution on respiratory health of children with chronic respiratory symptoms. *Am Rev Respir Dis*, 147, 118-124.

Schwartz, J. and Morris, R. (1995). Air pollution and hospital admissions for cardiovascular disease in Detroit, Michigan. *Am J Epidem*, 142, 23-35; *Am J Epidem*, 137, 701-705.

Seinfeld, J. H. and Pandis, S. N. (1998). *Atmospheric chemistry and physics.* New York: Wiley.

SEPA (1998). Study on mapping of SO_2 and acid rain control zones in China. Report No. 96301. State Environmental Protection Administration, Beijing, China, March 1998.

SEPA (2002). *Technical regulations on the coal burning SO_2 emission control.* Science and Technology Bureau of SEPA, Beijing, China.

Shandong Electricity Management Bureau (1999). *FGD plan for power plants in Shandong from 2000-2010.*

Sinton, J. E. and Fridley, D. G. (2000). What goes up: recent trends in China's energy consumption. *Energy Policy*, 28, 671-687

SPC (1998). *Phasing-out plan of small and low efficiency units by 2000.* State Power Corporation, Beijing, China.

SPC (2001). *The 10th five year plan and long-term plan on environmental protection of power generation.* State Power Corporation, Beijing, 2001.

Spadaro, J. V. (1999). *Quantifying the damages of airborne pollution: Impact models, sensitivity analyses and applications.* Ph.D. Thesis. Ecole des Mines de Paris, Centre d'Energétique, 60 boul. St. Michel, F75272, Paris, Cedex 06, France.

Spivakovsky, C. M., Logan, J. A., Montzka, S. A., Balkanski, Y. L., Foreman-Fowler, M., Jones, D. B. A.,
Horowitz, L. W., Fusco, A. C., Brenninkmeijer, C. A. M., Prather, M. J., Wofsy S. C. and McElroy, M. B. (2000). Three-dimensional climatological distribution of tropospheric OH: update and evaluation. *Journal of Geophysical Research – Atmospheres*, 105(D7), 8931-8980.

Streets, D. G. and Waldhoff, S. T. (2000). Present and future emissions of air pollutants in China: SO_2, NO_x, and CO. *Atmospheric Environment*, 34, 363-374.

Sun Quingrui and Wang Meirong (1997). Ammonia emissions and concentrations in the atmosphere over China. *Scientia Atmospherica Sinica*, 21 (5), Sept.1997, 590-598.

Susskind, J., Pirano, P., Lokke, L., Iredell, L. and Meht, A. (1997). Characteristics of the TOVS Pathfinder Path A dataset. *Bulletin of the American Meteorological Society,* 78 (7): 1449-1472.

Suutari, R., Amann, M., Cofala, J., Klimont, Z. and Schoepp, W. (2001). From economic activities to critical load exceedances in Europe - An uncertainty analysis of two scenarios of the RAINS Integrated Assessment Model. Interim Report IR-01-020, May 28, 2001. International Institute for Applied Systems Analysis (IIASA), Laxenburg, Austria.

Tol, R. S. J. and Downing, T. E. (2000). *The marginal costs of climate changing emissions.* Institute for Environmental Studies, Amsterdam, the Netherlands.

Touloumi, G., Samoli, E. and Katsouyanni, K. (1996). Daily mortality and 'winter type' air pollution in Athens, Greece - a time series analysis within the APHEA project. *J Epidem Comm Health*, 50 (suppl 1), 47-51.

Trukenmüller, A. and Friedrich, R. (1995). *Die Abbildung der großräumigen Verteilung, chemischen Umwandlung und Deposition von Luftschadstoffen mit dem Trajektorienmodell WTM.* Jahresbericht 1995. ALS Universität Stuttgart, pp. 93-108.

Tsinghua University (1998). *Study on the mapping of acid rain control zone and SO_2 control zone of China.* SEPA research report 96301, 1998.

UNEP (1996). Incorporation of environmental consideration in energy planning in the People's Republic of China. UNEP Report. Beijing: Chinese Environmental Science Press.

WHO (1999). Guidelines for Air Quality. World Health Organization. Retrieved from http://www.who.int/peh/air/Airqualitygd.htm.

Wordley, J., Walters, S., Ayres, J. G. (1997). Short term variations in hospital admissions and mortality and particulate air pollution. *Occup Environ Med.*, 4, 108-116.

World Bank (1994). *China: Issues and options in Greenhouse Gas emission control.* World Bank Report. Washington D.C., USA.

World Bank (1997). *Clear Water, Blue Skies: China's environment in the new century.* China 2020 series. World Bank Report. Washington D.C., USA.

Wu, Z. X. and Chen, W. Y. (2001). *Cleaner energy strategy of coal dominated multi-energies.* Beijing: Tsinghua University Press

Xu Songling (1997). Economic loss due to the ecological damages in China in 1985 and 1993. *Ecology and Economy* (in Chinese), No.4.

Xu, X. (1998). Air pollution and its health effects in urban China. In: M. B. McElroy, C. P. Nielson, P. Lydon Peter (Eds.), *Energizing China.* Harvard University Press.

Yee, L. W. (1998). *Study of economic aspects of ambient air pollution on health effects.* Final report R0036-3.98. Prepared by EHS Consultants Limited for Hong Kong Environmental Protection Department. Retrieved on 19 Oct 2001 from http://www.info.gov.hk/epd/air/ehs/.

Zhou Yuexian and Li Hong (1999). The costs of health effect due to air pollution in Luoyang, *Journal of Environment and Health*, 16 (2), 65-68.

NOTES

[1] All results on health and monetary damages cited here refer to year 1998. Data on air pollutant emissions from the power sector in Shandong for year 1999 indicate that policy measures (in particular use of coal with lower sulfur content), resulted in significantly lower emissions and to corresponding reductions of the associated external costs.

CHAPTER 10

ASSESSMENT OF SEVERE ACCIDENT RISKS

STEFAN HIRSCHBERG, PETER BURGHERR,
GERARD SPIEKERMAN, ERIK CAZZOLI, JIRINA VITAZEK
AND LULIAN CHENG

1. INTRODUCTION

This chapter deals with the Risk Assessment Task within CETP. This task addressed the comparative assessment of accidents risks associated with the various electricity supply options. The full account of the work carried out within this task is provided in the technical report (Hirschberg et al., 2003a).

A reasonably complete picture of the wide spectrum of health, environmental, and economic effects, characteristic for the energy supply systems of interest, must include a consideration of potential accidents as well as damages due to normal operations, and coverage of full energy chains. Severe accidents in particular are considered controversial. For purposes of this chapter, these include potential or actual accidents that represent a significant risk to people, property, and the environment and that may lead to serious consequences. They may occur at fixed installations where hazardous materials are stored and processed or during transportation of such materials by road, rail, pipelines, open sea, and inland waterways. Hazards to be considered include fires, explosions, structural collapses, and uncontrolled releases of toxic substances outside the boundaries of hazardous installations.

Since the results of the accident analysis have implications for external cost assessment we refer to the preceding chapter, focused on health and environmental impacts due to air pollution and to the corresponding technical report (Hirschberg et al., 2003b).

2. OBJECTIVES AND SCOPE

The overall goal for this activity within CETP was to provide a balanced perspective on the severe accident risks specific to China. The assessment addressed fossil energy sources (coal, oil, and gas), nuclear power, and on a lower level of detail, hydro power also.

While the focus of this chapter is on severe accidents, small accidents are also addressed in order to provide a broad perspective on the accident issue. Given that coal is and will remain the dominant energy source in China for a long time, accidents associated with coal were investigated in more detail than were those associated with other fossil energy sources (oil, natural gas, and LPG). The analyses were primarily based on historical accident records for fossil energy chains and to a lesser extent for hydro power. The availability of China-specific data on severe hydro accidents was a major problem. This limitation of hydro analysis is not of primary importance for the purpose of this work since significant water resources in Shandong are lacking. Nevertheless, even though the findings are relatively less robust than for fossil fuel accidents, hydro accidents are covered here because hydro power is a very important electricity supply option for China.

For the nuclear chain, a probabilistic approach was employed because, fortunately, relevant statistical experience to support the evaluation of severe accidents does not exist. China has an ambitious program for the expansion of nuclear power. There are currently no nuclear plants in Shandong but the local utility is seriously considering building them.

Coalbed methane and renewables such as biomass, tidal power, wind, solar, and geothermal resources contribute only marginally to China's present energy mix. Furthermore, the issue of severe accidents is less relevant for these energy sources. For this reason they have not been analyzed in this chapter.

In the present work the analysis of accident risks is not limited to power plants but covers full energy chains, including exploration, extraction, processing, storage, transport, and waste management since accidents can occur at any of these stages. The study includes analysis of the specific chains followed by a comparison of severe accident fatality rates associated with each chain.

The present work builds on an earlier major PSI effort to carry out comparative assessment of severe accidents in the energy sector in industrialized and developing countries (Hirschberg et al., 1998; Hirschberg et al., 2001). A major limitation in the previous work, which constitutes the most comprehensive comparison of severe accidents associated with energy chains, was the unsatisfactory (highly incomplete) coverage of accidents in the Chinese coal chain. One of the central achievements of the present study is that this topic is now adequately covered. For details concerning the procedures used for the establishment of the severe accident database, analytical treatment, uncertainties and limitations, etc., we refer to our earlier study.

The period of time covered in the historical evaluation is 1969-1999; the earlier PSI study covered the period 1969-1996 (Hirschberg et al., 1998). However, in some cases the quantitative analyses for China address a more limited time period because of the availability of reliable accident records.

The literature provides several definitions of the term *"severe accident"*. All definitions include various consequences or damage types (fatalities, injured persons, evacuees, people rendered homeless, land loss, or economic loss) and a minimum damage level for each type. The differences between the definitions concern both the set of specific consequence types considered and the damage threshold (Hirschberg et al., 1998).

The PSI database ENSAD (Energy-related Severe Accident Database) uses seven criteria to define a severe accident in the coal, oil, gas, hydro, or nuclear chain:

1) at least five fatalities or
2) at least ten injured or
3) at least 200 evacuees or
4) extensive ban on consumption of food or
5) releases of hydrocarbons exceeding 10'000 t[1] or
6) enforced clean-up of land and water over an area of at least 25 km^2 or
7) economic loss of at least five million 1996 USD.

Whenever any one of the above criteria is satisfied, the accident is considered to be severe. However, in this report the quantitative analysis of severe accidents primarily refers to the first criterion above; information related to the other criteria was in most cases not available.

3. INFORMATION SOURCES

To establish a statistical basis for the comparative assessment of accident risks associated with coal and other energy chains that was as complete as possible, the research team used a variety of commercial and non-commercial sources. Although scope, quality, and time periods covered differed among sources, they can be classified into four broad categories.

Major commercial and noncommercial databases

- ENSAD: Energy-related Severe Accident Database of PSI (Hirschberg et al., 1998). ENSAD was the primary source for data valid for OECD countries and non-OECD countries other than China. ENSAD integrates data from a large variety of sources.
- MHIDAS: Major Hazards Incidence Data Service of the UK Health and Safety Executive (HSE);
- HSELINE: Library and Information Services of the UK HSE;
- NIOSHTIC: Bibliographic database published by the US National Institute of Occupational Safety and Health;
- CISDOC: a product of the International Occupational Health and Safety Centre (CIS) of the International Labour Organisation (ILO); and
- Reuters Business Briefing.

The databases ENSAD, MHIDAS, HSELINE, NIOSHTIC and CISDOC are presented in detail elsewhere (Hirschberg et al., 1998). Reuters Business Briefing is a database of over 5000 specialist trade publications, business magazines, market research reports, the world's leading newspapers, and local and global newswires from Reuters and Dow Jones. The content of Reuters Business Briefing is primarily worldwide, focusing on Europe, the Middle East, and Asia. The database is available through the internet at http://factiva.com/.

Books, journals, and periodicals

Books consulted include:

– China Coal Industry Yearbook,
– China Petroleum Industry Yearbook,
– A Complete Work of Coal Mine Safety in China,
– China Occupational Safety and Health Yearbook,
– Accident Cases of Typical and Major Casualties on Labour Safety, and
– Health at Home and Abroad.

The *China Coal Industry Yearbook* is published by the China Coal Industry Publishing House in Beijing. It contains detailed information on accidents in coal mining especially for the years 1994-1999. Data on failures of large dams related to hydro power generation are hard to obtain and rather incomplete. Information used in this work originates primarily from official Chinese sources, databases established by the International Commission on Large Dams, and numerous publications and journals.

Journals and periodicals include:

– Mining Journal,
– Coal Age,
– Coal Outlook,
– Miners News,
– World Coal,
– Water Power & Dam Construction,
– International Journal on Hydro Power & Dams,
– Hydro World Review,
– Lloyd's Casualty Week, and
– SIGMA.

The journals *Engineering and Mining Journal*, and *Coal Age* are published monthly by Intertec Publishing. Both journals provide information about world-wide production, cost-effectiveness and safety for mines and mineral processing plants. *Coal Outlook* is published weekly by Financial Times Energy and gives coverage of coal markets specifically as affecting the USA, prices, contracts, transactions, mergers, acquisitions, transportation, utility deregulation, company profiles, and labor trends. *Miners News* is a bi-monthly (six issues annually) newspaper covering North American Mining Company International activities including hard rock, coal, and aggregates. It provides information particularly on mining, investments, safety and international developments in the mining sector. *World Coal* is a technical magazine, covering the global coal industry and providing total industry coverage. It is published monthly and provides business and technical information on important aspects of the international coal industry from longwall mining, to coal preparation, to transportation. *SIGMA* is published approximately eight times a year by Swiss Re company. A survey of the largest catastrophes and damages is provided each year in January. *Lloyd's Casualty Week* (LLP, formerly Lloyd's of London Press) is accessible online at http://www.llplimited.com.

National and international Chinese newspapers in English

- *China Daily* (Beijing),
- *Sichuan Daily*,
- *South China Morning Post*,
- *Beijing Evening News*,
- *China News Digest*,
- *China Daily* (published in Beijing, owned by the Chinese government),
- *Free China Journal* (newspaper covering China, published twice a week),
- *China Quarterly* (covers all aspects of modern China studies),
- *China Focus* (monthly newsletter published by Princeton University),
- *New York Times*.

Internet sources

These include, among many others:

- *MuziNet* (Muzi Lateline News),
- *Inside China Today*,
- *China E-News*,
- Xinhua News Service,
- BBC World Service,
- ABC News,
- CNN News, and the
- British Association for Immediate Care (BASICS).

MuziNet, founded in 1996, is an English-Chinese bilingual portal site on the World Wide Web. It was created and is maintained by Muzi Company, an internet content provider, with a strong commitment to act as an information gateway between the East and the West. The internet address of the Muzi Company is http://lateline.muzi.net/. *Inside China Today* offers a daily stream of news, analysis, commentary, and cultural information for China. It was created by an internet publishing company "European Internet Network" (EIN) which was launched in 1995 (http://www.einmedia.com). The internet address of *Inside China Today* is http://www.insidechina.com. *China E-News* is a special section of the Advanced International Studies Unit (AISU) web site dedicated to news, research, statistics and links relating to energy efficiency, environment, and the economy in China. The internet address is http://www.pnl.gov/china/aboutcen.htm. Xinhua News Service (http://www.xinhuanet.com/) is the state news agency of the People's Republic of China. BBC World News (http://news.bbc.co.uk/), ABC News (http://abcnews.go.com/) and CNN News (http://www.cnn.com/) all offer extensive internet sites. The British Association for Immediate Care (BASICS) maintains a database that currently contains details of more than 6350 disasters and incidents around the world.
The internet address is http://www.basedn.freeserve.co.uk/.

4. EVALUATIONS FOR SPECIFIC ENERGY CHAINS

Using the above sources, the risks to the public and the environment, associated with various energy systems that arise not only at the power plant stage but at all stages of energy chains, were evaluated. In general, an energy chain may comprise the following stages: exploration, extraction, transport, storage, power and/or heat generation, transmission, local distribution, waste treatment, and disposal. However, one should be aware that not all these stages are applicable to every energy chain.

For detailed information on accidents in the various energy chains that occurred world-wide we refer to Hirschberg et al. (1998). Some of this data will be used here as points of reference and/or as indicative values for China when considered applicable. The following sections detail and compare accident information for China's currently or potentially most important energy chains, especially as they pertain to Shandong province.

4.1 Coal chain

China is both the world's largest producer and consumer of coal (IEA, 2001). Chinese hard coal production is only slightly higher than that of the USA. However, China's hard coal consumption is substantially larger than in any other country, amounting to nearly half of the world's consumption.

4.1.1 Dominant hazards in the coal chain

Coal mining is by far the most hazardous stage of the coal chain. Records of the UK coal-mining industry show that since 1850 over 100'000 deaths occurred as a direct result of accidents (Clifton, 1992). The majority of underground fatalities have been historically attributable to four types of accidents (*Encyclopaedia Britannica,* 1972):

- Falls of roof or rock bursts,
- Haulage activities (truck, loader, sweeper, conveyor, train forklift),
- Toxic or explosive gas, dust explosions and fires, and
- Operating or moving machinery.

Surface coal mines, such as trip or opencast and auger operations, inherently represent a much lower degree of hazard to workmen than underground mining. Consequently, the death rates from surface mines are much lower. The worst accident in the history of coal mining happened in 1931 in Manchuria, resulting in about 3000 fatalities (Hirschberg et al., 1998).

Accidents in the transport stage of the coal chain are rare compared to other energy carriers such as oil, natural gas, or LPG, in which a large number of severe accidents have occurred at that stage (Hirschberg et al., 1998). Severe accidents occasionally occur when trains transporting coal collide with other vehicles transporting explosives or flammable substances.

The worst accident in terms of fatalities at the coal transport stage world-wide occurred in 1983 and involved a coal freighter carrying 27'000 tons of coal. The freighter capsized and sank in storm-battered seas off the coast of Chincoteague, Virginia, USA, resulting in the death of 33 of the 36 crewmen (Hirschberg et al., 1998).

Severe accidents in the power station stage with large number of fatalities are rare. However, the monetary losses can be large (Hirschberg et al. 1998). An investigation of past fatal and nonfatal accidents around the world showed that risks come from coal dust which can be ignited due to insufficient ventilation in storage vessels at the plant, overpressure in boilers, blocked water inflow to condensers of the power station, or tornados.

Coal refuse is a waste material generated in the mining and preparation of coal. It is often piled into huge slag tips on hillsides or used to build tailing dams that frequently lack the structural features of an engineered dam. Parts of these tips can slip down the hills and destroy villages and miners accommodations and tailing dams can breach causing flooding downstream. The largest accident worldwide within the coal waste storage stage occurred in 1982 when an avalanche of coal wastes killed 284 workers in China (Hirschberg et al., 1998).

Mine types. Given that severe accident risks in the coal chain are dominated by mining, some relevant characteristics of Chinese coal mines are salient. Several categories of mines can be distinguished in China's coal industry:

- There are 94 key coal mines or "super pits" that fall under the control of the administration of the State Coal Industry Administration (SCIA), formerly the Ministry of Coal Industry (*China Labour Bulletin*, 2000). Generally, they are supplied with relatively advanced technology and equipment.
- An additional 2500 local state-owned mines are also controlled by SCIA or local governments at the provincial, county, or prefecture level.
- Local collectively or individually owned township and village coal mines are mostly of small size. Although several definitions of small mines exist, they generally have less than 50 workers, an annual production between 15'000 and 250'000 tons, and low levels of mechanization (*China Labour Bulletin,* 2000).

In this chapter, a "large" mine is a large or medium-sized mine owned by a central or local government; a "small" mine is a collectively or privately owned village or town mine.

In November 1998, Vice Premier Wu Bangguo urged the closing of small coal mines that were operating illegally or that represented redundant or inefficient capacity to reduce overall production by about 250 Mt and to help eliminate the production of low quality coal and pollution. According to official sources, about 25'000 small coal mines were shut down by the end of 1999. As a result, total production decreased by approximately 200 Mt from 1998 to 1999 (*CCIY,* 1999; 2000). However, many of the closed mines reopened in the absence of constant government supervision because the areas in which they are located are often poor and people are willing to risk their lives for a job (*Times of India,* 2001).

Large mines bear the financial burdens of relatively high wages and a large number of retired staff to care for. In contrast, small mines employ cheap migrant work force, lack even basic safety standards, and thus have low overhead costs. Closing all small mines would destabilize local economics and create huge unemployment. Small mines in China provide labor for about three million people or roughly 75 percent of all employees in the coal sector. Finally, small mines generate substantial local purchasing power and lead to demand for locally sourced inputs (food, equipment, tools, housing) when they are available, or encourage their production (*China Labour Bulletin,* 2000).

Several factors restrict the development of large mines. The overstaffed management, the considerable amount of employees not directly involved in the mining work, and the burden of a large number of retired workers result in high production costs and reduced efficiency, thus decreasing overall competitiveness. Productivity in terms of coal produced per number of employees in China is 12 times lower than in South Africa and 25 times lower than in the USA.

The above factors, i.e. the structure of the Chinese mining industry and lack of efficiency contribute greatly to its poor safety record. Furthermore, in an official report the ministry acknowledged widespread violations of China's 1992 Mining Safety Law, but said federal authorities were ill-equipped to enforce safety standards. The government has estimated that the industry should have as many as 500'000 safety inspectors, but at present it has only a few thousands experts, responsible for inspecting more than 200'000 mines (BBC News Online, 2000).

4.1.2 Severe accidents in China's coal chain

The officially acknowledged number of fatalities in China's coal mines is higher than the total number of people killed annually in mining accidents anywhere else in the world (compare Hirschberg et al., 1998). It also exceeds the average number of people killed by floods in China, which every year sweep across vast swaths of territory populated by millions of people *(BBC News Online,* 2000). Furthermore, the government acknowledges that another 10'000 miners die of lung diseases every year in addition to those killed in accidents (*BBC News Online,* 1998; 2000).

Severe accidents as reflected in different information sources. For the period 1969-1999, a total of 1016 severe accidents with a total of 17'241 fatalities in the Chinese coal chain were identified from a variety of information sources. Figures 10.1 and 10.2 show a comparison of severe accidents and fatalities originally contained in ENSAD, additionally collected for CETP, and those finally used within the CETP framework. Incompleteness of the ENSAD data for China was earlier recognized to be a weak point. Both figures clearly demonstrate that ENSAD data alone strongly underestimate the real situation in China, and that no accurate assessment is possible without data from the China Coal Industry Yearbook. Furthermore, data for China before 1994 were of rather poor quality in terms of completeness and consistency. This shortcoming has been resolved thanks to the major improvements in reporting since 1994. Annual publication of the China Coal Industry Yearbook provides a reasonably complete and comprehensive picture of the number of severe accidents and fatalities. Because of inadequacies in the data for specific evaluations within the CETP framework (e.g., severe vs. smaller accidents; separation of accidents by criteria such as energy chain stages or causes), only data of the period 1994-1999 were considered; these represent 816 severe accidents with a total of 11'321 fatalities.

Severe accidents by year. Figures 10.3 and 10.4 show the number of accidents and fatalities separated by mine type from 1969-1999 that occurred in China's coal chain. Both figures exhibit two distinct peaks for 1995 and 1997 for large and small mines, respectively. The large increase in the number of accidents and fatalities since 1993 is probably due to improved reporting by the Chinese Ministry of Coal Industry.

The development of small mines took place in two steps. First, small mines began to develop with China's economic structural reform in the 1980s. Second, a much more dramatic increase followed upon the government's release of the indicated coal price in the early 1990s. Furthermore, the data support the premise that very poor safety standards in small mines result in higher accident and fatality rates.

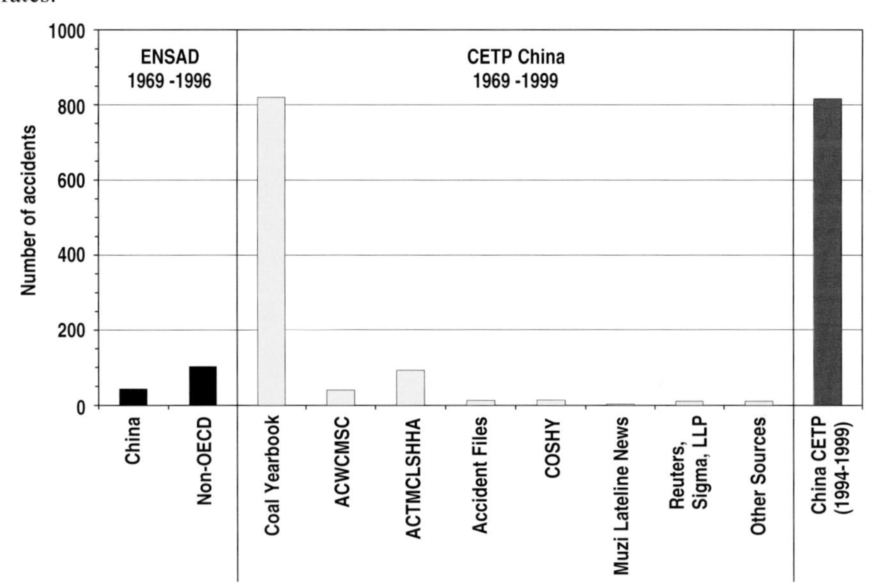

Figure 10.1 Cumulative number of severe accidents that occurred in the Chinese coal chain according to various sources. Black bars represent severe accidents contained in ENSAD. Yellow bars represent those additionally collected from various other sources during CETP. The red bar gives the number of accidents used for statistical evaluations within the CETP framework. Abbreviations used: ACWCMSC: A Complete Work of Coal Mine Safety in China; ACTMLSHA: Accident Cases of Typical and Major Casualties on Labour Safety and Health at Home and Abroad; COSHY: China Occupational Safety and Health Yearbook.

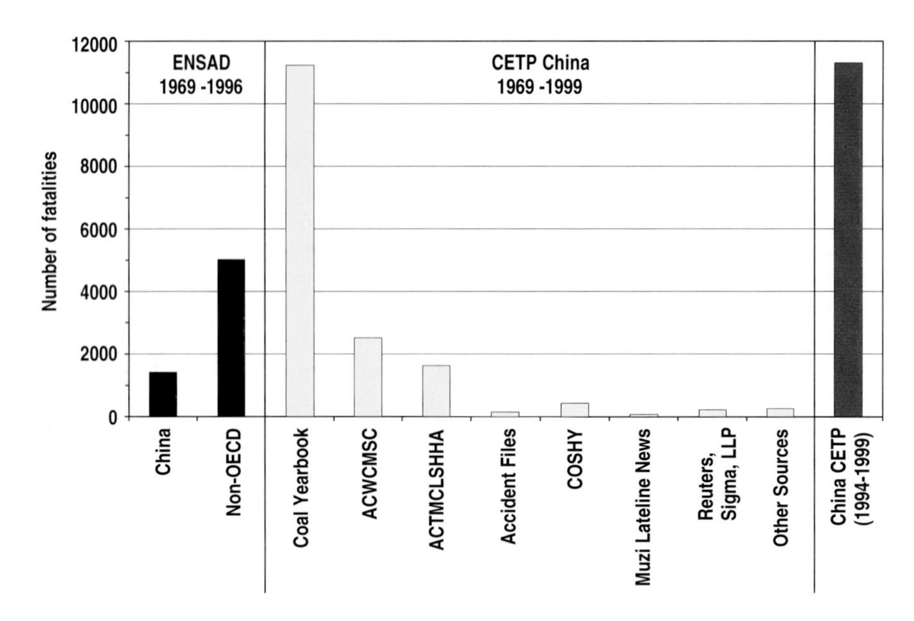

Figure 10.2 Cumulative number of severe accident fatalities that occurred in the Chinese coal chain according to various sources. Black bars represent fatalities from severe accidents contained in ENSAD. Yellow bars represent those additionally collected from various other sources during CETP. The red bar gives the number of fatalities used for statistical evaluations within the CETP framework. Abbreviations see Figure 10.1.

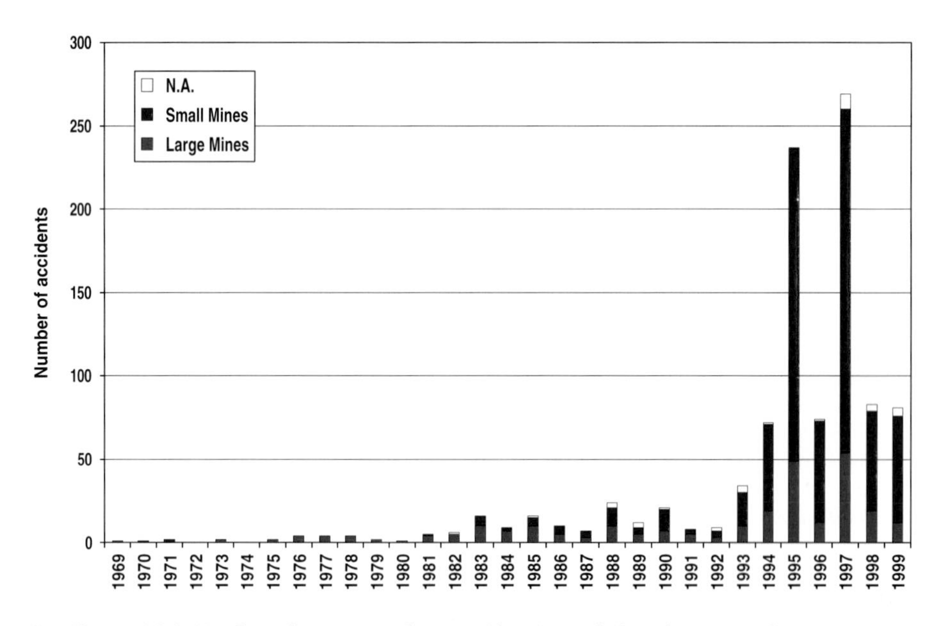

Figure 10.3 Number of severe accidents in China's coal chain by year and mine type.

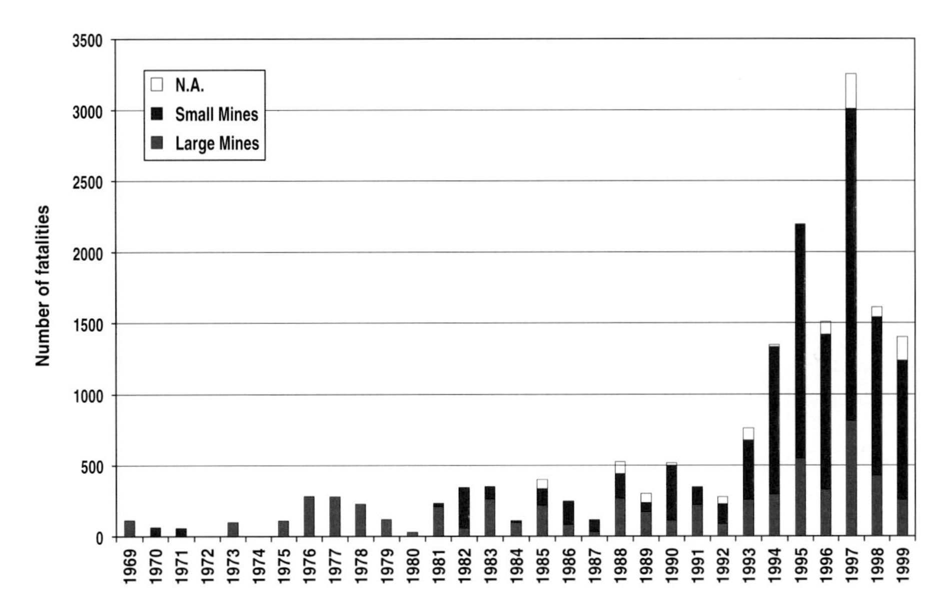

Figure 10.4 Number of severe accident fatalities in China's coal chain by year and mine type.

Severe versus smaller accidents. On average, accidents in China's coal chain resulted in 6200 fatalities per year for the period 1994-1999, of which severe accidents comprised 30 percent (Figure 10.5). Compared to the USA this is a much higher share. Averages per decade for the USA are given in Table 10.1. Values range from a minimum of 1.8 percent (1990-1999) to a maximum of 10.6 percent (1980-1989).

A detailed analysis of smaller accidents was not performed for three reasons. First, severe accidents are the focus of this chapter. Second, severe accidents are usually better documented than accidents with minor consequences; reports for severe accidents are likely to be more complete. Third, only cumulative data were available for smaller accidents which does not allow detailed statistical evaluations and comparisons.

Figures 10.6 and 10.7 show the number of fatalities per Mt of produced coal for severe and smaller accidents according to mine type. Overall patterns for severe and smaller accidents were quite similar, although variations were somewhat different. On average, fatality rates for small mines were about five times higher than for large mines. Furthermore, accident fatality rates were about 2.5 times higher for smaller accidents compared to severe accidents.

Finally, fatalities per Mt of coal in severe and smaller accidents by province were calculated (Figure 10.8). Values for individual provinces were only available for 1997-1999. Fatality rates were lowest in the large mines of provinces such as Shanxi, Shandong, Henan, Inner Mongolia, and Heilongjiang, the provinces that are most important for China's coal production. Correspondingly, provinces with low coal production (e.g., Zhejiang, Qinghai among others) had distinctly higher fatality rates. Differences in mechanization levels and safety standards may account for this finding, but lack of extensive data allows no validation of this hypothesis.

Severe accidents at different stages of the coal chain. Figures 10.9 and 10.10 show the number of accidents and fatalities according to the different stages in the coal chain for the years 1994-1999. The majority of accidents and fatalities occurred in the extraction stage. Exploration ranked second, though making a substantially lower contribution, while the transport stage contributed only marginally. For the period considered, no severe accidents were reported at the power station, heating, or waste storage and disposal stages. According to Hirschberg et al. (1998) severe accidents at these stages are rare world-wide. Table 10.2 lists the most severe accident for each coal chain stage that happened in China between 1969 and 1999.

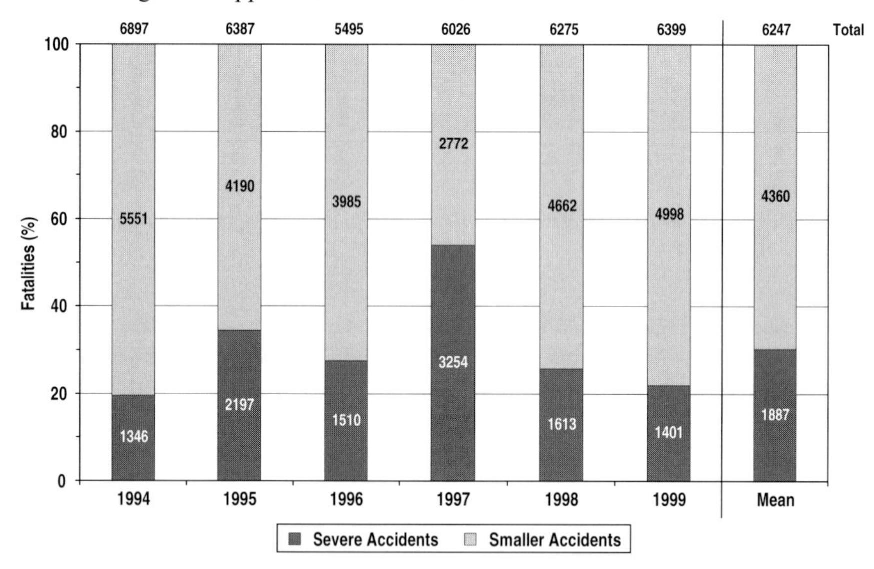

Figure 10.5 Shares and absolute number of fatalities in severe and smaller accidents that occurred in China's coal chain from 1994 to 1999. The mean value for this period is also given.

Table 10.1 Share of severe accident fatalities compared to all accidents in the US coal industry for the period 1950-1999.

Time period	Share of fatalities in severe accidents (%)
1950-1959	6.7
1960-1969	5.8
1970-1979	5.9
1980-1989	10.6
1990-1999	1.8

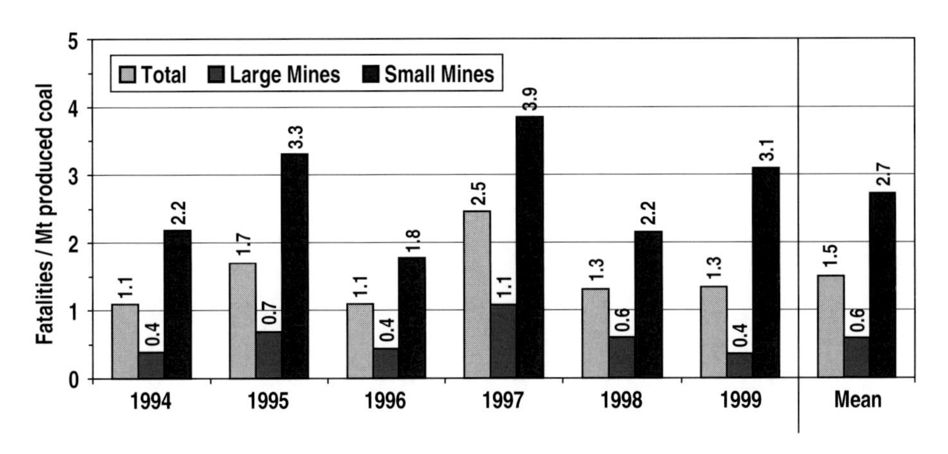

Figure 10.6 Fatalities per Mt of produced coal in severe accidents according to mine type for the years 1994-1999. The average value for this period is also given.

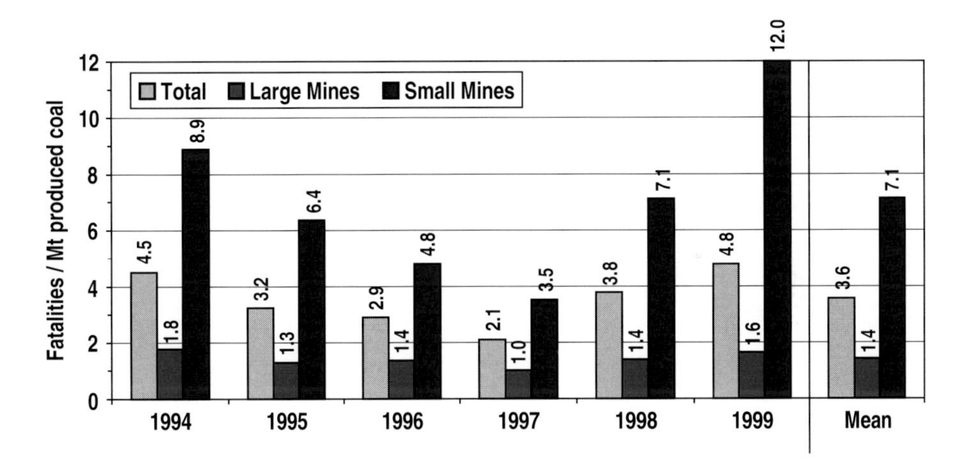

Figure 10.7 Fatalities per Mt of produced coal in smaller accidents alone according to mine type for the years 1994-1999. The average value for this period is also given.

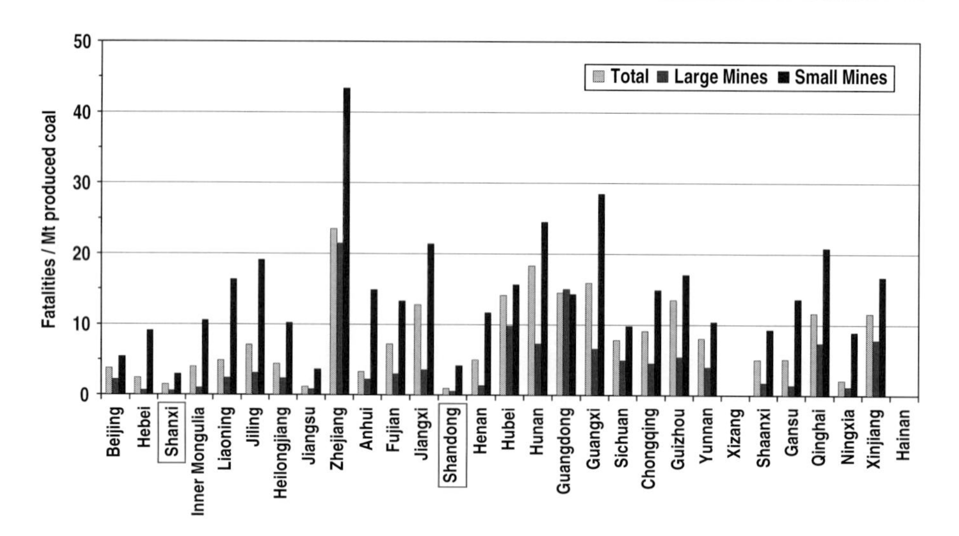

Figure 10.8 Fatalities per Mt produced coal in severe and smaller accidents according to mine type for individual provinces Values represent means for 1997-1999. Provinces marked by a frame are discussed in more detail below.

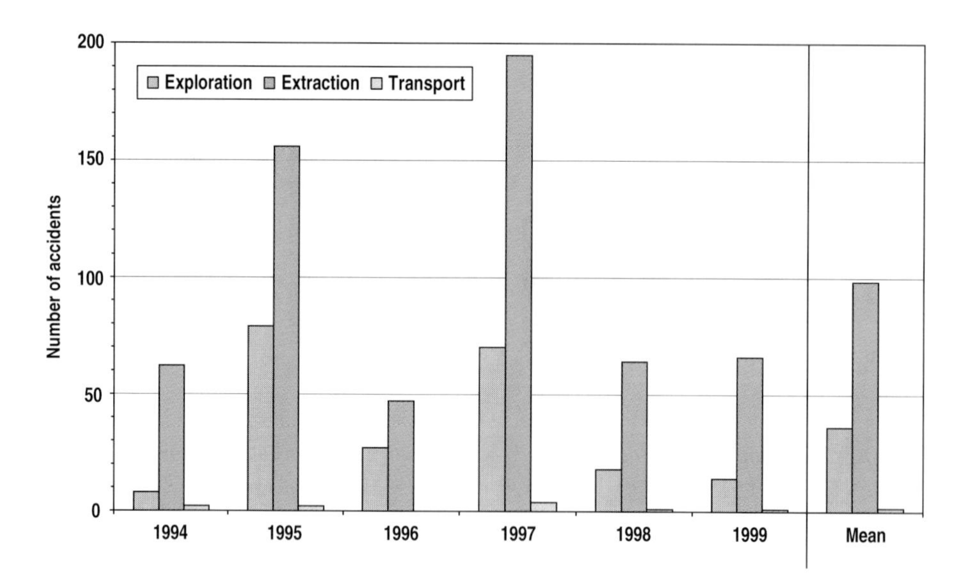

Figure 10.9 Number of accidents at each stage in the coal chain for the years 1994-1999. The average value for this period is also given.

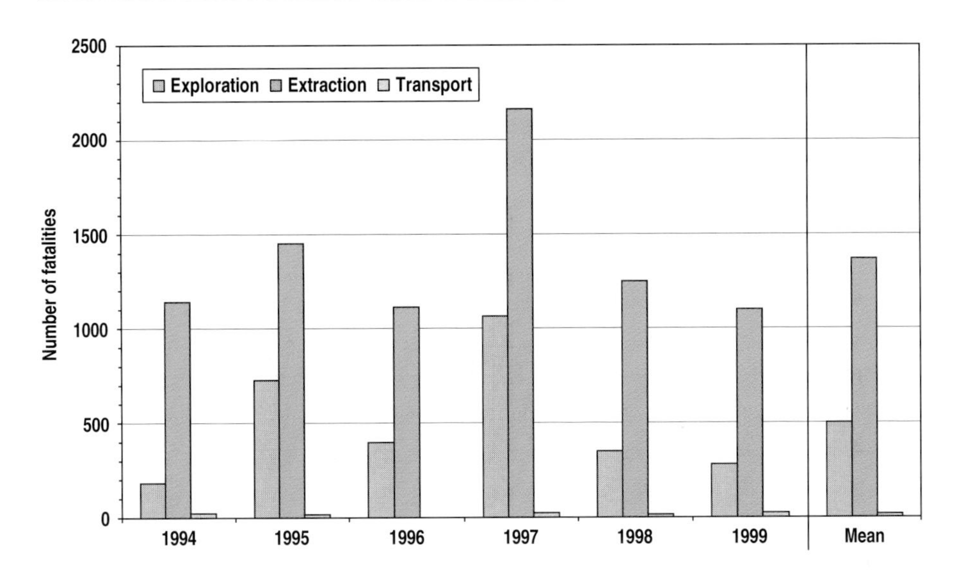

Figure 10.10 Number of fatalities at each stage of the coal chain for the years 1994-1999. The average value for this period is also given.

Table 10.2 The most severe accidents in each coal chain stage during 1969-1999.

Date	Location	Province	Energy chain stage	Number of Fatalities
21.04.1991	Hongtong	Shanxi	Exploration	148
24.12.1981	Pingdingshan	Henan	Extraction	133
31.10.1989	Bohai Bay	Offshore	Transport	30
10.03.1993	Beilungang	Zhejiang	Power Generation	20
1982	Unknown	Unknown	Waste Storage	284

General causes of severe accidents. Figures 10.11 and 10.12 show the number of accidents and fatalities from different causes for 1994-1999. Gas accidents during exploration and extraction dominated in terms of the number of accidents (80 percent) and fatalities (83 percent). Water hazard accidents ranked second, followed by roof, fire, and transport accidents. Electromechanical, blast, and other accidents were negligible. A study conducted in state-owned mines between 1988 and 1993 yielded similar results (Wang and Xu, 1996). On average, gas explosions caused 66 percent of all accidents, and 77 percent of the associated fatalities.

Detailed data on severe accidents makes it possible to assign sub-categories (i.e., specific causes) to each general cause. For example, fatalities in a gas accident could be due to gas asphyxiation (poisoning), gas (coal dust) explosion (ignition), or coal (rock) and gas outburst. Considering mine management and safety standards, the investigations yielding such data may help improve conditions specifically and in general. Relatively simple measures such as training of workers, adequate supervision, maintenance of equipment, and replacement of outdated equipment may have significant effects.

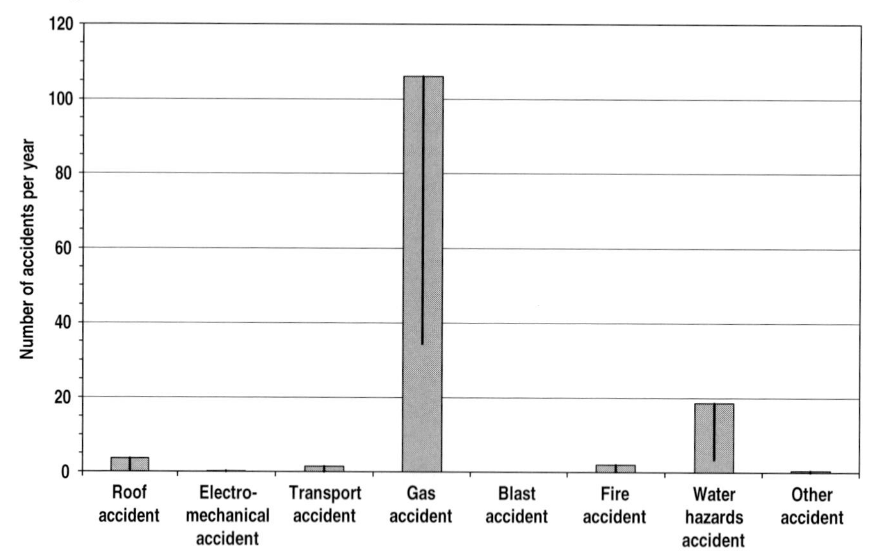

Figure 10.11 Average number of accidents per year attributed to different causes for the period 1994-1999. Dark lines within bars represent one standard deviation.

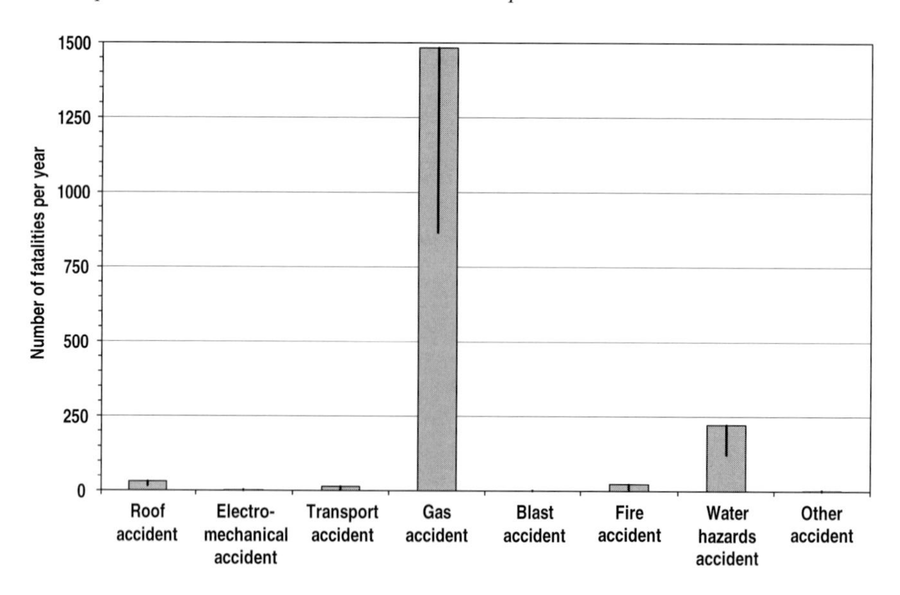

Figure 10.12 Average number of fatalities attributed to different causes for the period 1994-1999. Dark lines within bars represent one standard deviation.

Shandong versus Shanxi provinces. In 1999, Shanxi with 246 Mt and Shandong with 85 Mt were the two provinces making the largest contributions to China's overall coal production. Although production in Shanxi is only about three times higher than in Shandong, the number of accidents and fatalities were almost ten times higher in Shanxi in the period 1994-1999 (Figures 10.13 and 10.14).

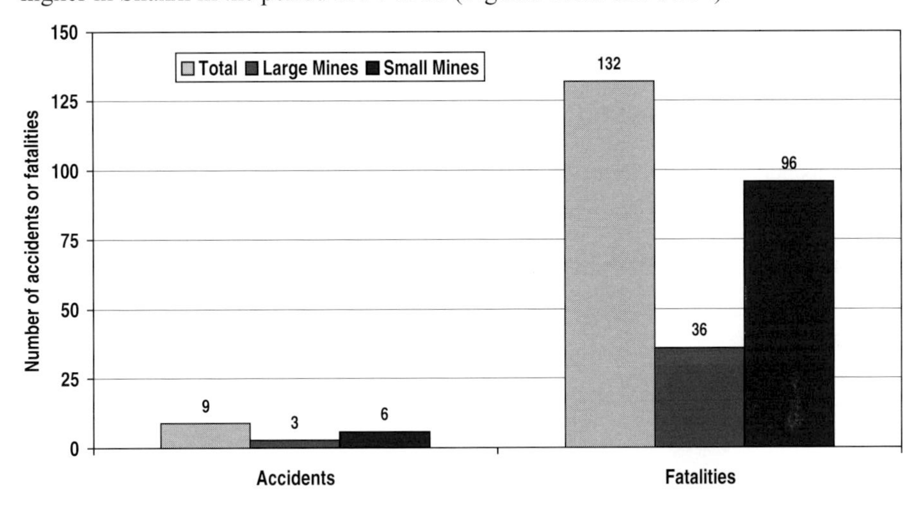

Figure 10.13 Number of severe accidents and fatalities according to mine type in Shandong province for the period 1994-1999.

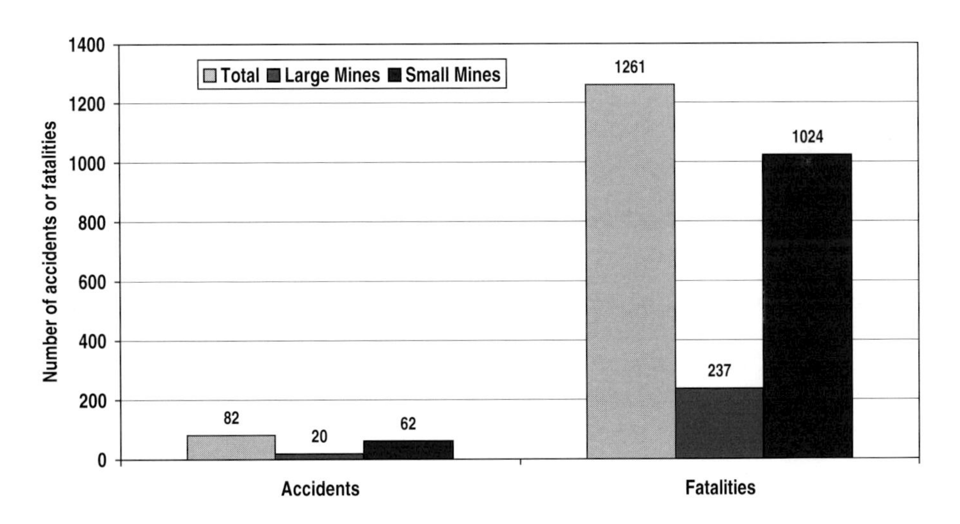

Figure 10.14 Number of severe accidents and fatalities according to mine type in Shanxi province for the period 1994-1999.

Both provinces followed the general pattern described before in terms of accident causation, with gas accidents clearly dominating (Figure 10.15). However, gas accidents contributed about 3.5 times more fatalities per Mt and yr in Shanxi than Shandong. This is most likely attributable to differences in fugitive coalbed methane (CBM) emissions between the two provinces. Coal mines in Shanxi are rich in CBM with 15 to 25 m^3 gas per ton of coal extracted, mines in Shandong contain only <1 to 3.6 m^3 gas per ton.

China has abundant CBM resources that provide a valuable resource. Additionally, extraction of CBM provides a potentially important safety measure to prevent explosions in coal mines, which could reduce the large number of gas accidents and contribute to lowering the number of accidents and fatalities.

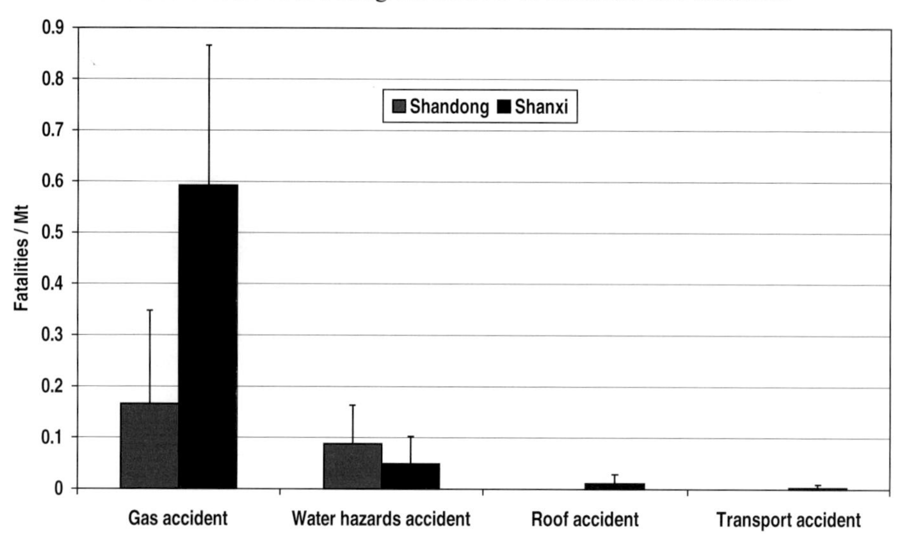

Figure 10.15 Average severe accident fatality rates per Mt according to different causes for Shandong and Shanxi for the period 1994-1999. Error bars indicate one standard deviation.

Comparison of severe accidents in coal mining and other industries. Health hazards need to be considered in virtually every industrial branch, from manufacturing, construction, and large-scale hydroelectric power plants, to oil, gas, and mining exploration. It is estimated that China suffers direct economic losses of several billions of yuans per year because of lost productivity, reduction in product quality, and increased retraining and medical expenses resulting from industrial injuries and illnesses. Toxic work environments are a serious concern. According to the central government, an estimated 33.8 million Chinese people work in factories that produce toxic substances. Exposed workers frequently encounter levels of toxicity that can have adverse human health effects. According to government statistics, in 1995, 18’160 workers died in industrial accidents in China's factories and mines; another 4879 died from various diseases caused by poisonous fumes and dusts; and 148 perished from acute poisoning. The mining industry accounted for 60 percent of the work-related fatalities. Furthermore, more than every third severe accident occurred in the coal industry.

4.2 Oil chain

Crude oil is defined as a naturally occurring mixture consisting predominantly of hydrocarbons; it exists in liquid phase in natural underground reservoirs and is recoverable as liquid at typical atmospheric conditions of pressure and temperature. For the purpose of this survey, crude oil also includes natural gas liquids, namely hydrocarbons that exist in the reservoir as natural gas but which are recovered as liquids in separators, field facilities, or gas-processing plants. In 1999, China ranked seventh among the world's largest oil producers and third in oil consumption (IEA, 2001).

4.2.1 Dominant hazards in the oil chain

During exploration and extraction, health impairment is likely to come in the form of mechanical injury as a result of accidents. Workers are exposed to hazardous situations resulting from maintenance activities, equipment failure, oil rig set-up, rig dismantling activities, and adverse weather conditions. In addition, high risk may be involved in geographical or seismographic exploration and testing which often require extensive aerial surveys and the use of explosives to determine geological formation profiles. Well blowouts that occur in the process of drilling, especially of deep wells, is a direct threat to the health and safety of workers.

The worst accident worldwide in the extraction stage happened on the oil rig "Piper Alpha" near the Scottish coast in the North Sea. Explosion and fires on the platform resulted in 167 fatalities and an insured damage of 2913 million USD at 2000 prices (SIGMA, 2001). The worst exploration accident occurred in 1980 when the flotel Alexander Kielland overturned off the coast of Norway with the loss of 123 lives (Hirschberg et al., 1998).

The long-distance transport and regional distribution stages involve distinctly different materials with different properties and potentials for incidents. The first stage involves the transportation of crude oil; the second deals with the transport of refined oil products. Every day about 119 billion liters of oil are being transported at sea (Cutter, 2001). Accidents with super-tankers on the sea are among the largest environmental catastrophes. But not all spills come from tankers. Some originate from storage tanks, pipelines, oil wells, tankers and vessels cleaning out tanks. The largest recorded spill was estimated at 455'000 t of crude oil and occurred on the coast of Mexico (in 1979) not from a tanker but from an oil platform (Ixtoc 1). Notably, in the notorious Exxon Valdez spill in Alaska in 1989 "only" 32'500 t of oil were released but the ecological and economic impacts were disastrous (Hirschberg et al., 1998).

The most severe accident in the transport to refinery stage occurred in 1987 when the Dona Paz ferry collided with the oil tanker Victor off the Coast of Mindoro (Philippines), causing between 3000 and 4375 fatalities (Hirschberg et al., 1998; Sigma, 2001). At the regional distribution stage, the collision of a Soviet fuel truck with another vehicle in the 2.7-km-long Salang tunnel in the northern part of

Afghanistan is the worst accident on record. Fire and noxious fumes following the explosion lead to the death of 2700 Soviet soldiers and Afghan civilians from burns and asphyxiation (Hirschberg et al., 1998).

The number of fatalities in accidents at refineries is much lower compared to the previously described stages. The two worst disasters occurred in 1983 in Teleajen (Romania) and 1988 in Shanghai (China) with more than 30 and 25 fatalities, respectively (Hirschberg et al., 1998).

At the power plant and heating stages only a few severe accidents, with relatively limited consequences occurred in the past.

4.2.2 Severe accidents in China's oil chain

Severe accidents by year. For the period 1969 to 1999, a total of 32 severe accidents with a total of 613 fatalities in the Chinese oil chain were identified. As for coal, Chinese information sources are of major importance in obtaining complete coverage as the Chinese-specific data in the ENSAD database are likely to be quite incomplete.

Figures 10.16 and 10.17 show the number of accidents and fatalities that occurred in China's oil chain in the period 1969-1999. The lack of data from 1969-1976 is most likely due to poor reporting, though it is possible that in 1985, 1986, and 1992 no severe accidents occurred. There were typically one or two severe accidents per year (Figure 10.16). Fatalities peaked in five years, namely 1979, 1980, 1983, 1990, and 1996 (Figure 10.17). These peaks were attributable to single large accidents and not to any particular accumulation in the annual number of accidents as can be seen from the comparison with Figure 10.16.

Severe accidents at different oil chain stages. Figures 10.18 and 10.19 show the number of severe accidents and fatalities according to different stages in the oil chain for the period 1969-1999. Half of all accidents occurred during transportation (i.e., transport to refinery or regional distribution), followed by the exploration and refinery stages (Figure 10.18). Considering fatalities, exploration ranks first with a share of about 40 percent, followed by regional distribution and transport to refinery (Figure 10.19).

General causes of severe accidents in the oil chain. Information on the general causes of severe accidents was not available. However, it has been reported that safety standards are relatively poor (Reuters, 1991). It appears that pressure to construct much-needed offshore infrastructure resulted in a lax control and application of safety standards. For instance, the captain of the oil barge "McDermott", involved in pipe-laying operations on the Huizhou 26-1 oilfield, ignored warnings of the impending typhoon Fred, which was predicted several days in advance. Survivors of the accident said that no warnings about the immediate danger were forthcoming until the barge started to capsize and sink. In particular, there was strong criticism that a special hyperbaric lifeboat which would have carried a decompression chamber for the divers working on the barge was not provided (Reuters, 1991).

Dangerous fires also result from inadequate safety standards. A circular issued by the Public Security Ministry Fire Department in 1997 stated that fires had caused substantial damages to the petrochemical industry throughout the country, i.e., refineries, oil tankers etc. because of lack of precautionary measures against fire, lax management, and serious disregard of rules and regulations (Reuters, 1997).

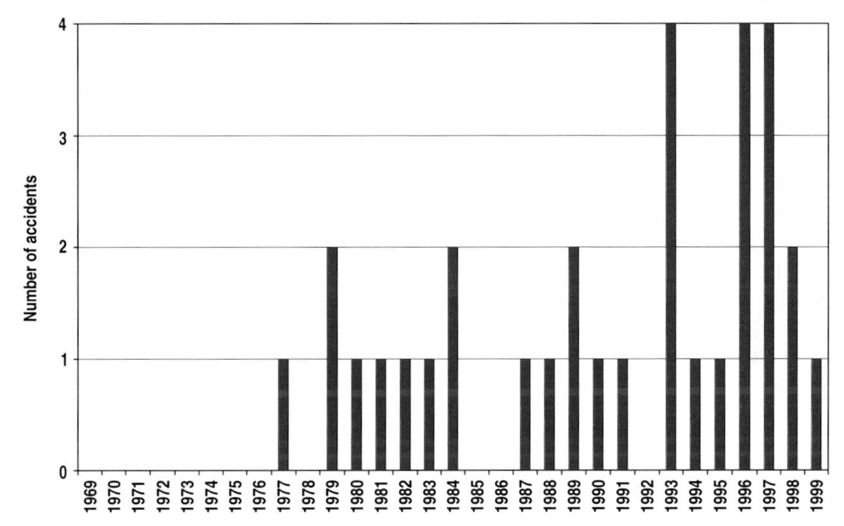

Figure 10.16 Number of severe accidents in China's oil chain in the period 1969-1999.

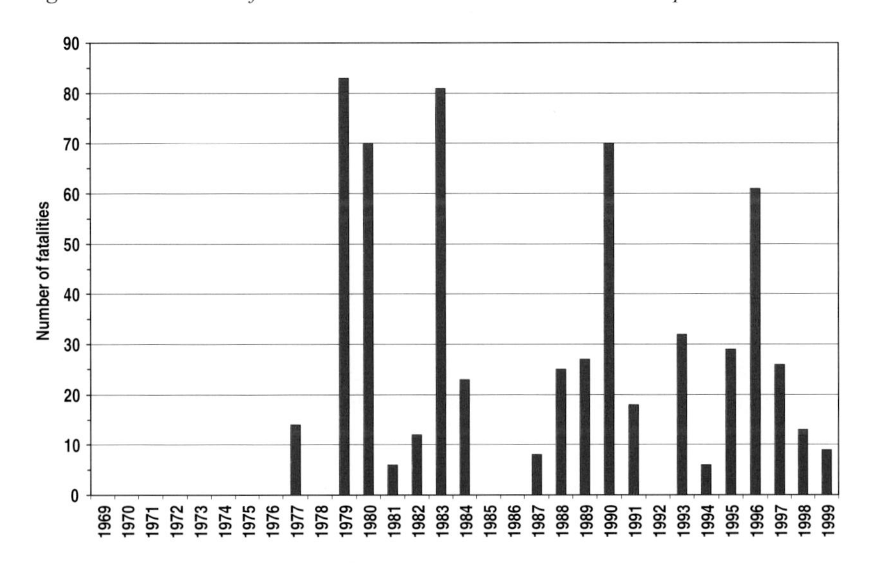

Figure 10.17 Number of severe accident fatalities in China's oil chain in the period 1969-1999.

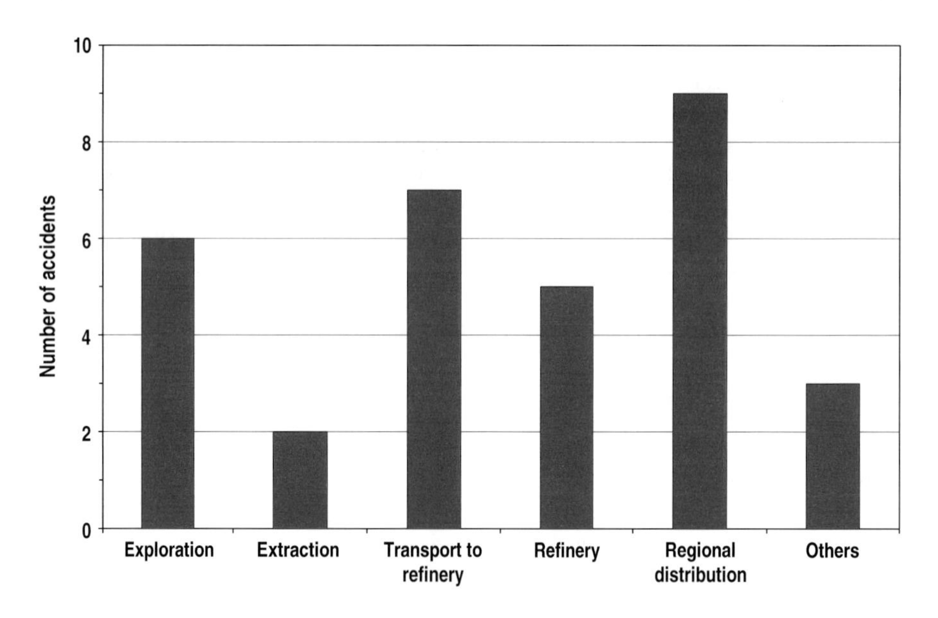

Figure 10.18 Number of severe accidents in China according to oil chain stages in the period 1969-1999.

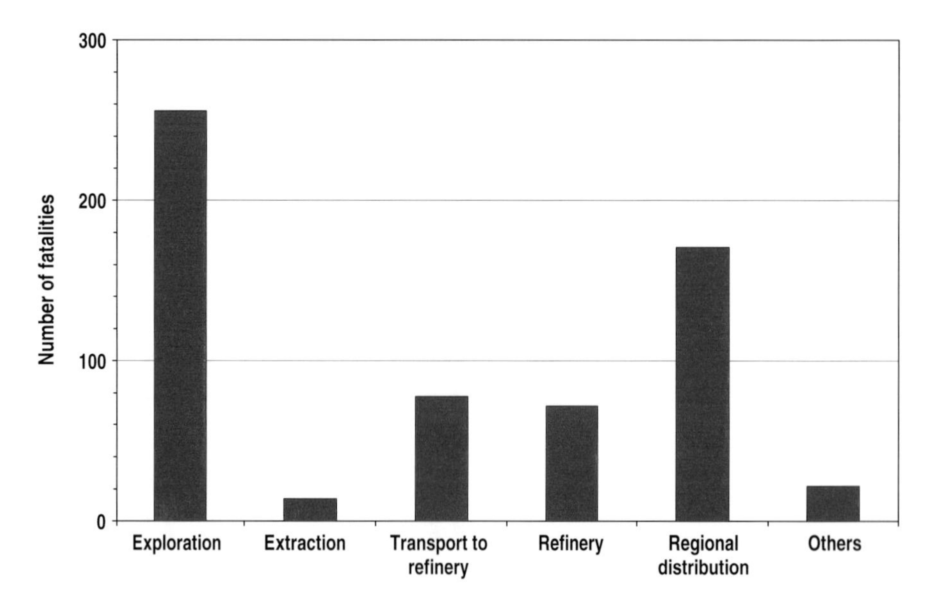

Figure 10.19 Number of severe accident fatalities in China related to oil chain stages in the period 1969-1999.

Oil spills. No major offshore and onshore oil (hydrocarbon) spills exceeding 25'000 t were reported for the period 1969-1999 in China (Table 10.3). The main reason could be that the amounts of hydrocarbons transported by ships, barges, lorries, or pipelines were relatively small. Table 10.3 indicates that most facilities transporting or storing hydrocarbons have capacities lower than 28'000 t. However, oil spills of relatively small quantities could be much more frequent and can also cause severe damages. In the South China Sea even small spills can have tremendous consequences because people in this densely populated coastal area depend largely on fish and oysters. For example in 1999, the collision of a tanker with a bulker carrier caused an oil spill of 150 t resulting in serious pollution and tonnes of dead fish (LLP, 1999).

4.3 Natural gas chain

China ranks only seventeenth in terms of gas production and consumption (IEA, 2001). However, between 1995 and 1999 there was a very substantial increase in consumption in China (by 26 percent), while it stabilized in USA and in Russia, the two largest producers and consumers of natural gas.

4.3.1 Dominant hazards in the natural gas chain

According to world-wide evaluation, severe accidents resulting in fatalities occur in the natural gas chain predominantly in long-distance transport (36.4 percent), local distribution (26.4 percent), and regional distribution (15.5 percent) stages while for the LPG chain the major contributors are regional distribution (64.7 percent) and long-distance transport" (10.9 percent) (Hirschberg et al., 1998).

4.3.2 Severe accidents in China's natural gas chain

The public is dangerously exposed to the consequences of gas accidents. The damages occur immediately and are local. In the case of natural gas the total number of fatalities world-wide does not exceed 250 in any year, with the worst accident (Tbilisi, Georgia, 1984) causing 100 deaths (Hirschberg et al., 1998). During the period 1969 to 1999, a total of eight severe accidents with a total of 201 fatalities were identified in the Chinese natural gas chain. This corresponds to roughly 17 percent of all accidents and 18 percent of all fatalities in severe accidents that occurred in the natural gas chain in non-OECD countries in the period considered.

For comparison with other energy chains, the number of accidents and fatalities per year are shown in Figures 10.20 and 10.21. The number of accidents per years ranged between zero and two with no reported accidents before 1989; the corresponding fatalities exhibit no particular trend and reflect only the severity of individual accidents. About 70 percent of all fatalities are attributable to two accidents in 1993, one in Shenzen (Guandong province) and one in Bahoe (Heilongjiang province), each causing 70 fatalities. More detailed analyses (e.g., separation by energy chain stages) were not performed because of the small size of the data basis.

Table 10.3 Chinese spills of hydrocarbons that caused major pollution in the period 1969-2000.

Date	Place/ Province	Unit(s)	Hydro-carbon	Cargo (t)	Spilled quantities (t)	Cause	Damages
13.06.1993	Shengli/ Shandong	Pipeline	crude oil	N.A.	6000	theft of a key valve in a pipeline	polluted farmland, huge traffic jam
23.10.1993	Nanjing/ Jinling	Storage tank	petrol	6000	<1000	human error	2 fatalities
08.01.1994	Yangtze River/ Zheijiang	Vessel	oil residue	500	< 500	N.A.	9 million USD
08.05.1994	Long Tau River	Vessel	fuel oil	1012	200	collision	considerable pollution
20.08.1995	Guangzhou/ Huangpu	Tanker	crude oil	28'000	150	N.A.	millions of USD
28.02.1996	Xiaoxi/ Fujian	Tanker	crude oil	57'000	N.A.	collision	considerable pollution
01.05.1996	Yantai	Tanker	oil	N.A.	N.A.	collision	considerable pollution
14.07.1996	Liuzhou/ Guangxi	2 vessels	diesel	5	5	drastic variation in temperature	monetary damage of vessels: 13'000 USD
17.07.1996	Huxi/ Liuzhou	2 vessels	diesel	30 (each)	< 30	collision	1 fatality, injured persons
11.11.1996	Island of Ketoy	Tanker	Fuel	400	< 400	N.A.	irretrievable damage to ecology of Kuril islands
29.12.1996	West of Hong Kong	Wooden barge	gasoline	60	60	N.A.	> 9 fatalities
1997	Zhanjiang, Guangzhou	Tanker	crude oil	N.A.	300	valve failure while loading	370'000 USD
07.06.1997	Nanjing/ Jinling	Tanker	crude oil	19'700	< 19'700	explosion	9 fatalities
24.09.1998	Jiangyou/ Sichuan	oil tank	crude oil	200	< 200	explosion	loss of at least 20 t of oil
13.11.1998	South China Sea	2 tankers	gasoline	N.A.	2000	collision	ruining fish grounds, 360'000 USD
24.03.1999	off Zhuhai (Guangdong) near Hong Kong	Tanker	fuel oil	1000	150	collision	serious pollution, tonnes of dead fish, 600 acres of fish farm destroyed, direct losses of 120'000 USD
16.07.1999	Zhanjiang/ Guangdong	Tanker	N.A.	N.A.	300	N.A.	N.A.
31.3.2000	Pearl River Delta	Tankers	crude oil	N.A.	200	collision	1.2 million USD

N.A. = not available

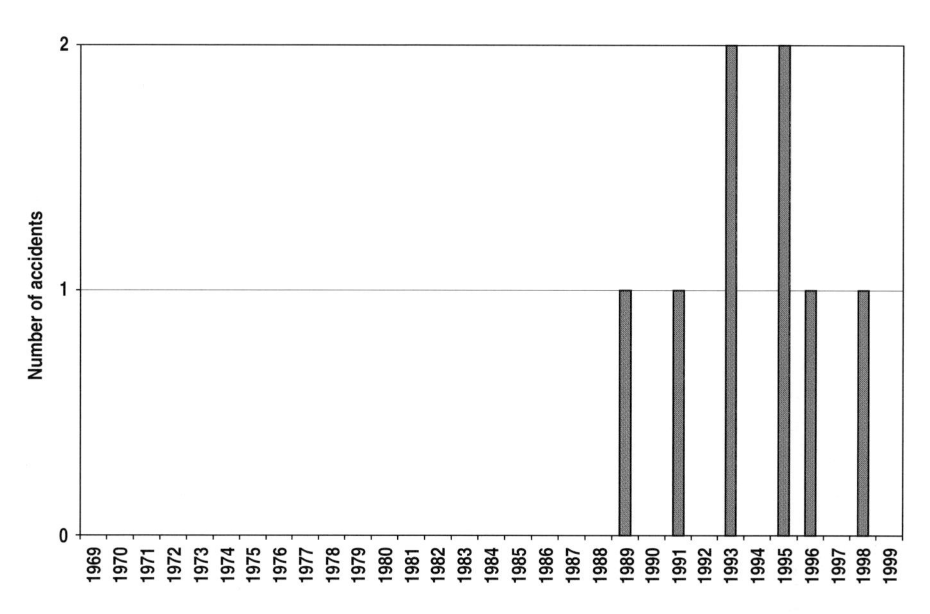

Figure 10.20 Number of severe accidents in China's natural gas chain in the period 1969-1999.

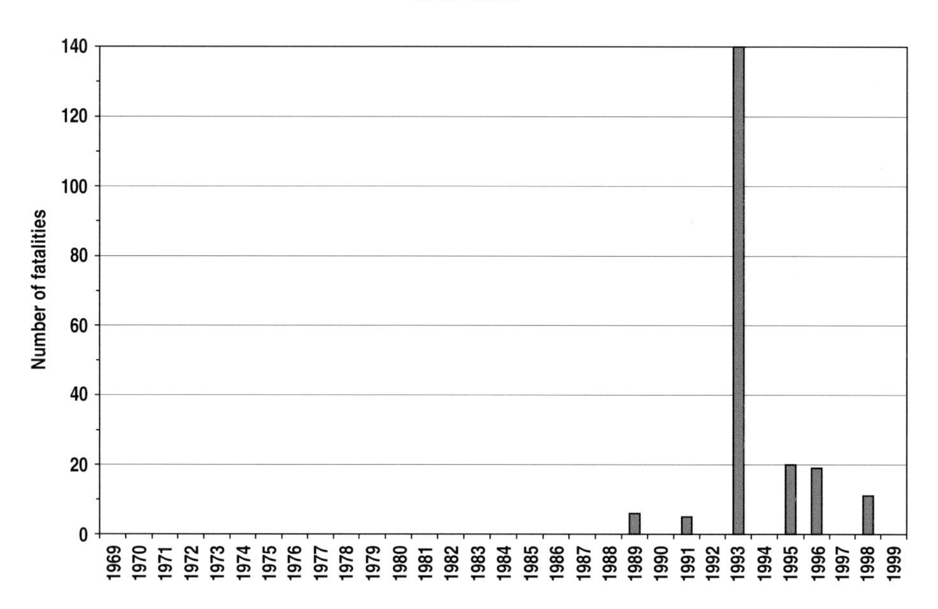

Figure 10.21 Number of severe accident fatalities in China's natural gas chain in the period 1969-1999.

4.4 Liquefied Petroleum Gas chain

Liquefied Petroleum Gas (LPG) is colorless, tasteless, nontoxic, and odorless, but it is also volatile (expands rapidly), heavier than air, and flammable. As the name suggests, it is partly a byproduct of the oil-refining process, but also occurs naturally in crude oil and natural gas fields. It is a mixture of light hydrocarbons which are gaseous at normal temperatures and pressures, and which liquefy readily at moderate pressures or reduced temperature. During the production process the gas is compressed, causing it to liquefy and remain in liquid form until the pressure has been removed. Its main constituents are propane and butane. For safety reasons, a pungent compound, ethyl mercaptan, is added to make any leaks easily detectable.

LPG is primarily used for heating and cooking, but is also an attractive automobile fuel among other uses. In contrast to natural gas no LPG is burned in power plants. China is the second largest producer of LPG after the USA, and the third largest consumer after the USA and Japan. Furthermore, the increase in production in China (+312 percent) is the second highest after Kuwait (+705 percent), and increase in consumption (+521 percent) is the highest worldwide for the period 1990-1999 (IEA, 2001).

World-wide historical experience shows that nearly 65 percent of all severe accidents in the LPG chain occurred at the regional distribution stage, over 13 percent in the heating stage, and about 11 percent during long-distance transport (Hirschberg et al., 1998). Very few severe natural gas and LPG accidents were recorded before 1970. Following that year the number ranged between one and seven annually. The worst accident (Asha-Ufa, Russia, 1989) resulted in 600 fatalities (Hirschberg et al., 1998).

For the period 1969 to 1999, a total of 11 severe accidents with a total of 291 fatalities were identified in the Chinese LPG chain. This amounted to about 28 percent of all accidents and about 11 percent of all fatalities in severe accidents that occurred in the LPG chain in non-OECD countries in the period considered.

For comparison with other energy chains, the number of accidents and fatalities per year are shown in Figures 10.22 and 10.23. The number of accidents per year ranged between zero and two with most accidents having occurred in the 1990s. Nevertheless, it would be speculative to conclude that the number of accidents and fatalities are increasing since the late 1990s because the data basis is rather small.

As was the case with natural gas, more detailed analyses (e.g., separation by energy chain stages) were not performed, given the limitations of the database.

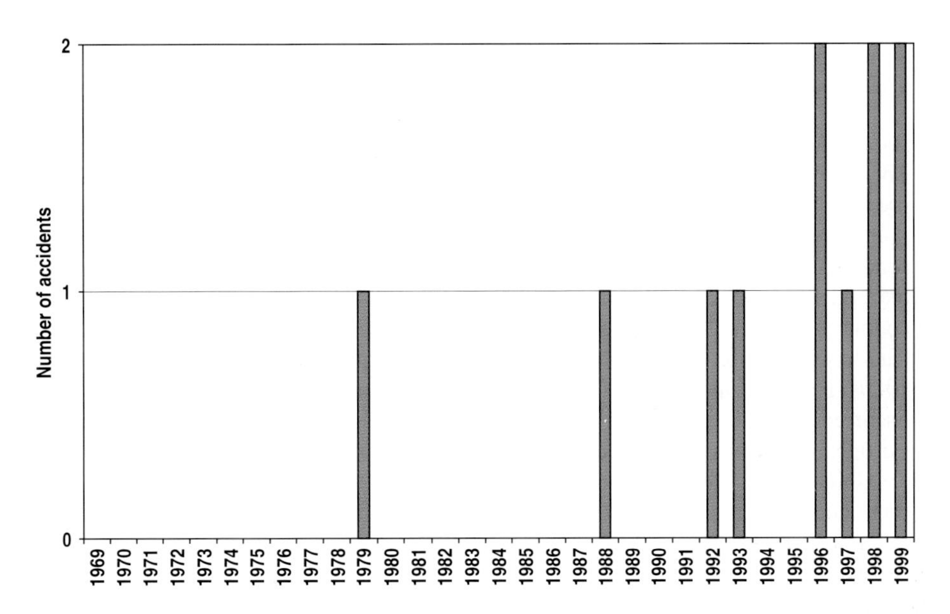

Figure 10.22 Number of severe accidents in China's LPG chain in the period 1969-1999.

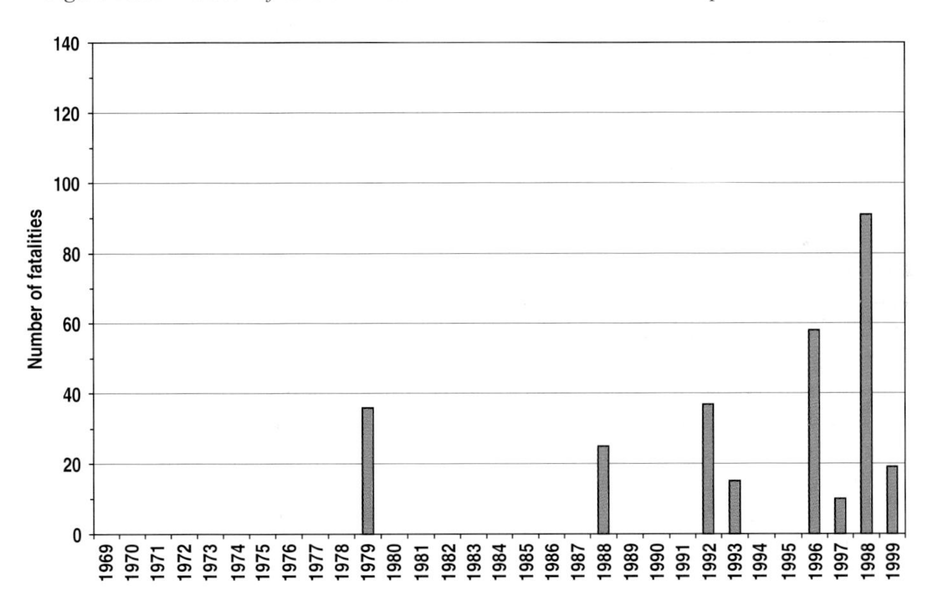

Figure 10.23 Number of severe accident fatalities in China's LPG chain in the period 1969-1999.

4.5 Hydro power

Use of dams for hydro power generation began at the end of the nineteenth century. The first hydroelectric power plant was built in 1882 in Appleton (Wisconsin) to provide 12.5 kW to light two paper mills and a home (National Renewable Energy Laboratory, 1998). In the twentieth century a rapid increase in large dam building occurred. By 1949, about 5000 large dams had been constructed worldwide, and by the end of the twentieth century there were over 45'000 large dams in over 140 countries (WCD, 2000). The period of economic growth following World War II lead to a huge rise in the global dam construction rate that lasted into the 1970s and 1980s. However, during the last two decades a decrease in the rate of dam building occurred because most technically attractive sites, especially in North America and Europe, are already developed (WCD, 2000).

China ranks fourth among the world's top hydro power producers, with production about 40 percent lower than that of Canada (IEA, 2001). Hydro power production in Canada and the USA remained at similar levels in the period 1990-1999, though it greatly increased in China (+161 percent) and Brazil (+142 percent). Currently, hydro power contributes about six percent to China's total primary energy, but it accounts for about 20 percent of China's electrical generation capacity, or almost all the electricity not generated by coal (SSB, 1999). By 2009, the Three Gorges Project (TGP) with its total installed capacity of 18.2 GW is expected to supply about three percent of China's power needs (Zhou et al., 2000). Once in operation, it will become the largest hydro power plant in the world.

Large dams have many positive but also some negative effects for which they have been occasionally criticized. For the discussion of technical and economic aspects, and environmental and social effects associated with dams in general and the Three Gorges Project (TGP) in particular, we refer to Hirschberg (2003a). The present discussion focuses on issue of severe accidents.

4.5.1 Classification of dams

For the purpose of this chapter, large dams are of primary interest. In the literature several definitions of *"large dams"* can be found. In the present discussion, he definition of a large dam follows the criteria established by the International Commission on Large Dams (ICOLD, 1995): A large dam is one with a height of 15 m or more from the foundation. However, if dams are between 1-15 m high and have a reservoir volume of more than three million m^3, they are also classified as large dams.

Dams can be classified according to either type or function (purpose). Six basic dam types can be distinguished. Dams of earth (Te) or rockfill (Er) type are embankment dams whose fill material is earth or rock. A gravity dam (Pg) is constructed of concrete and/or masonry and resists the pressure of impounded water through its own weight. An arch dam (Va) is built of concrete or masonry and resists the pressure of the water by having the form of a single arch often abutted by natural rock formations. Multi-arch dams (Mv) represent a variant of this type. A buttress dam (Cb) consists of a watertight part supported at intervals on the downstream side by a series of buttresses (walls perpendicular to the axis of the dam). For more detailed information see Hirschberg et al. (1998) and references therein.

Additionally, dams can be categorized by their function (purpose). Today, dams are primarily used for irrigation in agricultural production, electricity generation, and flood control. To a lesser extent, dams have been built to improve water supply and river transportation or for recreational purposes. Recent data show a trend towards multi-purpose dams. Furthermore, there is considerable variation in the functions served by large dams and these functions have changed over time (WCD, 2000). For example, the majority of large dams in Africa and Asia are for irrigation, but there is a growing interest in dams for flood protection and in pumped storage dams for power generation to meet peak demands in Asia.

Concerning severe accidents in the hydro power chain only dams used primarily for hydroelectric power generation are considered. Most hydro power plants are conventional in design, i.e., they use one-way water flow to generate electricity. Run-of-river plants use little, if any stored water to provide water flow through turbines. This type of plant exhibits significant fluctuations in power output due to seasonal changes in discharge of rivers, whereas storage plants have enough capacity to compensate for seasonal fluctuations in water flow and provide a constant supply of electricity throughout the year. In contrast to conventional hydro power plants, pumped storage plants reuse water. After water initially produces electricity, it flows from the turbines into a lower reservoir located below the dam. During off-peak hours (periods of low energy demand), some of the water is pumped into an upper reservoir and reused during periods of peak-demand.

4.5.2 Large dams and reservoirs in China

Dam construction in China has a long history. The most ancient reservoir, Shaopi, was built during Eastern Zhou Dynasty (598-591 BC) in Anhui province. It is a 10 m high earth dam that has been in regular operation up to the present (Zhang, 2000). The well-known Dujiangyan irrigation project, which supplied 800’000 hectares in China, is ca. 2200 years old (WCD, 2000; Zhang, 2000).

Dam construction using modern technology adopted from abroad began in the first half of the twentieth century. Before 1949 China had only 22 large dams (Zhang, 2000). Since 1950 dam construction has developed very fast. At present, an estimated 85’000 dams of all types are in operation in China, excluding small farm-scale irrigation and mini and micro hydro power units (WCD, 2000). China’s actual number of large dams may be around 22’000 - 46 percent of the world’s large dams (WCD, 2000). However, only 4434 large dams in China are contained in the voluntary World Register of Dams maintained by ICOLD (1998; 2000). In Figure 10.24, the number of large dams for the world’s top ten countries is given based on both WCD estimates and those provided in ICOLD’s database. The figure suggests that ICOLD data for China are highly incomplete considering the huge difference between estimated and registered large dams.

According to Zhang (2000) China had 17’526 large dams with a height of 15-30 m and 4578 dams over 30 m (including 32 higher than 100 m) by the end of 1999. An additional 320 large dams were under construction, of which 23 are higher than 100 m (WCD, 2000). Embankment dams comprise the majority in China (ICOLD,

2000). The number of concrete dams is gradually growing as dam heights steadily increase, but the new trend is toward concrete-faced rockfill dams and roller-compacted concrete (RCC) dams (Zhang, 2000). Figure 10.25 shows large dams in China according to purpose, based on ICOLD (2000) data. Irrigation dams (3982) dominate, followed by hydro power (1758) and flood control (1031) dams. The figure further shows that most dams, except those for irrigation, serve more than one purpose. Only 260, or six percent of China's large dams, are exclusively used for hydro power generation, whereas 1506, or 34 percent, are used for at least one other purpose besides hydro power.

4.5.3 Risks and failures of large dams

The International Commission on Large Dams (ICOLD, 1995) defines a dam failure as "collapse or movement of part of a dam or its foundations so that the dam cannot retain the stored water". This definition does not address partial dam failures. The dam is considered failed when all stored water is released.

Embankment dam failures can be grouped into three general categories, although the various types may often be interrelated in a complex manner. For example, uncontrolled seepage may weaken the soil and lead to a structural failure. The three types of failure are overtopping, seepage, and structural failures. Overtopping failures result from the erosive action of water on the embankment. Erosion is due to the uncontrolled flow of water over, around, and adjacent to the dam. Earth embankments are not designed to be overtopped, and therefore are particularly susceptible to erosion. Once erosion has begun during overtopping, it is almost impossible to stop. All earth dams have seepage resulting from water percolating slowly through the dam and its foundation. Seepage must, however, be controlled in both velocity and quantity. If uncontrolled, it can progressively erode soil from the embankment or its foundation, resulting in rapid failure of the dam. Structural failures can occur in either the embankment or the appurtenances. Structural failure of a spillway, lake drain, or other appurtenance may lead to failure of the embankment.

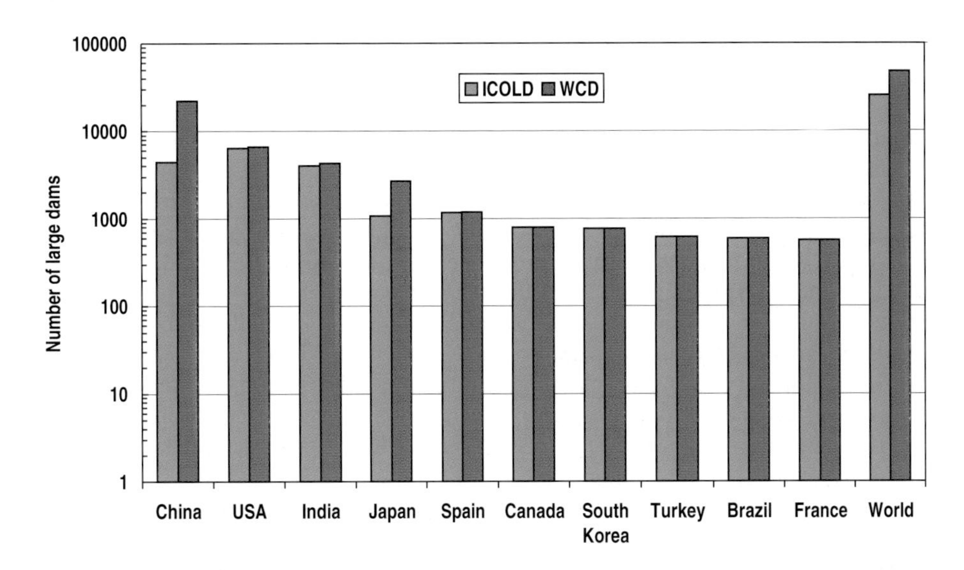

Figure 10.24 Top ten countries by number of large dams (WCD, 2000; ICOLD, 1998; 2000). For comparison, estimated numbers of large dams according to WCD and number of dams registered at ICOLD are shown.

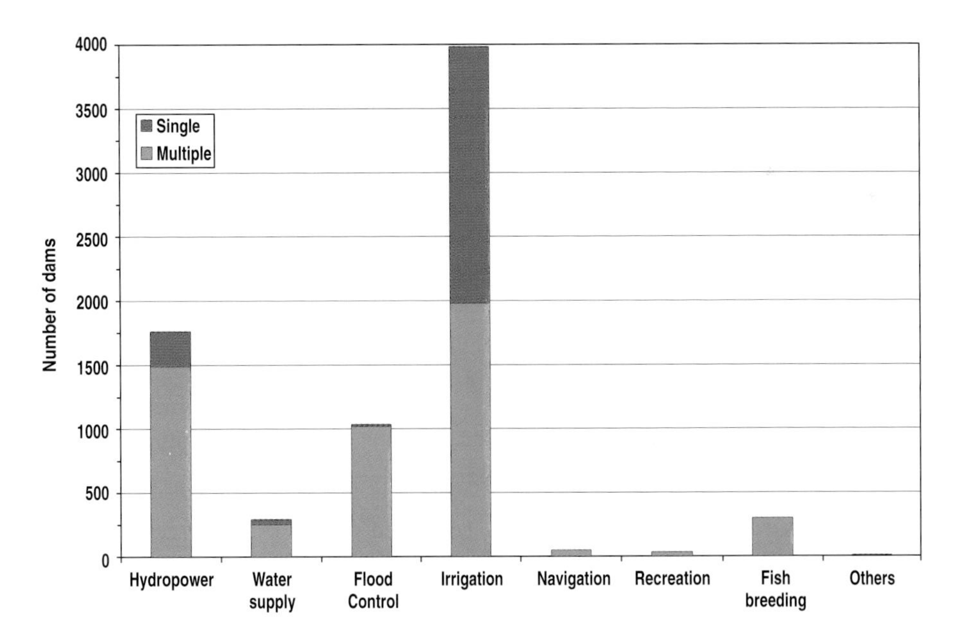

Figure 10.25 Large dams in China according to purpose, based on database by ICOLD (2000). Single = single purpose dam, Multiple = multiple purpose dam.

4.5.4 Severe accidents in China's hydro power chain

According to the Chinese Ministry of Water Resources, more than one third of China's estimated 85'000 dams are defective and must be repaired within the next decade to prevent disaster (*BBC News Online,* 1999; *Muzi Lateline News,* 1999). Costs of government plans to reinforce 33'000 dams by 2010 are estimated at about 33 billion yuan (3.9 billion USD). For example, the Foziling, Meishan, and Xianghogdian dams located on the Huaihe River in the western Anhui province are in bad condition after over 40 years of operation (Probe International, 1998). In 1995, a special inspection team identified some major problems at the Foziling dam, such as enlarged crevices, leaks in several areas, low strength, and dam body displacement, but no repairs have been made since then. Therefore the water has to be kept at low level to reduce the danger of a dam collapse due to a serious flood event. The Meishan and Xianghogdian dams are also in urgent need of repair.

The following summary of failures of dams in China is based on Zhang (1997) and additional information from *BBC News Online* (1999), ICOLD (1995), *Muzi Lateline News* (1998, 1999), Probe International (1998), Fu (1998) and Vogel (1998). By 1973, 40 percent or 4501 of the 10'000 Chinese reservoirs with capacities between 10'000 and 1'000'000 m^3 have been built below project specifications and were unable to control floods effectively. In the period 1950-1990, 3241 dams of all purposes had collapsed, of which 123 or 3.8 percent were large dams and 3118 or 96.2 percent were smaller dams. On average, China experienced 81 collapses per year, with the worst year being 1973, when 554 dams collapsed. The official death toll resulting from dam failures came to 9937, not including Banqiao and Shimantan collapses that accounted for 26'000 fatalities alone (Fu, 1998).

Table 4 shows failed dams of all purposes in China based on data from ICOLD (1995), Vogel (1998), official Chinese sources, and the journal *International Water Power & Dam Construction* (1991-1995). These dam-specific accident data are inconsistent with the overall dam collapse records provided above. Both, the number of specific accidents among all large dams and the number of specific accidents at hydro power dams appear rather low, suggesting this list could be highly incomplete.

- Considering that 40 percent of large Chinese dams are associated with hydro power purposes, roughly 50 of the 123 recognized collapses of large dams between 1950 and 1990 could be expected to have been hydro power accidents. Even when assuming that hydro power dams are safer than other dams, this contrasts sharply with the officially acknowledged three accidents, thus raising the question of underreporting.
- According to ICOLD data, 0.5 percent of dams built since 1950 (excluding China) have failed (WCD, 2000), compared to about four percent in China for the same time period (IRN, 1995-1999; Zhang, 1997).

Based on the available data no reasonably reliable evaluation of severe accidents is possible for hydro power, utilizing Chinese-specific data. Therefore, the pooled hydro power experience for non-OECD countries (including China) was used for comparative analyses.

Table 10.4 Severe accidents involving dams of all purposes in China.

Dam name River Province	Dam type	Purpose	Reservoir volume (10^6 m^3)	Construction year	Year of failure	Number of fatalities
Fushan Huai Shanxi	Earth	Block passage[(1)]	>10'000	516	516	10'000
Liaohe Huozehe Shanxi	N.A.	Irrigation	N.A.	N.A.	1973	29
Lijiazui N.A. Gansu	N.A.	Irrigation	N.A.	N.A.	1973	580
Banqiao Ru Henan	Earth	**Hydro power**	492	1956	1975	In total: 26'000
Shimantan Hong Henan	Earth	**Hydro power**	94.4	1952	1975	
Jishan &Wenchun Shanxi	N.A.	Irrigation	N.A.	N.A.	1977	30
Zianxinan I	N.A.	N.A.	N.A.	N.A.	1985	N.A.
Zianxinan II	N.A.	N.A.	N.A.	N.A.	1985	N.A.
  Anhui	N.A.	N.A.	N.A.	N.A.	1986	N.A.
Wujiangdu Wujiang Guizhou	Gravity Arch	**Hydro power**	N.A.	1985	1989	28
Hongqi [(2)] Shanghai	Earth	N.A.	N.A.	N.A.	1991	N.A.
Gouhou Qinghai	Earth/ Rockfill	Irrigation	3.3	1988	1993	300
Changping Changpinghe Shanxi	N.A.	Irrigation	N.A.	1982	1997	30

(1) The Fushan dam on the Huai River, was used to block passage across the Huai during an attack against the Wei Kingdom in AD 516.
(2) Deliberately breached to discharge water from Taihu lake, which was threatening to flood nearby cities.

The Banqiao and Shimantan [2] dams are located on the Ru River and Hong River in Henan province. These two earthfill dams were completed in 1956 and 1952, respectively. In August 1975, a typhoon created a maximum 24-hour rainfall of 1005 mm and a three-day rainfall of 1605 mm in Zhumadian Prefecture of Henan Province, leading to the world's most catastrophic dam failures. Within about one hour, floods overtopped the Banqiao and Shimantan dams causing their collapse. In total, more than 60 smaller dams broke because dikes and flood diversion projects further downstream could not resist such a deluge. More than one million hectares of land was flooded, over 100 km of the Beijing-Guangzhou railway line was damaged, countless of villages and small towns were submerged or partially covered, and millions of people lost their homes. Approximately 26'000 people (other sources state up to 85'000 people) were killed by the immediate flood waves from the failed dams, while a further 145'000 died of epidemics and famine during the ensuing weeks.

4.6 Nuclear chain

According to IAEA (WorldAtom Press Release, 2001; PRIS, 2002) China had three units in operation by the end of 2000 with a net capacity of 2167 MW_e; another eight units with 6420 MW_e were under construction. In 2000, nuclear power in China delivered 16 TW_eh to the grid, amounting to 1.2 percent of the country's total. A more detailed overview of the current status of China's installed and planned nuclear power capacity is given in Table 10.5. In addition, other sources report projects in Shandong (2 x 1000 MW), Zhejiang (Qinshan Phase IV 2 x 1000 MW and Sanmen 2 x 1000 MW), Fujian (2 x 1000 MW) and Guangdong (Yangjiang 6 x 1000 MW) (Nuclear Europe Worldscan, 2000).

Table 10.5 Installed and planned nuclear reactors in China. Op = operational, Uc = under construction, PWR = Pressurized Water Reactor.

Name	Type	Status	Province Location	Capacity (MWe)		Date Connected
				Net	Gross	
Guangdong 1	PWR	Op	Guangdong, Daya Bay	944	984	1993
Guangdong 2	PWR	Op	Guangdong, Daya Bay	944	984	1994
Lingao 1	PWR	Uc	Guangdong, Ling Ao	935	985	2002
Lingao 2	PWR	Uc	Guangdong, Ling Ao	935	985	2002
Qinshan 2-1	PWR	Uc	Zhejiang, Qinshan Phase II	610	642	2002
Qinshan 2-2	PWR	Uc	Zhejiang, Qinshan Phase II	610	642	2002
Qinshan 3-1	PWR	Uc	Zhejiang, Qinshan Phase III	665	728	2002
Qinshan 3-2	PWR	Uc	Zhejiang, Qinshan Phase III	665	728	2003
Qinshan 1	PWR	Uc	Zhejiang, Qinshan Phase I	279	300	1991
Tianwan 1	PWR	Uc	Jiangsu, Tianwan (Lianyungang)	1000	1060	2004
Tianwan 2	PWR	Uc	Jiangsu, Tianwan (Lianyungang)	1000	1060	2005

Source: PRIS, 2002

There are currently no nuclear plants in Shandong province but the local utility (Shandong Electric Power Corporation, or SEPCO) is seriously considering building them in the near future. Thus, nuclear energy is strongly represented in a number of scenarios analyzed within the CETP.

The analyses of severe accidents, carried out for fossil energy carriers, were primarily based on historical accident records. Particularly in the case of coal, the dominant energy carrier, the statistical basis is very rich. For nuclear energy there are fortunately no relevant statistical records. There is only one accident in the history of commercial nuclear power (Chernobyl in 1986), with severe consequences in terms of health effects that already occurred or are expected, damages to the environment, and economic losses. Since the design characteristics of the Chernobyl plant (reactor type and important safety features) show no parallels with the candidate plants that could be built in the future in Shandong province, the Chernobyl accident is considered to be irrelevant for the purpose of this work. It is necessary to use a probabilistic approach for the comparative evaluation of nuclear risks. For a detailed discussion of the Chernobyl accident and the associated risk comparisons we refer to Hirschberg et al. (1998).

4.6.1 Scope of analysis

The assessment concentrates on nuclear power plants as the risk of severe accident is greatest at the power generation stage within the nuclear energy chain. Based on probabilistic considerations, an assessment of accident risks associated with future nuclear plants in the Shandong province was carried out. A simplified approach to consequence analysis was applied in view of the budget constraints for the risk assessment task. The simplified approach is considered sufficiently accurate for the purpose of this study. For the application of a full scope approach to a mid-sized Swiss nuclear power plant we refer to Hirschberg and Cazzoli (1994) and Hirschberg et al. (1998).

Three alternative advanced designs as well as three designs representing currently operating plants were initially considered for the assessment in two locations on the opposite sides of the Shandong peninsula. The advanced designs are a KWU/EDF European Pressurized Reactor (EPR) (Eyink, 1997), an Advanced Boiling Water Reactor, ABWR 1000 (Jo et al., 1991), and a modernized Russian VVER-1000. Then three currently operating plants are a Siemens/KWU PWR (Pressurized Water Reactor), a BWR with MARK-III containment (ERI/HSK, 1996; 1999; HSK/KSA, 1996; 1999; SKI, 1999), and a VVER-1000 of old design (Mladý, 1999). The full set of probabilistic data (core damage frequencies, accident sequences, source terms) for the newer generation of plants is not public. However, total core damage frequency and makeup of accident sequences can be extrapolated from public information about the EPR and ABWR plants; also public information from risk assessments for similar, though less advanced plants, can be used as a surrogate. However, very little information is available in the public literature about probabilistic studies for VVERs of new designs and the relevant data (e.g. source terms) for older VVER designs are also quite scarce. For these reasons, only qualitative statements will be made about these types of plants.

A simplified evaluation was undertaken, documented in more detail in (Hirschberg et al., 2003a), leading to the approximate quantification of the expected risks of hypothetical nuclear accidents. The results are provided in terms of usual risk indicators: early, late, and latent fatalities, land contamination, maximum credible consequences, and the associated frequency-consequence curves. Formal uncertainty treatment is not feasible within the constraints of the present work, but some indication is provided, based on past evaluations of offsite risks.

The numerical assessment of the accident risks associated with future nuclear plants in Shandong province covers two types of reactors, i.e. BWR and PWR. Each of them was characterized by two Core Damage Frequencies (CDFs) differing by one order of magnitude, corresponding to current and advanced (evolutionary) designs. The plants were located at two sites (Qingdao and Yantai) in the Shandong province with different population density and land fraction relevant for the assessment of consequences of hypothetical accidents. Offsite consequences were assessed on the basis of extrapolation from known results.

4.6.2 Probabilistic methodology

The methodology for a simplified assessment of offsite consequences resulting from a severe accident, described in more detail in (Hirschberg et al., 2003a), is briefly presented in this section. A simplified model was chosen here as it is compatible with publicly available information while using appropriate elements from Probabilistic Safety Assessment (PSA). The full-scope PSA-based assessment of offsite risks for a given plant is a very complex process. For a detailed account of the overall PSA methodology we refer to USNRC (1989).

A full scope PSA consists of three levels of assessments. Level 1 deals with plant behavior following a disturbance (accident initiator). Systems behavior and interactions must be modeled, including operator interventions. This part of the study leads to an assessment of Core Damage Frequency (CDF) with associated uncertainties. The number of possible accident sequences can range into billions, but for each sequence the CDF is normally very small, so accident types are grouped by similarity in plant behavior into a finite number of Plant Damage States (PDSs). These are further studied in the Level 2 PSA, which deals with post core damage response by the plant. Level 2 considers severe accident phenomena, and for each PDS, end states of the containment and possible releases of radioactivity to the environment are evaluated.

These are normally reduced for the Level 3 assessment, which deals with offsite consequences, by considering the possible health effects of releases of radioactive substances within a manageable number of source terms. Consequences are then calculated using probabilistic codes, such as MACCS or COSYMA, and further site-specific uncertainties and characteristics such as weather and population are factored in. Evaluated consequences include immediate health effects, delayed health effects, and possibly economic consequences. Uncertainties are inherent in the assessment of health effects, especially the delayed ones, since cut-offs in doses leading to cancer may or may not be imposed. Even larger uncertainties are inherent in the assessment of economic effects, which depend on many factors, such as the value attributed to human life, health care costs, land costs, etc. It is normally found that uncertainties in CDF are small compared to uncertainties in post core damage behavior alone. Uncertainties due to site-specific information, including weather variability and economic costs, are normally not considered in the final results, and point value estimates are provided.

As described, and exemplified in Figure 10.26, the process requires many resources. Extensive work is required just to prepare the site-specific input to a consequence code. For this reason, simplified methods are necessary, given time and resource constraints. For the sake of calibration of consequences in relation to available PSA data (such as releases, population, weather), offsite risk measures for a power plant operating at 3600 MW_{th}, are used as the starting point for the simplified approach. Two hypothetical accidents, one with early containment rupture, and one with a late vented release, are considered. Data for two different sites is used for the dispersion and dose calculations, the first a European continental site with relatively large population density in the vicinity of the plant, and the other a US site with relatively low population density.

The calculations were performed for each of the nine normally defined (i.e., following PSA practices) radionuclide groups available as MACCS inputs. (the definition and core inventories used are described below), assuming that 100 percent of the group is released to the environment, without offsite countermeasures, and without cut-offs in effectiveness of doses in inducing cancers. The consequences were correlated then to the released activity, and factors (or effectiveness in causing consequences) were derived for each radionuclide group, i.e., consequences per Bq released. Consequences which were considered are early fatalities (i.e., deaths occurring because either inhalation or immersion in the passing cloud delivers doses larger than what is considered lethal, approximately three Gy); delayed cancer deaths due to doses from ingestion and inhalation while the cloud is passing; late cancer deaths due to doses incurred from ingestion of water and foodstuffs; and severely contaminated areas (which may be lost for up to 20 years or longer). Delayed cancer deaths together with late cancer deaths are normally called late fatalities, but due to the different pathways for exposure, they must be accounted for separately.

The calculations showed that the Xe group (noble gases) has essentially no influence on offsite consequences, with the exception of a background effect in late cancer fatalities (i.e., only released aerosol shows appreciable consequences). Moreover, the effect of the time of release (from scram, and start of radioactive decay) appears to have little influence on delayed health effects. This is because the calculations show that only long-lived radionuclides are relevant for late health effects. It was found that severe land contamination appears to be only due to the Cs and Sr groups, while the Ce group (which includes Pu) has practically no effect. Early health effects, on the other hand, are dominated by the I and Te groups which include, for the most part, short-lived radionuclides.

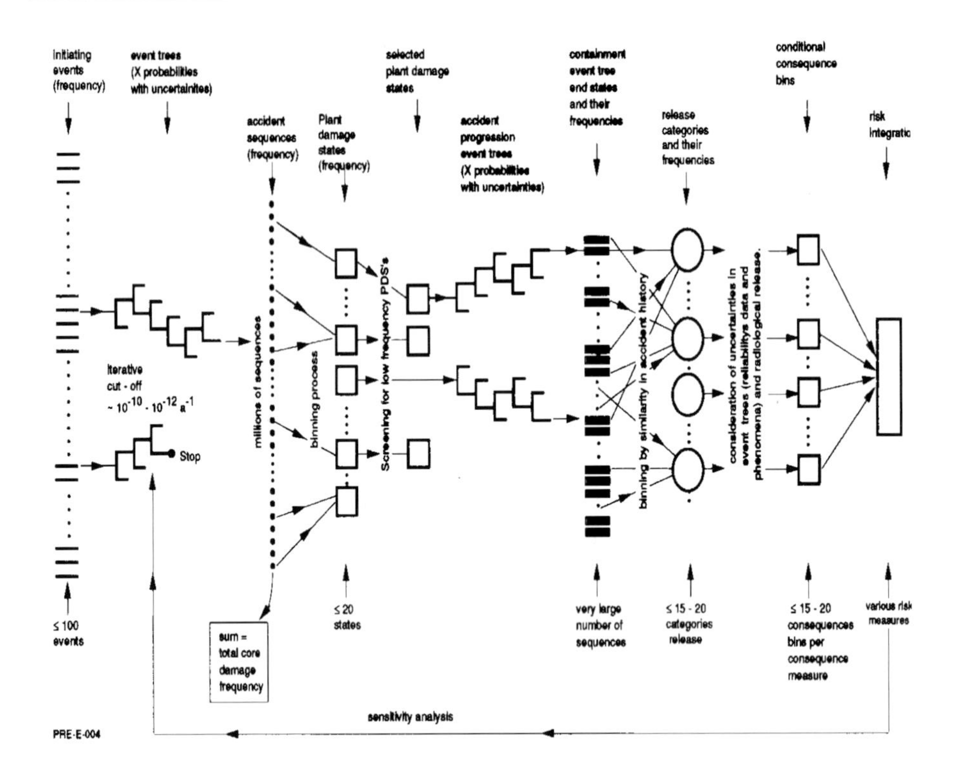

Figure 10.26 Exemplification of a Probabilistic Safety Assessment.
Source: Cazzoli et al. (1993).

In addition, the MACCS calculations showed that health effects are strongly correlated to the total population that can be affected, while the variability in weather from site to site appears to play a secondary role. It must be remembered that the full scope consequence code MACCS (Chanin et al., 1993) is a probabilistic tool, and calculations of cloud dispersion in first approximation do not include wind direction. The windrose, or real wind direction, is considered later in the calculations, when probabilities are assigned to possible consequences, according to overall frequency in wind direction, but since the code uses a Latin Hypercube Sampling (LHS) technique, i.e., with a very limited number of Monte Carlo samples, the probabilities assigned to a given weather and wind direction may be homogenized, and therefore, wind direction plays a small role in the calculations.

In fact, the input specifications includes 16 wind directions, and 8760 hours of weather data, while the LHS sampling chooses typically only about 100 weather types. It is for this reason that normal uncertainties in weather variability are not shown in this type of probabilistic results, and only point value estimates for offsite consequences are provided. The only alternative would be to perform deterministic calculations for each of the 8760 hours. This has been attempted, but it is still prohibitive in terms of resources. From the point of view of a probabilistic study, the results obtained with a code such as MACCS can be considered to be representative within factors of two or three. This does not include the uncertainties associated with source terms. The overall uncertainty band for the consequences typically corresponds to a factor of ten for high consequence releases.

As shown later, the coefficients derived from the various MACCS calculations (i.e., effects per Bq released) appear to be approximately consistent with the ratio of populations and land fractions regardless of weather variability.

Early fatalities can be extrapolated from one site to another from the ratio of population within eight to ten km. This is because the radioactive content in a passing cloud is effectively dispersed over a very large volume very quickly (the MACCS code uses a Gaussian dispersion model). The MACCS calculations show that acute mortality distance does not exceed 20 km under the worst possible weather conditions (i.e., with very small probability), since beyond this distance doses exceeding three Gy, i.e., the immediate mortality limit, are shown not to be delivered under any circumstance.

Delayed cancer deaths appear to be strongly correlated to the total population within 80 to 120 km. The MACCS calculations show that only a small background of delayed fatalities may occur beyond this distance. This, again, is due to cloud dispersion and dilution of activities, since the dose must be incurred via inhalation or submersion in the passing cloud. Cancer deaths occurring from ingestion (late deaths) are found to be correlated to the total population in the site considered. For both sites, a maximum distance of 800 km was considered; late deaths are considered proportional to the ratio of populations within 800 km. The calculations for the US site were extended to 1600 km, and the results in this case differ only by fractions of one percent with respect to the results to 800 km.

Finally, land contamination is assumed to be correlated to the ratio of land fractions to 120 km, even though the correlation was found to be weaker than the ones found for health effects. A distance of 120 km is assumed for this type of calculations, because the MACCS results show that the maximum distance to which land can be severely contaminated does not exceed 120 km for any of the radionuclide groups.

In conclusion, the results based on MACCS suggest that offsite consequences may be calculated approximately using activity of releases for each group of radionuclides, as assessed by Level-2 PSAs for the relevant source terms, and the ratios discussed above. For the assessment of early fatalities, the ratios for the early release sequence should be used, since they provide more conservative figures (as mentioned, early fatalities are mostly due to short-lived radionuclides). It should be emphasized that the results obtained using this simplified methodology should be viewed as order-of-magnitude results. However, uncertainties in probabilistic calculations themselves are normally very large, as discussed above.

4.6.3 Assumptions

The basic assumptions consider two different locations in the Shandong province, which were considered as possible sites for new NPPs and the different types of NPPs, as discussed earlier.

Sites and population. Two sites are considered for the present calculations (see Figure 10.27):

- A site around Yantai with relatively low population density to 20 km (the town includes about 200'000 citizens) and
- A site around Qingdao with high population density (the location chosen would account for about 2.2 million citizens within 20 km).

The precise location of NPPs was assumed to be about ten km from the city, and three different areas for consequence calculations around the planned NPPs were considered, according to the probabilistic methodology discussed above (see also Figure 10.27):

a. to eight km radius
b. to 100 km radius
c. to 800 km radius

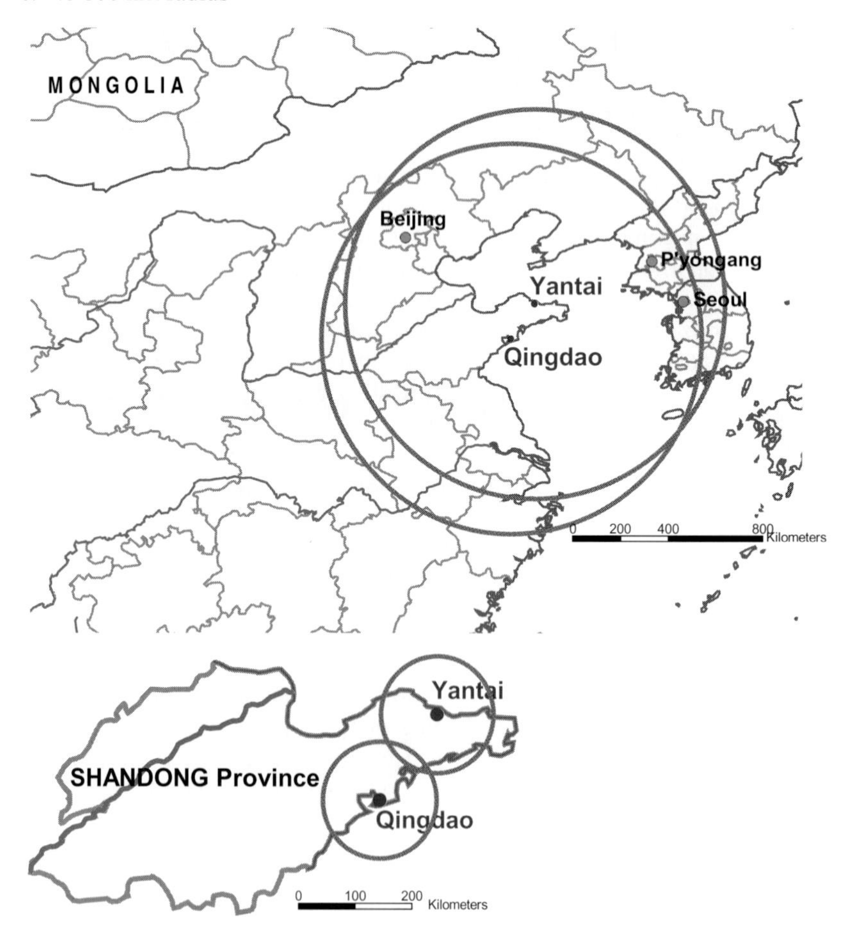

Figure 10.27 Maps with distances. Large and small circles have a radius of 800 km and 100 km, respectively. GIS Source: ESRI Data &Map CD.

Both the Yantai and Qingdao sites are located near the sea shore, so the number of people (in the region of eight km and 100 km radial areas from the sites) was assessed according to the ratio of land and water (Microsoft Encarta, 1999) (Figure 10.27, Table 10.6). For details of these calculations we refer to the technical report (Hirschberg et al., 2003a).

Table 10.6 Population in Qingdao and Yantai sites.

	100% circle area	**Land area Qingdao site**	**Population Qingdao site**	**Land area Yantai site**	**Population Yantai site**
to 8 km radius	201.06 km^2	100.53 km^2 (50% of circle)	~70’000	90.48 km^2 (45% of circle)	~50’000
to 100 km radius	31’415.92 km^2	15'707.96 km^2 (50% of circle)	~10.9 M	11’623.89 km^2 (37% of circle)	~6.4 M

Population to 800 km has been estimated at about 590 millions for the Qingdao site and 517 millions for the Yantai site. This includes affected population in South and North Korea.

Types of nuclear power plants considered. For western types of reactors two levels for Core Damage Frequencies (CDF) are used in the present work:

a. One corresponding to that of current NPPs (ERI/HSK, 1996; 1999; HSK/KSA, 1996; 1999; SKI, 1999), ranging from 2E-6/Ryr [3] to 5E-5/Ryr. In fact, results of PSAs for existing Swiss plants are used. The two plants, a BWR with MARK-III containment, and a Siemens/KWU PWR plant, have been assessed to have a total CDF of 4E-6/Ryr and 2E-6/Ryr, respectively, including some area (internal flooding and fires) and external events.

b. One order of magnitude lower than that for current NPPs to account for advanced features in the designs. Briefly, the new designs have many favorable severe accident progression features, such as less dependence on operator interventions, passive systems, stronger containments, etc. In this work CDF of 4E-7/Ryr and 2E-7/Ryr were used for ABWR and EPR, respectively.

For VVER-1000 type NPPs, published data (Mladý, 1999) indicates that, even for the most recently built plants, core damage frequency is of the order of 1E-4/Ryr, i.e., typically one order of magnitude higher than for plants of Western design. Extrapolating from this figure, since for the proposed designs (ABWR and EPR) CDFs are assumed to be one order of magnitude lower than for non-advanced Western reactors, it is plausible to assume that total CDF for advanced VVERs could be of the order of 1E-5/Ryr, i.e., still almost two orders of magnitude higher than the CDF used for the EPR design.

Table 10.7 shows design similarities and differences between the known or assumed features of the advanced designs, and of the plants, which have been used as reference. The combination of improved features as implemented in advanced designs reduces the expected core damage frequency from internal initiators, especially for the PWR type, which now would include some passive features. The advanced BWR, as opposed to the typical currently operating plants, would have extended containment systems, including a larger containment capacity, thus having only effects on the end states of the containment. The core damage profile is, however, assumed to be similar. It should be noted that the use of more modern equipment with a presumable lower failure rate, could reduce accident initiator frequency considerably and improve systems reliability. Finally, similarity of design would assure that severe accident progression would be similar to what is assessed for the operating plants used for reference, with exceptions that will be discussed later.

Table 10.7 Similarities and differences between basic and advanced types of NPPs.

Feature/NPP type	**BWR**	**ABWR**	**PWR**	**EPR**	**VVER**	**VVER advanced**
Thermal power (MWTh)	3600	3600	3000	3600	3600	3600
Containment venting	Yes	Yes	Yes	No	No	Not known
Additional Decay Heat Removal systems	Yes	Yes	Yes	Yes	No	No
AC/DC hydrogen igniters	Yes	Yes	No	Yes	No	Not known
Passive Emergency Core Cooling System (ECCS)	Yes	Yes	No	Yes	Yes	Not known
Improved Reactor Protection System (RPS) as protection against Anticipated Transient without Scram (ATWS) events	No	Yes	Not relevant for PWRs	Not relevant	Not relevant	Not relevant
Suppression pool	Yes	Yes	No	Yes	No	Not known
Containment capacity	< 5 bar	>> 5 bar	8-10 bar	> 10 bar	5-10 bar	Not known
Severe Accident Management (SAM) Measures	BWR Owners Group, to be implemented	As for BWR with MARK-III contain-ment, but with added cavity flooding	Siemens/ KWU defined, without hydrogen control devices	Siemens/ KWU defined, assuming hydrogen control devices	Not known	Not known
Core catcher	No	No	No	Yes	No	Not known

Table 10.8 shows the core inventories used for the present assessment. The inventories are taken from ERI/HSK (1996), and are based on ORIGEN calculations for an end-of-cycle core with MOX fuel, for the BWR plant. Inventories for a PWR core are slightly different, especially in the Sr-Ba content, but the difference is within the uncertainties generated by the ORIGEN code itself (about 20 to 30 percent).

Table 10.8 Initial core inventories, with 3600 MW_{th} operating power.

Radionuclide group	Initial Activity in Core ($x\ 10^{15}$ Bq)
Xe	1.6E4
I	3.1E4
Cs	1.1E3
Te	9.0E3
Sr	8.9E3
Ru	1.2E4
La	3.8E4
Ce	9.3E4
Ba	6.6E3
Mo	1.2E4

4.6.4 Source terms.

As a basis for the calculations of risks and consequences, source terms were taken from officially published references for both BWR and PWR NPP types (ERI/HSK, 1996; 1999; HSK/KSA, 1996; 1999; SKI, 1999). The source terms are shown in Tables 10.9 (PWR) and 10.10 (BWR). Activities of release are presented in Tables 10.11 (PWR) and 10.12 (BWR).

For advanced reactors both end states of containment and absolute release fractions are assumed to be the same as for European plants, mostly due to similarities in SAM systems, preventive systems, and design (especially containment capacity). Due to these similarities, accident progression following a given initiator, i.e., what follows post core damage and containment phenomena, is not expected to be different. However, for the EPR reactor type there are some major differences in the profile of frequency of containment end states because of the core catcher, which may effectively prevent basemat melt-through, and because almost no containment failure due to hydrogen combustion would be expected due to igniters and recombiners. Improved protection against containment isolation failure is also to be expected.

Radionuclide groups used in source terms (Tables 10.9 through 10.12) slightly differ from those shown in Table 10.8. Sr also includes the Ba group; the Mo group is included in the Ru group; and the La group in source terms consists of Ru, Ce, and La combined. This regrouping was necessary because of the abbreviated information shown in ERI/HSK (1999).

No reliable data on consequences for VVER plants has been published; however, it can be concluded that risks for a VVER design, located in the Shandong peninsula, are at least one order of magnitude higher than the consequences estimated for the BWR and PR plants, given a higher core damage profile.

Table 10.9 Source terms for PWR.

Release class	Xe	I	Cs	Te	Sr	La
RC1	9.10E-1	1.80E-01	1.80E-01	1.40E-01	1.70E-02	5.80E-03
RC2	9.60E-1	3.00E-02	3.00E-02	3.50E-02	1.00E-02	1.20E-03
RC3	9.50E-1	6.00E-03	6.00E-03	7.00E-03	2.70E-03	2.40E-04
RC4	1.70E-03	4.20E-06	4.20E-06	6.00E-06	1.60E-06	2.90E-07
RC5	5.00E-04	7.90E-06	7.90E-06	6.00E-06	7.00E-07	5.00E-07
RC6	3.30E-1	1.00E-01	1.00E-01	4.80E-02	1.00E-02	7.40E-03
RC7	3.70E-1	1.60E-01	1.60E-01	1.20E-01	1.00E-02	1.20E-03

Table 10.10 Source terms for BWR.

Release class	Xe	I	Cs	Te	Sr	La
RC1	7.20E-1	1.20E-01	1.10E-01	2.80E-02	1.70E-02	2.00E-04
RC2	7.30E-1	7.80E-03	7.60E-03	3.40E-03	2.20E-03	2.10E-05
RC3	7.40E-1	8.80E-02	8.00E-02	3.70E-02	3.70E-02	4.30E-04
RC4	7.40E-1	6.30E-04	2.80E-04	1.00E-04	7.80E-05	6.60E-07
RC5	7.40E-1	5.00E-03	2.90E-03	1.30E-03	1.00E-03	1.10E-05
RC6	7.40E-1	1.50E-03	1.40E-03	9.20E-04	4.80E-04	4.70E-06
RC7	7.80E-1	5.90E-02	5.70E-02	2.90E-02	1.90E-02	2.60E-04
RC8	7.40E-1	8.40E-05	3.80E-05	1.80E-05	1.40E-05	1.20E-07
RC9	7.40E-1	6.60E-03	4.00E-03	1.80E-03	1.40E-03	1.60E-05
RC10	7.40E-1	1.50E-03	1.50E-03	8.60E-04	4.50E-04	4.30E-06
RC11	7.40E-1	2.50E-02	2.60E-02	8.30E-03	4.10E-04	4.80E-06
RC12	7.30E-1	4.10E-05	1.80E-05	7.40E-06	3.50E-06	3.20E-08
RC13	7.40E-1	2.20E-03	1.20E-03	3.40E-04	6.60E-09	2.00E-10
RC14	3.60E-1	5.20E-05	5.30E-05	4.00E-05	4.10E-05	4.10E-07
RC16	3.70E-1	4.00E-06	2.40E-06	1.50E-06	1.70E-06	1.50E-08
RC18	7.80E-1	3.90E-06	4.00E-06	2.60E-06	2.70E-06	2.70E-08
RC21	1.30E-1	1.50E-06	5.40E-07	2.80E-07	2.30E-07	1.80E-09

Table 10.11 PWR releases in Bq.

Release class	Xe	I	Cs	Te	Sr	La
RC1	1.46E+19	5.58E+18	1.98E+17	1.26E+18	2.55E+17	9.28E+17
RC2	1.54E+19	9.30E+17	3.30E+16	3.15E+17	1.50E+17	1.92E+17
RC3	1.52E+19	1.86E+17	6.60E+15	6.30E+16	4.05E+16	3.84E+16
RC4	2.72E+16	1.30E+14	4.62E+12	5.40E+13	2.40E+13	4.64E+13
RC5	2.72E+16	2.50E+14	8.69E+12	5.40E+13	1.05E+13	8.00E+13
RC6	5.28E+18	3.10E+18	1.10E+17	4.32E+17	1.50E+17	1.18E+18
RC7	5.92E+18	4.96E+18	1.76E+17	1.08E+18	1.50E+17	1.92E+17

Table 10.12 BWR releases in Bq.

Release class	Xe	I	Cs	Te	Sr	La
RC1	1.152E+19	3.72E+18	1.21E+17	2.52E+17	1.462E+18	3.20E+16
RC2	1.168E+19	2.418E+17	8.36E+15	3.06E+16	1.892E+17	3.36E+15
RC3	1.184E+19	2.728E+18	8.8E+16	3.33E+17	3.182E+18	6.88E+16
RC4	1.184E+19	1.953E+16	3.08E+14	9.00E+14	6.708E+15	1.056E+14
RC5	1.184E+19	1.55E+17	3.19E+15	1.17E+16	8.6E+16	1.76E+15
RC6	1.184E+19	4.65E+16	1.54E+15	8.28E+15	4.128E+16	7.52E+14
RC7	1.248E+19	1.829E+18	6.27E+16	2.61E+17	1.634E+18	4.16E+16
RC8	1.184E+19	2.604E+15	4.18E+13	1.62E+14	1.204E+15	1.92E+13
RC9	1.184E+19	2.046E+17	4.40E+15	1.62E+16	1.204E+17	2.56E+15
RC10	1.184E+19	4.65E+16	1.65E+15	7.74E+15	3.87E+16	6.88E+14
RC11	1.184E+19	7.75E+17	2.86E+16	7.47E+16	3.526E+16	7.68E+14
RC12	1.168E+19	1.271E+15	1.98E+13	6.66E+13	3.01E+14	5.12E+12
RC13	1.184E+19	6.82E+16	1.32E+15	3.06E+15	5.676E+11	3.20E+10
RC14	5.76E+18	1.612E+15	5.83E+13	3.60E+14	3.526E+15	6.56E+13
RC16	5.92E+18	1.24E+14	2.64E+12	1.35E+13	1.462E+14	2.40E+12
RC18	1.248E+19	1.209E+14	4.40E+12	2.34E+13	2.322E+14	4.32E+12
RC21	2.08E+18	4.65E+13	5.94E+11	2.52E+12	1.978E+13	2.88E+11

4.6.5 Offsite consequences

Fatalities and land contamination. Offsite consequences were calculated using the simplified methodology, for the number of fatalities and area of severely contaminated land. Offsite consequences include fatalities and land contamination. Fatalities could continue to occur over an extended period. Early fatalities would occur shortly after exposure; the calculation involves early fatalities within an eight km radius of the source of the release. Based on earlier MACCS calculations, early fatalities would not be expected to occur further away than eight to ten km from the plant. Late fatalities result from exposure within a few days after the accident and would be manifested over a long period. These fatalities would result mostly from inhalation and immersion related to the passage of the radioactive cloud over an area. Population data up to 100 km from the site was used to assess the late fatalities; according to MACCS calculations the corresponding contributions beyond 100 km are negligible. Delayed fatalities would occur due to delayed exposure (long-term decay) in a long time period in the area in a radius up to 800 km from the source of the release. Latent fatalities include all those occurring within 60 years from the accident; these include the sum of late and delayed fatalities.

Land contamination due to releases after a hypothetical accident was calculated for an area up to 120 km of radius from release; beyond 120 km the cloud would be so diluted that high levels of ground activity would not be expected. "Heavy" land contamination means that the land cannot be decontaminated effectively for several years, as discussed below. For land contamination both interdicted and condemned areas were taken into account. In this chapter "interdicted" area is land which can be successfully decontaminated within 20 years and then resettled; a "condemned" area is land which cannot be decontaminated within 20 years.

Release sequences. Consequences were calculated for two specific release sequences with MACCS:

- The first starts at two hours from scram, is of short duration, and occurs at 100 m (stack elevation).
- The second starts 20 hours after scram, is of relatively long duration, and occurs at a height of ten meters.

One of the main assumptions leading to the calculation of hypothetical consequences for Shandong province was the fact, that the ratio of population *(p)* and fatalities *(f)* of two different sites can be considered to be approximately the same:

$$p1/p2 = f1/f2$$

Knowing the population and consequences for a reference site and the population for a different site makes it possible to calculate likely consequences for the other site.

The quotients q were used to calculate the consequences for the Qingdao and Yantai sites from known consequences for hypothetical accidents of European NPPs types.

$$q = p1/p2 = f1/f2 \quad \rightarrow \quad f1 = q*f2 \qquad \text{(see Table 10.13)}$$

Table 10.13 summarizes the results of population and land areas calculations as well the quotients (ratios) used for further consequence calculations. Calculations of consequences for reference sites were made with MACCS code, where two types of reference sites were considered: European with high population density, and US with lower population density.

Table 10.13 Quotients for consequence calculations.

	European site	US site	Quotient q for US/EU site	Qingdao	Quotient q for Qingdao *	Yantai	Quotient q for Yantai *
Population to 8 km	24'000	1400	0.06	70'000	2.92	50'000	2.08
Population to 100 km	5E+6	2.2E+6	0.44	1E+7	2.00	6.4E+6	1.28
Population to 800 km	226E+6	90E+6	0.40	590E+6	2.61	517E+6	2.29
Land fraction to 120 km	0.97	0.75	0.77	0.5	0.67	0.37	0.49

* For population, the ratio is taken to the European site because of much higher population density (closer to European), while for the land fraction the ratio is taken with respect to the US site, which, as the Shandong peninsula, is situated in the proximity of the ocean.

Weather. For these calculations, the weather was assumed to be uniform (i.e., without a preferred direction) as in the US site. According to the information from Hirschberg et al. (2003b) about the weather in the region, the prevailing wind direction for the sites considered is southwest, which could result in an added uncertainty of about 25 percent in the calculations. Average precipitation at both reference sites is approximately 75 cm/year, the same as the average precipitation in the whole of Shandong province. For results of the calculations of conditional consequences per Bq released we refer to the technical report Hirschberg et al. (2003a).

Offsite consequences for current reactors. In calculating the offsite consequences for current reactors core damage frequencies from SKI (1999) were primarily used to assess risks. Given that a plant's safety culture may have a decisive impact on the risk and that this aspect has not been addressed for China-specific conditions, this core damage frequency level is considered here to be more representative than the results for the advanced designs. Therefore, the frequencies of postulated accidents outlined here are ten times higher than what is normally assumed for advanced reactors.

Tables 10.14 to 10.18 show conditional consequences (i.e., without accounting for accident frequency) for the worst hypothetical accidents for the Qingdao site. For more detailed information for other release groups, different types of NPPs, and the (in relative terms somewhat more favorable) results for the Yantai site see Hirschberg et al. (2003a). Note that risk in these tables is not considered, just conditional consequences (i.e., consequences, given that such accidents would occur).

There are only a few early fatalities (Table 10.14). The high relative contribution of not only the iodine group but also of the La radionuclide group should be emphasized. This fact could be interesting from the point of view of countermeasures, which are in general focused on eliminating the influence of the iodine group.

As mentioned above, the La group is a significant contributor to the late fatalities due to early exposure (Table 10.15). Again for PWR reactors, the worst consequences seem to be from the release class RC1.

The main contributors to delayed fatalities due to delayed exposure seem to be according to the expectation Cs and Sr groups (Table 10.16). Table 10.17 provides the values for latent fatalities for the Qingdao site considering both early and delayed exposure. It summarizes the values from Tables 10.14 and 10.15 for late and delayed fatalities. It is evident that the most important contribution to the total latent fatalities is the delayed ones (i.e., those caused by late exposure).

Table 10.14 Early fatalities for the Qingdao site for the worst conditional consequences.

	Release class / NPP type	
Radionuclide group	**RC3 / BWR (Number of fatalities)**	**RC1 / PWR (Number of fatalities)**
Xe	0	0
I	11.2	22.9
Cs	0.14	0.39
Te	11.7	44.1
Sr	5.5	2.5
La	6.3	84.5
TOTAL	35	154

Table 10.15 Late fatalities (early exposure) for the Qingdao site for the worst conditional consequences.

	Release class / NPP type	
Radionuclide group	**RC3 / BWR (Number of fatalities)**	**RC1 / PWR (Number of fatalities)**
Xe	11.8	14.56
I	180.0	368.28
Cs	281.6	633.6
Te	226.4	856.8
Sr	1527	701.76
La	192.6	2598.4
TOTAL	2420	5173

Table 10.16 Delayed fatalities (delayed exposure) for the Qingdao site for the worst conditional consequences.

Radionuclide group	Release class / NPP type	
	RC3 / BWR (Number of fatalities)	**RC1 / PWR (Number of fatalities)**
Xe	0	0
I	190.96	390.6
Cs	7568	17’028
Te	113.22	428.4
Sr	7636.8	3508.8
La	64.672	872.32
TOTAL	15’574	22’228

Table 10.17 Latent fatalities (early and delayed exposure) for the Qingdao site for the worst conditional consequences.

Radionuclide group	Release class / NPP type	
	RC3 / BWR (Number of fatalities)	**RC1 / PWR (Number of fatalities)**
Xe	11.84	14.56
I	371	758.88
Cs	7849.6	17’661.6
Te	339.66	1285.2
Sr	9164.2	4210.5
La	257.3	3470.72
TOTAL	17’994	27’401

Table 10.18 Land contamination – land lost in km^2 for the Qingdao site for the worst conditional consequences.

Radionuclide group	Release class / NPP type	
	RC3 / BWR (Land lost in km^2)	**RC1 / PWR (Land lost in km^2)**
Xe	0	0
I	0	0
Cs	269.61	851.4
Te	0	0
Sr	2941.2	2631.6
La	0	0
TOTAL	3211	3483

Given the assumptions on source terms conditional consequences for advanced reactors would be the same as for current design plants. Needless to say, when risks are considered (i.e., the product of consequences and frequency of accidents), risk will be smaller for plants of newer design.

4.6.6 Main risk results

Risks of possible fatalities or land contamination are related to the frequency of core damage states and the conditional probability of containment failure modes (release categories, release bins), given particular plant damage states (accident classes) leading to the release after a hypothetical accident.

Average risk for a particular release category is understood to be the multiplication of the frequency of release per year with the absolute conditional consequences related to a given release bin. Total risk is the overall sum of all average risks. Conditional consequences from postulated severe accidents for PWRs and BWRs have been discussed in the previous section.

The comparison of risk results for particular consequences and the particular reactor types and particular sites are summarized in the Tables 10.19 and 10.20, and Figures 10.28 to 10.35. The results for current designs are shown first. Assuming that core damage frequency is ten times higher than for advanced designs, the associated risks (per Reactor*year, Ryr) are correspondingly higher, subject to additional differences when the indicators are normalized by $GW_e yr$ (depending on the power level of the various reference designs).

Table 10.19 Risk for the Qingdao and Yantai sites – current reactors.

Site / NPP Type	**Risk of early fatalities (number/$GW_e yr$)**		**Risk of latent fatalities (number/$GW_e yr$)**		**Risk of land contamination ($km^2/GW_e yr$)**	
	BWR	PWR	BWR	PWR	BWR	PWR
Qingdao	9.8E-6	2.5E-5	4.9E-3	6.1E-3	1.4E-3	6.7E-4
Yantai	7.0E-6	1.7E-5	4.2E-3	5.1E-3	1.1E-3	4.9E-4

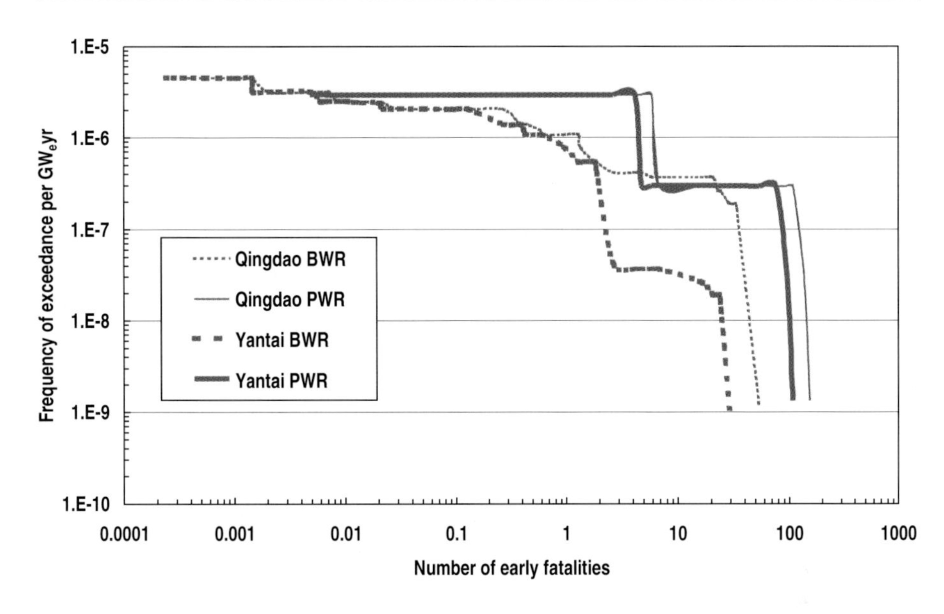

Figure 10.28 Frequency of exceedance of the number of early fatalities – current reactors.

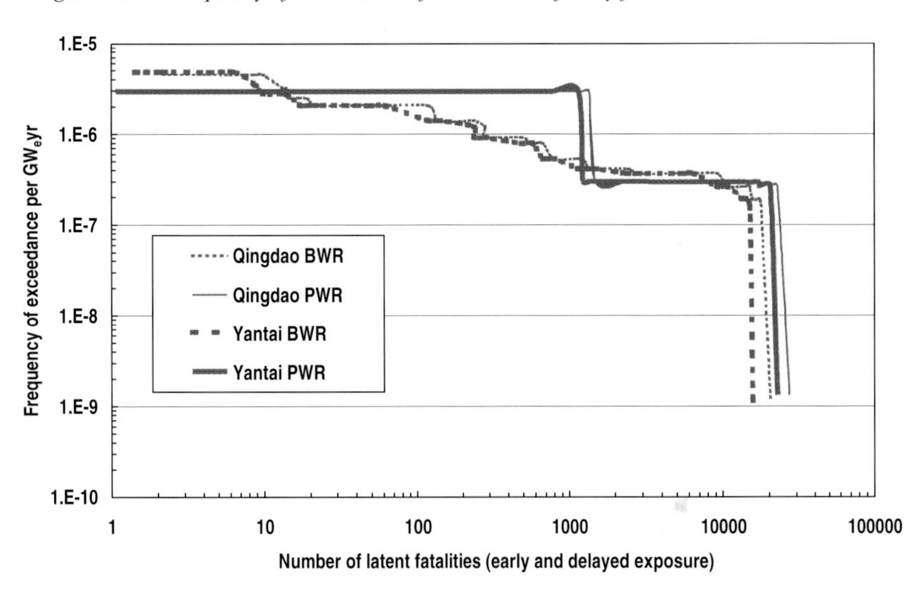

Figure 10.29 Frequency of exceedance of the number of latent fatalities – current reactors.

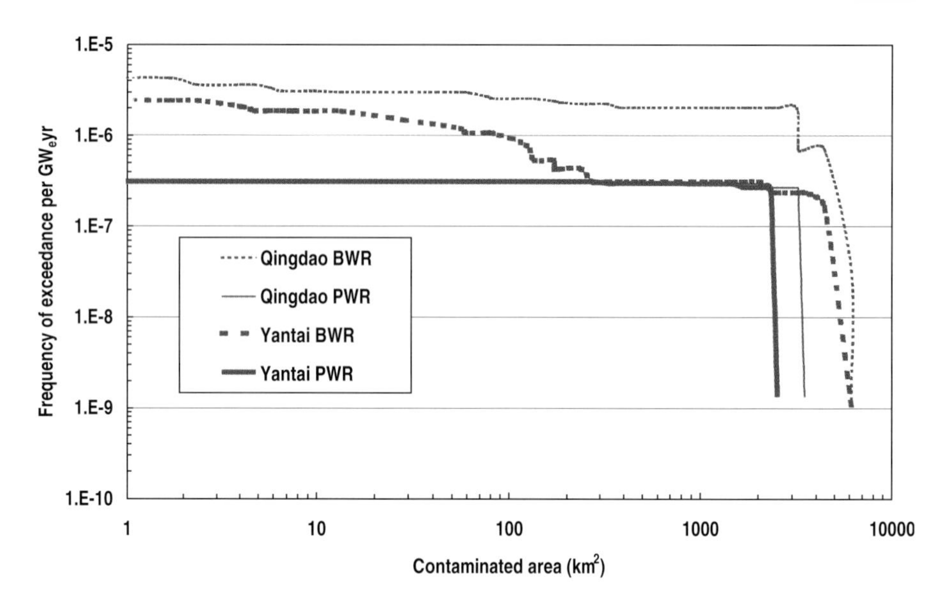

Figure 10.30 Frequency of exceedance for contaminated area – current reactors.

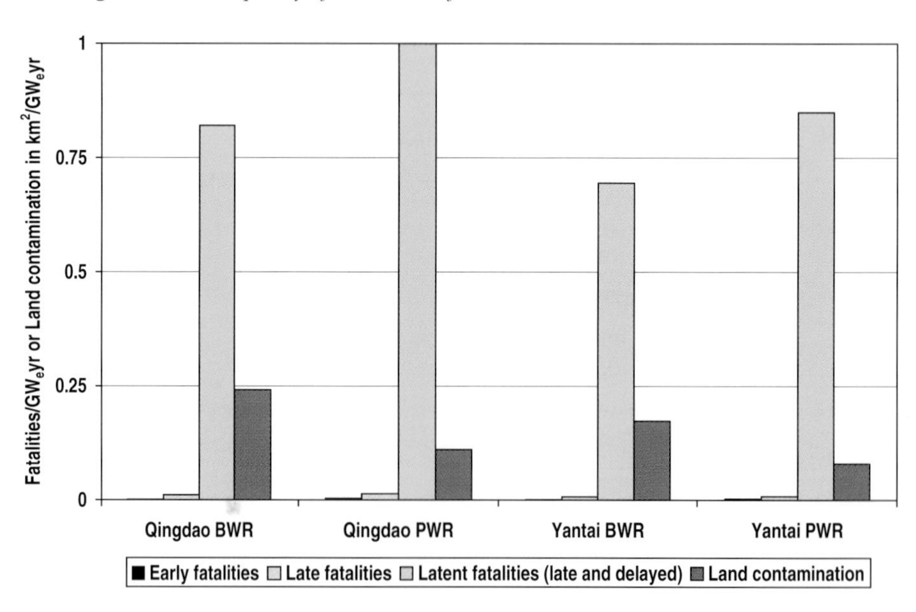

Figure 10.31 Comparison of risks normalized to the largest risk – current reactors (maximum value – Qingdao PWR, 6.1 E-3 GW_e yr).

Table 10.20 Risk for the Qingdao and Yantai sites – advanced reactors.

Site / NPP Type	Risk of early fatalities (number/GW_eyr)		Risk of latent fatalities (number/GW_eyr)		Risk of land contamination (km^2/GW_eyr)	
	ABWR	EPR	ABWR	EPR	ABWR	EPR
Qingdao	9.5 E-7	2.0 E-6	4.7 E-4	4.8 E-4	1.4 E-4	5.4 E-5
Yantai	6.7 E-7	1.4 E-6	4.1 E-4	4.1 E-4	1.0 E-4	3.9 E-5

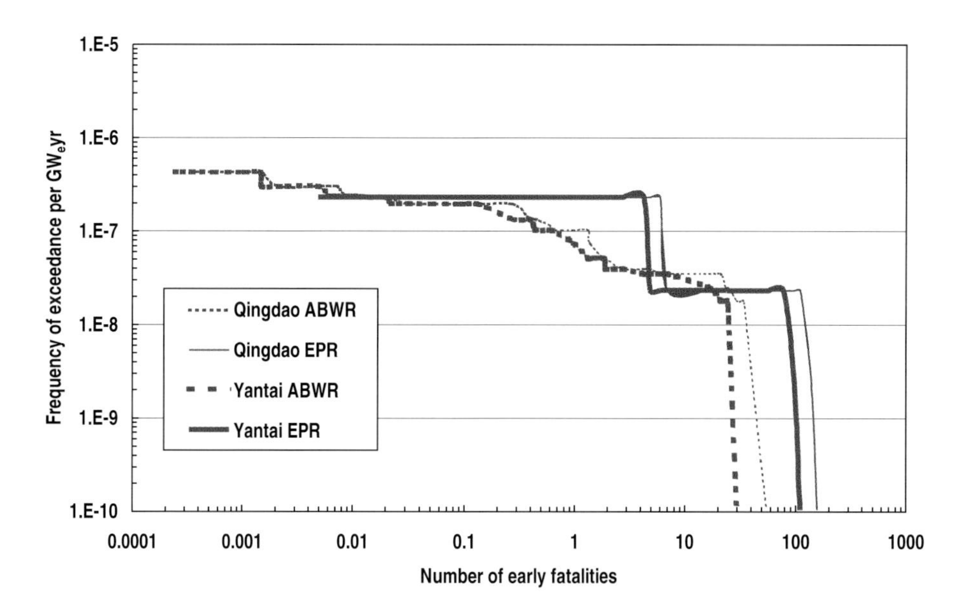

Figure 10.32 Frequency of exceedance for the number of early fatalities – advanced reactors.

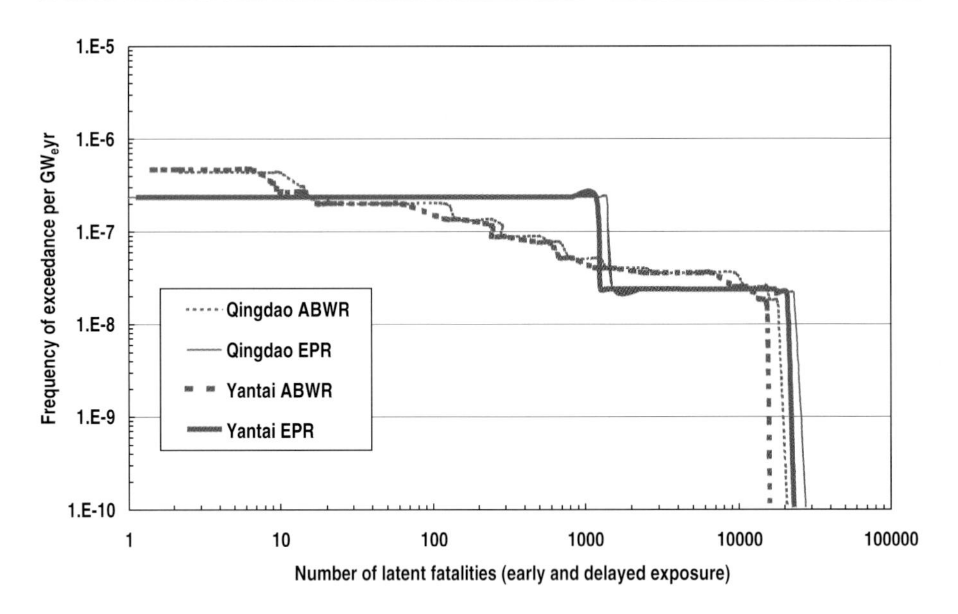

Figure 10.33 Frequency of exceedance for the number of latent fatalities – advanced reactors.

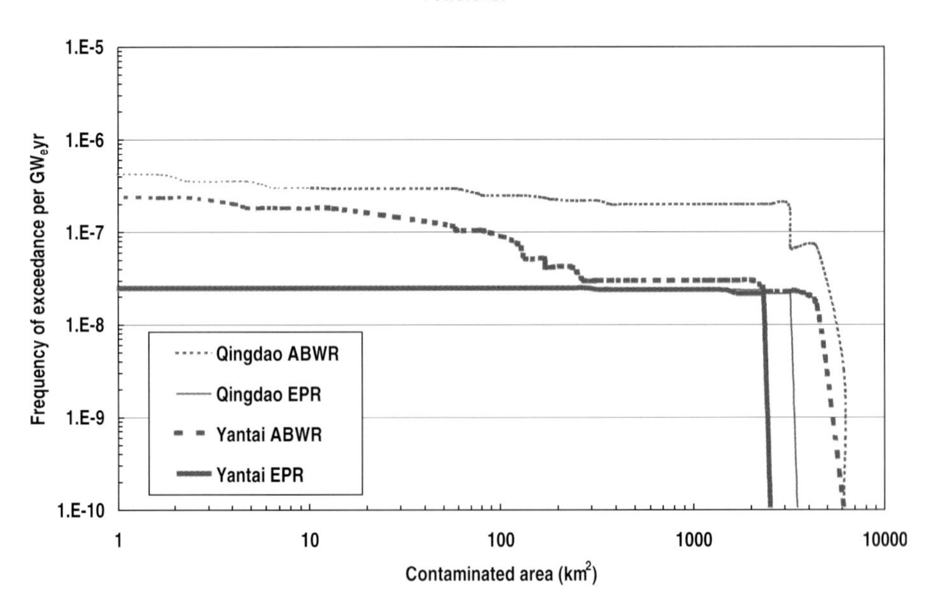

Figure 10.34 Frequency of exceedance for contaminated area in km^2 – advanced reactors.

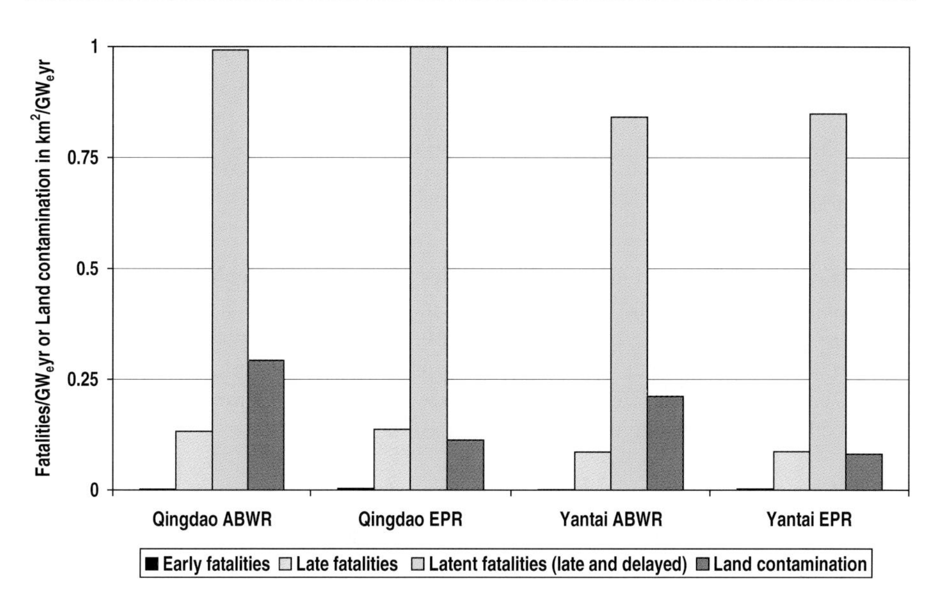

Figure 10.35 Comparison of risks normalized to the largest risk – advanced reactors (maximum value – Qingdao EPR, 4.8 E-4 GW_eyr).

As it is evident from Figures 10.31 and 10.35, for both current and advanced reactors, early fatalities are negligible compared to other risks. It should be noted that even early fatalities are overestimated in the simplified method used to extrapolate the results for fatalities using quotients. In reality the number of fatalities as a function of exposure is strongly exponential, once the fiftieth percentile of the lethal dose (known as LD-50), which is about three Gy, is reached. In the simplified method a linear extrapolation was used, which results in predicting possible fatalities even at low doses. Also, the risk of latent fatalities does not appear significant, given that it is calculated over a 60-year period. In this time, even 100'000 cancer deaths could not be statistically separated from normal cancer deaths (perhaps with the exception of thyroid cancers in the young population). In relative terms, the risk of land contamination is a major concern due to the related social consequences. Moreover, risk is found to be plant- and site-sensitive.

Influence of plant type. The risk of early fatalities for a BWR type reactor is about 100 smaller than for a PWR. It can be explained by the influence of the BWR pressure suppression pool, which is very effective in suppressing the release of aerosols, especially iodine. The difference in risk between BWR and PWR for latent fatalities is smaller because latent fatalities are caused mostly by long-lived radionuclides, and the release of these species is insensitive to plant design. On the other hand, results for land contamination appear more favorable for PWR type reactors. This is because, based on the results of MACCS calculations, the only contributors to land contamination are the Sr and Cs groups of radionuclides. The Sr-Ba group is mostly released late, ex-vessel, so the pressure suppression pool is less effective. Another reason can be the higher containment capacity in a PWR, so the frequency of relatively large and early releases of Sr and Cs for this type of reactor is lower.

In general it can be said that the differences in risk measures due to differences in plants are not decisive, i.e., risk measures for both BWR and PWR are comparable.

Influence of site. The results are in general less favorable for the Yantai site because of smaller population density and smaller land fraction in the area. The difference between risks attributable to difference in plant siting is comparable or somewhat smaller than the difference between risks caused by different plants designs, when plants belonging to the same generation are considered. Comparing current and advanced plants the design differences are dominant.

4.6.7 Uncertainties

The results presented here were obtained using simplified methods which involve uncertainties of different types for two different reactor designs, and include consideration of both currently operating plants and of advanced designs. In this section, sensitivity to some of the leading uncertainties in the assessment of offsite consequences is summarized. The calculations were carried out for advanced reactors but the conclusions would be the same for the current ones. Detailed information on the sensitivity calculations is provided in Hirschberg et al. (2003b).

Core inventories used in the calculations were assumed to be the same for all types of reactors. Assuming that the uncertainties of the ORIGEN code for core inventories depending on reactor design could be about ±20/30 percent, the uncertainty resulting from inventories alone can be as high. Calculations were repeated under these assumptions.

The calculations of weather influence did not consider the predominant wind direction in the Shandong peninsula. According to Hirschberg et al. (2003a), preferred wind direction is from the S/W quadrants towards N/E. This could add a source of uncertainty of 25 percent. Calculations for the Yantai site could be overestimated by 25 percent, while for the Qingdao site they could be underestimated by 25 percent.

Another source of uncertainties can occur from the population data. Though the most precise possible data for Shandong province was used (China Statistical Information Center, 2000) to estimate the population in the radial area of 800 km, the uncertainty may still be substantial. It was postulated that the uncertainty on this estimate could be as high as ±20 percent.

Source terms have large uncertainties in general, e.g. factors ±10-20 for small releases (fractions of iodine smaller than one percent), factors of ±2-3 for large releases (fractions of iodine larger than 10 percent). In the present sensitivity study, "large" release factors larger by two are used for large releases, and factors of 10 for small releases. Releases of noble gases have not been changed, but this has very little influence on risk.

Figure 10.36 shows a comparison of the risk measures for the sensitivity studies which have been considered. The figure shows that the increase of the source terms has the most significant influence on the average risk. To understand the differences in the source term sensitivity cases, it is important to know that early fatalities are driven by release of I, Te, and La; latent fatalities essentially by Cs; and land contamination by Cs and Sr. Thus, in the context of land contamination the higher sensitivity for ABWR can be explained by the much increased release of Sr inventory due to bypass of the pressure-suppression pool after vessel breach.

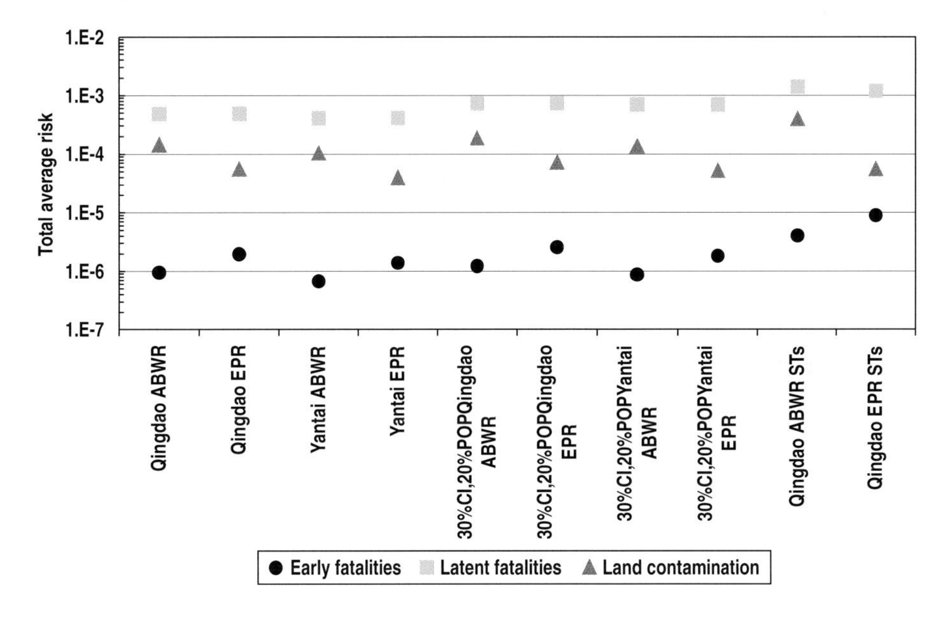

Figure 10.36 Comparison of risks for different sites and different plants and uncertainty results.

4.6.8 Conclusions on nuclear risk

In general it can be said that the sensitivity cases investigated were shown to be not of significant importance in relation to risk measures, with the exception of uncertainties in core damage frequency and source terms. Risk measures presented in the report are small, even lower than some US estimates (e.g. Park et al., 1991). The main reason for these results, as compared to earlier US studies, is that the NUREG-1150 analyses mostly relied on core damage frequencies, which were estimated according to technology which was available in the early 1980s, while the present results are based on studies performed with a much larger reliability data base and experience gained from earlier PSA studies. For instance, it may be remembered that estimated risks had been already reduced by a factor of about ten between the WASH-1400 and the NUREG-1150 studies.

Differences in risk between the designs of particular types of reactors only reflect known features, which are, for instance, less severe accident mitigative defenses in BWRs, but more preventive defenses and less capacity for containment of BWRs compared to PWRs, and the impact of pressure suppression pools in BWRs.

Differences in risk due to site characteristics are mostly due to different population density patterns at the considered sites but they still are approximately of the same order of magnitude. For this reason, the results can be considered not highly sensitive to assumptions on the population density in the Chinese sites.

The risk measures for all considered cases, including the sensitivity analyses, are small. The MACCS results indicate that weather effects could change the risk results by only about ±25 percent. Finally, even if total CDF were 100 times higher than assumed, risk measures would be still relatively small. However, it should be kept in mind that safety culture may have a decisive impact on risk and this aspect has not been addressed for China-specific conditions.

5. ENERGY CHAIN COMPARISONS

The data presented in Table 10.20 summarize the available information on severe accidents in various Chinese energy chains. Severe accidents in the coal chain are largely underestimated before 1994 due to restrictive reporting by Chinese authorities. The situation improved a lot with the official release of data and their annual publication in the *China Coal Industry Yearbook.* For this reason, coal accident statistics in the CETP framework are based on the period 1994-1999. In contrast, the China-specific database on severe accidents for oil, natural gas, LPG, and especially hydro power is much less extensive. Chinese data for these energy chains were pooled with those for non-OECD countries for comparative analysis purposes.

Table 10.20 Summary of the severe accident database used for CETP. Time periods considered are 1969-1999 for China and 1969-1996 for non-OECD countries without China and for OECD countries alone.

	China		**Non-OECD w/o China**		**OECD**	
Energy chain	**accidents**	**fatalities**	**accidents**	**fatalities**	**accidents**	**fatalities**
Coal Coal CETP[(a)]	1016 816	17'241 11'321	103	5023	43	1410
Oil	32	613	183	12'737	143	2627
Natural gas	8	201	38	920	45	536
LPG	11	291	27	2308	47	790
Hydro[(b)]	3	26'028	8	3896	1	14

(a) Coal CETP covers the period 1994-1999 as used for statistical evaluations of the Chinese coal chain.
(b) Banqiao and Shimantan dam failures in China caused 26'000 fatalities.

5.1 Aggregated indicators and frequency consequence curves

Based primarily on the historical evidence, comparisons among the different energy chains were carried out. They are focused on the experience in China, complemented by the records from non-OECD countries. Also for reference some evaluations for OECD countries are provided.

In view of the availability of extensive Chinese-specific statistical material for coal accidents in the period 1994-1999, and considerable differences between Chinese coal energy chain accident rates compared to other countries, within the CETP framework the Chinese data are considered as the only relevant representation of the risks associated with coal mines in China. With the possible exception of hydro power, there are no strong indications that the Chinese situation would be much different from the average in non-OECD countries.

The comparison covers aggregated indicators (i.e., fatality rates per GW_eyr) and frequency-consequence curves. It addresses fatality rates exclusively since no consistent data could be found for other accident indicators for all energy chains. Furthermore, some of the consequences are associated with only one energy carrier (e.g. oil spills) or are most pronounced for a specific chain (e.g. long-term land contamination). We refer to the appropriate sections of the preceding sections of this chapter for a detailed account of such issues.

Figure 10.37 shows immediate fatality rates for severe accidents in major energy chains for China, non-OECD countries without China, non-OECD countries with China, and OECD countries alone. Results for coal in China clearly demonstrate that the risk performance of small mines is much worse compared to large mines as reflected by an almost five times higher accident fatality rate. The Chinese severe accident fatality rate for the coal chain is about 6.5 per GW_eyr, i.e. about ten times higher than in non-OECD countries, and about 50 times higher than in OECD countries. The Chinese fatality rate for the oil chain is about 40 percent lower compared to non-OECD countries. This difference is primarily attributable to two very large accidents with 2700 (Afghanistan) and 3000 (Philippines) fatalities. If these two accidents were excluded, the non-OECD fatality rate would decrease to 0.47 per GW_eyr, similar to that of China. As observed for coal, the OECD fatality rate for oil is lowest. The natural gas, LPG, and hydro chains all exhibit distinctly lower fatality rates for OECD countries, compared to non-OECD countries (including China). The non-OECD fatality rate for the hydro chain is strongly dependent on whether the world's worst accident that occurred at Banqiao and Shimantan dams in 1975 in China is included or not in the evaluation.

Only immediate fatalities are covered here. Latent fatalities, of particular relevance for the Chernobyl accident, are not shown in the figure. The associated estimates and an extensive discussion of the related issues are provided in Hirschberg et al. (1998). We also refer to the results obtained for the types of nuclear power plants of more direct interest to China, generated using the Probabilistic Safety Assessment (PSA) approach and discussed in detail above. These results show that the risk of early fatalities associated with hypothetical accidents at the current type or advanced nuclear power plants at selected sites in Shandong is extremely low; risk of latent fatalities is higher but remains three to four orders of magnitude lower than the experience-based severe accident results for coal in China and hydro in non-OECD countries with China, and two to three and one to two orders of magnitude lower than severe accident fatality rates for oil and gas respectively in non-OECD countries with China. The predictive estimates for latent nuclear fatalities are of the same order as the experience-based results for hydro power in OECD countries.

Figure 10.38 gives the frequency-consequence curves for coal in China, non-OECD countries without China and OECD countries alone. Differences among the curves are very distinct with the Chinese curve ranking last with regard to risk performance. There is a large difference between the performance of large mines and small mines in China; the latter have quite shocking accident records. On the positive side, the maximum number of fatalities in any single accident was lower for China (114) than for other non-OECD (434) or OECD countries (272) for the periods considered. Nevertheless, accidents with more fatalities also occurred in the Chinese coal chain before and after the period 1994-1999. For example, an accident in 1982 caused 284 fatalities (exact location unknown), and one in 2000 resulted in 159 fatalities (Muchonggou mine, Guizhou).

Frequency-consequence curves for oil are shown for China, non-OECD countries with and without China and OECD countries alone (Figure 10.39). Although the data basis for severe accidents in the Chinese oil chain is rather small compared to non-OECD countries, deviations are much smaller than those found for coal. Therefore, the non-OECD experience is here considered to be also representative for China. In non-OECD countries, six accidents with more than 200 fatalities occurred, whereas in China the worst accident in 1983 in the South China Sea claimed 81 fatalities.

Figures 10.40 and 10.41 show frequency-consequence curves for the natural gas and LPG chains. The risk performance of China's natural gas chain appears to be worse compared to other non-OECD countries, but it is likely that this result is caused by very limited statistical experience. Therefore, it seems more appropriate to use non-OECD experience as representative for China also. Non-OECD experience is also considered to be representative for China's LPG chain.

Frequency-consequence curves for hydro power are given in Figure 10.42. Due to the generally small available data base, the Chinese hydro power experience is not shown separately. The most severe dam accident that occurred in the period examined was the failure at Banqiao and Shimantan dams in China resulting in 26'000 fatalities.

Frequency-consequence curves for the nuclear chain were based on Probabilistic Safety Assessment (PSA) and presented for the two sites Qingdao and Yantai (Figure 10.43).

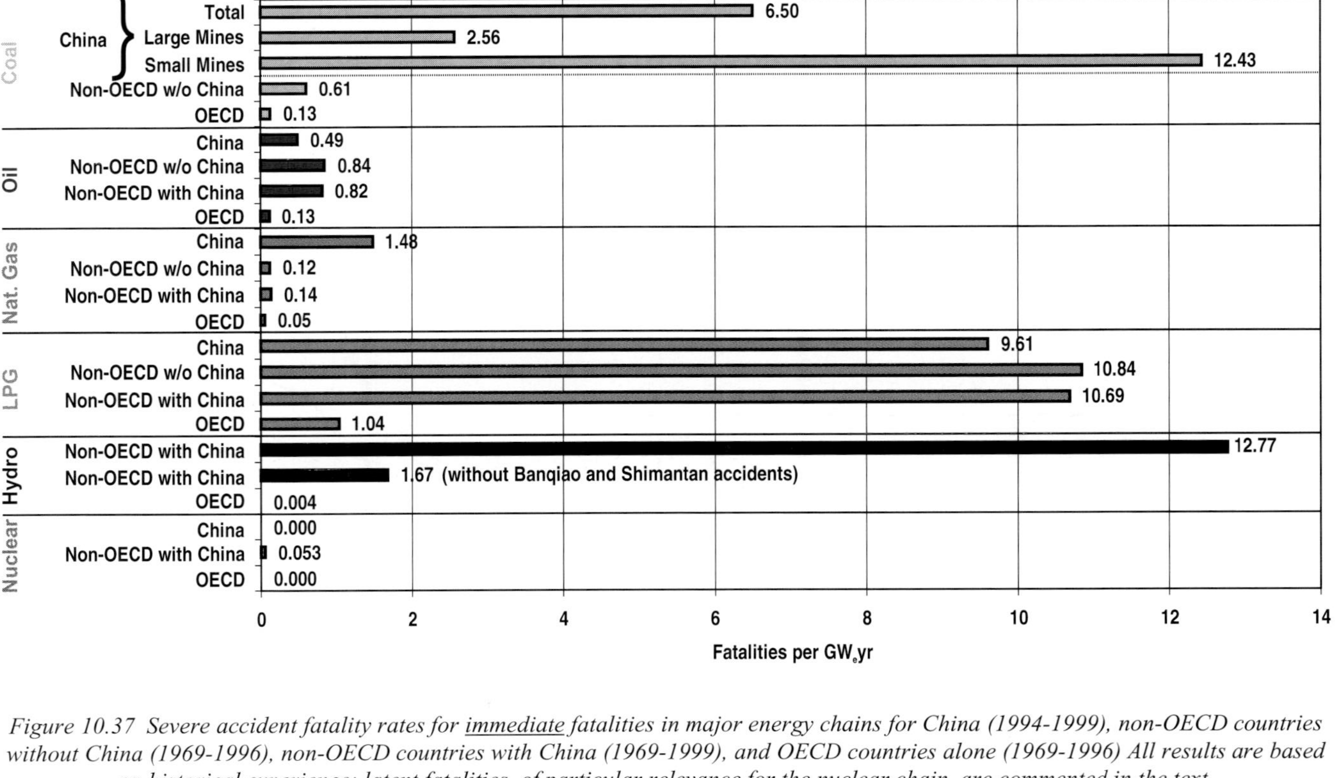

Figure 10.37 *Severe accident fatality rates for* <u>*immediate*</u> *fatalities in major energy chains for China (1994-1999), non-OECD countries without China (1969-1996), non-OECD countries with China (1969-1999), and OECD countries alone (1969-1996) All results are based on historical experience;* <u>*latent*</u> *fatalities, of particular relevance for the nuclear chain, are commented in the text.*

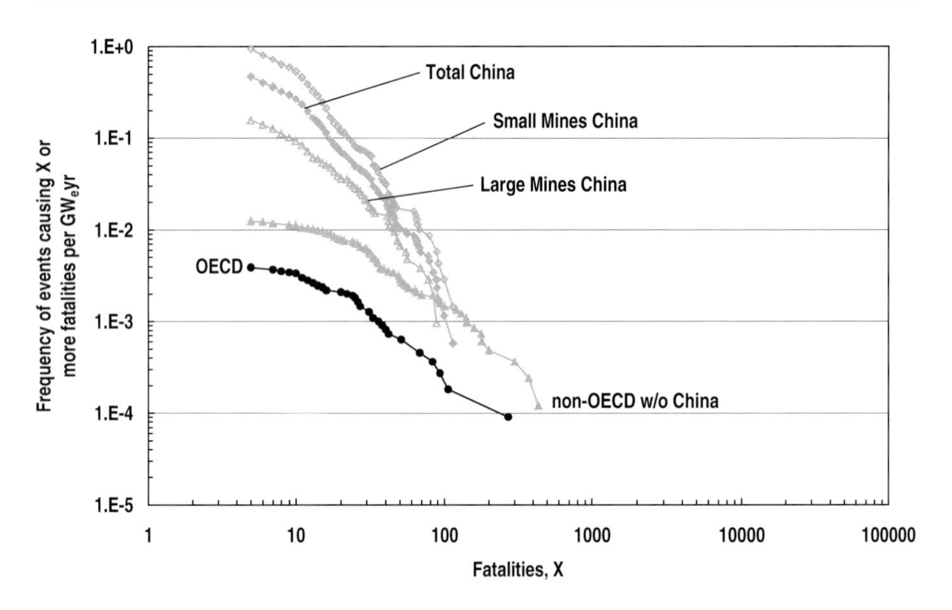

Figure 10.38 Frequency-consequence curves for the coal chain for China (1994-1999), non-OECD countries without China (1969-1996) and OECD countries alone (1969-1996).

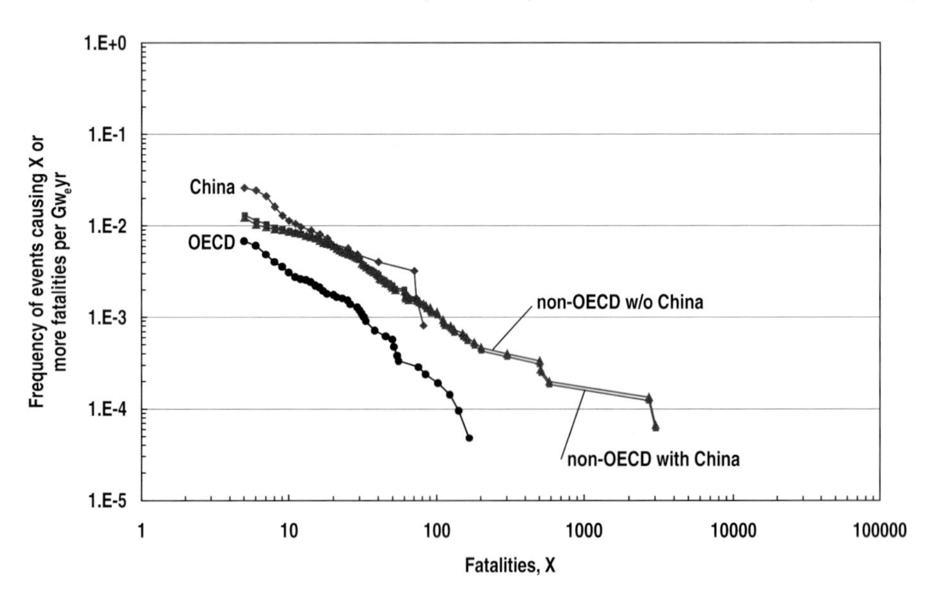

Figure 10.39 Frequency-consequence curves for the oil chain for China (CETP, 1969-1999), non-OECD countries without China (1969-1996), non-OECD countries with China (1969-1999), and OECD countries alone (1969-1996).

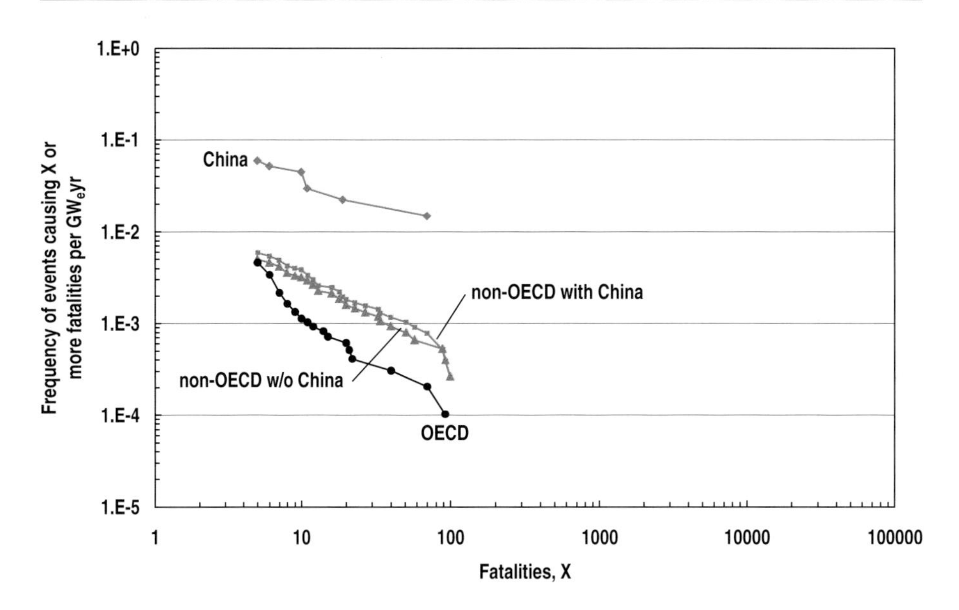

Figure 10.40 Frequency-consequence curves for the natural gas chain for China (CETP, 1969-1999), non-OECD countries without China (1969-1996), non-OECD countries with China (1969-1999), and OECD countries alone (1969-1996).

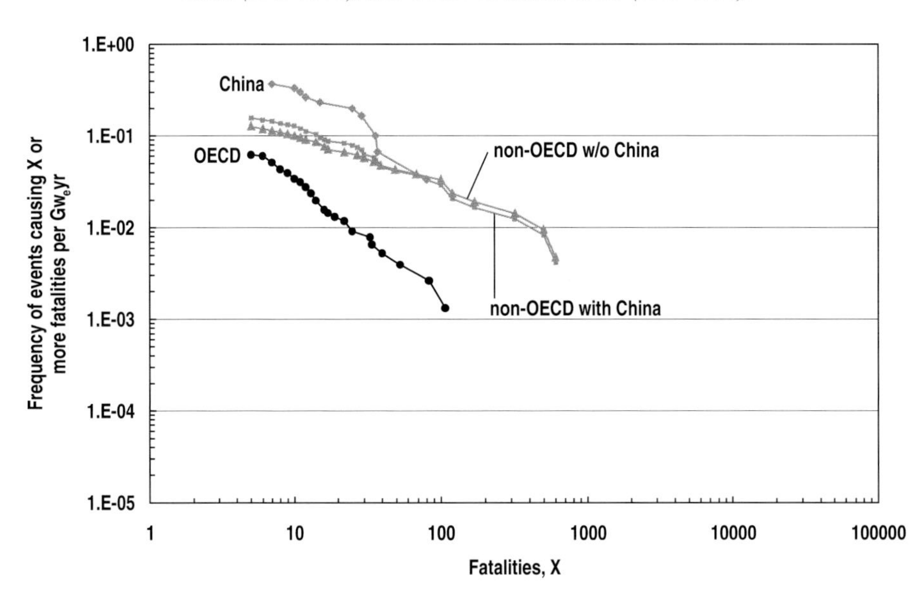

Figure 10.41 Frequency-consequence curves for the LPG chain for China (CETP, 1969-1999), non-OECD countries without China (1969-1996), non-OECD countries with China (1969-1999), and OECD countries alone (1969-1996).

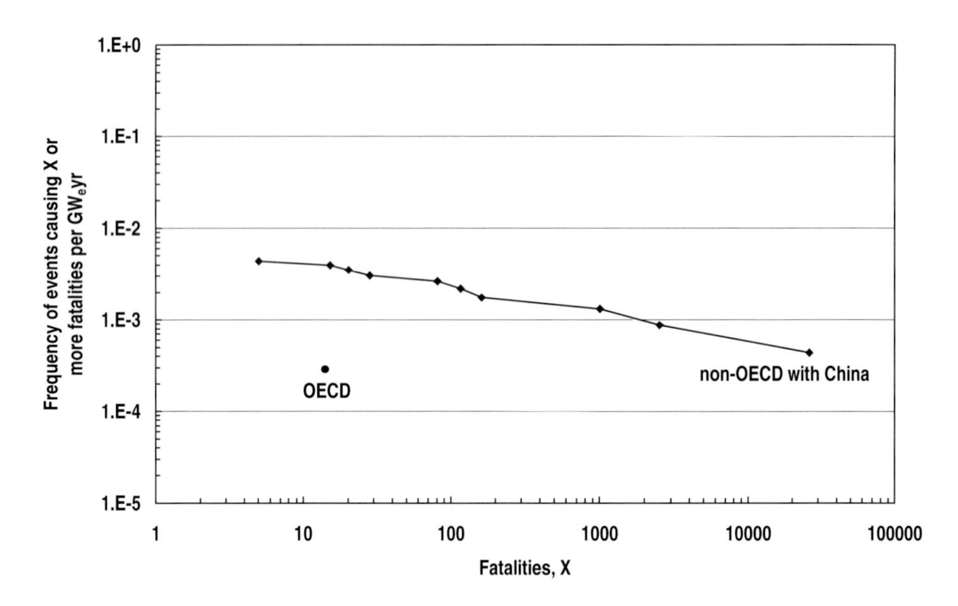

Figure 10.42 Frequency-consequence curves for the hydro power chain for non-OECD countries with China (1969-1999), and OECD countries alone (1969-1996).

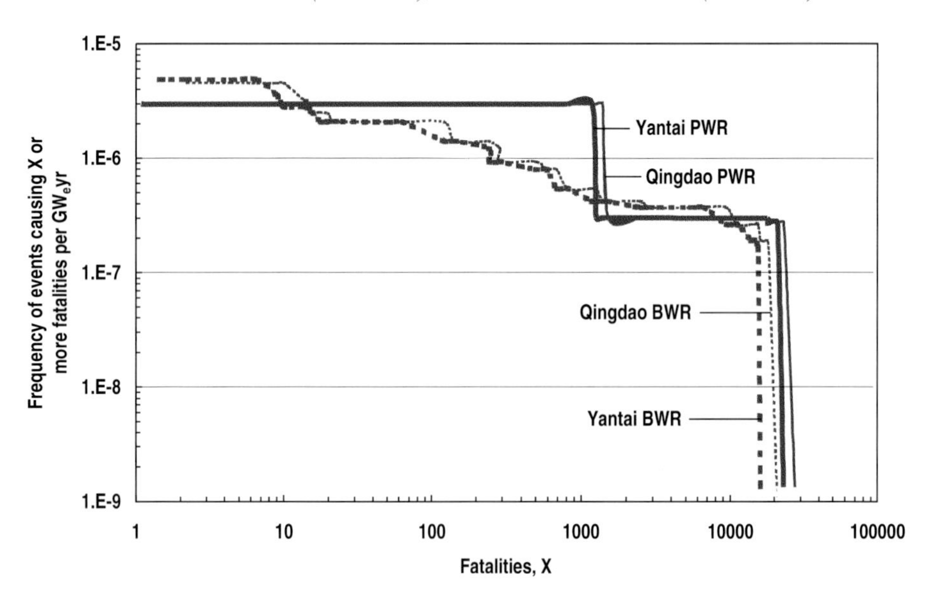

Figure 10.43 Frequency-consequence curves for nuclear power plants, based on Probabilistic Safety Assessment for two sites, Qingdao and Yantai, in Shandong province.

Figure 10.44 shows the final selection of frequency-consequence curves for comparison of the different Chinese energy chains. Among Chinese energy chains, coal exhibits the highest accident frequencies. However, the vast majority of severe coal accidents in China results in less than 100 fatalities. The natural gas chain shows a favorable performance with regard to accident frequencies and maximum fatalities, but natural gas is currently of relatively minor importance for China's electricity mix. Accident frequencies for the oil and hydro chains are also much lower than for the coal chain. However, maximum numbers of fatalities within the oil and hydro chains are respectively one and two orders of magnitude higher than for coal and natural gas chains. Finally, expectation values for severe accident fatality rates associated with hypothetical nuclear accidents are lowest among the relevant energy chains. The maximum credible consequences may be very large, i.e. comparable to the Banqiao and Shimantan dam accident that occurred in China in 1975.

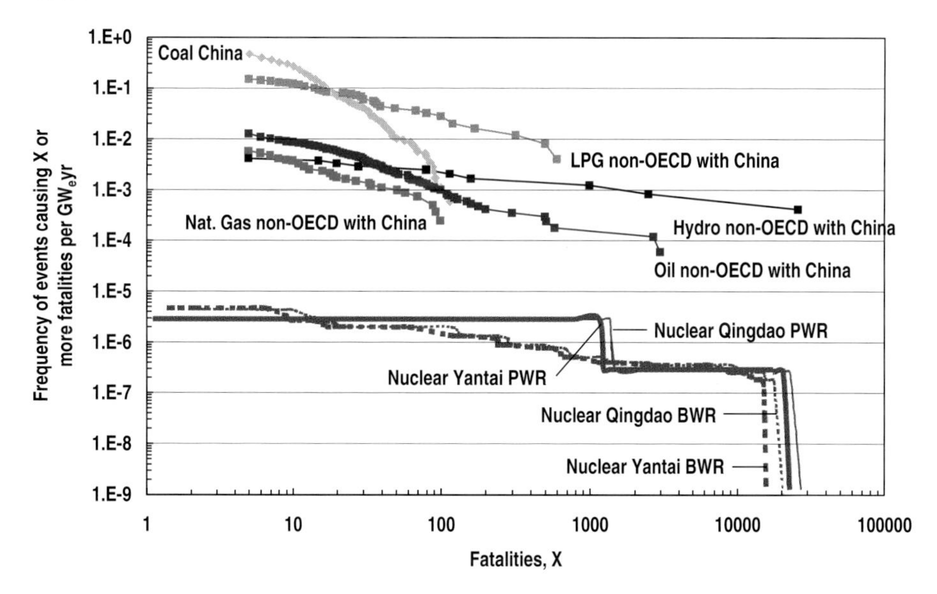

Figure 10.44 Comparison of severe accident records based on frequency-consequence curves for different energy chains in the period 1969-1999, except for coal (1994-1999). The results for the nuclear power plants are based on Probabilistic Safety Assessment and represent latent fatalities; the estimates for other chains are based on historical accidents and represent immediate fatalities.

5.2 Indicators for the future

Results of risk assessment were also used as part of the input for an overall multi-criteria decision analysis (MCDA) of selected Shandong electricity supply scenarios created using the Electric Sector Simulation (ESS). A modified set of indicators was chosen to meet the criteria definitions used in MCDA. Potential health impacts due to severe accidents were addressed using two sub-criteria. Expected risks under current conditions and for year 2020 were evaluated by means of aggregated indicators (Fatalities per $GW_e yr$). Additionally, maximum credible consequences were defined as the accident with the largest number of fatalities that ever occurred world-wide for a specific energy chain; for the nuclear case the Shandong-specific PSA-based results were used. Table 10.22 explains the assumptions on which the aggregated indicators for various energy chains were based. Hydro power is not included since its potential in Shandong province is insignificant.

Aggregated indicators for current conditions and 2020 are shown in Figure 10.45. Although the coal chain is predicted to improve by about 45 percent, it will still rank last in terms of expectation values. The expected fatality rate for the oil chain is in the same range as for coal, whereas no major changes in the natural gas chain are expected. For the nuclear chain current reactors are assumed as the reference technology for China; the estimates presented in this chapter reflect the design features of the plants analyzed as well as the physical conditions (primarily population density) around the selected reference sites but may not represent the possible negative impacts of different safety culture.

Figure 10.46 indicates the maximum credible consequences for each energy chain. Here the picture is somewhat reversed. Nuclear power has the lowest expected fatality rates but exhibits the potentially largest maximum credible consequences among the various energy chains. In contrast, accidents in the coal chain are more frequent but the number of maximum credible fatalities is much smaller. It should be noted that the basis for generating indicators for the maximum credible consequences is not as balanced as desirable due to lack of PSA studies for fossil energy chains. Such studies would probably generate higher values than those based on the historical experience though at a lower frequency level.

Table 10.22 Assumptions for selection of aggregated indicators for MCDA input.

	Current	**2020**
Coal	Weighted average based on coal output in Shandong and Shanxi corrected for contributions from large and small mines.	Performance of the Chinese coal chain is expected to become similar to non-OECD w/o China (1969-1996) experience. It is assumed that phasing out small, inefficient mines, characterized by very poor safety standards, will continue.
Oil	Non-OECD with China (1969-1999) experience is considered representative for China as the statistical basis in terms of China-specific records is relatively small.	The level corresponding to the current China-specific experience (1969-1999) appears as a meaningful reference for the future, particularly given the fact that the rate for non-OECD countries is driven by two very large and thus untypical accidents (Philippines, 1987 and Afghanistan, 1982).
Natural gas	Non-OECD with China (1969-1999).	Non-OECD w/o China (1969-1996).
Nuclear	There are currently no nuclear power plants in Shandong.	Current reactors (Qingdao site); no substantial change.

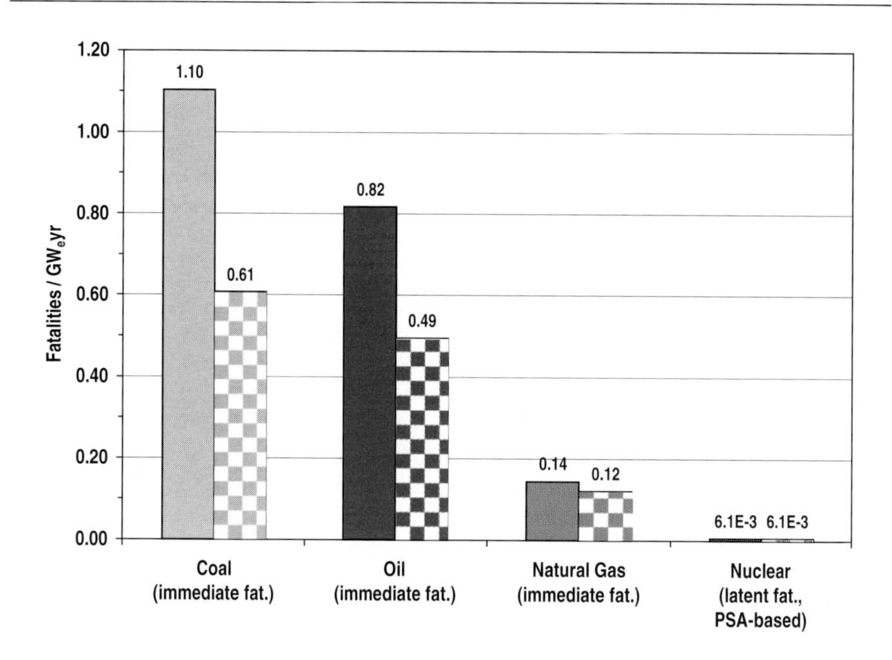

Figure 10.45 *Expected severe accident fatality rates for current conditions (solid bars) and in 2020 for (checkered bars) various energy chains.*

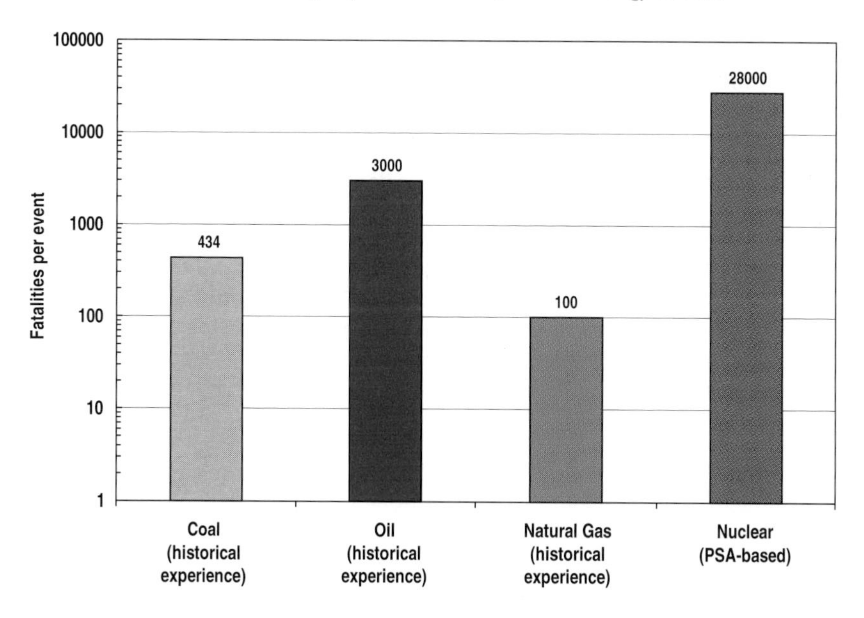

Figure 10.46 *Maximum credible consequences for various energy chains.*

6. CONCLUSIONS

Conclusions are provided here for the major energy chains individually and in comparison with each other. Risks associated with full energy chains were considered to the greatest extent possible unless it was clear that major risks are concentrated at one specific stage in the chain.

6.1 Specific chains

Coal chain

- The experience-based evaluation focused particularly on coal as the dominant energy chain in China. During the period 1994-1999 a total of 816 severe coal accidents occurred in China, with the total number of 11'321 fatalities.
- Every year about 6000 fatalities occur in Chinese mines due to small and severe accidents. Though severe accidents receive more attention than the small ones about two thirds of the accident-related fatalities is due to small accidents.
- More than every third severe industrial accident in China occurs in the coal industry.
- The Chinese severe accident fatality rate for the coal chain is about 6.5 per GW_eyr. On average, this is about ten times higher than in non-OECD countries and about 50 times higher than in OECD countries.
- Coal production in Shanxi province is not more than about three times higher than in Shandong but the number of accidents and fatalities for the period 1994-1999 was almost ten times higher in Shanxi.
- Small mines in China exhibit on average five times higher severe accident fatality rates than large mines. Closing small mines with unacceptably low safety standards would significantly reduce accident risks. The developments in the Chinese mining industry are currently moving in this direction and should continue.
- Additionally, extraction of coalbed methane (CBM) would greatly reduce the large number of gas accidents occurring in Chinese coal mines and thus potentially contribute to a reduction in the numbers of accidents and fatalities.
- The accident reports for the coal chain clearly show that the majority of severe accidents were due to bad mine management, lax safety standards by the mine owners, poor training, and lack of appropriate supervision by officials.

Oil chain

- During the period 1969-1999 a total of 32 severe oil chain accidents occurred in China, with total number of 613 fatalities.
- Due to the limited Chinese-specific experience the generic non-OECD experience is considered to be representative for China also.
- Half of the accidents occurred during transportation (i.e. transport to refinery or regional distribution), followed by the exploration and refinery stages. Considering fatalities, exploration ranks first with a share of about 40 percent, followed by regional distribution and transport to refinery.

- No major offshore or onshore oil spills exceeding 25'000 t were reported in China. However, many smaller oil spills that caused major environmental problems were identified. In the South China Sea even small spills can have serious social consequences due to the dependence of the population on fish.

Natural gas chain

- During the period 1969-1999 a total of eight severe accidents with a total of 201 fatalities occurred in the Chinese natural gas chain. The use of natural gas in China though currently increasing has been historically on a low level.
- Similar to oil the generic non-OECD experience is considered representative for China for the purpose of this chapter.

Hydro power

- Hydro power accident risks were not studied in detail. Hydro is very important for China but hydro resources in the Shandong province are low.
- Between 1950 and 1990 3241 dam failures occurred in China, of which 123 or 3.8 percent were at large dams and 3118 or 96.2 percent at medium and small dams. The consequences of these collapses have mostly not been reported.
- The official death toll resulting from dam failures is 9937, not including the failure of the Banqiao and Shimantan dams in 1975, which caused about 26'000 fatalities.
- Due to the clear indication that dam accidents are underreported, a reliable estimate of the Chinese-specific fatality rates from severe hydro power accidents is not feasible. Therefore, the pooled hydro power experience is used for comparative analyses in the present work. Progress towards more reliable Chinese-specific estimates is totally dependent on access to the relevant historical records. For the current major projects, such as Three Gorges, carrying out a dam-specific Probabilistic Safety Assessment (PSA) would be advisable.

Nuclear power

- Given the total lack of relevant historical accidents, PSA methodology utilizing a simplified approach to the assessment of consequences was employed to provide risk estimates for hypothetical nuclear accidents at two selected sites in Shandong. Typical current designs with good safety standards and advanced (evolutionary) designs were considered as reference plants.
- The estimated risks were found to be plant- and site-sensitive. The orders of magnitude for specific risk measures are, however, the same for almost all cross-comparisons. The only exception is the comparison of current and advanced designs, with the latter being one order of magnitude better from the risk point of view.
- Generally, early fatality risks are negligible compared to latent fatality rates. Latent fatality rates are significantly higher but remain at a low level in absolute terms. In relative terms, the risk of land contamination, though also low, is a major concern due to the related social consequences.
- A number of sensitivity cases showed that the estimated risk measures are relatively insensitive to moderate variations in a number of crucial parameters.

However, all results are subject to uncertainties associated with core damage frequency and source terms. When contemplating the results it should be kept in mind that they primarily reflect the good safety standard of the selected reference designs. Safety culture may have a decisive impact on risk once the plants are in operation; this aspect has not been addressed here for China-specific conditions.

6.2 Comparative aspects

- The comparison covers aggregated indicators (i.e., fatality rates per $GW_e yr$) and frequency-consequence curves. It addresses exclusively fatality rates since no consistent data could be found for other accident indicators for all energy chains. Furthermore, some of the consequences are associated with only one energy carrier (e.g. oil spills) or are most pronounced for a specific chain (e.g. long-term land contamination).
- Among the major energy chains relevant for electricity generation the coal chain exhibits the highest fatality rates when the aggregated indicators considered as most representative for the Chinese conditions are compared. The corresponding indicators for the oil chain are one order of magnitude lower and between one to two orders of magnitude lower for natural gas. Depending on the choice of reference and subject to the above-mentioned reservations that must be made regarding the underreporting of Chinese accidents, the severe accident fatality rate for hydro power in China is of the same order of magnitude as that for the coal chain. Probabilistic analysis indicates that risk of early fatalities associated with hypothetical accidents at the current type or advanced nuclear power plants at selected sites in Shandong is negligible compared to the immediate fatality rates for severe accidents in the other energy chains. The risk of latent fatalities is higher but remains three to four orders of magnitude lower than the experience-based severe accident results for coal in China and hydro in non-OECD countries with China, and two to three and one to two orders of magnitude lower than fatality rates for oil and gas respectively in non-OECD countries with China. The predictive estimates for latent nuclear fatalities are of the same order or lower than the experience-based results for hydro power in OECD countries. The results for nuclear are only valid under the assumptions that the nuclear power plants will be operated under stringent licensing requirements and in an environment characterized by safety cultures comparable to what is typical for OECD countries. Verifying the realism of these assumptions was beyond the scope of the present work.
- A somewhat different perspective is gained through the examination of frequency-consequence curves. Among Chinese energy chains, coal exhibits the highest accident frequencies. However, the vast majority of coal accidents in China results in less than 100 fatalities. The natural gas chain shows a favorable performance with regard to accident frequencies and maximum fatalities, but natural gas is currently of relatively minor importance for China's electricity mix. Accident frequencies in the oil and hydro chains are also much lower than for the coal chain. However, the maximum number of fatalities within the oil and hydro chains are respectively one and two orders of magnitude higher than for coal and

natural gas chains. Finally, the nuclear curves are at a very low frequency level, orders of magnitude below those for other chains. The maximum credible consequences due to hypothetical nuclear accidents may be very large, i.e. comparable to the Banqiao and Shimantan dam accident that occurred in China in 1975. The associated risk valuation is subject to stakeholder value judgments and has been pursued along with other economic, environmental, and social criteria in multi-criteria decision analysis carried out within CETP.

- Damage costs associated with severe accidents were not explicitly estimated within this project. The results expressed in terms of risk measures indicate that compared to OECD these damage costs are bound to be very high for the coal chain and probably also for hydro power. For Chinese conditions, however, they are of low significance when compared to the damage costs associated with air pollution.
- Diversification of the electricity mix, improvements in energy efficiency, energy conservation, and safety-promoting measures are key aspects for sustainable development in the energy sector, with significant gains in terms of reducing the number of severe accidents among other positive effects.

REFERENCES

BBC News Online (1998). *Coal-mine explosion in China kills 77.* Retrieved Jan 30, 1998, from http://news.bbc.co.uk/.

BBC News Online (1999). *Chinese dam disaster alert.* Retrieved Mar 23, 1999, from http://news.bbc.co.uk/hi/english/world/asia-pacific/newsid_301000/301476.stm#top.

BBC News Online (2000). *China's deadly mining industry.* Retrieved Sep 28, 2000, from http://news.bbc.co.uk/.

Cazzoli, E., Khatib-Rhabar, M., Schmocker, U. and Isaak, H. P. (1993). Approach to quantification of uncertainties in Probabilistic Safety Assessment. In P. Kafka and J. Wolf (Eds.), *Proceedings of the European Safety and Reliability Conference, ESREL '93* (pp. 899 - 909). Amsterdam, London: Elsevier.

CCIY (1997-2000). *China Coal Industry Yearbook.* Beijing: China Coal Industry Publishing House.

Chanin, D. I., Sprung, J. L., Ritchie, L. T. and Jow, H-N. (1993). *MELCOR Accident Consequence Code System (MACCS), User's Guide, Volume 1, NUREG/CR-4691, SAND86-1562.* February 1990 and updated 1993. Albuquerque, New Mexico: Sandia National Laboratories.

China Labour Bulletin (2000). Coal mining in China [Electronic Version]. Originally published in *China Labour Bulletin, 52,* Jan-Feb 2000. Retrieved Feb 28, 2000, from http://www.china-labour.org.hk/iso/.

China Statistical Information Center (2000). *Statistical communique of People's Republic of China on the 1999 national economic and social development.* All China Marketing Research Co., Ltd. 2000.

Clifton, J. (1992). The historical record of major accidents in the energy industries. *Applied Energy, 42*, 17-32.

Cutter Information Corp. (2001). *Oil Spill Intelligence Report* [Electronic version]. Retrieved from http://www.cutter.com/osir/index.html.

Encyclopaedia Britannica (1972). *1972 Britannica book of the year* [Electronic version]. Retrieved from http://www.eb.com/ (Online subscription). Chicago (IL): Encyclopædia Britannica Inc.

ERI/HSK (1996). *Regulatory evaluation of the Leibstadt Probabilistic Safety Assessment – part II: level 2*. ERI/HSK 95-303, HSK 13/444, Volume II, March 1996.

ERI/HSK (1999). *Regulatory evaluation of the Gösgen Probabilistic Safety Assessment – part II: level 2.* ERI/HSK 98-302, HSK 17/314, Volume II, April 1999.

Eyink, J. (Ed.) (1997). *Explanatory material on the European pressurised reactor.* AMM-BE-EJTs (96)-D16, NPI, June 1997.

Fu, S. (1998). A profile of dams in China. In: D. Qing, J. G. Thibodeau and P. Williams (Eds.), *The river dragon has come! The Three Gorges dam and the fate of China's Yangtze River and its people.* Armonk (NY, USA) & London (UK): Sharpe, M.E., Inc.

Hirschberg, S. and Cazzoli, E. (1994). Contribution of severe accidents to external costs of nuclear power. ENS Topical Meeting on PSA/PRA and Severe Accidents '94, 17-20 April 1994, Ljubljana, Slovenia.

Hirschberg, S., Spiekerman, G. and Dones, R. (1998). *Severe accidents in the energy sector.* PSI Report No. 98-16. Paul Scherrer Institute, Wuerenlingen and Villigen, Switzerland, November 1998.

Hirschberg, S., Spiekerman, G., Dones, R. and Burgherr, P. (2001). Comparison of severe accident risks in fossil, nuclear and hydro electricity generation (Invited paper), *Proceedings of EAE'2001, International Conference on Ecological Aspects of Electric Power Generation*, 14-16 November 2001, Warsaw, Poland.

Hirschberg, S., Burgherr, P., Spiekerman, G., Cazzoli, E., Vitazek, J. and Cheng, L. (2003a). *Comparative assessment of severe accidents in the Chinese energy sector.* PSI Report. Paul Scherrer Institute, Wuerenlingen and Villigen, Switzerland.

Hirschberg, S., Heck, T., Gantner, U., Lu, Y. Spadaro, J. V., Krewitt, W., Trukenmüller, A. and Zhao, Y. (2003b). *Environmental impact and external cost assessment in the China Energy Technology Program.* PSI Report. Paul Scherrer Institute, Wuerenlingen and Villigen, Switzerland.

HSK/KSA (1996). *Gutachten zum Gesuch des Kernkraftwerks Leibstadt um Leistungserhöhung auf 3600 MWth.* HSK 12/420, KSA 12/210, Würenlingen, März 1996.

HSK/KSA (1999). *Periodische Sicherheitsüberprüfung für das Kernkraftwerk Gösgen-Däniken - Zusammenfassung, Ergebnisse und Bewertung.* HSK 17/400, KSA 17/261, Würenlingen, November 1999.

ICOLD (1995). *Dam failures statistical analysis. ICOLD Bulletin, no. 99.* Paris: International Commission on Large Dams.

ICOLD (1998). *ICOLD world register of dams, computer database.* Paris: International Commission on Large Dams.

ICOLD (2000). *ICOLD world register of dams updating 2000 for China, computer database.* Paris: International Commission on Large Dams.

IEA (2001). *World energy statistics and balances of OECD countries (1960-1999) and non-OECD countries (1971-1999)*. Paris: International Energy Agency.

International Water Power & Dam Construction (1991-1995). World News. Issues Sep 1991, Sep/Oct 1993, May 1995.

IRN (1995-1999). *Dam Index*. Retrieved from http://www.irn.org/pubs/damindex.html.

Jo, J., Cazzoli, E., Tingle, A., Valtonen, K. and Pratt, W. T. (1991). *A Review of the advanced boiling water reactors probabilistic risk assessment.* NUREG/CR/5676P, BNL-NUREG-52276P, New York: Brookhaven National Laboratory, Upton, New York.

LLP (Lloyd's Casualty Archive) (1999). *DONG HAI (People's Republic of China)*. Retrieved Apr 8, 1999, from http://www.llplimited.com/casarchive/index.htm.

Microsoft Encarta (1999). *Microsoft Encarta World Atlas 98* (CD-ROM). Redmond: Microsoft Corp.

Mladý, O. (1999). *NPP Temelin safety analysis reports and PSA status. International Conference on the Strengthening of Nuclear Safety in Eastern Europe*. IAEA-CN-75, Vienna, Austria.

Muzi Lateline News (1998). *More than one third of China's dams "time bombs"*. Retrieved Mar 23, 1999, from http://lateline.muzi.net/.

Muzi Lateline News (1999). *Experts warn potential dam disasters from old dams*. Retrieved Mar 1, 1999, from http://lateline.muzi.net/.

National Renewable Energy Laboratory (1998). *Hydroelectric Power - turning water's mechanical energy into electricity*. Retrieved from http://www.nrel.gov/.

Nuclear Europe Worldscan (2000). China reconsiders new nuclear power project. *Nuclear Europe Worldscan 3-4/2000*, 23.

Park, C. et al. (1991) *Evaluation of severe accidents risks: Zion nuclear power plant*. NUREG/CR-4551, Brookhaven National Laboratory.

PRIS (2002). *IAEA's Power Reactor Information System*. Retrieved from http://www.iaea.org/programmes/a2/.

Probe International (1998). *Chinas old dams are in danger (by Zhang Kun, China Youth Daily)*. Retrieved from http://www.probeinternational.org/pi/3g/print.cfm?contentID=703.

Reuters (1991). *Editorial-China syndrome – loss of McDermott-operated barge renews calls for tighter safety standards* (Reuters Business Briefing). Retrieved Aug 19, 1991, from http://www.business.reuters.com/.

Reuters (1997). *China: crack down on fire safety after spate fatal blazes* (Reuters Business Briefing). Retrieved Aug 8, 1997, from http://www.business.reuters.com/.

Sigma (2001). *Natural catastrophes and man-made disasters in 2000: fewer insured losses despite huge floods, 2/2001*. Zurich: Swiss Re Company.

SKI (1999). *PSA InfoSys, prepared by SAC GmbH for SKI*. SKI report 97:17, updated December 1999.

SSB (State Statistical Bureau) (1999). *China Statistical Yearbook 1999* [Electronic version]. Retrieved from http://www.stats.gov.cn/yearbook/.

Times of India (2001a) *100 feared dead in Chinese mine blast*. Retrieved Jul 23, 2001, from http://www.timesofindia.com/articleshow.asp?catkey=314014066&art_id=1334683390&sType=1.

USNRC (1989). *Severe accidents risks: an assessment of five U.S. nuclear power plants*. NUREG-1150, US Nuclear Regulatory Commission, Washington DC.

Vogel, A. (1998). *Bibliography of the history of dam failu*res. Vienna, Austria: Data-Station for Dam Failures.

Wang, J. and Xu, J. (1996). Analysis on characteristics of fatal state-owned coal mine accidents since 1980. *China Coal, No.* 8, pp. 22-25 and 32. Coal Mine Safety Technology Training Centre (paper in Chinese).

WCD (2000). *Dams and development – a new framework for decision making; the report of the world commission on dams*. Retrieved from http://www.dams.org/report/.

WorldAtom Press Release (2001). *IAEA releases nuclear power statistics for 2000. PR 2001/07).* Retrieved May 3, 2001, from http://www.iaea.or.at/worldatom/Press/P_release/2001/prn0107.shtml

Zhang, H. (Ed.) (1997). *China flood and drought disaster, December 1997*. China Water Resources and Electric Power Press (in Chinese).

Zhang, L. (2000). *China social impacts of large dams*. Institute for Agricultural Economics, China, WCD Regional Consultation Paper.

Zhou, D., Yuan, G., Yingyi, S., Chandler, W. and Logan, J. (2000). *Developing countries and global climate change - electric power options in China.* Arlington, VA: Pew Center on Global Climate Change.

DSO (Dam Safety Office) (1999). *A procedure for estimating loss of life caused by dam failure*. DSO-99-06, US Department of Interior, Bureau of Reclamation, Denver, CO, USA. Retrieved from http://www.usbr.gov/research/dam_safety/documents/dso-99-06.pdf.

Human Rights Watch Asia (1995): The Three Gorges dam in China: forced resettlement, suppression of dissent and labor rights concerns. Human Rights Watch Asia Report, February 1995, vol. 7, no.2, Appendix III, pp.37-44.

Si, Y. (1998): The world's most catastrophic dam failures: the August 1975 collapse of the Banqiao and Shimantan dams. In: D. Qing, J. G. Thibodeau & P. Williams (Eds.): *The river dragon has come! The Three Gorges dam and the fate of China's Yangtze River and its people.* Armonk (NY, USA) & London (UK): Sharpe, M.E., Inc.

WCD China Country Review (2000): *Experience with dams in water and energy resource development in the People's Republic of China.* WCD Case Studies, Cape Town, South Africa. Retrieved from http://www.dams.org/report/.

World Rivers Review (1995): November 1995, vol. 10, no. 3. IRN, Berkeley, CA, USA. Retrieved from http://www.irn.org/pubs/wrr/index.shtml.

NOTES

[1] Other chemicals need to be considered on a case-by-case basis with view to their toxicity.

[2] Sources: DSO (1999), Human Rights Watch Asia (1995), Si (1998), WCD China Country Review (2000), World Rivers Review (1995)

[3] Ryr = Reactor*year

CHAPTER 11

MULTICRITERIA OUTPUT INTEGRATION ANALYSIS

PIERRE ANDRE HALDI AND JACQUES PICTET

CETP's relationship with the Alliance for Global Sustainability (AGS) implies that sustainability is a basic concern of the project's scenarios assessment process. Because sustainability is by nature a multidimensional concept, it requires the use of methodological approaches specifically designed to tackle decision-making problems of this type. Such a methodology should encompass multiple criteria, including economic, social and environmental issues, under sustainability constraints, and produce results that are transparent and easy for different stakeholders to understand.

This chapter first outlines the decision-making framework in China and then presents the foundation of the Multicriteria Decision Aiding (MCDA) methodology. It describes how this methodology has been used in the CETP project, how the weighting process was devised, the scenarios selected, and offers the MCDA analyses of the "performance matrix". Finally, the chapter discusses the scenarios rankings.

1. THE CONTEXTS OF THE MULTICRITERIA DECISION AIDING APPROACH IN THE CETP PROJECT

"The future is much more a matter of choice than a matter of destiny" (UNDP, 2000).

1.1 The Sustainability Context

Sustainability is the overriding general concern of AGS, and thus of the scenarios assessment and decision-making process in this project. According to Mrs. Gro Harlem Brundtland, chairwoman of the UN World Commission on Environment and Development, sustainable development is a process "*that meets the needs of the present without compromising the ability of future generations to meet their own needs*" (Brundtland, 1989). This deceptively simple definition hints at five underlying "core" principles:

1. respect for ecological integrity;
2. efficient use of natural, manufactured, and social capitals;
3. search for greater equity;

4. active participation of stakeholders in all decision making processes; and
5. environmental stewardship by all levels of decision-makers.

Although different, and not always converging, interpretations have been given of these basic principles, there is a general consensus on two important points:

a. sustainability is an intrinsically *multidimensional* concept, which goes beyond ecological concern alone to include also economical and societal issues (Figure 11.1);
b. the *active participation and involvement of the concerned "actors"* (stakeholders) is an absolute requirement for a sustainability assessment process.

This calls for the use of methodological approaches specifically designed to tackle decision making problems in a complex multidimensional / multistakeholders context. Policy choices made on this basis will be more acceptable to all parties concerned. Both adequate criteria (economic, social, and ecological) and a representative group of stakeholders are essential. This is particularly true in the power generation sector. Among the various forms of energy, electricity has gained particular importance because a) it can be produced from a variety of sources, b) it is easily transmitted to consumers, c) it may be transformed into other forms of useful energy (e.g. light or heat), d) it is, at least at the point of use, non polluting.

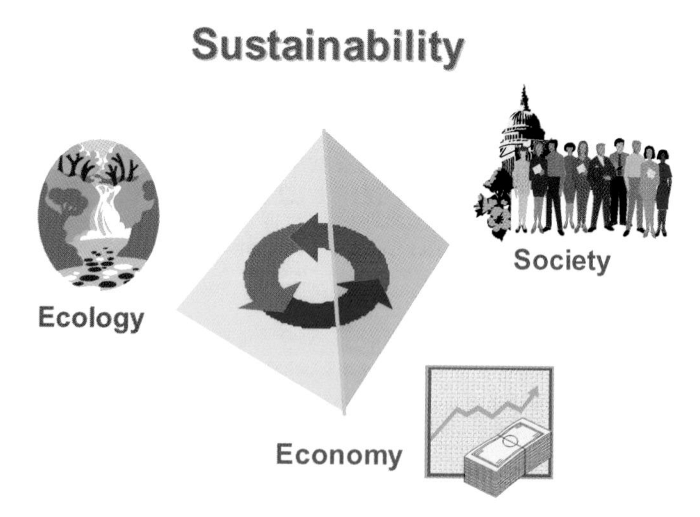

Figure 11.1 The three basic dimensions of the sustainability concept

Electricity is an unusual form of power in the sustainability context because energy saving and pollution mitigation rely largely upon it. It appears that the growth of electricity consumption will last longer than other forms of power and is the most resistant to energy demand reduction policies. The identification, design, selection, and finally implementation, of "sustainable electric power generation systems" are thus a key element to sustainable development.

Energy produced and used in ways that support human development over the long term, in all its social, economic, and environmental dimensions, is what is meant by the expression *sustainable energy*. In that sense, *sustainable power generation* does not refer simply to the search for a continuing supply of electricity, but to the production and use of this resource in ways that promote–or at least is compatible with–long-term human well-being and ecological balance.

However, it is not simple to alter the development course of energy systems in general, and of electricity-generating systems in particular. The picture is further complicated by the long lead times needed to implement a policy decision, the internationalization of constraints and consequences, and the impact of these consequences over subsequent human generations. Such a complex and long-term process requires major concerted efforts by governments, businesses, and members of civil society. Appropriate consensus-building approaches are also needed, because decision making in this domain has become far less straightforward than it used to be.

What is the "optimal" mix of energy sources for the supply of electricity? Who decides and how? How can selected policies be implemented? It appears that, while these questions have been successfully tackled in the past, they are now subject to extensive debate. The meanings of previously clear-cut concepts like "optimum" and "decision making" are no longer obvious.

In the past, "optimum" was taken to be more or less synonymous with "economic optimum" and "decision-making" was left entirely to the managers of electricity utilities themselves or, in some cases, political authorities. Although decisions made by the latter have so far satisfied society's needs remarkably well, they are now frequently contested and even blocked by opponents that vary from small but powerful groups to vast majorities of voters, from formally operating bodies to unofficial but influential public opinion makers.

Under today's circumstances, there is thus hardly any "optimal" solution in this field. Electric utilities or political authorities do not make decisions alone, and there are few well-defined, universally accepted procedures for discussion and decision making. In this context, multicriteria decision-aiding (MCDA) methods can be of great help as they can accommodate strong opposition to commonly accepted viewpoints. They avoid "optimization" of one (unique or synthesized) objective function and suggest "soft", "flexible" procedures rather than "hard", "rigid" mathematical algorithms; they do not authoritatively give *the* solution, but help decision-makers explore and clarify their own criteria. Consequently, these methods should favor move-forward consensus rather than fall-back compromise.

Because they are designed to be as realistic and practical as possible, thus avoiding the "black-box syndrome" of many "classical" mathematical approaches[1], MCDA methods are more adapted as "negotiating" tools in multidimensional / multistakeholders decision making processes, particularly when stakeholders with contrasting objectives are involved.

The multicriteria methodology proposed for the CETP project is based on the so-called "European School" of MCDA approaches. It makes use of tools such as ELECTRE or PROMETHEE (see § 11.7), which rely on syntheses by outranking relations supported by partial aggregation methods (as opposed to the usual complete aggregation techniques used in the more "classical" operations research approaches).

A main advantage of these methods is that they include both quantitative and qualitative criteria, as well as the inherent fuzziness associated with their evaluation, in a coherent and straightforward manner. Moreover, this kind of approach has an important and useful integrative role to play in a project like CETP, because it offers an appropriate methodological framework to gather and unify downstream the results of the various analyst teams' contributions as schematized in the following figure.

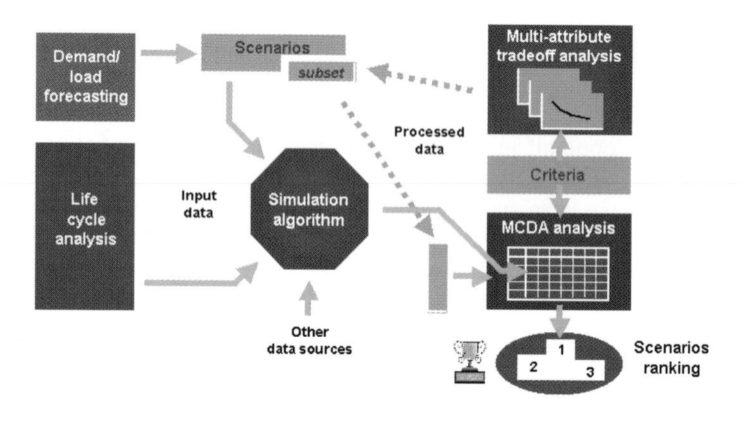

Figure 11.2 Integrative role of the MCDA procedure in the CETP project

1.2 The Decision-Making Context in China

One of the most original, but also most challenging, features of the CETP project is the application of analytical methods primarily developed in the context of Western countries (See for example: SESAMS, 2000) to a quite different sociopolitical environment, i.e. China. In the domain of decision making it is therefore of prime importance to have a closer look at how decisions of a technical / industrial nature are made in China before further presenting the intended MCDA analysis.

This section presents some general information initially gathered by EPFL-LASEN for another Chinese project. Although it concerns the specific approval procedures for a CCPP (Combined Cycle Power Plant), it should apply to other types of technical / industrial projects as well.

An overall view of the corresponding *planning* and "*call for tenders*" procedures is given in Figure 11.3.

Planning procedure *Call for tenders procedure*

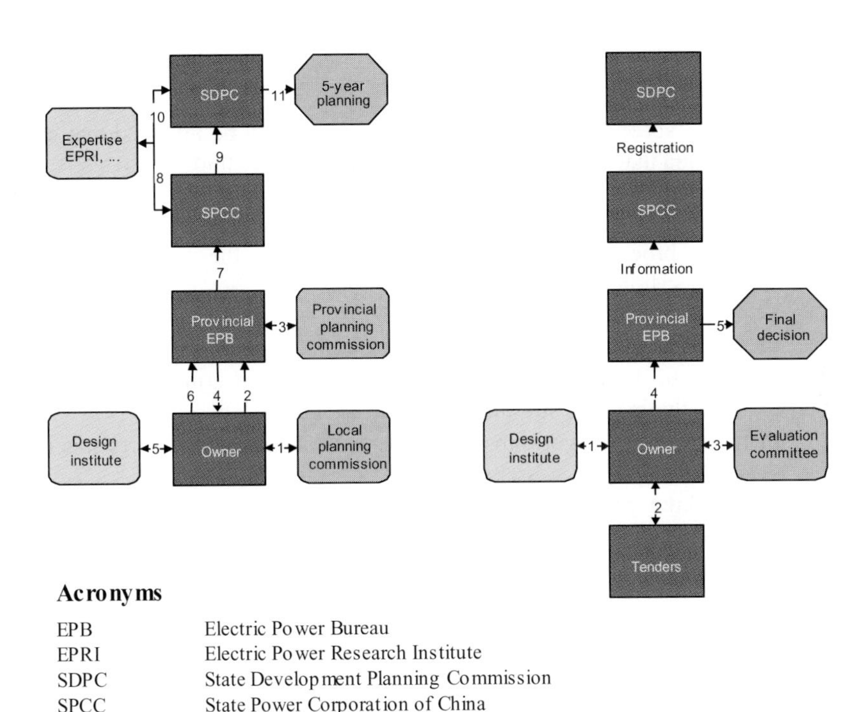

Figure 11.3 Approval procedures of technical/industrial projects in China

Both procedures involve three levels of administrative organization (local, provincial and national) and the concerned bodies at each level (planning commissions, electric power authorities, technical and economic experts, owner and, in the second case, tenders).

Both also follow essentially a bottom-up sequence (see numbers attached to the arrows in Figure 11.3.). The primary initiative falls within the province of the owner and the local authorities. The matter is then treated at the provincial level, where the final decision is made in the case of the "call for tenders" procedure (file conveyed upwards only for information and registration purposes). In the planning procedure, the formal decision is made at the national level, after consulting economic and technical experts, and must conform to the global objectives of the national five-year plan.

This figure shows that the decision-makers (central column) are the same ones in both procedures, even though they play different roles, especially at the national level as explained above.

In the "call for tenders" procedure, the evaluation committee is chaired by the provincial project manager and co-chaired by a SPCC (State Power Corporation of China) representative.

These flowcharts simplify a reality that can be considerably more complex and quite time-consuming (as demonstrated by the iterations, shown in Figure 11.3, between the owner and the provincial Electric Power Bureau in the planning procedure). Many secondary procedures (e.g. permits) may be involved that cannot be detailed here and that may vary from case to case.

These schematic diagrams show, however, that China has well structured and precise approval procedures. The CETP project should not address the formal decision making procedure itself, but rather deal with the content or general approach of the decision, i.e.:

- the selection of appropriate "decision objects" (scenarios);
- the basis of decision-making: broadening the set of criteria under scrutiny to correctly integrate sustainability issues (See § 1.1.);
- the mathematical algorithm used to partially aggregate the performances of the alternatives;
- the working procedures of the Stakeholder Advisory Group and other concerned groups;
- the technical and economic models: better assessment of the needs, more market-driven planning, etc.; and
- the international context: although China has recently been accepted as a new member of the World Trade Organization (WTO) and therefore must now comply with its Agreement on Government Procurement (AGP), adopted in Marrakech in 1994, it is probable and understandable that this country will need some time to adjust its procedures.

The MCDA procedures described in this chapter can provide the necessary means to improve the four first points by providing the basis for:

- the definition of suitable "decision objects",
- the joint use of criteria of various origins and quality,
- the application of a partial aggregation algorithm for ranking the scenarios, and
- the implementation of a suitable framework for group decision-making.

2. THE STAKEHOLDERS ADVISORY GROUP (SAG)

We have already mentioned (§ 1.1) that the active participation of the various bodies of concerned "actors" in decision-making processes of the kind considered in the CETP project is essential to its credibility and success. The involvement of ABB

China and other important Chinese partner institutions has been carefully cultivated from the very beginning of the study.

More specifically, a group of representatives of the principal Chinese bodies involved in strategic decision making for the electricity sector in China, and in Shandong province specifically, has been formed to assist the analyst teams in all the important phases of the project. This Chinese Stakeholders Advisory Group (SAG) provided direct inputs or assistance in (see figure 11.4):

- the collection of basic data and other fundamental information about the electricity sector in China and in Shandong province;
- the formulation of the various scenarios to be studied and of the underlying boundary conditions and assumptions;
- the selection of relevant criteria for assessing and comparing the performances of the above scenarios in the sustainability context adopted by the CETP project;
- the definition of the relative importance (so-called "weights") of these criteria in the decision-making process; and
- the assessment of the validity and accuracy of the scenario performance data used in the analysis.

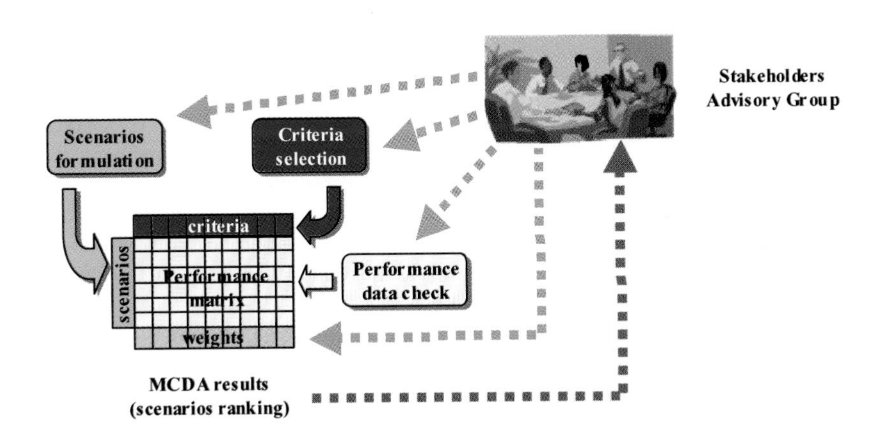

Figure 11.4 Involvement of the SAG in the MCDA process

The members of the SAG were selected and appointed by ABB project management, taking into account the general decision making structures presented in section 1.2 and the sustainability orientation of the CETP project. In addition to representing important elements in decision-making, SAG members also had to be genuinely interested in the project's objectives and outcomes, and be available to participate in a few common meetings. Forming an ad hoc advisory group fulfilling these basic requirements proved in practice to be markedly more difficult than initially foreseen. As a consequence, formation of the group was delayed; it became truly operational several months after the project was underway. Furthermore, interactions between the analysts and the SAG members, either as a group or individually, was very limited, mainly due to geographical, cultural and linguistic hurdles. This had a definite effect on the progress and completeness of the MCDA process.

The names and affiliations of the ten Chinese "stakeholder representatives" finally appointed to participate in the project are given in Table 11.1.

Table 11.1 Names and affiliations of the SAG members (Latin alphabetic order)

Stakeholder Name	Stakeholder Affiliation
Dr. Guo Risheng	Administrative Center for China's Agenda 21
Prof. Hu Zhenou	Chinese Academy of Sciences
Prof. Li Wenran	Shandong Environmental Protection Bureau
Mr. Shi Dinghuan	Ministry of Science and Technology
Prof. Wang Huijiong	Development Research Center of State Council
Mr. Xi Xiubin	Shandong Economy and Trade Commission
Prof. Zhan Zhonghui	State Power Corporation of China
Mr. Zhang Kun	State Environmental Protection Administration
Mr. Zhao Qingbo	Shandong Electric Power Research Institute
Prof. Zhou Dadi	Energy Research Institute

The fact that the group is formed of persons having different origins and backgrounds raised the question of how to reconcile, should the case arise, divergent opinions among these stakeholder representatives regarding the scenarios to be studied, the criteria to be considered or the relative importance to assign to these criteria. There are essentially two possible approaches to this problem (Belton, Pictet 1997): a) try to achieve a consensus among the group members on some unique, common solution, b) accept divergent individual opinions and try to find workable conclusions by the end of the process analysis.

After project management members debated this issue, they decided to obtain an agreement on a unique set of scenarios and criteria but to keep the individual weight sets given by each of the SAG members in the analysis.

3. THE "POTENTIAL ACTIONS" OR "DECISION SCENARIOS"

The first operation in any decision-making problem is, of course, to precisely define the "objects" of decision making. In the MCDA jargon, these "decision objects" are usually called "potential actions[2]" (a_i), the complete set of these actions being designated by A (A = $\{a_1, a_2, \ldots a_i, \ldots a_n\}$). The choice between these actions is open by definition, because if there are any constraints, they must either be accounted for as penalties or costs in the criteria, or be used to exclude unsuitable options. Thus, it may be said that identifying the potential actions means identifying all feasible options, regardless of the consequences.

The expression "potential action" is, however, unfamiliar to many people and could have been misunderstood by some CETP participants or readers of the present document. It has therefore been decided to call here these actions "decision scenarios". This designation emphasizes that they have direct links with other types of scenarios used in the CETP project, while being specifically designed to effectively constitute "decision objects" offered as choices to the stakeholders. More general-purpose scenarios contain aspects of the "environment" of the decision, which influence the selection of the actions and future consequences, but lie beyond the decision-makers' reach. In the CETP MCDA context, a potential action must be one that the decision-makers have the power to implement, or cause to be implemented, or, at least, to truly influence.

The aspects of the scenarios that are beyond the decision-makers' reach are directly related to the "futures" (according to the MIT terminology) defined in Chapter 6. In the MCDA context, they represent sets of possible but unpredictable and not decision-amenable "boundary conditions" that have direct effects on the performances of the scenarios (impact on the "performance matrix") and thus on the results of the multicriteria analysis. Various "futures" can be defined to cover the whole set of possible decision environments.

A comparison of the actions makes sense only for one "future" at a time. It is of little use to know that action a_i under "future" X is better than action a_j under "future" Y, as these situations are not directly comparable. For instance, let us consider two actions (to buy car A or car B), and two "futures" differing essentially by the price of gasoline (respectively p_I and p_{II} Yuan/liter). The decision-maker cannot use the information that car A in the p_I case is better than car B in the p_{II} case, because, at the time of the decision, he doesn't know what gasoline will cost in the future. However, if car A is better than car B for both prices, this information is relevant to the decision-maker's choice.

This "futures" related indetermination has to be taken into account in the sensitivity and robustness analyses of the MCDA results. Typically, one of the particularly plausible possible "futures" is chosen to define the baseline case and two more "extreme" ones are selected to make new analyses framing the results of the former on both sides. The MCDA conclusions are robust if the results of the baseline and "framing" analyses do not differ too much.

For practical reasons–distance, limited availability of the stakeholders, existing expertise, and the extensive experience of the MIT analyst's team–the selection of the decision scenarios (as well as more general purpose supply scenarios) has been primarily made by the latter, in constant consultation with colleagues from other concerned CETP teams and the project management staff. In this process, the SAG's (limited, but fundamental) role was essentially to comment on the suitability of the proposed scenarios from a Chinese viewpoint, to suggest desirable amendments, and to approve the final choices.

The deliverable is a list of scenarios, organized and named with acronyms reflecting their characteristics on specific aspects. As these scenarios have already been defined in Chapter 6 ("Electric Sector Simulation"), Table 11.2 only recapitulates those results.

Table 11.2 Definition of the decision scenarios (Description, se Chapter 6)

	Retire / Refire	**Retrofit FGD**	**Use clean coal**	**New gen. mix**	**FGD on new coal**	**Peak load mgt**	**End-use efficiency**
BOX-CENPAS	-	-	-	C	no	none	-
BOX-CONPAS	-	-	-	C	yes	none	-
BOX-CONPAM	-	-	old only	C	yes	none	10%
DOX-CONLAG	yes	-	old only	C	yes	yes	20%
BOX-LONLAM	-	-	old only	L	yes	yes	10%
DOX-MONLAM	yes	-	old only	M	yes	yes	10%
BOC-NONLAS	-	-	-	N	yes	yes	20%
BOX-NONLAM	-	-	old only	N	yes	yes	10%
BOX-NONLAG	-	-	old only	N	yes	yes	20%
DOX-TONLAG	yes	-	old only	T	yes	yes	20%
DUX-DONLAG	yes	yes	old only	D	yes	yes	20%
DUX-TONPAS	yes	yes	old only	T	yes	none	-

4. THE CRITERIA

The second operation to carry out in a MCDA study is to define the criteria. These should cover all the various aspects actually serving as a base for the comparison of the decision scenarios.

The definition of the criteria usually follows several steps. First, the various viewpoints, objectives, aptitudes or properties of the scenarios are listed as exhaustively as possible. This list is based on the discussions about the definition of the scenarios, ideally conducted, according to good MCDA practice, primarily with the members of the Stakeholders' Advisory Group.

At this stage, these aspects are not structured and form a "cloud". The next step is to put some order in that formless cloud by grouping them into clusters called "dimensions". Within each dimension, the aspects are organized further, avoiding redundancy as much as possible. At this stage, the criteria form a set designed by C (C = $\{1, 2, \ldots k, \ldots m\}$)[3].

The criteria definition task continues with the working out of the necessary characteristics of the criteria, i.e.:

- a name, to unequivocally identify the criterion;
- a definition, to clarify its meaning and avoid any misunderstanding;

- a type, depending of the kind of information available to assess the scenarios' performances:
  - *cardinal criterion*; characterized by figures that can be added, subtracted, multiplied or divided (e. g., a distance criterion is *cardinal* if the various distances can be defined in terms of meters (x [m] + y [m] = {x+y} [m]),
  - *ordinal criterion*; characterized by performances which can only be described in terms of order (e. g., a distance criterion is *ordinal* if the only possible statements are of the type "the distance for scenario A is longer than that for scenario B"),
  - *intermediate criterion*; an intermediate type in which performances can only be described in terms of relative differences (e. g., a distance criterion is *intermediate* if the statements are of the type "the distance difference between scenarios A and B is slightly longer than the distance difference between scenarios C and D);
- a scale, to have a common and well defined "metric" to measure the respective performances of the scenarios; this scale can be either continuous or discrete, depending on the nature of the "dimension" it is destined to assess;
- a direction of preference, to determine whether the criterion is to be maximized or minimized.

In the CETP project, criteria definition followed the same general process as did scenarios definition (See § 3.). It was moreover strongly influenced by the type of information that the different analyst teams judged realistically possible in framing the actual "project boundary conditions" (time and means).

The list of the finally retained criteria (names and definitions) is presented in Table 11.3.

Table 11.3 List of CETP MCDA-criteria

Nr	Name	Definition (unit, direction of preference)
1	**Economy Class**	
1.1	**Average Cost of Electric Service**	Inflation adjusted overall cost per unit of electricity produced by the mix of plants over the planning period (1999-yuan / kWh in 2020, min.)
1.2	**Total Electric Sector Investment**	Total amount of Chinese capital (inflation adjusted) that will have to be spent to implement the scenario (new plants, back-fitting of older plants, related infrastructures, etc.) (1999-yuan / kWh cumulated to 2020, min.)
1.3	**Fuel Transport Burden**	Increase of fuel transportation burden over the planning period ([tons·* km_{2020} - tons·* km_{2000}) / tons·* km_{2000}], %, min.)

2	**Health and Environment Class**	
2.1	**Global Warming**	Greenhouse gas emissions in terms of CO_2 equivalent for a 100 years time horizon (kg CO_2 eq. / kWh, min.)
2.2	**Public Health Impact (Air Pollution)**	Evaluation of the impacts of the major air pollutants on the public health; represented here by the resulting mortality (years of life lost / GWh in 2020, min.)
2.3	**Potential Health Impact due to Severe Accidents**	*The evaluation on this criterion results of the combination of the two sub-criteria mentioned below (–, min)*
2.3.1	*Expected Risk*	Expected fatalities that could result from potential severe accidents within considered energy chains (Expected number of fatalities / GWh, min.)
2.3.2	*Maximum Credible Consequences*	Maximum number of fatalities that could result from a credible accident in any part of a specific energy chain (Maximum number of fatalities / GWh, min.)
2.4	**Resource Consumption**	Consumption, by the technologies used, of non-renewable energetic resources related to their known recoverable resources world-wide, in tons (fast-breeders not considered) (–, min.)
2.5	**Wastes**	*The evaluation on this criterion results of the combination of the two sub-criteria mentioned below (–, min)*
2.5.1	*Amount of Wastes*	Overall evaluation of the quantitative burden of produced wastes (t / GWh, min.)
2.5.2	*Confinement Time of Critical Wastes*	Rough estimation (order of magnitude) of the necessary confinement time for critical wastes (years, min.)
2.6	**Land Use**	Overall evaluation of the surfaces and types of land used (surface degraded from one type to a lower environmental quality one; not associated to wastes) (km^2 / GWh, min.)
3	**Society Class**	
3.1	**Impact on Employment**	(Net) number multiplied by time and quality (represented by the salary level) of domestic direct Chinese jobs created by the scenario implementation over the planning period (yuan / GWh, max)
4	**Technology Class**	
4.1	**Maturity of Technologies**	Electricity production weighted according to the maturity of the technology used, relative to the actual production (%, max.)

The emphasis put on economic and environmental criteria reflects the general context of the project as well as the main concerns of the various Chinese "society circles" represented by the Stakeholders Advisory Group members.

5. THE AGGREGATION METHOD

5.1 The ELECTRE Family

ELECTRE methods have been developed by Bernard Roy and colleagues since the mid 1960s, first at SEMA, a large European consultant company, and then at LAMSADE, the laboratory Bernard Roy created at the beginning of the 1970s within the Dauphine University in Paris.

The various methods in the ELECTRE family share a common general approach but differ on two characteristics: the form of the results (See § 5.2.) and the type of criteria used (Table 11.4).

Table 11.4 Typology of the ELECTRE methods (Source: Maystre et al., 1994)

		Results		
Outranking	**Type of criterion (see § 5.4)**	**Selection**	**Assignment**	**Ranking**
Clear-cut	True criterion	ELECTRE I	–	ELECTRE II
Fuzzy	Pseudo-criterion	ELECTRE Is	ELECTRE Tri	ELECTRE III, IV

5.2 Forms of the results ("Problematics")

Most full-aggregation methods generate a clear-cut score for each action. Comparing the relative scores obtained by the different actions in this case determines the ranking of these actions, as well as the definition of the best (or the worst) ones, or the assignment of each action to predefined categories limited by hierarchically staged thresholds.

The partial-aggregation MCDA approach works differently. Each of the above "problematics" asks for specific methodological tools. There are therefore different methods attached to each of the following possible forms of results (Table 11.5). In this table, a fourth "problematic" is mentioned that simply describes the actions and/or their consequences in some formalized way.

Table 11.5 Possible forms of results in the partial-aggregation MCDA approach (Source: adapted from Roy, 1985)

Selection (α)	Selection of a subset of actions containing the actions not outranked by any other action of the subset.
Assignment (β)	Assignment of the actions to predefined categories.
Ranking (γ)	Ranking of the actions in decreasing or increasing order of preference.
Description (δ)	Description of the actions and/or their consequences in a systematic and formalized way or elaboration of a cognitive procedure.

5.3 General Approach

ELECTRE methods differ fundamentally from "classical" aggregation methods. They follow a two-step general procedure. In the first step, the potential actions are compared by pairs to evaluate the type of relationship that links them. (See below Table 11.6.)

In the second step, these relationships are analyzed to generate the desired results (according to § 5.2).

Table 6 defines the four basic types of relationship that can come out of the comparison of two actions at a global level, plus a fifth (derived) one that is central in the ELECTRE approach and explains why the methods of this type are called "outranking" methods.

Table 11.6 Basic and derived overall relationship types between two actions (Source: adapted from Roy, 1985)

Indifference (I)	There are clear reasons to consider the two actions as equivalent.
Strict preference (P)	There are clear reasons to prefer one action to the other.
Weak preference (Q)[4]	There are clear reasons to identify the relationship as intermediary to the ones mentioned above.
Incomparability (R)	There are clear reasons to consider the actions as too different to be compared.
Outranking (S)	There are clear reasons to consider the relationship to be either indifference, strict preference, or weak preference.

For a given pair of actions, the determination of the appropriate overall type of relationship depends on the answer given to the following question: "Does action a_i outrank action a_k?" (or, in other words: "is action a_i at least as good as action a_k?"). To test the hypothesis that a_i outranks a_k, two different tests are conducted:

- *Concordance test*: the larger the number of criteria in favor of a_i (meaning that the performance of a_i is at least as good as the performance of a_k for the considered criteria), the more credible the hypothesis;
- *Non-discordance test*: the smaller the extent of the maximal difference in favor of a_k (i.e. performance of a_k minus performance of a_i when the performance of a_i is worse than the performance of a_k), the more credible the hypothesis.

Combining the results of these two tests allows us to define a global credibility index for each pair of actions considered in the study, going from zero (no credibility at all) to one (maximum credibility). These indexes serve as a basis for the final comparison of the actions in one of the possible forms definable in the context of a MCDA analysis.

5.4 Types of criteria

The relationships between pairs of actions mentioned in Table 11.5 are defined at the global level (i.e. taking into account the relative performances for all the criteria). The same relationships–with the obvious exception of *incomparability*–can be used to compare the actions at the single-criterion level according to the quality of the available information on the respective action performances.

Traditionally, the information is assumed to be accurate enough to determine truly significant differences between the pairs; this situation corresponds to the so-called *true criterion* model in which the only possible relationships, at the criterion level, are:

- *(Strict) preference* (a_j P_j a_k) : a_i is clearly preferred to a_k on criterion j,
- *Indifference* (a_i I_j a_k) : a_i and a_k have exactly the same performance on criterion j.

However, if the information is not that precise, which is usually the case in practical situations, due to various sources of uncertainty, either "physical" or "mental", there is a case for other types of criteria (See Table 11.7.), using the above-mentioned relationships in combination with a third one:

- *Weak preference* (a_i Q_j a_k) : a_i is weakly preferred to a_k on criterion j.

In the *quasi-criterion* case, two actions may be termed indifferent if the difference $[g_j(a_i) - g_j(a_k)]$ between their respective performances for the considered criterion "j" remains small (rather than strictly equal as imposed in the true criterion case). This situation is modeled by delimiting an *indifference interval* (I_j) between $-q_j$ and q_j, with q_j being the *indifference threshold*. If, however, there is some hesitation between indifference (I_j) and strict preference (P_j), a *weak preference* (Q_j) interval can be added as in the *pre-criterion* (for "point" indifference) or in the *pseudo-criterion* (when indifference is as well defined on a finite interval) cases.

The transition from weak preference to strict preference requires the definition of a new threshold, the *strict preference threshold* p_j.

Table 11.7 The different types of criteria (Source: adapted from Schärlig, 1985)

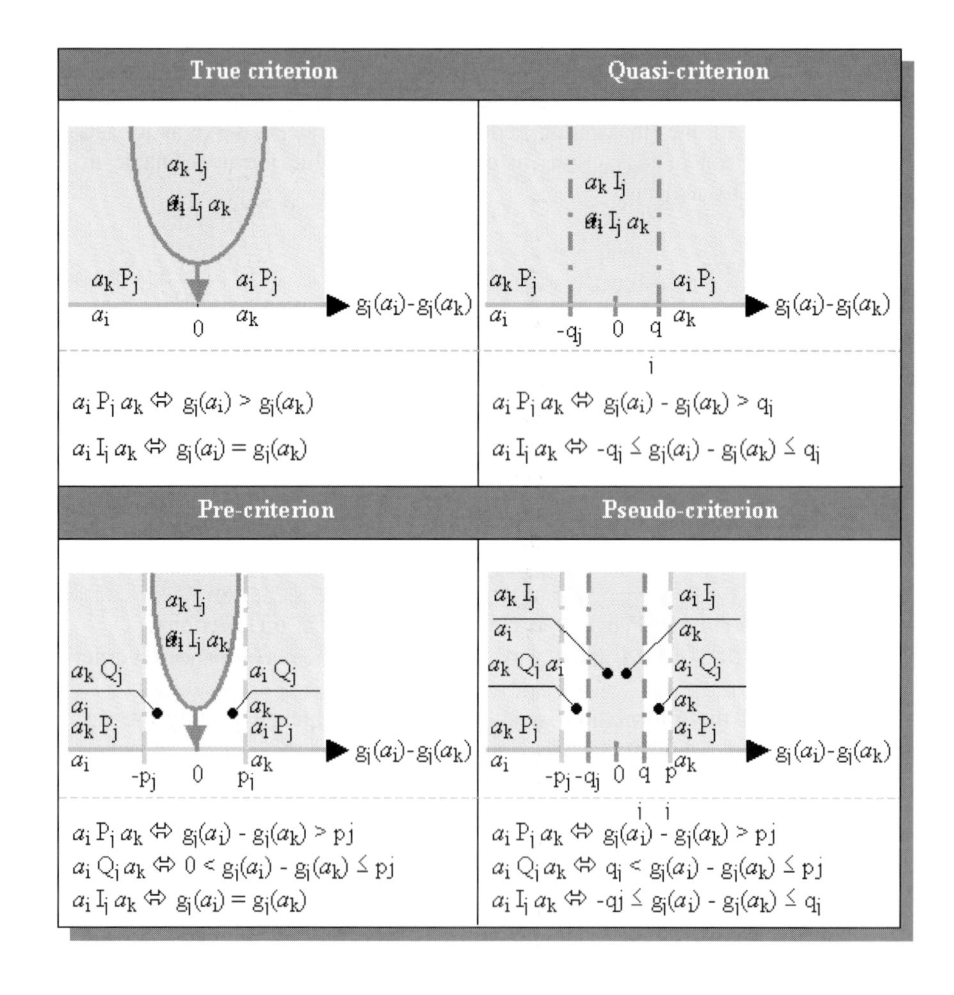

5.5 ELECTRE III

ELECTRE III (Roy, 1978) is a multicriteria method that generates a ranking of selected actions (See Table 11.6.). A short description of this method is given below. For more detailed information, the reader is referred to the existing literature (e.g.: Roy and Bouyssou, 1993; Maystre et al., 1994).

ELECTRE III considers pseudo-criteria and makes use of "fuzzy numbers" to quantify the results of the concordance and discordance tests mentioned in § 5.2. (See Figure 11.5.):

- *Concordance test*: If the difference of the actions' performances [$g_j(a_i)$ – $g_j(a_k)$] is smaller than (i.e. "to the left of") the "inverse" preference threshold -p_j, the concordance index for the criterion "j" is nil (action a_i is certainly worse than action a_k); if this difference is greater than (i.e. "to the right of") the "inverse" indifference threshold -q_j, the index is one (action a_i can be considered as "at least as good" as action a_k); in between, for simplicity's sake, the index increases linearly from zero to one[1]. Note that this means that ELECTRE III adopts a very "conservative" attitude regarding the rejection of the tested hypothesis ("a_i at least as good as a_k"). Concordance with this hypothesis is totally rejected (concordance index equal to zero) only when the difference in performances is definitely in favor of a_k rather than a_i.
- *Non-discordance test*: The upper limit at which there may be a possible strong discordance with the tested hypothesis occurs when the difference of the actions' performances [$g_j(a_i) - g_j(a_k)$] becomes smaller than the "inverse" strict preference threshold $-p_j$. However, it is only when this difference falls below some lower value (defined by a third threshold, the *veto threshold* -v_j) that the discordance can be called "absolute". Thus, above the "inverse" strict preference threshold, the discordance index is nil; below the veto threshold it takes the value one, and in between, again for simplicity's sake, it increases linearly from zero to one when the differences passes from $-p_j$ to $-v_j$. As above, this is a very conservative and prudent way of assessing some strong opposition to the truthfulness of the hypothesis "a_i at least as good as a_k".

When all the concordance indexes relative to a given pair of actions have been evaluated for each of the criteria, a *global concordance index* for the pair of actions under consideration is calculated by performing a weighted sum of these per-criterion indexes, taking into account the weight attributed to the different criteria by the Stakeholder Advisory Group members.

Concordance index: $c_j(a_i, a_k)$*

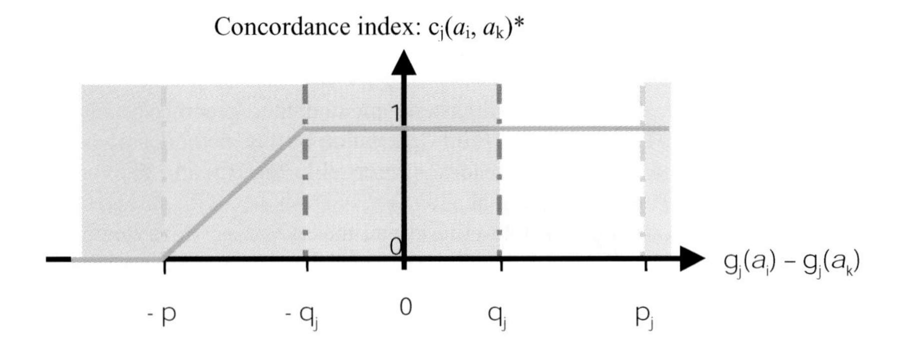

Discordance index: $d_j(a_i, a_k)$

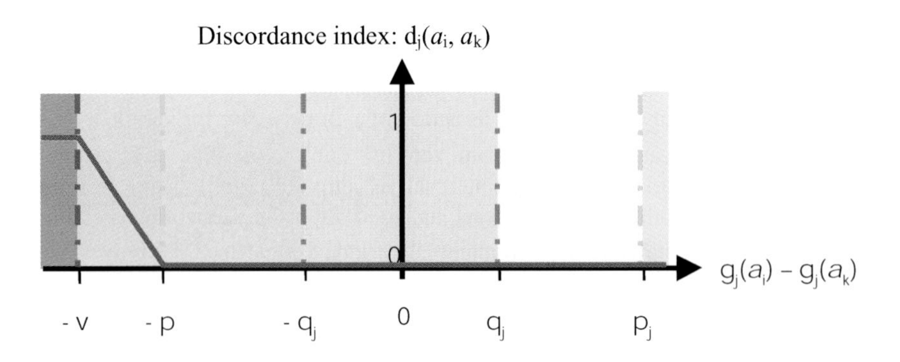

Figure 11.5 Definitions of the concordance and discordance indexes per criterion

Finally, a credibility index is calculated on the basis of the global preference index weakened by the discordance indexes. This is done in such a way that if even only a single discordance index is equal to one (unacceptable discordance with the tested hypothesis about a_i and a_k for the corresponding criterion), then the credibility index takes the value zero, whatever the value of the global concordance index (veto principle).

Note that all the above-defined indexes are normalized, i.e. they have a value between zero and one.

Obtaining a robust ranking of the selected actions out of the set of credibility indexes (one for each possible pairs of actions) is not an easy task, due to their fuzzy nature. The actions should be ranked beginning with those having the greatest possible credibility.

The algorithm used for this purpose in ELECTRE III is quite complex, being designed to take incomparability relationships into account[2]. To achieve this, a double calculation is necessary:

- *Descending distillation*: The calculation identifies the "best" action(s), putting it (them) at the highest rank; iteratively, the calculation then identifies, among the remaining actions, the "next best" action(s) and put it (them) in descending ranks; as incomparable actions are by definition not clearly "good", they tend to fall lower in this distillation.
- *Ascending distillation*: The calculation identifies the "worst" action(s), putting it (them) at the lowest rank; iteratively, the calculation then identifies, among the remaining actions, the "next worst" action(s) and put it (them) in ascending ranks; as incomparable actions are by definition not clearly "bad", they tend to rise in the ranking in this distillation.

This double calculation leads to two rankings, which can differ depending on how comparable the actions truly are. If all the actions are perfectly comparable, the two rankings are identical. Mixing them allows us to obtain what is called a "partial preorder":

- a preorder is a ranking allowing for ex æquo,
- it is partial because incomparability among actions is allowed (see § 8).

6. THE WEIGHTING PROCESS

As mentioned in the preceding section, the relative importance of the different criteria are assessed by the selected stakeholders, this information largely determines the definition of the credibility indexes used to rank the considered actions.

The need for such a weighting process is sometimes seen to be a weakness of the MCDA approach because of its perceived subjective aspect. In fact, almost all multiple criteria methods rely in one way or another on "weights" (even if the exact meaning of these differ from one method to another one). It is obvious that everyone has an opinion depending on training, personal experience, and value system, about the relative importance of the different criteria in the decision process. Ignoring this reality will make most comparative multicriteria analyses senseless.

However, experience has shown that it is not easy for stakeholders to assign direct weights that correctly reflect their opinions about the relative importance of the criteria. For this reason, the two-step weighting approach initially developed for the SESAMS Swiss case study (Haldi et al. 2002) was also used in the CETP project. In the first step, each stakeholder is asked to arrange in order of deceasing importance a set of cards (similar to playing cards) representing the different criteria. Each card mentions only the name of the considered criterion, to the exclusion of any other indicator likely to influence the stakeholder's judgment. In the second step, the stakeholder is asked to define the *relative* positioning of a criterion between its two immediate neighbors (the one ranked immediately above and the one ranked immediately below it in order of decreasing importance). These two types of information allow us to define a set of coupled algebraic equations linking the respective "weights" (unknowns), which can be solved by a matrix inversion algorithm to obtain "weights" or "importance levels" of the individual criteria.

The "weights" thus obtained in the framework of the CETP project are given in Table 11.7. To respect the wishes of the stakeholders, the individual results of the above-described process are given here anonymously. Because one of the stakeholders did not fully answer the questionnaire used to define the "weights", only nine results are presented.

Table 11.7 Results of the weighting process, expressed in relative (%) values

	SAG members								
Criteria	**A**	**B**	**C**	**D**	**E**	**F**	**G**	**H**	**I**
Average Cost of Electric Service	29	13	13	15	12	14	15	18	12
Total Electric Sector Investment	24	12	13	15	12	13	14	18	13
Fuel Transport Burden	21	12	12	14	12	13	6	12	9
Global Warming	2	10	2	7	7	4	6	4	1
Public Health Impact (Air pollution)	1	9	10	6	10	11	11	7	12
Potential Health Impact (due to Severe Accidents)	0	6	9	4	0	14	10	3	10
Resource Consumption	7	11	12	11	11	2	12	11	9
Wastes	0	9	6	3	11	7	8	4	8
Land Use	4	4	7	10	11	0	5	6	7
Impact on Employment	12	3	4	12	2	10	4	1	9
Maturity of Technologies	0	11	12	2	12	12	9	16	10

Although the SAG members had obvious differences of opinion regarding the relative importance of the selected criteria, it is clear that they seem principally preoccupied by the economic factors. "Health impacts" and "resource consumption" take second priority, but opinions on these are less unanimous than for the economic criteria. "Global warming" appears to be less important in the Chinese context than in Western countries. Finally, the importance of the "impact on employment" and the "maturity of technologies" criteria is clearly a matter of debate among the SAG members.

7. THE PERFORMANCE MATRIX

The *performance* or *evaluation* matrix is the basic information source for the MCDA analysis. It gives the respective "profiles" of the decision scenarios to be compared (rows of the matrix) with respect to the set of selected criteria (columns of the matrix). This is symbolically represented in Figure 11.6.

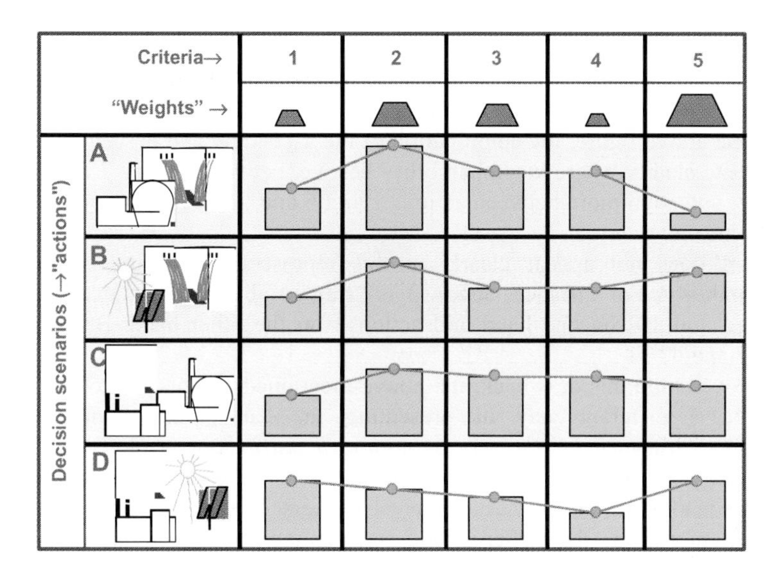

Figure 11.6 Symbolic representation of the performance matrix

The raw data come from the different studies presented in the preceding chapters by various CETP partners. These data were processed and adapted to the multicriteria analysis requirements by the PSI team and then delivered to the MCDA analysts in the form of a convenient Excel spreadsheet. In fact, there were three different performance matrices to analyze in the CETP project, one for each of the particular "futures" considered in the MCDA study (remember that the "futures" take into account "general" socioeconomic parameters that are beyond the influence of the decision-makers. (See Chapter 6.) These parameters include:

- FIB (*baseline case)*: moderate demand growth (5.11%/yr), baseline coal and natural gas costs;
- FAF: moderate demand growth (5.11%/yr), higher coal costs and low natural gas costs; and
- SUB: high demand growth (7.66%/yr), low coal costs and baseline natural gas costs.

As an example, the performance matrix corresponding to the baseline case (FIB) is given in Table 11.8.

8. ANALYSIS OF THE MCDA RESULTS

Each combination of the following elements–"futures", sets of weights, and ELECTRE III "control" parameters (thresholds)–generates a single result. As mentioned in §5, each result takes the form of a double ranking, which is used in combination to identify potential incomparability among the decision scenarios.

To make this point clearer, let us take an example developed in Pictet et al. (1994). The rankings considered in this example are presented in the "classical" way in Figure 11.7. In this figure the arrows indicate the direction of preference (e.g., in the descending-distillation order, action 4 is preferred to action 5); actions in the same column are indifferent (e.g., in the descending-distillation order, actions 8, 11 and 12).

In the above figure, the combination of the ascending and descending rankings shows quite clearly where incomparability lies, i.e. between actions 3 and 5, actions 2 and 12, and even more between actions 1 and 6 and the rest of the actions, except the extreme ones.

What does not appear clearly in this representation is the *extent* of this incomparability. For instance, action 3 is "framed" between the two best-ranked actions (4 and 10) on one hand and action 2 on the other hand. But what about actions 1 and 6?

To overcome this drawback, the above-mentioned authors (Pictet et al., 1994) developed a different way of presenting the "antagonist" rankings, called "Surmesure" (*Surface for the representation of outranking methods results*). The "Surmesure" presentation of the results of the preceding example is given in Figure 11.8. The reader can check that all the relative positions of the actions are indeed respected; moreover, the rather strong incomparability of actions 1 and 6 with respect to most actions, and their proximity to each other, are made more visible.

The analysis of the figure is rather simple:

- An action that is first in both rankings appears at the upper right corner.
- An action that is last in both rankings appears at the lower left corner.
- The nearer an action is to the diagonal, the more comparable it is.
- The farther an action is from the diagonal, the less comparable it is; moreover, if an action is positioned at the upper left or lower right corner it will be incomparable with most of the others.

"Surmesure" can moreover be used in a different way to display simultaneously different results. For each action, a "Surmesure" graph is drawn, on which the position of the action corresponding to the two rankings obtained for a given combination of elements (i.e. a "result") is marked by a dot. The cluster of dots thus obtained for the different results makes it possible to see at a glance how the considered action performs globally. (See Figure 11.9)

Table 11.8. CETP Performance Matrix used in the MCDA analysis (FIB baseline case example for 2020)

	Criteria / Scenarios	1.1 Average Cost of Electric Service	1.2. Total Electric Sector Investment	1.3. Fuel Transport Burden	2.1 Global Warming	2.2 Public Health Impact (Air Pollution) 10^{-4}	*2.3.1 Expected Risk [10^{-4}]*	*2.3.2 Maximum Credible Consequences*	2.4 Resource Consumption	*2.5.1 Amount of Wastes*	*2.5.2 Confinement Time of Critical Wastes*	2.6 Land Use	3.1 Impact on Employment	4.1 Maturity of Technologies *
1	BOC-CENPAS	0.307	0.057	6.419	1.05	4.58	2.11	434	0.140	152	1000	450	8553	1.000
2	BOC-CONPAS	0.326	0.058	6.755	1.05	2.37	2.15	434	0.142	154	1000	461	8553	1.000
3	BOX-CONPAM	0.297	0.057	6.457	0.91	2.12	1.84	434	0.129	129	1000	380	7437	1.000
4	DOX-CONLAG	0.283	0.059	5.132	0.76	1.99	1.57	434	0.116	113	1000	304	6320	1.000
5	BOX-LONLAM	0.298	0.057	5.330	0.88	2.04	1.73	434	0.126	133	1000	334	7438	1.000
6	DOX-MONLAM	0.309	0.056	6.061	0.87	2.41	1.77	434	0.129	128	1000	342	7252	1.000
7	BOC-NONLAS	0.327	0.058	4.847	0.82	2.10	1.72	28000	0.135	129	100000	356	7825	1.000
8	BOX-NONLAM	0.300	0.057	4.127	0.67	1.84	1.40	28000	0.124	103	100000	271	6709	1.000
9	BOX-NONLAG	0.289	0.061	3.214	0.53	1.57	1.12	28000	0.112	84	100000	206	5594	1.000
10	DOX-TONLAG	0.291	0.063	3.862	0.55	1.10	1.07	28000	0.111	82	100000	225	5594	1.000
11	DUX-DONLAG	0.292	0.061	3.472	0.54	1.37	1.13	28000	0.113	85	100000	216	5590	1.000
12	DUX-TONPAS	0.342	0.064	6.293	0.85	1.24	1.54	28000	0.131	119	100000	350	7824	1.000
	Criteria units (see Table 11.3)	yuan / kWh	yuan / kWh	%	kg CO_2 / kWh	YOLL / GWh	10^{-4} # fatal. / GWh	# fatal. / GWh	–	t / GWh	years	km^2/ GWh	yuan / GWh	–
	Sources, see chapter:	6	6	8	8	9	10	10	8	8	8	8	6	6

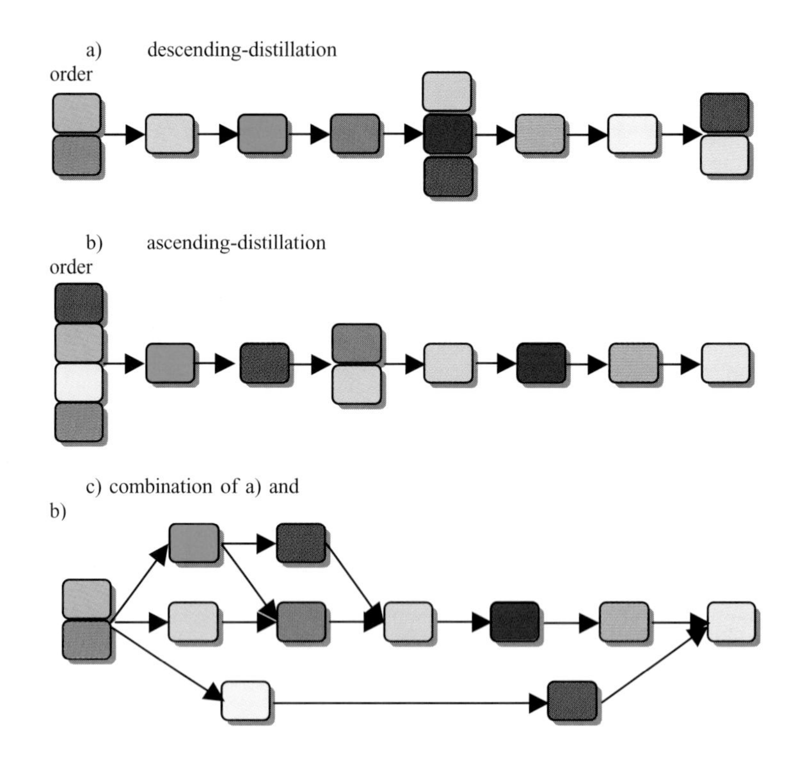

Figure 11.7. The two types of ranking and their combination ("classical" representation)

In the CETP case, the results are too numerous to be presented on only one graph per decision scenario. The presentation of the results will therefore be split according to:

- the "futures" (FIB, FAF, SUB), because it would make little sense to compare one decision scenario (= "action") in a "future" with one in a different "future";
- the possible use of the veto, which would reflect a stakeholder's position regarding the acceptance or rejection of large compensations when comparing the relative performances of the studied decision scenarios with respect to different criteria.

For each decision scenario and above-described "situation", each graph displays a series of dots representing the results obtained for the combinations of the nine sets of stakeholders' weights (A to I, see Table 7), three sets of criteria thresholds

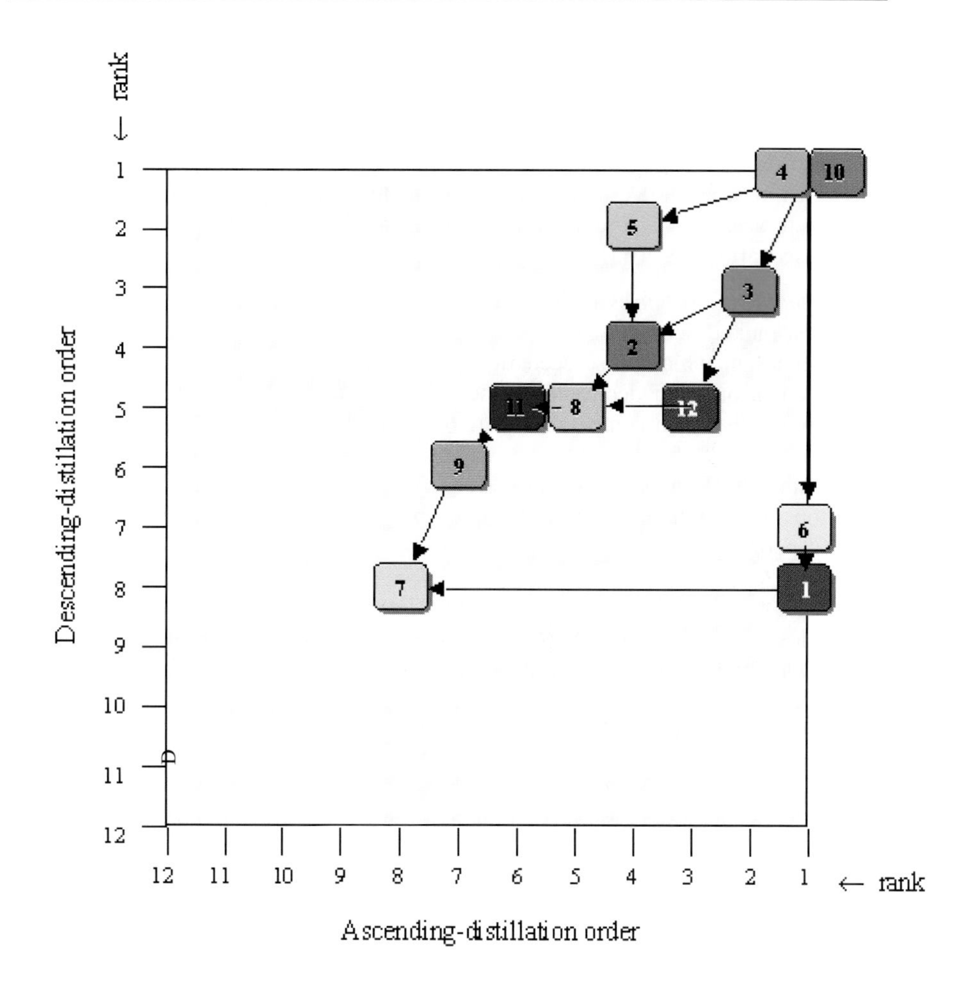

Figure 11.8 The "Surmesure" display of the results of Figure 11.7 example.

(indifference, strict preference and veto) and two sets of global thresholds (See Table 9)[1] This yields 54 dots (9×3×2) per graph. A short discussion of the various graphs is given below.

Table 11.9 Parameters used in the analysis.

	1st set	2nd set	3rd set
Criteria thresholds			
• indifference	25%	15%	35%
• preference	50%	40%	60%
• veto	90%	75%	99%
Global thresholds	0.3 - 0.15 λ	0.2 - 0.19 λ	–

FIB with veto (Figures 11.9 a and 11.9 b)
The group of scenarios 3, 4, 5, and 6 has good global rankings. If one considers only the occurrences in which these scenarios are ranked top in both orders, they can be classified in order of decreasing preference as follows: 4, 5, 6, 3. The other ones are less good; scenarios 1, 7 and 8 exhibit rather good comparability, while it is less so for the scenarios 2, 9, 10, 11 and 12.

FIB without veto (Figures 11.9 c and 11.9 d)
The best-ranked scenarios are here 4, 5, 9, 10 and 11. If one considers only the occurrences in which these scenarios are ranked top in both orders, they can be classified in order of decreasing preference as follows: 9, 10, 11, 4, 5. Scenarios exhibiting rather poor comparability are 6, 11 and 12.

FAF with veto (Figures 11.10 a and 11.10 b)
The top of the ranking is here occupied by only one scenario, 6. Scenarios that exhibit poor comparability are 5, 9, 10, 11 and 12 (particularly pronounced for the last two).
FAF without veto (Figures 11.10 c and 11.10 d)
Scenarios 6 and 11 have relatively good global rankings. Scenarios that are not easily comparable are 3, 4, 10 and 12.

SUB with veto (Figures 11.11 a and 11.10 b)
If one considers only the occurrences in which the scenarios are ranked top in both orders, they can be classified in order of decreasing preference as follows: 6 and 5. Scenarios that exhibit rather poor comparability are 2, 9, 10, 11 and 12.

SUB without veto (Figures 11.11 c and 11.11 d)
If one considers only the occurrences in which these scenarios are ranked top in both orders, they can be classified in order of decreasing preference as follows: 5, 4, 9, 10, 6 and 11. Only scenario 12 exhibits rather poor comparability.

Table 10 summarizes the results of the MCDA analysis, based on the number of occurrences in which a scenario is at the top of both rankings. The interpretation of this table must accommodate both the performances in terms of "Potential health impact due to severe accidents", and "Impact on employment" and the impact of the "futures" for each set of scenarios.

In terms of the performances, for two criteria ("Potential health impact due to severe accidents" and "Impact on employment"), there is a clear gap between scenarios 1 to 6, and scenarios 7 to 12. The veto effect emphasizes this separation as shown by the fact that no action of the second group appears in the row "With veto". (The veto prohibits them from outranking actions of the first group because of the above-mentioned gap.)

In terms of "futures" impacts the FIB "future" tends to suggest actions with average use of gas and coal. The FAF "future" favors scenarios emphasizing gas, but limiting coal (essentially action 6). The SUB "future" tends in the opposite direction, promoting scenarios that emphasize coal, while limiting gas.

Table 11.10 Best scenarios for the various cases.

	FIB	FAF	SUB
With veto	4, 5, 6, 3	6	6, 5
Without veto	9, 10,11, 4, 5	6, 11	5, 4, 9, 10, 6, 11

Altogether, scenario 6 (DOX-MONLAM) appears to be the best, being among the top ranked in five out of six cases. However, if the decision-makers think that the most probable "future" is FIB (second column of Table 11.10) they might also consider scenarios 4 (DOX-CONLAG) and 5 (BOX-LONLAM). Moreover, if the big gap observed for the two criteria "Potential health impact due to severe accidents" and "Impact on employment" is seen as not too significant (i.e. the veto option is not "activated", third row of Table 11.10), they might also consider scenarios 9 (BOX-NONLAG), 10 (DOX-TONLAG) and 11 (DUX-DONLAG). It is worth noting that the selected scenarios have a strong bias towards demand-side management. As the demand-side costs were assumed to be low (no analysis available), crediting conservation is a win-win situation. Therefore, scenarios not explicitly crediting demand-side management are a priori losers. The stakeholders reject the "dirty" coal scenario and favor clean coal technologies in the first place. Diversification of electricity supply is partially considered as attractive but the MCDA study indicates that the stakeholders have a more differentiated perspective on its benefits.

The above discussion of the CETP MCDA results should be considered to be only a rough outline of the conclusions that could be drawn from such a multicriteria analysis. Due to the important delays that occurred in the delivery of the decision scenario performance data (because of unforeseen difficulties in gathering the required basic information), and to meet the deadlines assigned to the program, it has unfortunately not been possible to carry out the iterative process and detailed sensitivity analyses that should be integral parts of a MCDA study. In particular, the first conclusions given above should have been presented and explained to the stakeholder advisory group so that its members could, if they desired, "correct" their initial positions regarding the importance they assigned to the different criteria; possibly modify the definitions of these criteria; or add or remove some scenarios. This however requires more time and availability of the concerned parties than could be achieved in the framework of the CETP program. Nevertheless, the main conclusions presented in this section appear sufficiently robust to not be significantly modified by a more thorough analysis.

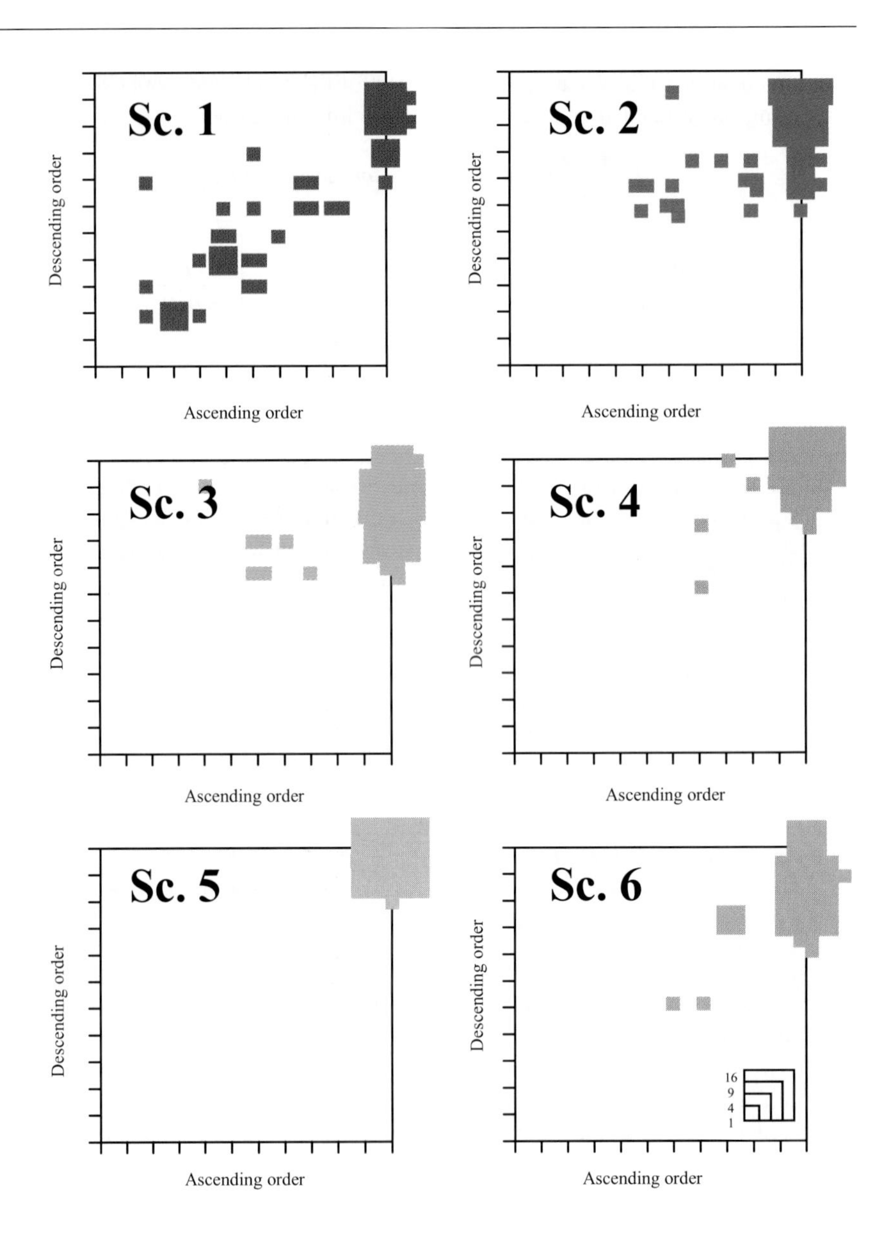

Figure 11.9 a "Surmesure" graphs for scenarios 1 to 6, FIB "future", with veto

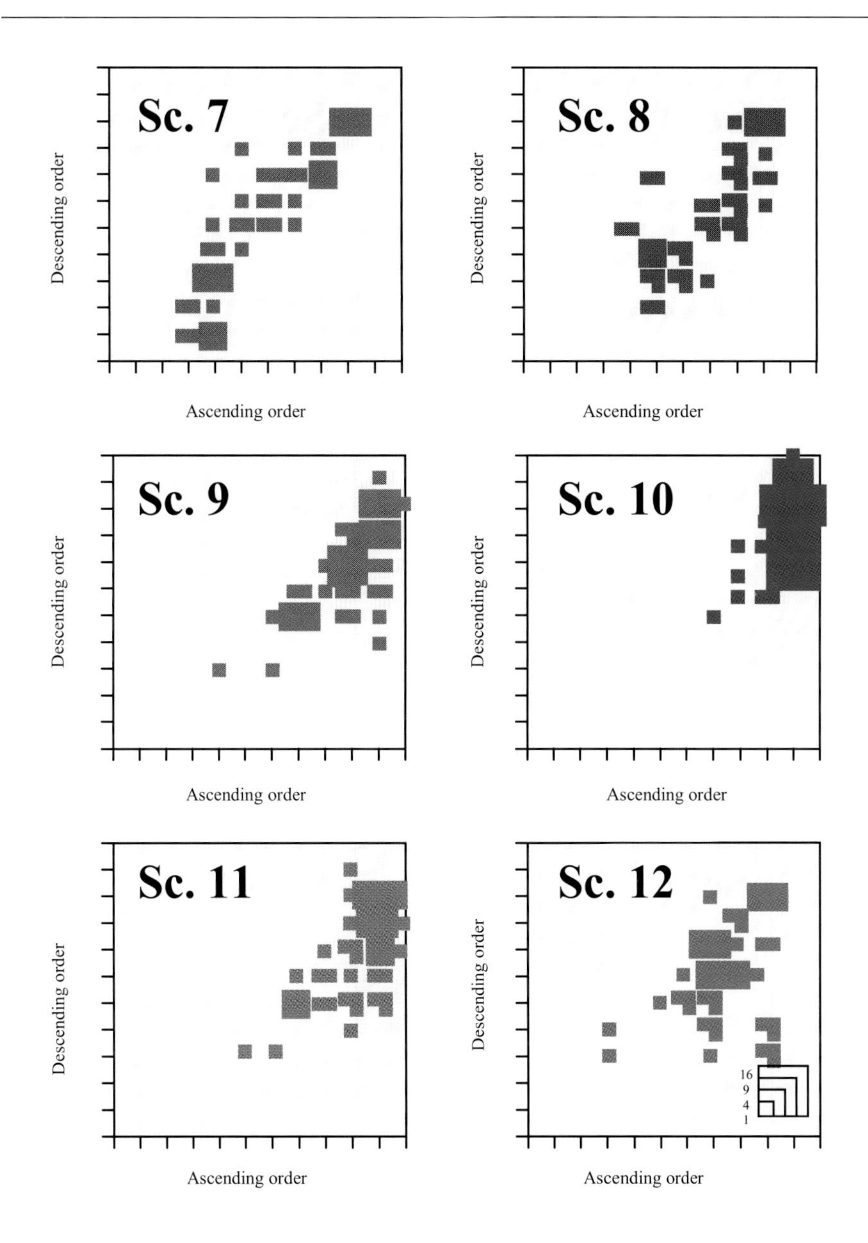

Figure 11.9 b "Surmesure" graphs for scenarios 7 to 12, FIB "future", with veto

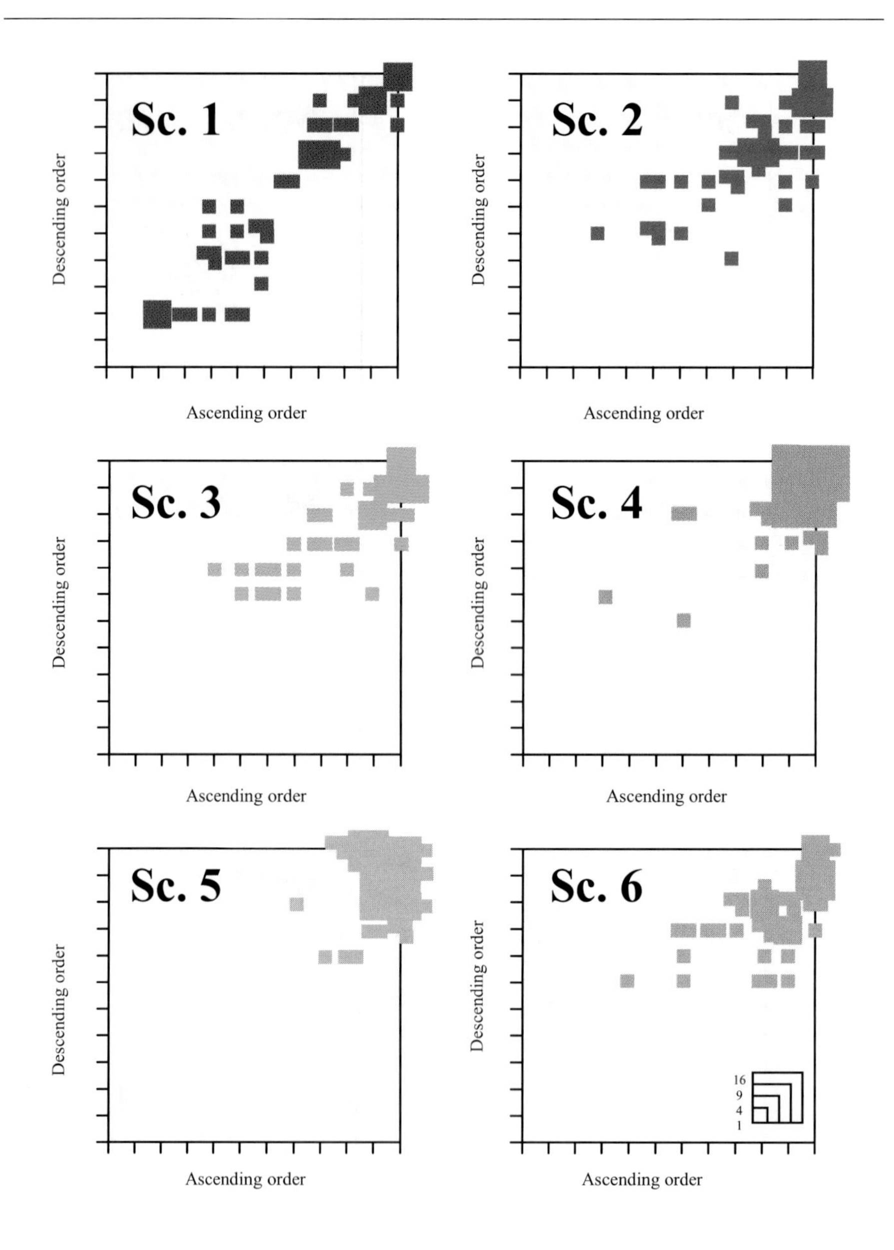

Figure 11.9 c "Surmesure" graphs for scenarios 1 to 6, FIB "future", without veto

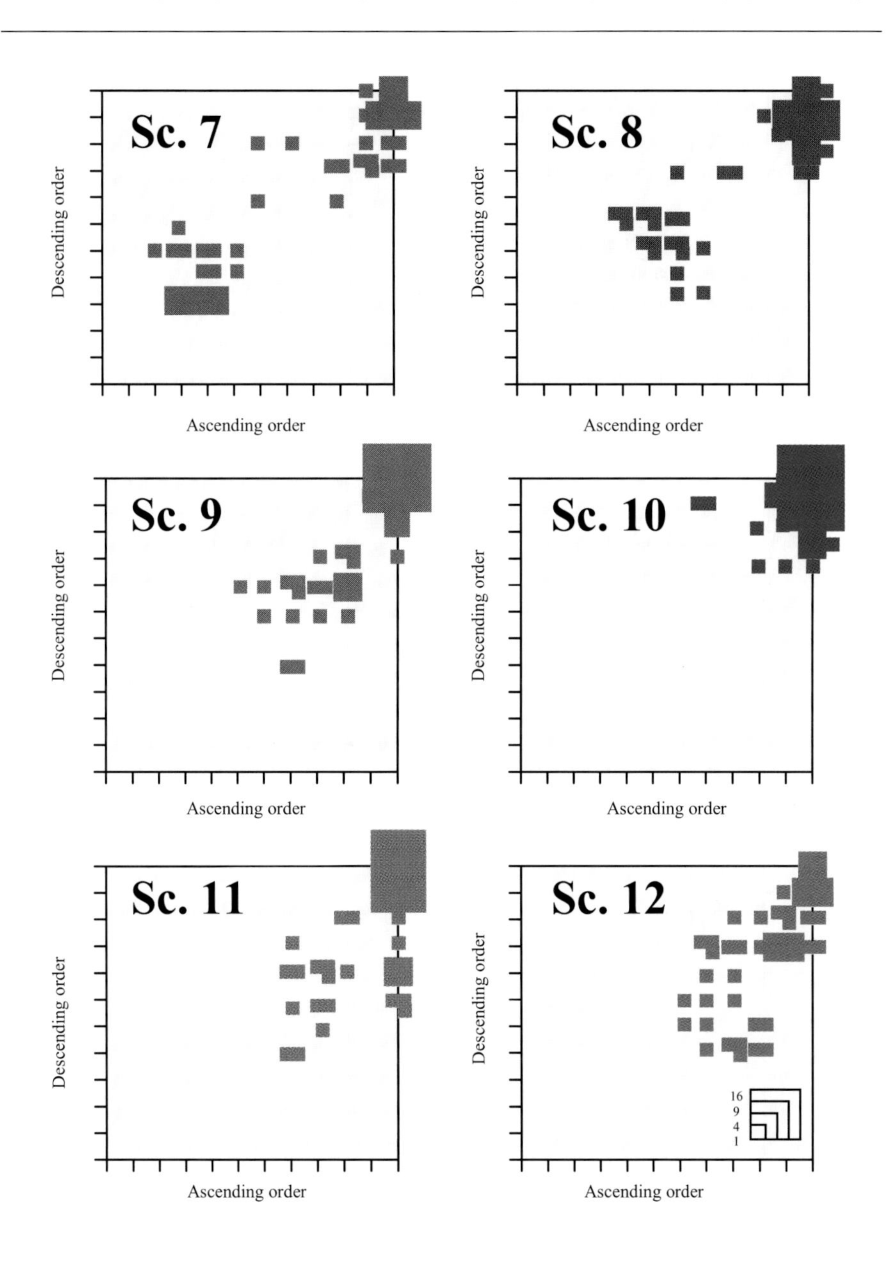

Figure 11.9 d "Surmesure" graphs for scenarios 7 to 12, FIB "future", without veto

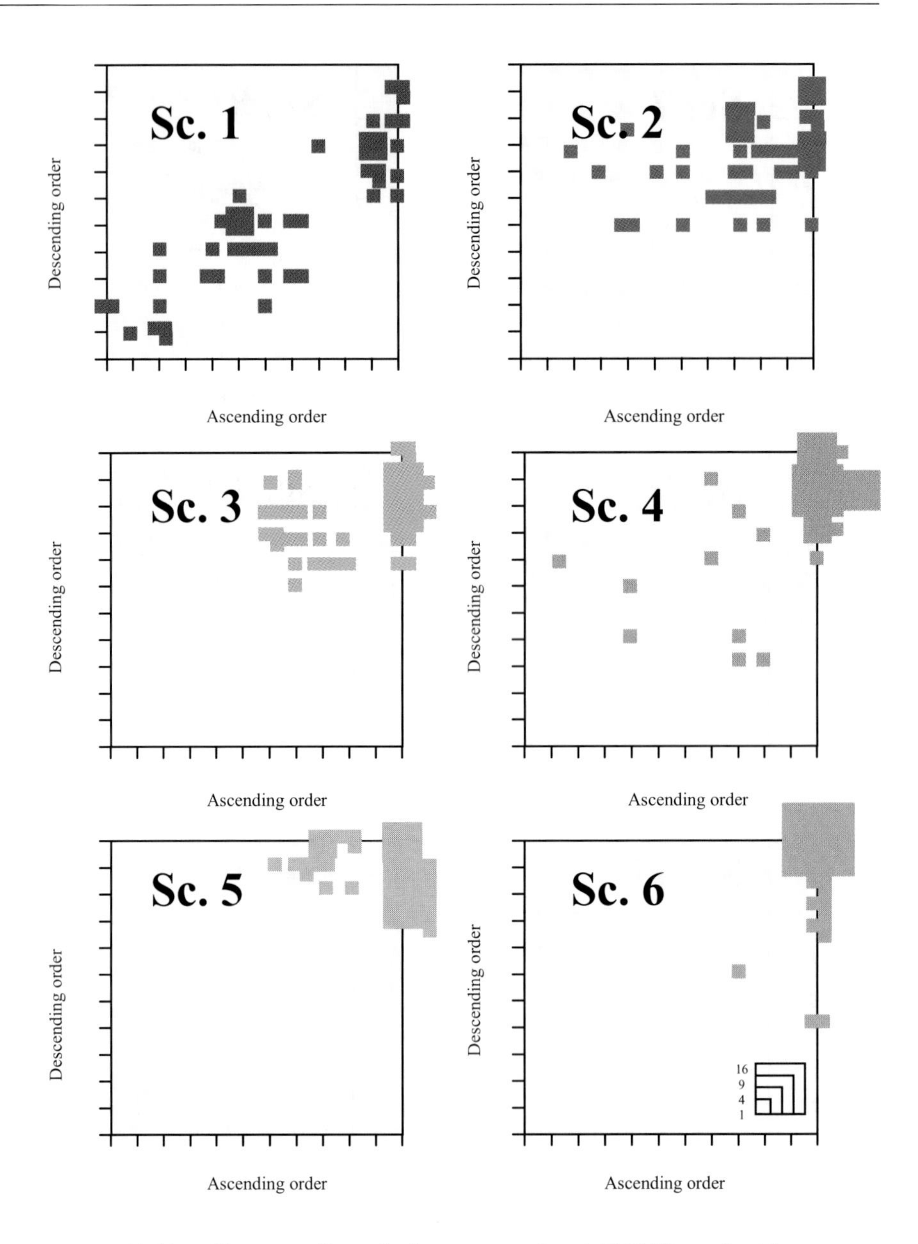

Figure 11.10 a "Surmesure" graphs for scenarios 1 to 6, FAF "future", with veto

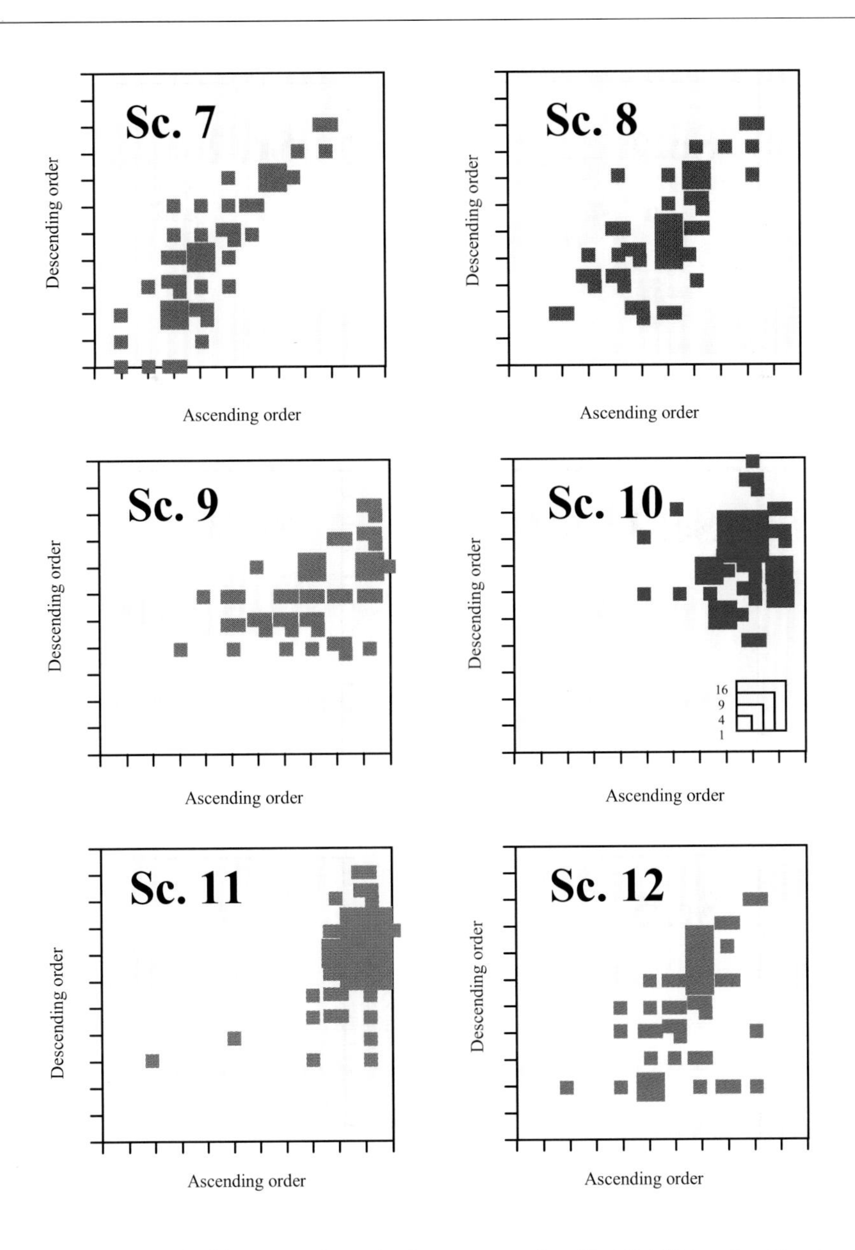

Figure 11.10 b "Surmesure" graphs for scenarios 7 to 12, FAF "future", with veto

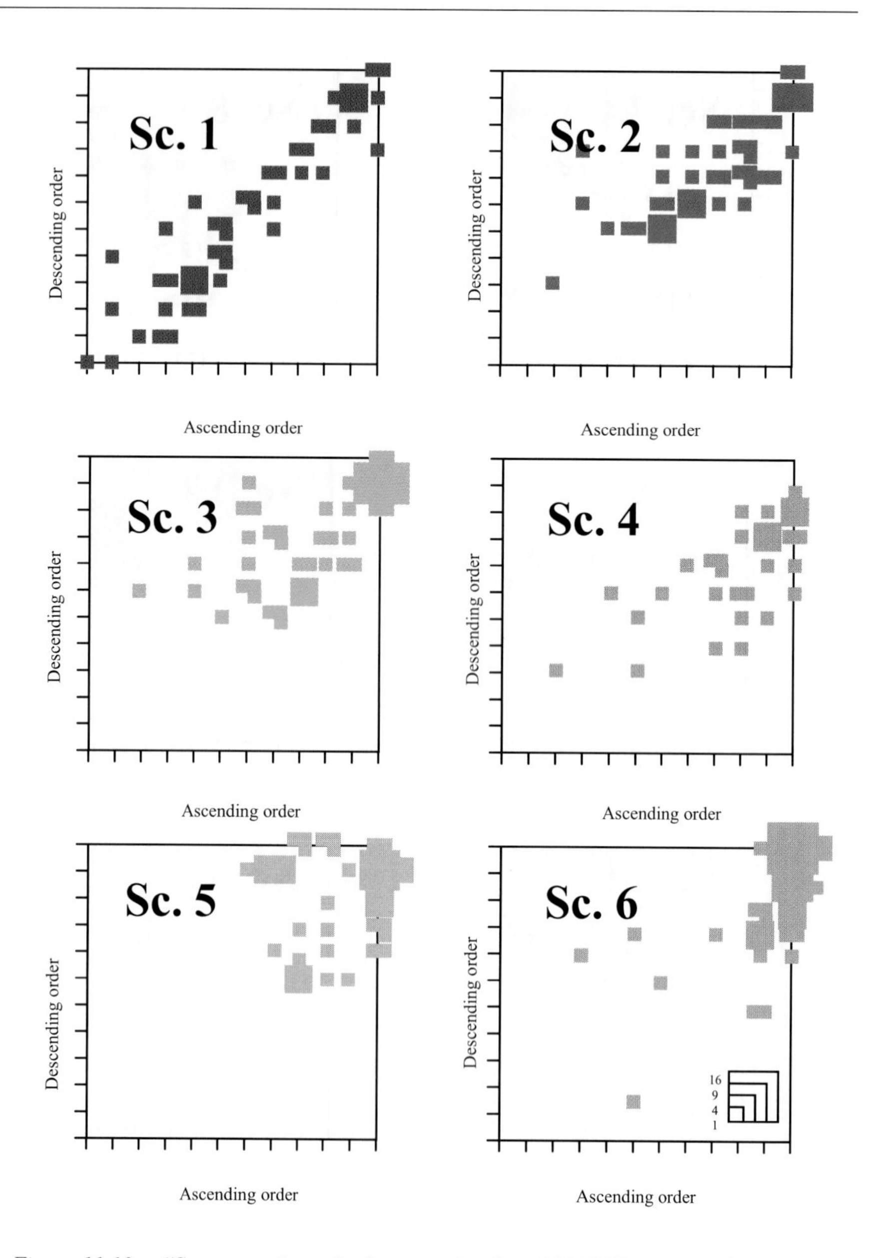

Figure 11.10 c "Surmesure" graphs for scenarios 1 to 6, FAF "future", without veto

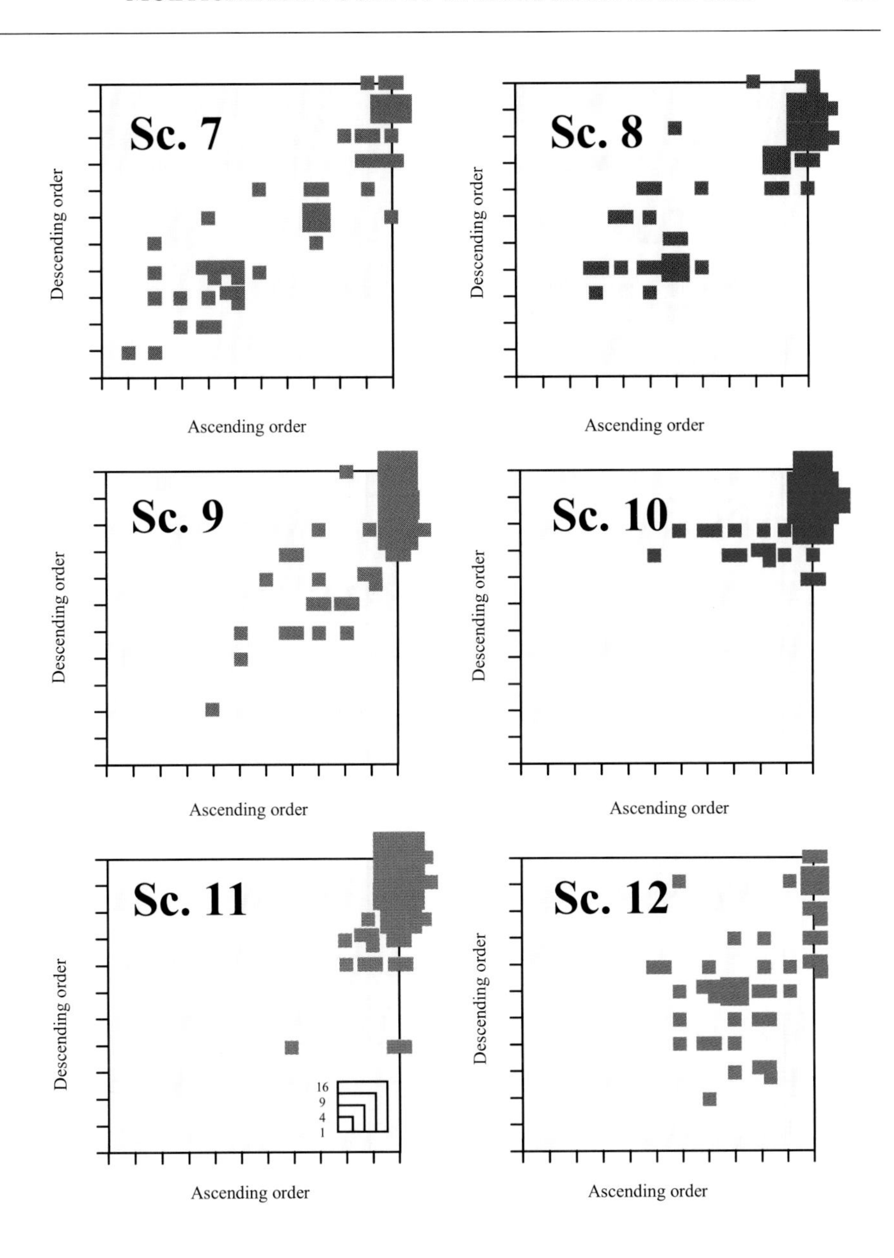

Figure 11.10 d "Surmesure" graphs for scenarios 7 to 12, FAF "future", without veto

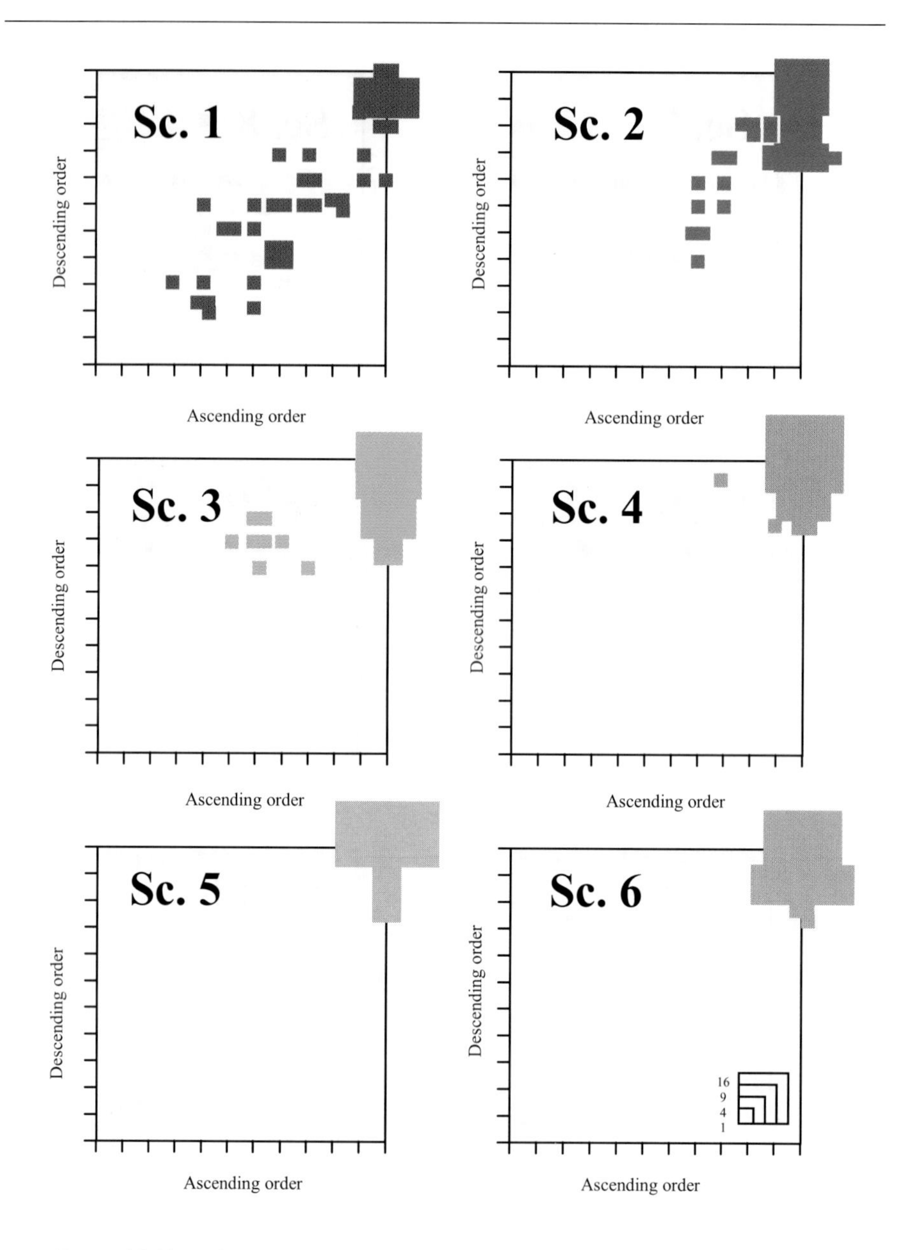

Figure 11.11 a "Surmesure" graphs for scenarios 1 to 6, SUB "future", with veto

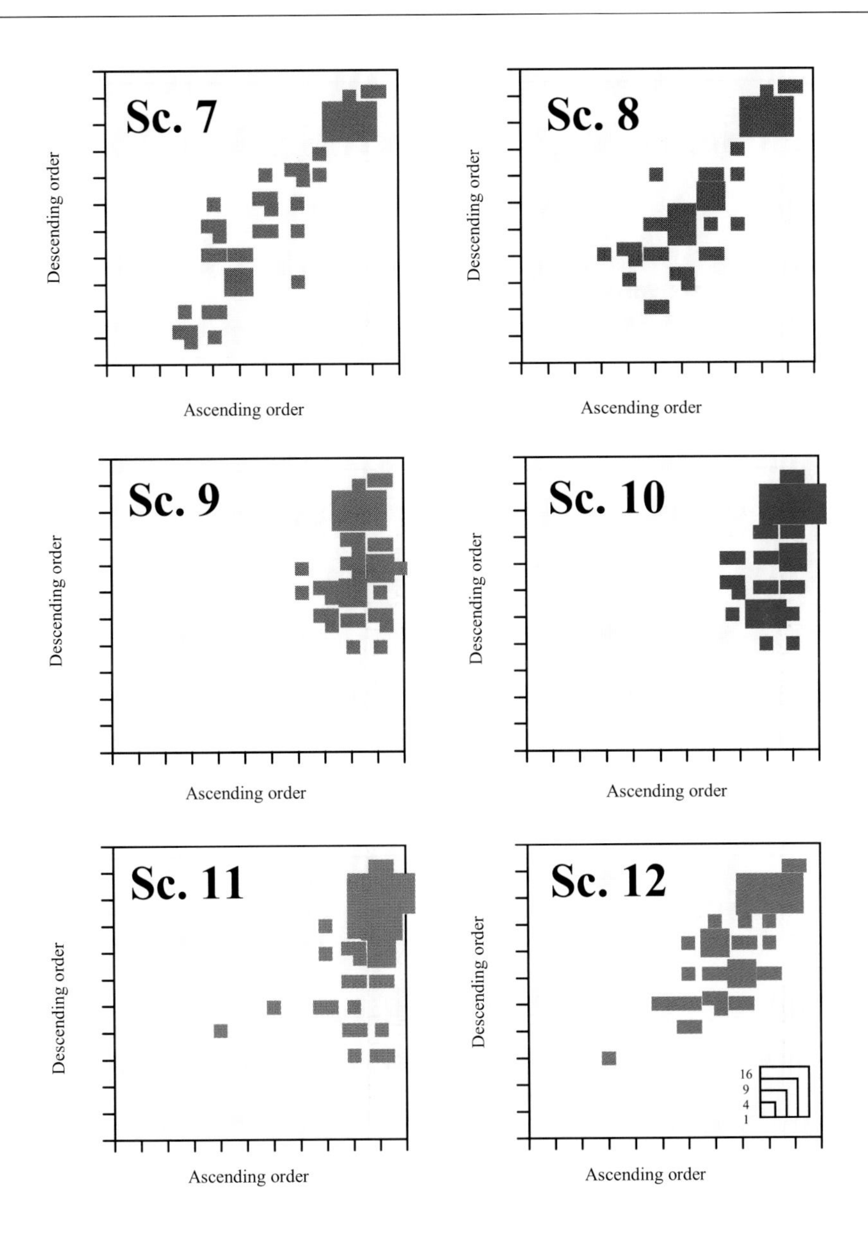

Figure 11.11 b "Surmesure" graphs for scenarios 7 to 12, SUB "future", with veto

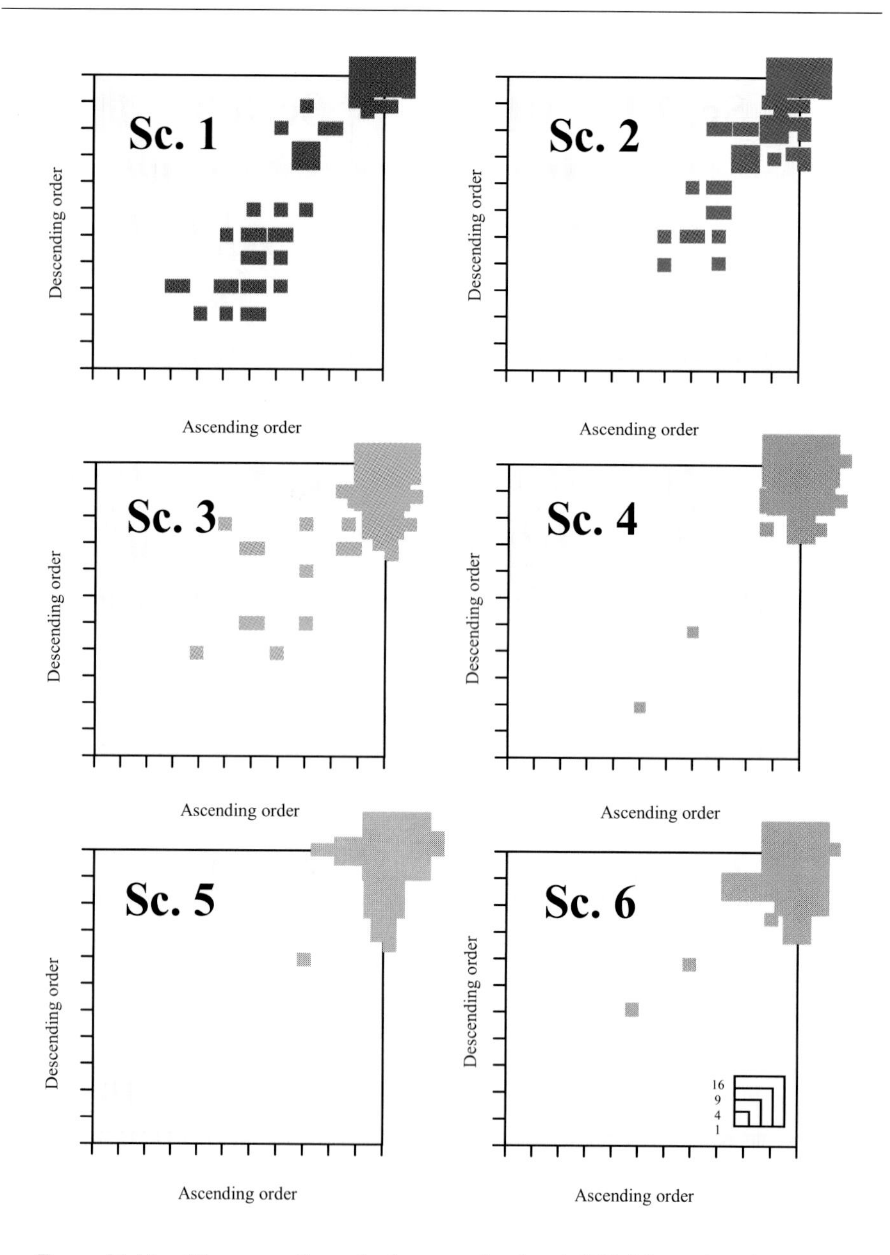

Figure 11.11 c "Surmesure" graphs for scenarios 1 to 6, SUB "future", without veto

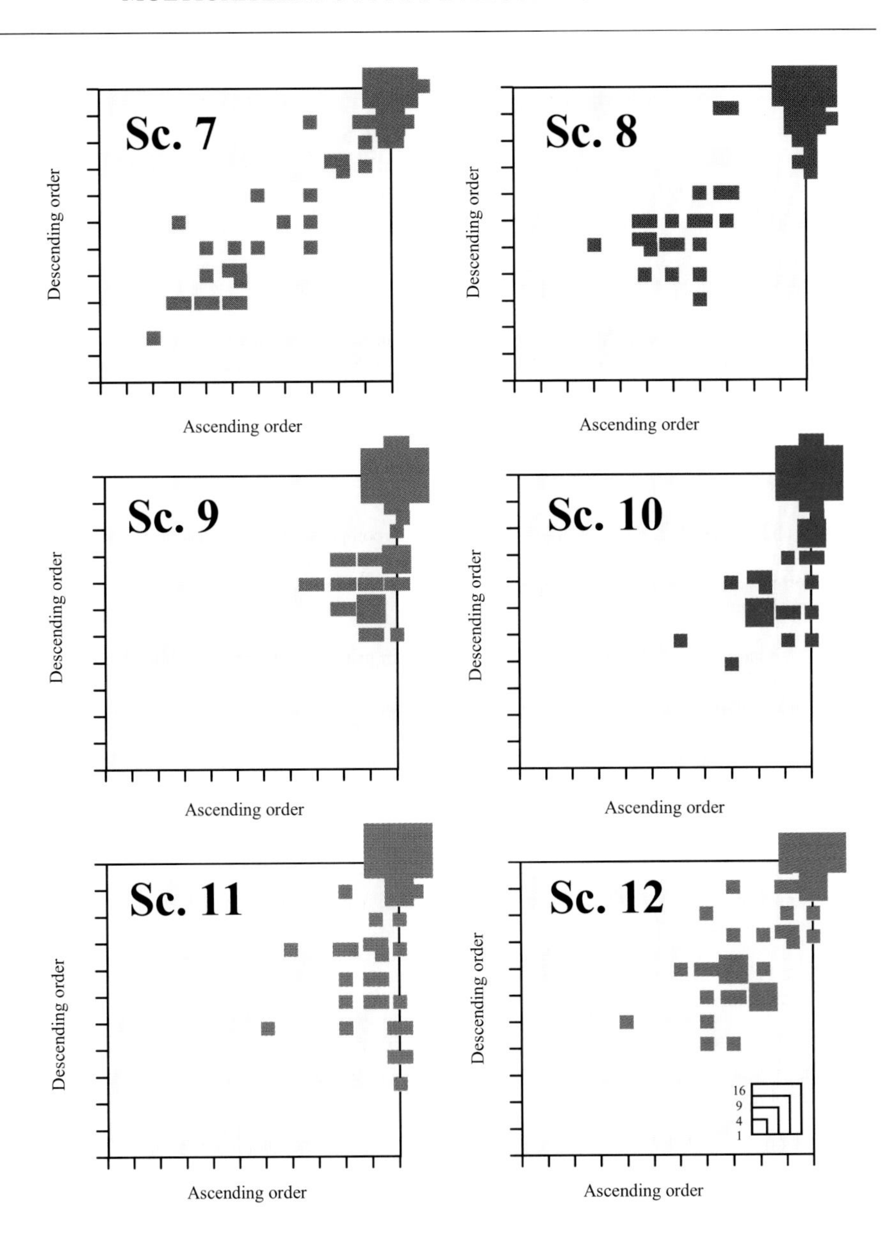

Figure 11.11 d "Surmesure" graphs for scenarios 7 to 12, SUB "future", without veto

REFERENCES

Belton V., Pictet J. (1997) A Framework for Group Decision Using a MCDA Model, Journal of Decision Systems 6 (3).

Brundtland, G.H. (1989) Global Change and our Common Future, Benjamin Franklin Lecture, Environment, Vol. 31, Washington D.C.

Haldi, P.-A. et al. (2002) Multi-criteria/multi-stakeholders comparative assessment of electricity generation scenarios in the sustainability context: a Swiss case study, International Journal of Sustainable Development, Vol. 5, Nos.1/2, Published by Interscience Enterprises Ltd., Oxford.

Maystre L.Y., Pictet J., Simos J. (1994) Méthodes multicritères Electre: description, conseils pratiques et cas d'application à la gestion environnementale, Presse polytechniques et universitaires romandes, Lausanne.

Pictet J, and Belton V. (2000) ACIDE: Analyse de la compensation et de l'incomparabilité dans la décision, in AMCD – Aide multicritère à la décision (Multiple criteria decision aiding), edited by Colorni A., Paruccini M. and Roy B., Joint research centre, EUR report 19808 EN, The European Commission.

Pictet J., Maystre L.Y. and Simos J. (1994) Surmesure. An instrument for presentation of results obtained by methods of the Electre and Promethee families, in Applying multiple criteria aid for decision to environmental management, edited by Paruccini M., Kluwer, Dordrecht.

Roy B. (1978) Electre III: Un algorithme de classement fondé sur une représentation floue des préférences en présence de critères multiples, Cahiers du CERO 20.

Roy B. (1985) Méthodologie multicritère d'aide à la décision, Economica, collection « Gestion », Paris (English version : Roy B. (1996) Multicriteria methodology for decision aiding, Kluwer, Dordrecht).

Roy B. and Bouyssou D. (1993) Aide multicritère d'aide à la décision : Méthodes et cas, Economica, collection « Gestion », Paris.

Schärlig A. (1985) Décider sur plusieurs critères. Panorama de l'aide à la décision multicritère, Presses Polytechniques Romandes, Lausanne.

SESAMS (2000) Strategic Electric Sector Assessment Methodology Under Sustainability Conditions, Final Research Report for the AGS SESAMS Project 1988-1999, MIT, Swiss Federal Institutes of Technology.

UNDP (2000) World Energy Assessment, United Nations Development Program, Bureau for Development Policy, New York.

Zaddeh L. (1978) Fuzzy sets as a basis for a theory of possibility, Fuzzy Sets and Systems 1.

NOTES

[1] It has been demonstrated indisputably that such mathematical models are not appropriate in all decision situations, particularly at a strategic level, where it is hardly possible to account in unambiguous mathematical terms for complex interactions between actors, constraints imposed by the environment, etc.

[2] The term "action" is preferred to "alternative" or "solution", also frequently used in this context, because "alternative" seems to imply that the actions are mutually exclusive. This is not necessarily true in a MCDA perspective and furthermore, to consider a potential action as a "solution" would anticipate the final conclusion of the analysis.

[3] An alternative to this essentially bottom-up approach consists in first defining the dimensions and then the criteria (top-down approach).

[4] Differences between the respective performances of the actions on the different criteria are not sufficiently marked to justify the strict preference, but important enough to definitely turn down the indifference option. The formulation "hesitation about the preference" would perhaps more accurately describe this idea.

[5] According to the fuzzy logic theory of Zaddeh (Zaddeh, 1978).

[6] This issue is central to MCDA. Most aggregation methods cannot account for incomparability, despite recent efforts to overcome this weakness (Pictet, Belton, 2000). Incomparability is quite natural in a multiple criteria context; it corresponds to the absence of clear and definitive evidence (because of "opposite" performances in different criteria) justifying either the indifference, strict preference or weak preference relationships.

[7] The definition of these global thresholds, related to the distillation process, will not be given here. For explanation, see Roy, Bouyssou (1993) and Maystre et al. (1994)

CHAPTER 12

COMPARISON AND INTEGRATION OF CETP TASKS

WARREN W. SCHENLER, STEFAN HIRSCHBERG, ROBERTO DONES AND YAM Y. LEE

1. INTRODUCTION

The CETP project has approached the problem of Shandong's electricity future in a way that has integrated the use of many different and complex methodologies within specific tasks and task interactions. These integration activities were conducted on different levels with varying degrees of implementation, and have included the following:

1. Parallel work by different analysis teams, with a regular exchange of information and the use of common basic assumptions and databases where possible (e.g. power plants characteristics). For example, the ESS, EEM, and ETM tasks all focused on the electric sector using different tools, which provided different perspectives for comparison and synergy. Some differences in data and assumptions were unavoidable due to differing models and task schedules.
2. Well-defined interactions between analysis tasks, with the use and transfer of consistent information between different analysis sectors. This included extending available models to use results generated in other task areas. For example, the LCA and EIA tasks were coupled to generate energy chain-specific impacts and damage costs. ESS scenarios were also coupled with attributes generated by LCA, EIA and RA for subsequent use in MCDA.
3. MCDA was applied to integrate many diverse decision factors, including factors that were subjective or not directly quantifiable.

This chapter addresses three primary topics. The first section discusses and compares the methodologies used by the different CETP tasks and their results. In particular, this chapter focuses on the comparison of the ESS, EEM, and ETM tasks. It presents issues related to comparing models, describes the individual models used, and discusses their different methods and assumptions, and how the results have been adjusted to put them on the most comparable basis. In the course of this analysis, a number of the results from the individual task chapters are presented and compared, although the integrated conclusions for the entire CETP project are presented in Chapter 13.

The second section of this chapter describes DVD that accompanies this book, which presents the entire CETP project in a flexible and interactive way. The DVD contains a short movie on CETP, general background information on CETP, Shandong and China, core information on individual tasks, methods, data and results and interactive tools for exploration of the task results. Thus, a range of formats is used to present different levels of detail so a wide range of users can explore the project to the depth they want. The section describes the DVD structure, and in particular the visualization and decision support tools.

The third and final section of this chapter presents a discussion of project management issues related to the technical cooperation between tasks, including interactions between stakeholders, researchers, the steering committee, the technical oversight committee and personnel exchanges between the different CETP institutions.

2. COMPARISON OF ELECTRIC SECTOR METHODOLOGIES

The three major CETP tasks that were used to analyze the electric sector in Shandong province were Electric Sector Simulation (ESS), Energy Economy Modeling (EEM), and Energy Transport Modeling (ETM). This section of Chapter 12 compares these three methodologies by discussing relevant modeling issues, briefly describes and compares the individual methodologies, discusses the comparability of task results, focusing on what subset of results should be used for comparison, and then describes how these results were adjusted to make the task results as comparable as possible.

2.1 Relevant Factors in Methodological Comparison

Choosing a methodology for electric sector analysis is a complex decision based on how the problem is perceived, what issues are considered important, how the issues are to be measured, what the appropriate level of detail and system boundaries are, and many other questions. A methodology is the entire structure of the analysis, including not just the core model, but also how the analysis is framed, data prepared, and post-model calculations performed. There is often a tendency to use the tools with which one is familiar, but this a simplistic way of saying that our training and experience influence both the choice of projects pursued and the methods chosen for analysis. Nonetheless, there are a number of issues that are relevant in discussing and comparing the electric sector methodologies used in CETP.

- Optimization v. Simulation – One of the primary distinctions between different methodologies is the difference between optimization and simulation models. Optimization models look for the 'best' solution, but the key question lies in how 'best' is defined, and how large is the solution-space over which the optimization searches or solves. Optimization implies that there is an 'optimization function' which combines all criteria of interest together into a single criterion. If the problem extends over a period of time (as in the electric sector planning problem), then an optimum strategy is determined, where the decision criterion is generally discounted to the present. This single criterion may be monetary (e.g. least cost or maximum profit), or it may be a utility function that combines criteria into a non-monetary measure of utility or 'goodness'. Optimization models are least controversial when the people who use the results can agree on the optimization function that is used, and the relative weights of the different decision criteria. Unfortunately this is often not the case in complex problems where different stakeholders have fundamentally different opinions on the relative weights of costs, environment, human health, and other factors.

Optimization models will find only a single optimum solution, either globally or locally if the solution-space is constrained. Constrained optima are not generally as 'good' as a global optimum, unless the constraints are not binding. On the other hand, optimization models generally supply the 'shadow price' or marginal cost of a constraint. This information can be very valuable. For example the shadow price of a binding emissions constraint is equivalent to the emissions tax required to make the economically optimum level of emissions just equal to the constraint.

It is customary to solve multiple scenarios by imposing different constraints that correspond to different strategies or uncertainties. For example, emissions constraints may be used to formulate specific scenarios. In problems that unfold over time, future uncertainties like load growth must be specified since it is necessary to assume what the future may be before it is possible to find the optimum solution.

Simulation models, on the other hand, do not generally produce the 'best' solution, or even generally define what is 'best' by using a single optimization function. Instead the system modeled is defined and allowed to evolve sequentially over time following structured rules. The solution algorithms are not generally as sophisticated as in optimization models, but it is often possible to incorporate more realistic modeling assumptions (e.g. non-linear behavior) that can often make optimization much more difficult. Although optimization models may be exercised for multiple scenarios, simulation models often have faster solution times and can be used to more fully explore non-optimum system behaviors.

Optimization and simulation are not mutually exclusive. For example the model used for the electric sector simulation optimizes system dispatch for least cost operation, but simulates many possible paths for system expansion or development. In the end, the choice depends upon on the modeler's purpose and philosophy.

- Endogenous v. Exogenous Factors – Any model designer must chose what factors to calculate inside the model (endogenously) and what factors must be supplied from outside the model (exogenously) in the form of assumptions or predictions. The choice of what should be included or excluded is key in focusing on what are the most important aspects of the problem being studied.

- System Boundaries - In addition to choosing what *factors* are calculated endogenously v. exogenously, a modeler must also specify system boundaries. The most obvious boundaries are geographic, and may depend upon provincial, national or airshed limits. Other boundaries can be sectoral, e.g. including only the electric sector v. the whole energy sector. Expanding system boundaries can sometimes reveal synergies by optimizing performance over a wider geographic area or the entire energy sector. Larger system boundaries also increase data requirements and solution times, often requiring a trade-off against the level of detail included.
- Single v. Multiple Scenarios – In an optimization model, a scenario is defined by what options are made available for inclusion by the model into the solution and what constraints are applied, while in a simulation model a scenario is more exogenously specified by combining one or more specific options or uncertainties. In either case, it is possible repetitively run the model in some (semi) automated way. In CETP, the optimization modeling philosophy has be to explore the optimum solutions for a number of constrained scenarios, while the simulation philosophy has been to analyze may different, non-optimum scenarios to find insights about electric sector system performance.
- Location of Costs v. Benefits – The costs and benefits of different strategies frequently occur in different locations. These may lie on different sides of the system boundaries, or affect different stakeholders. For example direct electricity generation costs may occur within the province, while external costs may occur outside the province (like health effects due to airborne pollution). Another example is the determination of utility costs (which include centralized generation, distribution and overhead), as opposed to societal costs that may also include distributed and renewable generation and energy savings costs.
- Combining Criteria – Different decision criteria may be combined either endogenously in an optimization function, or in the post-modeling combination of different results. This can include monetization of health and environmental externalities, but it can also include combining quantitative results (e.g. conversion factors for converting different greenhouse gases to CO_2 equivalents), and/or qualitative results (such as estimated risks due to future technology or fuel availability). As one specific example, the monetized value of pollution can be included in an optimization model to find the lowest total cost strategy. The uncertainties inherent in monetizing health and environmental damages are unfortunately high compared to direct generation costs, but such monetization is key to finding the 'real' cost of electricity, which is the main objective of CETP.

2.2 Discussion of Electric Sector Methodologies

Given these factors relevant to methodology evaluation, the models used by the three electric sector analysis tasks are briefly described below.

- ESS Methodology – The ESS task analysis was based on the repetitive use of a bottom-up, engineering based model that simulated least cost electric system dispatch. This means that the operation of the utility system was simulated, based upon minimizing dispatch cost (fuel costs plus variable operation and maintenance costs), while accounting for daily and seasonal variations in load and plant outages. Costs, generation, fuel use, and direct emissions were tracked by technology and location, but power transmission was not explicitly modeled. The modeling system boundaries for demand and generation were Shandong province, while allowing for fuel imports from Shanxi and other provinces. The modeling period was the 25 years from 2000 to 2024. The model was confined to the electric sector only, and projected electricity demand and fuel prices were exogenous assumptions. Strategies for future system expansion and operation were not optimized, but combined from a range of options, including plant retirement, fuel cleaning, retrofitted exhaust scrubbers, and a range of new generation technologies which were predominantly coal. A total of 18,144 scenarios were modeled and analyzed, while tracking over 200 primary and secondary attributes of system performance. Key results were combined with the results of the life cycle analysis task, and passed to the EIA and RA and MCDA tasks.
- EEM Methodology – The EEM methodology was based on an optimization approach, using the CRETM (China Regional Energy Technology Model) for all of China and the MARKAL-Macro model primarily for Shandong province. Both of these models are based on a linear programming core with a high level of technology-specific detail for a broad range of competing options. System expansion is optimized based upon minimizing generation costs, solving in 5-year periods, with generation growth from period to period limited by assumed constraints on the availability of financial capital. Within each modeling period, both the CRETM and MARKAL models are capable of optimizing the average load factor (as constrained by maximum availability), but this ability was only used in the MARKAL model, while in CRETM the capacity factors were endogenously specified for the sake of simplicity. Both models are based on the entire energy sector, instead of just the electric sector, so that fuel choices and new generation are optimized to meet the demand from competing technologies that provide end-use energy services. Primary energy transportation limits are recognized, but the model does not endogenously model electricity transmission. The overall level of demand for energy services (including electricity) is an exogenous input for these two models, but this reference demand can vary due to price changes or any constraints imposed. In addition, energy use technologies can compete with

energy supply technologies, and the load profiles of both must combine to meet the customers load profile. System boundaries are based on China, rather than just Shandong province, but results are tracked for individual regions, including Shandong. A total of 25 CRETM and 26 MARKAL scenarios were modeled by defining modeling constraints based on technology availability, emissions limits (or the equivalent emissions tax), and interest rate sensitivities. While a larger number of scenarios would have been possible using these models, the emphasis and philosophy of the EEM task focus on a more limited number of optimized scenarios and a few sensitivity analysis cases.

- ETM Methodology – The ETM task also used an optimization approach based on an underlying linear programming model, but unlike the other two electric sector tasks this methodology explicitly included the expansion of the existing transmission network, as well as the expansion of future generation capacity. The transmission modeling relied on simplified network flow, using an equivalent direct current formulation. Actual dispatch was not simulated, but the geographic distribution of generation was constrained by network flow limitations. Future demand was assumed to be geographically distributed proportional to current demand, and future transmission expansion was assumed to take place by strengthening existing transmission lines. The optimization function for this model is also based on minimizing costs, which include plant operation, construction of new generation and transmission capacity as well as the cost of fuel transportation to different locations within Shandong province. The model focuses on the electric sector, rather than the entire energy sector, and forecast demand is considered exogenous, rather than based on price elasticity. A total of 7 ETM scenarios were modeled.

Table 12.1 Comparison of energy modeling methodologies

Distinguishing Characteristics	**CETP Task**		
	ESS	EEM	ETM
Scope - Geographic	Shandong	China	Shandong
Scope - Sectoral	Electricity	Energy	Electricity
Scope - Criteria	Multi-criteria	Cost Minimization	Cost Minimization
Model Type	Simulation	Optimization	Optimization
Scenario Definition	Options, Uncertainties	Objective functions Constraints	Objective function, Constraints
Multiple Scenarios	Automated, easy	Sensitivity Analysis	Sensitivity Analysis

2.3 Comparison of Methodologies and Results

Comparison of Methodologies – Based on the issues and specific models described above, it is possible to make some specific observations and comparisons of the models used in the three electric sector analysis tasks.

- Simulation (ESS) v. Optimization (EEM, ETM) – The ESS and EEM methodologies are at the opposite ends of the simulation v. optimization spectrum within the CETP project, but this should be put within the context of a far wider range of energy models that exist. The ESS simulation describes system technologies and operation in more detail, and provides more detailed results based on economic dispatch. These are real strengths, which derive from the bottom-up approach. On the other hand the EEM models find optimum solutions based on a wider scope, extending system boundaries to other provinces and the whole energy sector. The ETM task lies somewhere in the middle. Although an optimization model, it is focused on Shandong and the electricity sector only.
  It is worthwhile noting that any optimization model will tend to produce "extreme" solutions. This is good rather than bad, since the optimum is by definition an extreme. But it does mean that since the model determines the system expansion strategy, the best option will dominate unless bounded. For example, if nuclear generation were favored economically, the model will make all new generation nuclear unless constrained by the ability to finance the capital cost and the need for peaking generators. This can be a valuable reference, since it tells the cost of constraints imposed in more 'realistic' scenarios. On the other hand, it is worthwhile comparing optimum solutions for a range of possible futures to create a balanced and robust portfolio.
  Optimization models will also tend to 'tip' from one optimum solution to another as the constraints that are applied shift. For example, if a CO_2 tax is reduced enough that nuclear is no longer favored, the generation mix may shift suddenly at a given tax level to be dominated by gas-fired generation. This means that balanced portfolio solutions are more likely to be based on the careful use of constraints.
  In the end, optimization is good in the sense that it maps out the extreme solution limits. However, since 'business-as-usual' strategies are often far from optimum, mapping out system performance over a range of constrained, intermediate and possibly more robust strategies is also important.
- Models Imply Options – The type of model used has a real impact on the types of options that can be modeled. This means that a simulation model will generally allow more detail on options related to individual plant technologies and especially the rules of system operation. On the other hand, optimization models (especially ones with a large amount of technical detail like CRETM and MARKAL) allow both a wide range of generic technologies, competition from non-electric options, the possibility of

economic price feedback and easier imposition of constraints (e.g. emissions caps).

- Number of Scenarios Possible – It is possible to automate an optimization model to run many scenarios, but in general due to greater run times, the types of scenarios considered, and the extreme solutions produced (see above) it is against the modeling philosophy to produce a really large number of scenarios that will map out the performance-space of the energy and/or electric sector.
- Realism in Detail – Simulation model have an edge in modeling detailed performance based on microeconomics. This is the reason that utilities use them for short-term, tactical planning studies. Explaining realistic simulation models to stakeholders can be easy when simple dispatch rules are implemented, but harder if advanced, multi-player game theory simulations are considered.
- Realism on the Large Scale – Conversely, optimization models have definite advantages in handling large-scale, macro-economic interactions. This is why they are used for long term, global, multi-region, multi-sector studies. The strengths, which are based on economic feedback (e.g. price elasticities), do offer their own challenges, such as the ability to reliably estimate key econometric parameters and the difficulty of communicating them to non-economist stakeholders. The simulation approach generally tries to counter its weakness in the macro-economic areas by sensitivity analysis (e.g. different demand growth assumptions), using the rationale that 'the forecast is always wrong', and that bracketing a reasonably wide range of uncertainties can reveal interesting results.

These relative strengths and weaknesses do partially explain why the detailed results from the ESS task were used as inputs to the EIA and MCDA tasks, but prior experience with model integration was also important. However, EEM task is better suited for mapping a smaller number of extreme, constrained solutions that better reflect macro-economic and non-electric interactions.

Comparability of Task Results – Based on the different approaches taken by the three electric sector modeling tasks, it is appropriate to ask which results should be chosen for comparison. Basically, the ESS, EEM and ETM models all produce approximately the same direct results, in particular costs and direct power plant emissions. However the greater detail and bottom-up dispatch simulation of the ESS task meant that it was used in combination with the LCA task to produce indirect environmental burdens used in turn by the EIA task to produce health and accident results for the MCDA task.

Choice of Key Results for Comparison – Because of the predominant concerns of cost and public health for Shandong, the primary result chosen for comparing results between the three electric sector modeling tasks was the tradeoff between cost and SO_2. Greenhouse gas and global warming concerns lead to selection of the cost v. CO_2 tradeoff being chosen as the secondary basis for comparison.

The three modeling efforts needed to make their own demand assumptions before the CETP demand forecasting task results became available. Each task also varied its demand forecasts somewhat differently for sensitivity analysis or bounding cases. For both of these reasons, the three tasks used demand forecasts that were sufficiently different to make total costs and total emissions non-comparable. To address this, the comparison was made on the basis of normalized cost and emissions (using a per kWh basis). This eliminates the differences due to different load growth assumptions. There is also an additional benefit. The ESS task included an analysis of reducing electricity demand by increases in end-use efficiency. By dividing the cost and emissions by the kWh that *would* have been used without any energy savings, the costs for supply and demand-side options are put on a comparable cost basis.

Table 12.2 Comparison of cost components by task

	ESS	CRETM	MARKAL	ETM	Comments/Actions
Cost Components					
Capital Costs - Old Generation	√	–	√	√	ESS included old plant capital costs based on base year revenue requirements. ETM based them on new unit costs & age (~0.5¢/kWh). ESS cost was added to CRETM for comparison (0.450 ¢/kWh).
Capital Costs - New Generation	√	√	√	√	CRETM, MARKAL higher. Difference apparently based on plant costs, amount of new capacity (effective reserve margin), and cost levelization v. depreciation.
Fixed Operation & Maintenance	√	√	√	√	
Variable Operation & Maintenance	√	√	√	√	
Fuel Costs	√	√	√	√	
Transmission & Distribution Costs (T&D)	T&D	0	T&D	T only	ESS and MARKAL T&D costs were excluded from comparison results. ETM cost is T only, negligible.
General & Administrative Costs (Overhead)	√	–	–	–	ESS overhead costs excluded from comparison results.
Demand Side Management	√	–	–	–	ESS includes DSM costs & adjusts kWh for comparison results. ESS DSM cost is uncertain, but would raise unit cost of only some scenarios.
Relevant Factors					
Metering Location (Busbar v. Meter)	B/M	B	B/M	M	Results from all groups adjusted to per kWh basis, measured at the customer's meter.
T&D Loss	6.20%	8%	7%	8%	
Fuel Price Growth	moderate, declining	flat, 5%	flat	flat	ESS base coal price forecast declines slightly in base year $.
Inflation Rate	<=5%	0%	0%	0%	All groups use base year $ & no discounting to calculate unit costs and emissions.
Discount Rate	10%	10%	10%	10%	ESS has lower effective real discount rate, which lowers effective cap costs.

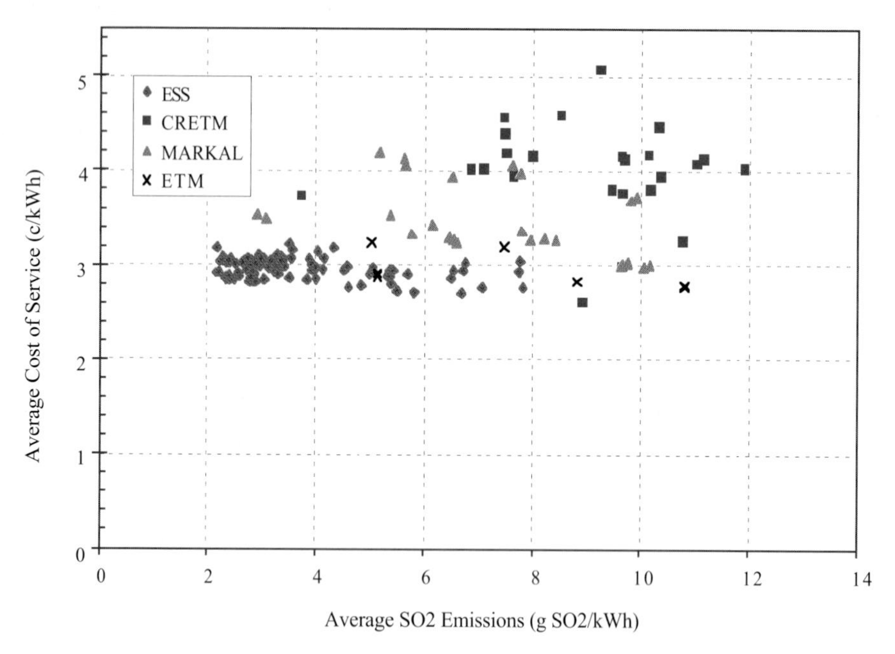

Figure 12.1 Comparison of cost v. SO_2 results

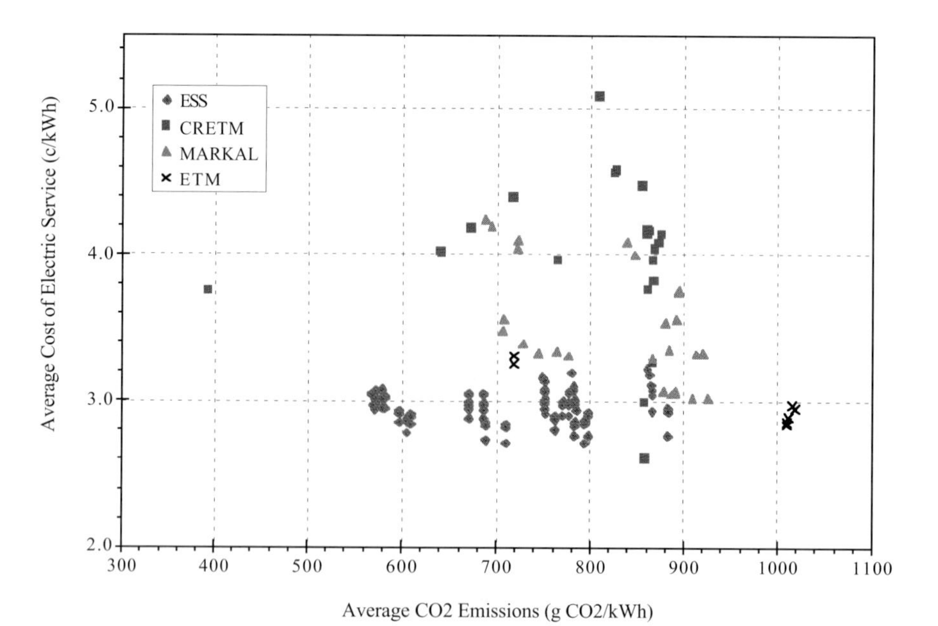

Figure 12.2 Comparison of CO_2 results

Adjusting Results for Comparison – The cost results for the three tasks contained slightly different cost components and were based on slightly different assumptions that needed to be adjusted before it was possible to compare them. Table 12.1 above shows the comparison of cost components and assumptions between the three tasks. This table also shows the steps taken to adjust the different results. For example, the ESS task included general and administrative utility costs so that its cost/kWh was basically what a customer would see on his bill, while the other two tasks focused on minimizing generation costs and did not include these overhead charges.

The cost v. SO_2 and cost v. CO_2 tradeoffs that show the comparative results of the three modeling tasks after the cost adjustments have been made are shown in Figures 12.1 and 12.2 above.

Discussion of Remaining Results – Even after adjusting for differences in the cost basis of the different task modeling results, there is still some variation in their results. These differences may be attributed to several factors, which include:

- Modeling Assumptions – Different models contain different assumptions. Some of these differences are structural, and are fundamentally based on the model paradigm. For example, in the ESS modeling plant dispatch is endogenous, and efficient new units can displace older, less efficient units. This reduces CO_2 emissions compared to the EEM models where unit load factors are fixed. These different modeling approaches also require different specific model inputs (e.g. EEM requires fixed load factors, and ESS does not), which means that it is impossible to make all inputs consistent across different tasks.
- Data Uncertainties - Even where the models have similar types of inputs, specific assumptions may differ, depending upon variations in data sources, original data quality and expert judgment. Each individual task chapter in this book describes key assumptions used in the modeling efforts. System-wide and macroeconomic data such as load growth and inflation and discount rates tend to have the largest impacts, while forecasts of future trends generally have the largest uncertainties.
- Model Propagation of Uncertainties – When a complex model is used, it may be very difficult to determine how uncertainty in one or more input assumptions may propagate through the model to produce uncertainty in the model results. Theoretical probabilistic methods exist, but they depend upon knowing the probability distributions of the data, which are harder to find than the best estimates for the data themselves, and the theory may not apply to the mathematical transformations used in the model. Sensitivity analyses frequently use varying input assumptions while observing the changes in results, but this is usually reserved for key assumptions and rarely done in a comprehensive way because the combinations of the different possible factors means that the number of model runs quickly becomes unmanageable.

In the end, it is unreasonable to expect that the results of research tasks using significantly different models, different types of input data, and different specific data assumptions will produce identical results. Of course, insofar as all the models attempt to describe reality (or at least the future), the results must be in the same ballpark. It is safe to say that this has been achieved, given that the model results agree within a range of about 25% for cost. There is a wider range for emissions results, which is largely due to the EEM models choosing extensive use of nuclear under certain constraints, while the ESS models used significant amounts of DSM. If these nuclear and DSM scenarios are excluded, the remaining scenarios agree much more, in the range of 800 g CO_2 /kWh. Life cycle analysis was not performed for the energy efficiency measures assumed for DSM scenarios, so emissions for these were also slightly lower for this reason.

It is also important to focus on the purpose of the results and the differences between relative and absolute errors. The real purpose of modeling is to study how the electric sector performs, trying to understand the performance of different strategies under different future circumstances. The goal is to aid stakeholders in their decision-making. If an error is systematically consistent and affects all strategies in a similar direction by roughly the same amounts, then it will not be likely to change any individual stakeholder's ranking of the strategies. This type of absolute error may (and should) be corrected when possible, but is not relevant to decision making. On the other hand, errors that affect the relative position or attractiveness of different strategies can be very relevant to ranking strategies and are most important to find and correct. This distinction between absolute and relative error is most important for the results produced within each task. When comparing between the task results, it is most important to see whether the conclusions drawn from each task (i.e. from the relative results of each task) are consistent across the three different tasks.

2.4 Comparison of Task Conclusions

Given the comparison of factors, models, assumptions and results that have already been discussed, this section presents and compares the major conclusions of the three electric sector modeling tasks. Detailed conclusions of these three tasks, and the results of other tasks have been presented in the individual chapters. Integrated conclusions of all tasks are presented in Chapter 13.

ESS Conclusions - The conclusions of the ESS task are based on the comparative results of over 18,000 scenarios, which focused on combining a wide range of options for the electric sector. The key conclusions of the ESS task include:

- Existing generation remains a large source of pollutant emissions, well into the future. For this reason, important options for old plants include retrofitting scrubbers, use of low sulfur coal and select retirement of old units.
- Significant cost-effective opportunities exist for reducing major power plant pollutants (SO_2, PM and NO_x), and reducing the growth of greenhouse gas emissions (CO_2). Pulverized coal units with flue gas

desulfurization are cost effective. Use of prepared coal improves plant efficiency and availability, and reduces PM and SO_2 at both old and new plants. Advanced coal technologies are a solid long-term option. A diverse mix of nuclear, natural gas combined-cycle, wind power and niche technologies should also be pursued.

- Implementing end-use efficiency programs and peak load management are cost-effective ways of reducing all emissions while supplying the same energy services.
- It is important to coordinate integrated strategies that combine old generation, new generation, fuel and demand-side options if cost-effective emissions reductions are to be achieved.
- While these conclusions are robust, the ESS study also showed places where uncertainty exists. For example:
- The impacts of government reforms in power and fuels markets still need to be addressed.
- We need to know more about changing consumption patterns, and the availability of renewable resources to develop these diversified strategies.

The relative strengths of the ESS task result from the detail of the bottom-up approach, the realism of assumptions about unit operation, and the large number of options which are combined, which allows the relative comparison of the effect of any one set of options, while leaving all others fixed. The relative weaknesses of the ESS task are that it focuses only on Shandong and the electricity sector, omitting fuel market competition between energy sectors and transmission related issues.

EEM Results – The results of the EEM task are based on two different models (CRETM and MARKAL), which were used to model a total of 51 scenarios. These scenarios focused on the least cost scenarios to meet certain emissions taxes or constraints, using a range of future technologies. The key conclusions of the EEM task include:

- Coal will remain the primary fuel.
- SO_2 reductions at modest cost also give only small CO_2 reductions.
- Significant CO_2 reductions are expensive.
- The cost of reducing emissions is less than cost damages due to pollution,
- SO_2 caps and/or trading permits are the most efficient way to control future emissions.
- Improvement of coal preparation, generation, scrubbers and advanced coal technologies is needed.
- Advanced coal technologies can be competitive when they are domestically built or when fuel prices are high.
- Fuel and technology diversification are important for China as a whole, including natural gas, hydro, nuclear, wind and small hydro resources.
- Nuclear is competitive on a direct cost basis with interest rates of 10% if construction times are less than five years and costs less than 1500 $/kW. They are also competitive if interest rates are less than 5%, or at greater than 5% with moderate emissions taxes.

- Interregional transmission reduces cost and local pollution.
- Supply-side efficiency needs further research and development.

The strengths of the EEM task models are their coverage of the whole energy sector and not just the electricity sector, their coverage of other regions of China and not just Shandong, and their inclusion of how price feedback influences demand and cross-fuel shifts, and their optimization of system expansion. Their weaknesses include a reduced level of detail on individual technologies and system dispatch, and a more limited number of scenarios. The range of emissions constraints modeled produced a wider range of production costs than in the other modeling tasks.

ETM Results – The results of the ETM task are based on a model that includes endogenous optimization of generation expansion, transmission expansion, and fuel source choices. The key conclusions of the ETM task include:

- Dirty coal remains cheapest option when only direct costs are considered.
- The current trial SO_2 tax rate of 0.2 yuan/kg-SO_2 ($50/t-S) is insufficient to control SOx emissions.
- Control zone policies as modeled were effective in shifting emissions. Overall emissions were not reduced but the shift can reduce damages in less sensitive locations.
- Flue gas desulfurization and low sulfur coal are effective and economic means for reducing SO_2, but not CO_2.
- Nuclear and gas are required for CO_2 control, and also help SO_2 emissions.
- Advanced clean coal technologies (IGCC and AFBC) were too expensive. Reduced costs would reduce CO_2 as well as SO_2 emissions.
- Natural Gas Combined Cycle units were not economic as the assumed gas price. Lower gas prices could reduce both SO_2 and CO_2.

The relative strength of the ETM model is in the way that it integrates consideration of both electricity and coal transportation. Its relative weaknesses are that it focuses on Shandong only, and has a reduced level of unit detail and system dispatch and a lower number of scenarios compared to the ESS model.

2.5 Comparison of Conclusions

Although the three CETP electricity-modeling tasks approach Shandong from different directions, they do agree on a number of conclusions:

- Coal will remain dominant. This is practically an input instead of an output, given China's fuel resources, but when *only* direct generation costs are considered then conventional technologies remain best.
- Reducing the emissions of major pollutants (SO_2, NO_x and PM) is cheaper than these emissions' external damages costs. All the electricity-modeling tasks agree that major emissions reductions can be achieved for

significantly less than the cost of externalities that are estimated by the EIA task.

- SO_2 reduction options for existing plants (i.e. flue gas desulfurization, low sulfur coal, coal preparation, and unit retirement) are better when combined. Some options are better than others, but even the best options may be constrained by available fuel, cost, or other factors.
- CO_2 control is expensive, but any technology that controls CO_2 also brings significant SO_2 reductions benefits for a marginal cost that is very low.
- Saving electricity reduces *all* emissions. The key questions are obtaining reliable cost estimates for electricity saving measures in China, and how to measure the real savings once such programs are in place. The best time to implement an electricity efficiency program is while demand growth is still high. The size of available electricity savings already dwarfs renewable resources, and will continue to grow.

Each modeling task offers its own special results and insights, but that their conclusions overlap and reinforce each other to the extent described in this book and this chapter only confirms what we have come to understand about the electricity sector in Shandong. These common conclusions on the electricity sector and the other CETP tasks are explored in more depth in Chapter 13.

3. DVD TOOL FOR DOCUMENTATION, EXPLORATION AND DECISION SUPPORT

3.1 Background

CETP has been composed of a large number of multi-disciplinary tasks, with relatively complex internal dependencies and interactions. The project produced a broad spectrum of deliverables including databases, analytical models, technical reports covering the sector-specific results, and this book summarizing the approaches used and the main achievements. The ambition was, however, to enable Chinese stakeholders and a wide range of other users a flexible access to the results and insights, including options for exploration directed by specific interests. This has been made possible through the extension of the integration framework already implemented within the CETP.

The analytic methods developed within CETP and implemented as software have been made available on a DVD. This product presents the various tasks for the benefit of the users and is itself a central contribution to the overall integration effort. The software has been developed by PSI in cooperation with external programming and multi-media specialists, and with the support and input provided by project partners.

3.2 Objectives and Targeted Users

The primary objectives of developing the DVD were:

- To present the approach, results and recommendations of CETP to decision-makers and other interested individuals and organizations in a clear and flexible way so they can readily grasp the material in relation to their own interests and needs. This includes high and medium level decision-makers (e.g. from electric utilities, plant suppliers and regulators), engineers, environmental experts, and academics.
- To allow high flexibility in using CETP results, including the possibility of analytical options that can extend results beyond the range of those obtained by CETP.

3.3 General Features

The following requirements for the DVD were established and achieved:

- User-friendliness and high graphic quality.
- Homogenous graphical user interface (GUI).
- Ability to present large amounts of data in different forms
- Platform independence, to the extent possible.
- Possible further expansion and implementation of other applications in the future.

All these requirements have been achieved by using Macromedia's "Flash" program to produce a product that can accessed using free and widely available web browsers such as Internet Explorer and Netscape Navigator.

3.4 Structure and Content

The information is presented in a hierarchical manner, allowing users to choose the level of detail most appropriate for their background and interests. Figure 12.3 shows an overview of the contents of the DVD, showing different levels of detail and interaction that are directly accessible as appropriate for different users. The components are then briefly described below. The software structure enables easy and flexible navigation within as well as between its various parts. Instructions are provided in the form of help and information text, facilitating the use of the software.

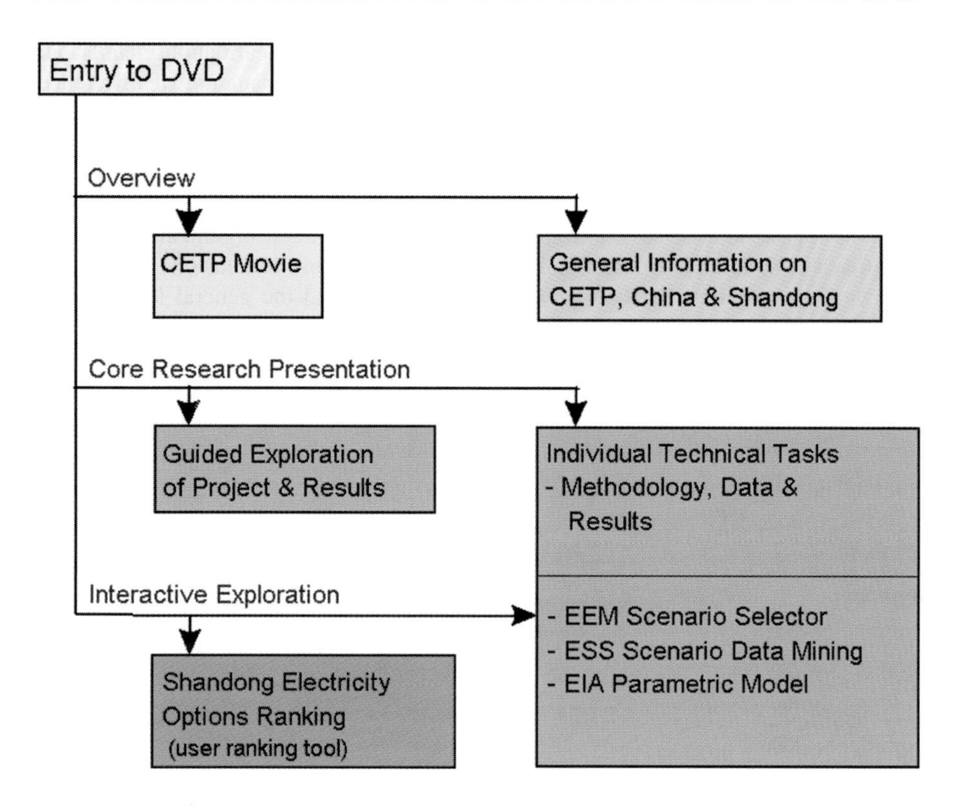

Figure 12.3 Contents of DVD by level of detail and interaction

The *Movie*, which is about 13 minutes long, serves as the primary introduction to the overall CETP. Any user of the DVD is expected to enter it by starting with the movie on at least the first occasion. In the case of high-level decision- and opinion-makers, non-specialists, or members of the general public it is possible that their direct exposure to CETP results will be limited to the information provided in the movie. The goal for this part of the DVD is to communicate in an accessible manner the scope, achievements and policy recommendations of the project through selected examples that use some technical, though non-specialist, terms. The movie also provides the rationale for CETP, illustrates the research process and shows some of the people who contributed to the work.

The *General Information* section has two major parts. First, the objectives, framework and organization of CETP are explained. Second, the basic features of energy and electricity supply in China and Shandong are sketched, including their environmental aspects.

The *Guided Exploration* gives a concise overview of the approaches used by the various tasks, a summary of the task-specific highlights and conclusions and a comparative perspective on results from selected modeling tasks. Finally, the overall conclusions and recommendations are summarized. This part of the DVD is intended for users who want to familiarize themselves with essentials of CETP from the technical point of view but who do not have the need to learn about details. The *Guided Tour* is also suitable for users interested in the explanation of differences between the EEM, ETM and ESS approaches as employed in the CETP.

The core of the DVD software is presented in the module on *Individual Technical Tasks*. All the technical tasks within CETP are covered with considerable detail. A common structure was adopted for task presentations, including menu bar buttons for sections on task introduction, task definition and organization, the task approach or methodology, task analysis, and the results and conclusions. Figure 12.4 shows an example of the task presentation structure and the general layout of the screen.

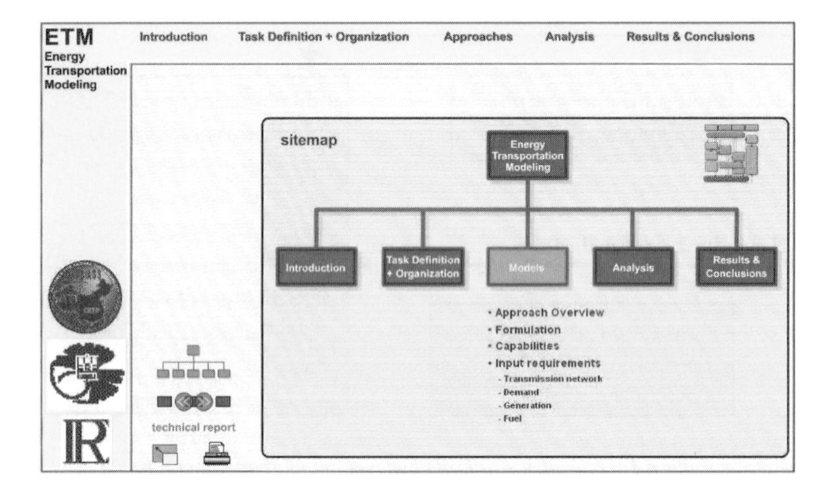

Figure 12.4 Example of task presentation structure (ETM task)

Some tasks also have special features for flexibility in the presentation of results or concerning extended analytical capabilities. One such feature in the EEM task is a scenario selector for choosing and comparing results for individual EEM scenarios. These scenarios are characterized by different values for driving factors, including environmental policies, economic growth rates and fuel prices for China or Shandong. The user can choose from a set of values for these options and then obtain the associated results. The process can be repeated for another scenario, and the results for two scenarios can be compared. For example, Figure 12.5 shows SO_2 emissions for China in the reference case, compared to the reference case with SO_2 and CO_2 caps added.

The ESS task contained too many scenarios, so it was not possible to use the Flash software to present the results in the same way as for EEM. Instead, another software package called Data Desk was used. This software is used for graphical analysis of statistical data. A structured presentation has been created and locked on the DVD, allowing the user to 'mine' the set of ESS results for the base or default

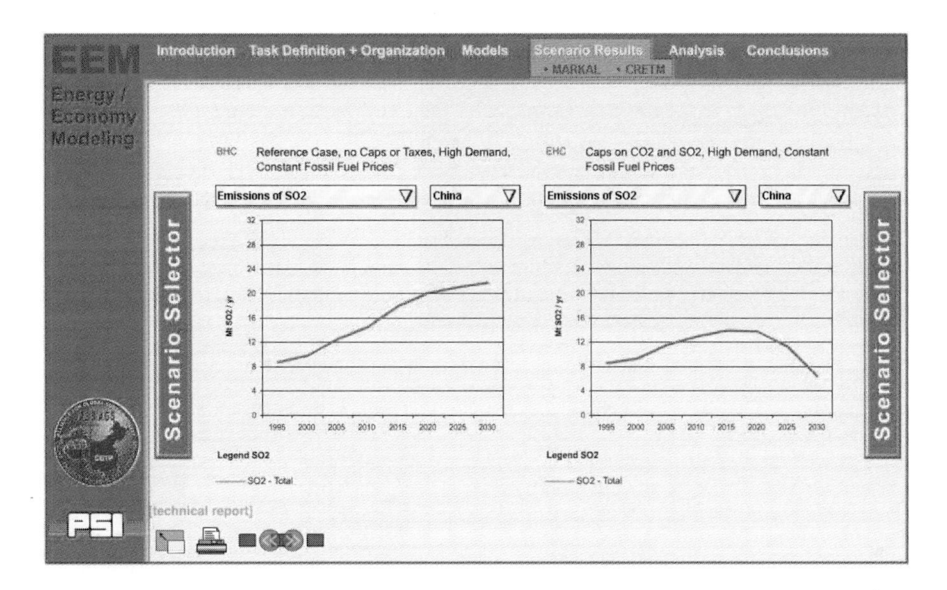

Figure 12.5 SO_2 emissions in China for scenarios with and without strong SO_2 and CO_2 constraints (EEM task)

future, which contains 1008 scenarios. The user first proceeds through a structured sequence of tutorial screens, and can then use buttons to interactively select subsets of results for a wide range of criteria. Most results are cumulative for the whole modeling period from 2000 to 2024, including cost measures, emissions, and external MCDA indicators and costs. Some time trajectory results are also available for a limited set of cost, emissions and generation measures. Figure 2.6 shows a picture of the DataDesk application interface structure.

The EIA task models the airborne transport of emissions and estimates the damages that they produce over a geographic region. Within the EIA task a tool was developed to allow an interactive simulation of impacts and the corresponding monetary damages for power plant sites in Shandong. For each site the user can select from several generation technologies, establish basic characteristics (power and load factor), and choose emission abatement levels for major pollutants. The direct plant emissions determined by these choices are then added to the upstream energy chain emissions and the total impacts are calculated, based on a parameterization of site-specific results. This allows the calculation of damage valuations within seconds, compared to hours needed by full-scale simulation using the EcoSense model, as described in Chapter 9. Figure 12.7 shows a comparison of two cases for the Laiwu site.

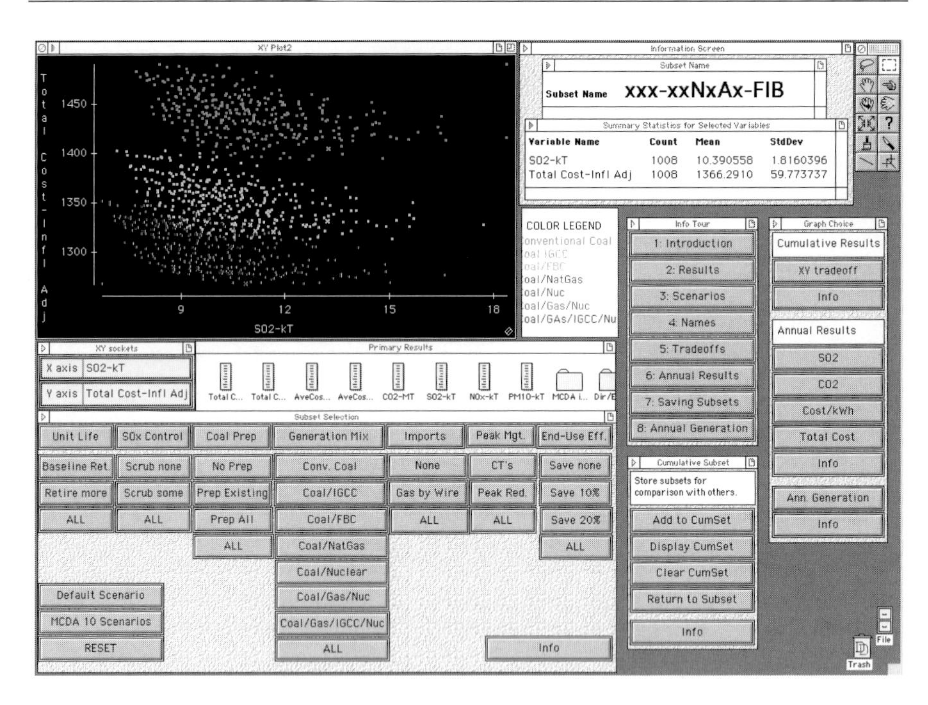

Figure 12.6 DataDesk tool for exploring tradeoff results for different criteria and subsets of scenarios (ESS task)

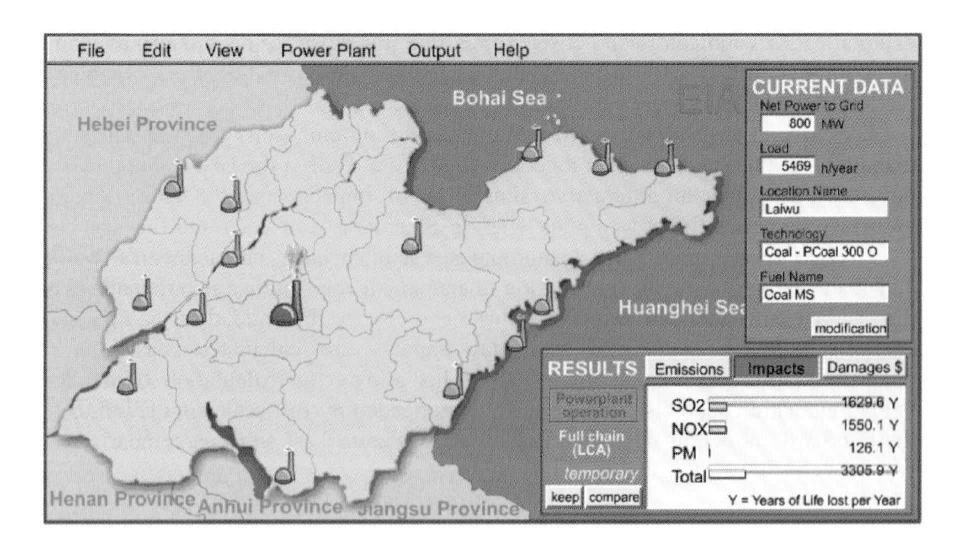

Figure 12.7 Mortality impacts from the Laiwu plant with and without scrubbers (EIA task, reference year 1998)

The MCDA task was designed to assist decision-makers by using multi-criteria decision analysis to produce scenario rankings based on stakeholder input for criteria weights. Figure 12.8 below shows a sample of ranking results.

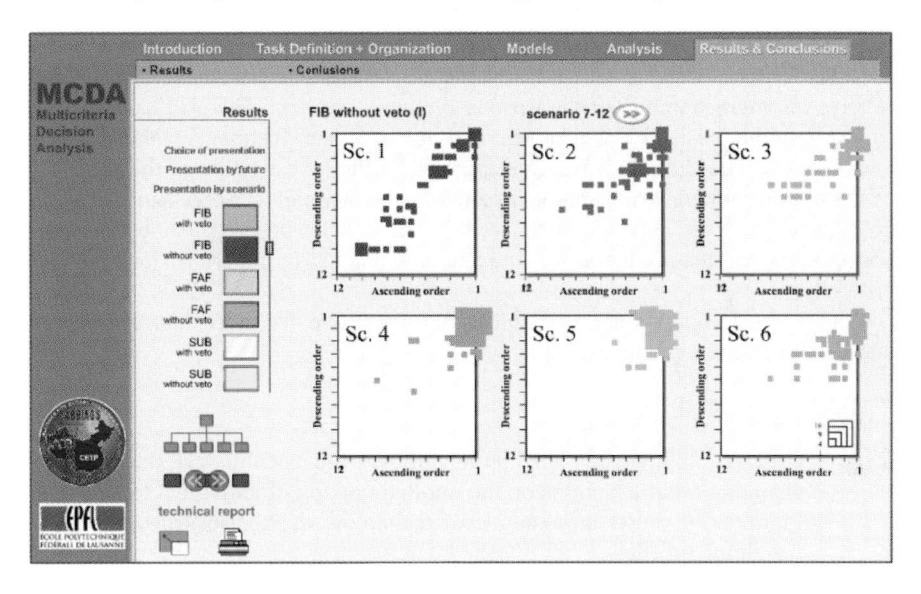

Figure 12.8 MCDA scenario ranking results

The *Shandong Electricity Options Ranking (SEOR)* module generates and presents ranking results using a simplified multi-criteria decision aiding approach in an interactive, graphical form. Numeric results for a range of criteria are combined with the users input on criteria weights and subjective indicators to produce a ranking of either single technologies or one or more mixes of individual technologies. By comparing this ranking with the users prior expectations, it is possible for the user to see whether his inputs (criterion weights) and outputs (scenario rankings) are consistent. The users can map the possible best options conditionally against different stakeholders preferences, based on their economic, social and political perspectives. A range of viewpoints expressed in the energy debate can be accommodated that may lead to different option rankings, with the hope that relatively robust alternatives can be identified.

The multi-criteria approach used to combine the cost, environment and social indicators on an aggregate level is based on a utility function that is a weighted average of the contributing criteria. This approach is less complex and formal than the one used by EPFL in the MCDA task (see Chapter 11), but it allows the *SEOR* to be interactive. The supply options represent the technologies of primary interest in the context of future electricity generation in Shandong, and their ranking performance using the weights supplied by the stakeholders was essentially consistent with the EPFL analysis. The main difference from the EPFL results (apart from the MCDA methodology) is that the EPFL results were based upon ESS scenario results that include plant dispatch simulation, while the SEOR results assume fixed capacity factors for plant operation. This was necessary for interactive operation on the DVD, but the capacity factors were fixed using typical system dispatch from the ESS task, rather than assuming baseload operation for each technology. This means that fixed costs per kWh depend critically on how many hours per year the plant is assumed to run.

The criteria weights assigned by the user ("the decision-maker") reflect the relative importance to him of the various evaluation criteria. The user has the option to explore the basis for the scores and the relevant features of the candidate technologies. The central step in this approach is the investigation of the robustness of the option rankings to changes in relative criteria weights. Some scores can also be changed and the corresponding sensitivities can be investigated, but this only concerns scores that are based on subjective evaluation (e.g. performance on social criteria).

Three basic options are available to the user, two for individual technologies and one for technology mixes:

Ranking Technologies:

- Criteria Preferences. This option allows the user to test the effects of changing criteria weights on the implied ranking of individual technologies. Figure 12.9 shows an example of technology ranking based on a specific weighting profile.
- Intuitive versus Structured. This option allows the user to test to what extent his initial, intuitive technology ranking matches the one based on the systematic approach using indicators generated by CETP in combination with user-specific weights. The software keeps track of ranking results based on different preference profiles, allowing the user to perform sensitivity analysis.

Ranking Technology Mixes :

- Mixes. This option allows the user to define up to three electricity supply mixes to meet a specified demand in the year 2020 and then perform the same type of evaluation as in the "Intuitive versus Structured" option. Figure 12.11 is an example of results that can be obtained using this option.

SEOR is a further development of PSI's "Energy Game", originally created for application in Switzerland. SEOR has been tailored for the specific characteristics of the electricity supply situation in Shandong and includes significant extensions of the scope and flexibility compared to the original software. Furthermore, state-of-the-art user interfaces were implemented based on Flash.

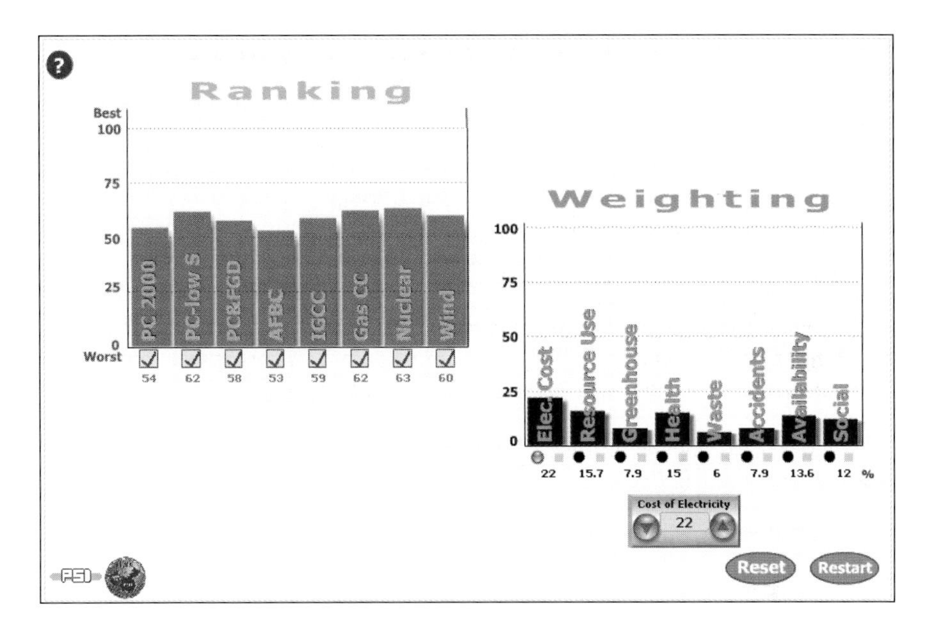

Figure 12.9 Technology ranking based on user-specific criteria weighting.

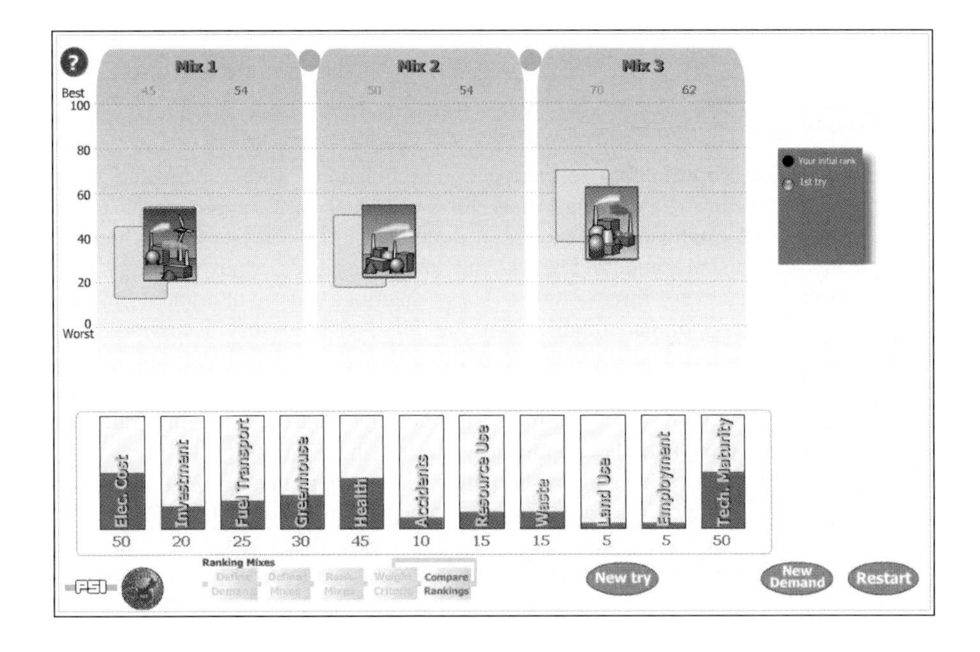

Figure 12.10 Calculated ranking (cards with pictures) versus intuitive ranking (blank cards) for three supply mixes using conventional coal, conventional and advanced coal, and a mix of conventional and advanced coal, nuclear and renewables.

4. PROGRAM MANAGEMENT FOR INTEGRATION

CETP has been a complex project and was organized by dividing it into twelve different tasks with a principal investigator (PI) responsible for each task (see chapter 2). Each PI had a clear plan of their task that was developed in the initial CETP proposal. The research team under each task was composed of scientists or engineers with knowledge, skills and competence to perform the particular task. Nevertheless the technical, geographic and cultural diversity, and the sheer scope of the project required effective management by the project's industry sponsor to ensure integrated results that were useful and on-schedule. In particular, some specific project challenges included:

- The scope and interdisciplinary nature of the research.
- Issues of data quality and availability.
- A tight schedule.
- Communication and language barriers.
- Differences in working styles between western and Chinese partners.
- Addressing Chinese stakeholder concerns and issues.
- Communicating results to stakeholders and the public

These challenges were primarily met by effective communications within the research teams, interactions with Chinese stakeholders, personnel exchanges and public outreach activities. This section addresses these management efforts to promote project coordination and integration.

4.1 Program Communications

For good project coordination, effective communications were established using three primary means. First, email lists for individual tasks and for all CETP participants were established. Second, a document-base was set up, which all participants could share and review via the Internet. Third, group meetings were held approximately every three months. These meetings mostly alternated between China and Switzerland, depending upon the relative need for discussions with stakeholder and technical discussions between researchers. The meetings allowed project management to track program progress and facilitated face-to-face discussion among the research partners. In the first phase of the program, the principal investigators had the opportunity during these meetings to present the methodology of their tasks and explain their data requirements to all the other researchers. This allowed the identification of common and specific data needed. Effective plans for collecting data were then developed with a better understanding of the relationship between the methodology and data required. In the second phase of the program, the group meetings allowed PIs to present their task progress and discuss related issues so that solutions could be found to resolve any problems. In the final phase of the program, the group meetings were used to discuss results, the writing of the book and technical reports, and the preparation of task presentations for the DVD. All these means of communication fostered teamwork and an understanding of each group's technical strengths and capabilities. Frequent interactions helped investigators retain an overall perspective, rather than focusing only on their own tasks.

Two oversight committees were also used to ensure appropriate management and technical quality. First, a steering committee provided project oversight, reviewing progress reports and providing decisions on key management issues. One major decision was to approve the development of the DVD in the later part of the program. This was a large challenge as the researchers were pressed to complete their technical research program, as well as helping in the development of the DVD, which has been described in the previous section of this chapter. Second, a technical advisory board was also set up to address technical quality concerns through meetings with PI's and review of specific issues.

4.2 Stakeholder Interaction

A key component of the CETP program was stakeholder interaction. This was to ensure that the program was designed and competed to meet the needs and expectations of the stakeholders or users. Early involvement of the stakeholders was important. SEPCO was a primary stakeholder, and also played an important role in providing modeling data. The Stakeholder Advisory Group had many functions, including identification of issues and criteria, review of scenario design, review of technical issues and data, providing criteria weights and receiving of results.

Communication with the stakeholders was a challenge due to the language barrier. Some of the stakeholders did not have sufficient knowledge of English for easy and reliable discussions, and most of the PI's did not speak Chinese. Translation of spoken presentations and visual aids was provided at major meetings, but discussion and feedback were still a problem. Some stakeholder discussion sessions were held just in Chinese so the stakeholders could communicate freely and express clearly their ideas and comments. The information obtained was then translated back to English for the CETP researchers. Translation was also done for some individual PI discussion with research partners, and individual interviews for eliciting stakeholder criteria weights for use in MCDA were conducted by Chinese partners. The language barrier and long travel distances were the major obstacles preventing more frequent interaction between the various CETP activities and the stakeholders. The language barrier also prevented setting up a web site for discussion with the stakeholders.

4.3 Technical Exchange and Outreach

One of the means of finding data and transferring the CETP methodology back to Chinese partners was the technical exchange between partner institutions. As mentioned above, frequent group meetings allowed technical exchanges between the partners in addition to communications by email and phone. However, the most effective way for technical exchanges to occur was for personnel from partner institutions to visit each other for short periods of two weeks to 2 months. A number of these exchanges took place during the course of CETP. Our Chinese industry partner SEPRI was able to take advantage of this technical exchange by sending personnel to different institutions to learn about the modeling methods used in this program and to build its expertise in these areas. This form of technology transfer was important if the CETP methods and not just the CETP results were to be transferred to China.

Outreach was also important for CETP, and went beyond informing the stakeholders about the program and its progress during the course of the program. In addition to stakeholder and researcher interactions, CETP has made a conscious outreach effort through press releases, TV coverage, press publications, the CETP website, the CETP movie, and of course the DVD and CETP book which you are holding. This chapter has discussed the integration of CETP task models and results, their presentation through the DVD and project management, but the real integration occurs when CETP results are used by decision-makers and the general public to influence the debate over sustainable energy and electricity use in China.

CHAPTER 13

CONCLUSIONS AND RECOMMENDATIONS

1. INTRODUCTION

The CETP has been a complex project, integrating a broad range of analyses into a comprehensive framework aimed at understanding electric sector planning and sustainability in China, and in particular in Shandong province. This book has described the component methodologies and results of the individual CETP tasks. This chapter integrates the conclusions that can be drawn from the different tasks, and presents them in a structured way. These conclusions start with observations about the CETP structure, and proceed to demand forecasting, the environmental, health and risk effects, and impacts, electric sector options and uncertainties, and recommendations.

2. THE ANALYTIC FRAMEWORK

The analytic framework of the CETP has been a significant achievement in itself. The complexity of the power planning problem has been addressed by combining the strengths of many different participants and approaches. This includes elements that are:

Multi-Disciplinary – CETP combined a wide range of academic disciplines, including specialists not just from energy systems and modeling, but also from the fields of economics, risk and environment.

Multi-Institution – Analysis team members representing a wide range of both Western and Chinese research institutions cooperated on CETP tasks, bringing a wealth of backgrounds and perspectives to the analysis.

Multi-Stakeholder – By assembling as diverse a group of Chinese stakeholders as possible, CETP addressed a range of issues that concern Chinese decision-makers, and used their input to interpret and rank results for their use.

Multi-Methodology – Because the scope of the CETP project, it was possible to have several tasks with different methodologies work in parallel to address the core questions of modeling future electric sector performance. The different approaches of these tasks allowed a wider perspective, based on the strengths of both individual results and common conclusions.

Integration – The CETP framework has succeeded in integrating analysis of the complete electricity chain, including demand, supply, direct and indirect environmental burdens, health impacts and accident risks. This includes not just modeling the future mix of electricity supply, but also life cycle analysis of generation technology chains, detailed analysis of major environmental externalities and risks, creation of the associated datasets required, and developing a wide range of specific tools, including custom software and a DVD for project presentation and exploration.

3. ENERGY AND ELECTRICITY DEMAND FORECASTING

Energy and electricity demand both fundamentally depend upon population and per capita economic activity. Population growth is less variable and easier to forecast than the economy, but demand growth remains inherently uncertain. Forecasting also depends upon the relative growth of the component sub-sectors of industrial, commercial, residential, and agricultural demand. Electricity prices have a feedback effect on demand, and the relative supply and demand for certain energy services also causes shifts between electricity and competing energy sources. The demand forecasting task for the CETP used two models to produce forecasts for energy and electricity demand, starting in the historical base year of 1995 and ending in 2020.

The MEDEE-S model was used to forecast total energy demand in Shandong. This is a bottom-up model that is based on economic activity, structural changes in the economy and energy efficiency, using detailed forecasts for individual economic sectors. Four major scenarios were modeled, including a "base scenario" using the best available data assumptions, a "lower GDP" scenario with one percent lower GDP growth per year, a "service sector" scenario where one percent of total demand shifts from the industrial to the service sector share of the economy by the year 2020, and an "inefficient" scenario where different sectoral energy efficiencies do not increase as much as expected. The forecast energy demand for these four scenarios increased from 44 Mtoe in 1995 to 85, 71, 84 and 95 Mtoe in 2020, respectively. These forecasts are equivalent to average energy growth rates of 2.7, 2.0, 2.6 and 3.2 percent per year, respectively. The future demand for higher quality energy carriers, including gas fuels (natural gas or LPG) and electricity grows at a significantly higher rate than the overall energy demand. Gas fuels and electricity are projected to increase from 6.0 percent and 14.5 percent of total energy demand in 1995 to 7.4 percent and 26.3 percent respectively in 2020.

The DEMELEC-PRO model was used to forecast electricity demand in Shandong. This a simulation model, based on a techno-economic description of different demand sectors with multiple levels. Four electricity modeling scenarios were again used, parallel to the energy scenarios, but with a somewhat increased level of electricity penetration. The electricity demand for the base forecast, lower GDP, increased service sector and lower efficiency scenarios rose from 94 TWh in 1995 to 262, 215, 259 and 318 TWh in 2020 respectively. These growth forecasts are equivalent to 5.2, 4.4, 5.1 and 6.0 percent per year, respectively. The base forecast is only slightly higher than the SEPCO forecast in the year 2010.

The shares of electricity use remained relatively fixed over the study period for the agricultural, construction and transportation sectors at six, one and one percent respectively. Meanwhile the industrial share dropped from 77 to about 65 percent, the service sector rose from 5 to about 13 percent and the residential sector rose from 11 to about 14 percent, which of course reflects the increased living standards related to economic growth and development. The shares of electricity demand for the agricultural, construction, service and transport sectors remained relatively fixed between the four scenarios in 2020 at about 6, 1, 13 and 1 percent, respectively. The main shifts were from the industrial to the residential sector. In the lower GDP case about three percent of demand shifted from industry to residential demand, and in the lower efficiency scenario about two to three percent of demand shifted in the other direction from household to industrial demand.

In both the energy and electricity modeling tasks, the base forecast and the three alternate scenario forecasts were relatively quite close. The chief intent was to produce the best forecast possible, and the three sensitivity cases represented shifts in assumptions that in the end did not produce vary large changes in the results.

The demand forecast task also highlighted the fact that energy efficiency is one of the most important factors in reducing demand and the corresponding environmental burdens and impacts linked to it. The demand task concludes that the available size of the energy savings available and the growth rates for energy and electricity mean that there is every reason to encourage efficiency by means of research and development, efficiency standards for new demand, retirement of old, inefficient technologies, and shifting industry to products and sectors that reflect both energy prices and market forces.

4. ENVIRONMENTAL DAMAGE, HEALTH AND ACCIDENT RISK

Environmental burdens and occupational risks depend both upon demand and the energy supply infrastructure. Public health depends not just upon emissions, but also upon the airborne transport of pollutants, their chemical transformations, and the density and geographic location of the affected population. These obvious steps are relatively simple to describe, but the bulk of the research is to understand the relationships and to quantify them. Using Life Cycle Assessment (LCA), Environmental Impact Assessment (EIA) and Risk Assessment (RA), results for China with emphasis on Shandong electricity sector include the following:

Overall Relative Performance – The LCA study shows that the environmental burdens of coal chains are generally the worst of all the energy chains considered for all selected indicators (Greenhouse Gases, SO_2, NO_x, particulates, solid waste, land use). Although the introduction of improved or new coal technologies should meaningfully reduce the harm to human health from major airborne pollutants, the balance of burdens per unit of electricity from the coal chain will still remain substantially higher than those associated with the natural gas or the nuclear chains.

The ranking of scenarios based on SO_2, NO_x and GHG emissions is dominated by direct power plant emissions and not greatly influenced by including upstream fuel chain emissions, especially for current dirty plants that are not refurbished. However LCA burdens do influence the ranking of scenarios for particulates, solid wastes, and land use.

Coal Chain SO_2 Emissions – Coal energy chain SO_2 emissions per kWh supplied to the Shandong high voltage grid are currently approximately three times greater than from chains associated with average W. European coal power plants in mid 90s. However, this factor may be up to 20 compared to the best European countries. The upstream chain contributed only about 5 percent but this would increase to about 10 percent if all coal comes from Shanxi. Total emission rates will dramatically decrease with better coal power plant technology. Scrubbers and advanced coal power plant technologies such as AFBC or IGCC should also allow increased use of local Shandong coal with relatively high sulfur content. This has the dual effect of reducing the amount of low-sulfur coal transported from the outer provinces, and freeing low-sulfur coal for distributed or diffuse applications where emissions control devices are much less economical than at large power plants. Increased attention should be paid to mine emissions, especially SO_2 and particulates from small mine-mouth power plants for self-supply of electricity. The shutdown of these power plants should be coupled with new, more efficient units that include mine waste and coal washing residues as fuel, and also use pollution control.

Coal Chain NO_x Emissions – NO_x emissions per kWh supplied from the coal energy-chain to the Shandong grid are currently approximately six times greater than from the W. European power mix in mid 90s. The upstream coal chain contributes only about 8 percent of the total emissions associated with Shandong power plants. The NO_x emission rates estimated for the MCDA scenarios in 2020, where the coal share always exceeds 60 percent of total supply, remain well above the average for W. Europe in 1995. NO_x emissions may become the dominant air pollution burden once particulates and sulfur emissions are reduced.

Coal Chain Greenhouse Gas Emissions – Coal chains relevant to China and to the mix of power plants in Shandong generate from 20 to 40 percent more Greenhouse Gases per kWh delivered to the grid than coal chains associated with Western European hard coal power plants. This rather broad range depends on whether or not coal fires in mines, coal seams and storage and waste heaps are included. GHG emissions from coal mining can be as high as 25 percent of the total. GHG emissions could be reduced by recovering methane at gassy mines, reducing coal fires and increasing efficiency throughout the coal chain.

Gas Pipeline Losses – GHG emissions for gas piped from gas fields near Shandong is comparable to future European conditions, but 20 percent higher if from distant fields in Xinjiang or Central Asia due to gas leaks and energy used for pumping. The LNG option and South Siberian pipeline gas are in the middle of this range.

Coal Chain Solid Wastes – Solid wastes from coal power plants should be recycled to the greatest extent possible, as limited by waste composition (for advanced technologies like AFBC and IGCC) and radioactivity content in the ash. Solid wastes from coal mining and coal washing with valuable heat content are already used for on-site energy conversion. Other mining waste is used in construction and to prevent subsidence by refilling shafts and tunnels. Conditioning and reclamation of waste heaps should be regularly practiced to decrease water pollution, reclaim land, and to prevent spontaneous fires.

Overall Health Impacts – Air pollution in China has a strongly negative feedback on the rate of economic growth. Health impacts dominate the assessed damages. Outdoor air pollution from all sectors in China resulted in about nine million years of life lost (YOLL) in 1998. This means that roughly one million Chinese die prematurely as a result of outdoor air pollution. The health damages caused by this pollution cost about 6-7 percent of GDP based on the willingness-to-pay method, which is comparable to current growth rates. The power sector contributes about one quarter of the total air pollution damage costs in China; Shandong's estimated share of the damage caused by the Chinese power sector is about 20 percent. Current power sector policies (in particular use of coal with lower sulfur content) already somewhat reduce emissions and their associated external costs.

Health Impacts of SO_2 – The major share of health damage is due to secondary particulates formed by chemical transformation of SO_2 and NO_x into sulfates and nitrates. Sulfates dominate damages, so SO_2 is the major contributor, followed by nitrates and primary particulates. The estimated physical impacts are sufficiently robust to be used for decision-making independently of their monetary value. For the reference Huangtai plant in Shandong, 1998 emissions caused about 25'000 YOLL per year. The average number of YOLL per tonne of SO_2 emitted in China is almost seven times higher than the average for the European Union; for Shandong the corresponding factor is about eleven. Impacts per tonne of pollutant emitted at different sites in Shandong vary by a factor of four, with the sites located on the coast having the lowest normalized impacts.

True Cost of Generation – The estimated average "true" cost of power generation from current power plants (including both internal and external costs for all relevant stages of their energy chains) amounts to 14.5 US cents per kWh in Shandong, including the highly uncertain global warming contribution of 1.7 US cents per kWh. The "true" cost is dominated by its external component (about 80 percent). The "rest of the chain" contribution is typically about 5 percent for the current coal chain. The parameterization of impacts and damage costs for the Shandong power sector in 1998 produced average external costs of 7057, 4579 and

5032 \$/tonne for the major pollutants SO_2, NO_x and PM, respectively, which dominated all other contributions.

Mortality Benefits of Alternative Technologies – Taking the Huangtai power plant in Shandong and its associated coal chain as a reference, the use of low sulfur coal reduces mortality (expressed in YOLL per GWh) by a factor of 1.7; and FGD with 95 percent SO_2 removal efficiency by a factor of 4.4. Replacement by AFBC gives a reduction factor of 8; by IGCC a factor of 13; by natural gas CC plant a factor of 52 and by nuclear power plant a factor of 63. The strong dependence of health damages on technology is also reflected in the corresponding reductions of the estimated "true" costs of electricity generated by these technologies.

Total Costs Justify Cleaner Technologies – The external costs of the "clean" Shandong electricity supply scenarios analyzed are typically a factor of three to four lower than for the "dirtiest" ones (depending on whether global warming damages are included or not). The total (internal plus external) costs of environmentally friendly strategies based on clean coal technologies and diversified supply (i.e. natural gas and nuclear power), are significantly lower than the total costs of "dirty" and non-sustainable strategies using conventional pulverized coal generation. Alternative monetization methods, considered inappropriate by the international scientific community, can lead to lower estimates of external costs. However, the damages avoided by cleaner technologies exceed the increase in internal costs in all the sensitivity cases analyzed, demonstrating the robustness of the findings. Reduction of major air pollutant emissions from electricity generation in China/Shandong, and the associated health and environmental damage is feasible, and economically and socially justified.

Land Area at Risk of Acidification – Critical load calculations demonstrate that the areas with the most sensitive ecosystems are unfortunately the ones most heavily exposed to SO_2 deposition. At present, about 25 percent of China is exposed to high acidification risks. Low levels of pollution control (the "Low Control" scenario in the EIA task) will lead to further deterioration with about 40 percent of China exceeding critical loads. If the current policy (the "Moderate Control" Scenario) is fully implemented, total SO_2 emissions in China in 2030 will be somewhat lower than in year 2000. Use of FGDs and increased use of low sulfur content coal at power plants are the most efficient measures for reducing emissions in the "Moderate Control" case. Full implementation of FGD for new coal power plants and extending control zones (beyond the two defined by the current policy) are the most important measures in the "High Control" Scenario. As a result, the emissions in 2030 drop by about one third below the current level and deposition in the most sensitive areas drops to about 30 percent. Thus, high levels of control would reduce areas at risk of exceeding critical loads to about 15 percent of China, in spite of the strong demand growth projected. For Shandong, the "High Control" Scenario with enhanced control of power plants and industry would reduce emissions from 2000 to 2030 by 45 percent, a larger relative reduction than for the whole of China.

Marginal Control Costs – For all China, the estimated additional annual control costs in 2020 are about 9 billion US$ for the "Moderate Control" scenario, and about 13 billion US$ per year for the "High Control" case, compared to the "Low Control" scenario. For Shandong, the corresponding annual estimates are 1.0 and 1.3 billion US$, respectively. Least-cost optimization of control measures was demonstrated by means of allocation of emissions reductions. As a result, it is possible to achieve the same target deposition for China as under the "Moderate Control" policy, but at 10 percent lower cost. Roughly the same cost reduction can be achieved in Shandong.

Coal Mine Accidents – When the best available datasets are combined and compared, the coal chain exhibits the highest accident fatality rates. About 6000 fatalities occur annually in Chinese coal mines due to small and severe accidents (five or more fatalities). Though severe accidents receive more attention, about two thirds of accident-related fatalities occur in small accidents. The Chinese severe accident fatality rate for the coal chain is about 6.5 per GW_eyr. On average, this is about ten times higher than in non-OECD countries, and about 50 times higher than in OECD countries. Small mines in China exhibit average severe accident fatality rates five times higher than large mines. Closing the most dangerous small mines could significantly reduce overall accident statistics. The Chinese mining industry is currently moving in this direction.

Oil and Gas Chain Accidents – Accident risk estimates based on other non-OECD countries appear to be representative for the oil and gas chains in China. Compared to coal, severe accident risks for the oil chain are one order of magnitude lower, and for the natural gas chain are between one to two orders of magnitude lower. The maximum credible number of fatalities per accident for the oil chains is one order of magnitude higher than for the coal and natural gas chains.

Hydro Chain Accidents – Hydropower accident risks were not studied in detail. Though hydro is very important for China, the hydropower resources in the Shandong province are low. Between 1950 and 1990, 3241 dam failures occurred in China, of which 123 or 3.8 percent were large dams and 3118 or 96.2 percent were medium and small dams. There is a rather clear indication that dam accidents and their consequences are underreported in China, so a reliable estimate of the Chinese-specific fatality rates from severe hydropower accidents is not feasible. Therefore, the pooled hydropower experience from non-OECD countries was used in the present work, indicating that the severe accident fatality rate for hydropower in China may be of the same order of magnitude as that for the coal chain. The maximum credible number of fatalities per accident within the hydro chain is two orders of magnitude higher than for coal and natural gas chains.

Nuclear Chain Accidents – Expected fatality rates associated with hypothetical nuclear accidents were based on Probabilistic Safety Assessment (PSA) due to the lack of relevant historical experience. The estimated risk of early fatalities associated with hypothetical accidents at current or advanced nuclear plants in

Shandong is negligible compared to the immediate fatality rates for severe accidents in the other energy chains. The risk of latent fatalities is higher but remains three to four orders of magnitude lower than the experience-based severe accident results for coal in China and hydro in non-OECD countries, while of the same order or lower than hydropower in OECD countries. The results for nuclear are only valid assuming that nuclear plants will be operated under stringent licensing requirements and in a safety culture comparable to OECD countries. The maximum credible consequences due to hypothetical nuclear accidents, though at an extremely low frequency level, may be very large, i.e. comparable to the Banqiao and Shimantan dams' accident that occurred in China in 1975 and resulted in 26,000 fatalities. The associated risk valuation is subject to stakeholder value judgments and has been pursued along with other economic, environmental, and social criteria in multi-criteria decision analysis carried out within CETP.

Severe Accident Costs – Damage costs due to severe accidents were not explicitly estimated. Risk measures results indicate that compared to OECD these damage costs are bound to be very high for the coal chain and probably also for early hydro plants. For Chinese conditions, however, they are of low significance when compared to the damage costs associated with air pollution.

5. ENERGY SUPPLY AND USE

Population, the demand for energy services, airborne transport of pollutants and impact mechanisms are basically beyond the control of electric sector decision-makers. However, there remain major options available for changing the relationship between energy demand and the external health and environmental costs and risks associated with energy use. Some options are predominantly technological, and are related to old generation capacity, new generation choices, and efficient energy use. Other options may be policies related to system operation, like emissions limits or taxes. This section explores the conclusions of the ESS, EEM and ETM tasks related to these options. Because of the way that the different tasks structure their scenarios, the ESS task options were formulated in terms of the old, new and energy saving technologies, while the EEM and ETM scenarios were composed using constraints like emissions taxes and caps, and sensitivities on load growth, fuel costs, interest rates and capital costs, which then determined the optimum choice of new technologies. Although the EEM and ETM tasks optimized/included generation from existing units, they did not include other options for old capacity or include demand side options, so these sections below discuss only ESS results.

Results presented below are in terms of cumulative total and average costs and emissions, summed over the study periods of each task. Cumulative differences between scenarios that appear small on a relative basis may represent large absolute differences that will continue beyond the end of the study period.

Old Generation Options – The existing generation base remains a large source of pollutant emissions well into the future. Results from the ESS task demonstrate that significant cost-effective strategies exist that can reduce power plant pollutants (SO_2, PM, NO_x) and reduce the rate of increase of greenhouse gases emission (CO_2). The most attractive scenarios consist of a combination of early retirement, coal preparation and FGD retrofits.

Although comparing changes of a single option against the base strategy in the base future is a limited way to view the rich set of ESS results, it does give some perspective. The base scenario had about 5% long-term growth in electricity demand and FGD on all new conventional coal units, no FGD retrofits, scheduled plant retirements, and not prepared coal. This produced a present value cost of 75 billion dollars (an average of 4.6 cents/kWh) and a cumulative total of 12.3 million tonnes of SO_2 emissions over 25 years. Compared to this, the range of single options for existing plants and their respective, cumulative SO_2 reductions include; FGD retrofits on some plants (-5 percent), retiring selected plants (-12 percent), using prepared coal at old plants (-12 percent), using prepared coal at all conventional coal plants (-17 percent), retirements and prepared coal at existing plants (-23 percent), and combining retirements, scrubbers and prepared coal at all conventional coal plants (-30 percent). These very significant SO_2 savings would be much smaller in relative terms if new generation was assumed in the worst case not to have FGD scrubbers, and come at a relatively narrow range of costs from 0.1 below to 0.5 percent above the base cost. It should also be noted that the prepared coal and retirement options achieve greater reductions in PM than SO_2 emissions. Although SO_2 emissions decline, CO_2 emissions for this set of scenarios which vary only by old generation options from the base case still increase from 0 to about 4 percent above the base case emissions of 3.0 billion tonnes (Gt). This is because average efficiency increases due to coal cleaning and plant retirements are overcome by the additional energy required for FGD scrubbers.

New Generation Options – Although the three electric sector modeling tasks differed in methodology and scenario formulation, they all included the same basic new generation technology options, and reached similar conclusions. China will continue to rely mainly on coal in the near future for power production and industrial energy use due to its abundance. However, over time new environmental regulations and policies will increase the share of other forms of power production such as hydro, natural gas, nuclear and renewables. Shandong has insignificant hydro resources and poorly assessed wind resources, so nuclear and natural gas remain as the largest options, but the attractiveness of these options depends critically on their capital costs (nuclear) and fuel costs (natural gas). Specific conclusions of the EEM and ETM tasks are further developed in the Policy Options section below.

All three tasks agree that new technologies can reduce SO_2 and other major pollutants for relatively low cost increases, far below the external costs of these pollutants. Advanced coal technologies such as IGCC therefore simultaneously offer major SO_2 and minor CO_2 reductions. For significant CO_2 emissions reductions it is necessary to pay a significantly higher premium for low or zero carbon generation technologies, particularly natural gas and nuclear power. Reducing the capital cost of the advanced coal and nuclear technologies is a key factor in making them more attractive. Research and development on increased efficiency, increased domestic Chinese manufacturing to decrease costs and international cooperation are all important factors in implementing these technologies. Without supply diversification and international cooperation, it will be difficult for China to sell CO_2 reductions to other countries, based on international protocols.

The dirtiest, least-cost solutions produced by the EEM and ETM tasks when no emissions caps or taxes were imposed were used as their basis for comparison with cleaner scenarios. In contrast, the ESS task reference scenario used conventional pulverized coal technology with FGD scrubbers on all new plants as the base option for new generation in Shandong. This assumption was based on stakeholder input, but it is arguable whether it will actually be implemented, so a worst-case scenario with no FGD scrubbers was also modeled. The ESS task analyzed eight other new generation technology options that were mixes of conventional coal, advanced coal, natural gas and nuclear generation technologies. The ESS base case had a present value cost of about 75 billion dollars, or 4.7 cents/kWh, with SO_2 emissions of 12.3 million tonnes and CO_2 emissions of 3.0 billion tonnes. The ESS worst case had a present value cost of 72 billion dollars, or 4.5 cents/kWh, with SO_2 emissions of 26.6 million tonnes and CO_2 emissions of 3.0 billion tonnes for the worst-case, no FGD scrubber scenario. The ESS worst case therefore already represents an increase in SO_2 emissions of 116 percent above the base case while saving only 0.2 cents/kWh or 4 percent. The other new technology mixes included natural gas combined cycle plants, nuclear power and natural gas, Atmospheric Fluidized Bed Combustion (AFBC), Integrated Gasification Combined Cycle (IGCC), a combination of nuclear, natural gas and IGCC, natural gas power imported by wire, and nuclear power. These alternatives have a range of emissions from 30 percent above to 9 percent below the base case, and come at a present value cost impact between 0.2 and 2.0 percent.

These changes are representative rather than average, but they tell several stories. First, while the possible SO_2 savings of the new technologies considered are significant, they are less than the savings from both the old capacity options discussed above and from implementing scrubbers on all new conventional coal units. The costs for new capacity options are also higher than for the old capacity options, even though the undiscounted average cost of electricity (cents/kWh) only rises one and a half percent.

Second, natural gas prices are very important. Power generated from natural gas and imported by transmission line was assumed to have "take or pay" gas contracts, making the plants effectively "must run," and also the most expensive generation option. In contrast, when natural gas plants are built in Shandong and dispatched by their cost they almost never run, with old, high sulfur units making up the difference. When gas is combined with new generation technologies this effect is decreased, but it still persists to some degree depending upon the new mix. Therefore, gas price forecasts are a key uncertainty for this technology. Unless natural gas prices are lower than the base forecast, the gas plants will not run enough to justify building them. Emissions taxes or permits trading as well as lower gas prices could alter this.

Third, the advanced coal technologies (IGCC and AFBC) are not that much better than the best pulverized coal plants with FGD, although the substantially higher thermal efficiency of IGCC made a bigger difference. They save only 4 to 8 percent on cumulative SO_2, emissions, at a cost increase of 1 to 2 percent. Because all of the option mixes contain at least some conventional coal generation, this emphasizes the need for scrubbers on all new conventional coal capacity.

Fourth, new technologies appear less cost-effective at SO_2 reductions than old technology options (in part due to the fact that the base case already assumed FGD scrubbers on all new conventional coal plants), but they have cumulative CO_2 reductions of up to 16 percent. The greatest savings came from the combined nuclear and advanced coal scenario, due to nuclear's zero air pollution and the higher efficiency of the advanced coal technologies.

Fifth, although the total difference in scenario costs can be as high as 1.5 billion dollars, this is a difference of only two percent and is totally surpassed by the external costs that can be avoided by the emissions savings. The ESS results may seem to deemphasize the role of new technologies in affecting future costs and emissions, but this is not entirely true for several reasons. First, the base case already included FGD on all new conventional coal plants, which has the largest single impact. Second, the cumulative rather than end-of-study-period results tend to de-emphasize differences between scenarios. Third, assumptions regarding technology and fuel availability were arguably realistic but limited new technology penetration. Fourth, the results do emphasize that the cleaner new technologies get, the more important old capacity and demand-side options become to create balanced strategies.

Demand Side Options – China has already significantly increased its energy efficiency, and continuing this improvement has a very large potential for environmental and health benefits. The ESS task asked "what if" electricity could be saved, over and above the normal market response due to cost or market barriers. Individual program measures were not identified, but the price of aggregate savings was estimated.

The demand side options analyzed in the ESS task included both increased end-use efficiency and peak load management relative to similar scenarios without them, and both effect system emissions in different and important ways. In contrast to old and new capacity options, energy efficiency and peak load management are both more effective at saving CO_2 than SO_2. Energy efficiency is also more effective than peak load management at saving both SO_2 and CO_2. Both these effects are based on least-cost dispatch and would change if emissions costs were included in the dispatch.

Increasing energy efficiency by ten or twenty percent relative to the base case reduces the need for new generation. Fewer new, clean and efficient plants means that older, dirtier and more inefficient plants run more, because they will be needed not just at peak loads, but also for intermediate and baseload service. The difference in SO_2 emissions between old and new technologies is large enough that SO_2 emissions only drop one or two percent. Peak load management also reduces the need for new generation, but without the same energy savings. This means that when peak load management is combined with end-use efficiency, SO_2 emissions can *increase* by about 16 to 20 percent. The impact on CO_2 emissions is much more favorable. New technologies are not much more efficient than old technologies, so end-use savings of 10 and 20 percent relative to the base case result in almost the same reductions in CO_2 emissions. Average emissions per amount of electrical service provided also drop, because the same services are provided with less electricity.

These results imply that end-use efficiency programs are more effective than peak load management in achieving emissions reductions. End-use efficiency alone saves some SO_2 and much more CO_2, while reducing the direct cost of electric service by between five and nine percent for the moderate and aggressive cases, respectively. This represents a real win-win situation. The real caveats lie in the difficulties of correctly estimating the price and effectiveness of end-use efficiency programs, and separating their effects from normal market responses that also increase efficiency. The assumed savings of 10 and 20 percent from the efficiency programs may not be available at the prices assumed. However, these prices could be approximately doubled before the cost savings were eliminated. At that point, demand side costs would equal supply side alternatives and the emissions savings would be achieved at no additional direct cost. So while efficiency is not as effective as the supply side options in SO_2 reductions, it is much more effective in CO_2 reductions and a very valuable component of a balanced energy strategy portfolio. The demand-side options were not analyzed using life cycle analysis in the same way as the supply side technologies. Their emissions savings are therefore slightly overestimated. It is clear from the supply-side analysis that the rest-of-chain emissions are dominated by direct power plant emissions for older fossil chains, although this shifts as clean coal technologies are considered. Demand-side options have no direct emissions, and only the rest-of-chain emissions for their manufacture and installation, so it seems clear that their savings will be almost equal to their avoided direct emissions.

It is worth noting that energy efficiency is not appropriate only in energy intensive sectors, but rather anywhere it is cheaper than energy supply and delivery. This means that for optimum efficiency, it is necessary that not just electricity and its more efficient end-use technologies be correctly priced, but also competing fuels and the technologies that provide competing energy services. Increased reliance on markets should not be limited to just one energy sector.

Policy Options – The EEM and ETM models included the same basic technologies as options for optimized future energy strategies, but their modeling scenarios were based on constraints and sensitivity analysis rather than fixed option combinations. This means that emissions taxes or caps for SO_2, CO_2 or both were combined with varying assumptions about demand growth, fuel prices, interest rates and capital costs. Because both the EEM and ETM tasks used optimization models, the minimum cost solutions were found for the imposed conditions. However some of the emissions caps or their equivalent taxes were sufficient to generate wider cost swings than in the ESS task.

The EEM task found that the cumulative discounted cost of building a more sustainable path for electricity generation in China increases by two to six percent between 2000 and 2030 for an annual discount rate of ten percent (the same as in ESS). Undiscounted costs increase significantly more (20 to 30 percent). These higher costs are still less than the damage costs caused by pollution in China without these future measures.

Specifically, the EEM task's MARKAL model of Shandong had a base, low-growth, constant fuel price (BLC) scenario that met a demand of about 275 TWh in 2020 with a generation mix dominated by conventional coal units (approximately 92 percent), with the rest from several other minor contributors. This scenario produced about 2.5 million tonnes of SO_2 and 230 million tonnes of CO_2 in 2020, at an average cost of about 3.4 cents/kWh. When a SO_2 tax of $1500/tonne was imposed (the SLC15 scenario), this shifted the generation mix to about 15 percent conventional coal without scrubbers, 43 percent conventional coal units with FGD scrubbers, and 34 percent advanced coal units. This change reduced SO_2 emissions by a factor of about five to roughly 0.5 million tonnes in 2020, while CO_2 emissions dropped slightly to about 220 million tonnes, at an average cost of about 4.2 cents/kWh. Finally, if emissions caps are applied which limit SO_2 emissions to 1.4 million tonnes and CO_2 emissions to 180 million tonnes in 2020 (the ELC scenario), then the generation mix shifts to about 48 percent conventional coal units, 9 percent conventional coal units with FGD scrubbers, 9 percent advanced coal units, 11 percent nuclear units, and 9 percent gas units, at an average price of about 4.0 cents/kWh.

The ETM task obtained similar specific results for Shandong province. Their business-as-usual (BAU) scenario with a generation mix in 2020 of about 23 percent old coal units without scrubbers, 69 percent new conventional coal units without scrubbers, and 8 percent new conventional coal units with FGD scrubbers to meet a demand of about 256 TWh of electricity. This scenario produced roughly 2.8 million tonnes of SO_2 and 258 million tonnes of CO_2 in 2020, at an average cost of about 2.8 cents/kWh. When the Total Sulfur Control (TSC) scenario was modeled, which capped total provincial SO_2 emissions at the 2000 level, this shifted the generation mix to about 15 percent old coal units without scrubbers, 29 percent new conventional coal units without scrubbers, and 48 percent new conventional coal units with FGD scrubbers. SO_2 emissions dropped to about 0.8 million tonnes in 2020, and CO_2 emissions rose slightly to about 261 million tonnes, while cost rose roughly 4 percent to about 2.9 cents/kWh. Finally, if local city SO_2 emissions are capped at their 2000 levels and provincial CO_2 emissions are capped at half their business-as-usual level (the TLSC scenarios), then the generation mix shifts to about 36 percent old and new conventional coal units without scrubbers, 7 percent new conventional coal units with FGD scrubbers, 40 percent nuclear units and 16 percent combined cycle gas turbine units. This produced SO_2 emissions of about 0.8 Mt and CO_2 emissions of about 129 Mt in 2020, with the average price rising roughly 17 percent to about 3.3 cents/kWh.

Both the EEM and ETM tasks agreed with the ESS task that SO_2 reductions were achievable for a modest investment in scrubbers, low sulfur coal and/or advanced clean coal technologies. Some reductions in CO_2 accompanied the SO_2

savings, depending basically on the net changes in efficiency of the new generation mix. In the MARKAL model of Shandong a tax of \$500/tonne of SO_2 was barely effective, but taxes of \$1000 and \$1500 per tonne reduced SO_2 emissions of about 750 thousand tonnes in the year 2030 by about one third and two thirds, respectively. However to obtain significant CO_2 reductions required significant investments in lower or non-carbon generation, which also reduced SO_2 at the same time. The scenario that capped CO_2 emissions produced a significant penetration of nuclear generation.

The more stringent the caps or taxes imposed, the more diversification of supply was required to meet them. This meant that in Shandong the current coal mix was shifted to include not just advanced coal technologies, but also natural gas, nuclear, wind, and other resources. Outside Shandong, a more diverse mix may also include hydropower. The diversification required to obtain these goals was observed to vary with interest rates, fuel costs and nuclear capital costs. Interest rates caused a shift between low capital cost technologies (e.g. natural gas) and capital intense technologies (hydro and nuclear). Although the sensitivity analysis on interest rates was not intended to represent an option, a policy offering artificially lower interest rates would constitute a direct subsidy. The penetration of nuclear at the base cost of \$1600/kW was about 5%, and only rose as high as 30 percent when the capital cost was reduced to around \$1200/kW, which was considered unrealistically low.

Both the EEM and ETM tasks brought perspectives to bear on transmission issues that were impossible in the ESS task. The CRETM model covered seven regions of China, including Shandong. According to its results, interregional transmission may be economically attractive and would reduce local Shandong pollution. Although this would presumably reduce local impacts in Shandong, the CRETM results were not coupled with the EIA task to estimate whether the overall health impacts would rise or fall.

The ETM task modeled both local and provincial control zones for Shandong, and found that the current trial sulfur tax rate of \$25/tonne SO_2 (\$50/tonne S) is insufficient to control SO_2 emissions. It also modeled the current SO_2 control zone policy and observed the resulting geographic shift in emissions, without any substantial reduction in overall emissions. Damages are presumably reduced by the shift to less sensitive locations, but this was not quantified. Fuel supply locations and new generation technologies were also observed to shift geographically, based on these policies. As in the ESS task, the assumed natural gas prices were too high for this option to compete economically, but the emissions benefits create a rationale for increasing pipeline capacity and lowering prices.

Putting It All Together – The results of all three tasks confirm the continuing role of coal, and also the need for a mix of options that combine and integrate options for old capacity, new capacity and energy savings. Some options like end-use efficiency appear to be clear winners, lowering both cost and emissions. Many other options save emissions for minimal increases in average cost, far below the costs of their externalities. The effects of fuel price and plant efficiency on the dispatch order that drives system behavior have been demonstrated by the lack of natural gas generation and the effect of peak load management on emissions. Taxes can alter system dispatch and expansion planning, but taxes need to be linked to the fuel or emission that actually causes the damage, and be set at a level that matches the external damage cost. Optimally, emissions taxes or permits might be regionally priced to match local variations in environmental and health damages.

Integration is not just a question of electricity supply and demand, but also of the supply and transportation networks that transport coal, gas, electric power and heat. The fuel supply chain needs to be an integrally coordinated part of the energy system, which means correct pricing of fuels, fuel preparation and treatments, and transportation. Rail networks, ships, pipelines and electric transmission and distribution networks all work together as part of an integrated system, if the right prices exist to coordinate them.

Finally, in emphasizing the need for integrated strategies, it should be remembered that strategies also need to be both robust and flexible. A good strategy that can meet many possible futures, and that is easy to change when needed is far better than a narrow optimum.

6. SUSTAINABILITY AND STAKEHOLDER PERSPECTIVES

Within CETP a wide spectrum of indicators were generated on electric sector system performance. Scenario rankings based on a single, different indicators do not generally agree with each other, and rankings based on multiple criteria depend upon personal preferences. Therefore decision-makers are interested in an overall ranking based on aggregating many individual indicators.

Total Cost – One method of aggregation often highly attractive to decision-makers is the monetization of scenario performance across the range of environmental, health and risk indicators. External costs can then be added to direct, internal costs to produce the total or "true" cost of electricity production. Some authors propose that total cost may serve as a measure of sustainability performance, since it reflects the aggregated economic and environmental efficiency of electric system performance. As already discussed, the total costs estimated by CETP show that an integrated Shandong electricity supply strategy based on reducing emissions from old plants, a diversified mix of new technologies including clean coal technologies and demand side efficiency has the cheapest total cost. Even more costly mixes of clean coal technologies, nuclear power and natural gas generation are still significantly cheaper in terms of total cost than dirtier scenarios. Based on the set of ESS results, the best scenarios can reduce total costs by slightly over 70 percent when compared to the worst scenario with no FGD scrubbers.

The use of total cost as an aggregate measure of system performance is not universally accepted. The balance between strictly economic and environmental components may shift or vary between developed and developing countries. Furthermore, the social dimension of sustainability is not explicitly included in the total cost approach. Health impacts are the dominant contributor to external, environmental costs, and may or may not be included as a social component in the evaluation. As another example, nuclear waste and severe accidents contribute only marginally to the external costs estimated, but nuclear power remains controversial in some countries. Another issue is whether all attributes important to stakeholders are adequately reflected in the total cost. For example, although fuel transport costs are included through fuel prices as part of the internal cost, extensive expansion of the transport infrastructure is not only a matter of cost. Finally, though some specific estimated impacts (such as mortality and morbidity) are considered robust, there may be no consensus on their monetization (and in particular, the value of reduced life spans measured by Years of Life Lost, or YOLLs).

MCDA and Stakeholder Inputs – The major alternative to the total cost approach is multi-criteria decision analysis (MCDA). The application of MCDA in CETP provided a framework that allowed the often-conflicting evaluation criteria to be addressed together, based on scenario results and CETP stakeholder preferences. The implementation of MCDA in CETP was limited in scope and resources, which could be extended in future follow-up work. Furthermore, the general validity of MCDA results depends upon how representative the preferences of the individual stakeholders are for their own organizations and for Chinese stakeholders in general. In spite of these practical and partially inherent limitations of MCDA, some rather clear patterns have been identified.

The CETP stakeholders did not subscribe to the idea that sustainability implies an equal weighting of the economic, environmental and social performance of the electric supply system. In relative terms, all stakeholders considered the economic component to be clearly the most important, followed by the environment. The social class of criteria was generally assigned relatively low weights, but some concerns were expressed regarding hypothetical severe accidents and long confinement times for nuclear waste.

MCDA and Total Cost Scenario Rankings – The results of the MCDA evaluation show that the common features of strategies favored by CETP stakeholders include; use of prepared coal in conventional coal plants, use of scrubbers for new conventional coal plants, extensive use of DSM, and in some cases also the select retirement of the worst existing coal plants. These strategies may be combined with natural gas, IGCC and/or nuclear energy. Natural gas was relatively more favored, with some differences in stakeholder opinions regarding nuclear power. One essential and positive observation was that the stakeholders clearly reject most non-sustainable strategy examined, i.e. no measures affecting current generation, no FGD scrubbers for new conventional coal plants, no diversification of supply and no DSM. This result is consistent with the total cost ranking where the overall benefit of FGD scrubbers was clearly demonstrated. The MCDA approach gives a somewhat lower ranking to diversification based on nuclear power. Apart from the reasons mentioned above, this was caused by the fact that most stakeholders assigned relatively low weights to health effects (typically around 10% of the total) that dominate external cost and are the largest benefits of nuclear power.

7. RECOMMENDATIONS

Based upon the conclusions that have been described above, what recommendations can be made for China, and in particular Shandong province that will improve the sustainability of their electric sectors? It is clear that some win-win strategies should be favored by all stakeholders. One example is the continued use of energy conservation, which reduces both direct costs and emissions. This conclusion is based on the assumed cost of energy efficiency, but even significantly higher conservation costs would yield emissions reductions for no increase in the cost of electricity. Other strategies present cost v. emissions tradeoffs that must be evaluated by stakeholders. Based on the CETP estimation of externality costs, the increased cost of many technology strategies is less than the associated reductions in emissions damage costs. Individual stakeholders may disagree on how to balance direct, internal costs against external costs that implicitly set values on human life. Nevertheless, based on stakeholder inputs and international scientific consensus, the analysis team feels able to make the following recommendations.

LONG-TERM ENERGY POLICY RECOMMENDATIONS FOR CHINA:

The combined modeling efforts of the CETP program have confirmed the idea that good energy policy for China as a whole will consist of a balanced portfolio of options that deal with old plants, new technologies and fuels, saving energy and improving markets. The specific mix of options in such balanced strategies may differ between provinces, but the general conclusions are clear.

Old Capacity – Even with China's current rates of high economic growth, parts of the old energy infrastructure will persist for decades to come. Although its market share will shrink, its inefficiency and pollution give it a disproportionate influence on future environmental and health burdens, damages, risks and costs. For this reason it is very important to improve or retire old infrastructure. This conclusion spans the energy spectrum from, dangerous mines to dirty power plants to inefficient industries. Specific examples of improvement include cleaning coal, retrofitting electrostatic precipitators to remove particulates, FGD scrubbers for SO_2 and retrofitting efficient energy use.

New Resources – Coal will remain China's most abundant fuel, but new technologies and alternate fuels will contribute to cleaner use and increased diversification. More efficient conventional generation with scrubbers and advanced coal plants will be complemented by natural gas, large hydro and nuclear capacity. Small hydro will remain a niche player, and the future role of wind power can only be determined once a detailed regional assessment of China's wind resources has been completed.

Saving Energy and Electricity – Energy conservation and demand side management options beyond normal, imperfect market-driven savings are of key importance because this resource is large, cost-competitive and non-polluting. Available energy savings dwarf renewable energy sources, and are cheaper, easier and faster to implement when the energy supply and demand infrastructure is growing quickly. Although costs of specific portfolios can be debated, this was the single biggest win-win option in CETP with both lower direct costs and lower pollution.

Market Allocation – All energy strategies require the allocation of resources between competing options. China expects that continued reforms will encourage efficiency, especially in the natural gas and electricity sector. Such competition should be encouraged as broadly as possible. Planning can play a complementary role with markets, particularly as the government seeks to make transactions more transparent, enforce patents, perform R&D where patents do not supply sufficient incentive, supply missing signals (such as internalizing external costs) and enforce contractual liabilities. Specific examples include competitive electricity markets, coal pricing by heat and sulfur content, pricing pollutants using emissions taxes or tradable emissions, competition between 'fuels' (e.g. gas v. electricity v. conservation), environmental externalities for pollution control, and the planning of gas, electric and rail transport networks. Energy efficiency belongs not just in energy intensive sectors, but wherever the cost is less than the cost for energy supply and delivery. Government support of energy sector development is important, but the market information contained in prices is key to both building and operating integrated energy systems.

SPECIFIC RECOMMENDATIONS FOR SHANDONG PROVINCE:

These general recommendations for China as a whole apply to Shandong with some specific emphases. Old electric generation capacity in Shandong has very high external costs, and total costs can be reduced by cleaner coal, select retirements, retrofitting scrubbers, efficient new and advanced coal plants, diversified supply and demand savings. However Shandong has basically no hydro and uncharacterized onshore and offshore wind resources, so fuel diversification is largely limited to natural gas and nuclear energy. Natural gas, if available at a competitive price, is especially suited to the increase in peaking capacity that is needed as Shandong's load shifts away from its current industrial emphasis and towards other sectors with higher daily peak loads. Shandong will need to import both natural gas and more coal in the future, as future demand for high quality energy in the forms of gas and electricity increase. Increased natural gas demand for electricity generation may provide the large initial demand necessary to justify investing in new pipeline capacity. However, when urban natural gas distribution networks become common other uses may compete with its use for power generation. These new gas uses will displace many dirty, distributed emissions sources, which are otherwise much harder to control than a central generation plant. Shandong will also benefit from its ongoing integration with the transmission grid beyond its current provincial bounds.

8. OVERALL REMARKS AND FUTURE OUTLOOK

CETP has been a successful story of cooperation between academia and industry. Its experience has reinforced the idea that a focused, industry-led effort with academic research expertise and stakeholder involvement is an effective and productive way of undertaking international multi-disciplinary research activities. In this age of globalization, international cooperation is especially important to ensure a sustainable growth of the world economy and better living standards for the developing countries. CETP is an excellent example of this kind of joint effort for sustainable development.

CETP has assembled a set of tools to investigate the impacts of power using Shandong Province in China as a case study. A number of specific conclusions and recommendations derived from the results of various research components have been presented. Overall highlights of the China Energy Technology Program that apply to the sustainable development of the electricity sector throughout China include:

- Reduction of major air pollutant emissions and the associated health and environmental damage, due to changes in how electricity is generated and used, is feasible, economic and socially justified.

- Cost-effective strategies have been identified that lead to large reductions of emissions of major air pollutants and substantially curtail the increase of greenhouse gas emissions.

- Improved and advanced power technologies, coal treatment, fuel diversification, and promotion of demand-side management are pre-requisites for a more sustainable electricity supply sector in China.

These results of the CETP project can be used by Chinese stakeholders and other interested parties immediately in electricity sector planning for Shandong Province and China. The DVD that accompanies this book provides easy access to the results of this study. For example, the environmental impact of airborne emissions from power plants at different sites in Shandong can be evaluated in only a few steps. The impact of different technologies can also be assessed quickly. Combined portfolios of technologies can be evaluated quantitatively in terms of costs and environmental impacts. The Shandong Electricity Options Ranking tool also allows experimentation with different economic, environmental and social preferences.

The CETP methodology is an integrated set of powerful tools for studying different issues related to power production and environmental impact assessment, which can be applied to different problems and locations around the world. Within China, these problems include deregulation, emissions controls under deregulation, regional surcharges for SO_2 permits based on actual damages, CO_2 markets and trading, and domestic power markets and trading. The focus should depend upon stakeholder priorities, but there is a need for increased integration, better efficiency costs and determining the scope of renewable energy. In other developed and developing countries, the CETP methodology can also provide tools for studying issues like deregulation, electrification, and other pathways for sustainable development that will allow higher productivity and standards of living. In short, the CETP methodology provides a platform for stakeholders to explore alternate paths and trade-offs towards sustainability.

APPENDIX A

STRUCTURE OF CETP

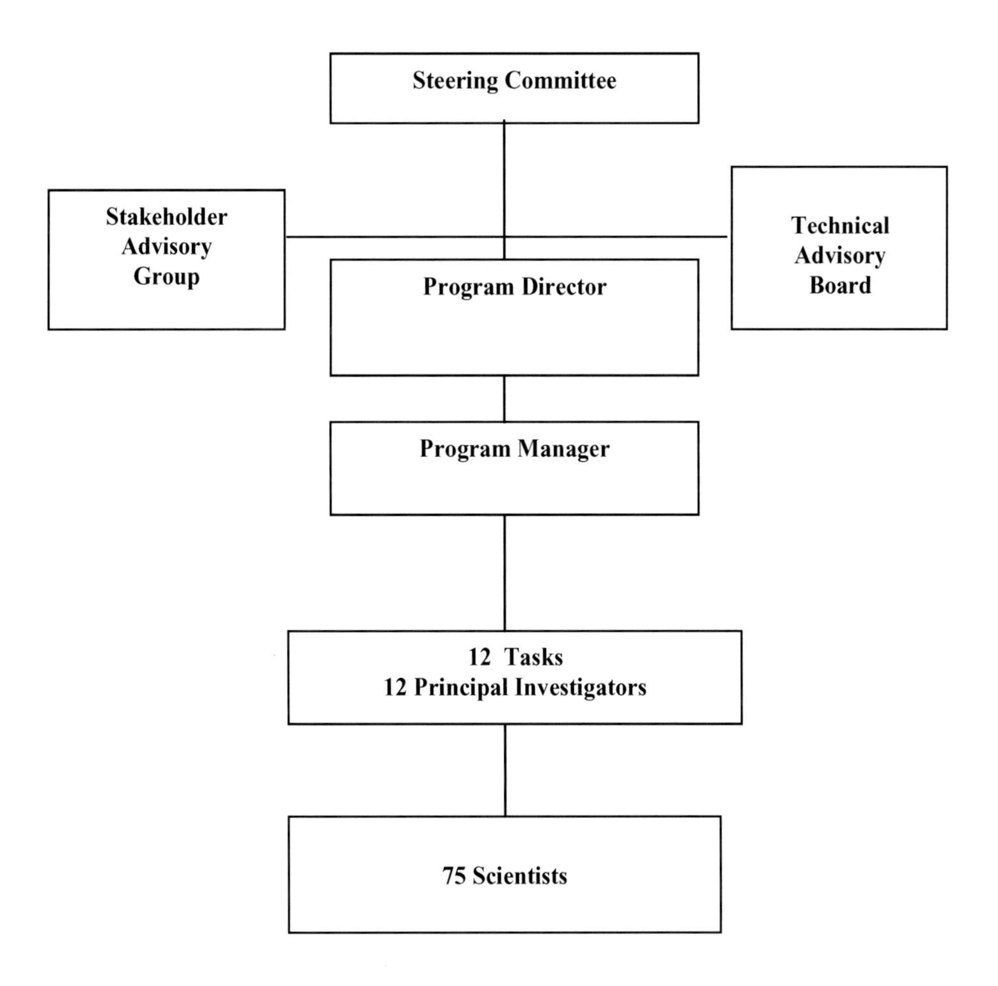

APPENDIX B

STEERING COMMITTEE

Markus Bayegan	Chairman Prof. Dr., Chief Technology Officer, Group R&D, ABB Ltd., Switzerland
Baldur Eliasson	Dr., Head, Energy & Global Change Dept., ABB Corporate Research, Baden-Dättwil, Switzerland
Paul T.P. Chan	Senior Vice President, Communication and Sustainability Affairs, ABB China, Beijing China
David H. Marks	James Mason Crafts Professor, Civil and Environment Engineering, Director, MIT Laboratory for Energy and The Environment, Cambridge, USA
Tomonori Matsuo	Professor, Director, Environment Science Center, The University of Tokyo
Shuichiro Asao	AGS Coordinator(Replaced Prof. Matsuo as SC member February 2001), The University of Tokyo
Bjorn Stigson	President, World Business Council for Sustainable Development, Geneva, Switzerland
Ulrich W. Suter	Professor, Polymer Chemistry, Dept. of Materials Institute of Polymers, ETH, Zurich, Switzerland Current Title: Vice President for Research and Industrial Relations

---

Roger Baud	Finances Financial Administration, Alliance for Global Sustainability, ETH, Zurich, Switzerland
Yam Y. Lee	Secretary Dr., Consultant, Energy & Global Change Dept., ABB Corporate Research, Baden-Dättwil, Switzerland

STAKEHOLDER ADVISORY GROUP

Risheng Guo	Deputy Director-General, the Administrative Center for China's Agenda 21, China
Zhenou Hu	Deputy Director, Office of External Financing Bureau of International Cooperation, Chinese Academy of Sciences, China
Wenran Li	Senior Consultant, former Vice Director, Environmental Protection Bureau of Shandong, China
Dinghuan Shi	Secretary General, Ministry of Science and Technology China
Huijiong Wang	Senior Economist, Senior Engineer, Vice President of Academic Committee, Development Research Center, State Council, China
Xiubin Xi	Shandong Economy and Trade Commission, China
Zhonghui Zhan	Vice Director General, Department of Science & Technology and Environmental Protection, State Power Corporation of China
Kun Zhang	Director General, China-Japan Center for Environmental Protection, State Environmental Protection Agency, China
Qingbo Zhao	Senior Engineer, President of Shandong Electric Power Research Institute, China
Dadi Zhou	Director General, Energy Research Institute, State Development Planning Commission, China

TECHNICAL ADVISORY BOARD

Jefferson W. Tester	Chairman H.P. Meissner Professor of Chemical Engineering, MIT Cambridge, USA
Baldur Eliasson	Dr., Head, Energy & Global Change Dept., ABB Corporate Research, Baden-Dättwil, Switzerland
Wolfgang Kröger	Prof. Dr., Head, Nuclear Energy and Safety Research Department, Paul Scherrer Institute, Switzerland

APPENDIX C

LIST OF CETP PARTICIPANTS

(Authors are marked with an *)

ABB Energy & Global Change Dept., Baden-Dättwil, Switzerland

Name	Function in CETP	Tasks
Baldur Eliasson*	Program Director	
.	Overall Management and Coordination	All
Yam Y. Lee*	Program Manager	All
Christopher Russo*	Principal Investigator	Database Development
Jiang Ma*	Coordination and Support	Data Collection

ABB China Ltd., Beijing, China

Name	Function in CETP	Tasks
Paul Chan	Advisory	Data Collection
Shiwen Zheng	Coordination China	Data Collection Support (Beijing)

ABB China Ltd., Jinan, China

Name	Function in CETP	Tasks
Tianpeng Liu (presently Shanghai)	Coordination China Support (Jinan)	Data Collection
Victoria Su	Coordination China Support (Jinan)	Data Collection

Paul Scherrer Institute, Villigen, Switzerland

Name	Function in CETP	Tasks
Stefan Hirschberg*	Principal Investigator	Risk Assessment
	Principal Investigator	Assessment of Environmental Impacts & External Costs
	Principal Investigator	DVD
Roberto Dones*	Principal Investigator	Life Cycle Assessment
	Technical Coordinator	DVD

Socrates Kypreos*	Principal Investigator	Energy Economy Modeling
Robert A. Krakowski*	Scientist	Energy Economy Modeling
Alexander Roeder*	Scientist	Energy Economy Modeling
Peter Burgherr*	Scientist	Risk Assessment
Thomas Heck*	Scientist	Environmental Impact Assessment

Swiss Federal Institute of Technology (EPFL), Laboratory of Energy Systems (LASEN), Lausanne, Switzerland

Name	**Function in CETP**	**Tasks**
Gerard Sarlos	Advisory	MCDA
Pierre-Andre Haldi*	Scientist	Output Integration (MCDA)
Jacques Pictet*	Scientist	Output Integration (MCDA)
(Current affiliation: Bureau d'Aide à la Décision Pictet & Bolliger)		
Edgard Gnansounou*	Scientist	Demand Forecasting
Jun Dong*	Doctoral student	Demand Forecasting

Swiss Federal Institute of Technology (ETH) Zurich, Switzerland

Name	**Function in CETP**	**Tasks**
Adrian V. Gheorghe*	Scientist	Electric Sector Simulation
Warren W. Schenler* (Current affiliation: PSI)	Scientist	Electric Sector Simulation

Stuttgart University, Stuttgart, Germany

Name	**Function in CETP**	**Tasks**
Wolfram Krewitt* (current affiliation: German Aerospace Center)	Scientist	Environmental Impact Assessment
Alfred Trukenmüller* (Current affiliation: Hamburg University)	Scientist	Environmental Impact Assessment

International Atomic Energy Agency, Vienna, Austria

Name	**Function in CETP**	**Tasks**
Joseph V. Spadaro*	Scientist	Environmental Impact Assessment

Cazzoli Consulting, Switzerland

Name	**Function in CETP**	**Tasks**
Erik Cazzoli*	Scientist	Risk Assessment

Vitty Consulting, Vitty, Slovakia

Name	Function in CETP	Tasks
Jirina Vitazek*	Scientist	Risk Assessment

Massachusetts Institute of Technology, Cambridge, USA

Name	Function in CETP	Tasks
Stephen R. Connors*	Principal Investigator	Electric Sector Simulation
Chia-Chin Cheng*	Doctoral student	Electric Sector Simulation

Energy Research Institute, State Development Planning Commission, Beijing China

Name	Function in CETP	Tasks
Dadi Zhou*	Principal Investigator	Demand Forecasting
Zhonghu Wu	Advisory	Demand Forecasting
Shixian Gao*	Scientist	Demand Forecasting
Xingshan Zhu	Scientist	Demand Forecasting
Xiaoli Liu	Scientist	Demand Forecasting
Xinmin Xue	Scientist	Demand Forecasting
Zhengming Su	Scientist	Demand Forecasting

China-Japan Friendship Center for Environmental Protection, State Environmental Protection Agency (SEPA), Beijing, China

Name	Function in CETP	Tasks
Kun Zhang	Advisory	Life Cycle Assessment Environmental Impact Assessment Risk Assessment

Policy research Center for Environment & Economy, State Environmental Protection Administration, Beijing, China

Name	Function in CETP	Tasks
Guang Xia	Advisory	Environmental Impact Assessment
Lulian Cheng*	Scientist	Risk Assessment
Yihong Zhao*	Scientist	Environmental Impact Assessment
Dianlin Li	Scientist	Environmental Impact Assessment
Xin Zhou*	Scientist	Life Cycle Assessment
Chunxiu Tian*	Scientist	Life Cycle Assessment
Xiangyang Xu	Scientist	Environmental

		Impact Assessment
Zhongping Zhou	Scientist	Environmental Impact Assessment
Yongqi Lu* (Tsinghua University)	Scientist	Environmental Impact Assessment

Shandong Electric Power Group Corp., Shandong, China

Name	**Function in CETP**	**Tasks**
Zhenlin Xu	Advisory	Data Collection
Jianmin Hu	Advisory	Data Collection

Shandong Electric Power Group Corp., Shandong Electric Power Research Institute, Jinan, Shandong, China

Name	**Function in CETP**	**Tasks**
Qingbo Zhao*	Advisory	Data Collection
Yong Xu*	Scientist	Data Collection
Baoguo Shan	Scientist	Data Collection
Jingwei Shen	Scientist	Data Collection
Huijun Zhang	Scientist	Data Collection

University of Tokyo, Tokyo, Japan

Name	**Function in CETP**	**Tasks**
Kenji Yamaji*	Principal Investigator	Energy Transportation Modeling
Takeo Imanaka* (CRIEPI)	Scientist	Energy Transportation Modeling

Tsinghua University, Global Climate Change Institute, Beijing, China

Name	**Function in CETP**	**Tasks**
Zhihong Wei*	Scientist	Energy Economy Modeling
Wenying Chen*	Scientist	Energy Economy Modeling

Participants who left their participating institutions during the course of this work:

	Institution/organization
Alain Bill	ABB, Switzerland
Bingzhang Xue*	ABB, Switzerland
Jean Gao	ABB China Ltd., Beijing, China
Urs Gantner*	PSI, Switzerland
Gerard Spiekerman*	PSI, Switzerland
Jennifer Barker	MIT, USA
Christopher Hansen*	MIT, USA
Nuko A. Iiiadis	MIT, USA

APPENDIX D
(CHAPTER 1)

TASK DESCRIPTION

1. Project Management

A main element of CETP is the program management by industry. A steering committee guides the program. An overall program coordinator is responsible for overseeing the running of the program, maintaining links to the steering committee and promoting outreach. A program manager is responsible for daily operation and coordination of the program. ABB was responsible for this task

.

2. Data Collection

The data collection task was responsible for coordinating and organizing the collection of data for different tasks from various sources and partners in CETP. ABB was responsible for this task

.

3. Database Development

In order to organize and disseminate the data collected efficiently to the project participants, a database has been developed. Along with a database, a website at http://www.cetp.ch was also developed. ABB was responsible for this task.

4. Demand Forecasting

To determine the strategy for energy planning, it is necessary to be able to forecast energy and electricity demand. The Energy Research Institute (ERI) of the SDPC of China is responsible for this task. They used the MEDEE-S model, which is a long-term, bottom-up final energy demand simulation model for developing countries in this study. EPFL worked closely with ERI and is responsible for performing an electricity demand and load forecasting using a bottom-up approach and the DEMELEC-PRO software.

Energy and Electricity Modeling Tasks

Tasks 5, 6 and 7 are three modeling activities relating to energy and electricity modeling. The approach is to investigate the energy mix options based on different economic and environmental constraints using both optimization and simulation techniques

5. Energy/Economy Modeling

The Energy Economics group of the Paul Scherrer Institute (PSI) was responsible for this task and worked closely with the Global Climate Change Institute of Tsinghua University. The Market Allocation Model (MARKAL) which is a process-oriented engineering model that describes all energy transformations from primary sources to energy services was used to study the Shandong Province. PSI has also developed a multi-regional optimization model called the China Regional Electricity Trade Model (CRETM). The CRETM was used to examine a set of energy-economic-environmental driven scenarios in order to quantify related policy implications.

6. Electric Sector Simulation

This task involved bottom-up, engineering-based modeling of the Chinese electricity sector in Shandong. The task was the responsibility of the AGREA research group at the MIT Energy Laboratory. The comparative analysis of alternative strategies, under various uncertainties, is a requirement in the search for robust, long-term technological strategies and related policies. MIT has engaged the stakeholders' interaction in this task and used a simulation dispatch model (EGEAS) to simulate both technological and operational alternatives. ETHZ works closely with MIT for integration of this task with other analytic initiatives in CETP.

7. Energy Transportation Modeling

The Energy Transportation Model developed at the University of Tokyo was used to simulate the electricity sector for Shandong Province in China. It is an optimization model used to study the least-cost expansion planning of the power system in Shandong Province taking into account the fossil fuel transportation to the power plants and the electricity transmission networks. It has investigated the configuration of power plants and electricity generation mix, power plant sites, expansion of transmission lines, coal flows and SO_2 and CO_2 emission control measures.

Environmental, health and safety related Tasks

PSI was responsible for three environmental/health/safety related tasks (Tasks 8, 9 and 10). In carrying out these tasks, PSI worked closely with the Policy Research Center for Environment and Economy of the State Environmental Protection Administration of China

8. Life Cycle Assessment

The Life Cycle Assessment (LCA) methodology was used to develop suitably complete environmental inventories for major present and future energy systems for the generation of electricity in Shandong. LCA provides input for the decision-making on energy choices by performing the full scope evaluation of the amounts of pollutants emitted into the environment from the entire chains of energy systems, including fuel mining, processing, power generation, transmission lines and waste management.

9. Environmental Impact Assessment

Two modeling approaches were used in this task. Firstly, the RAINS-Asia model developed by IIASA was used in this study to help decision-makers analyze future trends in emissions, estimate regional impacts of resulting deposition levels, and to evaluate costs and effectiveness of alternative mitigation options. Secondly, applying the 'impact pathway approach', the environmental external costs associated with the various energy chains were estimated. The basic tool used in this analysis was the model EcoSense developed by the University of Stuttgart. It was adapted to the Chinese conditions.

10. Risk Assessment

The objective of this task was to provide a balanced perspective on the severe accident risks specific for China. The assessment addressed fossil energy sources such as coal, oil and gas, nuclear power and hydropower. In addition to the power production step of these energy chains, whenever applicable, exploration, extraction, transports, processing, storage and waste disposal were also considered. Apart from focusing on the historical data, existing Probabilistic Safety Assessment approach was also used, particularly for the nuclear chain.

11. Integration, Decision Support and DVD

ABB was responsible for integrating the results from different tasks and developed a coherent product for this study. Integration was a continuous process and conducted at various levels.

One of the goals of CETP was to develop a user-friendly decision support-aiding tool. A set of scenarios of interests was chosen from the electric sector simulation for the multi-criteria Decision Aiding (MCDA) analysis. EPFL has chosen ELECTRE III , an MCDA tool, for use in CETP since the objective of CETP was not to choose a 'best' solution but to compare different options from a sustainability standpoint.

In order to allow the stakeholders and users to fully utilize the results obtained from this study, an integrated software tool, DVD, has also been developed. The wide spectrum of targeted users was reflected in the modular design of the software, which should satisfy the needs of users having different backgrounds and interests. This task was the responsibility of PSI and was based on the contributions from all Principal Investigators and their collaborators.

12. Outreach and Technical Exchange

Stakeholder involvement in CETP is essential to ensure that the results are applicable and will be used. The outreach activities were conducted to solicit inputs from stakeholders as well as inform the scientific community and various stakeholders about the progress of the program.

As CETP is an international cooperative program, technical exchange between the program participants played an important role in its activities. Technical exchange of personnel and visits to each other's institutions were encouraged as it strengthened the cooperative spirit, communication and technology transfer.

APPENDIX E
(CHAPTER 1)

ABB IN CHINA

ABB is one of the world's largest technology companies. Since ABB was created from the 1988 merger of Sweden's Asea and Switzerland's Brown Boveri, it has become the world leader in electrical engineering. In the last two years, ABB has pulled out of the capital-intensive and low-margin transportation and large-scale power generation businesses.

Currently, ABB is made up of two customer divisions responsible for serving end-users (ABB Website [2002] www.abb.com): The Power Technologies division serves electric, gas and water utilities as well as industrial and commercial customers, with a broad range of products, systems and services for power transmission, distribution and automation. PT produces transformers, switchgear, breakers, capacitors, cables and other products and technologies for high- and medium-voltage applications. The Automation Technologies division serves customers in the automotive, chemicals, consumer, electronics, life sciences, manufacturing, marine, metals, minerals, paper, petroleum, turbo-charging and utility industries.

ABB's friendship with China dates back to the beginning of last century, when it supplied the country with a steam boiler in 1907. Responding to China's open policy, ABB China established its headquarter in Beijing in 1994. In 1997, China selected ABB as the major supplier to the Three Gorges Project. In 2002, ABB China had 5500 employees working in 22 branches and sales offices in China. In addition, it had 26 joint ventures and wholly owned companies. The annual revenue of ABB China is about $650 million, which accounts for about one-third of ABB's revenue in Asia.

ABB Corporate Research has been actively supporting research work in the areas of coal combustion, pollution control, and greenhouse gas emission reduction in top universities in China since the mid-1990s. In the spirit of technology transfer, ABB has established Joint Research Laboratories with Tsinghua University in Beijing and Tianjin University in Tianjin. The Greenhouse Gas Laboratory of the Energy and Global Change Department of ABB Corporate Research in Switzerland, which has trained about ten visiting Chinese scientists over the last 5 years, was transferred to the Chemistry Department at Tsinghua University and to the Chemical Engineering Department at Tianjin University. This is a pioneering step for ABB in transferring technology and research knowledge to developing countries.

In addition, ABB has supported conferences, seminars, and training courses at a number of Chinese Universities including Fudan University in Shanghai, Tsinghua University in Beijing, Tianjin University in Tianjin, Jiaotong University in Xian and

Zhejiang University in Hangzhou. In 1995, ABB scholarships were established to support the nation's development of young talent at a number of top universities in China.

APPENDIX F
(CHAPTER 4)

ADDITIONAL TABLES OF DATA

Table 1. Physical value balance sheet of several types of energy of Shandong province in 1995 and 1999

Items	Coal (Mt)		Oil (Mt)		Gas (Mm^3)	
	1995	1999	1995	1999	1995	1999
Total energy available for consumption	**111.19**	**108.94**	**13.44**	**15.76**	**1285**	**352**
Storage at beginning of year	2.72	16.87		0.56		
Output	88.27	89.95	30.06	26.65	1285	733
Import from other province	23.62	25.02	0.08	0.6		
Import from outside				2.57		
Export to other province (-)	-3.36	-4.88	-13.36	-12.37		-381
Export to outside (-)	-0.06		-2.7	-1.88		
Storage at end of year		-18.02	-0.64	-0.37		
Processing and transformation input (-) output (+)	**-71.02**	**-63.27**	**-12.94**	**-14.96**	**-44**	**-18**
Thermal power	-34.87	-36.82	-0.37	-0.27	-39	-18
Heat supply	-4.74	-3.89			-5	
Coal washing and dressing	-30.13	-21.9				
Coking	-1.03	-0.47				
Petroleum refining			-12.57	-14.69		
Gas production	-0.25	-0.19				
Losses in transportation	**0.96**	**1.02**	**0.02**	**0.069**		
Total consumption	**39.16**	**44.61**	**0.52**	**0.73**	**1241**	**334**
Primary sector	1.77	5.03				
Secondary sector	29.45	33.92	0.51	0.72	641	88
- Industry	29.1	32.16	0.49	0.69	641	88
- Construction	0.35	1.76	0.02	0.03		
Tertiary sector	4.57	2.63	0.01	0.01		

- Transport, telecommunication, etc.	1.14	0.98	0.005	0.0083		
- Commerce	1.73	1.07	0.005	0.0017		
- Others	1.7	0.58				
Household sector	3.37	3.03			600	246
- Urban	1.99	2.27			600	246
- Rural	1.38	0.76				
Balance	**0.05**	**0.04**	**-0.04**	**0.001**	**0**	**0**

Source: Shandong Statistics Bureau

Table 2. Electricity balance sheet

Unit: TWh

Item	1985	1990	1995	1996	1997
Total energy available for consumption	**26.29**	**44.99**	**74.04**	**79.63**	**85.53**
Output	26.15	44.67	73.62	79.22	85.34
Thermal power	26.15	44.67	73.24	78.77	84.88
Heat supply			0.38	0.45	0.46
Import	0.15	0.32	0.42	0.43	0.2
Export (-)	0.014	0.004		0.02	0.011
Total consumption	**24.67**	**44.76**	**74.1**	**79.68**	**85.46**
Primary sector	2.22	3.45	4.28	4.62	4.91
Secondary sector	20.29	35.5	57.74	61.46	64.08
Industry	19.89	35.33	57.13	60.8	63.41
Construction	0.4	0.18	0.61	0.66	0.67
Tertiary sector	0.19	2.06	4.13	4.66	6.44
Transport, telecom, etc.	0.083	0.51	0.76	0.85	2.44
Commerce	0.018	0.53	1.19	1.31	1.5
Others	0.089	1.02	2.18	2.5	2.5
Household sector	1.97	3.75	7.95	8.94	10.03
Urban	1.1	2.1	3.62	4.15	4.62
Rural	0.87	1.65	4.33	4.79	5.41
Balance	**1.62**	**0.23**	**-0.06**	**-0.05**	**0.07**

Source: Shandong Statistics Bureau

Note: There are some differences in the electricity statistics derived from the data sources of the Shandong Statistics Bureau and SEPCO. In the statistics of the Shandong Statistics Bureau, losses of electricity in transmission and distribution are included in the figures for consumption.

Table 3. Electricity consumption, growth rate and structure

	Total	**Primary**	**Industry**	**Constructi on**	**Service**	**Trans, etc**	**House-hold**
Electricity consumption in 1995,TWh	74.076	4.281	57.096	0.612	3.376	0.757	7.954
Growth rate in 1985-1995, % p.a.	11.4	5.6	11.1	8.9	15.5	5	16.2
Electricity consumption in 2000,TWh	94.586	5.392	69.054	0.753	5.95	1.182	12.255
Growth rate in 1995-2000, % p.a.	4.8	3.66	3.76	4.07	11.95	8.07	8.61
Electricity structure in 1985, %	100	9.7	78	0.6	3.2	1.7	6.8
Electricity structure in 1995, %	100	5.78	77.08	0.82	4.56	1.02	10.74
Electricity structure in 2000, %	100	5.7	73	0.8	6.29	1.25	12.96

Table 4. Comparative Statistics among China, Shanghai and Shandong

In year 1995	**China**	**Shanghai**	**Shandong**
GDP (Billion Yuan)	5847.81	246.26	500.2
GDP per capita (Yuan/person by end of year)	4828	18928	5746
Economic structure (%):			
Primary sector	20.51	2.5	20.2
Secondary sector	48.8	57.3	47.4
Tertiary sector	30.69	40.2	32.4
Population (Million by end of year)	1211.21	13.01	87.05
Urban	351.74 (29%)	9.57 (73.6%)	21.7 (24.9%)
Rural	859.47(71%)	3.44 (26.4%)	65.35 (75.1%)
Electricity consumption (TWh)	1002.34	37.92	74.1
Electricity structure (%):			
Primary sector	5.81	3.14	5.78
Secondary sector	78.01	76.48	77.92
Tertiary sector	6.15	12.39	5.58
Household	10.03	7.99	10.72
Electricity growth (% p.a.) in 1986-1990:			
Total	8.64		11.1
Primary sector			8
Secondary sector			10.7
Tertiary sector			11.3
Household			18.4
Electricity growth in 1991-1995 (% p.a.):			
Total	10.16	8.33	12.3
Primary sector		10.6	2.3
Secondary sector		5.9	12.5
Tertiary sector		21.4	16.1
Household		17.8	16
Electricity growth in 1996 comparing to 1995 (%):			
Total	7.3	6.88	7.5
Primary sector	5.1	7.56	7.8
Secondary sector	5.78	3.62	6.4
Tertiary sector	11.3	21.49	12.7
Household	12.7	14.85	12.4
Electricity growth in 1997 comparing to 1996 (%):			
Total	4.4	6.14	5.4
Primary sector	5.6	-29	6.4
Secondary sector	2.9	5.52	4.3
Tertiary sector	10.5	15.76	6
Household	10.6	8.9	12.1
Electricity per capita (kWh/person by end of year)	827.6	2914	851.3
Electricity intensity (kWh/Yuan)			
Primary sector	0.049	0.193	0.042

Secondary sector	0.274	0.206	0.243
Tertiary sector	0.034	0.047	0.026

Table 5. Assumptions for households

	1995-2000	**2000-2005**	**2005-2010**	**2010-2015**	**2015-2020**	**1985-1995**
Population growth (% p.a.)	0.8	0.68	0.58	0.42	0.21	1.36
Year	2000	2005	2010	2015	2020	1995
Urbanization rate (% at end of period)	26	35	42	47	50	25
Family size (persons/household at end of period)						
Urban	3.1	3.06	3.04	3.03	3.02	3.19
Rural	3.9	3.8	3.65	3.4	3.23	4.07
	1995-2000	2000-2005	2005-2010	2010-2015	2015-2020	1985-1995
Growth of average electricity per household (% p.a.)						
Urban	7	5	3	2	2	3.79
Rural	6	5.5	5	4	4	16.41

Table 6. Assumptions for production sectors

	1995-2000	**2000-2005**	**2005-2010**	**2010-2015**	**2015-2020**	**1985-1995**
Economic growth (Baseline-E scenario), (% p.a.)	10	9	8	7.5	6.5	11.7
Lower economic growth (Lowgdp-E scenario), (% p.a.)	9	8	7	6.5	5.5	
	2000	2005	2010	2015	2020	1995
Economic structure (%) (Baseline-E scenario)						
Agriculture	16.5	14	12.4	11.5	11	20.2
Industry	43.8	44.2	43.6	42.8	41.6	42.2
Construction	3.7	2.8	2.2	1.7	1.4	5.2
Service	29.5	32	34	35.5	37	26.4
Transport, etc.	6.5	7	7.8	8.5	9	6
Changed economic structure (%) (Structure-E scenario)						
Agriculture	16.5	14	12.4	11.5	11	
Industry	42.8	43.2	42.6	41.8	40.6	
Construction	3.7	2.8	2.2	1.7	1.4	
Service	30.5	33	35	36.5	38	
Transport, etc.	6.5	7	7.8	8.5	9	
Quicker decrease of intensity year) (% p.a.) (Baseline-E scenario)						
Agriculture	0.2	0.2	0.1	-1	-1	0.9
Construction	-1	-1	-1	-1	-1	-1.6
Service	1	1	0.5	-1	-1	2.2
Transport, etc.	-5	-4	-4	-3	-3	-4.8
Slower decrease of intensity (% p.a.) (Inefficient-E scenario)						
Agriculture	0.7	0.7	0.6	-0.5	-0.5	
Construction	-0.5	-0.5	-0.5	-0.5	-0.5	
Service	1.5	1.5	1	-0.5	-0.5	
Transport, etc.	-4	-3	-3	-2	-2	

Table 7. Assumptions for intensity change in industry subsectors

	1995-2000	2000-2005	2005-2010	2010-2015	2015-2020
Quicker decrease of intensity (% p.a.) (Baseline-E scenario)					
Nonferrous	-5	-4.5	-3	-2	-1
Textile	-4.5	-4	-3	-2	-1
Chemical	-4.5	-4	-3	-2	-1
Energy	-3.5	-2.5	-2	-1	-1
Others	-5	-4.5	-3	-2	-1
Slower decrease of intensity (% p.a.)(Inefficient-E scenario)					
Nonferrous					
Textile	-4	-3.5	-2	-1	0
Chemical	-3.5	-3	-2	-1	0
Energy	-3.5	-3	-2	-1	0
Others	-2.5	-1.5	-1	-0.5	0
	-4	-3.5	-2	-1	0

Table 8. Assumptions for intensity change for EIPs in industry sector

	1995-2000	2000-2005	2005-2010	2010-2015	2015-2020
Quicker decrease of intensity (% p.a.) (Baseline-E scenario)	-5	-4.5	-3	-2	-1
Slower decrease of intensity (% p.a.)(Inefficient-E scenario)	-4	-3.5	-2	-1	0

Table 9. Energy demand forecast by sector and by fuel in the baseline scenario

Unit: ktoe

Industry						
	1995	2000	2005	2010	2015	2020
Fossil Fuels	26050.89	25435.39	27660.44	30187.5	34449.38	37762.48
Coal	16636.96	16017.58	17380.73	18992.49	21245.4	23122.72
Charcoal	2443.05	2775.23	3065.91	3453.26	4295.58	4685.55
Fuel Oil	4593.55	3922.63	3724.05	3580.83	3676.17	3771.67
LPG	266.09	347.72	522.83	689.78	894.84	1050.78
Gas	2111.24	2372.24	2966.92	3471.15	4337.38	5131.76
Motor Fuels	705.69	902.8	1197.26	1466.93	1744.41	1944.11
Electricity	4909.33	6379.00	8823.99	11105.71	13939.31	16427.56
TOTAL	31665.91	32717.19	37681.69	42760.14	50133.10	56134.15
Subsector of Industry	1995	2000	2005	2010	2015	2020
Fossil Fuels	18824.89	18039.06	19681.15	21575.21	24453.98	27145.15
Coal	12285.12	11760.44	12834.41	14091.42	15917.31	17640.47
Charcoal	330.92	556.52	832.97	1085.7	1378.09	1599.58
Fuel Oil	4082.88	3386.27	3125.94	2924.58	2906.8	2921.76
LPG	244.38	313.4	488.65	651.01	841.05	982.76
Gas	1881.59	2022.43	2399.19	2822.51	3410.72	4000.58
Motor Fuels	705.69	902.8	1197.26	1466.93	1744.41	1944.11
Electricity	4069.02	5443.01	7715.58	9726.97	12313.46	14489.43
TOTAL	23599.60	24384.87	28593.99	32769.11	38511.85	43578.69
TEXTILE	1995	2000	2005	2010	2015	2020
Motor Fuels	97.57	45.74	61.46	71.04	82.87	90.12
Electricity	367.03	321.77	423.87	496.56	588.65	726.54
Fossil Fuels	1586.56	687.01	691.08	781.46	902.88	1029.09
Coal	1499.66	507.45	502.31	563.40	657.71	675.74
Charcoal	1.67	2.28	2.72	3.39	4.08	4.25
Fuel Oil	8.57	39.20	42.05	50.83	56.04	165.36
LPG	76.66	138.08	144.00	163.84	185.05	183.74
TOTAL	2051.16	1054.52	1176.41	1349.05	1574.4	1845.76
CHEMICAL	1995	2000	2005	2010	2015	2020
Motor Fuels	55.85	67.24	72.88	81.14	87.94	89.12
Electricity	1084.94	1258.22	1569.06	1783.79	1970.84	2118.36
Fossil Fuels	10311.03	7892.22	6661.59	5791.11	5212.71	4695.46
Coal	5632.83	4545.80	3677.42	2984.44	2186.53	1538.17

Charcoal	153.87	107.95	108.64	115.73	122.59	128.52
Residual Fuel Oil	3256.35	2436.50	2137.42	1828.12	1736.06	1725.61
LPG	34.05	0.00	0.00	0.00	0.00	0.00
Gas	1233.93	801.98	738.10	862.82	1167.54	1303.17
TOTAL	11451.81	9217.68	8303.53	7656.04	7271.5	6902.93
ENERGY	1995	2000	2005	2010	2015	2020
Motor Fuels	417.14	632.57	880.04	1110.37	1351.83	1531.24
Electricity	1701.37	2694.27	4213.63	5646.76	7535.36	9023.07
Fossil Fuels	2317.12	3572.45	5120.65	6888.9	9064.2	11273.82
Coal	1328.46	1863.96	3099.33	4561.58	6473.56	8328.98
Charcoal	4.04	6.44	8.26	9.70	10.85	10.99
Fuel Oil	426.70	575.49	621.00	690.45	720.93	660.87
LPG	15.26	21.56	31.06	41.45	54.10	66.06
Gas	542.66	1105.00	1361.01	1585.72	1804.77	2206.93
TOTAL	4435.63	6899.29	10214.32	13646.03	17951.39	21828.13
OTHERS	1995	2000	2005	2010	2015	2020
Motor Fuels	113.97	144.27	169.85	191.23	208.6	220.29
Electricity	685.41	972.7	1305.26	1588.84	1984.71	2364.08
Fossil Fuels	3952.3	5225.07	6558.27	7458.55	8628.18	9505.25
Coal	3366.06	4276.76	5001.09	5467.52	6102.96	6633.03
Charcoal	132.84	387.59	661.35	904.10	1196.66	1411.58
Residual Fuel Oil	230.00	291.51	282.15	311.21	349.80	325.68
LPG	118.40	153.76	313.59	445.71	601.89	732.96
Gas	105.00	115.45	300.08	330.00	376.85	402.00
TOTAL	4751.68	6342.04	8033.37	9238.62	10821.48	12089.62
EIPs of Industry						
	1995	2000	2005	2010	2015	2020
Fossil Fuels	7226	7396.33	7979.28	8612.29	9995.4	10617.34
Coal	4351.84	4257.14	4546.32	4901.07	5328.09	5482.26
Charcoal	2112.14	2218.7	2232.94	2367.56	2917.49	3085.97
Fuel Oil	510.67	536.36	598.12	656.25	769.37	849.91
LPG	21.71	34.32	34.17	38.77	53.79	68.02
Gas	229.65	349.81	567.73	648.64	926.65	1131.19
Electricity	840.31	935.99	1108.41	1378.74	1625.85	1938.13
TOTAL	8066.31	8332.32	9087.69	9991.03	11621.25	12555.47
CEMENT	1995	2000	2005	2010	2015	2020
Electricity	144	175.5	193.79	224.61	249.71	282.31
Fossil Fuels	2645.85	2743.07	2865.77	2980.53	3096.46	3226.76

Coal	2352.5	2414.63	2498.32	2584.67	2645.25	2717.6
Fuel Oil	219.28	229.78	242.5	253.59	267.46	282.63
LPG	21.34	32.89	31.93	36.3	50.53	64.72
Gas	52.73	65.78	93.02	105.99	133.22	161.81
Total	2789.85	2918.58	3059.57	3205.14	3346.17	3509.07
GLASS	1995	2000	2005	2010	2015	2020
Electricity	229	256.9	294.2	325	357.9	388.7
Fossil Fuels	34.2	41.79	49.29	52.92	56.79	63.23
Coal	30.7	36.21	41.24	42.94	44.56	48.26
Fuel Oil	2.81	3.34	3.81	4.06	4.32	4.73
LPG	0.37	1.43	2.24	2.48	3.26	3.3
Gas	0.32	0.82	1.99	3.44	4.66	6.94
Total	263.2	298.69	343.49	377.92	414.7	451.93
PAPER	1995	2000	2005	2010	2015	2020
Electricity	164.15	193.5	276.02	441.35	587.75	750.54
Fossil Fuels	662.05	740.77	1020.96	1291.36	1685.4	2121.12
Coal	587.67	645.46	870.04	1092.5	1410.5	1756.81
Fuel Oil	74.38	82.15	113.1	144.08	183.36	235.1
Gas	0	13.17	37.82	54.77	91.54	129.22
Total	826.2	934.27	1296.98	1732.7	2273.15	2871.66
STEEL	1995	2000	2005	2010	2015	2020
Electricity	303.17	310.09	344.39	387.79	430.49	516.59
Fossil Fuels	3883.9	3870.69	4043.27	4287.48	5156.74	5206.23
Coal	1380.96	1160.85	1136.72	1180.95	1227.79	959.59
Charcoal	2112.14	2218.7	2232.94	2367.56	2917.49	3085.97
Fuel Oil	214.2	221.1	238.71	254.52	314.23	327.45
Gas	176.6	270.05	434.9	484.44	697.23	833.22
Total	4187.06	4180.78	4387.66	4675.27	5587.23	5722.81
Household Sector						
	1995	2000	2005	2010	2015	2020
Electricity	684.56	1051.13	1523.44	2099.51	2678.5	3097.16
Fossil Fuels	2518.37	2946.35	3447.52	3959.27	4445.73	4874.6
Coal	1688.22	2073.22	2482.78	2877.84	3193.51	3443.35
Kerosene	162.54	128.59	102.41	80.32	63.69	48.28
Gas	510.24	554.73	634.71	741.81	906.38	1080.83
LPG	157.37	189.81	227.61	259.3	282.15	302.14
TOTAL	3202.93	3997.48	4970.96	6058.78	7124.23	7971.76
CONSUMPTION BY USAGE						

Cooking	2510.39	2952.74	3465.54	3985.46	4483.71	4926.23
Lighting	114.54	164.96	222.12	281.48	327.02	359.75
Elect. Appliances	571.12	855.76	1217.27	1662.84	2095.57	2364.46
Air Conditioning	6.88	24.02	66.03	129	217.93	321.3
TOTAL	3202.93	3997.48	4970.96	6058.78	7124.23	7971.76
Service Sector						
	1995	2000	2005	2010	2015	2020
Electricity	290.53	412.42	556.82	722.32	914.29	1131.97
Fossil Fuels	2798.48	3052.32	3363.5	3729.21	4144.2	4601.54
Coal	1902.37	1837.83	1787.11	1747.53	1708.53	1678.11
Diesel	887.14	1201.38	1561.09	1961.68	2410.87	2893.28
Fuel Oil	8.96	11.26	13.14	15.06	16.62	18.2
Gas	0	1.85	2.16	4.94	8.18	11.95
TOTAL	3089.01	3464.74	3920.32	4451.54	5058.49	5733.5
Transport Sector						
	1995	2000	2005	2010	2015	2020
Petrol	1986.16	2454.75	3347.14	3964.38	4716.83	5411.7
Diesel	1128.49	1261.22	1638.31	2048.54	2501.78	3115.38
Electricity	6.34	7.26	9.14	11.49	15.77	21.35
Coal	76.36	65.74	39.32	0	0	0
Jet Fuel	67.8	86.83	106.34	127.39	151.93	178.96
TOTAL	3265.15	3875.79	5140.25	6151.8	7386.32	8727.39
TRANSPORT SECTOR						
	1995	2000	2005	2010	2015	2020
PASSENGER TRANSPORT						
DOMESTIC AIR TRANSPORT Jet Fuel	67.8	86.83	106.34	127.39	151.93	178.96
PUBLIC TRANSPORT	667.56	1019.51	1493.76	1871.23	2201.44	2473.55
ROAD,BUSES	570.62	912.55	1368.73	1712.04	2051.11	2334.66
Petrol	515.24	819.73	1223.12	1513.9	1765.65	1943.76
Diesel	55.38	92.82	145.61	198.14	285.46	390.9
RAIL	89.58	96.35	110.57	143.27	133.04	120.4
Coal	15.24	14.38	8.88	0	0	0
Diesel	74.34	81.96	101.46	142.63	131.71	118.36
Electricity	0	0.01	0.23	0.64	1.33	2.05
WATER (Diesel)	7.36	10.61	14.47	15.92	17.29	18.49
PRIVATE CARS TRANSPORT	82.94	226.58	413.27	591.94	801.16	1074.19
Petrol	77.59	204.24	350.4	455.13	572.51	738.5
Diesel	5.35	22.34	62.86	136.82	228.65	335.69

TOTAL	818.3	1332.92	2013.37	2590.57	3154.53	3726.7
Freight Transport						
	1995	2000	2005	2010	2015	2020
ROAD	1818.82	1915.6	2437.31	2816.64	3452.02	4198.04
Petrol	1393.33	1430.78	1773.62	1995.35	2378.68	2729.45
Diesel	425.5	484.82	663.69	821.29	1073.35	1468.59
RAIL	359.29	344.06	378.91	406.93	423.44	427.1
Coal	61.12	51.36	30.44	0	0	0
Diesel	298.16	292.68	347.7	405.12	419.21	419.84
Electricity	0.01	0.03	0.78	1.82	4.23	7.26
WATERWAYS (Diesel)	262.4	275.99	302.53	328.62	346.11	363.51
PIPE LINE (Electricity)	6.33	7.23	8.13	9.03	10.22	12.04
TOTAL	2446.84	2542.87	3126.88	3561.23	4231.79	5000.69
Agricultural Sector						
	1995	2000	2005	2010	2015	2020
MOTOR FUELS	919.53	1049.98	1190.72	1385.19	1540.56	1649.74
FOSSIL FUELS	887.19	1476.14	2238.96	2346.78	2415.18	2416.41
ELECTRICITY	369.22	489.94	734.78	1000.99	1239.1	1504.71
TOTAL	2175.94	3016.06	4164.46	4732.96	5194.84	5570.86
Construction Sector						
	1995	2000	2005	2010	2015	2020
MOTOR FUELS	101.45	136.13	182.06	240.34	290.38	364.05
ELECTRICITY (TWH)	0.6	0.71	0.86	1.02	1.16	1.46
TOTAL	152.9	197.27	255.59	327.84	390.26	489.46

Table 10. Forecast of electricity demand in the different scenarios

	Electricity (GWh)				
Scenarios	***2000***	***2005***	***2010***	***2015***	***2020***
Productive sectors					
Baseline-E					
Total	86677.53	110545.90	140102.00	178703.93	226101.39
Agriculture	*5639.08*	*7435.73*	*9725.38*	*12314.05*	*15346.87*
Industry	*72850.71*	*89596.05*	*109488.38*	*137070.01*	*171462.53*
Construction	*680.34*	*753.34*	*827.08*	*872.56*	*936.26*
Service	*6494.40*	*11392.16*	*18234.09*	*25992.73*	*35297.94*
Transport, etc.	*1013.00*	*1368.62*	*1827.06*	*2454.59*	*3057.80*
Lowgdp-E					
Total	83049.61	101062.05	122206.22	148719.32	179454.25
Agriculture	*5387.38*	*6783.89*	*8469.58*	*10234.39*	*12167.33*
Industry	*69839.96*	*81948.73*	*95545.61*	*114116.73*	*136135.39*
Construction	*649.97*	*687.30*	*720.29*	*725.20*	*742.29*
Service	*6204.52*	*10393.49*	*15879.60*	*21602.95*	*27984.96*
Transport, etc.	*967.78*	*1248.64*	*1591.14*	*2040.04*	*2424.29*
Structure-E					
Total	85466.01	109104.18	138375.00	176515.50	223260.19
Agriculture	*5639.08*	*7435.73*	*9725.38*	*12314.05*	*15346.87*
Industry	*71419.05*	*87798.33*	*107225.09*	*134149.38*	*167667.33*
Construction	*680.34*	*753.34*	*827.08*	*872.56*	*936.26*
Service	*6714.55*	*11748.17*	*18770.39*	*26724.92*	*36251.94*
Transport, etc.	1013.00	1368.62	1827.06	2454.59	3057.80
Inefficient-E					
Total	**90939.90**	**121464.79**	**160940.16**	**212989.50**	**281802.34**
Agriculture	5781.19	7815.22	10479.57	13607.47	17391.44
Industry	76736.82	99369.57	127794.64	166688.30	219408.17
Construction	697.69	792.26	892.01	965.05	1061.93
Service	6656.75	11968.86	19638.46	28708.77	39980.80
Transport, etc.	1067.45	1518.87	2135.48	3019.90	3960.00
Household	*2000*	*2005*	*2010*	*2015*	*2020*
Total	11883.54	17784.41	23961.61	29897.11	35942.01
Urban	5673.89	10215.94	14724.88	18639.45	22197.15
Rural	6209.65	7568.47	9236.73	11257.65	13744.86
Total	*2000*	*2005*	*2010*	*2015*	*2020*
Baseline-E	98561.07	128330.31	164063.60	208601.04	262043.40
Lowgdp-E	94933.15	118846.46	146167.83	178616.42	215396.26
Structure-E	97349.55	126888.59	162336.61	206412.61	259202.20
Inefficient-E	102823.44	139249.20	184901.77	242886.61	317744.35

APPENDIX G
(CHAPTER 7)

MATHEMATICAL DESCRIPTION OF THE MODEL

As mentioned above, ETM is a linear-programming (LP) model minimizing the total system cost, which is sum of generation, power transmission and fuel transportation cost. In this section, variables in the LP are written in lower case.

Constraints

1. Total capacity of generating plants

Total capacity of generating plants in the power system must cover the peak load in the system taking capacity reserve rate into account.

$$(1+CRR)PeakL_t \leq \sum_{n=1}^{N}\sum_{bg=1}^{BG} BGC_{n,bg,t} + \sum_{n=1}^{N}\sum_{cg=1}^{CG}\sum_{t'=1}^{t} cigc_{n,cg,t'} + \sum_{n=1}^{N}\sum_{ng=1}^{NG}\sum_{t'=1}^{t} nigc_{n,ng,t'} \quad (1)$$

where,

n (=1,...,N) : indices of node,

t (=1,...,T) : indices of period

c (=1,...,C) : indices of coal production site

bg (=1,...,BG) : indices of the coal-fired plant installed before the planning period (existing coal plant) at each node.

cg (=1,...,CG) : indices of coal-fired plant type installed in the planning periods (new coal plant)

ng (=1,...,NG): indices of plant type (other than coal-fired plant) installed in the planning periods (non-coal plant)

$BGC_{n,bg,t}$: capacity of existing coal plant.

$cgic_{n,cg,t}$: capacity of new coal plant installed in period t

$ngic_{n,ng,t}$: capacity of the non-coal plant installed in period t

$PeakL_t$: peak load (including loss on T&D)

CRR : capacity reserve rate

2. Existing plants constraints

2.1 Existing plants supplemented with FGD

All the existing plants are assumed to be the plants without FGD, and can be the candidates for installing FGD. The closing of the existing plant have to be considered for the FGD installation capacity. It's not applied only for existing plant but also for all the coal-fired plant, coal from any source can be used, changed, and mixed. And electrical energy output for instantaneous peak is set to be negligible in this model.

$$0 \le mbgc_{n,bg,t} \le BGC_{n,bg,t} \tag{2}$$

$$\begin{aligned} bgcm_{n,bg,t} &\ge mbgc_{n,bg,t} && for\ t = 1 \\ bgcm_{n,bg,t} &\ge mbgc_{n,bg,t} - mbgc_{n,bg,t-1} && for\ t = 2, \ldots, T \end{aligned} \tag{3}$$

$$\sum_{c=1}^{C} mbgo_{n,bg,c,t,h} \le mbgc_{n,bg,t} \tag{4}$$

$$\sum_{c=1}^{C}\sum_{h=1}^{H} \tau \cdot mbgo_{n,bg,c,t,h} \le CF \sum_{h=1}^{H} \tau \cdot mbgc_{n,bg,t} \tag{5}$$

where,
$bgcm_{n,bg,t}$: exisiting plant capacity modified (supplemented with FGD) in period t.
$mgbc_{n,bg,t}$: modified existing plant capacity.
h (=0,...,H) : indices of time zone (h=0: instantaneous peak, 1: peak, 2: middle, 3: off-peak)
$mbgo_{n,bg,c,t,h}$: output of modified existing plant using coal c
τ : width of time zones
CF : maximum capacity factor.

2.2 The other existing plants

$$\sum_{c=1}^{C} bgo_{n,bg,c,t,h} \le BGC_{n,bg,t} - mbgc_{n,bg,t} \tag{6}$$

$$\sum_{c=1}^{C}\sum_{h=1}^{H} \tau \cdot bgo_{n,bg,c,t,h} \le CF \sum_{h=1}^{H} \tau \cdot (BGC_{n,bg,t} - mbgc_{n,bg,t}) \tag{7}$$

where,
$bgo_{n,bg,c,t,h}$: output of existing plant (not modified) using coal c

3. New plants constraints

3.1 Pulverized Coal (PC) plants supplemented with FGD

PC plants, which constructed with FGD are distinguished in the model description by cg. FGD installation to PC plants w/o FGD is allowed one period after the installation of the plant.

$$\begin{aligned} cgcm_{n,t} &= 0 && for\ t = 1 \\ cgcm_{n,t} &\le \sum_{t'}^{t-1} cgic_{n,cg=[PC\cdot w/o\cdot FGD],t'} - \sum_{t'}^{t-1} cgcm_{n,t} && for\ t = 2, \ldots T \end{aligned} \tag{8}$$

$$\sum_{c=1}^{C} mcgo_{n,c,t,h} \le \sum_{t'=1}^{t} cgcm_{n,t'} \tag{9}$$

$$\sum_{c=1}^{C}\sum_{h=1}^{H}\tau\cdot mcgo_{n,c,t,h}\leq CF\sum_{h=1}^{H}\tau\cdot\sum_{t'=1}^{t}cgcm_{n,t'} \tag{10}$$

where,
$cgcm_{n,t}$: PC plant capacity modified (supplemented with FGD) in period t.
$mcgo_{n,c,t,h}$: output of modified PC plant using coal c

3.2 The other PC plants

for cg = PC w/o FGD,

$$\sum_{c=1}^{C}cgo_{n,cg,c,t,h}\leq\sum_{t'=1}^{t}cgic_{n,cg,t'}-\sum_{t'=1}^{t}cgcm_{n,t'} \tag{11}$$

$$\sum_{c=1}^{C}\sum_{h=1}^{H}\tau\cdot cgo_{n,cg,c,t,h}\leq CF\sum_{h=1}^{H}\tau\cdot\left(\sum_{t'=1}^{t}cgic_{n,cg,t'}-\sum_{t'=1}^{t}cgcm_{n,t'}\right) \tag{12}$$

where,
$cgo_{n,cg,c,t,h}$: output of PC plant using coal c

3.3 The other coal-fired plant

for cg≠ PC w/o FGD

$$\sum_{c=1}^{C}cgo_{n,cg,c,t,h}\leq\sum_{t'=1}^{t}cgic_{n,cg,t'} \tag{13}$$

$$\sum_{c=1}^{C}\sum_{h=1}^{H}\tau\cdot cgo_{n,cg,c,t,h}\leq CF\sum_{h=1}^{H}\tau\cdot\sum_{t'=1}^{t}cgic_{n,cg,t'} \tag{14}$$

3.4 Longkou coal limitation
As stated above, Longkou mine-mouth plant in Yantai can use the Longkou coal without rail or ship transportation, but the production amount from the Longkou mine is limited.

for c=Longkou, n=Yantai

$$\sum_{bg=1}^{BG}\sum_{h=1}^{H}\tau\cdot bgo_{n,bg,c,t,h}+\sum_{bg=1}^{BG}\sum_{h=1}^{H}\tau\cdot mbgo_{n,bg,c,t,h}+\sum_{cg=1}^{CG}\sum_{h=1}^{H}\tau\cdot cgo_{n,cg,c,t,h}+\sum_{h=1}^{H}\tau\cdot mcgo_{n,c,t,h}\leq LongkouMaxGWh \tag{15}$$

3.5 The other plants

$$ngo_{n,ng,t,h} \leq \sum_{t'=1}^{t} ngic_{n,ng,t'} \quad (16)$$

$$\sum_{h=1}^{H} \tau \cdot ngo_{n,ng,t,h} \leq CF \sum_{h=1}^{H} \tau \cdot \sum_{t'=1}^{t} ngic_{n,t'} \quad (17)$$

where,
$ngo_{n,cg,c,t,h}$: output of non-coal plants

3.6 Plants' construction constraints

$$ngic_{n,ng,t} = 0 \quad for\ t < Available\ period_{ng} \quad (18)$$

$$cgic_{n,cg,t} \geq LLcgic_{n,cg,t} \quad (19)$$

$Available\ period_{ng}$: the period generating technology *ng* become available.
$LLcgic_{n,cg,t}$: plants' installation lower bound (e.g. for the plants under construction)

4. Electric power transmission

4.1 Power flow equation

This is an equation by DC method, assuming no loss on the modeled transmission network. The left side of the equation is the net output from the node, and the right side is the total power flow to other nodes. This equation also represents the supply-demand balance in the power system.

$$\sum_{bg=1}^{BG} \sum_{c=1}^{C} bgo_{n,bg,c,t,h} + \sum_{bg=1}^{BG} \sum_{c=1}^{C} mbgo_{n,bg,c,t,h} + \sum_{cg=1}^{CG} \sum_{c=1}^{C} cgo_{n,cg,c,t,h} + \sum_{c=1}^{C} mcgo_{n,c,t,h}$$
$$+ \sum_{ng=1}^{NG} ngo_{n,ng,t,h} - L_{n,t,h} = \sum_{\substack{m=1 \\ m \neq n}}^{N} \frac{\delta_{n,t,h} - \delta_{m,t,h}}{X_{n,m}} \quad (20)$$

$L_{n,t,h}$: electricity load at node,
□$_{n,t,h}$: phase angle at node,

$X_{n,m}$: reactance of transmission line between node n and m (if there no transmission line (branch) between n and m, $1/X_{n,m} = 0$)

4.2 Capacity constraints of transmission lines

$$\left| \frac{\delta_{n,t,h} - \delta_{m,t,h}}{X_{n,m}} \right| \leq \gamma \cdot \left(BBC_b + \sum_{t'=1}^{t} ibc_{b,t'} \right) \tag{21}$$

b : indices of branch correspondence to a pair of nodes (e.g. n and m)
γ : maximum operation rate of transmission lines,
BBC_b : capacity of transmission line installed before the planning period,
$ibc_{b,t}$: transmission capacity installed in the planning periods.

5 Emission control

(1) Local SO_x Emission Control

$$nsulfur_{n,t} \leq nSEL_n \tag{22}$$

where,
$nsulfur_{n,t}$: SO_x emission at nodes (Mt-S)

$$= \sum_{bg=1}^{BG} \sum_{c=1}^{C} \sum_{h=1}^{H} \tau \cdot bSE_{n,bg,c} \cdot bgo_{n,bg,c,t,h} + \sum_{bg=1}^{BG} \sum_{c=1}^{C} \sum_{h=1}^{H} \tau \cdot mbSE_{n,bg,c} \cdot mbgo_{n,bg,c,t,h}$$
$$+ \sum_{cg=1}^{CG} \sum_{c=1}^{C} \sum_{h=1}^{H} \tau \cdot cSE_{cg,c} \cdot cgo_{n,cg,c,t,h} + \sum_{c=1}^{C} \sum_{h=1}^{H} \tau \cdot mcSE_c \cdot mcgo_{n,c,t,h}$$
$$+ \sum_{ng=1}^{NG} \sum_{h=1}^{H} \tau \cdot nSE_{ng} \cdot ngo_{n,ng,t,h} \tag{23}$$

$bSE_{n,bg,c}$, $mbSE_{n,bg,c}$, $cSE_{cg,c}$, $mcSE_c$, nSE_{ng} : SO_x emission coefficient (S-t/kWh)
$nSEL_n$: local (nodal) SO_x emission control level

(2) provincial Total SO_x Emission Control

$$\sum_{n=1}^{N} nsulfur_{n,t} \leq SEL \tag{24}$$

where,
SEL : provincial SO_x emission control level

(3) CO_2 Emission Control

$$\sum_{n=1}^{N}\sum_{bg=1}^{BG}\sum_{c=1}^{C}\sum_{h=1}^{H}\tau \cdot bCE_{n,bg,c} \cdot bgo_{n,bg,c,t,h} + \sum_{n=1}^{N}\sum_{bg=1}^{BG}\sum_{c=1}^{C}\sum_{h=1}^{H}\tau \cdot mbCE_{n,bg,c} \cdot mbgo_{n,bg,c,t,h}$$
$$+ \sum_{n=1}^{N}\sum_{cg=1}^{CG}\sum_{c=1}^{C}\sum_{h=1}^{H}\tau \cdot cCE_{cg,c} \cdot cgo_{n,cg,c,t,h} + \sum_{n=1}^{N}\sum_{c=1}^{C}\sum_{h=1}^{H}\tau \cdot mcCE_{c} \cdot mcgo_{n,c,t,h}$$
$$+ \sum_{n=1}^{N}\sum_{ng=1}^{NG}\sum_{h=1}^{H}\tau \cdot nCE_{ng} \cdot ngo_{n,ng,t,h} \leq CEL \tag{25}$$

where,
$bCE_{n,bg,c}$, $mbCE_{n,bg,c}$, $cCE_{cg,c}$, $mcCE_c$, nCE_{ng} : CO_2 emission coefficient (C-t/kWh)
CEL : CO_2 emission control level

(4) Sulfur dioxide Control Zone
In Sulfur dioxide control zone, new plants burning coal of which sulfur content is over 1% must build FGD, and existing plant burning coal of which sulfur content is over 1% must build FGD must be supplemented with FGD by 2010.

for $n \in$ SO_2 Control zone, cg=*PC w/o FGD*, *c* of sulfur contain $\geq$ 1%

$$cgo_{n,cg,c,t,h} \leq 0 \tag{26}$$

$$bgo_{n,ng,c,t,h} \leq 0 \text{ for } t \geq 2010 \tag{27}$$

Objective function

$$J = fc + tc + vc + cc + nc + tax \rightarrow \min$$

where,
fc : fuel cost,
tc : fuel(coal) transportation cost,
vc : variable O&M cost
cc : fixed cost of generating plants,
nc : fixed cost of transmission cost

stax : SO_x tax

$$
\begin{aligned}
fc = &\sum_{t=1}^{T} PVF_t \left(TVF \sum_{n=1}^{N} \sum_{bg=1}^{BG} \sum_{c=1}^{C} \sum_{h=1}^{H} \tau \cdot bFC_{n,bg,c} \cdot bgo_{n,bg,c,t,h} \right) \\
&+ \sum_{t=1}^{T} PVF_t \left(TVF \sum_{n=1}^{N} \sum_{bg=1}^{BG} \sum_{c=1}^{C} \sum_{h=1}^{H} \tau \cdot mbFC_{n,bg,c} \cdot mbgo_{n,bg,c,t,h} \right) \\
&+ \sum_{t=1}^{T} PVF_t \left(TVF \sum_{n=1}^{N} \sum_{cg=1}^{CG} \sum_{c=1}^{C} \sum_{h=1}^{H} \tau \cdot cFC_{cg,c} \cdot cgo_{n,cg,c,t,h} \right) \\
&+ \sum_{t=1}^{T} PVF_t \left(TVF \sum_{n=1}^{N} \sum_{c=1}^{C} \sum_{h=1}^{H} \tau \cdot mcFC_{c} \cdot mcgo_{n,c,t,h} \right) \\
&+ \sum_{t=1}^{T} PVF_t \left(TVF \sum_{n=1}^{N} \sum_{ng=1}^{NG} \sum_{h=1}^{H} \tau \cdot nFC_{ng} \cdot ngo_{n,ng,t,h} \right)
\end{aligned} \tag{28}
$$

$$
\begin{aligned}
tc = &\sum_{t=1}^{T} PVF_t \left(TVF \sum_{n=1}^{N} \sum_{bg=1}^{BG} \sum_{c=1}^{C} \sum_{h=1}^{H} \tau \cdot bTC_{n,bg,c} \cdot bgo_{n,bg,c,t,h} \right) \\
&+ \sum_{t=1}^{T} PVF_t \left(TVF \sum_{n=1}^{N} \sum_{bg=1}^{BG} \sum_{c=1}^{C} \sum_{h=1}^{H} \tau \cdot mbTC_{n,bg,c} \cdot mbgo_{n,bg,c,t,h} \right) \\
&+ \sum_{t=1}^{T} PVF_t \left(TVF \sum_{n=1}^{N} \sum_{cg=1}^{CG} \sum_{c=1}^{C} \sum_{h=1}^{H} \tau \cdot cTC_{n,cg,c} \cdot cgo_{n,cg,c,t,h} \right) \\
&+ \sum_{t=1}^{T} PVF_t \left(TVF \sum_{n=1}^{N} \sum_{c=1}^{C} \sum_{h=1}^{H} \tau \cdot mcTC_{n,c} \cdot mcgo_{n,c,t,h} \right)
\end{aligned} \tag{29}
$$

$$
\begin{aligned}
vc = &\sum_{t=1}^{T} PVF_t\left(TVF \sum_{n=1}^{N}\sum_{bg=1}^{BG}\sum_{c=1}^{C}\sum_{h=1}^{H} \tau \cdot bVOM \cdot bgo_{n,bg,c,t,h} \right) \\
&+ \sum_{t=1}^{T} PVF_t\left(TVF \sum_{n=1}^{N}\sum_{bg=1}^{BG}\sum_{c=1}^{C}\sum_{h=1}^{H} \tau \cdot mVOM_n \cdot mbgo_{n,bg,c,t,h} \right) \\
&+ \sum_{t=1}^{T} PVF_t\left(TVF \sum_{n=1}^{N}\sum_{cg=1}^{CG}\sum_{c=1}^{C}\sum_{h=1}^{H} \tau \cdot cVOM_{n,cg} \cdot cgo_{n,cg,c,t,h} \right) \\
&+ \sum_{t=1}^{T} PVF_t\left(TVF \sum_{n=1}^{N}\sum_{c=1}^{C}\sum_{h=1}^{H} \tau \cdot mVOM_n \cdot mcgo_{n,c,t,h} \right) \\
&+ \sum_{t=1}^{T} PVF_t\left(TVF \sum_{n=1}^{N}\sum_{ng=1}^{NG}\sum_{h=1}^{H} \tau \cdot nVOM_{ng} \cdot ngo_{n,ng,t,h} \right)
\end{aligned} \tag{30}
$$

$$
\begin{aligned}
cc = &\sum_{t=1}^{T} PVF_t\left(TVF \sum_{n=1}^{N}\sum_{bg=1}^{BG}\sum_{t'=1}^{t} mbARC_n \cdot bgcm_{n,bg,t'} \right) \\
&+ \sum_{t=1}^{T} PVF_t\left(TVF \sum_{n=1}^{N}\sum_{bg=1}^{BG} mbFOM \cdot mbgc_{n,bg,t} \right) \\
&+ \sum_{t=1}^{T} PVF_t\left(TVF \sum_{n=1}^{N}\sum_{cg=1}^{CG}\sum_{t'=1}^{t} cAFC_{n,cg} \cdot cgic_{n,cg,t'} \right) \\
&+ \sum_{t=1}^{T} PVF_t\left(TVF \sum_{n=1}^{N}\sum_{t'=1}^{t} mcAAFC_n \cdot cgcm_{n,t'} \right) \\
&+ \sum_{t=1}^{T} PVF_t\left(TVF \sum_{n=1}^{N}\sum_{ng=1}^{NG}\sum_{t'=1}^{t} nAFC_{ng} \cdot cgic_{n,ng,t'} \right)
\end{aligned} \tag{31}
$$

$$
nc = \sum_{t}^{T} PVF_t\left(TVF \sum_{b=1}^{B}\sum_{t'=1}^{t} ATC_b \cdot ibc_{n,b,t'} \right) \tag{32}
$$

$$
stax = \sum_{t}^{T} PVF_t\left(TVF \cdot \sum_{n \in SCzone} STAX \cdot nsulfur_{n,t} \right) \tag{33}
$$

where,

$$PVF_t = \left(\frac{1}{1+R}\right)^{I(t-1)}, \quad TVF = \sum_{y=1}^{I}\left(\frac{1}{1+R}\right)^{y-1}$$

$bFC_{n,bg,c}$, $mbFC_{n,bg,c}$, $cFC_{cg,c}$, $mcFC_c$, nFC_{ng} :fuel cost of generating plants ($/kWh)
*coal transportation cost is not included.
$bTC_{n,bg,c}$, $mbTC_{n,bg,c}$, $cTC_{n,cg,c}$, $mcTC_{n,c}$:coal transportation cost of generating plants ($/kWh)
$bVOM$, $mVOM_n$, $cVOM_{n,cg}$, $nVOM_{ng}$:variable O&M cost of generating plants ($/kWh),
$mbARC_n$: annual recovery cost for FGD installation of existing plant ($/kW/yr.)
$mbFOM$: fixed O&M cost of supplemented FGD in existing plant ($/kW/yr.)
$cAFC_{n,cg}$: annual fixed cost (annual recovery cost and fixed O&M cost) of new coal plant ($/kW/yr.)
$mcAAFC_n$: additional annual fixed cost for supplemented FGD in PC ($/kW/yr.)
$nAFC_{ng}$: annual fixed cost of non-coal plant ($/kW/yr.)
ATC_b : annual fixed cost of transmission line ($/kW/yr.),
$STAX$: Emission Charge of SO_x ($/t-S)
I : period interval
R : discount rate

ACRONYMS AND ABBREVIATIONS

A, AGI	Agricultural sector
ABB	Asea Brown Boveri
ABWR	Advanced Boiling Water Reactor
ACS	American Cancer Society
ACTMCLSHHA	Accident Cases of Typical and Major Casualties on Labor Safety and Health at Home and Abroad
ACWCMSC	A Complete Work of Coal Mine Safety in China
AFB	Atmospheric Fluidized Bed
AFBC	Atmospheric Fluidized Bed Coal Combustion power plant
AGP	Agreement on Government Procurement
AGRIMAT	Simplified impact assessment model included in the AirPacts package
AGS	Alliance for Global Sustainability
AirPacts:	A suite of simplified impact assessment programs
AIT	Asian Institute of Technology
ALWR	Advanced Light Water Reactor
AP1000	Advanced Pressurized water reactor 1000 MW
AP600	Advanced Pressurized water reactor 600 MW
APWR	Advanced Pressurized Water Reactor
AVLIS	Atomic Vapor Laser Isotope Separation
BASICS	British Association for Immediate Care. This database contains Details of more than 6350 disasters and incidents around the world.
BAU	Business As Usual (B, baseline, *etc.*)
BAU	Business As Usual (simulation case)
BCM	Billion Cubic Meter (natural gas)
BP	British Petroleum
BWR	Boiling Water Reactor
C	Commercial, Constant (see Scenario)
CACETC	Clean Air and Clean Energy Technology Cooperation
CANDU	Canadian Deuterium (pressurized heavy water nuclear reactor)
CB	Coal Bureau
CBM	Coalbed Methane

CC	Combined Cycle
CC	CO_2 Control (simulation case)
CCGT	Combined Cycle Gas Turbine
CCII	China Coal Information Institute
CCIY	China Coal Industry Yearbook
CCPP	Combined Cycle Power Plant
CDF	Core Damage Frequency
CENTREL	Regional group of Czech, Hungarian, Polish, and Slovak power companies
CETP	China Energy Technology Program
CFC	Chlorofluorocarbon
CHEM	Chemical sector
CI	Statistical Confidence Interval (usually, 68% or 95%)
CIESIN	Center for International Earth Science Information Network
CISDOC	International Occupational Health and Safety Centre Bibliographic Database by the International Labor Organization (ILO)
CLAB	Central interim storage facility for spent nuclear fuel, Sweden (Centralt Lager för Använt Bränsle)
CLSC	CO_2 and Local SO_x Control (simulation case)
CM	Construction Materials
CM	Coal Mine
CMD	Coal Mine District
CNEIC	China Nuclear Energy Industry Corp.
CO	Carbon monoxide
CO_2	Carbon dioxide
COD	Chemical Oxygen Demand
COE	Cost Of Electricity
CON	Conversion technologies
COSHY	China Occupational Safety and Health Yearbook
COSTCAP	Cost of Capital investment
COY	China Ocean Yearbook.
CPIY	China Petroleum Industry Yearbook
CRETM	China Regional Energy Trade Model
CRIEPI	Central Research Institute of the Electric Power Industry (Japan)
CT	Combustion Turbine

DC	Direct Current
$DELEC_t$	Electricity demand at time t
DEMELEC-PRO	DEM and of ELECtricity, a general model for electricity market study
DF	Demand Forecasting
DM	Demand
DOE	Department of Energy (US)
DOE/EIA	Department of Energy/Energy Information Administration
DOM	Domestic (coal-fired plants)
DRAG	Data Reliability Advisory Group
DSM	Demand Side Management
DSO	Dam Safety Office, Bureau of Reclamation, US Department of Interior
D_t	Demand at time t
E^3	Energy-Economy-Environment
EcoSense	Integrated Impact Assessment and External Cost Model (full impact pathway approach)
EE	Electrical Energy Generation
EEC	European Economic Community
EEM	Energy Economy Modeling
EGEAS	Electric Generation Expansion System
EIA	Environmental Impact Assessment
EIN	European Internet Network
EIP	Energy Intensive Sector
EIPs	Energy Intensive Products
EIS	Energy Intensive Product
ELASGDP	Income (GDP) elasticity of demand
ELASPRC	Price elasticity of demand
ELECTRE	Elimination and Choice Reflecting the Reality (Elimination Et Choix TRaduisant la REalité)
ENC	Total discounted energy cost
ENSAD	Energy-related Severe Accidents Database; this comprehensive database on severe accidents with emphasis on those associated with the energy sector has been established by the Paul Scherrer

	Institute, Switzerland.
EPA	Environmental Protection Agency (US)
EPB	Electric Power Bureau
EPFL	Federal Institute of Technology, Lausanne (Ecole Polytechnique Fédérale de Lausanne, Switzerland)
EPFL-LASEN	Laboratory of Energy Systems of EPFL, Ecole Polytechnique Federale de Lausanne (Switzerland)
EPR	European Pressurized Water Reactor
EPRI	Electric Power Research Institute (USA)
ERF	Exposure Response Function
ERI	Energy Research Institute (Beijing)
ES&H	Environment, Safety, and Health
ESBWR	European Simplified Boiling Water Reactor
ESP	Electrostatic Precipitator
ESS	Electric Sector Simulation
ETH	Federal Institute of Technology (Eidgenössische Technische Hochschule)
ETHZ	Federal Institute of Technology, Zurich (Eidgenössische Technische Hochschule Zürich)
ETL	Endogenous Technological Learning
ETM	Energy Transportation Modeling
EU	European Union
EUE	End-Use Efficiency
EXP	Export
F_{er}	Exposure response function
f_{er}	Exposure response function slope
FBC	Fluidized Bed Combustion coal power plant
FE	Final Energy demand
FeM	Ferrous Metals
FGD	Flue Gas Desulfurization
FYP	Five Year Plan
G&A	General and Administrative
GAMS	General Arithmetic Modeling System
GCC	Gas Combined Cycle power plant
GDP	Gross Domestic Product
GEN	electrical GENeration

GGDP	GDP Growth
GIS	Geographical Information System
GNP	Gross National Product
GHG, GHGs	Greenhouse Gas, Greenhouse Gases
GPCC	Global Precipitation Climatology Centre
GPCP	Global Precipitation Climatology Project
$GPRICE_t$	fuel price growth rate at time t
GT	GeoThermal
GT	Gas Turbine
GWP	Greenhouse Warming Potential
H, HH,	High, higher, *etc.*
HC:	Human Capital
HEU	Highly Enriched Uranium
HFC	Hydrofluorocarbons
HLW	High Level radioactive Waste
HM	Heavy Metal (nuclear spent fuel)
HSE	Health and Safety Executive (UK).
HSELINE	Library and Information Services of HSE
HTGR	High Temperature Gas Reactor
HTM	Harwell Trajectory Model, long-range atmospheric dispersion model for primary and secondary species
HYDRO	HYDROelectric
I	Health impact
I_{SUWM}	Simple Uniform World Model impact estimate
IAEA	International Atomic Energy Agency (UN)
ICOLD	International Commission on Large Dams
IEA	International Energy Agency (OECD, Paris)
IEPE	Institute of Energy Policy and Economics of the University of Grenoble, France
IGCC	Integrated Gasification Combined Cycle coal power plant
IIASA	International Institute for Applied Systems Studies
ILO	International Labor Organization
ILW	Intermediate Level radioactive Waste
IMP	Import
IND, I	INDustry

IPA	Impact Pathways Approach (or Analysis)
IPCC	Intergovernmental Panel on Climate Change (United Nations)
ISC	Industrial Source Complex Model
ISL	In-Situ Leaching (chemical mining technique)
ISO	International Organization for Standardization
k:	Pollutant depletion or atmospheric removal velocity (usually, in cm/s); with "p" subscript for primary pollutant and "s" subscript for secondary species
$k_{p \rightarrow s}$	Primary to secondary pollutant chemical transformation velocity
k_{uni}	Uniform World Model (uni subscript) depletion velocity
KWU/EDF	Kraftwerkunion/Electricité de France
L, LL, LLL	Low, Lower, *etc.*
LAMSADE	Laboratory for the Analysis and Modeling of Decision Aid Systems, Paris (Laboratoire d'Analyse et Modélisation des Systèmes pour l'Aide à la DEcision)
LASEN	Laboratory of Energy Systems (Laboratoire des Systèmes ENergétiques), Switzerland
LCA	Life Cycle Assessment
LCI	Life Cycle Inventory assessment
LCIA	Life Cycle Impact Assessment
LERF	Large Early Release Frequency
LHV	Lower Heating Value
LLP	Lloyd's Casualty Week; formerly Lloyd's of London Press
LLW	Low Level radioactive Waste
LNG	Liquefied Natural Gas
LP	Linear Programming
LPG	Liquefied Petroleum Gas
LSC	Local SO_x Control (simulation case)
LWR	Light Water Reactor
M:	Pollutant removal flux along the earth's surface
MARKAL	MARket Allocation model
max	maximum
MCDA	Multi-Criteria Decision Aid(ing)
MCDA	Multi-Criteria Decision Analysis
MED	Medium (capacity)

MEDEE-S	Modele d'Evaluation de la Demande En Energie
MHIDAS	Major Hazards Incidence Data Service of the UK Health and Safety Executive (HSE)
MIN	MINing
min	minimum
MIT	Massachusetts Institute of Technology, Cambridge, US
MOX	Mixed Oxide Fuel
MRC	Mixed-refrigerant cycle (natural gas liquefaction process)
NCAR	National Center for Atmospheric Research
NCEP	National Centers for Environmental Prediction
NEA	Nuclear Energy Agency (OECD)
NFeM	Non-ferrous Metals
NGCC	Natural Gas Combined-Cycle
NH_3	Ammonia
NIOSHTIC	National Institute of Occupational Safety and Health (USA)
NO_x	Nitrogen Oxides
NPIC	Nuclear Power Institute of China
NPP	Nuclear Power Plant
NPV	Net Present Value
NUCL, NE	Nuclear Energy
O&M	Operation and Maintenance
OECD	Organization for Economic Cooperation and Development
Q	Pollutant annual emissions rate
QUERI	Simplified impact assessment model included in the AirPacts package
PC	Pulverized Coal combustion power plant
PDS	Plant Damage State
PE	Primary Energy demand
PFBC	Pressurized Fluidized Bed Coal Combustion
PHWR	Pressurized Heavy Water Reactor (CANDU)
PI	Principal Investigator
PLM	Peak Load Management
PM10, PM_{10}	Particulate Matter of aerodynamic diameter of 10 microns (0.001 mm) or less

PM2.5, $PM_{2.5}$	Particulate Matter of aerodynamic diameter of 2.5 microns(0.001 mm) or less
PNNL	Pacific Northwest National Laboratory
POP	POPulations
PP	Power Plant
PPPGNP	Purchasing Power Parity Gross National Product
PRC	Processes
PRCEE	Policy Research Center for Environment and Economy, SEPA, China
PRIS	IAEA's Power Reactor Information System
PSA	Probabilistic Safety Assessment
PSI	Paul Scherrer Institute, Villigen, Switzerland
PSP	Pre-Specified Pathway Program
P_t	Price at time t
PV	PhotoVoltaic
PWR	Pressurized Water Reactor
QU	Qinghua University. Beijing, China (see TU)
R	Downwind distance from source or source-receptor position vector
R&D	Research and Development
RA	Risk Assessment
RAINS	Regional Air Pollution Information and Simulation
RC	Raw coal
RD&D	Research, Development, and Design
RES	Reference Energy System
RNW	Renewables
ROM	Run-of-mine coal
RUWM	Robust Uniform World Model
S	Sulfur
SAC	Shandong Aluminum Company
SAG	Chinese Stakeholders Advisory Group
SC	Steering Committee

Scenarios (Region)-(Policy)(Discount Rate)(Demand)(Fossil Fuel)-(Technology)

B	Business as usual (no specific emission controls); Baseline technology availabilities, costs, *etc.*
C	China (Region), Constant (fossil fuel prices)

E	Environmental (S + C) Policy
H	High demand, discount rate, Increasing fossil fuel prices
L	Endogenous Technological Learning (ETL)
S	Sulfur control, Shandong Province (Region)
S	Services
SCIA	State Coal Industry Administration, China
SCZ	SO_x Control Zones (simulation case)
SD	Shandong Province, China
SDPC	State Development Planning Commission, China
SEMA	French company, Paris (Société d'étude et de mathématiques appliquées)
SEPA	State Environmental Protection Administration, China
SEPCO	Shandong Electric Power Group Company, China
SEPRI	Shandong Electric Power Research Institute, China
SESAMS	Strategic Electric Sector Assessment Methodology Under Sustainability conditions
SIB	Siberia
SIGMA	Sigma is published approximately eight times a year by SwissRe's Economic Research & Consulting team based in Zurich, New York and Hong Kong.
SO_2	Sulfur Dioxide
SO_x	Sulfur Oxides
SPC, SPCC	State Power Corporation of China
SRC	Source
SSB	State Statistical Bureau of China
STK	Stocks
SUWM	Simple Uniform World Model
SWU	Separative Work Unit (for uranium enrichment)
SX	Shanxi Province
t	Time or tons (1 million grams)
T	Transportation sector
T&D	Transmission and Distribution
TCH	Technology (process, conversion, *etc.*)
TCM	Trillion Cubic Meter (natural gas)
TGP	Three Gorges Project
Trans,etc.	Transport, telecommunication and other sectors in the demand forecasting

TSC	Total SO_x Control (simulation case)
TSP	Total Suspended Particulates
TU	Tsinghua University, Beijing, China (see QU)
u (U)	Wind speed
UCTE	Union for the Coordination of Transmission of Electricity
UE	Useful Energy demand
UI	Uranium Institute (now World Nuclear Association)
UN	United Nations
UNDP	United Nations Development Program
UNSCEAR	United Nations Scientific Committee on the Effects of Atomic Radiation
URBAN	Simplified impact assessment model included in the AirPacts package
US	United States
USDOE	United States Department of Energy
USEPA	United States Environmental Protection Agency
UT	University of Tokyo, Tokyo, Japan
VLYL	Value of a Life Year Lost, economic cost of one year of life lost
VSL	Value of Statistical Life, amount of money society is willing to spend to avoid a premature death
VVER	Light water pressurized reactor of Russian design (Vodo-Vodyanoy Energeticheskiy Reactor)
WBCSD	World Business Council for Sustainable Development, Geneva, Switzerland
WCD	World Commission on Dams
WEC	World Energy Council, London
WNA	World Nuclear Association (former UI)
WTM	Windrose Trajectory Model, long-range atmospheric dispersion model for primary and secondary species
WTO	World Trade Organization, Geneva
WTP	Willingness To Pay to achieve an environmental benefit
WWW	World Wide Web
XJ	Xinjiang Province (Autonomous Region), China
YEARPP	YEARs Per Period
YOLL	Years Of Life Lost (reduction in life expectancy)
Y_t	GDP at time t
μ_g	Geometric median of lognormal distribution

ρ_{uni}	Uniform World Model (uni subscript) receptor density
ρ	Receptor density
σ_g	Geometric standard deviation of lognormal distribution

UNITS

a	annum, year
BCM	billion cubic meter = 10^9 cubic meter
Bq	1 Becquerel = amount of material which will produce 1 nuclear decay per second. The Becquerel is the more recent SI unit for radioactive source activity. The curie (Ci) is the old standard unit for measuring the activity of a given radioactive sample. It is equivalent to the activity of 1 gram of radium. 1 curie = 3.7 x 10^{10} Becquerels.
BYuan	Billion Yuan
EJ	ExaJoules = 10^{18} Joule
eq/ha/year:	Unit for critical loads, 1 eq/ha/year corresponds to 1.6 mg/m^2/year
g(CO_2-equiv.)	gram CO_2 equivalent, calculated using IPCC (2001) GWP
gce	gram coal equivalent (coal with LHV = 29.3 MJ_{th}/kg)
GJ	GigaJoule = 10^9 Joule
GW	GigaWatt = 10^9 W
GW_eyr	GigaWatt-electric-year; 1 GW_eyr = 8.76 x 10^9 kWh
GWh	GigaWatthour = 10^9 Wh
Gy	Gray; SI unit of absorbed radiation dose in terms of the energy actually deposited in the tissue. The Gray is defined as 1 joule of deposited energy per kilogram of tissue. The old SI unit is the rad. 1 Gy = 1 J/kg = 100 rad.
ha	hectare
J	1 J = 1 N m= 1 m^2 kg s^{-2}. The Joule is the SI unit of work or energy.
kA	1000 Ampere
keq/ha/year	1000 eq/ha/year
kgoe	kilogram oil equivalent = 42.1 MJ
km	kilometer
kt	kilotonne = 10^3 tonne, also kiloton
ktoe	kiloton oil equivalent = 42.1 TJ
kV	kilovolt
kW	kiloWatt = 10^3 W

kWh	kiloWatthour; 1 kWh = 3.6 MJ
kWh_{th}	kiloWatthourthermal
m	meter
mg	milligram = 10^{-3} gram
MJ	Megajoule = 10^6 Joule
MJ_{th}	Megajoule thermal = 10^6 Joule thermal
mm	millimeter
Mm^3	Million cubic meter
MSWU	Million Separative Work Unit = 10^6 SWU
Mt	Megatonne = 10^6 tonne, also megaton
Mtoe	Million ton oil equivalent = 42.1 PJ
MVA	MegaVolt-Ampere
MW	MegaWatt = 10^6 W
MWd_{th}	MegaWattday thermal
MWh	MegaWatthour = 10^6 Wh
MYuan	Million Yuan
nm	Nautical mile = 1852 m
Nm^3	Normal cubic meter
Ryr	Reactor*year
s	second
SWU	Separative Work Unit (for uranium enrichment)
t	tonne, metric ton (1 t = 1000 kg), also ton
tce	tonne coal equivalent = 29.3 GJ
TCM	Trillion Cubic Meter = 10^{12} cubic meter
TJ	10^{12} Joule
tkm	tonne-kilometer
toe	tonnes of oil equivalent = 42.1 GJ
TW	TeraWatt = 10^{12} W
TWh	TeraWatthour = 10^{12} Wh
Twhe, Tweh	TeraWatthour, electrical = 3.6 MJ
USD	US Dollar
W	Watt 1 W = 1 J/s = 1 $m^2 \cdot kg \cdot s^{-3}$

YOLL	Years Of Life Lost (reduction in life expectancy)
yr	Year
μg	microgram (10^{-6} grams)

64’000 or 64,000	64000	(sixty four thousand)
64.000	64	(sixty four)

INDEX

F

G

H

I

Q

R

S

T